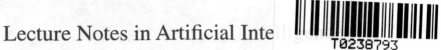

Lecture Notes in Artificial Inte

Subseries of Lecture Notes in Computer Science

LNAI Series Editors

Randy Goebel
University of Alberta, Edmonton, Canada
Yuzuru Tanaka
Hokkaido University, Sapporo, Japan
Wolfgang Wahlster
DFKI and Saarland University, Saarbrücken, Germany

LNAI Founding Series Editor

Joerg Siekmann
DFKI and Saarland University, Saarbrücken, Germany

De-Shuang Huang   Kang-Hyun Jo
Yong-Quan Zhou   Kyungsook Han (Eds.)

Intelligent Computing Theories and Technology

9th International Conference, ICIC 2013
Nanning, China, July 28-31, 2013
Proceedings

Springer

Volume Editors

De-Shuang Huang
Tongji University, Machine Learning and Systems Biology Laboratory
4800 Caoan Road, Shanghai 201804, China
E-mail: dshuang@tongji.edu.cn

Kang-Hyun Jo
University of Ulsan, School of Electrical Engineering
680-749 #7-413, San 29, Muger Dong, Ulsan, South Korea
E-mail: jkh2005@islab.ulsan.ac.kr

Yong-Quan Zhou
Guangxi University for Nationalities
Nanning, Guangxi 530006, China
E-mail: yongquanzhou@126.com

Kyungsook Han
Inha University, School of Computer Science and Engineering
Incheon 402-751, South Korea
E-mail: khan@inha.ac.kr

ISSN 0302-9743   e-ISSN 1611-3349
ISBN 978-3-642-39481-2   e-ISBN 978-3-642-39482-9
DOI 10.1007/978-3-642-39482-9
Springer Heidelberg Dordrecht London New York

Library of Congress Control Number: 2013942264

CR Subject Classification (1998): I.2, H.2.8, H.3-4, I.4-5, F.2, J.3

LNCS Sublibrary: SL 7 – Artificial Intelligence

Typesetting: Camera-ready by author, data conversion by Scientific Publishing Services, Chennai, India

Printed on acid-free paper

Springer is part of Springer Science+Business Media (www.springer.com)

Preface

The International Conference on Intelligent Computing (ICIC) was started to provide an annual forum dedicated to the emerging and challenging topics in artificial intelligence, machine learning, pattern recognition, image processing, bioinformatics, and computational biology. It aims to bring together researchers and practitioners from both academia and industry to share ideas, problems, and solutions related to the multifaceted aspects of intelligent computing.

ICIC 2013, held in Nanning, China, July 28–31, 2013, constituted the 9th International Conference on Intelligent Computing. It built upon the success of ICIC 2012, ICIC 2011, ICIC 2010, ICIC 2009, ICIC 2008, ICIC 2007, ICIC 2006, and ICIC 2005 that were held in Huangshan, Zhengzhou, Changsha, China, Ulsan, Korea, Shanghai, Qingdao, Kunming, and Hefei, China, respectively.

This year, the conference concentrated mainly on the theories and methodologies as well as the emerging applications of intelligent computing. Its aim was to unify the picture of contemporary intelligent computing techniques as an integral concept that highlights the trends in advanced computational intelligence and bridges theoretical research with applications. Therefore, the theme for this conference was "Advanced Intelligent Computing Technology and Applications". Papers focused on this theme were solicited, addressing theories, methodologies, and applications in science and technology.

ICIC 2013 received 561 submissions from 27 countries and regions. All papers went through a rigorous peer-review procedure and each paper received at least three review reports. Based on the review reports, the Program Committee finally selected 192 high-quality papers for presentation at ICIC 2013, included in three volumes of proceedings published by Springer: one volume of *Lecture Notes in Computer Science* (LNCS), one volume of *Lecture Notes in Artificial Intelligence* (LNAI), and one volume of *Communications in Computer and Information Science* (CCIS).

This volume of *Lecture Notes in Artificial Intelligence* (LNAI) includes 79 papers.

The organizers of ICIC 2013, including Tongji University and Guangxi University for Nationalities, made an enormous effort to ensure the success of the conference. We hereby would like to thank the members of the Program Committee and the referees for their collective effort in reviewing and soliciting the papers. We would like to thank Alfred Hofmann, executive editor from Springer, for his frank and helpful advice and guidance throughout and for his continuous support in publishing the proceedings. In particular, we would like to thank all the authors for contributing their papers. Without the high-quality submissions from the authors, the success of the conference would not have been possible.

Finally, we are especially grateful to the IEEE Computational Intelligence Society, the International Neural Network Society, and the National Science Foundation of China for their sponsorship.

May 2013   De-Shuang Huang
                                                   Kang-Hyun Jo
                                                   Yong-Quan Zhou
                                                   Kyungsook Han

ICIC 2013 Organization

General Co-chairs

De-Shuang Huang, China
Marios Polycarpou, Cyprus
Jin-Zhao Wu, China

Program Committee Co-chairs

Kang-Hyun Jo, Korea
Pei-Chann Chang, Taiwan, China

Organizing Committee Co-chairs

Yong-Quan Zhou, China
Bing Wang, China

Award Committee Co-chairs

Laurent Heutte, France
Phalguni Gupta, India

Publication Chair

Juan Carlos Figueroa, Colombia

Workshop/Special Session Chair

Vitoantonio Bevilacqua, Italy

Special Issue Chair

Michael Gromiha, India

Tutorial Chair

Luonan Chen, Japan

International Liaison Chair

Prashan Premaratne, Australia

Publicity Co-chairs

Kyungsook Han, Korea
Lei Zhang, China
Ling Wang, China
Valeriya Gribova, Russia

Exhibition Chair

Xing-Ming Zhao, China

Organizing Committee Members

Yong Huang, China
Yong Wang, China
Yuanbin Mo, China

Conference Secretary

Su-Ping Deng, China

Program Committee Members

Andrea Francesco Abate, Italy
Vasily Aristarkhov, Russian Federation
Costin Badica, Romania
Soumya Banerjee, India
Waqas Haider Khan Bangyal, Pakistan
Vitoantonio Bevilacqua, Italy
Shuhui Bi, China
Zhiming Cai, Macau
Chin-Chih Chang, Taiwan, China
Pei-Chann Chang, Taiwan, China
Guanling Chen, USA
Luonan Chen, Japan
Jingdong Chen, China
Songcan Chen, China
Weidong Chen, China

Xiyuan Chen, China
Yang Chen, China
Michal Choras, Poland
Angelo Ciaramella, Italy
Jose Alfredo F. Costa, Brazil
Mingcong Deng, Japan
Eng. Salvatore Distefano, Italy
Mariagrazia Dotoli, Italy
Haibin Duan, China
Hazem Elbakry, Egypt
Karim Faez, Iran
Jianbo Fan, China
Jianwen Fang, USA
Minrui Fei, China
Juan Carlos Figueroa, Colombia

Wai-Keung Fung, Canada
Jun-Ying Gan, China
Liang Gao, China
Xiao-Zhi Gao, Finland
Dunwei Gong, China
Valeriya Gribova, Russia
M. Michael Gromiha, India
Xingsheng Gu, China
Kayhan Gulez, Turkey
Phalguni Gupta, India
Fei Han, China
Kyungsook Han, Korea
Yong-Tao Hao, China
Jim Harkin, UK
Haibo He, USA
Jing Selena He, USA
Laurent Heutte, France
Wei-Chiang Hong, Taiwan, China
Yuexian Hou, China
Heyan Huang, China
Kun Huang, USA
Zhenkun Huang, China
Bora Peter Hung, Ireland
Chuleerat Jaruskulchai, Thailand
Umarani Jayaraman, India
Li Jia, China
Zhenran Jiang, China
Kang-Hyun Jo, Korea
Dong-Joong Kang, Korea
Sanggil Kang, Korea
Muhammad Khurram Khan,
    Saudi Arabia
Donald H. Kraft, USA
Harshit Kumar, Korea
Yoshinori Kuno, Japan
Takashi Kuremoto, Japan
Vincent C S Lee, Australia
Bo Li, China
Guo-Zheng Li, China
Kang Li, UK
Min Li, China
Shi-Hua Li, China
Xiaoou Li, Mexico
Honghuang Lin, USA
Chunmei Liu, USA

Ju Liu, China
Ke Lv, China
Jinwen Ma, China
Lorenzo Magnani, Italy
Xiandong Meng, USA
Tarik Veli Mumcu, Turkey
Roman Neruda, Czech Republic
Ken Nguyen, USA
Ben Niu, China
Yusuke Nojima, Japan
Sim-Heng Ong, Singapore
Francesco Pappalardo, Italy
Young B. Park, Korea
Surya Prakash, India
Prashan Premaratne, Australia
Seeja K.R., India
Ajita Rattani, Italy
Ivan Vladimir Meza Ruiz, Mexico
Angel D. Sappa, Spain
Li Shang, China
Fanhuai Shi, China
Jiatao Song, China
Stefano Squartini, Italy
Zhan-Li Sun, China
Evi Syukur, Australia
Naoyuki Tsuruta, Japan
Antonio E. Uva, Italy
Katya Rodriguez Vazquez, Mexico
Jun Wan, USA
Bing Wang, China
Lei Wang, China
Ling Wang, China
Shitong Wang, China
Wei Wang, China
Yijie Wang, China
Wei Wei, China
Zhi Wei, China
Qiong Wu, China
Xiaojun Wu, China
Yan Wu, China
Junfeng Xia, China
Shunren Xia, China
Yuanqing Xia, China
Liangjun Xie, USA
Bingji Xu, China

Hua Xu, USA
Shao Xu, Singapore
Zhenyu Xuan, USA
Tao Ye, China
Wen Yu, Mexico
Boyun Zhang, China
Lei Zhang, HongKong, China
Xiang Zhang, USA
Yi Zhang, China

Hongyong Zhao, China
Xing-Ming Zhao, China
Zhongming Zhao, USA
Bo-Jin Zheng, China
Chun-Hou Zheng, China
Fengfeng Zhou, China
Shuigeng Zhou, China
Li Zhuo, China

Reviewers

Kamlesh Tiwari
Aditya Nigam
Puneet Gupta
Somnath Dey
Aruna Tiwari
Erum Afzal
Gurkan Tuna
Ximo Torres
Chih-Chin Liu
Jianhung Chen
Ouyang Wen
Chun-Hsin Wang
Yea-Shung Huang
Huai-Jen Liu
James Chang
Wu Yu
Chyuan-Huei Yang
Yao-Hong Tsai
Cheng-Hsiung Chiang
Chia-Luen Yang
Wei Li
Xinhua Zhao
Yongcui Wang
Francesco Camastra
Antonino Staiano
Kyungsook Han
Francesca Nardone
Alessio Ferone
Antonio Maratea
Giorgio Gemignani
Mario Manzo
Shuhui Bi

Shengjun Wen
Aihui Wang
Shengjun Wen
Changan Jiang
Ni Bu
Guangyue Du
Lijun Mu
Yizhang Jiang
Zhenping Xie
Jun Wang
Pengjiang Qian
Jun Bo
Hong-hyun Kim
Hyeon-Gyu Min
One-Cue Kim
Xiaoming Wang
Zi Wang
Jekang Park
Zhu Teng
Baeguen Kwon
Jose A. Fernandez Leon
Saber Elsayed
Mohammad-Javad
    Mahmoodabadi
Amar Khoukhi
Hailei Zhang
Deepa Anand
Sepehr Attarchi
Siyu Xia
Yongfei Xiao
Xiaotu Ma
Jin-Xing Liu

Yan Cui
Guanglan Zhang
Jing Sun
Joaquín Dopazo
Miguel A Pujana
Wankou Yang
Xu Zhao
Chi-Tai Cheng
Shen Chong
Quan Yuan
Yunfei Wang
Zhanpeng Zhang
DungLe Tien
Hee-Jun Kang
Liu Yu
Kaijin Qiu
Chenghua Qu
Aravindan Chandrabose
Jayasudha John Suseela
Sudha Sadasivam
Minzhu Xie
Hongyun Zhang
Mai Son
Myeong-Jae Yi
Wooyoung Kim
Gang Chen
Jingli Wu
Qinghua Li
Lie Jie
Muhammad Amjad
Waqas Bangyal
Abdul Rauf

Durak-Ata Lutfiye
Aliahmed Adam
Kadir Erkan
Zhihua Wei
Elisano Pessa
Sara Dellantonio
Alfredo Pereira
Tomaso Vecchi
Assunta Zanetti
Wai-keung Fung
Guodong Zhao
Min Zheng
Aditya Nigam
Xiaoming Liu
Xin Xu
Yanfeng Zhang
Angelo Ciaramella
Saiful Islam
Mahdi Ezoji
Hamidreza Rashidy
  Kanan
Marjan Abdechiri
Saeed Mozaffari
Javad Haddadnia
Farzad Towhidkhah
Majid Ziaratban
Shaho Ghanei
Jinya Su
Minglei Tong
Shihong Ding
Yifeng Zhang
Yuan Xu
Xiying Wang
Haoqian Huang
Rui Song
Xinhua Tang
Xuefen Zhu
Zhijun Niu
Jun Lv
Muhammad Rashid
Muhammad Ramzan
Carlos Cubaque
Jairo Soriano
German Hernandez
Juan Carlos Figueroa

Hebert Lacey
Qing Liu
Tao Ran
Yan Yan
Chuang Ma
Xiaoxiao Ma
Jing Xu
Ke Huang
Prashan Premaratne
Geethan Mendiz
Sabooh Ajaz
Tsuyoshi Morimoto
Naoyuki Tsuruta
Shin Yatakahashi
Sakashi Maeda
Zhihua Wei
Chuleerat Jaruskulchai
Sheng-Yao Wang
Xiaolong Zheng
Huanyu Zheng
Ye Xu
Abdullah Bal
Xinwu Liang
Lei Hou
Tao He
Yong Wang
Zhixuan Wei
Xiao Wang
Jingchuan Wang
Qixin Wang
Xiutao Shi
Jibin Shi
Wenxi Zhang
Zhe Liu
Olesya Kazakova
Xiaoguang Li
Manabu Hashimoto
Yu Xue
Bin Song
Liang Zhao
Songsheng Zhou
Hung-Chi Su
Francesco Longo
Giovanni Merlino
Liang Zhao

Chao Xing
Lan Ma
Xiang Ding
Chunhui Zhang
Kitti Koonsanit
Duangmalai Klongdee
Atsushi Yamashita
Kazunori Onoguchi
Wencai Ma
Fausto Petrella
Diyi Chen
Sakthivel Ramasamy
Ryuzo Okada
Zhen Lu
Jida Huang
Wenting Han
Zhong-Ke Gao
Ning-De Jin
Fang-Fang Wang
Nobutaka Shimada
Huali Huang
Qiqi Duan
Kai Yin
Bojin Zheng
Hongrun Wu
Hironobu Fujiyoshi
Tarik Veli Mumcu
Liangbing Feng
Guangming Sun
Dingfei Ge
Wei Xiong
Yang Yang
Felix Albu
Mingyuan Jiu
SiowYong Low
Swanirbhar Majumder
Saameh Golzadeh
Saeed Jafarzadeh
Erik Marchi
Haciilhan
Tansalg
Li Liu
Ke Hu
Michele Scarpiniti
Danilo Comminiello

Min Zhao
Yuchou Chang
Saleh Mirheidari
Ye Bei
Xu Jie
Guohui Zhang
Hunny Mehrotra
Kamlesh Tiwari
Shingo Mabu
Kunikazu Kobayashi
Joaquín Torres-Sospedra
Takashi Kuremoto
Shouling Ji
Mingyuan Yan
Haojie Shen
Bingnan Li
Toshiaki Kondo
Yunqi Li
Ren Jun
Meng Han
Qixun Lan
Marius Brezovan
Amelia Badica
Sorin Ilie
Guoqin Mai
Miaomiao Zhao
Selena He
Michele Fiorentino
Francesco Ferrise
Giuseppe Carbone
Pierpaolo Valentini
Alfredo Liverani
Fabio Bruno
Francesca De Crescenzio
Bin Wang
Duwen Zhang
Ruofei Zhang
Leemon Baird
Zhiyong Zeng
Mike Collins
Yu Sun
Mukesh Tiwari
Gibran-Fuentes Pineda
Villatoro-Tello Esaú
Chao Shao

Xiaofang Gao
Prashan Premaratne
Hongjun Jia
Yehu Shen
Zhongjie Zhu
Tiantai Guo
Liya Ding
Dawen Xu
Jinhe Wang
Chun Chen
Anoosha Paruchuri
Angelo Riccio
Raffaele Montella
Giuseppe Vettigli
Tao Ye
Tower Gu
Xingjia Lu
Shaojing Fan
Chen Li
Qingfeng Li
Yong Lin
Mohebbi Keyvan
Lisbeth Rodríguez
Atsushi Shimada
Andrey Vavilin
Kaushik Deb
Liangxu Liu
Rina Su
Jie Sun
Hua Yu
Linhua Zhou
Nanli Zhu
Xiangyang Li
Dalong Li
Jiankun Sun
Xiangyu Wang
Jing Ge
Cong Cheng
Yue Zhao
James Jayaputera
Azis Ciayadi
Sotanto Sotanto
Rudy Ciayadi
Anush Himanshu
Simon Bernard

Laurent Heutte
James Liu
Changbo Zhao
Sheng Sun
Fanliang Bu
Shi-Jie Guan
Xinna Zheng
Jian Lin
Kevin Zhang
Changjun Hu
Kevin Zhang
Xin Hao
Chenbin Liu
Xiaoyin Xu
Jiayin Zhou
Mingyi Wang
Yinan Guo
Haibin Duan
He Jiang
Ming Zhang
Altshuler Yaniv
Donato Barone
Angelo Antonio Salatino
Xiaoling Zhang
Quanke Pan
Yu-Yan Han
Xiaoyan Sun
Ling Wang
Liang Gao
Xiangjuan Yao
Shujuan Jiang
Tian Tian
Changhai Nie
Yang Gao
Shang Li
Weili Wu
Yuhang Liu
Yinzhi Zhou
Haili Wang
Suparta Wayan
Ogaard Kirk
Samir Zeghlache
Yijian Liu
Wu-Yin Hui
Keming Xie

Yong-Wei Zhang
Kang Qi
Qin Xiao
Hongjun Tian
Jing Zhang
Xiangbo Qi
Yan Wang
Lijing Tan
Jing Liang
Eng.Marco Suma
Raffaele Carli
Fuhai Li
Lei Huang
Yunsheng Jiang
Shuyi Zhang
Yue Zhao
Marco Suma
Junfeng Qu
Ken Nguyen
Vladislavs Dovgalecs
Muhammad Rahman
Ferdinando Chiacchio
Surya Prakash
Yang Song
Xianxia Zhang
Dajundu
Kamlesh Tiwari
Sheng Ding
Yonghui Wu
Min Jiang
Liugui Zhong
Yichen Wang
Hua Zhu
Junfeng Luo
Chunhou Zheng
Chao Wu
Vasily Aristarkhov
Yinglei Song
Hui Li
Changan Jiang
Lin Yuan
Suping Deng
Zhiwei Ji
Yufeng Liu
Jun Zhang

Wei Jin
Buzhou Tang
Yaping Fang
Zhenyu Xuan
Ying Jiang
Min Zhao
Bo Sheng
Olivier Berder
Yunlong Zhao
Yu Zeng
Jing Deng
Jianxing Li
Shulin Wang
Jianqing Li
Bo Li
Akio Miyazaki
Peng Zhang
Dazhao Pan
Vlad Dovgalecs
Chen Fei
Xiaodi Li
Wenqiang Yang
Donguk Seo
Jingfei Li
Huabin Hong
Zhaoxi Wang
Xiujun Zhang
Angel Sappa
Stefano Squartini
Weili Guo
Lei Deng
Yang Kai
Qing Xia
Jinghua Yuan
Yushu Gao
Qiangfeng Zhang
Wei Xiong
Jair Cervantes
Guorong Cai
Zongyue Wang
Keling Li
Jianhua Zhang
Martin Pilat
Ondrej Kazik
Petra Vidnerová

Jakub Smid
Qiao Cai
Jin Xu
Zhen Ni
Fanhuai Shi
Jakub Smidbjunior
Jakub Smidmff
Smile Gu
Junjun Qu
Vitoantonio Bevilacqua
David Chen
Juan Li
Taeho Kim
Hyunuk Chae
Giangluca Percoco
Yongxu Zhu
Wei Wei
Chong Feng
Ying Qiu
Gumei Lin
Huisen Wang
Lanshen Guo
Surya Prakash
Shan-Xiu Yang
Qian Fu
Jian Guan
Fei Han
Nora Boumella
Xingsheng Gu
Chunjiang Zhang
Ji-Xiang Du
Fei Han
Miaomiao Zhao
Mary Thangakani
    Anthony
Sakthivel Ramasamy
Xiwei Tang
Jing Gu
Ling Wang
Stefanos Quartini
Yushu Gao
Songcan Chen
Ye Shuang

Table of Contents

Systems Biology and Computational Biology

Cognitive Science and Computational Neuroscience

Knowledge Discovery and Data Mining

Machine Learning Theory and Methods

Biomedical Informatics Theory and Methods

Complex Systems Theory and Methods

Natural Language Processing and Computational Linguistics

Fuzzy Theory and Models

Fuzzy Systems and Soft Computing

Particle Swarm Optimization and Niche Technology

Swarm Intelligence and Optimization

Unsupervised and Reinforcement Learning

Intelligent Computing in Bioinformatics

Intelligent Computing in Petri Nets/Transportation Systems

Intelligent Computing in Social Networking

Intelligent Computing in Network Software/Hardware

Intelligent Control and Automation

Intelligent Data Fusion and Information Security

Intelligent Sensor Networks

Intelligent Fault Diagnosis

Intelligent Computing in Signal Processing

Intelligent Computing in Pattern Recognition

Intelligent Computing in Biometrics Recognition

Intelligent Computing in Image Processing

Intelligent Computing in Computer Vision

Special Session on Biometrics System and Security for Intelligent Computing

Special Session on Bio-inspired Computing and Applications

Special Session on Intelligent Computing and Personalized Assisted Living

Computer Human Interaction Using Multiple Visual Cues and Intelligent Computing

Special Session on Protein and Gene Bioinformatics: Analysis, Algorithms and Applications

Research on Signaling Pathways Reconstruction by Integrating High Content RNAi Screening and Functional Gene Network

Zhu-Hong You[1], Zhong Ming[1], Liping Li[2], and Qiao-Ying Huang[1]

[1] College of Computer Science and Software Engineering, Shenzhen University
Shenzhen, Guangdong 518060, China
[2] The Institute of Soil and Water Conservation of Gansu
Lanzhou, Gansu 730020, China
zhyou@szu.edu.cn

Abstract. The relatively new technology of RNA interference (RNAi) can be used to suppress almost any gene on a genome wide scale. Fluorescence microscopy is an ever more advancing technology that allows for the visualization of cells in multidimensional fashion. The combination of these two techniques paired with automated image analysis is emerging as a powerful tool in system biology. It can be used to study the effects of gene knockdowns on cellular phenotypes thereby lightening shadow into the understanding of complex biological processes. In this paper we propose the use of high content screen (HCS) to derive a high quality functional gene network (FGN) for *Drosophila Melanogaster*. This approach is based on the characteristic patterns obtained from cell images under different gene knockdown conditions. We guarantee a high coverage of the resulting network by the further integration of a large set of heterogeneous genomic data. The integration of these diverse datasets is based on a linear support vector machine. The final network is analyzed and a signal transduction pathway for the mitogen-activated protein kinase (MAPK) pathway is extracted using an extended integer linear programming algorithm. We validate our results and demonstrate that the proposed method achieves full coverage of components deposited in KEGG for the MAPK pathway. Interestingly, we retrieved a set of additional candidate genes for this pathway (e.g. including *sev*, *tsl*, *hb*) that we suggest for future experimental validation.

Keywords: pathway reconstruction, high content screen, functional gene networks.

1 Introduction

The regulation and activation of transcription is one of the earliest and most critical steps in the expression of genes. In a signal transduction cascade information is transmitted from cell surface receptors to intracellular effectors that eventually activate this fundamental process [1]. The computational study of signal transduction pathways received lots of attention in the past years [2-4]. In this paper we are particularly interested in the mitogen-activated protein kinase (MAPK) path. This

D.-S. Huang et al. (Eds.): ICIC 2013, LNAI 7996, pp. 1–10, 2013.
© Springer-Verlag Berlin Heidelberg 2013

cascade is activated when growth factors bind to the cell surface receptors and plays an important role in several developmental processes[5]. It triggers a variety of intercellular responses including molecular interactions such as protein-protein interaction (PPI), protein-DNA interaction (PDI) or RNA-RNA interaction (RRI). Despite this variety, most computational analysis to data, focused on modeling PPI and genomics data only. An example integrating gene expression and PPI for network construction is given in [2, 3]. However, the performance of these approached is not optimal as these two data sources seem to be incomplete, in the sense that they do not cover all interactions.

An emerging tool to study cellular processes is high content screening (HCS) by automated fluorescence microscopy [6]. This approach has recently been used on a genome-wide scale to study signaling pathways in *Drosophila* BG-2 cell lines [7]. In these work the authors generated a large amount of cell images for different gene knockdown conditions. The analysis was based on the assumption that a cell phenotype depends on a complete signaling cascade. The interruption of such a pathway, at any location and gene, results in similar morphological cell formations. These shape information was translated into quantitative image descriptors which were further linked to statistical analysis tools to score cellular phenotypes. These features allow for a grouping of genes after phenotype and pathway and hence their integration into a functional gene network (FGN). However, such analysis is still in its infancy and bears great challenges. In this paper, we present a work that constructs a FGN to study signaling pathways using HCS.

We mentioned above that different types of genomic and proteomic data exists and the post-genomics era reveals ever more such data types. Most of these sets, in the one way or the other, also carry interaction information reflecting different aspects of gene associations. Therefore, one of the major challenges is to relate these data and obtain a systems level view on signaling relationships. Here we want to use these data in accordance to the HCS to propose a FGN and then extract the MAPK signaling from it. This network integrates information of HCS namely functional relationships between genes, PPI, microarray co-expressions, gene co-occurrence, genomic neighborhoods, gene fusion, text mining and gene homology and thus has a high coverage. We use a linear support vector machine (SVM) to assign linkage weights between gene pairs and ensures high-quality and minimal false positive rates. We study the MAPK pathway by an improved integer linear programming (IILP) algorithm and use the results to better understand its signal transduction. Further, we assessed the resulting pathways based on their functional enrichment in biological processes using GO terms, their P-value compared to its randomly obtained probability and their significance based on published literature. The validation demonstrates that our method achieves good performance. Additionally, fifteen candidate genes including *sev*, *tsl*, *hb* and many other genes were discovered. We show that these genes are involved in the MAPK signaling pathway and suggest their use for further experiment validation.

This paper has been structured as follows. Section 2 describes the materials and the methodology of our proposed algorithm. Firstly, the FGN is constructed by integrating HCS and heterogeneous genomic data using a linear SVM classifier. Then, we extract the MAPK signaling pathway from the FGN using an improved Integer Linear Programming (IILP) algorithm. Section 3 shows the evaluation method and the result of our proposed algorithm. Finally, the conclusion is drawn in Section 4.

2   Materials and Methodology

This section describes the datasets and algorithms used in more detail. We start with a discussion of the high-content screening (HCS) and the statistical methods used to prepare them for network integration. We also discuss the genomic and proteomic data used. Then we focus on the algorithms that we used to construct functional gene networks (FGN). Finally, we describe the algorithm used for signal pathway extraction from the obtained FGN.

2.1   High-Content Screening Data

The workflow that lead to the HCS data used here consist of several steps. At first a wetlab approach treats cells with specific dsRNA's to omit the function of those genes studied here. These cells are imaged and an automated computational approach is used to segment the cells from the background. The segmented cells are then classified and labeled into the following five groups Normal (N), Long Punctuate Actin (LPA), Cell Cycle Arrest (CCA), Rho family RNAi (Rho1) and rl-tear drop (T). In the final step a number of phenotype specific statistical properties are extracted as image descriptors. We want to look at this pipeline in more detail now.

2.1.1   Cell Culture and Image Acquisition

The use of dsRNA allows targeting and inhibiting the activity of specific genes and their later protein products. Using an image based approach the role of individual genes in the regulation of cell morphology can be systematically studied. Here we use Kc167 cells that were bathed in the presence of dsRNAs, targeting all known Drosophila protein kinases and phosphatases in 384-well plates (a detailed protocol is given in [8]). Following a five-day incubation period, the cells are fixed and stained with reagents in order to visualize the nuclear DNA (blue), polymerized F-actin (green), and α-tubulin (red). For each well, sixteen images from each of the three channels (blue, green and red) were acquired in automated fashion using an Evotec spinning-disk confocal.

2.1.2   Cell Image Segmentation

In order to analyze the morphology of all cells it is necessary to first delineate the boundaries of each individual cell. We utilize a two-step segmentation procedure first extracting the nuclei from the DNA channel and then the cell bodies from F-actin and α-tubulin channels [6]. The centers of the cell nuclei are first detected by a gradient vector field (GVF) based approach [9]. Using this information as seed points, the nuclei are fully segmented using a marker-controlled watershed algorithm. The steps are sufficient to obtain detailed shape and boundary information of the nuclei and cell bodies. We illustrated them in Fig. 1 (a) and (b).

2.1.3   Feature Extraction from the Segmented Cell Images

The identification of cellular phenotypes is basically a classification task depending on the right and rich set of descriptive image features. This is a standard problem and critical for most pattern recognition problems. In this study we capture the geometric and appearance properties using 211 morphology features belonging to five different image categories following [6]. The selected features include a total of 85 wavelet features, 10 geometric region features, 48 Zernike moments, 14 Haralick texture features [8] and a total of 54 shape descriptor features [6] .

A feature selection procedure is necessary to de-noise the dataset and to describe it in the most informative way. As the datasets and phenotype models are being updated adaptively, an unsupervised feature selection, not relying on phenotype labels, is used to supply a stable feature subset. It is based on iterative feature elimination using the k-nearest neighbors approach following [10]. Finally we selected an informative subset of twelve image features to quantify the segmented cells.

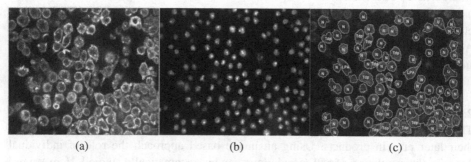

(a)   (b)   (c)

Fig. 1. Fluorescence images of Drosophila KC167 cells with three channels. The F-actin channel (a), the nuclear DNA (b) and the cellular phenotype classification results (c). In (c), the N, LPA, Tdp and U markers denote the phenotypes normal, long punctuate actin, teardrop and unclassified respectively.

2.1.4   Statistical Properties of Cells in Context of the Phenotypes

In order to generate statistical parameters able to adequately describe the cellular phenotypes, we extracted various types of parameters from the fluorescence labeled cells. In each phenotype group we use the mean of the cell number, area and perimeter. Additionally, the average cell number, area and perimeters corresponding to a single treatment are used. A total of 18 image descriptors were thus selected to represent the statistical properties of the cells in a HCS image. If two image descriptors for different gene knockdowns are similar we assume that they have a similar biology function. Furthermore, genes that act in the same SP are expected to result in similar cell morphology and thus will also have similar image descriptor patterns [7].

The HCS data is scored by Pearson correlation coefficients (PCC) calculated from the image descriptors between each gene pair. Therefore, 18 image descriptor values are used for a pair of genes and a PCC is always in the range [-1, 1]. A positive value means that the image descriptor of the two genes has a similar phenotype and a negative value indicates that opposite. Therefore we only keep PCC values greater than zero.

2.2   The Genomic and Proteomic Datasets

In the study we construct a high-coverage and high-quality FGN by integrating HCS with a number of genomic and proteomic datasets. These data sets, on the genomic or proteomic side, include PPI, gene expression microarray, gene co-occurrence, genomic neighborhoods, gene fusion, text mining and gene homology data.

PPI and gene expression profiles are the most widely used data source for biological network construction [3, 11]. PPI data can be modeled as a connectivity graph in which the nodes represent proteins and the edges represent physical interactions. In order to make use of gene expression profiles the PCC is calculated and used as a measure for the functional similarity of a pair of genes. In this work we compute PCCs for gene pairs in the microarray sets whenever a known PPI connection exists for their protein products. The resulting value is then used as individual score for these gene pairs. This set includes 6772 proteins with 19372 interactions. On contrast, gene expression profiles do not describe direct physical interaction but an inference on similar function. We integrate data from four microarray experiments with 73 conditions and 13046 genes obtained from the GEO database [12].

The genomic association between two genes means that they have similar function. As an example, a set of genes required for the same function often tends to be recovered in similar species, often localizes in close proximity on the genome or tend to be involved in gene-fusion events. Therefore we use gene co-occurrence (i.e. phylogenetic profiles), genomic neighborhood, gene fusion, text mining and gene homology data to grow the FGN. All of these data sets were obtained from the STRING database. Each of the prediction yields an individual score in the range [0, 1], which reflect the confidence that an interaction is true. In our work we directly use the individual score provided by STRING [13].

2.3   Functional Gene Network Construction

The power of a functional gene network rises and falls with the number and quality of datasets that can be integrated. However, most approaches in previous literature focus on the integration of PPI and gene expression profiles only [2, 14-17]. On contrast we propose to use high quality phenotype data from HCS, physical interactions from PPI obtained from BIND [13], inference information from microarrays obtained from GEO [12] and a list of genomic data obtained from the STRING database [18].

These data sources are integrated and combined scores for each functional association have to be calculated. In the study, we utilize a linear SVM to combine the scores of a total of eight individual data sources. The combined scores reflect the level of confidence in each gene pair interaction. Specifically, we use an unbounded linear SVM classifier with a real output value. Based on all individual datasets the input to the SVM consists of nearly 2862 gene interaction pairs. Of those, 580 gene pairs exist in at least one KEGG pathway and we select them as our gold standard for true positives (GSP). We defined our true negative gold standard (GSN) as the collection of pairs that are annotated in the KEGG but do not occur in the same pathway. In fact, this number is very large and to balance it with the GSP we selected only 580 gene pairs as GSN. For the GSP and GSN gene pairs, the final outputs are set to one and zero respectively, reflecting the probability of functional similarity between the gene pairs in GSP and GSN. The GSP and GSN gene sets are utilized to train the linear SVM and the experiment results show that about 87% of 10-fold cross validation accuracy has been achieved. The trained SVM is used to assign real valued scores to non-GSP and non-GSN gene pairs. These scores are further normalized into the range [0, 1]. Using this method we constructed a more accurate and higher coverage FGN, which we used to extract SP from in the next step.

3   Experiments and Results

We described above how HCS experiments can be utilized to infer gene function and in the construction of a FGN. In this section we discuss our results and compare our approach to a FGN constructed without image data.

3.1   Extracting Drosophila MAPK Signaling Pathways

We use the *Drosophila Melanogaster* MAPK signaling pathway, deposited in the KEGG database as the golden standard to validate our method. In the experiment we investigated the MAPKKK signaling pathway.

The MAPKKK pathway starts its signaling initiation at the membrane protein *Torso* and ends with the activation of the two transcription factors *Tll* and *Hkb* [19, 20]. We extend the integer linear programming algorithm (ILP) proposed in [3] by a weighting parameter α to our pathway detection problem. It is formulated as an optimization task that seeks to find a path starting from a membrane protein and ending at transcription factor. Given the starting membrane proteins, the ending TFs and a FGN, the IILP model is described as

$$\arg\min_{\{x_i, y_{ij}\}} (-\sum_{i=1}^{N}\sum_{j=1}^{N} W_{ij}^{\alpha} y_{ij} + \lambda \sum_{i=1}^{N}\sum_{j=1}^{N} y_{ij}) \tag{1}$$

subject to:

$$y_{ij} \le x_i, \quad y_{ij} \le x_j$$

$\sum_{j=1}^{N} y_{ij} \ge 1$,   if x_i is either a starting or ending node in SP;

$\sum_{j=1}^{N} y_{ij} \ge 2x_i$,   if x_i is an ordinary node in SP;

$x_i = 1$,   if x_i is a known node in SP;

$x_i \in \{0,1\}, i = 1, 2, ..., N$, $y_{ij} \in \{0,1\}, i, j = 1, 2, ..., N$

where x_i denotes whether the protein i is a component of the SP, y_{ij} denotes whether the edge $E(i, j)$ is a part of SP, W_{ij} is the weight of edge $E(i, j)$, λ is a positive constant parameter that is used to control the balance between the edge weight score and the SP size and N is the total number of nodes. The constraint $\sum_{j=1}^{N} y_{ij} \geq 2x_i$ is used to ensure that x_i has at least two linking edges if it is a component of the SP and the constraint $\sum_{j=1}^{N} y_{ij} \geq 1$ means that each starting node or ending node has at least one link to or from another node. These two constraints ensure that the components in the subnet are well connected. The constraints $y_{ij} \leq x_i$ and $y_{ij} \leq x_j$ means that if and only if proteins i and j are selected as the components of SP, the functional link $E(i,j)$ should be considered. The condition $x_i = 1$ means that the protein i is involved in the SP.

Therefore we have to select two important scalar parameters λ and α. The parameter α is used to give small weight a lower impact on determining the final SP. The parameter λ is used to control the length of the extracted SP. In our computations we use equation (2) to determine these parameters.

$$D = \frac{\sum_{i,j}^{N_1} W_{ij}^{\alpha}}{N_1} \tag{2}$$

As a result we set λ to 0.83 and α to 0.90 corresponding to the largest value of D. In Fig. 2 we illustrated the SP obtained by this method. When we compare this pathway to the one given in KEGG, we can see that it covers its entire components. Additionally it shows components that do not exist in the corresponding KEGG pathway, which can be used to guide the future experimental pathway analysis.

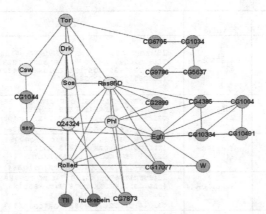

Fig. 2. The obtained MAPKKK SP extracted from the FGN. The green, red and yellow circles present the component nodes in the KEGG database. The green circle denotes the starting node and the red circle denotes the ending node of the pathway. The blue circles denote additional component nodes found by our method.

We employ two measurements, a P-value statistic and the GO term enrichment, to evaluate the function similarity of proteins assigned to the predicted SP. The p-value is based on the probability of a predicted SP to be found in a randomly weighted network. We compared the extracted SP with a random network and obtained a P-value that is less than 0.004, indicating a high significant. The GO enrichment (P_{GO})

reflects the likelihood of randomly observing a given enrichment level in the GO category that one finds particularly interesting. For the functional enrichment we choose five GO terms that are related to signal transduction, the results are shown in Table 1. These values show good agreement and similar biological function. These measurements suggest that our methodology is well suited to extract a signaling pathway from a FGN. We illustrate the extracted SP in Fig. 2. A comparison with the network stored in KEGG reveals that all components of the MAPKKK SP were successfully detected. Moreover, several other components are recovered, including the proteins CG1044 (*dos*), *sev*, CG6705(*tsl*), CG1034(*bcd*), CG9786(*hb*), CG5637(*nos*). In brief, the protein CG1044 is involved in the anti-apoptotic PI3K/Akt and the mitogenic Ras/MAPK pathways[21], sev regulates signal transduction events in the Ras/MAPK cascade [22], CG1034 is involved in the Torso signal transduction pathway[23], CG9786 is one of representative members of the first three levels of a signaling hierarchy which determines the segmented body plan [24], CG5637 is involved in the nitric oxide SP in Drosophila Melanogaster Malpighian tubules [25]. Other components detected by our method also relate to signal transduction. This suggests that our model is more complete that the KEEG pathway and potential candidates might be used for further experimental validation.

Table 1. The PGO value of functional enrichment for the MAPKKK pathway in Drosophila found by IILP based on the FGN. N1 denotes the number of annotated proteins to a certain Go term in the obtained signaling network; N2 denotes the number of annotated proteins to a certain Go term in the whole fruitfly genome.

GO term	P_{GO} value	N1	N2
transmembrane receptor protein tyrosine kinase signaling pathway	6.8790E-37	20 (EGFR|CG6705|CG17077|CSW|CG9786|TLL|CG1004|DRK|CG7873|RAS85D|CG2899|SEV|CG10491|SOS|PHL|CG1044|Q24324|CG4385|TOR|CG10334)	106
cell surface receptor linked signal transduction	2.9360E-23	20(EGFR|CG6705|CG17077|CSW|CG9786|TLL|CG1004|DRK|CG7873|RAS85D|CG2899|SEV|CG10491|SOS|PHL|CG1044|Q24324|CG4385|TOR|CG10334)	475
enzyme linked receptor protein signaling pathway	2.4814E-34	20(EGFR|CG6705|CG17077|CSW|CG9786|TLL|CG1004|DRK|CG7873|RAS85D|CG2899|SEV|CG10491|SOS|PHL|CG1044|Q24324|CG4385|TOR|CG10334)	139
Signal transduction	4.1611E-20	21(EGFR|CG6705|CG17077|CSW|CG9786|TLL|CG1004|DRK|CG7873|RAS85D|W|CG2899|SEV|CG10491|SOS|PHL|CG1044|Q24324|CG4385|TOR|CG10334)	842
cell communication	2.8712E-18	21(EGFR|CG6705|CG17077|CSW|CG9786|TLL|CG1004|DRK|CG7873|RAS85D|W|CG2899|SEV|CG10491|SOS|PHL|CG1044|Q24324|CG4385|TOR|CG10334)	1030

4   Discussion and Conclusions

Signal transduction pathways are perceived to be so central to biological processes that it is becoming an increasingly important research area in both fundamental and clinical perspective. Also with recent advances in fluorescence microscopy imaging techniques and technologies of gene knock down by RNA interference (RNAi), high-content screening (HCS) has emerged as a powerful approach to systematically identify all parts of complex biological processes. In this study, we first integrate high-content screening and other publicly available multi-modality genomic data together to construct a high-coverage and high-quality functional gene network using advanced machine learning method. Then we propose a novel computational model to discover MAPK signaling pathways in Drosophila Melanogaster based on improved integrated linear programming algorithm. The signaling pathways discovered are the optimal sub-graph with maximum weight value and minimum edges in the constructed functional gene network. The validation demonstrates that our method can discover signaling pathways which consist of major components of the MAPK pathway deposited in KEGG database. Several additional gene candidates were found by our method, which makes our model more complete. These potential candidates associated with the MAPK signaling pathway can be used for further experiment validation.

Acknowledgments. This work is supported in part by the National Science Foundation of China, under Grants 61102119, 61170077, 71001072, 61272339, 61272333, 61001185, in part by the China Postdoctoral Science Foundation, under Grant 2012M520929.

References

1. Baudot, J.B., Angelelli, G.A., et al.: Defining a Modular Signalling network from the fly interactome. BMC Syst. Biol. 2(45) (2008)
2. You, Z.H., Yin, Z., Han, K., Huang, D.S., Zhou, X.B.: A semi-supervised learning approach to predict synthetic genetic interactions by combining functional and topological properties of functional gene network. BMC Bioinformatics 11(343) (2010)
3. Zhao, X.M., Wang, R.S., Chen, L.: Uncovering signal transduction networks from high-throughput data by integer linear programming. Nucleic Acids Res. 36(9) (2008)
4. Scott, J., Ideker, T., Karp, R.M.: Efficient algorithms for detecting signaling pathways in protein interaction networks. J. Comput. Biol. 13(2), 133–144 (2006)
5. Aoki, K.F., Kanehisa, M.: Using the KEGG database resource. Curr Protoc Bioinformatics, ch.1, pp. 1– 12 (2005)
6. Li, F., Zhou, X.B., Ma, J.: An automated feedback system with the hybrid model of scoring and classification for solving over-segmentation problems in RNAi high content screening. Journal of Microscopy-Oxford 226(2), 121–132 (2007)
7. Bakal, C., Aach, J., Church, G.: Quantitative morphological signatures define local signaling networks regulating cell morphology. Science 316(5832), 1753–1756 (2007)
8. Yin, Z., Zhou, X.B., Bakal, C., et al.: Using iterative cluster merging with improved gap statistics to perform online phenotype discovery in the context of high-throughput RNAi screens. BMC Bioinformatics 9(264) (2008)
9. Perrimon, N., Mathey-Prevot, B.: Applications of high-throughput RNA interference screens to problems in cell and developmental biology. Genetics 175(1), 7–16 (2007)

10. Lee, I., Lehner, B., Crombie, C.: A single gene network accurately predicts phenotypic effects of gene perturbation in Caenorhabditis elegans. Nature Genetics 40(2), 181–188 (2008)
11. You, Z.H., Lei, Y.K., Huang, D.S., Zhou, X.B.: Using manifold embedding for assessing and predicting protein interactions from high-throughput experimental data. Bioinformatics 26(21), 2744–2751 (2010)
12. Edgar, R., Domrachev, M., Lash, A.E.: Gene Expression Omnibus: NCBI gene expression and hybridization array data repository. Nucleic Acids Res. 30(1), 207–210 (2002)
13. Maraziotis, I.A., Dimitrakopoulou, K., Bezerianos, A.: Growing functional modules from a seed protein via integration of protein interaction and gene expression data. BMC Bioinformatics 8, 408 (2007)
14. Mering, C., Jensen, L.J., Kuhn, M.: STRING 7 - recent developments in the integration and prediction of protein interactions. Nucleic Acids Research 35, D358–D362 (2007)
15. You, Z.H., Lei, Y.K., Zhu, L., Xia, J.F., Wang, B.: Prediction of protein-protein interactions from amino acid sequences with ensemble extreme learning machines and principal component analysis. BMC Bioinformatics 14(S10) (2013)
16. Lei, Y.K., You, Z.H., Ji, Z., Zhu, L., Huang, D.S.: Assessing and predicting protein interactions by combining manifold embedding with multiple information integration. BMC Bioinformatics 13(S3) (2012)
17. Zheng, C.H., Huang, D.S., Zhang, L., Kong, X.Z.: Tumor clustering using non-negative matrix factorization with gene selection. IEEE Transactions on Information Technology in Biomedicine 13(4), 599–607 (2009)
18. Bader, G.D., Betel, D., Hogue, C.W.: BIND: the Biomolecular Interaction Network Database. Nucleic Acids Res. 31(1), 248–250 (2003)
19. Igaki, T., Kanda, H., Yamamoto-Goto, Y.: Eiger, a TNF superfamily ligand that triggers the Drosophila JNK pathway. EMBO J. 21(12), 3009–3018 (2002)
20. Lim, Y.M., Nishizawa, K., Nishi, Y., et al.: Genetic analysis of rolled, which encodes a Drosophila mitogen-activated protein kinase. Genetics 153(2), 763–771 (1999)
21. Maus, M., Medgyesi, D., Kovesdi, D.: Grb2 associated binder 2 couples B-cell receptor to cell survival. Cell Signal 21(2), 220–227 (2009)
22. Sawamoto, K., Okabe, M., Tanimura, T.: The Drosophila secreted protein Argos regulates signal transduction in the Ras/MAPK pathway. Dev. Biol. 178(1), 13–22 (1996)
23. Janody, F., Sturny, R., Catala, F.: Phosphorylation of bicoid on MAP-kinase sites: contribution to its interaction with the torso pathway. Development 127(2), 279–289 (2000)
24. Spirov, A.V., Holloway, D.M.: Making the body plan: precision in the genetic hierarchy of Drosophila embryo segmentation. Silico Biol. 3(1-2), 89–100 (2003)
25. Davies, S.A., Stewart, E.J., Huesmann, G.R.: Neuropeptide stimulation of the nitric oxide signaling pathway in Drosophila melanogaster Malpighian tubules. Am J. Physiol. 273(2 Pt 2), R823–R827 (1997)
26. Linghu, B., Snitkin, E.S., Holloway, D.T.: High-precision high-coverage functional inference from integrated data sources. BMC Bioinformatics 9(119) (2008)

A Preliminary Study of Memory Functions in Unaffected First-Degree Relatives of Schizophrenia

Xiao-Yan Cao[1,2], Zhi Li[1], and Raymond C.K. Chan[1,*]

[1] Neuropsychology and Applied Cognitive Neuroscience Laboratory, Key Laboratory of Mental Health, Institute of Psychology, Chinese Academy of Sciences, Beijing, China
{caoxy,rckchan}@psych.ac.cn, moonlightforever@126.com
[2] University of Chinese Academy of Sciences, Beijing, China

Abstract. Schizophrenia is a neuropsychiatric disorder with etiologies caused by both genetic and environmental factors. However, very few studies have been done to examine the differential pattern of working memory dysfunction in individuals at risk for schizophrenia. The current study aimed to examine the different modalities of working memory performances in the first-degree relatives of patients with schizophrenia. Results showed that unaffected first-degree relatives characterized by high but not low schizotypal traits demonstrated significantly poorer performances in the verbal 2-back tasks, the immediate and delayed recall of logical memory compared to healthy controls. These preliminary findings suggest memory function impairment was more closely associated with schizotypal traits in unaffected first-degree relatives of schizophrenia patients.

Keywords: unaffected first-degree relatives, schizophrenia, memory, working memory.

1 Introduction

Schizophrenia is a neuropsychiatric disorder with etiologies caused by both genetic and environmental factors [1-4]. It is a spectrum of disorders covering both patients with psychotic symptoms fulfilling the diagnostic criteria for schizophrenia and those with mentally at-risk for psychoses [5, 6]. Substantial evidence has suggested that these at-risk individuals also demonstrate similar impairments with their psychotic probands, including attention, memory and executive functions [7-14].

Working memory has been considered to be one of the core features of cognitive impairments in schizophrenia [15-23]. This kind of impairment has been demonstrated in individuals with prodromal symptoms [24-26] and unaffected siblings of patients with schizophrenia [27-29]. Recent studies also suggest working memory is a potential endophenotype for schizophrenia [30-34].

Theoretically, Baddeley and Hitch (1974) have put forward a multi-componental concept of working memory comprising the central executive system, and two slave

---

* Corresponding author.

D.-S. Huang et al. (Eds.): ICIC 2013, LNAI 7996, pp. 11–19, 2013.

subsystems subserving for two independent route, namely the visuo-spatial sketchpad, and the phonological loop. Empirical findings from neuroimaging also showed that verbal task mainly evoked activation of left hemisphere [35-38] while spatial materials activated the right hemisphere of right-handers [39-43].

Recent meta-analysis of working memory in schizophrenia suggested that there may be differential effects of impairment of different modalities of working memory in this clinical group, with larger effect size demonstrated in spatial modalities than verbal modalities [18]. However, very few studies have been done to examine the differential pattern of working memory dysfunction in individuals at risk for schizophrenia. The current study aimed to examine the different modalities of working memory performances in the unaffected first-degree relatives of patients with schizophrenia. Given the similarity of neuropsychological dysfunctions demonstrated in biological relatives of patients with schizophrenia, it was hypothesized that the unaffected first-degree relatives of schizophrenia would demonstrate similar deficits in different components of working memory as compared to healthy controls.

2   Materials and Methods

2.1   Participants

Thirty-one unaffected first-degree relatives (including parents, siblings or off-springs) of patients with schizophrenia were recruited from the Mental Health Center, Shantou University, Shantou, Beijing Anding Hospital, Beijing Hui-long-guan Hospital, and the Institute of Mental Health of Peking University. All the relatives were interviewed by experienced psychiatrists to ascertain they did not suffer from any psychiatric illness, and had no history of neurological disorders and substance abuse. Thirty-one healthy volunteers were also recruited from the community. A semi-structured interview was conducted by a trained research assistant to ascertain that the volunteers had no family history of psychiatric and neurological disorders. All the participants were administered the Schizotypal Personality Questionnaire (SPQ) [44,45] to reflect the tendency of schizotypal personality trait. IQ was estimated by the short-form of the Chinese version of the Wechsler Adult Intelligence Scale-Revised (WAIS-R) [46].

This current study was approved by the local ethical committees of the related hospitals stated above. Written consent was obtained from each participant before the administration of the test and questionnaires.

2.2   Tasks

The verbal 2-back [47] and the visuo-spatial 2-back [48] tasks were applied in the present study to capture the updating ability in visuo-spatial domain. The participants' correct response rate and correct response time were recorded. The Chinese version of the Letter–Number Span Test[49] was also applied to assess the participants' working memory function that the total items and the longest item were recorded. Moreover, the Logical Memory and Visual Reproduction subtests from the Chinese version [50] of the Wechsler Memory Scale-Revised [51] were also administered to all the participants.

2.3   Data Analysis

Chi-square and one-way analysis of variance (ANOVA) were used to examine the differences of demographics between relatives of schizophrenia and healthy controls. Then multivariate analysis of covariance （MANCOVA） controlling for age and education was used to examine the main effect of factor group and potential interaction of independent factors. Subsequent one-way ANOVA was conducted to further examine the exact differences observed between relatives and controls. To explore the possible impact of schizotypal personality trait upon working memory in the unaffected first-degree relatives of schizophrenia, they were further classified into two subgroups according to the the median split of the SPQ score. Cohen's d values (low,0.2~0.3; medium,0.5; high,0.8 and above) and partial Eta-squared (η_p^2, low, 0.01; medium, 0.06; high, 0.14 and above) were calculated to estimate effect size and the extent of differences found between groups.

3   Results

Table 1 summarizes the demographic information of the participants. There were no significant differences found between the first-degree relatives of schizophrenia patients and healthy controls in age, number of years of education, gender proportion, and IQ estimates. Results from MANCOVA (Table 2) controlling for age and education showed that relatives of schizophrenia patients performed significantly poorer than the healthy controls in the correct response rate (F $_{(1,61)}$=4.65, p=0.035, η_p^2=0.077) and reaction time (F $_{(1,61)}$=7.64, p=0.008, η_p^2=0.120) in the verbal 2-back task. Moreover, the first-degree relatives also showed significantly poorer performances in both the immediate (F $_{(1,61)}$=10.86, p=0.002, η_p^2=0.162) and delayed (F $_{(1,61)}$=8.75, p=0.005, η_p^2=0.135) scores of logical memory than healthy controls. A check on the Cohen's d also indicated a range of medium to large effect sizes between the two (0.4~0.8).

Table 1. Demographics of relatives of schizophrenia and healthy controls

	HC (N = 31)		REL (N=31)		χ^2/t	p
	Mean	SD	Mean	SD		
Gender (M/F)	20/11		11/18		3.175	0.075
Age	38.55	10.14	42.90	12.17	-1.53	0.131
Education	11.45	2.61	11.94	2.78	-0.71	0.482
IQ _estimate	104.90	13.06	104.97	12.99	-0.02	0.985

Note: F=females, M=males; HC= healthy controls, REL= relatives of schizophrenia

Table 2. Differences over memory function between relatives of schizophrenia and healthy controls

	HC (n=31)	REL (n=31)	$F_{(1,61)}$	P	η_p^2	Cohen's d
	Mean (SD)	Mean (SD)				
Verbal 2-back Task						
Correct Response Rate	0.44 (0.2)	0.36 (0.17)	4.65	0.035	0.077	0.414
Mean Reaction Time	587.64 (218.23)	775.6 (240.53)	7.64	0.008	0.12	-0.818
Spatial 2-back Task						
Correct Response Rate	0.56 (0.17)	0.56 (0.22)	0.06	0.809	0.001	0.018
Mean Reaction Time	1048.2 (226.07)	1051.32 (206.07)	0.02	0.887	0	-0.014
Letter-Number Span						
Longest span passed	5.65 (1.2)	5.38 (1.18)	1.86	0.178	0.032	0.224
Total Scores	13.74 (3.4)	12.76 (2.89)	0.79	0.378	0.014	0.312
Logical Memory						
Immediate recall	13.03 (3.95)	10.14 (3.86)	10.86	0.002	0.162	0.741
Delayed recall	10.74 (4.08)	8.24 (3.5)	8.75	0.005	0.135	0.658
Visual Memory						
Immediate recall	22.45 (3.12)	23.07 (1.73)	1.07	0.305	0.019	0.245
Delayed recall	21.87 (3.95)	22.45 (2.5)	0.33	0.566	0.006	0.175

Note: HC for healthy controls, REL for relatives of schizophrenia

Table 3 shows the demographic summary of the subdivision of the first-degree relatives into the high-SPQ group (n=15) and low-SPQ group (n=17) with a median split of for the total SPQ (score of 17).

Table 3. Demographics of healthy controls and two subtypes of relatives of schizophrenia according to SPQ scores

	HC (N = 31)	High-SPQ REL (N=14)	Low-SPQ REL (N=17)	χ^2/F	P
	Mean (SD)	Mean (SD)	Mean (SD)		
Gender (M/F)	20/11	6/8	7/10	3.183	0.204
Age	38.55 (10.14)	41.53 (13.78)	44.57 (10.15)	1.44	0.244
Education	11.45 (2.61)	12.00 (2.32)	11.86 (3.35)	0.26	0.775
IQ _estimate	104.90 (13.06)	108.59 (11.25)	100.57 (13.99)	1.5	0.231

Note: HC for healthy controls, REL for relatives of schizophrenia

Results from MANCOVA (Table 4) showed that there were significant differences found between the reaction time (F $_{(2,61)}$=9.83, p<0.0001, η_p^2=0.263) of verbal 2-back, immediate as well as delayed logical memory (F $_{(2,61)}$=5.52 and 4.48; p=0.007 and 0.016, η_p^2=0.167 and 0.14). Further paired comparisons by Bonferroni correction found that the high-SPQ relatives performed significantly poorer than healthy controls over these variables. It was also noted that there was a trend of significant among the three groups in the correct response of the verbal 2-back (F $_{(2,61)}$=2.84, p=0.08, η_p^2=0.088), with a relatively large effect size observe between the high-SPQ relatives and controls (Cohen's d 0.798).

Table 4. Differences over memory function among healthy controls and two subtypes of relatives of schizophrenia

	HC	REL		F $_{(2,61)}$	p	η_p^2	p_value (Cohen's d) for paired comparisons		
		High-SPQ	Low-SPQ				HC vs. High-SPQ	HC vs. Low-SPQ	Low-SPQ vs. High-SPQ
	(N=31)	(N=14)	(N=17)						
	Mean (SD)	Mean (SD)	Mean (SD)						
Verbal 2-back Task									
Correct Response Rate	0.44 (0.2)	0.3 (0.15)	0.4 (0.17)	2.64	0.08	0.088	0.545 0.095 (0.798)	(0.175)	1 (0.667)
Mean Reaction Time	587.64 (218.23)	916.42 (178.12)	616.6 (238.5)	9.83	<0.001	0.263	<0.001 (1.651)	1 (0.127)	0.006 (1.424)
Spatial 2-back Task									
Correct Response Rate	0.56 (0.17)	0.5 (0.21)	0.6 (0.23)	0.38	0.687	0.014	1 (0.324)	1 (0.161)	1 (0.432)
Mean Reaction Time	1048.2 (226.07)	986.86 (230.48)	1084.01 (194.44)	0.79	0.459	0.028	1 (0.269)	1 (0.17)	0.651 (0.456)
LNS task									
Longest span passed	5.65 (1.2)	4.93 (0.92)	5.75 (1.24)	1.36	0.265	0.047	0.38 (0.672)	1 (0.086) 0.927	0.514 (0.754)
Total Scores	13.74 (3.4)	12 (3.09)	13.31 (2.57)	0.93	0.399	0.033	0.712 (0.537)	(0.143)	1 (0.462)
Logical Memory									
Immediate recall	13.03 (3.95)	9.14 (4.44)	10.88 (3.1)	5.52	0.007	0.167	0.055 0.014 (0.926)	(0.608)	1 (0.453)
Delayed recall	10.74 (4.08)	7.29 (3.29)	8.88 (3.59)	4.48	0.016	0.14	0.029 (0.932)	0.106 (0.485)	1 (0.461)
Visual Memory									
Immediate recall	22.45 (3.12)	22.93 (1.59)	23.06 (1.91)	0.64	0.531	0.023	0.823 (0.193)	1 (0.236)	1 (0.076)
Delayed recall	21.87 (3.95)	21.93 (2.73)	22.44 (2.94)	0.19	0.828	0.007	1 (0.017)	1 (0.163)	1 (0.179)

Note: HC for healthy controls, REL for relatives of schizophrenia

4   Discussion

This study showed that there were significant differences found between the unaffected first-degree relatives and healthy controls in the verbal 2-back task, and the immediate and delayed logical memory. These findings were particularly demonstrated in the relatives characterized by high schizotypal trait. The findings are in general consistent with the existing literature concerning the working memory function in the unaffected first-degree relatives with schizophrenia [52-57]. Our current study did not show that the unaffected first-degree relatives of schizophrenia as a whole demonstrated significant spatial working memory deficits as compared to healthy controls. At a first glance, these findings seem to be inconsistent with the existing literature concerning the working memory deficits observed in unaffected first-degree relatives of schizophrenia [15]. However, we found that these kinds of deficits, especially the verbal working memory and semantic memory, were only demonstrated in the relatives associated with higher schizotypal traits. These findings highlight the importance of schizotypy contributing to the cognitive impairments in genetically at-risk individuals for schizophrenia.

The current study has a number of methodological limitations. First, we only recruited a relatively small sample size that might have limited the power of discriminating the true differences of memory functions found between the participants. Second, although we attempted to measure the different modalities of working memory function in our current study, we only adopted a narrow range of tests to capture the verbal and visuo-spatial modalities of working memory. These relatively simple behavioral tasks might not be sensitive enough to detect any differences demonstrated in at-risk individuals. Future study should recruit a larger sample size with a wider range of tests to cover different modalities of working and semantic memory functions in unaffected first-degree relatives. Target participants should extend to the patients with schizophrenia. Neuroimaging or electrophysiological paradigms may be more sensitive to detect such a subtle impairment in at-risk individuals for schizophrenia.

Acknowledgement. This work was supported a grant from the Key Laboratory of Mental Health, Institute of Psychology, Chinese Academy of Sciences. The authors would like to acknowledge the staff of the Mental Health Center, Shantou University, Beijing Anding Hospital, Beijing Hui-long-guan Hospital, and the Institute of Mental Health of Peking University for their recruitment of clinical cases.

References

1. Derks, E.M., Allardyce, J., Boks, M.P., Vermunt, J.K., Hijman, R., Ophoff, R.A., et al.: Kraepelin Was Right: A Latent Class Analysis of Symptom Dimensions in Patients and Controls. Schizophrenia Bulletin (2010)
2. McGrath, J.A., Avramopoulos, D., Lasseter, V.K., Wolyniec, P.S., Fallin, M.D., Liang, K.Y., et al.: Familiality of novel factorial dimensions of schizophrenia. Arch Gen Psychiatry 66(6), 591–600 (2009)
3. Tsuang, M.T., Stone, W., Faraone, S.V.: Genes, environment and schizophrenia. The British Journal of Psychiatry 178(40), 18–24 (2001)
4. Tsuang, M.T., Stone, W.S., Faraone, S.V.: Schizophrenia: a review of genetic studies. Harv. Rev. Psychiatry 7(4), 185–207 (1999)
5. Cadenhead, K.S., Perry, W., Shafer, K., Braff, D.L.: Cognitive functions in schizotypal personality disorder. Schizophr Res. 37(2), 123–132 (1999)

6. Tsuang, M.T., Faraone, S.V.: Genetic transmission of negative ad positive symptoms in the biological relatives of schizophrenics. In: Marneos, A., Tsuang, M.T., Andersen, N. (eds.) Positive vs Negative Schizophrenia, pp. 265–291. Springer, New York (1991)

7. Bora, E., Yucel, M., Pantelis, C.: Cognitive impairment in schizophrenia and affective psychoses: implications for DSM-V criteria and beyond. Schizophr Bull. 36(1), 36–42 (2010)

8. Eastvold, A.D., Heaton, R.K., Cadenhead, K.S.: Neurocognitive deficits in the (putative) prodrome and first episode of psychosis. Schizophr Res. 93(1-3), 266–277 (2007)

9. Heinrichs, R.W., Zakzanis, K.K.: Neurocognitive deficit in schizophrenia: a quantitative review of the evidence. Neuropsychology 12(3), 426–445 (1998)

10. Husted, J.A., Lim, S., Chow, E.W., Greenwood, C., Bassett, A.S.: Heritability of neurocognitive traits in familial schizophrenia. Am. J. Med. Genet B Neuropsychiatr Genet. 150B(6), 845–853 (2009)

11. Irani, F., Kalkstein, S., Moberg, E.A., Moberg, P.J.: Neuropsychological Performance in Older Patients With Schizophrenia: A Meta-Analysis of Cross-sectional and Longitudinal Studies. Schizophrenia Bulletin (2010)

12. Jahshan, C., Heaton, R.K., Golshan, S., Cadenhead, K.S.: Course of neurocognitive deficits in the prodrome and first episode of schizophrenia. Neuropsychology 24(1), 109–120 (2010)

13. Roitman, S.E., Mitropoulou, V., Keefe, R.S., Silverman, J.M., Serby, M., Harvey, P.D., et al.: Visuospatial working memory in schizotypal personality disorder patients. Schizophr Res. 41(3), 447–455 (2000)

14. Shamsi, S., Lau, A., Lencz, T., Burdick, K.E., DeRosse, P., Brenner, R., et al.: Cognitive and symptomatic predictors of functional disability in schizophrenia. Schizophr Res. 126(1-3), 257–264 (2011)

15. Forbes, N.F., Carrick, L.A., McIntosh, A.M., Lawrie, S.M.: Working memory in schizophrenia: a meta-analysis. Psychol. Med. 39(6), 889–905 (2009)

16. Giersch, A., van Assche, M., Huron, C., Luck, D.: Visuo-perceptual organization and working memory in patients with schizophrenia. Neuropsychologia 49(3), 435–443 (2011)

17. Goldman-Rakic, P.S.: Working memory dysfunction in schizophrenia. J. Neuropsychiatry Clin Neurosci. 6(4), 348–357 (1994)

18. Lee, J., Park, S.: Working memory impairments in schizophrenia: a meta-analysis. J. Abnorm Psychol. 114(4), 599–611 (2005)

19. Quee, P.J., Eling, P.A., van der Heijden, F.M., Hildebrandt, H.: Working memory in schizophrenia: a systematic study of specific modalities and processes. Psychiatry Res. 185(1-2), 54–59 (2011)

20. Silver, H., Feldman, P., Bilker, W., Gur, R.C.: Working memory deficit as a core neuropsychological dysfunction in schizophrenia. Am. J. Psychiatry 160(10), 1809–1816 (2003)

21. White, T., Schmidt, M., Karatekin, C.: Verbal and visuospatial working memory development and deficits in children and adolescents with schizophrenia. Early Interv. Psychiatry 4(4), 305–313 (2010)

22. White, T., Schmidt, M., Kim, D.I., Calhoun, V.D.: Disrupted functional brain connectivity during verbal working memory in children and adolescents with schizophrenia. Cereb Cortex 21(3), 510–518 (2011)

23. Zanello, A., Curtis, L., Badan Ba, M., Merlo, M.C.G.: Working memory impairments in first-episode psychosis and chronic schizophrenia. Psychiatry Res. 165(1-2), 10–18 (2009)

24. Hambrecht, M., Lammertink, M., Klosterkotter, J., Matuschek, E., Pukrop, R.: Subjective and objective neuropsychological abnormalities in a psychosis prodrome clinic. Br. J. Psychiatry Suppl. 43, S30–S37 (2002)

25. Lencz, T., Smith, C.W., McLaughlin, D., Auther, A., Nakayama, E., Hovey, L., et al.: Generalized and specific neurocognitive deficits in prodromal schizophrenia. Biol. Psychiatry 59(9), 863–871 (2006)
26. Wood, S.J., Pantelis, C., Proffitt, T., Phillips, L.J., Stuart, G.W., Buchanan, J.A., et al.: Spatial working memory ability is a marker of risk-for-psychosis. Psychol. Med. 33(7), 1239–1247 (2003)
27. Bachman, P., Kim, J., Yee, C.M., Therman, S., Manninen, M., Lonnqvist, J., et al.: Efficiency of working memory encoding in twins discordant for schizophrenia. Psychiatry Res. 174(2), 97–104 (2009)
28. Barrantes-Vidal, N., Aguilera, M., Campanera, S., Fatjó-Vilas, M., Guitart, M., Miret, S., et al.: Working memory in siblings of schizophrenia patients. Schizophrenia Research 95(1-3), 70–75 (2007)
29. Delawalla, Z., Csernansky, J.G., Barch, D.M.: Prefrontal cortex function in nonpsychotic siblings of individuals with schizophrenia. Biol. Psychiatry 63(5), 490–497 (2008)
30. Glahn, D.C., Therman, S., Manninen, M., Huttunen, M., Kaprio, J., Lonnqvist, J., et al.: Spatial working memory as an endophenotype for schizophrenia. Biol. Psychiatry 53(7), 624–626 (2003)
31. Greenwood, T.A., Lazzeroni, L.C., Murray, S.S., Cadenhead, K.S., Calkins, M.E., Dobie, D.J., et al.: Analysis of 94 Candidate Genes and 12 Endophenotypes for Schizophrenia From the Consortium on the Genetics of Schizophrenia. Am. J. Psychiatry 10050723 (2011) appi.ajp.2011.10050723
32. Gur, R.E., Calkins, M.E., Gur, R.C., Horan, W.P., Nuechterlein, K.H., Seidman, L.J., et al.: The Consortium on the Genetics of Schizophrenia: neurocognitive endophenotypes. Schizophr Bull 33(1), 49–68 (2007)
33. Gur, R.E., Nimgaonkar, V.L., Almasy, L., Calkins, M.E., Ragland, J.D., Pogue-Geile, M.F., et al.: Neurocognitive endophenotypes in a multiplex multigenerational family study of schizophrenia. Am. J. Psychiatry 164(5), 813–819 (2007)
34. Hill, S.K., Harris, M.S., Herbener, E.S., Pavuluri, M., Sweeney, J.A.: Neurocognitive allied phenotypes for schizophrenia and bipolar disorder. Schizophr Bull 34(4), 743–759 (2008)
35. Fletcher, P.C., Henson, R.N.: Frontal lobes and human memory: insights from functional neuroimaging. Brain 124(Pt 5), 849–881 (2001)
36. Norman, J.: Two visual systems and two theories of perception: An attempt to reconcile the constructivist and ecological approaches. Behav Brain Sci. 25(1), 73–96 (2002) discussion -144
37. Öztekin, I., Davachi, L., McElree, B.: Are representations in working memory distinct from representations in long-term memory? Neural evidence in support of a single store. Psychological Science 21(8), 1123–1133 (2010)
38. Smith, E.E., Jonides, J.: Storage and executive processes in the frontal lobes. Science 283(5408), 1657–1661 (1999)
39. D'Esposito, M., Postle, B.R., Rypma, B.: Prefrontal cortical contributions to working memory: evidence from event-related fMRI studies. Exp. Brain Res. 133(1), 3–11 (2000)
40. Jonides, J., Smith, E.E., Koeppe, R.A., Awh, E., Minoshima, S., Mintun, M.A.: Spatial working memory in humans as revealed by PET. Nature 363(6430), 623–625 (1993)
41. Paulesu, E., Frith, C.D., Frackowiak, R.S.: The neural correlates of the verbal component of working memory. Nature 362(6418), 342–345 (1993)
42. Sala, J.B., Rama, P., Courtney, S.M.: Functional topography of a distributed neural system for spatial and nonspatial information maintenance in working memory. Neuropsychologia 41(3), 341–356 (2003)

43. Smith, E.E., Jonides, J., Koeppe, R.A.: Dissociating verbal and spatial working memory using PET. Cereb Cortex 6(1), 11–20 (1996)
44. Chen, W.J., Hsiao, C.K., Lin, C.C.: Schizotypy in community samples: the three-factor structure and correlation with sustained attention. J. Abnorm Psychol. 106(4), 649–654 (1997)
45. Raine, A.: The SPQ: a scale for the assessment of schizotypal personality based on DSM-III-R criteria. Schizophr Bull 17(4), 555–564 (1991)
46. Gong, Y.X.: Manual of Wechsler Adult Intelligence Scale—Chinese Version. Chinese Map Press, Changsha (1992)
47. Callicott, J.H., Ramsey, N.F., Tallent, K., Bertolino, A., Knable, M.B., Coppola, R., et al.: Functional magnetic resonance imaging brain mapping in psychiatry: methodological issues illustrated in a study of working memory in schizophrenia. Neuropsychopharmacology 18(3), 186–196 (1998)
48. Aronen, E.T., Vuontela, V., Steenari, M.R., Salmi, J., Carlson, S.: Working memory, psychiatric symptoms, and academic performance at school. Neurobiol Learn Mem. 83(1), 33–42 (2005)
49. Chan, R.C., Wang, Y., Deng, Y., Zhang, Y., Yiao, X., Zhang, C.: The development of a Chinese equivalence version of letter-number span test. Clin Neuropsychol. 22(1), 112–121 (2008)
50. Gong, Y.X., Jiang, D.W., Deng, J.L., Dai, Z.S.: Manual of Wechsler Memory Scale-Chinese Vesion. Hunan Medical College Press, Changsha (1989)
51. Wechsler, D.: Wechsler Memory Scale Manual. Psychological Corp, New York (1987)
52. Conklin, H.M., Curtis, C.E., Calkins, M.E., Iacono, W.G.: Working memory functioning in schizophrenia patients and their first-degree relatives: cognitive functioning shedding light on etiology. Neuropsychologia 43(6), 930–942 (2005)
53. Egeland, J., Sundet, K., Rund, B.R., Asbjornsen, A., Hugdahl, K., Landro, N.I., et al.: Sensitivity and specificity of memory dysfunction in schizophrenia: a comparison with major depression. Journal of Clinical and Experimental Neuropsychology 25(1), 79–93 (2003)
54. MacDonald III, A.W., Thermenos, H.W., Barch, D.M., Seidman, L.J.: Imaging genetic liability to schizophrenia: systematic review of FMRI studies of patients' nonpsychotic relatives. Schizophr Bull 35(6), 1142–1162 (2009)
55. Choi, J.S., Park, J.Y., Jung, M.H., Jang, J.H., Kang, D.H., Jung, W.H., et al.: Phase-Specific Brain Change of Spatial Working Memory Processing in Genetic and Ultra-High Risk Groups of Schizophrenia. Schizophr Bull (2011)
56. O'Connor, M., Harris, J.M., McIntosh, A.M., Owens, D.G., Lawrie, S.M., Johnstone, E.C.: Specific cognitive deficits in a group at genetic high risk of schizophrenia. Psychol Med. 39(10), 1649–1655 (2009)
57. Park, S., Holzman, P.S., Goldman-Rakic, P.S.: Spatial working memory deficits in the relatives of schizophrenic patients. Arch Gen. Psychiatry 52(10), 821–828 (1995)

Classifying Aging Genes into DNA Repair or Non-DNA Repair-Related Categories

Yaping Fang[1], Xinkun Wang[2,3], Elias K. Michaelis[2,3], and Jianwen Fang[1,*]

[1] Applied Bioinformatics Laboratory, The University of Kansas, Lawrence, KS 66047, USA
{ypfang,jwfang}@ku.edu
[2] Department of Pharmacology and Toxicology,
[3] Higuchi Biosciences Center, The University of Kansas, Lawrence, KS 66047, USA
{xwang,emichaelis}@ku.edu

Abstract. The elderly population in almost every country is growing faster than ever before. However, our knowledge about the aging process is still limited despite decades of studies on this topic. In this report, we focus on the gradual accumulation of DNA damage in cells, which is a key aspect of the aging process and one that underlies age-dependent functional decline in cells, tissues, and organs. To achieve the goal of discriminating DNA-repair from non-DNA-repair genes among currently known genes related to human aging, four machine learning methods were employed: Decision Trees, Naïve Bayes, Support Vector Machine, and Random Forest (RF). Among the four methods, the RF algorithm achieved a total accuracy (ACC) of 97.32% and an area under receiver operating characteristic (AUC) of 0.98. These estimates were based on 18 selected attributes, including 10 Gene Ontology and 8 Protein-Protein Interaction (PPI) attributes. A predictive model built with only 15 PPI attributes achieved performance levels of ACC= 96.56% and AUC=0.95. Systems biology analyses showed that the features of these attributes were related to cancer, genetic, developmental, and neurological disorders, as well as DNA replication/recombination/repair, cell cycle, cell death, and cell function maintenance. The results of this study indicate that genes indicative of aging may be successfully classified into DNA repair and non-DNA repair genes and such successful classification may help identify pathways and biomarkers that are important to the aging process.

Keywords: Aging, DNA-repair, Random Forest, Classification, Feature selection.

1 Introduction

The significant gains in life expectancy experienced by human populations over the past decades have led to a rapid increase in the aged in many countries. As a result, aging is increasingly becoming a major burden to family and society in these countries [1, 2] because the aging process is related to increases in cardiovascular disease, type 2 diabetes, pulmonary disease [1], cancer [2], and many other diseases.

---

* Corresponding author.

D.-S. Huang et al. (Eds.): ICIC 2013, LNAI 7996, pp. 20–29, 2013.
© Springer-Verlag Berlin Heidelberg 2013

Despite intensive research in the past [3], the specific mechanisms of aging are far from being fully understood. One of the prevailing theories is that aging is a consequence of accumulation of unrepaired DNA damage [4]. Although DNA damage occurs frequently in our cells due to external (e.g., ionizing radiation and genotoxic drugs) and internal factors (e.g., reactive oxygen species), there is an effective DNA-repair system that has evolved so that it compensates for the damage to DNA [5]. Yet, this repair system may not be able to repair all DNA damages. The accumulation of non-repaired DNA damages during the aging process leads to an up-regulation in the expression of DNA repair genes in order to combat the detrimental consequences of accumulating DNA lesions [6]. Therefore, genes associated with the repair of DNA damage play important roles in the aging process. Of course, among the long list of genes that have been identified to be aging-related [7-9], only a small portion is estimated to be related to DNA repair. The task of identifying DNA repair genes from a long list of genes that are related to many aspects of the aging process is not trivial. To our knowledge there is currently no report that describes the use of bioinformatics techniques in order to identify or classify DNA repair genes among all other genes related to aging. A related report was published recently by Freitas and colleagues [7] on constructing models that can divide DNA repair genes into aging-related and non-aging-related categories using Decision Tree (DT) and Naïve Bayes (NB) algorithms. The biggest area under receiver operating characteristic (AUC) of there models were 0.826 for their NB model with 137 attributes [7].

In this study, we attempted to build models that would sub-classify aging-related genes into DNA repair and non-DNA-repair genes. We used the previously employed algorithms of J48, NB, and SVM [7, 10], plus the algorithm Random Forest (RF) [11-14]. RF also provides a convenient and often good approach to feature selection.

In the following sections, we first describe the procedure of data collection and feature generation. The most significant features that contribute to the predictive models were identified based on the rank order of the RF importance. A network was constructed based on PPI attributes and most of the selected attributes have important roles in both biology and disease.

2   Methods

2.1   Gene List Collection

A list of 261 human aging-related genes were retrieved from GenAge (Build 15, http://genomics.senescence.info/genes/) [9], among which 35 are annotated as DNA repair genes by Wood et al. [15, 16]. All 261 aging-related genes were grouped into two classes, one of which contains 226 non-DNA repair genes (i.e., negative samples) and the other includes 35 DNA repair related aging genes (positive samples).

2.2   GO Attributes

GO data (version 1.1.2393) were downloaded from the Gene Ontology Project site (http://www.geneontology.org/). The Biological Process (BP) annotations of all aging-related genes were extracted. Each BP term was treated as a feature of aging-related genes and labeled as 1 or 0 based on whether there is "is a" relationship between aging genes and this BP term. This procedure produced 2,189 binary attributes.

2.3   PPI Attributes

Two PPI datasets were used to generate the PPI attributes: the Human Protein Reference Database (HPRD, http://www.hprd.org/) [17] and the Biological General Repository for Interaction Datasets (BIOGRID, http://thebiogrid.org/) [18]. We filtered both datasets using the following criteria: (1) human proteins or genes; (2) for a PPI pair, at least one protein was the product of an aging-related gene; and, (3) gene interactions were experimentally determined. All PPI data sets filtered according to these criteria were merged to generate a list of PPIs that involved aging-related genes. Overall, there were 3,811 proteins that interacted with the aging-related genes. Each of these proteins was treated as a feature of aging-related genes which is labeled as 1 if an aging gene (or its product) interacts with it or 0 otherwise.

2.4   Data mining Algorithms

J48, NB, SVM, and RF algorithms were used in the present study. J48 is an open source Java implementation of the C4.5 algorithm for generating decision trees [19]. NB is a classification algorithm based on the Bayes' theorem of conditional probability with a prerequisite assumption that all attributes are independent of each other [7]. SVM uses a kernel function to perform nonlinear mapping by projecting descriptors onto a high-dimensional feature space. In the multi-dimensional space, the best hyper-plane is the one that can maximize the margin between different classes.

The RF algorithm is an ensemble approach that utilizes many independent decision trees to perform classification or regression [20]. In this study, the R implementation of the algorithm was used to rank all attributes (http://cran.r-project.org/web/packages/randomForest/index.html). The Gini index is used to measure the relative importance of each attribute [21]. Each RF model developed in the present study included 5,000 trees.

2.5   Performance Evaluation

A standard 5-fold cross-validation was used to assess the performance of different classifiers. The Receiver Operating Characteristic (ROC) curve, Matthew's correlation coefficient (MCC), and total accuracy (ACC) were used to evaluate the performance of different classifiers. The ROC curve is a graph of the true-positive rate (sensitivity) against the false-positive rate (1-specificity). The area under ROC curve (AUC) is a trade-off between sensitivity and specificity. An AUC of 1 represents a perfect prediction model while an AUC of 0.5 represents a random prediction.

3   Results

3.1   Analysis of Attributes

There were 6,000 attributes for all aging-related genes, including 2,189 GO terms and 3,811 PPI attributes. The occurrences of these attributes vary and some of them are quite low. Most GO attributes appear only once and the occurrences of most PPI attributes are fewer than 10. The number of GO attributes with 3 or more occurrences is 602 and the number of PPI attributes with 10 or more occurrences is 587.

3.2   Performance of Different Classification Algorithms

Several different cutoff values of occurrences were tested in filtering attributes that were then employed to build models based on the four algorithms. We selected cutoff values between 1 and 20 for GO attributes. And, for the PPI attributes, the cutoff values were set at between 1 and 200 with an interval of 10. The AUC's of these models that were calculated on the basis of different cutoff values are shown in Figure 1.

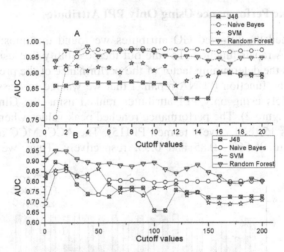

Fig. 1. The influence of cutoff values on the performance of different algorithms. Panel A is for GO term attributes and B for PPI attributes.

Figure 1 shows that RF is the most robust algorithm. The best performance was achieved at a cutoff value of 4 for NB and RF models. It can also be seen that the best performance was achieved at the cutoff value of 10. Thus, we combined the GO attributes at the cutoff value of 4 with the PPI attributes at the cutoff value of 10.

The above analysis indicated that some attributes may make little contribution to the performance of the models. Thus we used feature selection methods to select attributes important to DNA repair as it related aging. The J48 algorithm chose 6 attributes (JC6), including 4 GO and 2 PPI attributes. A standard 5-fold cross-validation based on RF was used to rank all attributes. The average values and standard deviation of the Gini importance were calculated. All attributes were ranked and the top 25 features (Top25) were selected, including 12 GO and 13 PPI features. The 13 PPI attributes are referred to in subsequent sections as the Combined13. Of the 25 attributes, only 4 GO and none of the PPI attributes were shared by both JC6 and Top25.

3.3   Performance of Selected Attributes

To evaluate the performance of the two feature selection algorithms, we built J48 and RF models using selected features (Table 1). Table 1 clearly shows that selected features can be used to classify aging-related genes into DNA-repair and non-DNA-repair ones. Overall, the RF algorithm outperformed the J48 algorithm.

Table 1. Performance of selected features using J48 and RF

Feature set	#Attributes	J48		RF	
		ACC	AUC	ACC	AUC
Top25	25	94.25	0.88	96.17	0.99
JC6	6	96.17	0.87	97.32	0.94

3.4   Algorithmic Performance Using Only PPI Attributes

When we examined the selected GO attributes we found that most of them were related to the DNA repair function. To rule out the possibility that using these cyclic GO attributes was the determining factor of the performance of the predictors, as they describe exactly the function ("DNA repair") that we wanted to classify, we built a series of RF models using only PPI attributes ranked using the Gini index that is available in RF (Figure 2). The performance reached peak values when the number of PPI attributes was 15 (collectively termed PPI15). The ACC, MCC and AUC of 15 PPI attributes were 96.56%, 0.85 and 0.95, respectively. Thus, we could classify

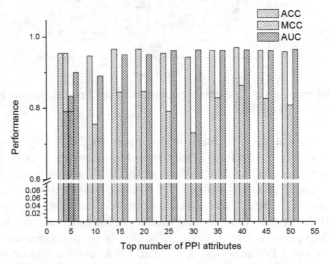

Fig. 2. Performance of the RF models using different numbers of PPI attributes ranked using the Gini importance

aging-related genes into either DNA-repair or non-DNA-repair genes based on as few as 15 PPI attributes. The name and sub-cellular localization of each of the corresponding proteins are listed in Table 2. It should be noted that most of the proteins shown in Table 2 are nuclear proteins, in accordance with their potential role as DNA-repair genes.

Table 2. Selected PPI attributes and their respective proteins and localizations

Attributes	Symbol	EntrezProtein Name	Location
472[b]	ATM	ataxia telangiectasia mutated	Nucleus
4361[b]	MRE11A	MRE11 meiotic recombination 11 homolog A (S. cerevisiae)	Nucleus
3308[b]	HSPA4	heat shock 70kDa protein 4	Cytoplasm
545[ab]	ATR	ataxia telangiectasia and Rad3 related	Nucleus
641[ab]	BLM	Bloom syndrome, RecQ helicase-like	Nucleus
2074[ab]	ERCC6	excision repair cross-complementing rodent repair deficiency, complementation group 6	Nucleus
9656[ab]	MDC1	mediator of DNA-damage checkpoint 1	Nucleus
4436[ab]	MSH2	mutS homolog 2, colon cancer, nonpolyposis type 1 (E. coli)	Nucleus
5111[ab]	PCNA	proliferating cell nuclear antigen	Nucleus
6117[ab]	RPA1	replication protein A1, 70kDa	Nucleus
6742[ab]	SSBP1	single-stranded DNA binding protein 1	Cytoplasm
7486[ab]	WRN	Werner syndrome, RecQ helicase-like	Nucleus
7515[ab]	XRCC1	X-ray repair complementing defective repair in Chinese hamster cells 1	Nucleus
2068[a]	ERCC2	excision repair cross-complementing rodent repair deficiency, complementation group 2	Nucleus
2189[a]	FANCG	Fanconi anemia, complementation group G	Nucleus
5428[a]	POLG	polymerase (DNA directed), gamma	Cytoplasm
56949[a]	XAB2	XPA binding protein 2	Nucleus
11098[a]	PRSS23	protease, serine, 23	Extracellular Space

a: PPI15 attributes; b: Combined13 attributes; ab: attributes in both Combined13 and PPI15.

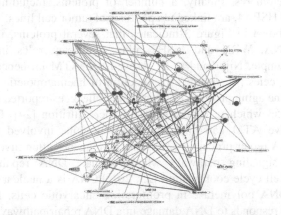

Fig. 3. Network of interactions of proteins with the PPI attributes described in the text. Common attributes in both Combined13 and PPI15 are filled red. Unique attributes for PPI15 are magenta and unique attributes for Combined13 are green. Pink links represent proteins whose function is repair of double-stranded DNA breaks. Cyan lines represent check point control and DNA damage check point related proteins. Green lines represent repair of DNA related proteins. Orange lines represent cell cycle control related proteins and red lines represent apoptosis related proteins. Blue lines represent maintenance of telomeres related proteins.

4   Discussion

4.1   Biological Implications of Selected Attributes

Among the top 25 predictive attributes identified in the present study, 13 were PPI attributes and of those, 10 attributes were shared with the PPI15 group of attributes. The net result was that there were 3 unique attributes under the Combined13 group and 5 under the PPI15 group, for a total of 18 PPI attributes represented under these two categories (Table 3). To investigate the biological functions of these 18 attributes, we used Ingenuity Pathway Analysis (IPA) (http://www.ingenuity.com/) to perform a systems biology analysis. Only a single biological network was formed which included these 18 proteins and their interaction partners (Fig. 3). The PPI attributes of the 18 proteins were related to several important biological functions. Nine proteins, namely BLM, ATM, RPA1, WRN, MSH2, MRE11A, ERCC6, PCNA and ATR, are involved in cell cycle control. Some proteins in the network shown in Figure 3 (BLM, ERCC2, FANCG, MSH2, ATM, MRE11A and ATR) play a role in check-point control and DNA damage check. For example, ATM is a serine-threonine protein kinase that is activated upon detection of double-stranded DNA breaks, whereas ATR is a serine-threonine kinase that responds to single-stranded DNA breaks, and both proteins function as cell cycle check-point controls [22, 23]. It is important to note that mutations in ATM and ATR lead to multiple system dysfunction, including the nervous and immune systems, altered responses to UV radiation and genotoxic stress, and increased probability of cancer development.

DNA repair-related proteins among those shown in Table 2 included BLM, WRN, MSH2, XRCC1, RPA1, MRE11A, ATM, ATR, PCNA, ERCC6, ERCC2 and XAB2. Four proteins, BLM, WRN, MRE11A and Mre11, are associated with the maintenance of telomeres. Finally, a number of proteins, including BLM, ATM, RPA1, PCNA and HSPA4, are linked to apoptosis of tumor cell lines.

The network shown in Figure 3 indicates that several proteins, including BLM, ATM, ATR, PCNA, WRN and MSH2, play significant roles in a number of pathways. For example, Nijnikand et al. showed that ATM influences the levels of reactive oxygen species [4], restricts the function of haematopoietic stem cells, and has an impact on the aging process. Lombard and colleagues reported that ATM/ATR kinases activate p53 which in turn induces telomere attrition [24]. It has also been reported that active ATM can bind to various proteins involved in check point responses and DNA repair processes [24]. Check points are often involved in cellular surveillance and signaling pathways which regulate DNA repair, chromosome metabolism, and cell-cycle coordination [25, 26]. PCNA is a nuclear antigen protein associated with DNA polymerase in proliferating eukaryotic cells. It functions via ubiquitination and responds to DNA damage in a DNA repair pathway [27, 28]. WRN protein is related to Werner's syndrome, a condition of premature aging. The protein WRN has both an ATP-dependent 3'—5' DNA helicase function as well as a 3'—5' exonuclease activity and thus, together with other proteins, including PCNA [29], is involved in double stranded DNA repair [30]. Furthermore, the interaction of WRN with the RecQ helicase-like protein is important in maintaining normal genomic stability during the aging process, and defects in the WRN protein disrupt such interactions and accelerate aging [31, 32].

The cell functions, diseases and physiological processes in which the 18 PPI attributes participate were analyzed. Most of the 18 PPI attributes are related to cancer and genetic disorders and this finding fits well with the suggestion made by others that the aging process shares some common factors with cancer and genetic disorders. Among the 18 PPI attributes we identified, 16 were related to DNA replication, recombination, and repair functions, 13 to cell cycle, and 12 to cell death processes.

5   Conclusions

In the present study we showed that a set of aging-related genes could be classified with high degree of confidence into either the DNA-repair or the non-DNA-repair categories. We also demonstrated that a small number of selected attributes were sufficient to build classification models for distinguishing DNA-repair genes from non-DNA-repair ones. Interestingly, many interaction partners of aging-related DNA-repair genes are also involved in cancer and other genetic disorders, including immune system and neurological diseases, consistent with previous findings that the aging process shares some common properties with cancer and other human diseases. We believe that the identification of these select genes that are associated with both aging and DNA repair processes will be important in guiding future research into aging-related pathways and biomarkers.

Acknowledgements. This work was supported in part by the National Institutes of Health (NIH) Grant P01 AG12993 (PI: E. Michaelis).

References

1. Sahin, E., Depinho, R.A.: Linking functional decline of telomeres, mitochondria and stem cells during ageing. Nature 464, 520–528 (2010)
2. Finkel, T., Serrano, M., Blasco, M.A.: The common biology of cancer and ageing. Nature 448, 767–774 (2007)
3. Tse, M.T.: Brain ageing: a fine balance. Nat. Rev. Neurosci. 13, 222 (2012)
4. Nijnik, A., Woodbine, L., Marchetti, C., Dawson, S., Lambe, T., Liu, C., Rodrigues, N.P., Crockford, T.L., Cabuy, E., Vindigni, A., Enver, T., Bell, J.I., Slijepcevic, P., Goodnow, C.C., Jeggo, P.A., Cornall, R.J.: DNA repair is limiting for haematopoietic stem cells during ageing. Nature 447, 686–690 (2007)
5. Thoms, K.M., Baesecke, J., Emmert, B., Hermann, J., Roedling, T., Laspe, P., Leibeling, D., Truemper, L., Emmert, S.: Functional DNA repair system analysis in haematopoietic progenitor cells using host cell reactivation. Scand J. Clin Lab Invest. 67, 580–588 (2007)
6. Lu, T., Pan, Y., Kao, S.Y., Li, C., Kohane, I., Chan, J., Yankner, B.A.: Gene regulation and DNA damage in the ageing human brain. Nature 429, 883–891 (2004)
7. Freitas, A.A., Vasieva, O., de Magalhaes, J.P.: A data mining approach for classifying DNA repair genes into ageing-related or non-ageing-related. BMC Genomics 12, 27 (2011)
8. Kenyon, C.J.: The genetics of ageing. Nature 464, 504–512 (2010)

9. de Magalhaes, J.P., Budovsky, A., Lehmann, G., Costa, J., Li, Y., Fraifeld, V., Church, G.M.: The Human Ageing Genomic Resources: online databases and tools for biogerontologists. Aging Cell 8, 65–72 (2009)

10. Jiang, H., Ching, W.K.: Classifying DNA repair genes by kernel-based support vector machines. Bioinformation 7, 257–263 (2011)

11. Fang, J.W., Dong, Y.H., Williams, T.D., Lushington, G.H.: Feature selection in validating mass spectrometry database search results. J. Bioinform Comput. Biol. 6, 223–240 (2008)

12. Wang, L., Yang, M.Q., Yang, J.Y.: Prediction of DNA-binding residues from protein sequence information using random forests. BMC Genomics 10(suppl. 1), S1 (2009)

13. Sikic, M., Tomic, S., Vlahovicek, K.: Prediction of protein-protein interaction sites in sequences and 3D structures by random forests. PLoS Comput. Biol. 5, e1000278 (2009)

14. Li, Y., Fang, Y., Fang, J.: Predicting Residue-Residue Contacts Using Random Forest Models. Bioinformatics 27, 3379–3384 (2011)

15. Wood, R.D., Mitchell, M., Sgouros, J., Lindahl, T.: Human DNA repair genes. Science 291, 1284–1289 (2001)

16. Wood, R.D., Mitchell, M., Lindahl, T.: Human DNA repair genes, 2005. Mutat Res. 577, 275–283 (2005)

17. Keshava Prasad, T.S., Goel, R., Kandasamy, K., Keerthikumar, S., Kumar, S., Mathivanan, S., Telikicherla, D., Raju, R., Shafreen, B., Venugopal, A., Balakrishnan, L., Marimuthu, A., Banerjee, S., Somanathan, D.S., Sebastian, A., Rani, S., Ray, S., Harrys Kishore, C.J., Kanth, S., Ahmed, M., Kashyap, M.K., Mohmood, R., Ramachandra, Y.L., Krishna, V., Rahiman, B.A., Mohan, S., Ranganathan, P., Ramabadran, S., Chaerkady, R., Pandey, A.: Human Protein Reference Database–2009 update. Nucleic Acids Research 37, D767–D772 (2009)

18. Stark, C., Breitkreutz, B.J., Chatr-Aryamontri, A., Boucher, L., Oughtred, R., Livstone, M.S., Nixon, J., Van Auken, K., Wang, X., Shi, X., Reguly, T., Rust, J.M., Winter, A., Dolinski, K., Tyers, M.: The BioGRID Interaction Database: 2011 update. Nucleic Acids Research 39, D698–D704 (2011)

19. Hall, M., Frank, E., Holmes, G., Pfahringer, B., Reutemann, P., Witten, I.H.: The WEKA Data Mining Software: An Update. SIGKDD Explorations 11 (2009)

20. Breiman, L.: Random Forests. Machine Learning 45, 5–32 (2001)

21. Li, Y., Fang, J.: Distance-dependent statistical potentials for discriminating thermophilic and mesophilic proteins. Biochem Biophys Res. Commun. 396, 736–741 (2010)

22. Lim, D.-S., Kim, S.-T., Xu, B., Maser, R.S., Lin, J., Petrini, J.H.J., Kastan, M.B.: ATM phosphorylates p95/nbs1 in an S-phase checkpoint pathway. Nature 404, 613–617 (2000)

23. Falck, J., Coates, J., Jackson, S.P.: Conserved modes of recruitment of ATM, ATR and DNA-PKcs to sites of DNA damage. Nature 434, 605–611 (2005)

24. Lombard, D.B., Chua, K.F., Mostoslavsky, R., Franco, S., Gostissa, M., Alt, F.W.: DNA repair, genome stability, and aging. Cell 120, 497–512 (2005)

25. Branzei, D., Foiani, M.: Regulation of DNA repair throughout the cell cycle. Nat. Rev. Mol. Cell Biol. 9, 297–308 (2008)

26. Bartek, J., Lukas, J.: DNA damage checkpoints: from initiation to recovery or adaptation. Curr. Opin. Cell Biol. 19, 238–245 (2007)

27. Zlatanou, A., Despras, E., Braz-Petta, T., Boubakour-Azzouz, I., Pouvelle, C., Stewart, G.S., Nakajima, S., Yasui, A., Ishchenko, A.A., Kannouche, P.L.: The hMsh2-hMsh6 complex acts in concert with monoubiquitinated PCNA and Pol eta in response to oxidative DNA damage in human cells. Mol Cell 43, 649–662 (2011)

28. Aggarwal, M., Sommers, J.A., Shoemaker, R.H., Brosh Jr., R.M.: Inhibition of helicase activity by a small molecule impairs Werner syndrome helicase (WRN) function in the cellular response to DNA damage or replication stress. Proceedings of the National Academy of Sciences of the United States of America 108, 1525–1530 (2011)
29. Rodrı, X., Guez-López, A.M., Jackson, D.A., Nehlin, J.O., Iborra, F., Warren, A.V., Cox, L.S.: Characterisation of the interaction between WRN, the helicase/exonuclease defective in progeroid Werner's syndrome, and an essential replication factor, PCNA. Mechanisms of Ageing and Development 124, 167–174 (2003)
30. Chen, L., Huang, S., Lee, L., Davalos, A., Schiestl, R.H., Campisi, J., Oshima, J.: WRN, the protein deficient in Werner syndrome, plays a critical structural role in optimizing DNA repair. Aging Cell 2, 191–199 (2003)
31. Hasty, P., Vijg, J.: Accelerating aging by mouse reverse genetics: a rational approach to understanding longevity. Aging Cell 3, 55–65 (2004)
32. Multani, A.S., Chang, S.: WRN at telomeres: implications for aging and cancer. J. Cell Sci. 120, 713–721 (2007)

Data Mining with Ant Colony Algorithms

Ilaim Costa Junior

Fluminense Federal University, Institute of Computing, Niterói, Brazil
ilaim@ic.uff.br

Abstract. The Ant-Miner algorithm, Ant-Miner2, Ant-Miner3 and Taco-Miner have an excellent performance in classification tasks, what can be seen in literature. These algorithms are inspired on the behavior of real ant colonies and some data mining concepts as well as principles. This paper presents a new algorithm based on Ant Colony whose experiments comparing with the others suggest superiority.

Keywords: Rule discovery, data mining, computational intelligence, ant colony algorithm, multi-agent systems.

1   Introduction

For some time, the data generated by the various organizations of the world have been collected and stored on magnetic media. As time passes and technology advances, it becomes possible to store increasingly significant volumes of various types of information. However, this has a cost: the difficulty of efficiently analyze all that is stored is also growing.

One of the main solutions for this problem is the Data Mining (DM), which is a part of the Knowledge Discovery in Database (KDD).

Data mining is an interdisciplinary field that relates primarily techniques of machine learning, statistics, computational intelligence, database, and others. Your goal is, in essence, extract knowledge - or, more specifically, structural patterns - from data as a tool to help explain them and make predictions using the same.

There are several DM tasks, including classification, regression, clustering, dependence modeling, etc [17]. Each of these tasks can be regarded as a kind of problem to be solved. So the first step is to decide what kind of problem will be address to choose the right algorithm to solution.

In this paper, we propose a new ant colony optimization algorithm for the classification task of Data M.

In the remainder of this work there are: section 2 Ant Colony Optimization, section 3 Ant-Miner algorithm, section 4 Density-Based Heuristic Algorithm – Ant-Miner 2, section 5 Ant-Miner 3, section 6 Taco-Miner, section 7 modifications of the Ant-Miner – the new algorithm and in section 8 we have the experiments and conclusions.

D.-S. Huang et al. (Eds.): ICIC 2013, LNAI 7996, pp. 30–38, 2013.

2 Ant Colony Optimization

From a mathematical model we would like to simulate the foraging behavior of an ant colony. Initially it was proposed for solving combinatorial optimization problems, as can be seen in [1] and [2]. By prove promising, efficient, has also been used to solve another kind of problems.

The ants behave as extremely simple agents. They have a small ability to process data and things around them and also to exchange information. As a result, it is essential the shortest path between the nest and the food source, because this way they reduce the risks.

By modifying your environment depositing pheromone, they generate a system of indirect communication, which necessarily influence their behaviors leading to choosing the best path to follow. So, it will converge to the shortest path without any hierarchy or visual communication.

This paper proposes modifications to the Ant-Miner algorithm [3]. These are based on the mathematical model of foraging mechanism proposed in [4] and suggest an improved performance in the discovery of classification rules, it is a new algorithm.

3 The Ant-Miner Algorithm

The Ant-Miner is proposed and detailed in [3] and [5] whose goal is to discover classification rules whose form is: IF <term1 AND term2 AND...> THEN   <class>.

The algorithm seeks an ordered list of classification rules aimed at covering nearly all training cases. In the beginning the list is empty. The algorithm can be seen below.

```
TrainingSet = {all training cases};
DiscoveredRuleList = [ ];/*rule list is initialized with na empty list*/
WHILE (TrainingSet > Max_uncovered_cases) DO
    t = 1; /* ant index */
    j = 1; /*convergence test index */
    Initialize all trails with the same amount of pheromone;
    REPEAT
        Ant t starts with an empty rule and incrementally constructs a
        classification rule Rt by adding one terma at a time to the
        current rule;
        Prune rule Rt;
        Update the pheromone of all trails by increasing pheromone in
        the trail followed by ant t (proportional to the quality of Rt)
        and decreasing pheromone in the other trails (simulating
        pheromone evaporation)
        IF Rt is equal to Rt -1 /*update convergence test*/
            THEN j = j + 1;
            ELSE j = 1;
        END-IF
```

```
      t = t + 1;
  UNTIL (i ≥ No_of_ants) OR (j ≥ No_rules_converg)
  Chose the Best rule R_melhor among all rules R_t constructed by all the
  ants;
  Add rule R_melhor to DiscoveredRuleList;
  TrainingSet = TrainingSet - {set of cases correctly covered by
      R_melhor };
```

END-WHILE.
[The Ant-Miner Algorithm [3]]

Some steps of the algorithm are seen in more detail below.

3.1 Pheromone Initialization

The following formula gives the same amount of all the pheromone trails.

$$\tau_{ij}(t = 0) = \frac{1}{\displaystyle\sum_{i=1}^{a} b_i} \tag{1}$$

Where a is the total number of attributes and b_i is the number of values in the domain of the attribute i.

3.2 Rules Construction

Equation 2 generates a probability term ij to be chosen and added to the rule part of an ant.

$$P_{ij}(t) = \frac{\tau_{ij}(t).\eta_{ij}}{\displaystyle\sum_{i}^{a}\sum_{j}^{b_i}\tau_{ij}(t).\eta_{ij}}, \forall i \in I \tag{2}$$

Where:

- η_{ij} is an heuristic function value dependent on the problem for the term ij;

- τ_{ij} is the amount of pheromone associated with the term ij at a time t, corresponding to the amount of pheromone currently available at the position ij currently available at the position ij in the track being followed   by the ant;

- a is the total number of attributes;

- b_i is the total number of values in the i-th field attributes;

- I is the set of attributes that are not used by the ants.

3.3   Heuristic Function

The major role of the heuristic function is to decide what change will take place, taking into account the value of pheromone. In Ant-Miner it's based on Information Theory [9]. The function is given by the following equation:

$$
\eta_{ij} = \frac{\log_2 k - H(W \mid A_i = V_{ij})}{\sum_{m=1}^{a} x_m X \left(\sum_{n=1}^{b_m} \log_2 k - H(W \mid A_m = V_{mn}) \right)}
\tag{3}
$$

where:

- a is the total number of attributes;
- X_m receives value 1 if the attribute A_i has not been used by this ant and the value 0 otherwise;
- b_m is the number of values in the do m-*th* attribute.

3.4   Pruning Rule

The act of pruning a rule can bring various benefits such as removing irrelevant terms. The quality of the rule is calculated using the following equation:

$$
Q = \left(\frac{VP}{VP + FN} \right) X \left(\frac{VN}{FP + VN} \right)
\tag{4}
$$

Where

- VP is the number of cases   that is provided by the class rule;
- FP is the number of cases that have different class to class provided by the rule;
- FN is the number of cases that are not covered by the rule, but that have a class provided by rule;
- VN is the number of cases that are not covered by the rule and do not have the class provide by rule;

3.5   Pheromone Update Rule

After construction of each rule by each ant is made the pheromone update by the equation:

$$
\tau_{ij}(t+1) = \tau_{ij}(t) + \tau_{ij}(t) x Q, \forall ij \in R
\tag{5}
$$

where:

- R is the set of terms occurring in the rule constructed by the ant at time t;
- $0 \le Q \le 1$. The higher the Q value, the better the quality of the rule.

4   The Density Based Heuristic – The Ant Miner 2

In [12] authors think that Ant Colony Optimization Algorithm dos not need accurate information in the heuristic value since the idea of the pheromone should compensate the small potential errors in the heuristic values. In other words, a simpler heuristic value may do the job as well as the complex one. As a result, is proposed the equation below:

$$\eta_{ij} = \frac{majority_{classT_{ij}}}{T_{ij}} \tag{6}$$

where:

   $majority\_classT_{ij}$ is the majority class in partition T_{ij}.

This simple heuristic produces equivalent results to the entropy function of the Ant-Miner. The experimental results is in conclusion section.

5   Ant Miner 3

The Ant-Miner 3 algorithm presents a new method for pheromone update and new state transition rules that improve the classification accuracy [13].

5.1   Pheromone Update Method

After an ant constructs a rule, the amount of pheromone associated with each term that appears in construction of the rule is updated with the following equation, and the unused terms is updated by standardization.

   Note that Q varies in the range [0,1]. The greatest Q is the largest amount of pheromone associated with each used term. Otherwise, if Q is very small (near zero), the level of pheromone associated with each used term will decrease.

$$\tau_{ij}(t) = (1 - \rho).\tau_{ij}(t-1) + (1 - \frac{1}{1+Q}).\tau_{ij}(t-1) \tag{7}$$

where:

   — ρ is the evaporation rate of pheromone;
   — Q is the quality of the constructed rule.

ρ is the evaporation rate which controls how fast evaporation is the pheromone of the old paths. This parameter controls the influence of the history of the current pheromone trails [14]. In this method, a high value of ρ represents a fast evaporation and vice - versa. In the experiments it was set at 0.1.

5.2   Transition Rule

Ants behave as agents cooperating in a colony. These intelligent agents live in an environment without global knowledge, but can benefit from the pheromone updating [15]. The pheromone deposited on the edges in an ant colony system plays the role of a distributed long term memory. This memory is not stored within the individual ants, but distributed by the edges of the route of a solution of a given problem, which creates an indirect form of communication. This helps the operation of prior knowledge, but increases the likelihood choice of terms belonging to previously discovered rules according to equation (2), which inhibits the ants to exhibit alternative forms along the operation. To improve operation is applied the transition rule shown below:

If $q_1 \leq \varphi$

loop

if $q_2 \leq \sum P_{ij}$ where $j \varepsilon J_i$

then choose the term ij

end-loop
else
choose the term ij com maior P_{ij}

Fig. 1. The transition rule

As can be seen in [13], this method has a performance slightly higher than Ant-Miner Algorithm.

6   Threshold Ant Colony Optimization Miner (Taco – Miner)

In this algorithm a new form of construction rule is introduced [16]. The information gained is mainly used to check the credibility of the term being selected for the antecedent part of the rule. If the information gained of the selected term is less than the threshold of 0.6, then it will be rejected from inclusion otherwise will be included, fig 2. The threshold 0.6 is a empirical criterion and studies are still needed to arrive at

REPEAT

Choose the first term

if (gain of information $\geq$ 0.6) THEN

add the term to the ruleset

ELSE

reject the term to be included

UNTIL NO MORE TERMS

Fig. 2. Procedure of rule construction [16]

an exact value. Thus only the fittest terms are provided in the antecedent of the rule and consequently with rules varying size are obtained.

As can be seen in [16] this method has outperformed the Ant-Miner algorithm and several other algorithms for generating classification rules.

7   The Proposed Method

In this section we will present a new proposal for the Ant-Miner algorithm, which aims to improve its performance.

The proposal is based on [4]. In this, the way to how to choose the trail in which ants must pass are more consistent and therefore generates better results.

Initially it will be considered the following verbal description: "if a mass forager arrives at a fork in a chemical recruitment trail, the probability that it takes the left branch is all the greater as there is more trail pheromone on it than on the right one". So the probability of choosing the path is [4]:

$$P(D) = \frac{(1 + \tanh(qD))}{2} \tag{8}$$

where:

- D represents the difference of concentration of pheromone in the left branch minus the concentration of pheromone in the right branch, denoted by $D = L - R$ and q is the parameter that determines the shape of curve.
- In [4] was suggested value for $q=0.016$.

This proposal presents two possible options for choosing the path to be followed by the ant:

1. The probability of the ant choose a ij term depends only on the concentration difference between the ij pheromone term and the other values of the attribute domain i. This is called a local strategy and will be given by:

$$P_{ij} = \frac{1 + \tanh\left(q\left(2\tau_{ij}(t) - \sum \tau_{ij}(t)\right)\right)}{b_i} \tag{9}$$

2. To avoid obtaining only optimal local solutions by ants, the probability of choice of the term ij is based on the pheromone concentration difference between the term ij and all other values in the domain of all attributes that have not been used by the ant. This is called a global strategy and will be given by:

$$P_{ij} = \frac{1 + \tanh\left(q\left(2\tau_{ij}(t) - \sum x_i * \sum \tau_{ij}(t)\right)\right)}{\sum x_i b_i} \tag{10}$$

Note that, rather than one, we have two different versions of probability formula and, therefore, two versions of the algorithm. We call them AMP1 and AMP2.

8   Computational Results and Discussions

In these experiments we used seven public domain datasets repositories from the University of California [10]. They are Ljubjana Breast Cancer, Wisconsin Breast Cancer, Tic-tat-toe, Dermatology, Hepatitis, Cleveland Heart Disease e Diabetes.

The collection of the results was performed by running each version of the program ten times with each data set, using cross-validation. We calculated the arithmetic average of the accuracy achieved in each of the executions. Tests can be seen in the table 1 below. Note that there are six algorithms: Ant-Miner, Ant-Miner2, Ant-Miner3, Taco-Miner and our versions AMP1 and AMP2.

Table 1. Experiments and Results

Data Set	Ant-Miner	Ant-Miner2	Ant-Miner3	Taco-Miner	AM 1b	AM 2a
Ljubljana	73,60% +/- 2,57%	73,60% +/- 2,57%	71,62% +/- 1,68 %	72,60% +/- 2,55%	**75,00% +/- 3,15%**	**74,38% +/- 3,09%**
Wisconsin	91,85% +/- 1,00%	91,85% +/- 1,00%	91,75 +/- 2,55%	91,23% +/- 1,00%	**92,02% +/- 1,17%**	**92,41% +/- 1,10%**
Tic-tac-toe	72,11% +/- 1,86%	72,11% +/- 1,86%	70,23 +/- 2,35 %	72,11% +/- 1,86%	70,25% +/- 1,69%	69,53% +/- 1,76%
Dermatology	95,28% +/- 1,18%	95,28% +/- 1,18%	88,35 +/- 1,98%	84,86% +/- 1,28%	**95,32% +/- 1,15%**	**95,53% +/- 0,91%**
Hepatitis	83,67% +/- 2,91%	83,67% +/- 2,91%	81,26 +/- 1,22%	86,58 +/- 1,97%	79,17% +/- 3,51%	79,00% +/- 3,26%
Cleveland	77,99%+/- 2,38%	77,99%+/- 2,38%	75,88 +/- 2,08%	81,25 +/- 2,36%	**78,06% +/- 2,67%**	**78,86% +/- 1,99%**
Diabetes	67,16%+/- 1,72%	67,16%+/- 1,72%	68,55 +/- 1,95%	77,22 +/- 1,25 %	**67,33% +/- 1,46%**	**66,32% +/- 1,67%**

In all experiments, the parameters were Min_cases_per_rule=5 and Max_uncovere_cases=10.

Note the superiority of the modified algorithm, especially in the bold lines.

In our future studies and works we will are trying new heuristic functions to maximize the predictive accuracy of the classification rules discovered.

References

1. Dorigo, M., Di Caro, G.: The ant colony optimization meta-heuristic. In: New Ideas in Optimization, pp. 11–32. McGraw Hill, London (1999)
2. Dorigo, M., Di Caro, G., Gambardella, L.M.: Ant algorithms for discrete optimization. Artificial Life 5(2), 137–172 (1999)
3. Parpinelli, R.S., Lopes, H.S., Freitas, A.A.: Data Mining with an Ant Colony Optimization Algorithm. IEEE Transactions on Evolutionary Computing 6(4) (2002)
4. Rozin, V., Margaliot, M.: The Fuzzy Ant. IEEE Computational Intelligence Magazine 2(4) (2007)
5. Parpinelli, R.S.: Um Algoritmo Baseado em Colônias de Formigas para Classificação e, Data Mining. Dissertação de Mestrado, UTFPR, Curitiba (2001) (in Portuguese)
6. Chen, M.S., Han, J., Yu, P.S.: Data mining: an overview from database perspective. Proceedings of the IEEE Transactions on Knowledge and Data Engineering, 866–883 (1996)
7. Quinlan, J.R.: C4.5: Programs for Machine Learning. Morgan Kaufmann (1993)
8. Clark, P., Neblett, T.: The CN2 induction algorithm. Machine Learning 3, 261–283 (1989)
9. Cover, T.M., Thomas, J.A.: Elements of Information Theory. John Wiley & Sons, New York (1991)
10. Frank, A., Asuncion, A.: UCI Machine Learning Repository. University of California, School of Information and Computer Science, Irvine, CA (2010), http://archive.ics.uci.edu/ml
11. Quinlan, J.R.: C4.5: Programs for Machine Learning. Morgan Kaufmann (1993)
12. Liu, B., Abbass, H.A., Mckay, B.: Density-Based Heuristic for Rule Discovery with Ant-Miner. In: Australia-Japan Workshop on Intelligent and Evolutionary Systems (2002)
13. Liu, B., Abbass, H.A., Mckay, B.: Classification Rule Discovery with Ant Colony Optimization. In: IAT 2003, International Conference on Intelligent Agent Technology (2003)
14. Schools, L., Naudts, B.: Ant Colonies are Good at Solving Constraint Satisfaction Problems. In: Proceedings of the Congress on Evolutionary Computation, vol. 2, pp. 1190–1195 (2000)
15. Sun, R., Tatsumi, S., Zhao, G.: Multiagent Reinforcement Learning Method with An Improved Ant Colony Systems. In: Proceedings of the 2001 IEEE International Conference on Systems, Man and Cybernetics, vol. 3, pp. 1612–1617 (2001)
16. Thangavel, K., Jaganathan, P.: Rule Minig Algorithm with a New Ant Colony Optimization Algorithm. In: IEEE International Conference on Computational Intelligence and Multimedia Applications (2007)
17. Fayyad, U.M., Piatetsky-Shapiro, G., Smyth, P.: From data mining to knowledge discovery: An overview. In: Fayyad, U., Piatetsky-Shapiro, G., Smyth, P., Uthurusamy, R. (eds.) Advances in Knowledge Discovery & Data Mining, pp. 1–34. MIT Press, Cambridge (1996)

Content-Based Diversifying Leaf Image Retrieval

Sheng-Ping Zhu, Ji-Xiang Du, and Chuan-Min Zhai

Department of Computer Science and Technology, Huaqiao University, Xiamen 361021
{jxdu77,cmzhai}@gmail.com

Abstract. In recent years, content-based image retrieval achieved continuous development, the main goal so far has been to retrieve similar objects for a given query, and only the relevance is cared in retrieval system, so many duplicate or near duplicate documents retrieved in response to a query. For efficient content-based image retrieval, we propose the Content-based Diversifying Leaf Image Retrieval in this paper. In order to make the retrieval results have relevance and diversity, we extract leaf image feature and use the relevance feedback technique based of SVM and the AP clustering algorithm. We also proposed a new evaluation function - Maximal Scatter Diversity (MSD) static evaluation function. Experimental results show that our approach can achieve good performance with improving the diversity of the retrieval results without reduction of their relevance.

Keywords: diversity retrieval, relevance feedback, AP clustering, Maximal Scatter Diversity.

1 Introduction

Content-based image retrieval is perhaps the most researched and most famous application of multimedia retrieval, and it applied in many domains, such as plant image retrieval and it get well development. Since the shape of leaves is one of important features for charactering various plants, the study of leaf image retrieval will be an important step for plant identification. In 2000, Z. Wang presented an efficient two-step approach of using a shape characterization function called centroid-contour distance curve and the object eccentricity (or elongation) for leaf image retrieval [1]. In the subsequent work, he presents the combination of different shape based feature sets using fuzzy integral for leaf image retrieval [2],[3], Experimental results on 1400 leaf images from 140 plants show that the proposed approach can achieve a better retrieval performance.

There also in lies a serious challenge that is often referred to as the semantic gap. For instance, we instantly recognize a house in a drawing of a house even in its simplest form, but we can't extract the effect feature from the image. The semantic gap is defined in the context of image retrieval in [4] as "the lack of coincidence between the information that one can extract from the visual data and the interpretation that the same data have for a user in a given situation". To narrow down the semantic gap, relevance feedback is widely researched these years, and the

D.-S. Huang et al. (Eds.): ICIC 2013, LNAI 7996, pp. 39–46, 2013.

relevance feedback technique based of SVM is reached a better retrieval performance [5], so we use it in this paper to make the retrieval result more relevance.

In the other side, image search results are usually displayed in a ranked list, it may exist problem in this ranking. To solve this problem, we propose to create a visually diverse ranking of the image search results, through clustering of the images based on their visual characteristics [6],[7]. To organize the display of the image search results, a cluster representative is shown to the user. This approach guarantees that the user will be presented a visually diverse set of images. In this way, we also can remove the duplicate or near duplicate document retrieved in a query, to make the results not only relevance, but diverse as well.

Therefore, in this paper we will now extend our interest toward obtaining diverse results. Not only should the results be relevant, the top of the ranking should in fact reflect the diversity of the relevant objects that are present in the collection, we test on leaf images, and extract leaf image feature and use the relevance feedback technique based of SVM and the AP clustering algorithm, experimental results show our approach achieve good performance. We also proposed a new evaluation function - Maximal Scatter Diversity (MSD) static evaluation function.

The rest of this paper is organized as follows. In section 2, we briefly discuss some related works .The proposed new evaluation function - Maximal Scatter Diversity (MSD) is described in section 3. In section 4, we present the experiment results on leaf image database and analyze the experimental results. Finally some conclusions are drawn in section 5.

2   Related Works

2.1   Image Description

From all of the visual properties, shape is the mandatory property that carries the identity of visual stimuli, It can be well described the contour of the leaf. In the following, we present the shape descriptors used in our experiments: GIST descriptor [8], Pyramid of histograms of oriented gradients (PHOG) [9].

GIST Descriptor: Orientation histograms are computed on a 4×4 grid over the entire image. We extract at 4 scales with 8 orientation bins respectively.

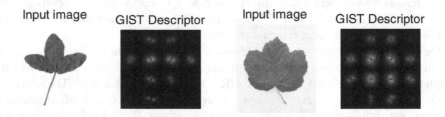

Fig. 1. GIST descriptor

PHOG Descriptor: The PHOG descriptor first extracts canny edges. It, then, quantizes the gradient orientation on the edges ($0°$ to $180°$) into 20 bins. Three spatial pyramid levels are used (1×1, 2×2, 4×4). Each level is used in an independent kernel.

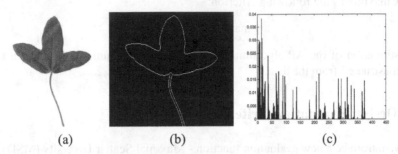

       (a)   (b)   (c)

Fig. 2. PHOG descriptor, (a) Original image, (b) Gradient magnitude figure, (c) Histogram of Orientated Gradients

2.2 Relevance Feedback

To narrow down the gap between low-level features and high-level concepts, relevance feedback (RF) initially developed in text retrieval was introduced into CBIR during mid-1990's and has been shown to provide dramatic performance boost in retrieval systems [5][10]. The main idea of RF is to let users guide the system. During retrieval process, the user interacts with the system and rates the retrieved images, according to his/her subjective judgment.

As a core machine learning technology, SVM has not only strong theoretical foundations but also excellent empirical successes. SVM has also been introduced into CBIR as a powerful RF tool, and performs fairly well in the systems that use global representation [11]. The classifiers for the Low-level image features were obtained based on a standard non-linear binary SVM using LIBSVM [12].

2.3 Affinity Propagation

In brief, the AP algorithm propagates two kinds of information between two data points [13]: the "responsibility" r (i, k) sent from data point i to data point k, which reflects how well k serves as the exemplar of i considering other potential exemplars for i, and the "availability" a (i, k) sent from data point k to data point i, which reflects how appropriate i chooses k as its exemplar considering other potential points that may choose k as their exemplar. The information are updated in an iterative way as

$$r(i,k) \leftarrow s(i,k) - \max_{k' s.t. k' \neq k}\{a(i,k') + s(i,k')\} \tag{1}$$

$$a(i,k) \leftarrow \min\{0, r(k,k) + \sum_{i' s.t. i' \in \{i,k\}} \max\{0, r(i',k)\}\} \tag{2}$$

The self-availability is updated in a slightly different way as

$$a(k,k) \leftarrow \sum_{i \, s.t. \, i \neq k} \max(0, r(i', k)) \tag{3}$$

Upon convergence, the exemplar for each data point x_i is chosen as $e(x_i) = x_k$ where k maximizes the following criterion:

$$\arg\max_k (a(i,k) + r(i,k)) \tag{4}$$

The justification of the AP algorithm roots from the max-sum algorithm in a factor graph constructed from the data.

3   Design Evaluation Criteria

Here, we introduce a new evaluation function - Maximal Scatter Diversity (MSD). The top of the ranking should in fact reflect the diversity of the relevant objects that are present in the collection. So our evaluation index must be different with the other evaluation measure, We take full advantage of the training samples of category labels to express the first level's relevance and proposed a new evaluation function - Maximal Scatter Diversity (MSD) static evaluation function.

Given a collection of images $\Phi = \{I_1, I_2, ..., I_n\}$, we denote the binary relevance label of I_1 with respect to the given query as $r(I_i)$, $r(I_i) = 1$, if I_1 is relevant; otherwise $r(I_i) = 0$. Let N_t be the number of true relevant images in the set Φ.

$$MSD = \frac{1}{N_t} \sum_{i=1}^{N_t} r(I_i) \cdot d(I_i, \overline{I}) \qquad \text{Where,} \qquad \overline{I} = \frac{\sum_{i=1}^{N_t} [r(I_i) \cdot I_i]}{\sum_{i=1}^{N_t} r(I_i)} \tag{5}$$

$$\text{Or} \quad MSD = \frac{1}{N_t} \sum_{i=1}^{N_t} r(I_i) \cdot [d(I_i, \overline{I}) - II]^2 \quad \text{Where,} \quad II = \frac{\sum_{i=1}^{N_t} [r(I_i) \cdot d(I_i, \overline{I})]}{\sum_{i=1}^{N_t} r(I_i)} \tag{6}$$

And $d(\cdot)$ is a distance between two images, we can use Euclidean distance or cosine distance. The evaluation index MSD in the expression has level of relevance with the degree of diversity; it can ensure the relevance of the retrieved results, to improve the evaluation of the diversity. And it also different with the evaluation measure MMR, MSD evaluation function don't have variable parameters, so it is more stability.

4   Experimental Results

The process we adopt to implement the image retrieval in Image CLEF 2012 Plant [14] is overall shown in Fig.3, which can also be depicted by the following steps:

1) Extract the visual features from the all image data;
2) Image retrieval rank without clustering based on image features;
3) Use relevance feedback technique based of SVM in first-ranking image retrieval results;
4) Use AP clustering algorithm after SVM classify on the relevant images;
5) Select images in each category and re-rank to ensure the retrieval results diversity.

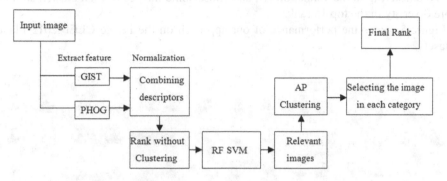

Fig. 3. Framework of our image retrieval system

Images are represented by visual feature descriptors. Retrieval consists in finding the nearest neighbors in high-dimensional descriptor space. To combine GIST feature vectors and PHOG feature vectors, each of them should be normalized.

To evaluate the performance of the proposed approach, we test on the retrieval of particular leaves on 220 classes we self-built image database and the Image CLEF 2012 Plant dataset. Our database contains more than 30,000 pieces of Anhui plant leaves and two types of image content are considered: Scan, and Photograph (unconstrained leaf with natural background).Figure 4 shows the performance of our approach on we self-built dataset.

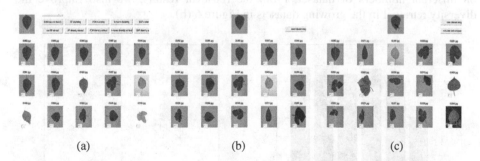

(a)   (b)   (c)

Fig. 4. diversifying retrieval result, (a) First retrieval result, (b) SVM relevance feedback retrieval result, (c) AP clustering retrieval result

From Figure.4 we can see: (a) the top 15 output of an image retrieval system on the topic of 'mulberry', it is mostly based on the similarity between the retrieval and sample images, so there are more duplicate images retrieved, and some irrelevant images are also retrieved in the list. When we chose the relevant images as positive samples, the others as the negative samples to construct the validation set, and use the relevance feedback technique based of SVM, as (b) show the top 15 output are all relevance images, but it is also more duplicate images ranked in the list. As (c) when we use AP clustering in the validation set, and chose some images in each cluster, so it is more diversity in the top 15 rank.

Figure 5 shows the performance of our approach on the Image CLEF 2012 Plant dataset.

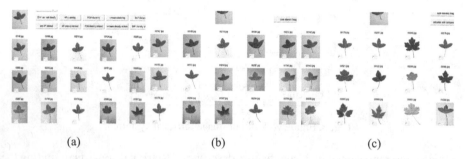

| (a) | (b) | (c) |

Fig. 5. diversifying retrieval result, (a) First retrieval result, (b) SVM relevance feedback retrieval result, (c) AP clustering retrieval result

Figure 5 shows an example of the list of the top 15 output of an image retrieval system on the topic of 'maple'. We can see Figure.5 (a) and (b). All of the top 15 results are highly relevant, but the leaves in (c) are not only relevant, but diversity as well. We use formula (4) calculate the different of retrieval results' MSD in Figure.6 (a). Using different clustering algorithms to guarantee the retrieval result more diversity, we find that AP clustering has more stably performance. To evaluate large-scale search, we test on different numbers of dataset. From the retrieval results, we also improve the diversity retrieval in the growing datasets in Figure.6 (b).

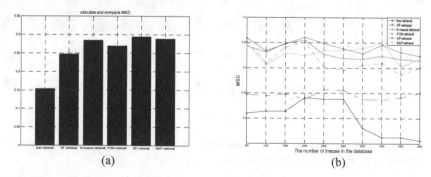

| (a) | (b) |

Fig. 6. (a) Top results of different ranking methods of MSD, (b) Performance on the different numbers of dataset

In order to eliminate the differences of different category image to affect the diversity retrieval result, we test on different category of leaf image. According to the different retrieval results, we calculate the MSD and Mean Maximal Scatter Diversity (MMSD).

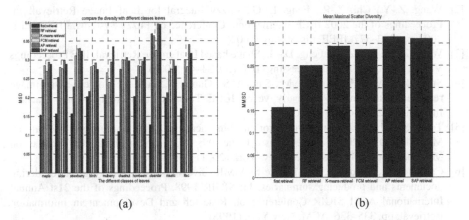

(a)   (b)

Fig. 7. (a) different types of image retrieval result MSD, (b) Mean Maximal Scatter Diversity

From the retrieval results we can find there is a low diversity and exist some false retrieving image in the first retrieval. It will remove the false retrieving image by using the relevance feedback technique based of SVM, so the RF retrieval improves the relevance and diversity in the results. When we use the clustering algorithm, the diversity performance can be further improved according to our experiments, and AP clustering has better performance.

5   Conclusion

The approach that proposed in our paper can achieve good performance with improving the diversity of the retrieval results without reduction of their relevance. It is used in leaf image retrieval can improve agricultural information development. Finally, as we have done throughout this thesis, Content-based Diversifying Image Retrieval should be done further investigate in other domains, the accuracy and diversity need to be improved by other methods or new techniques.

Acknowledgments. This work was supported by the grants of the National Science Foundation of China (Nos. 61175121&61102163), the Program for New Century Excellent Talents in University (No.NCET-10-0117), the grants of the National Science Foundation of Fujian Province (No.2011J01349, 11136006), the Program for Excellent Youth Talents in University of Fujian Province (No.JA10006), the Fundamental Research Funds for the Central Universities of Huaqiao University (No.JB-SJ1003, 09HZR15).

References

[1] Wang, Z.-Y., Chi, Z.-R., Feng, D.-G., Wang, Q.: Leaf Image Retrieval with Shape Features. In: Laurini, R. (ed.) VISUAL 2000. LNCS, vol. 1929, pp. 477–487. Springer, Heidelberg (2000)

[2] Wang, Z.-Y., Chi, Z.-R., Feng, D.-G.: Fuzzy Integral for Leaf Image Retrieval. In: Proceedings of the 2002 IEEE International Conference on Fuzzy Systems, FUZZ-IEEE 2002, pp. 372–377. IEEE, Piscataway (2002)

[3] Wang, Z.-Y., Chi, Z.-R., Feng, D.-G.: Shape based Leaf Image Retrieval. IEE Proceedings Vision, Image and Signal Processing- IET 150, 34–43 (2003)

[4] Smeulders, A.W.M., Worring, M., Santini, S., Gupta, A., Jain, R.: Content-based image retrieval at the end of the early years. IEEE Trans. Pattern Analysis and Machine Intelligence 22, 1349–1380 (2000)

[5] Tong, S., Chang, E.: Support vector machine active learning for image retrieval. In: MULTIMEDIA 2001, Proceedings of the 9th ACM International Conference on Multimedia, pp. 107–118. ACM, New York (2001)

[6] Carbonell, J., Goldstein, J.: The use of MMR, diversity-based reranking for reordering documents and producing summaries. In: SIGIR 1998, Proceedings of the 21st Annual International ACM SIGIR Conference on Research and Development in Information Retrieval, pp. 335–336. ACM, New York (1998)

[7] Zhao, Z.-Q., Glotin, H.: Diversifying Image Retrieval by Affinity Propagation Clustering on Visual Manifolds. IEEE Mutimedia 16, 34–43 (2009)

[8] Oliva, A., Torralba, A.: Modeling the Shape of the Scene: A Holistic Representation of the Spatial Envelope. International Journal of Computer Vision 42, 145–175 (2001)

[9] Bosch, A., Zisserman, A., Munoz, X.: Representing shape with a spatial pyramid kernel. In: CIVR 2007, Proceedings of the 6th ACM International Conference on Image and Video Retrieval, pp. 401–408. ACM, New York (2007)

[10] Zhou, X.S., Huang, T.S.: Relevance feedback in image retrieval: a comprehensive review. Multimedia Systems 8, 536–544 (2003)

[11] Zhang, L., Lin, F.-Z., Zhang, B.: Support Vector Machine Learning for Image Retrieval. In: Proceedings of 2001 International Conference on Image Processing, vol. 2, pp. 721–724. IEEE Computer Society, Washington (2001)

[12] Chang, C.C., Lin, C.J.: LIBSVM: a library for support vector machines. ACM Transactions on Intelligent Systems and Technology (TIST) 2(3), 27 (2011)

[13] Frey, B.J., Dueck, D.: Clustering by passing messages between data points. Science 315, 972–976 (2007)

[14] Plant Identification (2012), http://www.imageclef.org/2013/plant

Binary Coded Output Support Vector Machine

Tao Ye and Xuefeng Zhu

College of Automation Science and Engineering,
South China University of Technology, Guangzhou, P.R. China
yetao@scut.edu.cn, xfzhu@scut.edu.cn

Abstract. To solve multi-class classification problems for large-scale datasets, the authors propose a coded output support vector machine (COSVM) by introducing the idea of information coding. The COSVM is built based on the support vector regression (SVR) machine that is implemented by the sequential minimal optimization (SMO) algorithm. The paper first introduces the soft ε-tube SVR's basic principles, next gives the idea and procedure of the SMO algorithm, and then illustrates the COSVM's topology. For studying the parameters impact on the binary COSVM's performance, we perform two experiments with the Character Trajectories dataset, in which output labels are coded with the binary number system. And some useful results are obtained in these experiments. The final section gives a conclusion and further research ideas.

Keywords: Support vector machine (SVM), binary coded output, classification, regression, number system.

1 Introduction

The support vector machine (SVM) is a machine learning method developed in the frame of statistical learning theory (SLT) [16,17]. To improve the generalization capability (GC), the SVM method applies the structural risk minimization (SRM) induction principle, rather than the empirical risk minimization (ERM) induction principle. Back to the mid-1960s, Vapnik and Chervonenkis had started basic research on the SLT and published several important papers in Russian. In their papers, the Vapnik-Chervonenkis (VC) dimension was invented to evaluate the complexity expressing ability of a set of functions [18]. With the VC-dimension, they further proposed the SRM principle, and pointed out the SRM principle outdoes the ERM principle for a limited sample set (especially the small sample set). Until 1992, Boser, Guyon and Vapnik published the significant paper, "A training algorithm for optimal margin classifiers" [1], which makes the SLT be widely recognized in the AI area. They creatively applied the SRM principle, convex quadratic programming (CQP) duality theory, kernel function (KF) theory to build an SVM. After that, researchers developed a series of SVM algorithms, including hard margin support vector classifier (SVC), soft margin SVC [2,3], hard ε-tube support vector regression (SVR) [4,5], and soft ε-tube SVR machine [14]. Furthermore, Suykens proposed the least squares SVM [15], and Schölkopf *et al.* developed the ν-SVM algorithm [11].

D.-S. Huang et al. (Eds.): ICIC 2013, LNAI 7996, pp. 47–55, 2013.

The SVM method replaces the artificial neural network (ANN) method in many applications because of its excellent GC. To improve the training efficiency of large-scale learning problems, Cortes and Vapnik designed a chunking algorithm [3], and Osuna *et al.* proposed a decomposition algorithm [9]. Based on Osuna's idea of dataset decomposing, Platt devised the sequential minimal optimization (SMO) algorithm [10]. Joachims utilized the steepest gradient descent method to select working sets for the decomposition algorithm [7]. Most machine learning problems are the multi-class classification problems. Deng and Tian summarize methods of construct-ing a multi-class SVC [5]. Dietterich and Bakiri applied the error-correcting output codes to decision tree (DT) C4.5 for solving multi-class classification problems [6]. Zhu and Dai used 2-class SVCs as decision nodes (non-leaf nodes) of a binary DT to design a multi-class SVC, and presented two types of DT topology [20]. We design a coded output SVM (COSVM) based on the SVR machine.

2   The Soft ε-SVR Machine and SMO Algorithm

Most real-life classification problems are the multi-class classification problems. To solve such problems, Platt proposed the SMO classification algorithm in 1998, and Smola generalized his idea to the regression algorithm [13]. The paper proposes the COSVM algorithm to solve the multi-class classification problem. The algorithm is easier to achieve parallel implementation (PI) with multiple MCUs, which can greatly improve the algorithm training efficiency. Since the SMO algorithm has higher training efficiency, the paper applies it to realize the ε-SVR machine. The SMO regression algorithm is given in the following.

The SMO algorithm is a special case of the decomposition algorithm, whose work-ing set contains only 2 samples. The algorithm includes two phases, i.e., selecting a working set and solving a 2-dimension CQP problem. As for the normal regression dual problem [4], both α_i and α_i^* are greater than or equal to 0, and there must be a 0 among them. So one can use α_i to replace $(\alpha_i - \alpha_i^*)$ in the problem and obtain another version of the regression dual problem as follows

$$\underset{\alpha \in \Re^\ell}{\text{maximize}} \quad \sum_{i=1}^{\ell} \alpha_i y_i - \varepsilon \sum_{i=1}^{\ell} |\alpha_i| - \frac{1}{2} \sum_{i,j=1}^{\ell} \alpha_i \alpha_j K(x_i, x_j)$$

$$\text{subject to} \quad \sum_{i=1}^{\ell} \alpha_i = 0; -C \le \alpha_i \le C, (i = 1, \cdots, \ell) \tag{1}$$

And its regression function is $f(x) = \sum_{i=1}^{\ell} \alpha_i K(x_i, x) + b$. The SMO algorithm is derived from problem (1). Firstly, one heuristically selects a pair of dual variables, denoted by $(\alpha_{s1}, \alpha_{s2})$. To keep other dual variables fixed and consider equality con-straint $\sum_{i=1}^{\ell} \alpha_i = 0$, one can solve the 2-dimension CQP problem on working set $(\alpha_{s1}, \alpha_{s2})$ and obtain α_{s2}'s unclipped solution as below

$$\alpha_{s2}^{\text{new,unc}} = \alpha_{s2}^{\text{old}} + \frac{1}{\kappa} \{ (E_1 - E_2) - \varepsilon[\text{sgn}(\alpha_{s2}^{\text{new}}) - \text{sgn}(\alpha_{s1}^{\text{new}})] \} \tag{2}$$

where $\kappa \triangleq K_{11} + K_{22} - 2K_{12}$, $K_{ij} \triangleq K(\boldsymbol{x}_{si}, \boldsymbol{x}_{sj})$, and $E_i \triangleq f(\boldsymbol{x}_{si}) - y_{si}$, $(i, j = 1, 2)$. Considering inequality constraint $-C \le \alpha_i \le C$, one can clip the unclipped solution $\alpha_{s2}^{\text{new,unc}}$ to obtain the feasible solution as follows

$$\alpha_{s2}^{\text{new}} = U, \text{if } \alpha_{s2}^{\text{new,unc}} < U \left| \alpha_{s2}^{\text{new,unc}}, \text{if } U \le \alpha_{s2}^{\text{new,unc}} \le V \right| V, \text{if } \alpha_{s2}^{\text{new,unc}} > V \qquad (3)$$

where lower limit $U = \max(C_U^2, \alpha_{s1}^{\text{old}} + \alpha_{s2}^{\text{old}} - C_V^1)$, upper limit $V = \min(C_V^2, \alpha_{s2}^{\text{old}} + \alpha_{s2}^{\text{old}} - C_U^1)$. If $\alpha_{si}^{\text{old}} < 0, (i = 1, 2)$, then $C_U^i = -C, C_V^i = 0$; if $\alpha_{si}^{\text{old}} \ge 0, (i = 1, 2)$, then $C_U^i = 0$, $C_V^i = C$. By using α_{s2}^{new}, one can compute α_{s1}^{new} as below

$$\alpha_{s1}^{\text{new}} = \alpha_{s1}^{\text{old}} + \alpha_{s2}^{\text{old}} - \alpha_{s2}^{\text{new}} \qquad (4)$$

With the equation condition of KKT conditions, bias b can be computed as below

$$\begin{cases} b_{s1}^{\text{new}} = -E_1 - \varepsilon \cdot \text{sgn}(\alpha_{s1}^{\text{new}}) - (\alpha_{s1}^{\text{new}} - \alpha_{s1}^{\text{old}})K_{11} - (\alpha_{s2}^{\text{new}} - \alpha_{s2}^{\text{old}})K_{21} + b^{\text{old}}, \text{if } \left| \alpha_{s1}^{\text{new}} \right| \in (0, C) \\ b_{s2}^{\text{new}} = -E_2 - \varepsilon \cdot \text{sgn}(\alpha_{s2}^{\text{new}}) - (\alpha_{s1}^{\text{new}} - \alpha_{s1}^{\text{old}})K_{12} - (\alpha_{s2}^{\text{new}} - \alpha_{s2}^{\text{old}})K_{22} + b^{\text{old}}, \text{if } \left| \alpha_{s2}^{\text{new}} \right| \in (0, C) \end{cases}$$
$$(5)$$

If $\left| \alpha_{si}^{\text{new}} \right| \in (0, C)$, let $b^{\text{new}} = b_{si}^{\text{new}}$; if both fall within $(0, C)$, let $b^{\text{new}} = b_{s1}^{\text{new}}$ or b_{s2}^{new}; and if both fall outside $(0, C)$, let $b^{\text{new}} = (b_{s1}^{\text{new}} + b_{s2}^{\text{new}})/2$. As bias b is obtained by an iterative method, its numerical stability can be guaranteed.

Selecting a working set $(\alpha_{s1}, \alpha_{s2})$ is a prerequisite to implement the 2-dimension CQP in the SMO algorithm. The heuristic search is used to select the working set for improving the efficiency. The sample that seriously violates KKT conditions is selected as the first sample, and the second sample is the one that maximizes $\left| (E_1 - E_2) - \varepsilon[\text{sgn}(\alpha_{s2}^{\text{new}}) - \text{sgn}(\alpha_{s1}^{\text{new}})] \right|$. KKT conditions are used as the stopping criteria, i.e., the algorithm iterates until all training samples satisfy KKT conditions.

3   Coded Output Support Vector Machine

We can build a COSVM by using the SVR machines as basic units. Selecting a number system (NS) and designing a code mapping are two key steps before training the COSVM's SVR members.

The topology of the d-digit COSVM classifier is shown in Figure 1, where $[\boldsymbol{x}]_i$ is the i-th component of sample $\boldsymbol{x}$, $y_{:i}^d$ the i-th digit of output number coded with the d-digit NS, and α_j^i the dual variable corresponding to sample

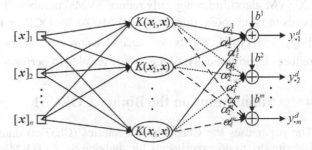

Fig. 1. Topology of the d-digit COSVM classifier

x_j of the i-th SVR member. Output digits of each SVR need to be rounded. The SVM is very similar to the generalized RBF network in topology. But the core ideas of two algorithms are essentially different: First, the former uses the SRM principle, while the latter uses the ERM principle. Second, the former acquires a unique optimal solution by solving a CQP problem, while the latter possibly falls into a local minimum point due to applying error back-propagation (BP) algorithm. The regularization term of the SRM principle can reduce the impact of large-error samples, especially, outlier samples. The ERM principle only pursues the smallest LF value without controlling the VC-dimension of function sets, which makes the ANN overfit noises and errors contained in the dataset. Differing from the strong coupling of the BP-based multi-output MLP's, the COSVM shows a weak coupling among SVR members. There is only an order relation among output coding digits, so each SVR member can be trained by itself. The COSVM's code coupling feature makes it easy to be implemented in parallel.

According to Vapnik's SRM theory, increasing margin Δ can reduce VC-dimension h, and then reduce confidence interval $\Gamma(h/\ell)$ of the structural risk, which will improve the GC of a learning machine. Margin comparing makes sense only if it is discussed within hyperspheres with the same radius (say the unit hypersphere). The code distance (CD) is defined as the normalized distance between two consecutive digits in a number system (NS). Thus the CD of the d-digit NS is $1/(d-1)$. For example, the CD of the binary, decimal, and natural NS is 1, 1/9, and $1/(M-1)$, respectively. Obviously, a large CD makes for a large margin Δ. The binary coding has the largest CD and most likely achieves a good GC, but it needs the largest number of output nodes. The increase of output nodes means the increase of computing time or resources (such as MCUs and memories used in a PI system). The multiple 2-class SVCs integrating algorithm is suitable for PI. Let the category number be $M > 2$, the "one-vs-rest" method needs to construct M 2-class SVCs, and the "one-vs-one" method needs $M(M-1)/2$ 2-class SVCs. The SVM DT algorithm needs $(M-1)$ 2-class SVCs. Supposing the i-th category has $\overline{\ell}_i$ samples, so the dataset has $\sum_{i=1}^{M} \overline{\ell}_i$ samples in total, which shows the learning problem scale increases rapidly with M increasing. When M is rather large, say the primary Chinese Character (CC) library having 3755 CCs (i.e. $M = 3755$), it is very time-consuming to train so many SVCs. It makes the PI unfeasible due to using too much MCUs. The COSVM algorithm can greatly reduce SVMs' number. If taking the d-digit NS, one needs to build about $[\log_d M]+1$ SVMs. As for OCR on the primary CC library, only $[\log_2 3755]+1=12$ SVMs are needed using the binary coding. The training time reduces logarithmically, so it is totally feasible to perform the PI.

4   Simulations on the Binary COSVM

The paper uses the Character Trajectories (ChaTra) dataset, downloaded from UCI website [8], to do experiments for studying the COSVM's characteristics. The dataset collected 2858 English letters, which belong to 20 types of lowercases with a single pen-down segment (except for 'f', 'i', 'j', 'k', 't', and 'x'). Each sample contains 3 time-series, i.e., x-coordinate, y-coordinate, and pen-tip force of the handwriting trajectory. Seventy samples are extracted from each character to compose the experimental dataset. The first 50 samples of each character are used to compose the training dataset, and the last 20 samples are used to compose the testing dataset. There are 20 characters in total, so the dataset contains 1400 samples. Each time signal can be expanded into a Fourier orthogonal function series on interval [0,1] as the form $S_r(t) = a_0 + \sum_{k=1}^{\infty} a_k \sqrt{2} \sin(2k\pi t) + \sum_{k=1}^{\infty} b_k \sqrt{2} \cos(2k\pi t)$, where a_0, a_k, b_k are Fourier expansion coefficients of the signal. Each sample contains 3 time-series, from which $3(2r+1)$ low-frequency features are extracted to form an input vector [19]. By taking $r = 2$, 15 features can be extracted from each sample to make a 15-dimension vector. One can attain the feature dataset to do the same operation on all samples. For easily referring to, the training dataset and testing dataset are denoted by $\mathcal{D}_{trn}^d$ and $\mathcal{D}_{tst}^d$, respectively, where d means the output label is coded with the d-digit NS. Two datasets $\{(\mathcal{D}_{trn}^M, \mathcal{D}_{tst}^M), (\mathcal{D}_{trn}^2, \mathcal{D}_{tst}^2)\}$ are obtained by coding output labels with the natural NS and binary NS. They are applied to do two experiments for studying the parameters impact on a binary COSVM classifier's performance. Two micro-computers are used in the experiments, whose configurations are dual-core CPU 2.10GHz, 2.0GB RAM (PC1), and dual-core CPU 2.00GHz, 1.0GB RAM (PC2).

For comparing performance between the binary COSVM and natural COSVM, experimental results of the natural COSVM are also given in Table 1 and 2. Only one output node is needed, and the experiments are run on PC1. Table 1 shows that the running time rises in the general trend with parameter C increasing. And the accuracy reaches the maximal value 16% at $C = 40$. Table 2 shows that the accuracy reaches the maximal value 43.5% at $\sigma = 0.4$, where the running time is quite large (45800 sec). There is no obvious advantage in contrast to the QP algorithm.

4.1   Member Parameters Optimizing Identically for the Binary COSVM

As to dataset $(\mathcal{D}_{trn}^2, \mathcal{D}_{tst}^2)$, five output nodes are needed. In this experiment, the SMO algorithm is still applied and is run on PC2. The parameter C and σ's impact on the binary COSVM's performance is given in Table 3 and 4, respectively. Table 3 shows that the accuracy increases slightly and then decreases slowly with parameter C increasing, and achieves the maximum 89.25% at $C = 6$. The COSVM has 5 SVR members, and there are differences in their running time. The 5th member needs most running time because it needs to learn the most complex decision function. Table 4 shows that the accuracy appears an upward trend, with parameter σ decreasing, and reaches the maximum 91.25% at $\sigma = 0.3$. Each SVR member's running time firstly rises and then declines. The experiment on parameter ε is not performed. According to the theory, reducing ε will make SV numbers increase, and then make running time rise. As for the binary NS dataset, the best accuracy achieves 91.25%, and the corresponding total running time (9968 sec) is only 13% of the natural NS dataset's. And if the PI is applied, the running time (4215 sec) is only 5.5% of the

Table 1. Parameter C's impact on the performance in the natural coded dataset ($\varepsilon = 0.30, \sigma = 6.0$)

Parameter C	10	20	30	40	50	60	70	80	90	100
Accuracy (%)	11.25	13.75	14	**16**	15.25	14.75	14.5	15.25	12.5	13.75
Training Time (s)	373.4	406.8	471.3	627.3	869.0	711.4	828.8	1310	1339	2042

Table 2. Parameter σ's impact on the performance in the natural coded dataset ($\varepsilon = 0.30, C = 40$)

Parameter σ	0.2	0.4	0.6	0.8	1.0	2.0	3.0	4.0	5.0	6.0
Accuracy (%)	33.25	**43.5**	35.75	35.25	30.25	19.25	14.25	15	17.5	15.25
Training Time (s)	13288	45800	64589	66607	25978	7807	2717	1503	1097	599.4

Table 3. Parameter C's impact on the performance in the binary coded dataset ($\varepsilon = 0.30, \sigma = 1.0$)

Parameter C	3	6	8	10	15	20	25
Accuracy (%)	88.5	**89.25**	88.5	88	87.25	87.75	87
Training Time (s)	2637,2860, 4293,4842, 9192	2246,4645, 4904,4142, 8223	3713,6372, 5852,7368, 13001	2969,8009, 4318,4722, 11489	4734,7004, 4201,4601, 13463	4982,10190, 7916,9348, 18919	7964,11167, 5550,5381, 13174

Table 4. Parameter σ's impact on the performance in the binary coded dataset ($\varepsilon = 0.30, C = 10$)

Parameter σ	0.3	0.5	1.0	2.0	4.0	6.0	8.0
Accuracy (%)	**91.25**	90	88	83	60.25	50.5	42.75
Training Time (s)	1004,1168, 1378,2203, 4215	1074,1826, 3344,4028, 5831	2969,8009, 4318,4722, 11489	2790,4114, 5808,5236, 9230	2771,2808, 4582,2983, 3219	918,1003, 2804,1092, 1246	478.6,427.7, 1149,1071, 802.6

natural NS dataset's. It is worth noting that five SVR members take the same parameter set. But if each member's parameter set is optimized respectively, the accuracy has the possibility of further improving. The formula of accuracy is given as below

$$Ar = \frac{number(y = \hat{y})}{\ell_{tst}} \times 100\% \tag{6}$$

where $number(\cdot)$ is a function counts the number of true category label codes matching predicted label codes (binary code), and ℓ_{tst} denotes the test data-set size.

4.2 Member Parameters Optimizing Respectively for the Binary COSVM

The SMO algorithm is run on PC2 in this experiment. Each SVR member's parameter-set (ε, σ, C) is optimized respectively. By taking $\varepsilon = 0.3$, Table 5 gives the

Table 5. Parameter (σ, C) 's impact on the performance in binary coded dataset ($\varepsilon = 0.30$)

σ	Performance	C					
		3	6	8	10	15	20
0.3	Accuracy (%)	**97.25**, 94.5, <u>**98.25**</u>, **98.25**, <u>95.75</u>	**97.25**, 94.25, <u>**98.25**</u>, **98.0**, 95.5	**97.25**, 94.5, <u>**98.25**</u>, **98.0**, 95.5	**97.25**, 94.25, <u>**98.25**</u>, **98.0**, 95.5	**97.25**, 94.25, <u>**98.25**</u>, **98.0**, 95.5	**97.25**, 94.25, <u>**98.25**</u>, **98.0**, 95.5
	Training Time (s)	952.4, 1055, 1784, 1781, 3623	970.4, 1134, 1670, 2204, 3198	1010, 1379, 2029, 2035, 3118	990.1, 938.7, 2090, 1956, 4288	1024, 1008, 1496, 1636, 3140	962.2, 1090, 1346, 1968, 3736
0.5	Accuracy (%)	97.25, <u>**96.75**</u>, 97.75, 98.25, 95.25	97.25, **95.75**, 97.75, <u>**98.75**</u>, 95.25	97.25, **96.0**, 98.0, 98.25, 95.25	97.0, 95.75, 98.0, 98.0, 95.5	97.0, 95.75, 98.0, 97.5, 95.5	97.0, 95.5, 98.0, 97.5, 95.5
	Training Time (s)	1277, 1781, 1631, 1372, 8135	1077, 1288, 1577, 2522, 6744	1113, 1595, 1726, 2444, 4424	1206, 1575, 2527, 3056, 4753	1011, 1382, 1267, 2000, 6027	1351, 1502, 1530, 2176, 4000
1.0	Accuracy (%)	<u>97.5</u>, 96.25, 97.75, 97.75, 94.75	97.25, 96.0, 97.5, 97.25, 95.5	97.0, 95.75, 97.25, 97.0, <u>95.75</u>	97.0, 96.25, 97.25, 96.75, 94.75	97.25, 96.0, 97.0, 96.5, 93.75	<u>97.5</u>, 96.0, 97.25, 96.75, 94.25
	Training Time (s)	1782, 2813, 4358, 5725, 8001	2922, 3108, 5005, 5232, 10847	2552, 5535, 6394, 5020, 10457	3586, 5031, 6896, 4460, 8815	4512, 8444, 5271, 5595, 11972	5715, 8765, 7729, 5393, 13145
2.0	Accuracy (%)	95.5, 92.25, 94.5, 95.5, 89.25	96.25, 93.25, 96.75, 96.25, 90.75	96.5, 94.0, 96.5, 96.25, 91.25	96.5, 94.0, 96.5, 95.75, 91.75	97.0, 94.5, 96.5, 95.5, 91.25	<u>97.5</u>, 94.0, 96.5, 94.5, 91.5
	Training Time (s)	1816, 1476, 2015, 3081, 5200	1830, 2619, 5475, 4785, 3592	3009, 3154, 4535, 4358, 5883	4053, 4067, 4816, 4819, 9447	4752, 6807, 6902, 7034, 9721	3769, 8981, 8396, 7745, 9550

parameter-set (σ, C) 's impact on each member' performance. Each SVR member has 24 different optional parameter-sets. Maximal accuracy rates of each member are marked with underlined numbers. There are three parameter-sets achieving the maximal accuracy 97.5% for SVR1, one parameter-set achieving the maximum 96.75% for SVR2, six parameter-sets achieving the maximum 98.25% for SVR3, one parameter-set achieving the maximum 98.75% for SVR4, and two parameter-sets achieving the maximum 95.75% for SVR5. One has to search for the binary COSVM's maximal accuracy in a space having 24^5=7962624 parameter-sets. The work needs 346 sec (about 5.77 min), and finds the maximal accuracy is 92.25%. There are 6×3×6×7×1=756 member combinations that can achieve this maximal accuracy. That is, there are six optimal parameter-sets for SVR1, three optimal parameter-sets for SVR2, six optimal parameter-sets for SVR3, seven optimal parameter-sets for SVR4, and one optimal parameter-set for SVR5. Accuracy rates related to these parameter-sets are marked with bold numbers in Table 5. The results show that each SVR member's optimal parameter-sets for the binary COSVM are not necessarily the optimal parameter-sets for each individual SVR. It is because that there exists slight inconsistency among SVR members' misrecognized samples. Although lots of work is done, the maximal accuracy increases by only 1% in comparison with that of the uniform parameter-set binary COSVM.

Table 6. Abbreviation list for academic terms

Abbreviation	Full name	Abbreviation	Full name
ANN	Artificial Neural Network	NS	Number System
BP	Back-Propagation	PI	Parallel Implementation
CC	Chinese Character	RBF	Radial Basic Function
CD	Code Distance	SLT	Statistical Learning Theory
COSVM	Coded Output Support Vector Machine	SMO	Sequential Minimal Optimization
CQP	Convex Quadratic Programming	SRM	Structural Risk Minimization
DT	Decision Tree	SVC	Support Vector Classifier
ERM	Empirical Risk Minimization	SVM	Support Vector Machine
GC	Generalization Capability	SVR	Support Vector Regression
KF	Kernel Function	VC	Vapnik-Chervonenkis Dimension

5   Conclusion

This paper proposes a COSVM by coding category label with a number system (NS), which only needs to train about $[\log_d M]+1$ sub-SVRs. It presents a logarithmic relationship between the sub-SVR number and category number M. Moreover, the SVR members of a smaller d NS COSVM are relatively simple, so their training time can be reduced significantly by using the SMO algorithm. Experimental results on the ChaTra dataset show that the binary COSVM does far outperform the natural COSVM on the accuracy and time cost indexes. The binary COSVM needs more sub-SVR members than the natural COSVM, but whose total time cost is still less than that of the natural COSVM if using the SMO algorithm. Furthermore, the time cost can be reduced exponentially by using the PI algorithm. The parameters-optimizing respectively binary COSVM slightly outdoes the parameters-optimizing identically binary COSVM. The follow-up research work is how the code mapping impact on the generalization capability of a COSVM classifier? Doing systematic comparative experiments is also a meaningful research work.

Acknowledgements. This work is partially supported by the Natural Science Foundation of Guangdong Province (China) under Grant no. 10451064101004837. The authors gratefully acknowledge anonymous reviewers for their constructive comments. The first author would also like to thank his family.

References

1. Boser, B.E., Guyon, I.M., Vapnik, V.N.: A training algorithm for optimal margin classifiers. In: Haussler, D. (ed.) 5th Annual ACM Workshop on Computational Learning Theory, pp. 144–152. ACM Press, Pittsburgh (1992)
2. Burges, C.J.C.: A tutorial on support vector machines for pattern recognition. Data Mining and Knowledge Discovery 2(2), 121–167 (1998)
3. Cortes, C., Vapnik, V.N.: Support vector networks. Machine Learning 20, 273–297 (1995)
4. Cristianini, N., Shawe-Taylor, J.: An Introduction to Support Vector Machines. Cambridge University Press, Cambridge (2000)
5. Deng, N.Y., Tian, Y.J.: A New Method in Data Mining: Support Vector Machine. China Science Press, Beijing (2004) (in Chinese)
6. Dietterich, T.G., Bakiri, G.: Solving multi-class learning problems via error-correcting output codes. Journal of Artificial Intelligent Research 2, 263–286 (1995)
7. Joachims, T.: Making large-scale SVM learning practical. In: Schölkopf, B., Burges, C.J.C., Smola, A.J. (eds.) Advances in Kernel Methods - Support Vector Learning, pp. 169–184. MIT Press, Cambridge (1999)
8. Newman, D.J., Hettich, S., Blake, C.L., et al.: UCI Repository of machine learning databases. University of California, Irvine (1998),
   http://www.ics.uci.edu/~mlearn/MLRepository.html
9. Osuna, E., Freund, R., Girosi, F.: An improved training algorithm for support vector machines. In: Principe, J., Gile, L., Morgan, N. (eds.) Neural Networks for Signal Processing VII - The 1997 IEEE Workshop, pp. 276–285. IEEE Press, Piscataway (1997)
10. Platt, J.C.: Sequential minimal optimization: A fast algorithm for training support vector machines. Technical report, MSR-TR-98-14, Microsoft Research (1998)
11. Schölkopf, B., Smola, A.J., Williamson, R., et al.: New support vector algorithms. Neural Computation 12, 1207–1245 (2000)
12. Smola, A.J., Schölkopf, B., Müller, K.R.: General cost functions for support vector regression. In: Downs, T., Frean, M., Gallagher, M. (eds.) 9th Australian Conference on Neural Network, pp. 79–83. University of Queensland, Brisbane (1998)
13. Smola, A.J.: Learning with Kernels. PhD dissertation. Technische Universität Berlin, Berlin (1998)
14. Smola, A.J., Schölkopf, B.: A tutorial on support vector regression. Statistics and Computing 14, 199–222 (2004)
15. Suykens, J.A.K., Vandewalle, J.: Least squares support vector machine classifiers. Neural Processing Letters 9(3), 293–300 (1999)
16. Vapnik, V.N.: Statistical Learning Theory. John Wiley & Sons, New York (1998)
17. Vapnik, V.N.: An overview of statistical learning theory. IEEE Transactions on Neural Networks 10(5), 988–999 (1999)
18. Vapnik, V.N., Chervonenkis, A.: On the uniform convergence of relative frequencies of events to their probabilities. Theory of Probability and its Applications 16(2), 264–280 (1971)
19. Ye, T., Zhu, X.F.: The bridge relating process neural networks and traditional neural networks. Neurocomputing 74(6), 906–915 (2011)
20. Zhu, Y.P., Dai, R.W.: Text classifier based on SVM decision tree. Pattern Recognition & Artificial Intelligence 18(4), 412–416 (2005) (in Chinese)

Edge Multi-scale Markov Random Field Model Based Medical Image Segmentation in Wavelet Domain[*]

Wenjing Tang[1,2], Caiming Zhang[1], and Hailin Zou[2]

[1] School of Computer Science and Technology, Shandong University, Jinan, China
`twj_tang@126.com, czhang@sdu.edu.cn`
[2] School of Information and Electrical Engineering, Ludong University, Yantai, China
`zhl_8655@sina.com`

Abstract. The segmentation algorithms based on MRF often exist edge block effect, and have low operation efficiency by modeling the whole image. To solve the problems the image segmentation algorithm using edge multiscale domain hierarchical Markov model is presented. It views an edge as an observable data series, the image characteristic field is built on a series of edge extracted by wavelet transform, and the label field MRF model based on the edge is established to integrate the scale interaction in the model, then the image segmentation is obtained. The test images and medical images are experimented, and the results show that compared with the WMSRF algorithm, the proposed algorithm can not only distinguish effectively different regions, but also retain the edge information very well, and improve the efficiency. Both the visual effects and evaluation parameters illustrate the effectiveness of the proposed algorithm.

Keywords: medical image segmentation, MRF, wavelet, edge.

1   Introduction

Image segmentation is an important means of medical image processing, and is crucial to a variety of medical applications including image enhancement and reconstruction, surgical planning, disease classification, data storage and compression, and 3D visualization. Medical images, however, are particularly difficult to be segmented due to the diversity of imaging conditions and the complexity of imaging content. At present, to tackle the difficult problems of image segmentation, researchers have proposed a variety of methods, such as the segmentation based on edge [1], region [2], statistics, fuzzy theory [3-5], and active contour model[6,7], and so on.

---

[*] Project supported by the National Nature Science Foundation of China(Nos 61020106001, 60933008, 60903109, 61170161), the Nature Science Foundation of Shandong Province(Nos.ZR2011G0001, ZR2012FQ029) , and Nature Science Foundation of Ludong University(No. LY2010014)

D.-S. Huang et al. (Eds.): ICIC 2013, LNAI 7996, pp. 56–63, 2013.

For the segmentation based on statiscal methodology, Markov Random Fields(MRFs) are widely used that constitute a powerful tool to incorporate spatial local interactions with a global context[8,9]. In general, to solve different problems image data is modeled as the two MRFs (label and characteristic field), which can take spacial interactions of pixels into account, and obtain the segmentations in a probabilistic framework based on the Bayesian theory.

According to whether to use the image information of different scales, image segmentation algorithms are divided into two categories: single scale and muti-scale MRFs modeling segmentation. The former brings great improvement in image segmentation, but causes the over-segmentation or under segmentation because of being unable to obtain accurate segmentation characteristics. In order to reduce the computational complexity, capture the image structural information with different resolution, and reduce the dependence on the initial segmentation, multi-scale random field(MSRF) is proposed by Bouman etc[10]. In MSRF model the interactions of label field between scales are constructed with quad-tree structure or pyramid structure, and hierarchical MRF model is designed, which makes the model calculation avoid the complicated iterative process. But its characteristic field is still built in image space, non-directional and non-redundant, so it is hard to describe nonstationarity of image. The wavelet transform can obtain the singular of image in different resolution, and express the nonstationary of image, so some scholars have proposed a segmentation algorithm based on wavelet domain multiscale random field model(WMSRF)[11]. In WMSRF, its characteristic field is modeled on a series of wavelet domain, and has different vector at different scale. It significantly improves the regional, but there is a problem of insufficient statistics of local spatial information. The segmentation algorithms which are used to model the whole image based on MRF have boundary block effect and low efficiency. To solve these problems, edge multi-scale markov random field model based image segmentation is presented. Its general idea is: the image edge can be made use of dividing the image into different regions. And the wavelet transform with directional, non-redundant, multi-resolution features provides a clear mathematical framework for multi-scale. Therefore, the image characteristic field is constructed on a series of edges extracted by wavelet transform, and the label field model based on edge is also established considering interscale and intrascale markovianity. Then the parameters are estimated, the likelihood of characteristic field of the current scale and its segmentation result are computed, and the segmentation result at the coarser scale is transferred to result at the next fine scale, thus the final segmentation result is achieved. The whole framework of the paper is shown in figure 1.In the experiments, we use medical images to test our method.

The rest of the paper is organized as follows: Section 2 presents the multi-scale markov random field model based on edge. Section 3 descripts the image segmentation algorithm in detail using the model. Then, the experiments of the segmentation and some discussions are made in section 4. We conclude this paper with section 5.

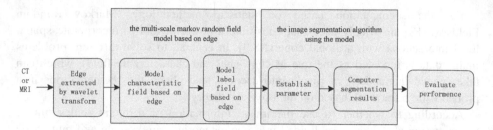

Fig. 1. The whole framework of the paper

2   Edge Multiscale Markov Random Field Model

2.1   Edge Multiscale MRF Model of Characteristic Field

The image edge is its signal discontinuity and singularity. To construct a model, we must first extract the edge using wavelet transform, i.e., the modulus maxima of wavelet coefficients. Image edge extraction by wavelet transformation has been discussed in detail in the literature [12], which can not be repeated here.

Given the image y defined on the grid S, $J-1$ layer wavelet decompositions on y together with the original image form J scales expression of the image multi-resolution. Each scale is represented by the corresponding layer number $n(1 \le n \le J-1)$. Let Wedge denote the edge extracted by the above method at each scale. The edge of the four bands (LL, LH, HL and HH) is contained in the lowest resolution, while the edge of three bands (LH, HL and HH) is contained in other resolutions. These edges of different bands are composed of eigenvector, then eigenvectors of each resolution form multi-scale characteristic field, expressed as $\text{Wedge} = \left\{ \text{Wedge}^{(0)}, Wedge^{(1)}, ..., Wedge^{(J-1)} \right\}$, where the eigenvector of scale $n = J-1$ is $\text{Wedge}^{(n)} = \left[\omega_{ij}^{LL,n}, \omega_{ij}^{LH,n}, \omega_{ij}^{HL,n}, \omega_{ij}^{HH,n} \right]$, and the eigenvector of scale $n(1 \le n < J-1)$ is $Wedge^{(n)} = \left[\omega_{ij}^{LH,n}, \omega_{ij}^{HL,n}, \omega_{ij}^{HH,n} \right]$. It can be seen that the characteristic sequences from the coarser scale to fine scale form the first order markovian chain with the order of the scale or resolution.

Wavelet coefficients distribution is a kind of ditribution with high peak and long heavy tail. Therefore, the characteristic field of edge is modeled by Gauss-MRF. The formula is as below:

$$P\left(Wedge \mid D\right) = \prod_{n=0}^{J-1} f\left(Wedge^{(n)} \mid D^{(n)}\right) = \prod_{n=0}^{J-1} \prod_{(i,j)\in Sedge} f\left(\omega_{ij}^{(n)} \mid d_{ij}^{(n)} = m, \omega_{\eta_{ij}}^{(n)}\right)$$

$$= \prod_{n=0}^{J-1} \prod_{(i,j)\in Sedge} \frac{1}{\left(\sqrt{2\pi}\right)^{r} \left|\sum_{m}^{(n)}\right|^{1/2}} \exp\left\{-\frac{1}{2}\left(e_{m,ij}^{(n)}\right)'\left(\sum_{m}^{(n)}\right)^{-1}\left(e_{m,ij}^{(n)}\right)\right\}$$

$$(1)$$

Where D represents the image edge label field, Sedge is the set of edge pixels, r is the dimension of the wavelet coefficients vector, $e_{m,ij}^{(n)}$ denotes Gaussian noise of zero-mean, and $\sum_{m}^{(n)}$ denotes conditional covariance matrix of class m.

2.2   Edge Multi-scale MRF Model of Label Field

For different resolution the region of characteristics field is different, so the correspondent label field is established on the multi-resolution edge, i.e., the set of label field is

$$D = \left\{ D^{(0)}, D^{(1)}, ..., D^{(J-1)} \right\}$$

where $D^{(n)}$ denotes the label field at scale n.

Similar with the characteristic field based on edge, the label sequences of edge from large scale to small scale also form the first order markovian chain with the order of the scale or resolution. Therefore, the label field of edge is modeled by MSRF based on the following important assumptions:

Assumption 1. The reality of the label field at the scale n depends only on the reality at the scale n+1, i.e., $P\left(D^{(n)} \mid D^{(l)}, l > n\right) = P\left(D^{(n)} \mid D^{(n+1)}\right)$。

Assumption 2. Label of each edge pixel depends on the label set of neighborhood location at the next coarser scale.

Relying on these assumptions, edge markovian representation across scales in the MSRF model can be considered that the label of node s depends on not only its parent node $\rho(s)$, but also three brothers $u(s)$ of $\rho(s)$, so the pyramid model is applied to describe it, i.e.,

$$P\left(d^{(n)} \mid d^{(n+1)}\right) = \prod_{s \in Sedge} P\left(d_s^{(n)} = m \mid d_{\rho(s)}^{(n+1)} = i, d_{u(s)}^{(n+1)} = \{j, k\}\right)$$

$$= \prod_{s \in Sedge} \left[\frac{\theta}{7}\left(3\delta_{m,i} + 2\delta_{m,j} + 2\delta_{m,k}\right) + \frac{1-\theta}{|L|} \right]$$

(2)

where m, i, j, $k \in L$ are class labels, $|L|$ is the number of edge label,

$$\delta_{m,k} = \begin{cases} 0\,(m \neq k) \\ 1\,(m = k) \end{cases}$$, parameter θ denotes the possibility of the same label between

s and $\rho(s)$, also known as the interaction across scales.

The segmentation algorithms based on MRF are parameterized. Image segmentation algorithms using Bayesian decision[13] use the similarity of different features of the image and the model of known parameters to label different regions, and obtain the segmentation results at different scales. With the theorem, combined with equation (1), (2), the parameters involved in multi-scale MRF model based on edge are: ① covariance matrix $\sum_m^{(n)}$ and mean vector $e_{m,ij}^{(n)}$ in edge characteristic field; ② the interaction parameter θ across scales in edge label field model. These parameters are defined as a parameter vector, denoted by Θ, to be estimated by the EM(Expectation Maximization)algorithm.

3   Image Segmentation Using Edge Multiscale MRF Model in Wavelet Domain

In multi-resolution image segmentation, every pixel at coarser scales is behalf on a large region, and every pixel at finer scales is behalf on a small region in the original resolution image. This illustrates that the segmentation error at different scales is different to the total cost function. So Bouman etc. proposed the SMAP (sequential maximum a posterior)[9]. It introduces different penalty factor at different scales, and assigns different weights to segmentation errors at different scales. The greater are the segmentation errors on the final cost function at the coarser scales, the smaller are those at the finer scales. This is similar to human visual process. Using the non-iterative algorithm of the SMAP estimator, we can formally express the image segmentation as follows:

$$\hat{d}^{(J-1)} = \arg\max \log f\left(Wedge^{(J-1)} \mid d^{(J-1)}\right) \tag{3}$$

$$\hat{d}^{(n)} = \arg\max \left\{ \log f\left(Wedge^{(n)} \mid d^{(n)}\right) + \log P\left(d^{(n)} \mid d^{(n+1)}\right) \right\} \tag{4}$$

The segmentation result at the largest scale is estimated by equation (3) using maximizing the likelihood. After obtaining the segmentation result at next finer scale, we can calculate the segmentation result at the current scale by equation (4) that takes markovian dependency between different scales and the likelihood at the current scale into account. The process is repeated until the segmentation result at the original image is obtained. The steps of segmentation are described as follows:

Step1. Set the initial interaction in scales;

Step2. Calculate the likelihood of the current scale with the interaction in scales;

Step3. Computer the segmentation result at scale J-1 by equation (3) ;

Step4. For n from J-2 to 0,

(i)Calculate the multi-scale likelihood of the characteristic field at the current scale with the interaction parameters between scales;

(ii)Correct the new interaction parameters in scales ;

(iii)Obtain the segmentation result at the current scale by equation (4);

Step5. Repeat Step2~Step4, the algorithm is over.

It can be seen that during the segmentation process only a bottom-up likelihood calculation process and a top-down segmentation process are needed, and no complex iterative calculation is required.

4   Experiments

To demonstrate the proposed segmentation algorithm in this paper, the medical images are used to test it. We select Haar wavelet and carry out three-level decomposition. Figure 2 shows the testing medical image whose size is 512×512 and their segmentation results with the proposed algorithm and WMSRF introduced in [11].

From figure 2, both algorithms can get good region in visual, but the proposed algorithm in this paper is better in details. And compared with WMSRF, our results have lower boundary block effect, and is closer to the real segmentation results. Furthermore, the lesion in the diagram (b) can be also detected by the proposed algorithm, which can provide help for subsequent image analysis and understanding.

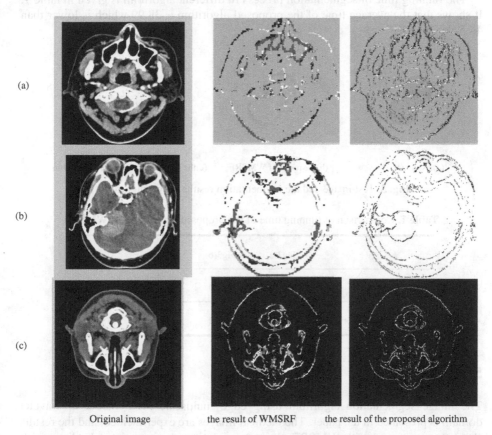

    Original image   the result of WMSRF   the result of the proposed algorithm

Fig. 2. Medical images and their segmentation results

From the quantitative point of view in order to verify the performance of our proposed algorithm, we adopt the test image shown in figure 3(a) whose size is 256×256, carry on three-level decomposition using Haar wavelet, and is compared with the WMSRF algorithm. The segmentation results using WMSRF and the proposed algorithm are respectively shown in figure3(b) and (c). Edge detection rate is calculated to compare two algorithms, which is defined as the proportion of the true edge in the segmentation result, and the formula is

$$P = \frac{E_a}{E_{real}} \tag{5}$$

Where E_a is the number of correct edge pixels in the segmentation result, and E_{real} is the number of the true edge pixels. Obviously, the higher is the edge detection, the better is the algorithm performance. The edge detection of the proposed algorithm is 93.5%, while the edge detection of the WMSRF algorithm is 88.27%. Apparently the proposed algorithm is better, which is consistent with the visual effects.

The running time of segmentation process in different algorithm is given in table 1. It shows that the average time of the proposed algorithm is 38.99 which is lower than 44.51 of WMSRF.

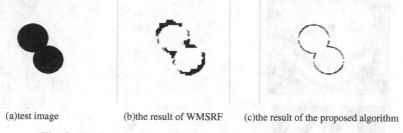

(a)test image   (b)the result of WMSRF   (c)the result of the proposed algorithm

Fig. 3. Test image and the segmentation results of two algorithms

Table 1. Comparison of running time of the proposed algorithm and WMSRF

	Fig3(a)	Fig3(b)	Fig3(c)	Average
WMSRF	44.48	48.70	40.36	44.51
The proposed algorithm	38.67	41.72	36.59	38.99

5   Conclusion

The image segmentation algorithm using edge multiscale MRF model in wavelet domain is presented in the paper. The medical images are experimented, and the results show that compared with WMSRF, the proposed algorithm can not only distinguish effectively different regions, but also retain the edge information very well, improve the efficiency, and provide help for the subsequent target identification and detection.

References

1. Qin, X.-J., Du, Y.-C., Zhang, S.-Q., et al.: Boundary Information Based C_V Model Method for Medical Image Segmentation. Journal of Chinese Computer Systems 32(5), 972–977 (2011) (in Chinese)
2. Sun, R.-M.: A New Region Growth Method for Medical Image Segment. Journal of Dalian Jiaotong University 31(2), 91–94 (2010) (in Chinese)
3. Ji, Z.-X., Sun, Q.-S., Xia, D.-S.: A framework with modified fast FCM for brain MR images segmentation. Pattern Recognition 44, 999–1013 (2011)
4. Li, B.N., Chui, C.K., Chang, S., et al.: Integrating spatial fuzzy clustering with level set methods for automated medical image segmentation. Computers in Biology and Medicine 41, 1–10 (2011)
5. Tang, W., Zhang, C., Zhang, X., et al.: Medical Image Segmentation Based on Improved FCM. Journal of Computational Information System 8(2), 1–8 (2012)
6. He, L., Peng, Z., Everding, B., et al.: A comparative study of deformable contour methods on medical image segmentation. Image and Vision Computing 26, 141–163 (2008)
7. Truc, P.T.H., Kim, T.S., Lee, S., et al.: Homogeneity and density distance-driven active contours for medical image segmentation. Computers in Biology and Medicine 41, 292–301 (2011)
8. Geman, S., Geman, D.: Stochastic relaxation Gibbs distribution and the Bayesian restoration of images. IEEE Transactions on Pattern Analysis and Machine Intelligence 16, 721–741 (1984)
9. Li, S.Z.: Markov Random Field Modeling in Image Analysis. Springer, Berlin (2001)
10. Bouman, C.A., Shapiro, M.: A Multiscale Random Field Model for Bayesian Image Segmentation. IEEE Transaction on Image Processing 3(2), 162–178 (1994)
11. Liu, G., Ma, G., Wang, L., et al.: Image modeling and segmentation in wavelet domain based on Markov Random Field——Matlab Environment. Science Press, Beijing (2010)
12. Zhang, Y.: Image Segmentation. Science Press, Beijing (2001)
13. Choi, H., Baraniuk, R.G.: Multiscale image segmentation using wavelet-domain hidden Markov models. IEEE Transactions on Image Processing 10(9), 1309–1321 (2001)

Model of Opinion Interactions Base on Evolutionary Game in Social Network[*]

Li Liu[1], Yuanzhuo Wang[2], and Songtao Liu[3]

[1] School of Automation & Electrical Engineering,
University of Science and Technology Beijing, Beijing 100083, China
[2] Institute of Computing Technology, Chinese Academy of Sciences, Beijing, 100190, China
[3] Beijing AVIC Information Technology Company, Aviation Industry Corporation of China
Liuli@ustb.edu.cn, wangyuanzhuo@ict.ac.cn

Abstract. Due to convenient information change at individual level in recent years, social network has become popular platform of information dissemination, more precisely, micro-blog has attracted a lot of attention, thus modeling the real opinion interactions on micro-blog is important. This paper is devoted to apply evolutionary game models to reveal some basic rules and features of opinion interactions in micro-blog. We have divided the users into three different kinds of player with respective payoff. Then the evolutionary game model simulates a group of individuals participating in a discussion, aiming at persuading opponent into an agreement. Our results show some characteristics consistent with facts, and also offer possible explanations for the emergence of some real features.

Keywords: evolutionary games, replicator dynamics, opinion interactions, social network.

1 Introduction

Direct, frequent, and low-cost information exchange at individual level is allowed with widespread of Internet, attendant to the extensive use of blogs, bulletin boards, forums and micro-blog [1][2], all of these information dissemination platform are part of social network. With the development of social network service, networked individuals online become a large group, which is usually called "virtual community" in the complex network science. Information disseminates through these networked users can make public opinion and behavior [3], and rumors can spread astoundingly fast via social networks[4]. How to understand and analyze the rules of information dissemination in social network is one of the most challenging issues.

Tracing back to its fundamental, it is opinion interactions between networked information users online that promote the propagation of information. Thus it will be of great theoretical and practical value to model the process of opinion interactions

---

[*] This work is supported by the State High-Tech Development Plan (No.2013AA01A601), and National Natural Science Foundation of China(No.61173008).

D.-S. Huang et al. (Eds.): ICIC 2013, LNAI 7996, pp. 64–72, 2013.

between social network users, in order to analyze, even predict dissemination of information and network structure changes in future work. Socialists research more on the analysis of user's information behavior long time ago, but it remains on the qualitative level. As the evolution of user's information behavior, more and more scientists pay more attention on this issue, considering game relationship between users, influence of network community structure on behavior evolution, dynamic mechanism of maintaining group stability and so on. However, we still don't have a system of models and methods to solve this issue.

Stochastic models are widely used to describe network information behavior. Rao et.al propose a hybrid Markov model to analyze user's information seeking behavior using data mining, which is able to predict wanted categories[5]. Chung et.al propose a Markov model to analyze human behavior based on video stream[6].Rafael propose a model to describe human behavior when infectious disease outbreak[7]. Stochastic petri nets are also important direction of this area[8][9], mainly in attack action and security analysis. These models describe and analyze user's behavior on individual level, not reflecting the evolution over all networked behavior properties.

Thus, models concerning evolution of user's behavior and community structure appeared in recent years, represented by complex network and related fields. Social network is a huge network with intricate connections, so it is unpredictable and may appear chaotic state, complex network can uncover some macroscopic properties with analysis of network topology from overall situation. Community detection is helpful when try to analyze connection of structure and evolution. Xu et.al analyze efficiency of local control of virus in complex network with different topologies[10]. Shen et.al [11] point out that online social network have different underlying mechanisms with the real-world. Akihiko et.al present the feasibility of evolutionary game applied to analyze social system diversity and evolution, and interaction within the system[12], and it has been a common method in behavior analysis, like economic behavior[13], human behavior[14], network resource allocation[15] and so on.

In this paper, we combine opinion interaction with evolutionary game theory and present a framework considering that people may care about several aspects of their benefits. Three different kinds of users with various characteristic and payoffs are defined, which effect on individuals' viewpoint. In order to illustrate our model, calculation of evolutionary stable strategy is proposed in the paper using replicator dynamics, also an simulation is provided from which we get some appropriate laws that may help us to understand how opinion interactions evolution during the process. Besides, we try to explain the emergence of cooperation in social network, which is common result of user's opinion evolution in several other scenes.

2   General Features of Social Network Users

Microblog is a new form of communication through which users can describe their current status in short posts distributed by instant messages, mobile phones, email or the Web. SinaWeibo, a popular microblogging tool in China has seen a lot of growth since it has been launched in 2009. A lot of researchers are exploring explanations to opinion formation and evolution by modeling and simulating opinion interactions from an individual level.

Opinions in this paper are defined as user's view values about some certain information, a significant part of information diffusion are shown with the form of opinion interactions, through which users accept new messages, express their own views so that more users get to know this information. And in the model, we simply assume that users' opinions are presented by a number between 0 and 1. Numbers between 0 and 1 represent users' view values about certain information from negative to positive.

The model describes the opinion interactions via micro-blog, and how can users choose to cooperate based on selfish assumptions. Focus of this paper are how players choose and update their own strategies according to some rules, analyzing influences that different kind of users have brought to group. In the model, we suppose that the human being have bounded rational, which means that they prefer to choose strategy bringing themselves the best benefit through a trial-and-error way, though they may deviate from the best choice in reality.

Based on theoretical studies of opinion interactions in SinaWeibo, this paper focuses on studying how different kinds of online discussants interact according to the same guideline. The model studies three typical kinds of people online. According to Louis[16], there're three types of user accounts on SinaWeibo, regular user accounts, verified user accounts and the expert (star) user accounts. From the same paper, we can see that a large percentage of trends in SinaWeibo are actually due to artificial inflation by some fake users. Thus, we redefined three kinds of users on the basis of their patterns of behavior. Opinion leaders are usually famous people in a certain field, whose comments are more influential than normal ones. They tend to devote more time to investigate and think over the event, once taking part in the discussion, they would follow the event closely for a period of time. The number of their followers is always much larger than the users they would follow. Chatters are such kind of people: they may be new to a group, they enjoy commenting on everything and taking part in discussion actively or simply for some purpose. They comment on things casually without any special attention on a typical event, thus less influential. Their number of followers is exactly the opposite to the opinion leaders. The third type is General people, who are likely to stay in some fixed groups, unlikely to communicate with people out of the group. They would consider the issue for a while, but no comments on them. Their number of followers and people they are following tend to stay in a balance.

3   The Evolutionary Game of Opinion Interactions

Interactions between individuals are described as games involved several players, under certain conditions, players analyze, select and implement strategies according to information they have, then get responding benefits. Evolutionary game theory doesn't require players to be fully rational, we assume the user progressively update their behavior to respond to the current policy environment according to replicator dynamics, which describe a replicator process through ordinary differential equations.

Table 1. Payoffs for different users

Opinion Leader	Opponent's Strategy	
	Keep	*Agree*
Keep	0	$p_1V+q_1U-r_1W+1/d$
Agree	$-p_1V-q_1U-r_1W+1/d$	$p_1V-q_1U-r_1W+1/d$
Chatter	Opponent's Strategy	
	Keep	*Agree*
Keep	Q	$p_2V+q_2U-r_2W+1/d$
Agree	$-p_2V-q_2U-r_2W+1/d$	$p_2V-q_2U-r_2W+1/d$
General People	Opponent's Strategy	
	Keep	*Agree*
Keep	$-r_3W$	$p_3V+q_3U-r_3W+1/d$
Agree	$-p_3V-q_3U-r_3W+1/d$	$p_3V-q_3U-r_3W+1/d$

3.1 Basic Conditions of the Model

The model describe the phenomenon that people are likely to persuade each other rather than change their own opinion. Firstly, consider a game with three strategies, keep its own mind, agree with opponent and totally change its mind. Obviously, totally change its mind to agree to opponent is a opponent's strictly dominant strategy, a rational individual won't choose it since no matter what strategy opponent choose, the benefit individual gets from strategy "change" is lowest. Thus, such a strategy will be eliminated during evolution, strategy "change" will not be discussed in the rest of the paper.

Users are divided into three different kinds of people with respective payoffs, but they have the same strategy set. There're two strategies, agree with opponent or keep its own mind. According to the analysis of different users, payoffs of them are as following.

p_iV stands for the benefit user will get when opponents agree with its opinion, q_iU represents benefits it gets from holding its own opinion, and r_iW is the loss of participating. d is the opinion differences between two players, so $1/d$ is the extra earnings from compromise, since we have made the assumption of d is a number between 0 and 1, thus, the maximum value of d is 1. As what can be seen in TABLE I, the smaller d is, the more earnings user will get when they choose to agree, which means the closer two users' opinions are, the more attractive to agree with each other. According to the analysis before, such assumptions are made, $p_2>p_1>p_3$, $q_1>q_3>q_2$, $r_3>r_1$ and $r_2=0$ because they're eager to participate.

3.2 Game Solution and Analysis

Replicator dynamics are used in the paper to analysis this problem, which is able to provide the evolution rate towards the equilibrium at a certain time point, with current and average earnings as inputs.

Since three different kinds of users are proposed above, we will take opinion leaders for an example of analysis, analyzing evolutionary process with replicator dynamics. The evolutionary process is a repeated game with bounded rationality, during which individuals keep adjusting strategy to pursuit the best. It is assumed that x stands for the probability of user 1 choosing keep strategy, then 1-x will be the probability of agree strategy, y and 1-y respectively represent the probability of user 2 choosing keep and agree.

The expected profit of user 1 choosing strategy keep is equation (1), expected profit of user 1 choosing strategy agree is equation (2), then the total expected profits of user 1 will be equation (3), based on replicator dynamics, grown rate over time of x is equation (4). In accordance of similar calculation we can conclude with equation (5). Similarly, when it comes to chatter and general people, we can get similar results. Three kinds of users lead to six different opinion interaction conditions.

- An opinion leader meets an opinion leader.
- An opinion leader meets a chatter.
- An opinion leader meets a general person.
- A chatter meets a chatter.
- A chatter meets a general person.
- General people interaction.

When dx/dt = dy/dt =0, five local equilibrium points can be draw, (0,0), (0,1), (1,0), (1,1) and $(-2q_1U+1/d)/(p_1V-q_1U+r_1W),(2p_1V+1/d)/(p_1V-q_1U+r_1W))$ among which only (0,0) and (1,1) are evolution stable equilibrium.

$$E_1^1 = y \cdot 0 + (1-y)(p_1V + q_1U - r_1W + 1/d) \tag{1}$$

$$E_1^2 = y(-p_1V - q_1U - r_1W + 1/d) + (1-y)(p_1V - q_1U - r_1W + 1/d) \tag{2}$$

$$E_1 = xE_1^1 + (1-x)E_1^2 \tag{3}$$

$$dx/dt = x(E_1^1 - E_1) \tag{4}$$

$$dy/dt = x(E_2^1 - E_2) \tag{5}$$

4   Simulation Results

In order to focus on the parameters we are interested, simple assumptions of model parameters are made here. The payoff value of opponent changing opinion, V=10; payoff value of holding itself opinion, U=8; energy loss of taking part in a discussion, W=5. Besides, Q---profit of chatters get involved and d---difference between two users opinions are changeable. Also we made $p_1=1/4$, $p_2=1/2$, $p_3=1/8$, $q_1=1/3$, $q_2=1/9$, $q_3=1/6$, $r_1=1/10$, $r_3=1/5$, in accordance with the relationship of $p_2 > p_1 > p_3$, $q_1 > q_3 > q_2$, $r_3 > r_1$ and $r_2 = 0$.

(0,0) stands for strategy combination of (agree, agree), (1,1) represents (keep, keep), (0,1) stands for (agree, keep) and (1,0) is (keep, agree).

When 1/d = 1, payoff users get from strategy keep is much more than benefit of strategy agree, so that both of them would like to stay keep, but if this is really happen, none of them will get any profit; if they have made a compromise to (agree, agree), then they will get better earnings. But, if one of them betray, which means the user A stays in keep when its opponent B trust A will choose agree, then A will get much higher benefit than B. This is what happens in Fig.1, even know a possibility of betrayal, B still choose to trust A will choose agree, so B stick to agree strategy, which shows a strong cooperative characteristic in selfish human beings.

This phenomenon shows more clearly in Fig.2, where exists two stable strategies when two chatters debate, (keep, keep) and (keep, agree), in such cases two keeps will bring them the best earnings, but still, a part of users choose to stay in agree. The diagram down in Fig.2 shows that when the start value of y is smaller, which will lead user2 to trust user 1 will choose to cooperate, so user 2 is more likely to choose agree.

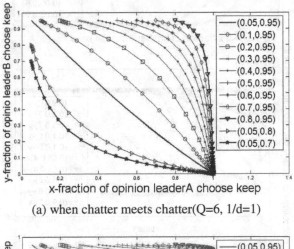

(a) when chatter meets chatter(Q=6, 1/d=1)

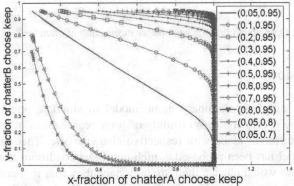

(b) when opinion leader meets opinion leader (1/d = 1)

Fig. 1. Opinion evolution

In Fig.2, strategy probability distribution over time is shown, what reveals the cooperative tendency more clearly. A part of the users who end in the strategy of keep their own opinion, in the evolution process, they are tend to choose agree with their opponents before realizing that the payoff of keep strategy is much more than agree. The smaller the initial state of x is, the more likely y would end in the value of 1, which corresponding to the reality of when A pretend to be cooperative, user B will be deceived to stay in cooperation since it believes in A.

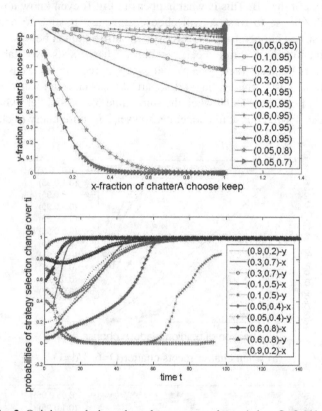

Fig. 2. Opinion evolution when chatter meets chatter(when Q=8,1/d=1)

5   Conclusion

In this paper, a simple evolutionary game model to simulate opinion interactions among micro-blog users is built. With study of some real features of user, we divided them into three different kinds with respective characteristic. Then, an evolutionary game theory model has been proposed, which considers conditions that users are not totally rational and would change their strategy once they find it is not good enough. Replicator equation are used to describe the trial and error process. Simulation results of the model offer some possible interpretations to some real facts: A strong cooperative characteristic exists in selfish human beings. In conditions when likely to be betrayed by opponents, still a big part of users stay in the agree strategy. Our results shows that opinion leaders in micro-blog are a group of people who will stick their own mind strongly, chatters are easily persuaded since they may just want certain information to be known; general people stay in the neutral zone. For further research, put in more real features of users through statistical results to carry classification of users more compelling. And structure characteristic should be concerned, a combination of complex network and evolutionary theory applied to social network is also necessary.

References

[1] Ding, F., Liu, Y.: Influences of Fanatics and Chatters on Information Diffusion on the Internet. International Conference on Natural Computation, 301–306 (2009)

[2] Ding, F., Liu, Y.: Modeling Opinion Interactions in a BBS Community. The European Physical Journal B 78, 245–252 (2010)

[3] Nekovee, M., Moreno, Y., Bianconi, G., Marsili, M.: Theory of rumor spreading in complex social networks. Physica A: Statistical Mechanics and its Applications 374(1), 457–470 (2007)

[4] Kostka, J., Oswald, Y.A., Wattenhofer, R.: Word of Mouth: Rumor Dissemination in Social Networks. In: Shvartsman, A.A., Felber, P. (eds.) SIROCCO 2008. LNCS, vol. 5058, pp. 185–196. Springer, Heidelberg (2008)

[5] MaheswaraRao, V.V.R., ValliKumari, V.: An Efficient Hybrid Successive Markov Model for Predicting Web User Usage Behavior using Web Usage Mining. International Journal of Data Engineering (IJDE) 1(5), 43–62 (2011)

[6] Chung, P.C., Liu, C.D.: Adaily behavior enabled hiddenMarkovmodel for human behaviorunderstanding. The Journal of Pattern Recognition Society 41, 1572–1580 (2008)

[7] Brune, R.: A Stochastic Model forPanic Behavior in Disease Dynamics. Angefertigtim InstitutfürNichtlineareDynamikder Georg-August-Universitätzu Göttingen (2008)

[8] Wang, Y.Z., Lin, C., Ungsunan, P.D.: Modeling and Survivability Analysis of Service Composition Using Stochastic Petri Nets. Journal of Supercomputing 56(1), 79–105 (2011)

[9] Wang, Y.Z., Yu, M., Li, J.Y., Meng, K., Lin, C., Cheng, X.Q.: Stochastic Game Net and Applications in Security Analysis for Enterprise Network. International Journal of Information Security

[10] Xu, D., Li, X., Wang, X.F.: An investigation on local area control of virus spreading in complex networks. Acta Physical Sinica 56(03), 1313–1317 (2007)

[11] Shen, H.W., Cheng, X.Q., Fang, B.X.: Covariance, correlation matrix and the multiscale community structure of networks. Physical Review E 82, 016114 (2010)

[12] Hu, H.B., Wang, K., Xu, L., Wang, X.: Analysis of Online Social Networks Based on Complex Network Theory. Complex Systems and Complexity Science 5(2), 1–14 (2008)

[13] Akihiko, M.: On Cultural Evolution: Social Norms, Rational Behavior and Evolutionary Game Theory. Journal of the Japanese and International Economies 10, 262–294 (1996)

[14] Xu, Z.Y., Li, C.D.: Evolutionary Game Analysis on Attitudinal Commitment Behavior in Business Relationships. In: 2009 International Conference on Information Management, Innovation Management and Industrial Engineering, pp. 245–248 (2009)

[15] Wang, D., Shen, Y.Z.: The Evolutionary Game Analysis of Mine Worker's Behavior of Violating Rules. In: Proceedings of 2010 IEEE the 17th International Conference on Industrial Engineering and Engineering Management, vol. 2, pp. 1120–1123 (2010)

[16] Chen, Y.: Cooperative Peer-to-Peer Streaming: An Evolutionary Game-Theoretic Approach. IEEE Transactions on Circuits and Systems for Video Technology, 1–12 (2010)

[17] Louis, L.Y., Sitaram, A., Bernardo, A., Huberman: Artificial Inflation: The True Story of Trends in Sina Weibo, Social Computing Lab, HP Labs,
http://www.hpl.hp.com/research/scl/papers/chinatrends/weibospam.pdf

A Possibilistic Query Translation Approach for Cross-Language Information Retrieval

Wiem Ben Romdhane[1], Bilel Elayeb[1,2], Ibrahim Bounhas[3], Fabrice Evrard[4], and Narjès Bellamine Ben Saoud[1]

[1] RIADI Research Laboratory, ENSI, Manouba University 2010, Tunisia
br.wiem@yahoo.fr, Narjes.Bellamine@ensi.rnu.tn
[2] Emirates College of Technology, P.O. Box: 41009, Abu Dhabi, United Arab Emirates
Bilel.Elayeb@riadi.rnu.tn
[3] LISI Lab. of computer science for industrial systems, ISD, Manouba University 2010, Tunisia
Bounhas.Ibrahim@yahoo.fr
[4] IRIT-ENSEEIHT, 02 Rue Camichel, 31071 Toulouse Cedex 7, France
Fabrice.Evrard@enseeiht.fr

Abstract. In this paper, we explore several statistical methods to find solutions to the problem of query translation ambiguity. Indeed, we propose and compare a new possibilistic approach for query translation derived from a probabilistic one, by applying a classical probability-possibility transformation of probability distributions, which introduces a certain tolerance in the selection of word translations. Finally, the best words are selected based on a similarity measure. The experiments are performed on CLEF-2003 French-English CLIR collection, which allowed us to test the effectiveness of the possibilistic approach.

Keywords: Cross-Language Information Retrieval (CLIR), Query Translation, Possibilistic Approach.

1   Introduction

With the huge expansion of documents in several languages on the Web and the increasing desire of non-native speakers of the English language to be able to retrieve documents in their own languages, the need for Cross-Language Information Retrieval (CLIR) System has become increasingly important in recent years. In fact, in the CLIR task, either the documents or the queries are translated. However, the majority of approaches focus on query translation, because document translation is computationally expensive. There are three main approaches to CLIR: Dictionary-based methods, parallel or comparable corpora-based methods, and machine translation methods.

The *Dictionary-based methods* [16][14] are the general approaches for CLIR when no commercial MT system with a recognized reputation is available. Several information retrieval systems (IRS) have used the so-called "bag-of-words" architectures, in which documents and queries are decayed into a set of words (or phrases) during an indexing

D.-S. Huang et al. (Eds.): ICIC 2013, LNAI 7996, pp. 73–82, 2013.

procedure. Therefore, queries can be simply translated by replacing every query term with its corresponding translations existing in a bilingual term list or a bilingual dictionary. Nevertheless, dictionary-based methods suffer from several difficulties such as: i) no translation of non-existing specific words in the used dictionary; ii) the addition of irrelevant information caused by the intrinsically ambiguities of the dictionary; iii) the decreasing of the effectiveness due to the disappointment to translate multiword expressions. To reduce ambiguity, one may adopt a corpus-based approach.

In *corpus-based methods* [17], a set of multilingual terms extracted from parallel or comparable corpora is exploited. Approaches based statistical/probabilistic method on parallel text written in multiple languages with the intention of selecting the correct word translation provides a good performance, but they suffer from many drawbacks. Firstly, the translation association created among the parallel words in the text is generally domain restricted, which means that accuracy decreases outside the domain. Secondly, parallel texts in different pairs of languages, are not always available.

In *machine translation (MT) techniques* [5][13], the main aim is to analyze the context of the query before translating its words. In fact, syntactic and semantic ambiguities are the principal problems decreasing MT performance. Besides, MT-based approaches suffer from several others limits decreasing the effectiveness of CLIR. Firstly, MT systems have serious difficulties to appropriately generate the syntactic and semantic analysis of the source text. Secondly, full linguistic analysis is computationally expensive, which decreases search performance.

In fact, query translation approaches need training and matching models which compute the similarities (or the relevance) between words and their translations. Existing models for query translation in CLIR are based on poor, uncertain and imprecise data. While probabilistic models are unable to deal with such type of data, possibility theory applies naturally to this kind of problems [8]. Thus, we propose a possibilistic approach for query translation derived from a probabilistic one using a probability/possibility transformation [6]. This approach begins with a query analysis step, then a lexical analysis step, and finally the selection of the best translation using different similarity measures.

This paper is organized as follows. Section 2 details our approach which is experimented in section 3. In section 4, we conclude our work and give some directions for future research.

2   The Proposed Approach

We propose a new possibilistic approach for query translation in CLIR. The proposed approach is an extension of a probabilistic model proposed by [12] into a possibilistic framework, using an existing probability/possibility transformation method [6]. In this approach we used a greedy algorithm to choose the best translation [12]. The calculation of similarity between the terms and the cohesion of a term x with a set X of other terms are two essential steps before selecting the best term translation. In our case, we used the *EMMI* weighting measure [15] to estimate the probabilistic similarity between terms. Then, we extended it to a possibilistic framework (*EMMI-POSS*) using an existing probability/possibility transformation [6]. We briefly recall in the follow-ing this transformation and we detail the three main steps of our query translation process, which are summarized in figure 1.

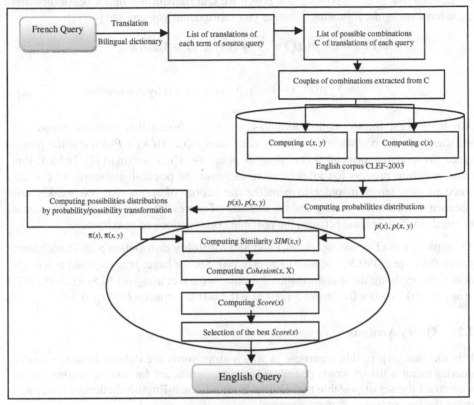

Fig. 1. Overview of query translation process

Formally the similarity between the terms x and y is given by formula (1). Howev-er, the cohesion of a term x with a set X of other words is the maximum similarity of this term with each term of the set, as given by formula (4).

$$SIM\,(x,\,y) = p(x,\,y) \times \log_2(\frac{p(x,\,y)}{p(x) \times p(y)}) \tag{1}$$

$$p(x,\,y) = \frac{c(x,\,y)}{c(x)} + \frac{c(x,\,y)}{c(y)} \tag{2}$$

$$p(x) = \frac{c(x)}{\sum_x c(x)}\,;\;p(y) = \frac{c(y)}{\sum_y c(y)} \tag{3}$$

$$Cohesion\,(x,\,X) = \underset{y \in X}{Max}(SIM\,(x,\,y)) \tag{4}$$

Where $c(x,\,y)$ is the frequency that the term x and the term y co-occur in the same sentences in the collection. $c(x)$ is the number of occurrences of term x in the collection.

2.1   Probability/Possibility Transformation

Given the universe of discourse $\Omega = \{\omega_1, \omega_2,..., \omega_n\}$ and a probability distribution p on Ω, such that $p(\omega_1) \geq p(\omega_2) \geq ...\geq p(\omega_n)$), we can transform p into a possibility distribution π using the following formulas (for more detail you can see [6][7]):

$$\pi(\omega_i) = i * p(\omega_i) + \sum_{j=i+1}^{n} p(\omega_j), \forall i = 1,...,n \tag{5}$$

$$\sum_{j=1}^{n} p(\omega_j) = 1 \quad \text{and} \quad p(\omega_{n+1}) = 0 \text{ by convention.} \tag{6}$$

Among several transformation formulas, we have chosen this formula, because it satisfies both the probability/possibility consistency (i.e. $\Pi(A) \geq P(A)$) and the preference preservation principles (i.e. $p(\omega_i) > p(\omega_j) \Leftrightarrow \pi(\omega_i) > \pi(\omega_j)$ [7]). Indeed, this transformation process has allowed us to increase the possibilistic scores of coexistence of two terms in order to penalize the scores of terms that are weakly co-occurring. In fact, the penalty and the increase of scores are proportional to the power of words to discriminate between the possible combinations of coexistence.

Example: Let $\Omega = \{\omega_1, \omega_2, \omega_3, \omega_4\}$ and a probability distribution p on Ω such that: $p(\omega_1)=0.2$; $p(\omega_2)=0.5$; $p(\omega_3)=0.3$; $p(\omega_4)=0$. So, we have: $p(\omega_2) > p(\omega_3) > p(\omega_1) > p(\omega_4)$. By applying the transformation formula, we have: $\pi(\omega_2)=(1*0.5)+(0.3+0,2)=1$; $\pi(\omega_1) = (3*0.2) + 0 = 0.6$; $\pi(\omega_3) = (2*0.3) + 0.2 = 0.8$; $\pi(\omega_4) = (4*0) + 0 = 0$.

2.2   Query Analysis

It is the first step in this approach, in which stop words are deleted from the source queries using a list of words considered as non-significant for source queries. Then, we extract the set of possible translations from a French-English dictionary generated using the free online dictionary Reverso[1].

2.3   Lexical Analysis

The step of lemmatization aims to find the canonical form of a word, so that different grammatical forms or variations are considered as instances of the same word. We applied the process of lemmatization on the test collection and queries before their translations. This reduction mechanism gives better results for the following matching phase.

2.4   Selection of Best Translation

It is the main step in our approach. Indeed, selecting the best translation among several ones existing in the bilingual dictionary is summarized as follows. The suitable translations of source query terms co-occur in the target language documents contrary to incorrect translations one. Consequently, we select for each set of the source query

---

[1] http://www.reverso.net/text_translation.aspx?lang=FR

terms the best translation term, which frequently co-occurs with other translation terms in the target language. However, it is computationally very costly to identify such an optimal set. For that reason, we take advantage from an approximate Greedy algorithm as used in [12].We briefly summarize in the following the main principle of this algorithm. Firstly, and using the bilingual dictionary, we select a set T_i of translation terms for each of the n source query terms $\{f_1,...,f_n\}$. Secondly, we compute the cohesion of every term in each set T_i with the other sets of translation terms. The best translation in each T_i has the maximum degree of cohesion. Finally, the target query $\{e_1,...,e_n\}$ is composed of the best terms from every translation set.

Cohesion is based on the similarity between the terms. We transform the weighting measure *EMMI* to a possibilistic one (*EMMI-POSS*), which is successfully used to estimate similarity among terms. However, the measure *EMMI-POSS* does not take into account the distance between words. In fact, we observe that the local context is more important for the selection of translation. If two words appear in the same document, but in two remote locations, it is unlikely to be strongly dependent. Therefore, a distance factor was added by [12] in computing word similarity.

2.5   Illustrative Example

Let us consider the following French source query Q: $\{L'Union\ Européenne\ et\ les\ Pays\ Baltes\}$. Indeed, we have a set of possible translations for each term in the query Q from the used dictionary. The term "*union*" has two possible translations (*union, unity*), the term "*Européenne*" has the unique translation (*european*), the term "*pays*" has two possible translations (*country, land*) and the term "*Baltes*" has the unique translation (*Baltic*). In fact, Given the source query Q, we generate the set of possible translations combinations from a bilingual dictionary. In this example, there are 4 possible translation combinations (cf. table 1). The best translation is which has the greater possibilistic score. Table 2 and give detail of calculus.

Table 1. Translation combinations for $\{L'Union\ Européenne\ et\ les\ Pays\ Baltes\}$

	Translation Combinations
1	union AND european AND country AND Baltic
2	union AND european AND land AND Baltic
3	unity AND european AND country AND Baltic
4	unity AND european AND land AND Baltic

Probability values in table 2 and 3 are very low comparing to the possibility ones. So, they have a poor discriminative effect in the selection of the suitable translation. Consequently, we risk having very close probabilistic similarity scores, in which ambiguity translation cannot be correctly resolved.   Moreover, the selected English translation of the given French source query {union, européenne, pays, balte} is the target Enlish query {unity, european, country, baltic}. We remark here that the suitable translation of the name phrase (NP) "union européenne" is not "European unity" but "European union". Consequently, we mainly need to identify the NP in the source query and translate them before translating one-word terms.

Table 2. Possibilistic similarity scores for the different pairs of words (x, y)

Pairs of words (x, y)	C(x, y)	P(x, y)	π(x, y)	SIM(x, y)
union-european	1317	0.2642	9.3428	34.0590
union - country	209	0.0442	4.9091	12.9894
union - baltic	6	0.0583	5.5669	43.4782
european - country	535	0.0894	6.6168	20.0428
european -baltic	5	0.0485	5.1314	39.2299
country –baltic	8	0.0872	6.5532	51.9453
union – european	1317	0.2642	9.3428	34.0590
union - land	1	0.0077	1.5429	2.2342
union - baltic	6	0.0583	5.5669	43.4782
european-land	51	0.0134	2.3534	4.7296
european -baltic	5	0.0485	5.1314	39.2299
land - baltic	1	0.0097	1.8719	12.3381
unity - european	15	0.0380	4.5408	25.4750
unity - country	10	0.0282	3.8353	20.3097
unity -baltic	0	0.0	0.0	0.0
european - country	535	0.0894	6.6168	20.0428
european - baltic	5	0.0485	5.1314	39.2299
country - baltic	8	0.0872	6.5532	51.9453
unity -european	15	0.0380	4.5408	25.4750
unity - land	1	0.0024	0.5520	1.6403
unity-baltic	0	0.0	0.0	0.0
european-land	51	0.0134	2.3534	4.7296
european - baltic	5	0.0485	5.1314	39.2299
land - baltic	1	0.0097	1.8719	12.3381

Table 3. The final possibilistic score of each possible translation

Possible translations of word x	C(x)	P(x)	π(x)	Score (x)
union	9894	0.0148	0.8499	14.0363
unity	455	0.0006	0.1058	**50.9500**
european	11121	0.0165	0.8784	**119.0682**
country	13469	0.0204	0.9228	**103.8907**
land	5592	0.0083	0.6653	24.6762
baltic	108	0.0001	0.0291	**186.5989**

3   Experimental Evaluation

Our experiments are performed through our possibilistic information Retrieval System [10], and implemented using the platform Terrier[2]. It provides many existing matching models such as OKAPI and a new possibilistic matching model proposed by [11]. We propose and compare here our results using these two matching model in order to study the generic character of our approach.

---

[2] http://terrier.org/

Experiments are achieved using a subset of the collection CLEF-2003. This part includes articles published during 1995 in the newspaper "*Glasgow Herald*". This collection consists of 56472 documents and 54 queries, forming 154 MB. We only take into account the part <title> of the test queries, because it contains several isolate words, which are suitable to experiment our approach. However, we plan to consider other part of queries such as <description> and <narrative>, in which the context is relevant in the translation process. To evaluate our possibilistic approach, we compare our results to some existing probabilistic similarity measures such as T-score (*TS*) [3], Log Likelihood Ratio (*LLR*) score, [9], Dice Factor (*DF*) [16] and Mutual Information (*MI*) [4].

Table 4 contains statistics on two elements u and v which are in this case, the components of an expression. O_{11} is the number of co-occurrences of u with v. O_{12} is the number of occurrences of u with an element other than v, etc.

Table 4. The contingency table

	$t_1 = v$	$t_1 \neq v$
$t_2 = u$	O_{11}	O_{12}
$t_2 \neq u$	O_{21}	O_{22}

We have also:

$$R_1 = O_{11} + O_{12} \tag{7}$$

$$R_2 = O_{21} + O_{22} \tag{8}$$

$$C_1 = O_{11} + O_{21} \tag{9}$$

$$C_2 = O_{12} + O_{22} \tag{10}$$

$$N = R_1 + R_2 = C_1 + C_2. \tag{11}$$

We also calculate the expected frequency of collocation as follows:

$$E_{11} = (R_1 * C_1)/N \tag{12}$$

The *LLR, MI, TS* and *DF* score are calculated as follows:

$$\text{LLR}(u, v) = -2 \log\left(\frac{L(O_{11}, C_1, r) * L(O_{12}, C_2, r)}{L(O_{11}, C_1, r_1) * L(O_{12}, C_2, r_2)} \right) \tag{13}$$

$$L(k, n, r) = k^r * (1 - r)^{n-k} \; where: r = R_1/N, r_1 = O_{11}/C_1, r_2 = O_{12}/C_2 \tag{14}$$

$$MI = \log_2(O_{11}/E_{11}) \tag{15}$$

$$TS = \frac{(O_{11} - E_{11})}{\sqrt{O_{11}}} \tag{16}$$

$$DF = 2 * \frac{O_{11}}{R_1 + C_1} \tag{17}$$

The proposed approach is assessed using the mean average precision (MAP) as a performance measure. The formula of computing the MAP is the following:

$$MAP = \frac{1}{N}\sum_{j=1}^{N}\frac{1}{Q_j}\sum_{i=1}^{Q_j}P(doc_i)$$ (18)

Where:
Q_j: The number of relevant documents for query j;
N: The number of queries;
$P(doc_i)$: The precision at the ith relevant document.

Moreover, we compare the possibilistic approach (*EMMI-POSS*) both to monolingual IR task and to others probabilistic similarity measures, using OKAPI and Possibilistic matching models (Figure 2 and 3, respectively). In fact, we used the precision (Y-axe) over 11 points of recall in the X-axe (0.0, 0.1, ..., 1.0) to draw all recall-precision curves.

Using OKAPI (figure 2) or the possibilistic (figure 3) matching model, results in both figures show that the possibilistic query translation approach has the closest recall-precision curve to the Monolingual task, which confirm its effectiveness comparing to other probabilistic approaches. Indeed, the discriminative character of the possibilistic approach improves its ability to solve the problem of query translation ambiguity and consequently enhance its efficiency.

On the other hand, the mean average precision of *EMMI-POSS* (0.23) is very close to that obtained for the Monolingual (0.24) and that obtained for *TS* (0.21). The *LLR* metric has the worst result (0.15). These results are also confirmed using the possibilistic matching model. Indeed, results in figure 3 prove that the mean average precision of *EMMI-POSS* (0.165) is very close to that obtained for the Monolingual (0.17). The *LLR* metric stays also the worst one with 0.104.

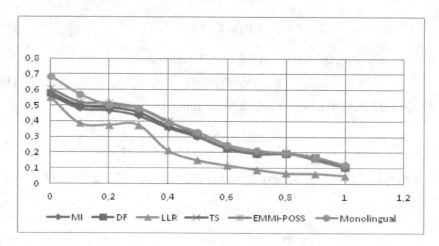

Fig. 2. Recall-Precision curves of Monolingual vs. All similarity measures (OKAPI)

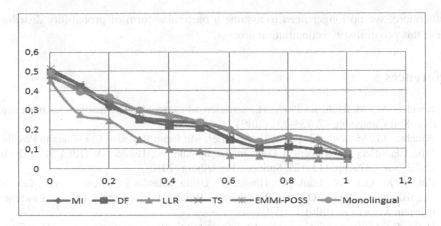

Fig. 3. Recall-Precision curves of Monolingual vs. All similarity measures (Possibilistic)

In fact our approach for CLIR has some drawbacks such us: (i) the limited coverage of dictionary and; (ii) The complexity of the algorithm allowing to choose the suitable translation among the set of the translations proposed by the dictionary. To overcome these limitations, we exploited the cohesion between a given query term and its possible translations in the training corpus and a particular similarity score measure to select the suitable translation of each query term. However, the results were mainly influenced by the specific properties of the used document collection.

4   Conclusion

In this paper, we presented a possibilistic query translation approach based on the cohesion between the translations of words. This approach is based on probability/possibility transformation improving discrimination in the selection of suitable translation. Besides, this transformation did not increase the complexity such as in [1][2]. We have tested and compared several similarity scores to improve query translation based dictionaries in CLIR.

The idea of applying possibility theory to query translation is identical to the use of probabilities in the Bayesian probability model. In fact, it is necessary to evaluate many parameters, a task that cannot be compatible with poor data. The problem of accurately estimating probability distributions for probabilistic query translation is important for the accurate calculation of the probability distribution of translations. However, due to the use of the product to combine probability values (which are frequently small), the probability estimation error may have a significant effect on the final estimation. This contrasts with the possibility distributions which are less sensitive to imprecise estimation for several reasons. Indeed, a possibility distribution can be considered representative of a family of probability distributions corresponding to imprecise probabilities, which are more reasonable in the case of insufficient data (such as the case when some words do not exist in the bilingual dictionary).

Furthermore, we no longer need to assume a particular form of probability distribution in this possibilistic reconciliation process.

References

1. Bounhas, M., Mellouli, K., Prade, H., Serrurier, M.: Possibilistic classifiers for numerical data. Soft Computing 17, 733–751 (2013)
2. Bounhas, M., Mellouli, K., Prade, H., Serrurier, M.: From Bayesian Classifiers to Possibilistic Classifiers for Numerical Data. In: Deshpande, A., Hunter, A. (eds.) SUM 2010. LNCS, vol. 6379, pp. 112–125. Springer, Heidelberg (2010)
3. Church, K., Gale, W., Hanks, P., Hindle, D.: Using statistics in lexical analysis. Lexical Acquisition: Exploiting On-Line Resources to Build a Lexicon, pp. 115–164. Lawrence Erlbaum Associates, Hillsdale (1991)
4. Daille, B.: Approche mixte pour l'extraction de terminologie : statistique lexicale et filtres linguistiques. Ph.D. Thesis, University of Paris 7 (1994) (in French)
5. Mavaluru, D., Shriram, R., Banu, W.A.: Ensemble Approach for Cross Language Information Retrieval. In: Gelbukh, A. (ed.) CICLing 2012, Part II. LNCS, vol. 7182, pp. 274–285. Springer, Heidelberg (2012)
6. Dubois, D., Prade, H.: Unfair coins and necessity measures: Towards a possibilistic interpretation of histograms. Fuzzy Sets and Systems 10, 15–20 (1985)
7. Dubois, D., Prade, H., Sandri, S.: On Possibility/Probability transformation. Fuzzy Logic: State of the Art, 103–112 (1993)
8. Dubois, D., Prade, H.: Possibility Theory: An Approach to computerized Processing of Uncertainty. Plenum Press, New York (1994)
9. Dunning, T.: Accurate Methods for the Statistics of Surprise and Coincidence. Computational Linguistics 19, 61–74 (1994)
10. Elayeb, B., Bounhas, I., Ben Khiroun, O., Evrard, F., Bellamine Ben Saoud, N.: Towards a Possibilistic Information Retrieval System Using Semantic Query Expansion. International Journal of Intelligent Information Technologies 7, 1–25 (2011)
11. Elayeb, B., Evrard, F., Zaghdoud, M., Ben Ahmed, M.: Towards an Intelligent Possibilistic Web Information Retrieval using Multiagent System. The Interactive Technology and Smart Education, Special issue: New learning support systems 6, 40–59 (2009)
12. Gao, J., Nie, J.Y., Xun, E., Zhang, J., Zhou, M., Huang, C.: Improving Query Translation for Cross-Language Information Retrieval using Statistical Models. In: Proceedings of SIGIR 2001, New Orleans, Louisiana, USA, pp. 9–12 (2001)
13. Iswarya, P., Radha, V.: Cross Language Text Retrieval: A Review. International Journal Of Engineering Research and Applications 2, 1036–1043 (2012)
14. Mallamma, V.R., Hanumanthappa, M.: Dictionary Based Word Translation in CLIR Using Cohesion Method. In: INDIACom-(2012) ISSN 0973-7529, ISBN 978-93-80544-03-8
15. Rijsbergen, V.: Information Retrieval. Butterworths, Londres (1979)
16. Smadja, F., Mckeown, K.R., Hatzivassiloglou, V.: Translating collocations for bilingual lexicons: a statistical approach. Computational Linguistics 22, 1–38 (1996)
17. Vitaly, K., Yannis, H.: Accurate Query Translation For Japanese-English Cross-Language Information Retrieval. In: International Conference on Pervasive and Embedded Computing and Communication Systems, pp. 214–219 (2012)

Three Kinds of Negation of Fuzzy Knowledge and Their Base of Logic

Zhenghua Pan

School of Science, Jiangnan University, Wuxi, 214122, China
panzh@jiangnan.edu.cn

Abstract. Negative information plays an important role in fuzzy knowledge representation and reasoning. This paper distinguish between contradictory negative relation and opposite negative relation for the fuzzy information, a characteristic of fuzzy information is discovered that if a pair of opposite concepts are fuzzy concepts, then there must exists a "medium" fuzzy concept between them; conversely, if there is a medium fuzzy concept between the two opposite concepts, then opposite concepts must be fuzzy concepts. We thus consider that negation of fuzzy information include contradictory negation, opposite negation and medium negation. In order to provide a logical basis for three kinds of negation in fuzzy information, we propose a fuzzy propositional logic with contradictory negation, opposite negation and medium negation (FLCOM), discussed some interesting properties of FLCOM, presented a semantic interpretation of FLCOM, and proved the reliability theorem.

Keywords: fuzzy knowledge, fuzzy propositional logic, contradictory negation, opposite negation, medium negation.

1 Introduction

Negation in knowledge processing is an important notion, especially in fuzzy knowledge processing. The concept of negation plays a special role in non-classical logics and in knowledge representation formalisms, in which the negative information has been taking into account par with positive information. Some scholars suggested that uncertain information processing needed different negations in various domains. For all of FS (fuzzy sets)[1], RS(rough sets)[2] and the various extensions of FS such as IFS(intuitionistic fuzzy sets)[3], the negation $\neg$ is defined as $\neg x = 1 - x$. In Hájek's BL(basic logic)[4], the negation $\neg$ is defined as $\neg x = x \rightarrow 0$. In Wang's fuzzy propositional calculus system $\mathcal{L}$ [5], the negation $\neg$ is still defined as $\neg x = 1 - x$. In other words, there is only a sort of negation in these theories, the negation is simply different forms of expression in definition.

However, the some of scholars have cognized that negation is not a clean concept from a logical point of view in knowledge representation and knowledge reasoning, in database query languages (such as *SQL*), in production rule systems (such as *CLIPS Jess*), in natural language processing and in semantic web and so on, there are different

D.-S. Huang et al. (Eds.): ICIC 2013, LNAI 7996, pp. 83–93, 2013.

negations[6-13]. Wagner et al proposed that there are (at least) two kinds of negation in above domains, a *weak negation* expressing non-truth (in the sense of "she doesn't like snow" or "he doesn't trust you"), and a *strong negation* expressing explicit falsity (in the sense of "she dislikes snow" or "he distrusts you")[6,7,8]. Ferré introduced an epistemic extension for the concept of negation, considered that there are *extensional negation intentional negation* in logical concept analysis and natural language [9]. Kaneiwa proposed that "description logic ALC_{-} with *classical negation* and *strong negation*, the classical negation $\neg$ represents the negation of a statement, the strong negation ~ may be more suitable for expressing explicit negative information (or negative facts), in other worlds, ~ indicates information that is directly opposite and exclusive to a statement rather than its complementary negation [10]. Since 2005, Pan et al introduced an epistemic extension for the concept of negation, considered that negative relations in knowledge should differentiate the contradictory relation and the opposite relation, and discovered a characteristic of fuzzy information that is if two opposite concepts are fuzzy concepts, there must exists a "medium" fuzzy concept between them, vice versa, thus proposed that negations of fuzzy knowledge include the contradictory negative relation, the opposite negative relation and the medium negative relation, and described these relations using the medium logic[11-15]. In order to provide a set basis to contradictory negation, opposite negation and medium negation in fuzzy information, Pan presented a fuzzy sets with contradictory negation, opposite negation and medium negation (FSCOM)[16,17].

This paper presents a new fuzzy propositional logic system with contradictory negation, opposite negation and medium negation (FLCOM) corresponding to FSCOM, discussed some interesting properties of FLCOM, presented a semantic interpretation and proved the reliability theorem for FLCOM.

2   Three Kinds of Negative Relation in Fuzzy Concept

In the formal logic, the relation between concept A and concept B is relation between the extension of A and the extension of B, it included consistent relation and inconsistent relation. The inconsistent relation means that extension of A and extension of B is disjoint. For example, *white* and *nonwhite*, *conductor* and *nonconductor*, and so on. The inconsistent relation between concept A and B is showed by the following figure (Fig.1) (the rectangle denotes extension of a concept):

Fig. 1. Extensions of concept A and B have nothing in common when the relation between A and B is inconsistent relation

A concept and its negation are subconcepts of one *genus concept*, the sum of extensions of all the subconcepts is equal to the extension of genus concept. Since Aristotle, the inconsistent relation had distinguished as contradictory relation and opposite relation, so the relation between a concept and its negation should include *contradictory negative relation* and *opposite negative relation*.

Due to fuzziness of concept means that extension of concept is vague, we consider that relationship between a fuzzy concept and its negation take on the following three cases, in which F denotes a fuzzy concept, F_c denotes the contradictory negation of F, F_o denotes the opposite negation of F, and $E(x)$ denotes the extension of concept x.

(1) *Contradictory negative relation in Fuzzy Concept (CFC)*

Character of CFC: "the borderline between $E(F)$ and $E(F_c)$ is vague, the sum of $E(F)$ and $E(F_c)$ is equal to the extension of genus concept."

For example, *"young people"* and its contradictory negation *"non-young people"* are fuzzy subconcepts of the genus concept *"people"*, the relation between *young people* and *non-young people* is CFC, the borderline between $E(young\ people)$ and $E(non\text{-}young\ people)$ is vague, and the sum of $E(young\ people)$ and $E(non\text{-}young\ people)$ is equal $E(people)$. *"quick"* and its contradictory negation *"not quick"* are fuzzy subconcepts of the genus concept *"velocity"*, the relation between *"quick"* and *"not quick"* is CFC, the borderline between $E(quick)$ and $E(not\ quick)$ is vague, and the sum of $E(quick)$ and $E(not\ quick)$ is equal to $E(velocity)$, and so on. The character of CFC is showed by the following figure (Fig. 2):

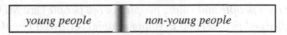

Fig. 2. Borderline between $E(young\ people)$ and $E(non\text{-}young\ people)$ is vague, and the sum of $E(young\ people)$ and $E(non\text{-}young\ people)$ is equal to $E(people)$.

(2) *Opposite negative relation in Fuzzy Concept (OFC)*

Character of OFC: "the ambit between $E(F)$ and $E(F_o)$ is vague, the sum of $E(F)$ and $E(F_o)$ less than the extension of genus concept."

For example, *"young people"* and its opposite negation *"old people"* are fuzzy subconcepts of the genus concept *"people"*, the relation between *young people* and *old people* is OFC, the ambit between $E(young\ people)$ and $E(old\ people)$ is vague, and the sum of $E(young\ people)$ and $E(old\ people)$ less than $E(people)$. *"quick velocity"* and its opposite negation *"slow velocity"* are fuzzy subconcepts of the genus concept *"velocity"*, the relation between *"quick velocity"* and *"slow velocity"* is OFC, the ambit between $E(quick\ velocity)$ and $E(slow\ velocity)$ is vague, and the sum of $E(quick\ velocity)$ and $E(slow\ velocity)$ less than $E(velocity)$, and so on. The character of OFC is showed by the following figure (Fig. 3):

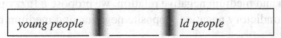

Fig. 3. Ambit between $E(young\ people)$ and $E(old\ people)$ is vague, and the sum of $E(young\ people)$ and $E(old\ people)$ less than $E(people)$.

In real life, a lot of opposite concepts taken on a character: there is a transitional *"medium concept"* between the two opposite concepts, the medium concept represents an transition from one to another one for two opposite concepts. For instance, *"zero"* between positive number and negative number, *"semiconductor"* between conductor and nonconductor, "*dawn*" between daylight and night. We discover that a pair of opposite concepts and their medium concept take on following characteristic after we investigated large numbers of instances:

if a pair of opposite concepts are fuzzy concepts, then there is *medium fuzzy concept* between them. Conversely, if there is a *medium fuzzy concept* between the two opposite concepts, then opposite concepts must be fuzzy concepts.

A pair of fuzzy opposite concepts and its medium concept are subconcepts of one genus concept, and the sum of extensions of all the subconcepts is equal to the extension of generic concept. We thus consider that relation of opposite concepts and its medium concept is a negative relation, this negative relation is called as *medium negative relation* in following depiction. For the convenience of depiction, the following F_m denotes the medium negation of fuzzy concept F.

(3) Medium negative relation in Fuzzy Concepts (MFC)

Character of MFC: "borderlines between $E(F_m)$ and $E(F)$ (or $E(F_o)$) are vague, the sum of $E(F_m)$, $E(F_o)$ and $E(F)$ is equal to the extension of genus concept."

For example, "*middleaged people*" between opposite concepts "*young people*" and "*old people*" is a medium fuzzy concept of one genus concept "*people*", the relations between *middleaged people* and *young people* (or *old people*) are MFC, the borderlines between $E(middleaged people)$ and $E(young people)$ (or $E(old people)$) are vague, and the sum of $E(middleaged people)$, $E(young people)$ and $E(old people)$ is equal to $E(people)$. "*dawn*" between "*daylight*" and "*night*" is medium fuzzy concept of one genus concept "*day*", the relations between *dawn* and *daylight* (or *night*) are MFC, the borderlines between $E(dawn)$ and $E(daylight)$ (or $E(night)$) are vague, and the sum of $E(dawn)$, $E(daylight)$ and $E(night)$ is equal to $E(day)$. The character of MFC is showed by the following figure (Fig. 4):

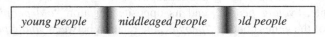

| *young people* | *niddleaged people* | *old people* |

Fig. 4. Borderlines between $E(dawn)$ and $E(night)$ (or $E(dawn)$) are vague, and the sum of $E(young people)$, $E(old people)$ and $E(middleaged people)$ is equal to $E(people)$

3   Base of Logic for Three Kinds of Negative Relation

In order to provide a logical basis for the contradictory negative relation, opposite negative relation and medium negative relation, we propose a fuzzy propositional logic system with contradictory negation, opposite negation and medium negation.

3.1   Fuzzy Propositional Logic System FLCOM

Definition 1. Let S be non-empty set, element in S is called atom proposition or atom formula, the formal symbols "¬", "⫪", "~", "→", "∧" and "∨", parentheses "(" and ")" are connectives, in which ¬, ⫪, ~, →, ∧ and ∨ denote contradictory negation, opposite negation, medium negation, implication, conjunction and disjunction. And

(I):
    (a)   for any $A \in S$, A is fuzzy formula, formula for short;
    (b)   if A and B are formulas then $\neg A$, $\daleth A$, $\sim A$, $(A \rightarrow B)$, $(A \vee B)$ and $(A \wedge B)$ are formulas;
    (c)   formulas are built from (a) and (b), all the formulas constituted a set $\mathfrak{I}(S)$, for short $\mathfrak{I}$.

(II): the following formulas in $\mathfrak{I}$ are called axioms:
    (A1) $A \rightarrow (B \rightarrow A)$
    (A2) $(A \rightarrow (A \rightarrow B)) \rightarrow (A \rightarrow B)$
    (A3) $(A \rightarrow B) \rightarrow ((B \rightarrow C) \rightarrow (A \rightarrow C))$
    (M_1) $(A \rightarrow \neg B) \rightarrow (B \rightarrow \neg A)$
    (M_2) $(A \rightarrow \daleth B) \rightarrow (B \rightarrow \daleth A)$
    (H) $\neg A \rightarrow (A \rightarrow B)$
    (C) $((A \rightarrow \neg A) \rightarrow B) \rightarrow ((A \rightarrow B) \rightarrow B)$
    ($\vee_1$) $A \rightarrow A \vee B$
    ($\vee_2$) $B \rightarrow A \vee B$
    ($\wedge_1$) $A \wedge B \rightarrow A$
    ($\wedge_2$) $A \wedge B \rightarrow B$
    ($Y_\daleth$) $\daleth A \rightarrow \neg A \wedge \neg \sim A$
    ($Y_\sim$) $\sim A \rightarrow \neg A \wedge \neg \daleth A$

(III): the deduction rule MP(modus ponens): deduce B from $A \rightarrow B$ and A.

(I), (II) and (III) compose a formal system, which is called *fuzzy propositional Logic system with Contradictory negation, Opposite negation and Medium negation*, FLCOM for short.

In [16], we have proved that contradictory negation $A^{\neg}$ of a fuzzy set A have the following relations with the opposite negation $A^{\daleth}$ and the medium negation $A^{\sim}$ in FSCOM :

$$A^{\neg} = A^{\daleth} \cup A^{\sim}$$

Corresponding to this relation, the formulas $\neg A$ have relations with $\daleth A$ and $\sim A$ in FLCOM can be defined as follows.

Definition 2. In FLCOM,

$$\neg A = \daleth A \vee \sim A. \tag{1}$$

Remark 3.1. We could proven that (i) the some of axioms in FLCOM are not independent, (ii) the some of connectives in FLCOM are not independent, connectives set $\{\neg, \daleth, \sim, \vee, \wedge, \rightarrow\}$ is reduced to $\{\daleth, \sim, \vee, \rightarrow\}$, moreover, $\{\neg, \daleth, \sim, \vee, \wedge, \rightarrow\}$ is reduced to $\{\daleth, \sim, \rightarrow\}$ if $\vee$ and $\wedge$ are defined by $A \vee B = \daleth A \rightarrow B$ and $A \wedge B = \daleth (A \rightarrow \daleth B)$ in FLCOM, well then $\daleth$, $\sim$ and $\rightarrow$ are original connectives.

Remark 3.2. By reason that $A^{\daleth} = A^{\neg} \cap A^{\neg\neg}$ and $A^{\sim} = A^{\neg} \cap A^{\neg}$ had been proven in FSCOM [16], $Y_\daleth$ and $Y_\sim$ are taken as axioms of FLCOM.

Remark 3.3. $A \rightarrow B$ is not logical equivalent with $\neg A \vee B$ and $\daleth A \vee B$.

3.2   Properties and Meaning of FLCOM

By reason that negations of fuzzy concept is distinguished as contradictory negation, the opposite negation and the medium negation in FLCOM, we can prove the following particular properties of FLCOM, in which some properties is the same as other fuzzy logics.

Theorem 1. In FLCOM,

[1] $\vdash A \to A$

[2] $\vdash ((A \to B) \to C) \to (B \to C)$

[3] $\vdash A \to ((A \to B) \to (C \to B))$

[4] $\vdash A \to ((A \to B) \to B)$

[5] $\vdash (((A \to B) \to B) \to C) \to (A \to C)$

[6] $\vdash (A \to (B \to C)) \to (B \to (A \to C))$

[7]* $\vdash (B \to C) \to ((A \to B) \to (A \to C))$

[8] $\vdash ((A \to B) \to (A \to (A \to C))) \to ((A \to B) \to (A \to C))$

[9] $\vdash (B \to (A \to C)) \to ((A \to B) \to (A \to C))$

[10] $\vdash (A \to (B \to C)) \to ((A \to B) \to (A \to C))$

Theorem 2. In FLCOM,

[11] $\vdash \neg A \to (A \to \neg(B \to B))$

[12] $\vdash B \to ((A \to \neg B) \to \neg A)$

[13] $\vdash (A \to \neg B) \to ((A \to B) \to \neg A)$

[14]* $\vdash (A \to B) \to ((A \to \neg B) \to \neg A)$ *(Reductio ad Absurdum)*

[15]* $\vdash (A \to B) \to (\neg B \to \neg A)$

[16] $\vdash (A \to \neg A) \to \neg A$

[17]* $\vdash A \to \neg\neg A$

[18]* $\vdash (\neg A \to \neg B) \to (B \to A)$

[19] $\vdash \neg\neg A \to (\neg\neg A \to A)$

[20]* $\vdash \neg\neg A \to A$

[21]* $\vdash (\neg A \to B) \to ((\neg A \to \neg B) \to A)$ *(Reduction to Absurdity)*

The theorem 1 and the theorem 2 showed that FLCOM keep many basal properties of fuzzy logic, such as the *Reductio ad Absurdum* ([14]*) and the *Reduction to Absurdity* ([21]*), as well as [7]*, [15]*, [17]*, [18]* and [20]*.

Theorem 3. In FLCOM,

[1⁰] $\vdash \rightleftharpoons A \to (A \to \rightleftharpoons (B \to B))$

[2⁰] $\vdash B \to ((A \to \rightleftharpoons B) \to \rightleftharpoons A)$

[3⁰] $\vdash (A \to \rightleftharpoons B) \to (A \to (B \to \rightleftharpoons (B \to B)))$

[4⁰] $\vdash (A \to B) \to ((A \to \rightleftharpoons B) \to \rightleftharpoons A)$ *(new Reductio ad Absurdum)*

[5⁰] $\vdash (A \to B) \to (\rightleftharpoons B \to \rightleftharpoons A)$

[6⁰] $\vdash (A \to \rightleftharpoons A) \to \rightleftharpoons A$

[7⁰] $\vdash A \to \rightleftharpoons\rightleftharpoons A$

[8⁰] $\vdash (\rightleftharpoons A \to \rightleftharpoons B) \to (B \to A)$

[9⁰] $\vdash \rightleftharpoons\rightleftharpoons A \to (\rightleftharpoons\rightleftharpoons A \to A)$

[10⁰] $\vdash \rightleftharpoons\rightleftharpoons A \to A$

The theorem 3 showed that FLCOM take on some particular properties, in which not only kept the reductio ad absurdum ([14]*), but also add a *new reductio ad absurdum* ([4^0]), thus FLCOM has extended the meaning of the reductio ad absurdum.

Theorem 4. In FLCOM,

[11^0]	$\vdash \neg(A \wedge \neg A)$	(*law of contradiction*)
[12^0]	$\vdash \neg(\daleth A \wedge \sim A)$	(*new law of contradiction 1*)
[13^0]	$\vdash \neg(A \wedge \sim A)$	(*new law of contradiction 2*)
[14^0]	$\vdash \neg(A \wedge \daleth A)$	(*new law of contradiction 3*)

In respect that negation is distinguished as the contradictory negation, the opposite negation and the medium negation in FLCOM, thus the accepted meaning of "negation" should be extended. The theorem 4 reflected such expansion, that is, in addition to kept the law of contradiction in fuzzy logics ([11^0]), added three new law of contradiction ([12^0], [13^0], [14^0]).

Definition 3. The proof of a formal theorem A in FLCOM is finite formulas sequence $A_1, A_2, \ldots, A_n$, where A_k (k = 1, 2, …, n) is an axiom, or formula which be deduced from A_i and A_j(i < k, j < k) by MP, and A_n = A.

3.3 An Interpretation of FLCOM

Definition 4 (λ-evaluation). Let $\lambda \in (0, 1)$. Mapping $\partial : \mathfrak{J} \rightarrow [0, 1]$ is called a λ-evaluation if

$$\text{(a)} \quad \partial(A) + \partial(\daleth A) = 1 \tag{2}$$

(b) $\partial(\sim A) =$

$$\begin{cases} \lambda - \dfrac{2\lambda - 1}{1 - \lambda}(\partial(A) - \lambda), & \text{when } \lambda \geq \tfrac{1}{2} \text{ and } \partial(A) \in (\lambda, 1] & (3) \\[2ex] \lambda - \dfrac{2\lambda - 1}{1 - \lambda}\partial(A), & \text{when } \lambda \geq \tfrac{1}{2} \text{ and } \partial(A) \in [0, 1-\lambda) & (4) \\[2ex] 1 - \dfrac{1 - 2\lambda}{\lambda}(\partial(A) + \lambda - 1) - \lambda, & \text{when } \lambda \leq \tfrac{1}{2} \text{ and } \partial(A) \in (1-\lambda, 1] & (5) \\[2ex] 1 - \dfrac{1 - 2\lambda}{\lambda}\partial(A) - \lambda, & \text{when } \lambda \leq \tfrac{1}{2} \text{ and } \partial(A) \in [0, \lambda) & (6) \\[2ex] \partial(A), & \text{otherwise} & (7) \end{cases}$$

$$\text{(c)} \quad \partial(A \vee B) = \max(\partial(A), \partial(B)), \quad \partial(A \wedge B) = \min(\partial(A), \partial(B)) \tag{8}$$

$$\text{(d)} \quad \partial(A \rightarrow B) = \mathfrak{R}(\partial(A), \partial(B)), \text{ where } \mathfrak{R}: [0, 1]^2 \rightarrow [0, 1] \text{ is a binary function.} \tag{9}$$

The meaning of (b) in definition 4, and $\partial(A)$ relationships with $\partial(\daleth A)$ and $\partial(\sim A)$ in [0, 1] can be showed by the following figures (Fig.5, Fig.6), in which the symbols "•" and "○" denote the closed terminal and the open terminal in figure, respectively.

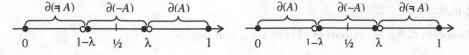

Fig. 5. $\partial(A)$ has relationships with $\partial(\daleth A)$ and $\partial(\sim A)$ in [0, 1] when $\lambda \geq \tfrac{1}{2}$.

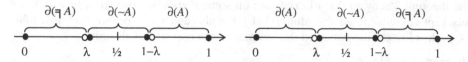

Fig. 6. $\partial(A)$ has relationships with $\partial(\daleth A)$ and $\partial(\sim A)$ in $[0, 1]$ when $\lambda \le \frac{1}{2}$

According to the definition 4, the following fact can be easily proven.

Proposition 1. Let A be a formula in FLCOM, $\lambda > \frac{1}{2}$. Then

$$\partial(A) \ge \partial(\sim A) \ge \partial(\daleth A) \text{ when } \partial(A) \in (\lambda, 1], \tag{10}$$

$$\partial(\daleth A) \ge \partial(\sim A) \ge \partial(A) \text{ when } \partial(A) \in [0, 1-\lambda). \tag{11}$$

Proposition 2. Let A be a formula in FLCOM, $\lambda \le \frac{1}{2}$. Then

$$\partial(A) \ge \partial(\sim A) \ge \partial(\daleth A) \text{ when } \partial(A) \in (1-\lambda, 1], \tag{12}$$

$$\partial(\daleth A) \ge \partial(\sim A) \ge \partial(A) \text{ when } \partial(A) \in [0, \lambda). \tag{13}$$

Proposition 3.

$$\lambda \ge \partial(\sim A) \ge 1-\lambda \text{ when } \lambda \ge \frac{1}{2}, \ 1-\lambda \ge \partial(\sim A) \ge \lambda \text{ when } \lambda \le \frac{1}{2}. \tag{14}$$

Definition 5 (λ-tautology). Let Γ be λ-evaluation set of $\mathfrak{I}$, $\forall A \in \mathfrak{I}$. A is called as Γ-tautology if $\xi(A) = 1$ for each $\xi \in \Gamma$, A is called as λ-tautology if $\xi(A) \ge \lambda$ for each $\xi \in \Gamma$.

In order to proving all the axioms are λ-tautologies in FLCOM, we require confirm the function $\mathfrak{R}$ in (9).

Definition 6. Let a, b $\in [0, 1]$, $\mathfrak{R}_o$: $[0, 1]^2 \to [0, 1]$ is a binary function $\mathfrak{R}$ if

$$\mathfrak{R}_o(a, b) = 1 \text{ when } a \le b, \tag{15}$$

$$\mathfrak{R}_o(a, b) = \max(1-a, b) \text{ when } a > b. \tag{16}$$

We prove easily the following property of $\mathfrak{R}_o$:
Property 1. Let a, b, c $\in [0, 1]$. Then

$$a \ge b, \text{ if and only if } \mathfrak{R}_o(a, c) \le \mathfrak{R}_o(b, c), \tag{17}$$

$$a \ge b, \text{ if and only if } \mathfrak{R}_o(c, a) \ge \mathfrak{R}_o(c, b). \tag{18}$$

Theorem 5. If A and $A \to B$ are λ-tautologies then B is λ-tautology.

Proof: Let $\mathfrak{R} = \mathfrak{R}_o$, A and $A \to B$ are λ-tautologies. If B is not λ-tautology, there is a λ-evaluation $\beta \in \Gamma$ such that $\beta(B) < \lambda$ according to the definition 5, but $\beta(A) \ge \lambda$, and $\beta(A \to B) = \mathfrak{R}_o(\beta(A), \beta(B)) \ge \lambda$ by (9). Due to $\beta(A) > \beta(B)$ and (16), $\mathfrak{R}_o(\beta(A), \beta(B)) = \max(1-\beta(A), \beta(B)) < \lambda$, which in contradiction to $\mathfrak{R}_o(\beta(A), \beta(B)) \ge \lambda$. Therefore, B is λ-tautology.

Theorem 6. All the axioms in FLCOM are λ-tautologies.

Proof: For any $\upsilon \in \Gamma$, a, b and c denote $\upsilon(A)$, $\upsilon(B)$ and $\upsilon(C)$ respectively, let $\Re = \Re_o$. Then the axioms (A_1), (A_2) and (A_3) can be denoted as following by (9) and the definition 6:

(A_1): $\Re(a, \Re(b, a)) \geq \lambda$

(A_2): $\Re(\Re(a, \Re(a, b)), \Re(a, b)) \geq \lambda$

(A_3): $\Re(\Re(a, b), \Re(\Re(b, c), \Re(a, c))) \geq \lambda$

For the (A_1), according to the definition 6, if $a \leq b$ then $\Re(a, \Re(b, a)) = \Re(a, \max(1-b, a))$, in which if $1-b>a$ then $\Re(a, \max(1-b, a)) = \Re(a, 1-b) = 1 \geq \lambda$, and if $1-b \leq a$ then $\Re(a, \max(1-b, a)) = \Re(a, a) = 1 \geq \lambda$. If $a > b$ then $\Re(a, \Re(b, a)) = \Re(a, 1) = 1 \geq \lambda$. Therefore, the axiom (A_1) is λ-tautology.

For the (A_2), we require only prove to $\Re(a, \Re(a, b)) \leq \Re(a, b)$ according to the definition 6. Because of $\Re(a, \Re(a, b)) \leq \Re(a, b)$ when $a \leq b$, so prove only to $\Re(a, \Re(a, b)) \leq \Re(a, b)$ when $a > b$.

Suppose $\Re(a, \Re(a, b)) > \Re(a, b)$ when $a > b$. According to (18), there is $\Re(a, b) > b$ that is $\Re(a, b) = \max(1-a, b) > b$, so $\Re(a, b) = 1-a > b$, hence obtain $\Re(a, 1-a) > 1-a$ by hypothesis. But when $a > 1-a$, $\Re(a, 1-a) = \max(1-a, 1-a)$ by (16), i.e. $1-a > 1-a$, which is incompatible. Thus, $\Re(a, \Re(a, b)) \leq \Re(a, b)$ holds. Therefore, the axiom (A_2) is λ-tautology.

For the (A_3), so as prove $\Re(a, b) \leq \Re(\Re(b, c), \Re(a, c))$ according to the definition 6.

Suppose $\Re(a, b) > \Re(\Re(b, c), \Re(a, c))$. Then $\Re(\Re(b, c), \Re(a, c)) \neq 1$, according to the definition 6, there is

$$\Re(b, c) > \Re(a, c). \qquad (*)$$

By (17), there is $a > b$. Because $\Re(a, c) \neq 1$, there is $a > c$ by (17), then there is $\Re(a, c) = \max(1-a, c)$ by (16), which is substituted into (*), obtain $\Re(b, c) > \max(1-a, c)$, that is $\Re(b, c) > \max(1-a, c)$ whether $b > c$ or $b \leq c$. When $b \leq c$, there is $\Re(b, c) = 1$ by (15), so, $\Re(b, c) = 1 > \max(1-a, c)$ holds. When $b > c$, there is $\Re(b, c) = \max(1-b, c)$ by (16), due to $a > b$, that is $1-a<1-b$, so $\Re(b, c) = \max(1-b, c) > \max(1-a, c)$, in which if $1-b \leq c$, then $\max(1-b, c) = c > \max(1-a, c) = c$, which is incompatible. Thus, $R(b, c) > \max(1-a, c)$ when $b > c$ and $1-b > c$. Using $a > b$, $b > c$ and $1-b > c$ substitute in hypothesis $\Re(a, b) > \Re(\Re(b, c), \Re(a, c))$, there is $\max(1-a, b) > \max(1-\max(1-b, c), \max(1-a, c)) = \max(b, \max(1-a, c))$ according to (17). However, when $1-a < b$, there is $\max(1-a, b) = b > \max(b, \max(1-a, c)) = b$, which is incompatible. When $1-a \geq b$, there is $\max(1-a, b) = 1-a > \max(b, \max(1-a, c)) = 1-a$, which is also incompatible. So, the hypothesis $\Re(a, b) > \Re(\Re(b, c), \Re(a, c))$ not hold. Therefore, (A_3) is λ-tautology.

Similarly, we can proved that other axioms (M)-(Y-) are λ-tautologies.

Theorem 7 (*Reliability*). Any provable formula in FLCOM is λ-tautology.

Proof: Let A be a provable formula. According to the definition 3, the proof of A is a formulas sequence $A_1, A_2, ..., A_n$, we thus induce to length n of the sequence $A_1, A_2, ..., A_n$.

(i) When n = 1, A_1(namely A) is an axiom in FLCOM according to the definition 3, so A (namely A_1) is λ-tautology by the theorem 6.

(ii) When k < n, suppose that A_1, A_2, ..., A_k are λ-tautologies, we prove that A_n is λ-tautology when n = k. According to the definition 3, there are two cases: (a) A_n is axiom in FLCOM. (b) A_n(namely A) is a formula which be deduced from A_i and A_j(i < n, j < n) by MP.

In the case (a), proof as (i). In the case (b), then A_i and A_j are certain to take the form of B and $B{\rightarrow}A$. By the above hypothesis, B and $B{\rightarrow}A$ are λ-tautologies, thus A (namely A_n) is a λ-tautology according to the theorem 5.

Therefore, according to the induction principle, any provable formula in FLCOM is λ-tautology.

The facts mentioned above which FLCOM taken on the reliability theorem for $\Re$ = $\Re_o$. For the other $\Re$, we can prove that FLCOM not always take on the reliability theorem. Such as, for the Łukasiewicz's $\Re_{\text{Łu}}$ ($\Re_{\text{Łu}}(a, b) = 1$ when $a \le b$, $\Re_{\text{Łu}}(a, b) =$ min(1, 1 − a + b) when a > b), holding the reliability theorem, but for the Gődel's $\Re_G$ ($\Re_G(a, b) = 1$ when $a \le b$, $\Re_G(a, b) = b$ when a > b), the Gaines-Recher's $\Re_{GR}$ ($\Re_{GR}(a, b)$ = 1 when $a \le b$, $\Re_{GR}(a, b) = 0$ when a > b) and the Mamdani's ($\Re_M(a, b) = 1$ when $a \le b$, $\Re_M(a, b) = $ min(a, b) when a > b), no holding the reliability theorem.

Acknowledgments. This work was supported by the National Natural Science Foundation of China (60973156) and the Fundamental Research Funds for the Central Universities(JUSRP51317B).

References

1. Zadeh, L.A.: Fuzzy Sets. Information and Control 8(3), 338–353 (1965)
2. Pawlak, Z.: Rough Sets. International Journal of Computer and Information Sciences 11, 341–356 (1982)
3. Atanassov, K.: Intuitionistic Fuzzy Sets. Fuzzy Sets and Systems 20(1), 87–96 (1986)
4. Hájek, P.: Metamathematics of Fuzzy Logic. Kluwer Academic Publishers, Dordrecht (1998)
5. Wang, G.J.: A fuzzy propositional calculus system. Bulletin of Science 42(10), 1041–1045 (1997)
6. Wagner, G.: Web Rules Need Two Kinds of Negation. In: Bry, F., Henze, N., Małuszyński, J. (eds.) PPSWR 2003. LNCS, vol. 2901, pp. 33–50. Springer, Heidelberg (2003)
7. Herre, H., Jaspars, J., Wagner, G.: Partial Logics with Two Kinds of Negation As a Foundation For Knowledge-Based Reasoning. In: Gabbay, D., Wansing, H. (eds.) What Is Negation, pp. 121–159 (1999)
8. Analyti, A., Antoniou, G., Wagner, G.: Negation and Negative Information in the W3C Resource Description Framework. Annals of Mathematics, Computing & Teleinformatics 1(2), 25–34 (2004)
9. Ferré, S.: Negation, Opposition, and Possibility in Logical Concept Analysis. In: Missaoui, R., Schmidt, J. (eds.) Formal Concept Analysis. LNCS (LNAI), vol. 3874, pp. 130–145. Springer, Heidelberg (2006)

10. Kaneiwa, K.: Description Logic with Contraries, Cntradictories, and Subcontraries. New Generation Computing 25(4), 443–468 (2007)
11. Zhenghua, P.: A New Cognition and Processing on Contradictory Knowledge. In: Fifth International Conference on Machine Learning and Cybernetics, pp. 1532–1537 (2006)
12. Pan, Z.H., Zhang, S.L.: Differentiation and Processing on Contradictory Relation and Opposite Relation in Knowledge. In: Fourth International Conference on Fuzzy Systems and Knowledge Discovery, pp. 334–338. IEEE Computer Society Press (2007)
13. Pan, Z.H.: Five Kinds of Contradictory Relations and Opposite Relations in Inconsistent Knowledge. In: Fourth International Conference on Fuzzy Systems and Knowledge Discovery, vol. 4, pp. 761–766. IEEE Computer Society Press (2007)
14. Pan, Z.H.: A Logic Description on Different Negation Relation in Knowledge. In: Huang, D.-S., Wunsch II, D.C., Levine, D.S., Jo, K.-H. (eds.) ICIC 2008. LNCS (LNAI), vol. 5227, pp. 815–823. Springer, Heidelberg (2008)
15. Pan, Z.H., Wang, C., Zhang, L.: Three Kinds of Negations in Fuzzy Knowledge and Their Applications to Decision Making in Financial Investment. In: Pan, J.-S., Chen, S.-M., Nguyen, N.T. (eds.) ICCCI 2010, Part II. LNCS (LNAI), vol. 6422, pp. 391–401. Springer, Heidelberg (2010)
16. Pan, Z.H.: Fuzzy Set with Three Kinds of Negations in Fuzzy Knowledge Processing. In: Ninth International Conference on Machine Learning and Cybernetics, vol. 5, pp. 2730–2735 (2010)
17. Pan, Z.H., Yang, L., Xu, J.: Fuzzy Set with Three Kinds of Negations and Its Applications in Fuzzy Decision Making. In: Deng, H., Miao, D., Lei, J., Wang, F.L. (eds.) AICI 2011, Part I. LNCS (LNAI), vol. 7002, pp. 533–542. Springer, Heidelberg (2011)

Experimental Teaching Quality Evaluation Practice Based on AHP-Fuzzy Comprehensive Evaluation Model

Yinjuan Huang and Liujia Huang

College of Information Science and Engineering,
Guangxi Key Laboratory of Hybrid Computational and IC Design Analysis,
Guangxi University for Nationalities, Nanning, 530006, China
huangyinjuan_1@126.com

Abstract. In this thesis, we use the integration method of AHP and fuzzy comprehensive evaluation as the evaluation model for the experimental teaching evaluation system. First, we build a hierarchy model and calculate the weigh of evaluation factor by AHP, and then execute hierarchical the evaluation and obtain the total evaluation results by fuzzy comprehensive evaluation. Finally, we compare the evaluation results with the machine test scores of the final grade examination, students' evaluation of classroom teaching evaluation. The result reflects the effectiveness, maneuverability, and fairness of the model.

Keywords: Experimental teaching, Quality evaluation, AHP, Fuzzy comprehensive evaluation model.

1 Introduction

In recent years, due to the experimental teaching special role in the culture of science and engineering students' innovative spirit and practical ability, it has been more attention of educators. At present, for the teaching theory, many universities have established a set of relatively complete theory of teaching quality evaluation and monitoring system. But in the study of the experimental teaching quality evaluation, people focus on the construction of the evaluation system, and most use a single method to evaluate, it easily bring the deviation for the evaluation results due to the limitations of the method, this makes the evaluation results unconvincing [1-4]. Although the fuzzy comprehensive evaluation is used in [1], [2], [3], but it is under the given weight, the evaluation cannot resolve the index weight reasonably, and can not reflect the fuzziness of the evaluation. In this paper, we make full use of the respective advantages of AHP and fuzzy comprehensive evaluation method, and integrate their as the method of experimental teaching evaluation system, which can be a good solution to the above problem.

D.-S. Huang et al. (Eds.): ICIC 2013, LNAI 7996, pp. 94–100, 2013.

2   AHP-Fuzzy Comprehensive Evaluation Model

In view of systematic, hierarchical, fuzzy characteristics of the experimental teaching evaluation, and the easiness of operation and realization at the same time, we take into account AHP, which can well settle the weight coefficient of the experimental teaching quality evaluation system, and Fuzzy comprehensive evaluation method can solve the vagueness of remark in evaluation process. Thus we chose AHP-Fuzzy comprehensive evaluation model for the experimental teaching evaluation. It integrated AHP [5] and Fuzzy comprehensive evaluation method [6]. At first, it built hierarchy evaluation index system by AHP, and calculated the weight for each layer indicator. Secondly it carried the evaluations for the different levels of fuzzy comprehensive evaluation. And finally it has given the overall evaluation results. This integration makes full use of the advantages of two methods, which incarnates a combination of qualitative and quantitative, and can quantify the fuzziness factors. It is simple and easy to understand and operate. The basic steps of the method are as follows:

(1) Use AHP to analysis problem, and build the hierarchical model of comprehensive evaluation.

(2) Determine the evaluation factor set $U = \{u_1, u_2, \cdots, u_m\}$ and the remark set $V = \{v_1, v_2, \cdots, v_t\}$.

(3) Determine the weight of factors.

(4) Establish the first layer fuzzy evaluation. Quantify the evaluated objects from each evaluation factor. Calculate the frequency distribution r_{ij}, which is the judge result from valuation factor to remark and construct the fuzzy evaluation matrix $R_k = (r_{ij})_{m_k \times t}$. Select the appropriate fuzzy operator which compose W_k and R_k to obtain the fuzzy comprehensive evaluation result vector $B_k = W_k \circ R_k$, and normalized it.

(5) Execute the secondary layer fuzzy evaluation. Construct the total fuzzy evaluation matrix $R = (B_1, B_2, \cdots, B_n)^T$, and select the appropriate fuzzy operator to calculate the total fuzzy comprehensive evaluation result vector $B = W \circ R$, and normalized it to obtain the fuzzy comprehensive evaluation result vector $(b_1, b_2, \cdots\cdots, b_t)$.

(6) Determine the results of the evaluation. If $b_k = \max\{b_1, b_2, \cdots\cdots, b_t\}$, then the remark of the total fuzzy comprehensive evaluation result is v_k according to the principle of maximum degree of membership.

3   Experimental Teaching Quality Evaluation Practice

In constructing the model of the teaching quality evaluation system of teachers' practical teaching experiment, it requires that it not only reflects the actual teaching dynamic of teachers, but also base on the experimental teaching reform and development law of laboratory construction, real effectively reflect and evaluation of teachers teaching, reflects the goal of the cultivation of students' creative consciousness and practical ability.

3.1   Building the Hierarchical Model of Comprehensive Evaluation

According to the curriculum design, referencing some of the established experimental teaching evaluation system [1-4], and taking into account the evaluation of easy operation and easy realization, we discuss with teachers of basic computer course teaching staff to determine the index system. The whole system is divided into 3 terms of the first layer indicators: experimental preparation, experimental procedure and experimental effect, and 12 terms of the secondary layer indicators (fig.1.).

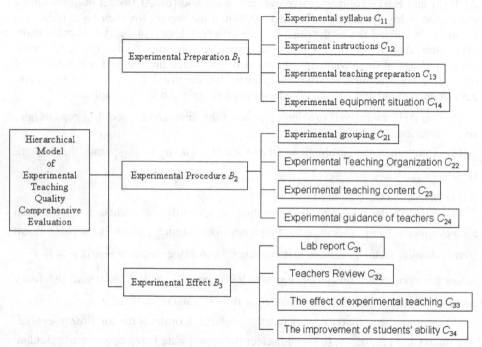

Fig. 1. Hierarchical Model of Experimental Teaching Quality Comprehensive Evaluation

3.2   Executing the First Layer Fuzzy Evaluation

According to the previously established hierarchical model, we obtain the factor set:

$$A = \{B_1, B_2, B_3\}, \; B_1 = \{C_{11}, C_{12}, C_{13}, C_{14}\},$$
$$B_2 = \{C_{21}, C_{22}, C_{23}, C_{24}\}, \; B_3 = \{C_{31}, C_{32}, C_{33}, C_{34}\},$$

and the remark set

$$V = \{v_1, v_2, v_3, v_4\} = \{excellent, \; good, \; qualified, \; unqualified\},$$

which corresponding to the grade of $\{A, B, C, D\}$ in practice.

3.3   Determining the Weight of Factors

According to the previously established hierarchical model, teachers working in basic computer course teaching team vote all evaluation factors based on the 1-9 scale method by pairwise comparisons, and construct judgment matrix by using the majority principle, there list the vote result of the first layer (table 1).

Table 1. Judgment matrix of comprehensive evaluation

Comprehensive Evaluation A	B_1	B_2	B_3
Experimental Preparation B_1	1	3	5
Experimental Procedure B_2	1/3	1	2
Experimental Effect B_3	1/5	1/2	1

We calculate the consistency index $CR = 0.0036 < 0.1$, which through the consistency test, and obtain weight $W = (0.0345, 0.1592, 0.2452)$.

Similarly, we can obtain the weight of the experimental preparation B_1, the experimental procedure B_2, the experimental effect B_3, which through the consistency test, are

$$W_1 = (0.4668, 0.2776, 0.0953, 0.1603);$$
$$W_2 = (0.0700, 0.1899, 0.2885, 0.4516);$$
$$W_3 = (0.1142, 0.1358, 0.2797, 0.4704).$$

3.4   Executing the First Layer Fuzzy Evaluation

The teaching team consisted of 10 teachers evaluated four teachers through collective attended a lecture. We count the data to construct r_{ij}. Here, we list only part of the evaluation results of teacher 1 (table 2).

Table 2. The evaluation results of teacher 1

The first layer indicator	The secondary layer indicator	Excellent	Good	Qualified	Unqualified
Experimental preparations B_1	C_{11}	6	4	0	0
	C_{12}	10	0	0	0
	C_{13}	9	1	0	0
	C_{14}	8	2	0	0

Thus, we obtain the first layer fuzzy evaluation matrix of teacher 1:

$$R_1 = \begin{pmatrix} 0.6 & 0.4 & 0 & 0 \\ 1 & 0 & 0 & 0 \\ 0.9 & 0.1 & 0 & 0 \\ 0.8 & 0.2 & 0 & 0 \end{pmatrix}, R_2 = \begin{pmatrix} 0.5 & 0.4 & 0.1 & 0 \\ 0.5 & 0.5 & 0 & 0 \\ 0.5 & 0.4 & 0.1 & 0 \\ 0.5 & 0.4 & 0.1 & 0 \end{pmatrix}, R_3 = \begin{pmatrix} 0.2 & 0.8 & 0 & 0 \\ 0.3 & 0.7 & 0 & 0 \\ 0.1 & 0.3 & 0.5 & 0.1 \\ 0.7 & 0.3 & 0 & 0 \end{pmatrix}.$$

We select ordinary composite operator to calculate, and obtain the fuzzy comprehensive evaluation results vector

$$B_1 = W_1 \circ R_1 = (0.7717, 0.2283, 0, 0),$$
$$B_2 = W_2 \circ R_2 = (0.5, 0.419, 0.081, 0),$$
$$B_3 = W_3 \circ R_3 = (0.4208, 0.4674, 0.0839, 0.028).$$

3.5   Executing the Secondary Layer Fuzzy Evaluation

We calculate the composition by matlab to obtain the total fuzzy comprehensive evaluation result vector

$$B = W \circ R = W \circ (B_1, B_2, B_3)^T = (0.4771, 0.431, 0.0763, 0.01567).$$

Thus according to the principle of maximum degree of membership, we have $\max B = 0.4771$, so the results of the fuzzy comprehensive evaluation result of teacher 1 is excellent.

Similarly, we obtain the total fuzzy comprehensive evaluation result vector of other teachers:

$$(0.3614, 0.4193, 0.1932, 0.0261), (0.5926, 0.3932, 0.0142, 0), (0.7265, 0.22735, 0, 0).$$

3.6   Analyzing the Result

We compare the total fuzzy comprehensive evaluation results with the machine test scores of the final grade examination, students' evaluation of classroom teaching evaluation at table 3.

Although the classroom teaching and experiment teaching is different, but from the above table it can be seen that the consistency of results between the experimental teaching quality evaluation, the pass rate of the final grade examination, and the students' evaluation of teaching classroom teaching evaluation is good. The validity, operability and fairness of the evaluation results show that this evaluation model is feasible. Also, we use AHP method to determine the hierarchical model and to determine the weight. The final result is only grade, which fully embodies the fuzzy of evaluation.

Table 3. Comparison of the evaluation results and others

Teacher	Experimental teaching evaluation	Pass rate of the grade examination	Excellent rate of the grade examination	Students' evaluation of classroom teaching evaluation
Teacher 1	Excellent	97.56%	43.90%	92.68
Teacher 2	Good	77.11%	25.30%	89.39
Teacher 3	Excellent	92.68%	53.66%	93.33
Teacher 4	Excellent	89.19%	14.86%	92.46

4   Conclusions

In this paper, we apply the AHP-Fuzzy comprehensive evaluation model to practice experimental teaching quality evaluation. First, we use AHP to determine the hierarchical model of experimental teaching quality comprehensive evaluation for teachers and calculate the weight of factors. Second, we execute fuzzy comprehensive evaluation to obtain the evaluation result. The operation is easier more and the results is more objective, also reflect the fuzzy of evaluation.

Acknowledgements. The work was supported by Guangxi Higher Education Teaching Reform Projects (No. 2013JGB138), Scientific Research Projects of Guangxi Province Education Department (No. 200911MS72), and Higher Education Reform Projects of Guangxi University for Nationalities (No. 2011XJGC12).

References

1. Chen, H.: Construction and Application of College Practice Teaching Quality Evaluation Model. Research and Exploration in Laboratory 28(6), 159–161 (2009)
2. Xi, L.P., Qi, X.H., Dong, H.R.: Application of Fuzzy Comprehensive Evaluation Theory in Estimation of University Experiment Teaching Quality. Experiment Science & Technology 6(6), 29–31 (2008)
3. Chen, M.H.: Method for the Quality Assessment of Experiment Teaching based on the Fuzzy Comprehensive Evaluation. Laboratory Science 14(6), 228–230 (2011)
4. Gao, Y.P., Zhou, M., Guo, X.J.: He Donggang: Application of Artificial Neural Network in Teaching Quality Evaluation System of Basic Experiment. Journal of Liaoning Normal University (Natural Science Edition) 31(4), 433–435 (2008)
5. Saaty, T.L.: How to make a decision: the analytic hierarchy process. Interfaces 24(6), 19–43 (1994)
6. Du, D., Pang, Q.H.: Modern Comprehensive Evaluation Method and Case Selection. Tsinghua University Press, Beijing (2009)

7. Li, Y.L., Gao, Z.G., Han, Y.L.: The Determination of Weight Value and the Choice of Composite Operators in Fuzzy Comprehensive Evaluation. Computer Engineering and Applications 23, 38–42 (2006)
8. Chen, M.H.: Method for the Quality Assessment of Experiment Teaching based on the Fuzzy Comprehensive Evaluation. Laboratory Sciense 14(6), 228–230 (2011)
9. Chen, G.H., Chen, Y.T., Li, M.J.: Research on the Combination Evaluation System. Journal of Fudan University (Natural Science) 42(5), 667–672 (2003)
10. He, F.B.: Comprehensive Evaluation Method MATLAB Implementation. China Social Sciences Press, Beijing (2010)

Behavior of the Soft Constraints Method Applied to Interval Type-2 Fuzzy Linear Programming Problems

Juan C. Figueroa-García[1,2] and Germán Hernández[2]

[1] Universidad Distrital Francisco José de Caldas, Bogotá – Colombia
jcfigueroag@udistrital.edu.co
[2]Universidad Nacional de Colombia, Sede Bogotá
gjhernandezp@gmail.com

Abstract. This paper presents some considerations when applying the Zimmermann soft constraints method to linear programming with Interval Type-2 fuzzy constraints. A descriptive study of the behavior of the method is performed using an example with an explanation of the obtained results.

1   Introduction

Figueroa [2–6] and [7] proposed the use of Interval Type-2 fuzzy sets (IT2FS) in Linear Programming (LP) problems to deal with linguistic uncertainty on the constraints of an LP problem. The main idea is to apply the Zimmermann soft constraints method [16] in a problem where its constraints include IT2FS with linear membership functions, which are useful to address the opinion of multiple experts in problems where the constraints of the LP model are flexible, as financial, production planning and related problems.

In this paper we provide a descriptive study about some of the possible results of the soft constraints method when using IT2FS. As IT2FSs are composed by an infinite amount of fuzzy sets, an analyst can take many possible choices of fuzzy sets before applying the soft constraints method, so there is a need for analyzing some key properties of some fuzzy sets embedded into an IT2FS in order to visualize how the soft constraints method changes its behavior.

The paper is organized as follows: a first section introduces the scope of this paper and presents the Soft Constraints method. Section two shows some basic concepts of IT2FS as constraints of an LP. Section three presents the methodology used to describe the behavior of the soft constraints when using IT2FS. In Section four, an illustrative example is presented and finally in section five some concluding remarks of the work are shown.

1.1   The Soft Constraints Method

The classical way to model of a fuzzy LP problem (FLP) with fuzzy constraints is a matrix A which solves i row inequalities with fuzzy values represented by a vector

D.-S. Huang et al. (Eds.): ICIC 2013, LNAI 7996, pp. 101–109, 2013.

called $\breve{b}$, where its optimal solution x^* is found as a function of $c'x$. Its mathematical representation is:

$$\max z = c'x + c_0$$
$$s.t.$$
$$Ax \underset{\sim}{\leqslant} \breve{b} \tag{1}$$
$$x \geqslant 0$$

where $x \in \mathbb{R}^n, c_0 \in \mathbb{R}, \breve{b} \in \mathcal{F}(S)$ and A is an $(n \times m)$ matrix, $A_{n \times m} \in \mathbb{R}^{n \times m}$. $\mathcal{F}(S)$ is the the set of all fuzzy sets.

The most used method to solve LP problems with fuzzy constraints was proposed by Zimmermann [16] and [17] who defined a fuzzy set of solutions $\breve{z}(x^*)$ to find a joint α-cut to $\breve{z}(x^*)$ and $\breve{b}$, which is summarized next:

Algorithm 1. [Zimmermann's Soft Constraints method]

1. Calculate an inferior bound called Z $Minimum$ $(\underline{z})$ by solving a LP model with $\underline{b}$.
1. Calculate a superior bound called Z $Maximum$ $(\overline{z})$ by solving a LP model with $\overline{b}$.
2. Define a fuzzy set $\breve{z}(x^*)$ with bounds $\underline{z}$ and $\overline{z}$, and linear membership function. This set represents the degree that any feasible solution has regarding the optimization objective:
3. If the objective is to minimize, then its membership function[1] is:

$$\mu_{\breve{z}}(x; \underline{z}, \overline{z}) = \begin{cases} 1, & c'x \leqslant \underline{z} \\ \dfrac{\overline{z} - c'x}{\overline{z} - \underline{z}}, & \underline{z} \leqslant c'x \leqslant \overline{z} \\ 0, & c'x \geqslant \overline{z} \end{cases} \tag{2}$$

— Thus, solve the following LP model:

$$\max \{\alpha\}$$
$$s.t.$$
$$c'x + \alpha(\overline{z} - \underline{z}) = \overline{z} \tag{3}$$
$$Ax - \alpha(\overline{b} - \underline{b}) \geqslant \underline{b}$$
$$x \geqslant 0$$

where $\alpha \in [0,1]$ is the overall satisfaction degree of all fuzzy sets, $x \in \mathbb{R}^n$, $\underline{z}, \overline{z}$ and $c_0 \in \mathbb{R}, \underline{b}, \overline{b} \in \mathcal{F}(S)$, and A is an $(n \times m)$ matrix, $A_{n \times m} \in \mathbb{R}^{n \times m}$.

---

[1] For a maximization problem, its membership function is defined as the complement of $\mu_{\breve{z}}(x; \underline{z}, \overline{z})$.

2   FLP with Interval Type-2 Fuzzy Constraints

Interval Type-2 fuzzy sets allow to model linguistic uncertainty, i.e. the uncertainty about concepts and linguistic labels of a fuzzy set. Many current decision making problems are related to uncertainty about the parameters of the problem and the way how the experts perceive and handle it. In this way, ITFSs can be used to represent the uncertainty generated by those multiple opinions and perceptions about the same problems and their parameters.

Mendel (See [13, 15, 14, 10, 8, 9]), and Melgarejo (See [11, 12]) provided formal definitions of IT2FS, and Figueroa [2–4, 6] proposed an extension of the FLP which includes linguistic uncertainty using IT2FS, called Interval Type-2 Fuzzy Linear Programming (IT2FLP).

An IT2 fuzzy constraint $\tilde{b}$ is characterized by a type-2 membership function $\mu_{\tilde{b}}(x,u))$, so we have that

$$\tilde{b} = \left\{((b,u), \mu_{\tilde{b}}(b,u)) | \forall b \in \mathbb{R}, \forall u \in J_b \subseteq [0,1]\right\} \tag{4}$$

Now, $\tilde{b}$ is an IT2FS bounded by two membership functions namely *Lower* and *Upper* primary membership functions, denoted by $\underline{\mu}_{\tilde{b}}(x)$ with parameters $\underline{b}^\Delta$ and $\underline{b}^\nabla$ and $\bar{\mu}_{\tilde{b}}(x)$ with parameters $\bar{b}^\Delta$ and $\bar{b}^\nabla$ respectively. In a mathematical for, an IT2 fuzzy constraint is a set $\tilde{b}$ defined on the closed interval $\tilde{b}_i \in [\underline{b}_i, \bar{b}_i]$, $\{\underline{b}_i, \bar{b}_i\} \in \mathbb{R}$ and $i \in \mathbb{N}_m$. The membership function (see Figure 1) which represents $\tilde{b}$ is:

$$\tilde{b}_i = \int_{b_i \in \mathbb{R}} \left[\int_{u \in J_{b_i}} 1/u \right] /b_i, \quad i \in \mathbb{N}_m, J_{b_i} \subseteq [0,1] \tag{5}$$

The Footprint of Uncertainty *(FOU)* of each set $\tilde{b}$ can be represented by two distances called Δ and ∇. Then Δ is defined as the distance between $\underline{b}^\Delta$ and $\bar{b}^\Delta$, $\Delta = \bar{b}^\Delta - \underline{b}^\Delta$ and ∇ is defined as the distance between $\underline{b}^\nabla$ and $\bar{b}^\nabla$, $\nabla = \bar{b}^\nabla - \underline{b}^\nabla$. Note that this interval can be represented as the union of an infinite amount of fuzzy sets. In other words:

$$FOU(\tilde{b}) = \bigcup_{b \in \mathbb{R}} \left[{}^\alpha\bar{b}, {}^\alpha\underline{b} \right] \forall \alpha \in [0,1], u \in J_b \tag{6}$$

where J_b is the interval of memberships for which a particular value b is defined.

2.1   The IT2FLP Model

In the same way as presented in (1), an IT2FLP can be defined as follows:

$$\max z = c'x + c_0$$

$$s.t.$$

$$Ax \precsim \tilde{b} \tag{7}$$

$$x \geqslant 0$$

where $x \in \mathbb{R}^n, c_0 \in \mathbb{R}, \tilde{b} \in \mathcal{F}(S)$ and A is an *(n x m)* matrix, $A_{n \times m} \in \mathbb{R}^{n \times m}$. Note that $\tilde{b}$ is an IT2 FS defined by its two primary membership functions $\mu_{\underline{b}}(x)$ and $\mu_{\bar{b}}(x)$. The lower membership function for $\precsim$ is:

$$\mu_{\underline{b}}(x; \underline{b}^\Delta, \underline{b}^\nabla) = \begin{cases} 1, & x \leqslant \underline{b}^\Delta \\ \dfrac{\underline{b}^\nabla - x}{\underline{b}^\nabla - \underline{b}^\Delta}, & \underline{b}^\Delta \leqslant x \leqslant \underline{b}^\nabla \\ 0, & x \geqslant \underline{b}^\nabla \end{cases} \tag{8}$$

and its upper membership function is:

$$\mu_{\bar{b}}(x; \bar{b}^\Delta, \bar{b}^\nabla) = \begin{cases} 1, & x \leqslant \bar{b}^\Delta \\ \dfrac{\bar{b}^\nabla - x}{\bar{b}^\nabla - \bar{b}^\Delta}, & \bar{b}^\Delta \leqslant x \leqslant \bar{b}^\nabla \\ 0, & x \geqslant \bar{b}^\nabla \end{cases} \tag{9}$$

A graphical representation of $\tilde{b}$ is shown in Figure 1.

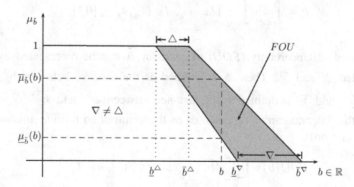

Fig. 1. Interval Type-2 fuzzy constraint with Joint Uncertain Δ & ∇

In this figure, the interval between $\mu_{\bar{b}}$ and $\mu_{\underline{b}}$ is the FOU of $\tilde{b}$. Note that there is an infinite amount of fuzzy sets embedded into the FOU of $\tilde{b}$, so an infinite amount of soft constraints models can be defined and solved.

3   Methodology of the Study

Now we have defined what an IT2FLP is (See (7), (8) and (9)), we performed an experiment by computing optimal solutions for some selected combinations of fuzzy sets using two key points $\underline{b}$ and $\overline{b}$ enclosed into Δ and ∇ respectively. The following are the selected combinations of parameters for computing $\overline{z}^*$:

Pessimistic approach: This approach uses $\mu_{\underline{b}}$ defined in (8), namely $\underline{b}^\Delta$ and $\underline{b}^\nabla$, based on the idea of having the minimum possible availability of resources.

Optimistic approach: This approach uses $\mu_{\overline{b}}$ defined in (9), namely $\overline{b}^\Delta$ and $\overline{b}^\nabla$, based on the idea of having the maximum possible availability of resources.

min-max approach: This approach uses $\underline{b}^\Delta$ and $\overline{b}^\nabla$, based on the idea of having extreme availability of resources.

max-min approach: This approach uses $\overline{b}^\Delta$ and $\underline{b}^\nabla$, based on the idea of having extreme   availability of resources.

Incremental approach: This approach divides Δ and ∇ into a set of proportional values from $\underline{b}^\Delta$ to $\overline{b}^\Delta$ and $\underline{b}^\nabla$ to $\overline{b}^\nabla$ using a value of $\delta \in [0,1]$, as follows:

$$\underline{b} = \underline{b}^\Delta + \Delta * \delta \tag{10}$$

$$\overline{b} = \underline{b}^\nabla + \nabla * \delta \tag{11}$$

Note that the values of $\delta = 0$ and $\delta = 1$ are equivalent to the pessimistic and optimistic approach.

For all those selected combinations of Δ and ∇ we use the soft constraints method (See Algorithm 1) and return the optimal values of z^*, x^* and α^*. This method is based on the the Bellman-Zadeh fuzzy decision making principle which looks for the maximum intersection among fuzzy goals and constraints as optimal decision (See Bellman and Zadeh [1]). In this way, the following are some key aspects to be analyzed:

— The behavior of $z^*, \underline{z}, \overline{z}$ and α^*.

— The minimum and maximum values of α^*, that is $\inf\{\alpha\}$ and $\sup\{\alpha\}$.

— The behavior of α for the optimistic and pessimistic approaches.

In next section, we present an application example for which we collect and analyze their results.

4   Application Example

In this example, the problem has a set of optimal solutions $\tilde{z}^*$ which is the set of all possible optimal solutions of the problem (1). All crisp optimal solutions coming from $\tilde{b}$ are embedded into $\tilde{z}^*$, so all choices of $\tilde{z}^*$ are embedded into $\tilde{z}^*$ as well. For further details about $\tilde{z}^*$ see Figueroa [4], Figueroa and Hernández [6], and Figueroa et.al. [7]. Basically, we solve a problem of four variables and five fuzzy constraints using the soft constraints method, using the following data:

$$
A = \begin{bmatrix} 2 & 3 & 3 & 2 \\ 3 & 1 & 2 & 3 \\ 4 & 5 & 2 & 6 \\ 3 & 2 & 4 & 4 \\ 3 & 6 & 4 & 5 \end{bmatrix} ; \check{b} = \begin{bmatrix} 20 \\ 24 \\ 18 \\ 20 \\ 32 \end{bmatrix} ; \hat{b} = \begin{bmatrix} 24 \\ 30 \\ 23 \\ 26 \\ 40 \end{bmatrix} ; \overline{b} = \begin{bmatrix} 22 \\ 30 \\ 20 \\ 24 \\ 40 \end{bmatrix} ; \overline{b} = \begin{bmatrix} 22 \\ 30 \\ 20 \\ 24 \\ 40 \end{bmatrix} ; c = \begin{bmatrix} 4 \\ 5 \\ 3 \\ 5 \end{bmatrix}
$$

Now, we have selected 9 equally distributed values of δ, $\delta = \{0.1, 0.2, \cdots, 0.9\}$ using (10) and (11) for which we computed the mixed approach. Note that each of problem have a value of $\underline{z}, \overline{z}$ and z^*. The results are summarized as follows.

Table 1. Optimization results for different values of α

Value	α^*	z^*	$\underline{z}$	$\overline{z}$
Pessimistic	0.5	25,125	22	28,25
Optimistic	0.5149	29,377	25	33,5
min − max	0.5194	27,973	22	33,5
max − min	0.5	26,625	25	28,25
$\delta = 0.1$	0.5	25,650	22,3	29
$\delta = 0.2$	0.5072	26,124	22,6	29,5474
$\delta = 0.3$	0.515	26,586	22,9	30,0579
$\delta = 0.4$	0.5208	27,037	23,2	30,5684
$\delta = 0.5$	0.5189	27,433	23,5	31,0789
$\delta = 0.6$	0.5171	27,828	23,8	31,5895
$\delta = 0.7$	0.5154	28,223	24,1	32,1
$\delta = 0.8$	0.5138	28,619	24,4	32,6105
$\delta = 0.9$	0.5138	29,003	24,7	33,075

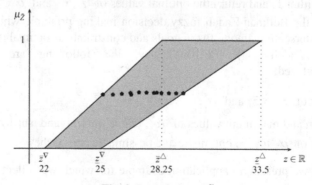

Fig. 2. Boundaries of $\tilde{z}$

Using the algorithm shown in Section 1, we can compute $\tilde{z}(x^*)$ which is displayed in Figure 2.

4.1   Analysis of the Results

The obtained results show us the following:

- The largest and smallest values of α were achieved by the min$-$max and max$-$min approach respectively. This means that using extreme points of the FOU of $\tilde{b}$ seem to lead to extreme points of α.
- At a first glance, the behavior of $\underline{z}, \overline{z}$ and α^* seem not to be proportional to δ. In this way we compute the variation rate, namely θ as

$$^\theta\alpha_i = \alpha_i - \alpha_{i-1} \ \forall\, i \in \delta$$
$$^\theta z_i^* = z_i^* - z_{i-1}^* \ \forall\, i \in \delta$$
$$^\theta \underline{z}_i = \underline{z}_i - \underline{z}_{i-1} \ \forall\, i \in \delta$$
$$^\theta \overline{z}_i = \overline{z}_i - \overline{z}_{i-1} \ \forall\, i \in \delta$$

The obtained results are:

Table 2. Variation rate θ for $\alpha, z^*, \underline{z}$ and $\overline{z}$

Value	$^\theta\alpha^*$	$^\theta z^*$	$^\theta \underline{z}$	$^\theta \overline{z}$
$\delta = 0.1$	0	0.525	0.3	0.750
$\delta = 0.2$	0.0072	0.474	0.3	0.547
$\delta = 0.3$	0.0078	0.463	0.3	0.511
$\delta = 0.4$	0.0058	0.451	0.3	0.511
$\delta = 0.5$	-0.0019	0.395	0.3	0.511
$\delta = 0.6$	-0.0018	0.395	0.3	0.511
$\delta = 0.7$	-0.0017	0.395	0.3	0.511
$\delta = 0.8$	-0.0016	0.395	0.3	0.511
$\delta = 0.9$	0	0.385	0.3	0.465
$\delta = 1$	0.0011	0.374	0.3	0.425

Table 2 shows that the behavior of α, z^* and $\overline{z}$ is not linearly incremental, so we can see that even when $\underline{z}$ is linearly incremented, the remaining results has no a proportional variation rate. This leads us to think that the soft constraints method has no a linear behavior, even when it is an LP problem itself.

- On the other hand, the interaction between $\tilde{z}(x^*)$ and $\tilde{b}$ is not proportional to δ, so the Bellmann-Zadeh when applying the soft constraints method leads to nonlinear results.
- There is no any α less than 0.5, so it seems that α has minimum and maximum boundaries. It is clear that $\alpha \leqslant 0.5$, so the soft constraints method accomplishes the

Bellman-Zadeh fuzzy decision making principle, using LP methods but achieving nonlinear results.

— The shape of $\tilde{b}$ affects the solution and the behavior of α. Different configurations of $\mu_{\underline{b}}$ and $\mu_{\overline{b}}$ lead to different values of α^*, so its behavior is a function of the FOU of $\tilde{b}$. The effect of having multiples shapes into the FOU of $\tilde{b}$ is to have nonlinear increments of $\alpha, z^*, \underline{z}$ and $\overline{z}$.

— Note that if we compute linear increments on $\mu_{\underline{b}}$ instead of proportional ones, the results should be different than using $\mu_{\overline{b}}$, since their shapes are different. For instance, we encourage the reader to compute $\alpha, z^*, \underline{z}$ and $\overline{z}$ using linear increments of $\mu_{\underline{b}}$, and compare to the results of having linear increments of $\mu_{\overline{b}}$.

5   Concluding Remarks

Some considerations about the Zimmermann's soft constraints method applied to uncertain RHS are done. The method does not show linear increments in their results when some linear shapes of $\mu_{\underline{b}}$ and $\mu_{\overline{b}}$ are used to define the FOU of $\tilde{b}$.

The application example shows us that when applying the soft constraints method to different configurations of fuzzy sets, its results can vary from linearity. This leads us to point out that even when the soft constraints uses LP methods, their results could not be linear, so the analyst should solve the problem and then analyze its results.

A sensibility analysis should be performed in order to get more information about the problem. This could be useful for computing the boundaries of $\alpha, z^*, \underline{z}$ and $\overline{z}$.

Finally, the main concluding remark is that FLP problems can achieve nonlinear results, no matter what kind of shapes for $\mu_{\underline{b}}$ and $\mu_{\overline{b}}$ are being used. In this way, the reader should not assume any linear (or linearly incremented) behavior of $\alpha, z^*, \underline{z}$ and $\overline{z}$ before solving the problem.

References

1. Bellman, R.E., Zadeh, L.A.: Decision-making in a fuzzy environment. Management Science 17(1), 141–164 (1970)
2. Figueroa, J.C.: Linear programming with interval type-2 fuzzy right hand side parameters. In: 2008 Annual Meeting of the IEEE North American Fuzzy Information Processing Society (NAFIPS) (2008)
3. Figueroa, J.C.: Solving fuzzy linear programming problems with interval type-2 RHS. In: 2009 Conference on Systems, Man and Cybernetics, pp. 1–6. IEEE (2009)
4. Figueroa, J.C.: Interval type-2 fuzzy linear programming: Uncertain constraints. In: IEEE Symposium Series on Computational Intelligence, pp. 1–6. IEEE (2011)

5. Figueroa, J.C.: A general model for linear programming with interval type-2 fuzzy technological coefficients. In: 2012 Annual Meeting of the North American Fuzzy Information Processing Society (NAFIPS), pp. 1–6. IEEE (2012)
6. Figueroa-García, J.C., Hernandez, G.: Computing Optimal Solutions of a Linear Programming Problem with Interval Type-2 Fuzzy Constraints. In: Corchado, E., Snášel, V., Abraham, A., Woźniak, M., Graña, M., Cho, S.-B. (eds.) HAIS 2012, Part III. LNCS, vol. 7208, pp. 567–576. Springer, Heidelberg (2012)
7. Figueroa, J.C., Kalenatic, D., Lopez, C.A.: Multi-period mixed production planning with uncertain demands: Fuzzy and interval fuzzy sets approach. Fuzzy Sets and Systems 206(1), 21–38 (2012)
8. Karnik, N.N., Mendel, J.M.: Operations on type-2 fuzzy sets. Fuzzy Sets and Systems 122, 327–348 (2001)
9. Karnik, N.N., Mendel, J.M., Liang, Q.: Type-2 fuzzy logic systems. Fuzzy Sets and Systems 17(10), 643–658 (1999)
10. Liang, Q., Mendel, J.M.: Interval type-2 fuzzy logic systems:. Theory and design. IEEE Transactions on Fuzzy Systems 8(5), 535–550 (2000)
11. Melgarejo, M.: Implementing Interval Type-2 Fuzzy processors. IEEE Computational Intelligence Magazine 2(1), 63–71 (2007)
12. Melgarejo, M.A.: A Fast Recursive Method to compute the Generalized Centroid of an Interval Type-2 Fuzzy Set. In: Annual Meeting of the North American Fuzzy Information Processing Society (NAFIPS), pp. 190–194. IEEE (2007)
13. Mendel, J.M.: Uncertain rule-based fuzzy logic systems: Introduction and new directions. Prentice Hall (1994)
14. Mendel, J.M., John, R.I.: Type-2 fuzzy sets made simple. IEEE Transactions on Fuzzy Systems 10(2), 117–127 (2002)
15. Mendel, J.M., John, R.I., Liu, F.: Interval type-2 fuzzy logic systems made simple. IEEE Transactions on Fuzzy Systems 14(6), 808–821 (2006)
16. Zimmermann, H.J.: Fuzzy programming and Linear Programming with several objective functions. Fuzzy Sets and Systems 1(1), 45–55 (1978)
17. Zimmermann, H.J., Fullér, R.: Fuzzy Reasoning for solving fuzzy Mathematical Programming Problems. Fuzzy Sets and Systems 60(1), 121–133 (1993)

Surprise Simulation Using Fuzzy Logic

Rui Qiao[1,2], Xiuqin Zhong[3], Shihan Yang[4], and Heng He[1,2]

[1] College of Computer Science and Technology,
Wuhan University of Science and Technology, Wuhan, China
freeqr@gmail.com
[2] Intelligent Information Processing and Real-time Industrial System
Hubei Province Key Laboratory, Wuhan, China
[3] College of Computer Science and Engineering,
University of Electronic Science and Technology of China, Chengdu, China
[4] College of Information Sciences and Engineering,
Guangxi University for Nationalities, Nanning, China

Abstract. Emotional agents are useful to variety of computer application. This paper focuses on the emotion surprise. Surprise is the automatic reaction to a mismatch, which plays an important role in the behaviors of intelligent agents. We represent psychology theories of surprise through a fuzzy inference system, as fuzzy logic helps to capture the fuzzy and complex nature of emotions. The system takes four factors related to surprise as inputs and the degree of surprise as output. We put forward fuzzy inference rules and reasoning parameters for the system. The results of the inference can be used in single or multi-agent system simulation.

Keywords: Surprise, fuzzy inference, emotion simulation, agent.

1   Introduction

Emotional agents are useful to variety of computer applications, which are modeled and developed by many researchers recently. This paper focuses on the emotion surprise[1]. Surprise is the automatic reaction to a mismatch. It is a felt reaction/response of alert and arousal due to an inconsistency (discrepancy, mismatch, non-assimilation, lack of integration) between an incoming input and our previous knowledge, in particular an actual prediction or a potential prediction [5].

According to [11,29], surprise plays an important role in the cognitive activities of intelligent agents, especially in attention focusing [7, 15, 17, 19], learning [20] and creativity [2,25]. Article [15,11,29] holds that surprise has two main functions, the one informational and the other motivational: it informs the individual about the occurrence of a schema-discrepancy, and it provides an initial impetus for the exploration of the unexpected event. Thereby, surprise promotes both immediate adaptive actions to the

---

[1] Some authors do not consider surprise as an emotion. However, whether surprise is an emotion is not the point of this paper.

D.-S. Huang et al. (Eds.): ICIC 2013, LNAI 7996, pp. 110–119, 2013.

unexpected event and the prediction, control and effective dealings with future occurrences of the event. Experiencing surprise has also some effects on human's other behaviors, for example, expression through facial expressions [3].

Papers [11, 9, 10, 8] proposed a surprise-based agent architecture and showed some experiments' results. The author computed the intensity of surprise by the degree of unexpectedness of events.

Since fuzzy logic helps to capture the fuzzy and complex nature of emotions, a lot of research work uses fuzzy rules to explore the capability of fuzzy logic in modeling emotions, such as [12, 13, 1, 24, 16, 4, 28].

Following the above-mentioned work to extrapolate psychologically grounded models of emotion through fuzzy logic systems, we represent psychology theories of surprise through a fuzzy logic system, which is implemented using Matlab. The rest of the paper is organized as follows: Section2 presents the fuzzy inference system. Section3 describes the fuzzy reasoning process. Section4 takes a case study.

2   Fuzzy Inference System

Fuzzy inference is the process of formulating the mapping from a given input to an output using fuzzy logic [27]. Fuzzy inference system(FIS) as major unit of a fuzzy logic system uses IF-THEN rules, given by, IF antecedent, THEN consequent [21]. The FIS formulates suitable rules, and based upon the rules the decision is made.

We adopt Mandani fuzzy inference method [21, 27] for our system. Mandani   is the most commonly seen fuzzy methodology, whose rules' form is more intuitive than Sugeno fuzzy inference method. The surprise fuzzy inference system is shown in Fig. 1. It has four inputs, one output, and eighty-one rules, which will be elaborated in this section.

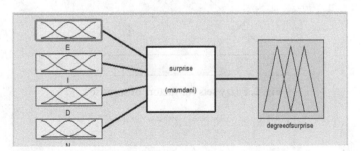

Fig. 1. Surprise inference system

2.1   Fuzzy Sets

A fuzzy set [26] is a pair (U, μ) where U is a crisp set and μ is a function denoted by $\mu : U \rightarrow [0,1]$. For each $x \in U$, the value $\mu(x)$ is called the grade of membership of x in (U, μ).

For a finite set $U = \{x_1,...,x_n\}$, the fuzzy set (U,μ) is often denoted by $\{\mu(x_1)/x_1,...,\mu(x_n)/x_n\}$. Let $x \in U$, then x is called not included in the fuzzy set (U,μ) if $\mu(x) = 0$, x is called fully included if $\mu(x) = 1$, and x is called a fuzzy member if $0 < \mu(x) < 1$. The function μ is called the membership function of the fuzzy set (U,μ).

Article [14] summarized the main factors affecting surprise intensity. A number of experiments in [22] show that expectancy disconfirmation is indeed an important factor for surprise. The experimental results shown in [6] shows in a number of experiments unexpected events that are seen as more important by a subject are experienced as more surprising. Article [18] shows that an unexpected event is seen as less surprising if the surprised person is offered a reasonable explanation that more or less justifies the occurrence of the surprising event. Several experiments [23] show that amongst other factors, events that are familiar are less surprising.

In our work, the four factors, i.e. expectation disconfirmation(E), importance of observed event(I), difficulty of explaining/fitting it in schema(D), novelty(N) of surprise emotion are all divided into three levels. We design three fuzzy sets i.e. high(H), middle(M) and low(L) for the levels. The intensity of surprise is divided into three levels. We design three fuzzy sets: very surprise(V), surprise(S) and little surprise(L) for the levels.

We take the four factors as inputs of the inference system, and the intensity of surprise as output. The membership functions for both inputs and output adopt the "trimf" function, which are shown in Fig. 2 and Fig. 3 respectively.

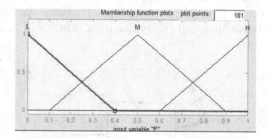

Fig. 2. Fuzzy sets for factors of surprise

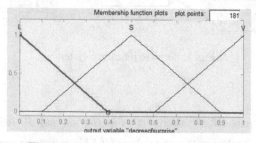

Fig. 3. Fuzzy sets for intensity of surprise

2.2   Fuzzy Rules

Fuzzy rules are IF-THEN rules to capture the relationship between inputs and outputs. According to the psychology theories[22, 6, 18, 23], we propose fuzzy rules of the emotion surprise. There are eighty-one fuzzy rules for the fuzzy system. In order to save space, only fourteen of them are listed in Table 1.

In Table 1, E,I,D,N represent expectation disconfirmation, importance of observed event, difficulty of explaining, and novelty respectively. H,M,L denote high, middle, and low degrees, while V,S,L denote very surprise, surprise, and little surprise respectively. Each rule occupies a row. For example, the first row in the table means: For an agent, if expectation disconfirmation, importance of observed event, difficulty of explaining, and novelty respectively are all high, then the agent is very surprise.

Table 1. Fuzzy rules for surprise

E	I	D	N	intensity of surprise
H	H	H	H	V
H	H	H	M	V
H	H	H	L	V
H	M	M	M	S
H	M	M	L	S
H	M	L	H	S
H	M	L	M	S
H	M	L	L	S
L	H	L	L	L
L	M	H	H	S
L	M	H	M	S
L	L	L	H	L
L	L	L	M	L
L	L	L	L	L

3   Fuzzy Reasoning

The fuzzy inference system applies Matlab to implement reasoning process. Matlab is an advanced interactive software package specially designed for scientific and engineering computation, which has been proven to be a very flexible and usable tool for solving problems in many areas. There exist large set of toolbox including Fuzzy logic toolbox, or collections of functions and procedures, available as part of the Matlab package [21].

The steps of fuzzy reasoning performed by FISs in Fuzzy logic toolbox [21,27] are: 1 fuzzification; 2 combine inputs(applying fuzzy operator); 3 implication; 4 combine outputs(aggregation); 5 defuzzification.

Step1: fuzzification. The first step is to take the crisp inputs and determine the degree to which they belong to each of the appropriate fuzzy sets via membership functions. The input is always a crisp numerical value limited to the universe of discourse of the input variable, and the output is a fuzzy degree of membership in the qualifying linguistic set. Designed uncertainty enables us to carry on the computation in those areas that are not clearly defined by crisp values.

Step2: combine inputs(applying fuzzy operator). If the antecedent of a given rule has more than one part, the fuzzy operator is applied to obtain one number that represents the result of the antecedent for that rule. The input to the fuzzy operator is two or more membership values from fuzzified input variables. The output is a single truth value. We choose method "min" for AND operator among the methods min and prod. There are only AND operators and do not exist OR operators in our fuzzy rules.

Step3: implication. Implication is used to obtain the membership function degree from antecedent of a rule to consequent of a rule. A consequent is a fuzzy set represented by a membership function, which weights appropriately the linguistic characteristics that are attributed to it. The consequent is reshaped using a function associated with the antecedent a single number. The input for the implication process is a single number given by the antecedent, and the output is a fuzzy set. Implication is implemented for each rule. We choose method "prod" for implication process among the methods min and prod.

Step4: combine outputs(aggregation). Aggregation is the process by which the fuzzy sets that represent the outputs of each rule are combined into a single fuzzy set. The input of the aggregation process is the list of truncated output functions returned by the implication process for each rule. The output of the aggregation process is one fuzzy set for each output variable. We adopt method "sum" for aggregation process among the methods max, probor and sum.

Step5: defuzzification. When the FIS is used as a controller, it is necessary to have a crisp output. Therefore in this case defuzzification method is adopted to best extract a crisp value that best represents a fuzzy set [21]. As much as fuzziness helps the rule evaluation during the intermediate steps, the final desired output for each variable is generally a single number. The input for the defuzzification process is the aggregate output fuzzy set, and the output is a single number. Defuzzification methods include max-membership principle, centroid method, weighted average method, mean-max membership. We adopts the "centroid" method, which is the most efficient and used defuzzification method. The method determines the center of the area of the combined membership functions.

There are six surfaces in three dimensions space, fixing two inputs, adopting "N-D","I-N","I-E","I-D","E-N" and "E-D" as the other two inputs respectively, and applying intensity of surprise as output. Take "I-E" surface as a representation shown in Fig. 4.

The Matlab code constructing the surprise fuzzy inference system is shown in Appendix A.

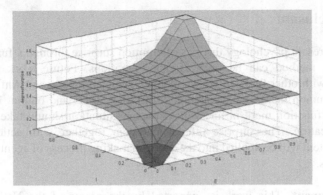

Fig. 4. I E surface

4   Case Study

Here is a scene. An agent gets lost in mountains. It is at the foot of a mountain, hungry and thirsty. It decides to climb over the mountain to search food and water on the other side of the mountain. When it climbs over the mountain, it finds some objects on the other side. The objects represent different degrees of stimulating factors sequence, which next stimulate different intensities of surprise. Table 2 shows the crisp inputs and output of the surprise fuzzy inference system in the situations. The function in Matlab fuzzy logic toolbox calculating output from inputs is shown in Appendix B.

The varying intensities of surprise result in different behaviors. Not only emotion generation but also action selections can be modeled by fuzzy rules. In this case, the relationship between the surprise degrees and behaviors can be described by fuzzy rules: If the agent is very surprise, then actions are cry and run away; If the agent is surprise, then action is cry; If the agent is little surprise, then there's only an facial expression. The fuzzy inference system for actions selection is omitted in this paper.

The scene can occur in a game or combat simulation, where agents make decisions strengthened by emotions.

Table 2. The other side of the mountain

Object	E	I	D	N	intensity of surprise
Monster	0.9	0.8	0.9	0.9	0.8700
Farmland	0.1	0.9	0.1	0.1	0.1300
Orchards	0.1	0.9	0.1	0.1	0.1300
River	0.1	0.9	0.1	0.1	0.1300
Lake	0.1	0.9	0.1	0.1	0.1300
Sea	0.9	0.9	0.5	0.1	0.5000
Mountain	0.1	0.8	0.1	0.1	0.1300
Buildings	0.5	0.8	0.1	0.1	0.4245

5   Conclusion

We have represent psychology theories of emotion surprise through a fuzzy inference system. The method can be used to other emotions. Future work will make a comparison with other fuzzy inference method, such as Sugeno and Anfis. This work takes the common membership function "trimf" for inputs and output. However, the membership function more in line with the actual situation should take shape in the actual application. The surprise inference system together with actions selection inference system will be applied into the decision making process of agents in uncertain environments.

Acknowledgment. This work is supported by the grants of the National Natural Science Foundation of China(61273303) and the Young Scientists Foundation of Wuhan University of Science and Technology (2013xz012).

References

1. Albert van der Heide, G.T., Sanchez, D.: Computational Models of Affect and Fuzzy Logic. In: European Society for Fuzzy Logic and Technology, pp. 620–627 (2011)
2. Boden, M.: Creativity and Unpredictability. SEHR 4(2), 123–139 (1995)
3. Ekman, F.W., Friesen, W.V.: Unmasking The Face: A Guide to Recognizing Emotions from Facial Clues (2003)
4. El-Nasr, M.S.: Modeling Emotions Dynamics in Intelligent Agents. PhD Thesis (1998)
5. Emiliano Lorini, C.C.: The Cognitive Structure of Surprise Looking for Basic Principles. An International Review of Philosophy 26(1), 133–149 (2007)
6. Gendolla, G.H.E., Koller, M.: Surprise And Motivation of Causal Search: How Are They Affected by Outcome Vlaence and Importance? Motivation and Emotion 25(4), 327–349 (2001)
7. Izard, C.: The psychology of emotions (1991)
8. Macedo, L., Cardoso, A.: Sceune - surprise/curiosity-based Exploration of Uncertain and Unknown Environments. In: 26th Annual Conference of The Cognitive Science Society, pp. 73–81 (2001)
9. Macedo, L., Cardoso, A.: Using Surprise To Create Products That Get The Attention of Other Agents. In: AAAI Fall Symposium Emotional and Intelligent II: The Tangled Knot of Social Cognition, pp. 79–84 (2001)
10. Macedo, L., Cardoso, A.: Towards Artificial Forms of Surprise And Curiosity. In: Proceedings of the European Conference on Cognitive Science, pp. 139–144 (2000)
11. Macedo, L., Cardoso, A.: Modeling Forms of Surprise in An Artificial Agent. In: Proceedings of The 23rd Annual Conference of The Cognitive Science Society, pp. 588–593 (2001)
12. El-Nasr, M.S., Yen, J., Ioerger, T.R.: Flame. Flame-fuzzy Logic Adaptive Model of Emotions. Autonomous Agents and Multi-Agent Systems 3(3), 219–257 (2000)
13. Deng, R.-R., Meng, X.Y., Wang, Z.-L.: An affective model based on fuzzy theory. Journal of Communication and Computer 5(2), 7–11 (2008)
14. Merk, R.: A Computational Model on Surprise and Its Effects on Agent Behaviou. In: Simulated Environments. Technical report, National Aerospace Laboratory NLR (2010)

15. Meyer, W.U., Reisenzein, R., Schützwohl, A.: Towards A Process Analysis of Emotions: The Case of Surprise. Motivation and Emotion 21, 251–274 (1997)
16. Vlad, O.P., Fukuda, T.: Model Based Emotional Status Simulation. In: Proceedings of the IEEE Int. Workshop on Robot and Human Interactive Communication, pp. 93–98 (2002)
17. Ortony, A., Partridge, D.: Surprisingness and Expectation Failure: What's The Difference? In: International Joint Conference on Articial Intelligence, pp. 106–108 (1987)
18. Maguire, R., Keane, M.T.: Surprise: Disconfirmed Expectations or Representation-fit. In: Proceedings of the 28th Annual Conference of the Cognitive Science Society, pp. 1765–1770 (2006)
19. Reisenzein, R.: The Subjective Experience of Surprise. The message within: The Role of Subjective Experience in Social Cognition and Behavior, 262–279 (2000)
20. Schank, R.: Explanation Patterns: Understanding Mechanically and Creatively. Lawrence Erlbaum Associates, Hillsdale (1986)
21. Sivanandam, S.N.: Introduction to Fuzzy Logic Using Matlab. Springer (2007)
22. Stiensmeier-Pelster, J., Martini, A., Reisenzein, R.: The Role of Surprise in The Attribution Process. Cognition and Emotion 9(1), 5–31 (1995)
23. Teigen, K.H., Keren, G.: Surprises: Low Probabilities or High Contrasts? Cognition 87(2), 55–71 (2002)
24. Dang, T.H.H., Letellier-Zarshenas, S., Duhaut, D.: Comparison of Recent Architectures of emotions. In: International Conference on Control. Automation. Robotics and Vision., pp. 1976–1981 (2008)
25. Williams, M.: Aesthetics and The Explication of Surprise. Languages of Design 3, 145–157 (1996)
26. Zadeh, L.A.: Fuzzy set. Information and Control 8(3), 338–353 (1965)
27. Zadeh, L.A.: Fuzzy Logic Toolbox Users Guide (1995)
28. PourMohammadBagher, L.: Intelligent Agent System Simulation Using Fear Emotion. World Academy of Science, Engineering and Technology 48 (2008)
29. Macedo, L., Cardoso, A., Reisenzein, R.: A Surprise-based Agent Architecture, Cybernetics and Systems, Trappl, R (ed.), vol. 2 (2006)

Appendix: Programs

A. Surprise fuzzy inference system

```
a=newfis('surprise','mamdani', 'min','max','prod','sum','centroid');
        a.input(1).name='E';
        a.input(1).range=[0 1];
        a.input(1).mf(1).name='L';
        a.input(1).mf(1).type='trimf';
        a.input(1).mf(1).params=[-0.4 0 0.4];
        a.input(1).mf(2).name='M';
        a.input(1).mf(2).type='trimf';
        a.input(1).mf(2).params=[0.1 0.5 0.9];
        a.input(1).mf(3).name='H';
        a.input(1).mf(3).type='trimf';
        a.input(1).mf(3).params=[0.6 1 1.4];
        a.input(2).name='T';
```

```
a.input(2).range=[0 1];
a.input(2).mf(1).name='L';
a.input(2).mf(1).type='trimf';
a.input(2).mf(1).params=[-0.4 0 0.4];
a.input(2).mf(2).name='M';
a.input(2).mf(2).type='trimf';
a.input(2).mf(2).params=[0.1 0.5 0.9];
a.input(2).mf(3).name='H';
a.input(2).mf(3).type='trimf';
a.input(2).mf(3).params=[0.6 1 1.4];
a.input(3).name='D';
a.input(3).range=[0 1];
a.input(3).mf(1).name='L';
a.input(3).mf(1).type='trimf';
a.input(3).mf(1).params=[-0.4 0 0.4];
a.input(3).mf(2).name='M';
a.input(3).mf(2).type='trimf';
a.input(3).mf(2).params=[0.1 0.5 0.9];
a.input(3).mf(3).name='H';
a.input(3).mf(3).type='trimf';
a.input(3).mf(3).params=[0.6 1 1.4];
a.input(4).name='M';
a.input(4).range=[0 1];
a.input(4).mf(1).name='L';
a.input(4).mf(1).type='trimf';
a.input(4).mf(1).params=[-0.4 0 0.4];
a.input(4).mf(2).name='M';
a.input(4).mf(2).type='trimf';
a.input(4).mf(2).params=[0.1 0.5 0.9];
a.input(4).mf(3).name='H';
a.input(4).mf(3).type='trimf';
a.input(4).mf(3).params=[0.6 1 1.4];
a.output(1).name='degreeofsurprise';
a.output(1).range=[0 1];
a.output(1).mf(1).name='L';
a.output(1).mf(1).type='trimf';
a.output(1).mf(1).params=[-0.4 0 0.4];
a.output(1).mf(2).name='S';
a.output(1).mf(2).type='trimf';
a.output(1).mf(2).params=[0.1 0.5 0.9];
a.output(1).mf(3).name='V';
a.output(1).mf(3).type='trimf';
a.output(1).mf(3).params=[0.6 1 1.4];
```

```
    ruleList=[
3 3 3 3 3 1 1; 3 3 3 2 3 1 1; 3 3 3 1 3 1 1; 3 3 2 3 3 1 1; 3 3 2 2 3 1 1;
3 3 2 1 2 1 1; 3 3 1 3 3 1 1; 3 3 1 2 2 1 1; 3 3 1 1 2 1 1; 3 2 3 3 3 1 1;
3 2 3 2 3 1 1; 3 2 3 1 2 1 1; 3 2 2 3 3 1 1; 3 2 2 2 2 1 1; 3 2 2 1 2 1 1;
3 2 1 3 2 1 1; 3 2 1 2 2 1 1; 3 2 1 1 2 1 1; 3 1 3 3 3 1 1; 3 1 3 2 2 1 1;
3 1 3 1 2 1 1; 3 1 2 3 2 1 1; 3 1 2 2 2 1 1; 3 1 2 1 2 1 1; 3 1 1 3 2 1 1;
3 1 1 2 2 1 1; 3 1 1 1 1 1 1; 2 3 3 3 3 1 1; 2 3 3 2 3 1 1; 2 3 3 1 2 1 1;
2 3 2 3 3 1 1; 2 3 2 2 2 1 1; 2 3 2 1 2 1 1; 2 3 1 3 2 1 1; 2 3 1 2 2 1 1;
2 3 1 1 2 1 1; 2 2 3 3 3 1 1; 2 2 3 2 2 1 1; 2 2 3 1 2 1 1; 2 2 2 3 2 1 1;
2 2 2 2 2 1 1; 2 2 2 1 2 1 1; 2 2 1 3 2 1 1; 2 2 1 2 2 1 1; 2 2 1 1 1 1 1;
2 1 3 3 2 1 1; 2 1 3 2 2 1 1; 2 1 3 1 2 1 1; 2 1 2 3 2 1 1; 2 1 2 2 2 1 1;
2 1 2 1 1 1 1; 2 1 1 3 2 1 1; 2 1 1 2 1 1 1; 2 1 1 1 1 1 1; 1 3 3 3 3 1 1;
1 3 3 2 2 1 1; 1 3 3 1 2 1 1; 1 3 2 3 2 1 1; 1 3 2 2 2 1 1; 1 3 2 1 2 1 1;
1 3 1 3 2 1 1; 1 3 1 2 2 1 1; 1 3 1 1 1 1 1; 1 2 3 3 2 1 1; 1 2 3 2 2 1 1;
1 2 3 1 2 1 1; 1 2 2 3 2 1 1; 1 2 2 2 2 1 1; 1 2 2 1 1 1 1; 1 2 1 3 2 1 1;
1 2 1 2 1 1 1; 1 2 1 1 1 1 1; 1 1 3 3 2 1 1; 1 1 3 2 2 1 1; 1 1 3 1 1 1 1;
1 1 2 3 2 1 1; 1 1 2 2 1 1 1; 1 1 2 1 1 1 1; 1 1 1 3 1 1 1; 1 1 1 2 1 1 1;
1 1 1 1 1 1 1;
];
a=addrule(a,ruleList);
writefis(a,'surprise.fis');
```

B. Case study

```
a=readfis('surprise.fis');
out=evalfis([0.9 0.8 0.9 0.9;0.1 0.9 0.1 0.1;
0.1 0.9 0.1 0.1;0.1 0.9 0.1 0.1;
0.1 0.9 0.1 0.1;0.9 0.9 0.5 0.1;
0.1 0.8 0.1 0.1;0.5 0.8 0.1 0.1],a);
out =
0.8700
0.1300
0.1300
0.1300
0.1300
0.5000
0.1300
0.4245
```

Colon Cell Image Segmentation Based on Level Set and Kernel-Based Fuzzy Clustering

Amin Gharipour and Alan Wee-Chung Liew

School of Information and Communication Technology
Gold Coast campus, Griffith University, QLD4222, Australia
amin.gharipour@griffithuni.edu.au, a.liew@griffith.edu.au

Abstract. This paper presents an integration framework for image segmentation. The proposed method is based on Fuzzy c-means clustering (FCM) and level set method. In this framework, firstly Chan and Vese's level set method (CV) and Bayes classifier based on mixture of density models are utilized to find a prior membership value for each pixel. Then, a supervised kernel based fuzzy c-means clustering (SKFCM) algorithm assisted by prior membership values is developed for final segmentation.

The performance of our approach has been evaluated using high-throughput fluorescence microscopy colon cancer cell images, which are commonly used for the study of many normal and neoplastic procedures. The experimental results show the superiority of the proposed clustering algorithm in comparison with several existing techniques.

Keywords: Image segmentation; fuzzy c-means; kernel based fuzzy c-means; supervised kernel based fuzzy c-means; Chan-Vese method; Bayes classifier.

1   Introduction

In recent years, although different methods for the segmentation of cell nuclei in fluorescence microscopy images have been proposed, it is still considered as one of the critical but challenging tasks in the analysis of the function of cell in fluorescence microscopy images. Typically, image segmentation techniques based on fuzzy clustering and level set methods are two important techniques utilized for segmenting images into different non-overlapping, constituent regions. Whereas level set methods impose constraints on boundary smoothness, fuzzy clustering segmentation is based on spectral characteristics of individual pixel.

The level set technique for image segmentation was initially proposed by Osher and Sethian [1]. Based on the idea of Osher and Sethian, Malladi et al. [2] suggested a front-propagation method for shape modeling. Later, Chan and Vese [3] proposed CV method using an energy minimization technique and incorporated region-based information in the energy functional as an extra constraint. The basic idea behind their method was to look for a specific partition of an image into two parts, one characterizing the objects to be identified, and the second one characterizing the background. The active contour was represented by the boundary among these two parts.

D.-S. Huang et al. (Eds.): ICIC 2013, LNAI 7996, pp. 120–129, 2013.

As a powerful data clustering algorithm, FCM [4, 5] reveals the original structure of the image data and classifies pixels into clusters based on a similarity measure. This allows a segmentation of groups of pixels into homogenous region during image segmentation [6,7,8]. This clustering is acquired by iterative minimization of a cost function that is influenced by the Euclidean distance of the pixels to the cluster prototypes. One of the disadvantages of FCM is that Euclidean distance can only be used to detect 'spherical' clusters. To overcome this drawback, many modified FCM algorithms based on the kernel methods (KFCM) have been suggested to handle non-spherical clusters [9].

Whereas CV method focuses on geometrical continuity of segmentation boundaries, FCM aims to analyze spectral properties of image pixels and does not consider topological information. Since level set method and FCM use different types of information, it is possible that they yield different segmentation results. Nevertheless, the topological information captured in the CV method and the pixel's spectral properties captured by FCM are complementary in nature and consider them together would improve the segmentation result. Hence, an integration strategy is employed in this work to use the rich information offered by both methods. The main contribution of this paper is the introduction of an integration framework for KFCM and CV. The performance improvement compared to some well-known techniques using cell images is reported.

The organization of this paper is as follows. Firstly, a brief overview of CV level set based approach and FCM is given. Next, the proposed integration procedure is addressed in detail. Experiments are then discussed in Section 4. Finally, Section 5 concludes the paper and highlights possible future work.

2   Previous Works

2.1   Chan–Vese Model

Let Ω denotes the image domain, $C \subset \Omega$ denotes the smooth, closed segmenting curve, $I_0 : \Omega \to R$ is a given image, I denotes the piecewise smooth estimation of I_0 with discontinuities among C, $|C|$ denotes the length of C. μ and v denote constant with positive values. The general form for the Mumford–Shah energy functional can be written as follows:

$$E^{MS}(I,C) = \int_{\Omega} |I_0(x,y) - I(x,y)|^2 \, dxdy + \mu \int_{\Omega \backslash C} |\nabla I(x,y)|^2 \, dxdy + v.|C| \qquad (1)$$

The CV method solves (1) through minimizing the following energy functional:

$$E^{CV}(c_1,c_2,C) = \lambda_1. \int_{inside(c)} |I_0(x,y) - c_1|^2 \, dxdy + \lambda_2. \int_{outside(C)} |I_0(x,y) - c_2|^2 \, dxdy + |C| \qquad (2)$$

where λ_1, λ_2 and μ denote positive constants. c_1, c_2 are the intensity averages of I_0 inside C and outside C.

In the CV method, the level set method which replaces the unknown curve C by the level set function $\emptyset(x,y)$ is utilized to minimize (2). Therefore (2) can be reformulated in terms of level set function $\emptyset(x,y)$ as follows:

$$E^{CV}\left(c_1,c_2,\emptyset(x,y)\right)=$$

$$\lambda_1.\iint_\Omega\left|I_0(x,y)-c_1\right|^2 H_\varepsilon\left(\emptyset(x,y)\right)dxdy+\lambda_2.\iint_\Omega\left|I_0(x,y)-c_2\right|^2(1-H_\varepsilon(\emptyset(x,y))dxdy$$

$$+\mu.\oint_\Omega\delta_\varepsilon\left(\emptyset(x,y)\right)|\nabla\emptyset(x,y)|dxdy$$

(3)

where $H_\varepsilon(z)$ and $\delta_\varepsilon(z)$ denote the regularized versions of the Heaviside function $H(z)$ and $\delta(z)$ specified by $H(z)=\begin{cases}1 \text{ if } z\geq 0\\0 \text{ if } z\leq 0\end{cases}$ and $\delta(z)=\dfrac{\partial H(z)}{\partial z}$, respectively.

The existence of minimizer for (3) has been considered in [10,11]. Using Euler–Lagrange method, the level set equation for evolution procedure is given by:

$$\frac{\partial\emptyset}{\partial t}=\delta_\varepsilon(\emptyset)\left[\mu \text{ div}\left(\frac{\nabla\emptyset}{|\nabla\emptyset|}\right)-\lambda_1(I_0-c_1)^2+\lambda_2(I_0-c_2)^2\right]$$

(4)

where c_1 and c_2 can be, respectively, updated as follows:

$$c_1(\emptyset)=\frac{\int_\Omega I_0(x,y)H_\varepsilon(\emptyset(x,y))dxdy}{\int_\Omega H_\varepsilon(\emptyset(x,y))dxdy}\quad c_2(\emptyset)=\frac{\int_\Omega I_0(x,y)(1-H_\varepsilon(\emptyset(x,y)))dxdy}{\int_\Omega(1-H_\varepsilon(\emptyset(x,y)))dxdy}$$

(5)

2.2   Kernel-Based Fuzzy c-Means Clustering Algorithm

The FCM algorithm, as a variation of the standard k-means clustering, was first proposed by Dunn [4], and further developed by Bezdek [5]. FCM aims to partition $\xi=(\xi_1,\ldots,\xi_n)$ i.e., a set of voxel or pixel locations in Ω into η clusters, that are defined by prototypes $V=(v_1,\ldots,v_\eta)$. With fuzzy clustering, each ξ_j is a member of all partitions simultaneously, but with different membership grades. The FCM algorithm achieves the final solution by solving:

$$\text{Minimize } J_{FCM}(U,V)=\sum_{i=1}^\eta\sum_{j=1}^n u_{ij}^m\|\xi_j-v_i\|^2$$

$$\text{subject to } U\in\mathcal{M}, m\in[1,\infty],$$

(6)

Where $\mathcal{M}=\left\{U=\left[u_{ij}\right]_{\substack{i=1,\ldots,\eta \\ j=1,\ldots,n}}\middle| u_{ij}\in[0,1],\ \sum_{i=1}^{\eta}u_{ij}=1\right\}$

The parameter m affects the fuzziness of the clusters. As can be seen from (6) every pixel has the same weight in the image's data set. Fuzzy clustering under constraint (6) is called probabilistic clustering and u_{ij} is the posterior probability $p(\eta_i \mid \xi_j)$ [12]. Alternating optimization, which alternates between optimizations of $\tilde{J}_{FCM}(U\mid\dot{v})$ over U with fixed $\dot{v}$ and $\tilde{J}_{FCM}(v\mid\dot{U})$ over v with fixed U converges to a local minimizer or a saddle point of J_{FCM} [13].

Now we replace the Euclidean distance in (6) by the distance function $\|\Theta(\xi_j)-\Theta(v_i)\|^2$ where Θ denotes a map on feature space. Therefore (6) can be rewritten as follows:

$$J_{KFCM}(U,V)=\sum_{i=1}^{2}\sum_{j=1}^{n}u_{ij}^{m}\|\Theta(\xi_j)-\Theta(v_i)\|^2 \tag{7}$$

In terms of inner product, the distance function can be expressed as follows:

$$\|\Theta(\xi_j)-\Theta(v_i)\|^2=<\Theta(\xi_j),\Theta(\xi_j)>-2<\Theta(\xi_j),\Theta(v_i)>+<\Theta(v_i),\Theta(v_i)> \tag{8}$$

We set $<\Theta(\xi_j),\Theta(v_i)>=K(\xi_j,v_i)$. Hence (8) can be formulated as follows:

$$\|\Theta(\xi_j)-\Theta(v_i)\|^2=K(\xi_j,\xi_j)-2K(\xi_j,v_i)+K(v_i,v_i) \tag{9}$$

By adopting the hyperbolic tangent function, $K(x,y)=1-\tanh\left(-\|x-y\|^2/\sigma^2\right)$, as a kernel function, we can define $d(x,y)$ as the following:

$$d(\xi_j,v_i)\triangleq\|\Theta(\xi_j)-\Theta(v_i)\|=\sqrt{2(1-k(\xi_j,v_i))} \tag{10}$$

It's easy to prove that $\forall x\neq y, d(x,y)=d(y,x)>0$ and $d(x,x)=0$. Also we have: $d(x,y)=\|\Theta(x)-\Theta(y)\|\leq\|\Theta(x)-\Theta(z)\|+\|\Theta(z)-\Theta(y)\|=d(x,z)+d(z,y)$

Hence, $d(x,y)$ is a metric.

According to (10), (7) can be rewritten as follows:

$$J_{KFCM}(U,V)=\sum_{i=1}^{2}\sum_{j=1}^{n}u_{ij}^{m}d(\xi_j,v_i)^2 \tag{11}$$

3   Integration Strategy

In this section the integration strategy is discussed. First, the Bayes classifier based on mixture of density models is utilized to find prior membership degrees of each pixel.

Then, the novel supervised kernel-based fuzzy c-means clustering algorithm (SKFCM) is presented to implement approximated prior membership degrees.

3.1   Bayes Classifier Based on Mixture of Density Models

For every pixel ξ , we suppose that $\xi \in \mathcal{C}_i \Leftrightarrow p(\mathcal{C}_i|\xi) \geq p(\mathcal{C}_j|\xi)$ where $\mathcal{C}_1 = \{\xi|\varnothing(x,y) \geq 0\}$ and $\mathcal{C}_2 = \Omega / \mathcal{C}_1$.

We can implement Bayes classifier based on mixture of density models [14] for finding the class-conditional densities. Therefore, $p(\mathcal{C}_i|\xi)$ can be calculated as follows:

$$p(\mathcal{C}_i|\xi) = \sum_{l=1}^{R} p(r_l|\xi) P(\mathcal{C}_i|r_l) \text{ for } i=1,2 \qquad (12)$$

where $R > 2$, $p(r_l|\xi)$ and $P(\mathcal{C}_i|r_l)$ denote the number of mixture models, the posteriori probability of ξ derived from r_lth local model and the prior probability of this model represents the class $\mathcal{C}_i$, respectively.

Based on the Bayesian rule, $P(r_l|\xi)$ can be written as follows:

$$p(r_i|\xi) = \frac{p(\xi|r_i) P(r_i)}{\sum_{j=1}^{R} p(\xi|r_j) P(r_j)}$$

Hence the posteriori class probability can be formulated as follows:

$$p(\mathcal{C}_i|\xi) = \frac{p(\xi|\mathcal{C}_i) P(\mathcal{C}_i)}{P(\xi)} = \sum_{l=1}^{R} \frac{p(\xi|r_l) P(r_l)}{\sum_{j=1}^{R} p(\xi|r_j) P(r_j)} P(\mathcal{C}_i|r_l) = \sum_{l=1}^{R} \frac{p(\xi|r_l) P(r_l) P(\mathcal{C}_i|r_l)}{P(\xi)} \qquad (13)$$

in which $p(\xi|r_i) = \frac{1}{|2\pi F_i|^{\frac{n}{2}}} \exp\left(-\frac{1}{2}(\xi_j - \mu_i)^T F_i^{-1}(\xi_j - \mu_i)\right)$

where μ_i and F_i represent the mean value and the covariance matrix of pixels in r_i , respectively.

3.2   Supervised Kernel-Based Fuzzy c-Means Clustering Algorithm

Prior membership degree given by (13) is denoted as $\aleph_{ij}$ and expressed as follows:

$$\aleph = \left\{ \left[\aleph_{ij} \right]_{\substack{i=1,2 \\ j=1,\dots,n}} \middle| \aleph_{ij} = p(\mathcal{C}_i|\xi_j) \right\} \qquad (14)$$

To improve the performance of KFCM, we incorporate $\aleph$, as a component of supervised learning, into modified KFCM objective function as follows:

$$J_{SKFCM}(u,v) = \sum_{i=1}^{2}\sum_{j=1}^{n}u_{ij}^{m}d(\xi_j,v_i)^2 + \gamma u_{ij}^{m}\sum_{\varsigma\epsilon\varsigma_i}\aleph_{\varsigma j}^{m} \tag{15}$$

in which $\varsigma_i = \{\{1,2\}\setminus i\}$ and γ is a control parameter which keeps a balance between the original KFCM and CV method in the optimization procedure.

To find the optimal values of v_i and u_{ij} considering the constraint (6), we introduce a Lagrange multiplier l and minimize the following Lagrange function $\mathcal{L}$:

$$\mathcal{L}(U,V,l) = \sum_{i=1}^{2}\sum_{j=1}^{n}u_{ij}^{m}d(x,y)^2 + \gamma u_{ij}^{m}\sum_{\varsigma\epsilon\varsigma_i}\aleph_{\varsigma j}^{m} - l(\sum_{i=1}^{\eta}u_{ij}-1)$$

Assume that u_{kj} is the membership degree of ξ_j belonging to the cluster η_k whose centroid is v_k. The stationary point of the optimized functional can be obtained as

$$\left(u_{kj},v_k,l\right) \text{ if and only if } \frac{\partial\mathcal{L}}{\partial l}=0, \quad \frac{\partial\mathcal{L}}{\partial v_k}=0 \text{ and } \frac{\partial\mathcal{L}}{\partial u_{kj}}=0.$$

Taking these derivatives returns the relationships as follows:

$$\frac{\partial\mathcal{L}}{\partial l} = \sum_{i=1}^{2}u_{kj}-1 = 0$$

The optimal solution for v_k is given as follows:

$$v_k = \frac{\sum_{j=1}^{n}u_{kj}^{m}\left(1-(K(\xi_j,v_k)-1)^2\right)\xi_j}{\sum_{j=1}^{n}u_{kj}^{m}\left(1-(K(\xi_j,v_k)-1)^2\right)} \tag{16}$$

Similarly, we take the derivative of $\mathcal{L}(u_{kj})$ with respect to u_{ij} and fixed it to zero:

$$\frac{\partial\mathcal{L}}{\partial u_{kj}} = 2mu_{kj}^{m-1}(1-k(\xi_j,v_k)) + \gamma mu_{kj}^{m-1}\sum_{\varsigma\epsilon\varsigma_k}\aleph_{\varsigma j}^{m} - l = 0$$

thus

$$2mu_{kj}^{m-1}(1-k(\xi_j,v_k)) + \gamma mu_{kj}^{m-1}\sum_{\varsigma\epsilon\varsigma_k}\aleph_{\varsigma j}^{m} = l$$

We set m=2, thus

$$u_{kj} = \left(\frac{l}{2\left(2(1-k(\xi_j,v_k))+\gamma\sum_{\varsigma\epsilon\varsigma_k}\aleph_{\varsigma j}^{2}\right)}\right)$$

$u_{kj} \in \mathcal{M}$ indicates that $\quad 1 = \dfrac{l}{2} \displaystyle\sum_{i=1}^{2} \left(\dfrac{1}{\left(2(1-k(\xi_j, v_k)) + \gamma \sum_{\varsigma \in \varsigma_i} \aleph_{\varsigma j}^{2}\right)} \right)$

Hence

$$l = \dfrac{2}{\displaystyle\sum_{i=1}^{2} \left(\dfrac{1}{\left(2(1-k(\xi_j, v_k)) + \gamma \sum_{\varsigma \in \varsigma_i} \aleph_{\varsigma j}^{2}\right)} \right)}$$

Therefore, the optimal solution of u_{kj} is formulated as follows:

$$u_{kj} = \dfrac{1}{\displaystyle\sum_{i=1}^{2} \left(\dfrac{\left(2(1-k(\xi_j, v_k)) + \gamma \sum_{\varsigma \in \varsigma_k} \aleph_{\varsigma j}^{2}\right)}{\left(2(1-k(\xi_j, v_i)) + \gamma \sum_{\varsigma \in \varsigma_i} \aleph_{\varsigma j}^{2}\right)} \right)} \tag{17}$$

4   Experimental Results

We tested our algorithm on the human HT29 colon cancer cells with a size of 512 x 512 pixels from image set BBBC008v1 [15,16] which were stained cell nuclei and have ground truth to show the segmentation performance of the proposed technique. An example of original image, ground truth and segmentation result based on the proposed method are shown in Fig. 1. The proposed method is compared with FCM, CV, SFCM [17], SFLS [18] and region-scalable fitting energy (RSFE) [19]. To evaluate the performance of our algorithm, we calculate the recognition error rate as follows:

$$\text{recognition error rate}\,(\%) = \dfrac{T_{cell} + T_{background}}{n} \times 100 \tag{18}$$

where n indicates the total number of pixels in the given image, T_{cell} and $T_{background}$ are the number of pixels incorrectly classified as the cell and background, respectively.

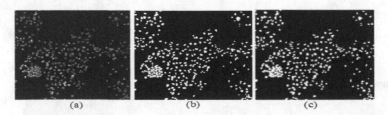

Fig. 1. Example of (a) Original image, (b) Ground truth and (c) Segmentation result

Table 1. Recognition error rate for different segmentation methods

Image index	human HT29 colon cancer cells						human HT29 colon cancer cells with 5% Gaussian noise					
	FCM	CV	SFCM	RSFE	SFLS	Proposed Method (SKFCM)	FCM	CV	SFCM	RSFE	SFLS	Proposed Method (SKFCM)
1	2.43	4.03	3.2	2.1	3	1.07	4.68	5.03	4.2	3.68	4.1	2.07
2	2.39	3.21	3.34	2.02	3.07	2.01	5.39	4.53	4.14	3.92	3.87	2.91
3	1.64	3.12	2.17	1.41	2.05	0.91	4.64	3.52	3.07	3.21	3.15	2.66
4	0.94	1.08	1.55	1.05	1.41	0.57	1.94	2.08	1.75	1.82	1.71	1.29
5	3.2	5.56	7.6	3.01	8.34	2.31	7.54	7.71	6.86	4.32	7.34	3.77
6	5.37	3.055	3.32	2.89	3.04	1.78	6.46	4.53	5.17	3.94	4.12	2.79
7	1.63	2.01	2.23	2.22	2.07	1.14	2.91	3.04	2.65	2.39	2.53	1.97
8	1.16	1.45	1.7	1.57	1.71	0.67	1.92	1.85	1.77	1.87	1.8	1.07
9	1.19	1.39	1.59	1.4	1.5	0.81	2.38	1.99	1.83	1.8	1.81	1.31
10	5.6	2.18	2.24	2.17	2.24	1.17	5.6	4.89	4.91	4.7	4.81	2.76
11	1.15	1.23	1.33	1.18	1.33	0.49	3.13	2.8	2.54	2.42	2.51	1.69
12	1.47	1.5	1.77	1.45	1.79	1.41	2.5	2.29	2.47	2.55	2.33	2.19

The recognition error rates for both normal and noisy image are reported in Table1. Fig.3 also shows the segmentation results (error rate) based on the different segmentation methods and noise levels. In order to evaluate the robustness of the various methods on BBBC008v1, we plot the standard deviation. Fig. 2 shows the standard deviation (error bar) of the compared approaches. As can be seen in Fig. 2 the proposed algorithm has better average segmentation accuracy and small standard error compared to the five other existing algorithms.

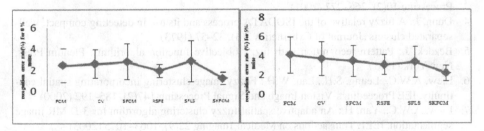

Fig. 2. The standard deviation (error bar) of the various methods on BBBC008v1

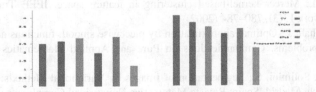

Fig. 3. The average of error rate for 0% and 5% Gaussian noise on BBBC008v1

5   Conclusion

In this study, we proposed an image segmentation algorithm based on combining KFCM and CV for the segmentation of the colon cancer cells images. We compared the results of our algorithm with the results of FCM, CV, SFCM, SFLS and RSFE, and show that our algorithm performed significantly better than these methods in terms of segmentation accuracy.

There are two critical issues related to our method. One is the selection of suitable kernel for the specific image. In practice, the performance of kernel-based methods is influenced by the selection of the kernel function. While solutions for estimating the optimal kernel function for supervised tasks can be found in the literatures, the challenge remains an open problem for unsupervised tasks. Secondly, a potential drawback of the Bayes classifier used in the integration process is that for each $\mathfrak{C}_i$ many characteristics of the class are considered which may not be useful to differentiate the classes.

Acknowledgement. This work was supported by the Australian Research Council (ARC) Discovery Grant DP1097059 and the Griffith International Postgraduate Research Scholarship (GUIPRS).

References

1. Osher, S., Sethian, J.A.: Fronts propagating with curvature-dependent speed: Algorithms based on Hamilton–Jacobi Formulation. Journal of Computational Physics 79, 12–49
2. Malladi, R., Vemuri, B.C.: Shape modeling with front propagation: A level set approach. IEEE Transaction on Pattern Analysis and Machine Intelligence 17(2), 158–175 (1995)
3. Chan, T.F., Vese, L.A.: Active Contours without edges. IEEE Transactions on Image Processing 10(2), 266–277 (2001)
4. Dunn, J.: A fuzzy relative of the ISODATA process and its use in detecting compact, well-separated clusters. Journal of Cybernetics 3(3), 32–57 (1973)
5. Bezdek, J.: Pattern recognition with fuzzy objective function algorithms. Plenum Press, New York (1981)
6. Liew, A.W.C., Leung, S.H., Lau, W.H.: Fuzzy image clustering incorporating spatial continuity. IEE Proceedings-Vision Image and Signal Processing 147(2), 185–192 (2000)
7. Liew, A.W.C., Yan, H.: An adaptive spatial fuzzy clustering algorithm for 3-D MR image segmentation. IEEE Transactions on Medical Imaging 22(9), 1063–1075 (2003)
8. Liew, A.W.C., Leung, S.H., Lau, W.H.: Segmentation of color lip images by spatial fuzzy clustering. IEEE Transactions on Fuzzy Systems 11(4), 542–549 (2003)
9. Girolami, M.: Mercer kernel-based clustering in feature space. IEEE Transactions on Neural Network 13(3), 780–784 (2002)
10. Mumford, Shah, J.: Optimal approximation by piecewise smooth functions and associated variational problems. Communications on Pure and Applied Mathematics 42, 577–685 (1989)
11. Morel, J.M., Solimini, S.: Segmentation of Images by Variational Methods: A Constructive Approach. Madrid, Spain: Revista Matematica Universidad Complutense de Madrid 1, 169–182 (1988)

12. Pal, N.R., Pal, K., Keller, J.M., Bezdek, J.: A possibilistic fuzzy c-means clustering algorithm. IEEE Transactions on Fuzzy Systems 13(4), 517–530 (2005)
13. Höppner, F., Klawonn, F.: A contribution to convergence theory of fuzzy c-means and derivatives. IEEE Transactions on Fuzzy Systems 11(5), 682–694 (2003)
14. Abonyi, J., Szeifert, F.: Supervised fuzzy clustering for the identification of fuzzy classifiers. Pattern Recognition Letters 24, 2195–2207 (2003)
15. http://www.broadinstitute.org/bbbc
16. Carpenter, A.E., Jones, T.R., Lamprecht, M.R., Clarke, C., Kang, I.H., Friman, O., Guertin, D.A., Chang, J.H., Lindquist, R.A., Moffat, J., Golland, P., Sabatini, D.M.: CellProfiler: image analysis software for identifying and quantifying cell phenotypes. Genome Biology 7, R100 (2006)
17. Chuang, K.S., Wu, J., Chen, T.J.: Fuzzy c-means clustering with spatial information for image segmentation. Computerized Medical Imaging and Graphics 30, 9–156 (2000)
18. Li, B.N., Chui, C.K., Chang, S., Ong, S.H.: Integrating spatial fuzzy clustering and level set methods for automated medical image segmentation. Computers in Biology and Medicine 41, 1–10 (2011)
19. Li, C., Kao, C., Gore, J., Ding, Z.: Minimization of region-scalable fitting energy for image segmentation. IEEE Transactions on Image Processing 17, 1940–1949 (2008)

High Dimensional Problem Based on Elite-Grouped Adaptive Particle Swarm Optimization

Haiping Yu[*] and Xueyan Li

Faculty of Information Engineering, City College Wuhan University of Science and Technology, Wuhan, China
seapingyu@gmail.com

Abstract. Particle swarm optimization is a new globe optimization algorithm based on swarm intelligent search. It is a simple and efficient optimization algorithm. Therefore, this algorithm is widely used in solving the most complex problems. However, particle swarm optimization is easy to fall into local minima, defects and poor precision. As a result, an improved particle swarm optimization algorithm is proposed to deal with multi-modal function optimization in high dimension problems. Elite particles and bad particles are differentiated from the swarm in the initial iteration steps, bad particles are replaced with the same number of middle particles generated by mutating bad particles and elite particles. Therefore, the diversity of particle has been increased. In order to avoid the particles falling into the local optimum, the direction of the particles changes in accordance with a certain probability in the latter part of the iteration. The results of the simulation and comparison show that the improved PSO algorithm named EGAPSO is verified to be feasible and effective.

Keywords: particle swarm optimization, high dimensional problem, elite-grouped.

1 Introduction

Particle swarm optimization is a swarm intelligence technique developed by Kennedy and Eberhart in 1995[1], it is inspired by the paradigm of birds flocking. In recent years, particle swarm optimization has been used increasingly as an effective technique for solving complex and difficult optimization problems, such as multi-objective optimization, training neural network and emergent system identification [2-4]. However, it suffers from premature convergence especially in high dimensional problem, and it is hard to escape from the local optimization, these shortcoming has restricted the wide application of PSO. As a result, accelerating convergence rate and avoiding local optima have become the most important and appealing goals in PSO research[5].

---

[*] Corresponding author.

D.-S. Huang et al. (Eds.): ICIC 2013, LNAI 7996, pp. 130–136, 2013.

Therefore, it is extremely significant to solve these problems mentioned above. In recent years, in order to solve the problem of multimodal function, various new versions have appeared, for example, In 2011, A Hybrid PSO-BFGS Strategy for Global Optimization of Multimodal Functions has been proposed in references[6]. new approaches have been presented from using Cauchy and Gaussian to produce random numbers to updating the velocity equation of standard particle swarm optimization[7],a hybrid niching PSO enhanced with recombination-replacement crowding strategy has been proposed in references[8]. Genetic Algorithm with adaptive elitist-population strategies has been proposed in references [9]. and so on. In order to avoid convergence and local optima, in this paper, we put forward a new particle swarm optimization called elite-grouped adaptive particle swarm optimization(EGAPSO). The new algorithm has been validated using six benchmark functions. The simulation results demonstrate that the new method is successful to resolve complex numerical function optimization problems.

This paper is organized as follows: section 2 describes the standard PSO algorithm, section 3 proposes the improved PSO algorithm and describes it in detail, section 4 illustrates the improved PSO using four benchmark functions is given to illustrate the improved PSO, and section 5 concludes our work and the future work to do.

2   Review of Standard Particle Swarm Optimization

The particle swarm optimization is one of the most Swarm intelligences, and it is inspired by the paradigm of birds or fish, and this algorithm has been introduced by James Kennedy and Russell Eberhart in 1995. It has achieved a lot of successes in real-world applications, including electric power system[10],automatic system[11], image processing techniques[12][13] etc. The standard PSO is a randomly optimal algorithm, and it's principle is easy. The algorithm principle description is as follows:

The Detailed steps are as follows: the population initialized by random particles, and calculating their fitness values, finding the personal best value, and global-best value, and the iteration, every particle update their velocity using formula (1), and their position using formula (2). Each time if a particle finds a better position than the previously found best position; its location is stored in memory. In other words ,the algorithm works on the social behavior of particles in the swarm. As a result, it finds the global-best position by simply adjusting their own position toward its own best location and toward the best particle of the entire swarm at each step.

$$v_{ij}(t+1) = wv_{ij}(t) + c_1 r_1 (pbest_{ij}(t) - x_{ij}(t)) + c_2 r_2 (gbest_{gj}(t) - x_{ij}(t)) \qquad (1)$$

$$x_{ij}(t+1) = x_{ij}(t) + v_{ij}(t+1) \qquad (2)$$

Where t indicates the iteration number is the inertia weight,r1and r2 are two random vectors range 0 to 1,i =1,2,3,...N,N is the warm size ,j=1,2,3,...D,D means the dimension of Searching place. v indicates the velocity of each particle, x means the position of each particle. We call this contains the inertia weight particle swarm optimization for IWPSO

The velocity vector of each particle is affected by their personal experiences and their global particle factors, and the position of each particle in the next iteration is calculated by adding its velocity vector to its current position.

In 2002, a constriction named λ coefficient is used to prevent each particle from exploring too far away in the range of min an max, since λ applies a Suppression effect to the oscillation size of a particle over time. This method constricted PSO suggested by Clerc and Kennedy is used with λ set it to 0.7298, according to the formula [14]

$$\lambda = \frac{2}{\left|2 - \varphi - \sqrt{\varphi^2 - 4\varphi}\right|}, \; where \; \varphi = \varphi_1 + \varphi_2 \tag{3}$$

3   The Elite-Grouped Adaptive Particle Swarm Optimization

3.1   The Principle of Improved Algorithm

PSO algorithm is a population-based algorithm which means a lot of particles exploit the search space. Each particle moves with an adaptable velocity within the search space, and retains in its memory the best position it ever encountered.

Therefore, in the beginning of the optimization process, some particle is relatively close to the optimal solution, but other particles is far away from the optimal solution, these particles go a lot of detours. We call the former elite particles, called the latter as bad particles. The standard PSO algorithm is a fast convergence and simple swarm algorithm, however, it is easy to fall into local optimal value. As a result, how to effectively solve the above problems is the focus of this paper.

In order to reserved the prototype of the particle swarm algorithm, improved algorithm called elite-grouped adaptive particle swarm optimization(EGAPSO) is proposed in two aspects. Test functions have been tested by the EGA PSO algorithm, and the results show the EGAPSO can verify the effectiveness of the algorithm. First, in the initial stage of the iterative, each particle will be identified a elite particle or a bad particle according to its fitness value, in which the number of elite particle and bad ones is the same. And the same number of middle particles will be obtained through Elite particles and bad particles mutation. Bad particles are replaced with the same number of middle particles, which expand the diversity of the population. Second, in the latter part of the iterative , in order to avoid particle into a local optima , the direction of some particles will be changed according to a certain proportion.

3.2   The Description of Algorithm

According to the introduction of algorithm, procedure has described in the previous section.

Step 1: Initializing the swarm:

a. initializing the size of particle, initializing the global optimum(gbest) and the personal optimum(pbest), first step set gbest=pbest.

b. set the learning factor c1 and c2.

Step2: Calculating the fitness value of each particle.

identifying elite particles and the same number of bad particles, calculating the middle particles, and bad particles are replaced with the same number of middle particles. transfer step4.

Step3: Group elite particles

Elite number of particles is divided into several groups, and each group will be composed of a new swarm.

Step4:Update velocity

Compare the particle of current fitness value with its own individual fitness value, if the current fitness value is better than individual's, then set the best personal value to the current fitness value, and set the location of individual extreme position for the current.

Compare each particle's fitness value and global extreme, if the particle's fitness value is more than the global fitness value, then modify the global extreme value for the current particle's fitness value . On the contrary, according to the Metropolis criterion, accepting worse solution as a certain probability.

Step5: update the particle position and velocity, the formula as above (1)(2).

Step6: If the Condition is true, the algorithm terminates, otherwise the transfer step2.

4   Simulation Results

This section conducts numerical experiments to verify the performance of the proposed method named EGAPSO problems. Firstly, the benchmark functions which have been commonly used in previous researches are described. Then the performance of the proposed method and the specific algorithms for comparison are tested on these functions, and experimental results are reported and discussed.

A . The test Functions[15]

(1) Rastrigin function

$$f_1(x) = \sum_{i=1}^{n} [x_i^2 - 10\cos(2\pi x_i + 10)], \, x_i \in [-5.12, 5.12] \tag{4}$$

(2) Spherical function

$$f_2(x) = \sum_{i=1}^{n} x_i^2, \, x_i \in [-5.12, 5.12] \tag{5}$$

(3) Griewank function

$$f_3(x) = \frac{1}{4000}\sum_{i=1}^{n} x_i^2 - \prod_{i=1}^{n}\cos(\frac{x_i}{\sqrt{i}})+1, \quad x_i \in [-600,600] \tag{6}$$

(4) Ackley Function

$$f_4(x) = 20 + e - 20e^{-\frac{1}{5}\sqrt{\frac{1}{n}\sum_{i=1}^{n}x_i^2}} - e^{\frac{1}{n}\sum \cos(2\pi x_i)}, \quad x_i \in [-32,32] \tag{7}$$

The above four functions are all multimodal function. In each role domain, when it is minimized, the only global optimum is x* = (0, 0, . . .,0) and f(x*) = 0.

B. The Experimental Settings

Detailed settings is as shown in Table 1.

Table 1. The Experimental Settings

Algorithm	dim	size	c_1	c_2	w_{max}/w_{min}	ite_{rmax}
PSO	20	50	2	2	0.9/0.4	10000
IWPSO	20	50	2.05	2.05	0.95/0.4	10000
GAPSO	20	50	2	2	0.9/0.4	10000
The improved PSO(EGAPSO)	20	50	2	2	0.9/0.4	10000

where dim is the dimension of the function, size is the definition domain, and Fmin means the minimum value of the function.c1 and c2 mean cognitive coefficient; wmax/wmin mean the maximum and minimum values of the inertia weight.

The improved PSO(EGAPSO) includes 25 elite particles and 25 bad particles. Bad particles are replaced with 25 middle particles by mutating elite particles and bad particles randomly. At the same time, 25 elite particles are divided into three groups in this example.

Table 2. The Result of The Experiment Comparing With Other PSOs

Function	f_1	f_2	f_3	f_4
PSO	5.008366	2.4409171e-10	0.01788601	0.02041571
IWPSO	18.776353	8.99103103e-12	0.20775331	3.08875532
GAPSO	4.010037	**9.61331974e-97**	0.008439601	4.02175591e-15
The improved PSO(EGAPSO)	**1.998339**	5.87003377e-78	**0.00544611**	**3.00954412e-15**

C. The Result of the Experiment

The results are shown in Table 2, Table 2 shows the average optimal fitness using the four PSOs over 30 independent runs respectively, and the best results are marked in boldface. On the multimodal functions, the experimental results of the four PSOs are analyzed as follows: it is clear our proposed PSO offers the best.

5 Conclusion

In this paper, a comparison between four optimization techniques, four test functions have been used to compare between these methods. Simulation results show that our proposed algorithm effectively solves the PSO algorithm in the presence of local optimum, stagnation and shock, and effectively improve the efficiency of the algorithm. But the number of grouped optimization of the elite particles number of needs further research in the near future.

References

1. Eberhart, R., Kennedy, J.: A new optimizer using particle swarm theory. In: Proc. 6th Int. Symp Micromach. Hum. SCI. Nagoya, Japan, pp. 39–43 (1995)
2. Wang, Y.F., Zhang, Y.F.: A PSO-based Multi-objective Optimization Approach to the Integration of Process Planning and Scheduling. In: 2010 8th IEEE International Conference on Control and Automation, pp. 614–619 (2010)
3. Hu, X., Eberhart, R.: Multi-objective optimization using dynamic neighborhood particle swarm optimization. In: Congress on evolutionary computation (CEC), vol. 2, pp. 1677–1681. IEEE Service Center, Piscataway (2002)
4. Sun, Y., Zhang, W.: Design of Neural Network Gain Scheduling Flight Control Law Using a Modified PSO Algorithm Based on Immune Clone Principle. In: 2009 Second International Conference on Intelligent Computation Technology and Automation, pp. 259–263 (2009)
5. Ho, S.Y., Lin, H.S., Liauh, W.H., Ho, S.J.: OPSO:orthogonal particle swarm optimization and its application to task assignment problems. IEEE Trans. Syst, Man, Cybern. A, Syst. Humans 38(2), 288–298 (2008)
6. Li, S., Tan, M., Kwok, J.T.-Y.: A Hybrid PSO-BFGS Strategy for Global Optimization of Multimodal Functions. IEEE Transactions on Systems, Man, and Cybernetics—Part B: Cybernetics 41(4) (August 2011)
7. Coelho, L.S., Krohling, R.A.: Predictive controller tuning using modified particle swarm optimization based on Cauchy and Gaussian distributions. In: Proceedings of the VI Brazilian Conference on Neural Networks, Sao Paulo, Brazil (June 2003) (in Portuguese)
8. Li, M., Lin, D., Kou, J.: A hybrid niching PSO enhanced with recombination-replacement crowding strategy for multimodal function optimization. Applied Soft Computing 12, 975–987 (2012)
9. Liang, Y., Leung, K.-S.: Genetic Algorithm with adaptive elitist-population strategies for multimodal function optimization. Applied Soft Computing 11, 2017–2034 (2011)
10. Zhang, W., Liu, Y.: Adaptive particle swarm optimization for reactive power and voltage control in power systems. In: Wang, L., Chen, K., S. Ong, Y. (eds.) ICNC 2005. LNCS, vol. 3612, pp. 449–452. Springer, Heidelberg (2005)

11. Su, C.-T., Wong, J.-T.: Designing MIMO controller by neuro-traveling particle swarm optimizer approach. Expert System with Applications 32, 848–855 (2007)
12. Yi, W., Yao, M., Jiang, Z.: Fuzzy particle swarm optimization clustering and its application to image clustering. In: Zhuang, Y.-T., Yang, S.-Q., Rui, Y., He, Q. (eds.) PCM 2006. LNCS, vol. 4261, pp. 459–467. Springer, Heidelberg (2006)
13. Jiao, W., Liu, G., Liu, D.: Elite Particle Swarm Optimization with Mutation. In: 7th Intl. Conf. on Sys. Simulation and Scientific Computing, pp. 800–803 (2008)
14. Li, X.: Niching Without Niching Parameters: Particle Swarm Optimization Using a Ring Topology. IEEE Transactions on Evolutionary Computation, 150–169 (February 2010)
15. Norouzzadeh, M.S.: Plowing PSO: A Novel Approach to Effectively Initializing Particle Swarm Optimization. In: 2010 3rd IEEE International Conference on Computer Science and Information Technology (ICCSIT), pp. 705–708 (2010)

An Improved Particle Swarm Optimization Algorithm with Quadratic Interpolation

Fengli Zhou and Haiping Yu

Faculty of Information Engineering, City College Wuhan University of Science and Technology, Wuhan 430083, China
thinkview@163.com, yhp0308@yahoo.com.cn

Abstract. In order to overcome the problems of premature convergence frequently in Particle Swarm Optimization(PSO), an improved PSO is proposed(IPSO). After the update of the particle velocity and position, two positions from set of the current personal best position are closed at random. A new position is produced by the quadratic interpolation given through three positions, i.e., global best position and two other positions. The current personal best position and the global best position are updated by comparing with the new position. Simulation experimental results of six classic benchmark functions indicate that the new algorithm greatly improves the searching efficiency and the convergence rate of PSO.

Keywords: Particle Swarm Optimization, Convergence Speed, Quadratic Interpolation, Global Optimization.

1 Introduction

Particle Swarm Optimization (PSO) is one of the evolutionary computational techniques proposed by Kennedy and Eberhart in 1995[1]. PSO has many advantages of fast convergence speed, few parameters setting and simple implementation, so it has attracted much attention from researchers around the world and uses widely in recent years, such as function optimization, neural networks trainings, pattern classification, fuzzy system control and other engineering fields[2-5]. Existing PSO can easily fall into local optimal solution, to overcome the limitation, a centroid particle swarm optimization algorithm is developed by Y. S. Wang and J. L. Li[11]. This method sees the optimal position of individual history as a particle with mass and updates the current particle's best position by solving their centroid position. Y. C. Chi proposes an improved particle swarm algorithm based on niche[12], this algorithm can determine outlier in the best location of individual history according to the number of niche, then it uses cross and selection operator to update all particles which history best value is inferior to the outlier value. These two algorithms update the speed and position of particles through standard PSO, and then update the individual history optimal position and global optimal position by secondary operating on the current individual history optimal position, they greatly improve the basic performance and enhance the

D.-S. Huang et al. (Eds.): ICIC 2013, LNAI 7996, pp. 137–144, 2013.

robustness of PSO. This paper presents an improved particle swarm algorithm, it can generate global optimal position and individual history optimal position based on standard PSO, and then randomly selects two individual history optimal position, after that uses these two history optimal position and global optimal position to generate a new location by quadratic interpolation. We can compare it with the individual history optimal position of *i-th* particle, if it is better that the history optimal position is replaced, otherwise unchanged. In order to verify the effective of the algorithm, we select six typical test functions and compare it with other particle swarm algorithm, the results show that it can improve the convergence speed and enhance the performance of global optimization.

2   Standard Particle Swarm Optimization Algorithm

The standard particle swarm algorithm is an optimization algorithm based on populations, individual in the particle swarm is called particle[6-9]. Assume that a certain group size is N in the D-dimensional search space, the position of its *i-th* particle's t-generation in this search space is $X_i^t = (x_{i1}^t, x_{i2}^t, \cdots, x_{iD}^t)$, $i = 1,2,\cdots N$, velocity is $V_i^t = (v_{i1}^t, v_{i2}^t, \cdots, v_{iD}^t)$, individual history optimal position is $P_i^t = (p_{i1}^t, p_{i2}^t, \cdots, p_{iD}^t)$, global optimal position of t-generation is $P_g^t = (p_{g1}^t, p_{g2}^t, \cdots, p_{gD}^t)$, the *i-th* particle changes its velocity and position according to the following equations[1][2]:

$$V_i^{t+1} = \omega \times V_i^t + c_1 \times r_1 \times (P_i^t - X_i^t) + c_2 \times r_2 \times (P_g^t - X_i^t) \tag{1}$$

$$X_i^{t+1} = X_i^t + V_i^{t+1} \tag{2}$$

Where c_1 and c_2 are constants and are known as acceleration coefficients; r_1, r_2 are random values in the range of (0, 1); ω denotes the inertia weight factor, maximum ω_{max} reduces to ω_{min} linearly as the iteration proceeds, that is the equation[3]:

$$\omega = \omega_{max} - \frac{(\omega_{max} - \omega_{min}) \times iter}{max\_iter} \tag{3}$$

Where *iter* is the current iteration frequency, max_*iter* is the max iteration frequency.

3   Improved Particle Swarm Optimization Algorithm

The PSO algorithm has better global exploration ability in the early, but in the later slow convergence speed and poor solution precision will lead to lower optimization ability. Because of PSO algorithm's poor local searching ability, we can locally search the individual history optimal position and global optimal position to improve the convergence speed of the algorithm. Therefore a quadratic interpolation operator is introduced into the PSO algorithm, as shown in equation[4]. The point generated by equation[4] is the minimum point through quadric surface of P_i, P_j, P_g three points in

D-dimensional space, so equation[4] has local search ability and can improve the convergence speed of algorithm. And equation[4] will always chooses the global optimal position produced by each iteration in order to seek better global optimal position. This paper presents an improved particle swarm optimization(IPSO) based on the above reasons.

Assume that the position of *i-th* particle's t-generation in this search space is $X_i^t = (x_{i1}^t, x_{i2}^t, \cdots, x_{iD}^t)$, $i = 1,2,\cdots N$, velocity is $V_i^t = (v_{i1}^t, v_{i2}^t, \cdots, v_{iD}^t)$, individual history optimal position is $P_i^t = (p_{i1}^t, p_{i2}^t, \cdots, p_{iD}^t)$, global optimal position of t-generation is $P_g^t = (p_{g1}^t, p_{g2}^t, \cdots, p_{gD}^t)$, selecting two positions P_l, P_j from the N individual history best positions randomly, $l, j \neq g$.

$$q_{ik} = \frac{1}{2} \times \frac{(p_{jk}^2 - p_{gk}^2) \times f(p_l) + (p_{gk}^2 - p_{lk}^2) \times f(p_j) + (p_{lk}^2 - p_{jk}^2) \times f(p_g)}{[(p_{jk} - p_{gk}) \times f(p_l) + (p_{gk} - p_{lk}) \times f(p_j) + (p_{lk} - p_{jk}) \times f(p_g)] + e}, \quad k = 1,2,\cdots,D \quad (4)$$

Where f is the objective function, e is a very small positive and can prevent the divisor to be zero, make $Q_i = (q_{i1}, q_{i2}, \cdots, q_{iD})$, $i = 1,2,\cdots N$.

$$P_i = \begin{cases} Q_i, f(Q_i) < f(P_i) \\ P_i, f(Q_i) \geq f(P_i) \end{cases}, \quad i = 1,2,\cdots N \quad (5)$$

The IPSO algorithm step can be described as follows:

Step1 Randomly initialize the particle's location $X_i^0 = (x_{i1}^0, x_{i2}^0, \cdots, x_{iD}^0)$ and velocity $V_i^0 = (v_{i1}^0, v_{i2}^0, \cdots, v_{iD}^0)$, $i = 1,2,\cdots N$, calculate the fitness value of the particle(take the objective function in this paper), determine the individual optimal position in history $P_i = X_i^0$ and the global optimal position P_g. Make t=0.

Step2 Update particle optimal position in history P_i and global optimal position P_g according to the equations[4][5].

Step3 Update particle velocity and position according to the equations[1][2].

Step4 Calculate fitness values of particles, determine current global optimal position P_g of the particle swarm and each individual optimal position P_i in history.

Step5 If algorithm termination condition is satisfied, then stop, otherwise make t=t+1, go to step2.

4 Experiment and Analysis

The experiment selects six typical test functions to evaluate the performance of the ISPO algorithm proposed in this paper, and compares the performance with NCSPSO[10]、CSPO[11]、LinWPSO[12] algorithm. Six test functions are Shpere、Griewank、Quadric、Hartman、Rosenbrock、Goldstein-Price, specific parameters are shown in Table1.

Table 1. Test Functions

Test Functions	Dimension	Search Range	Optimal Value
Shpere(f_1)	30	[-50,50]	0
Griewank(f_2)	30	[-300,300]	0
Quadric(f_3)	30	[-100,100]	0
Rosenbrock(f_4)	10	[-30,30]	0
Hartman(f_5)	3	[0,1]	-3.862 78
Goldstein-Price(f_6)	2	[-2,2]	3

Parameter settings are as follows: learning factor is set to $c_1 = c_2 = 2$, maximum weight is chosen as $\omega_{max} = 0.9$ and minimum weight is chosen as $\omega_{min} = 0.4$ in all the algorithms. The population scale of CPSO and LinWPSO algorithms is set to 80 in the functions Quadric、Rosenbrock(experiments prove that population scale taken to 80 is better than 40), the rest of the population scales are taken to 40. The NCSPSO algorithm niche radius is taken as 0.8, crossover probability is chosen as $CR = 0.8$.

Table 2. Comparison the Results of Test Functions by Four Algorithms

Function	Algorithm	Best Value	Worst Value	Average Optimal Value	Variance	Average Time/s
f_1	LinWPSO	0.005 560	0.296 599	0.113 432	0.005 033	0.776 009
	NCSPSO	9.485935E-006	0.001 506	2.603 382E-004	1.613 424-007	4.354 207
	CPSO	0.002 832	0.071 480	0.025 589	2.886 035E-004	1.862 705
	IPSO	0	8.783 183E-006	1.537 853E-006	7.895 045E-012	0.089 845
f_2	LinWPSO	0.001 268	0.040 110	0.011 853	5.826 110E-005	1.780 896
	NCSPSO	8.522 636E-006	2.615 096E-005	1.076 528E-005	1.207 820E-011	5.131 467
	CPSO	1.602 590E-004	0.003 561	0.001 662	7.339 640E-007	4.176 582
	IPSO	0	9.980 363E-006	2.872 900E-006	1.450 315E-011	0.141 107
f_3	LinWPSO	0.003 225	0.459 828	0.119 370	0.007 651	3.262 293
	NCSPSO	0.003 288	0.382 635	0.085 456	0.009 242	5.567 015
	CPSO	0.030 585	0.206 831	0.109 037	0.001 826	4.000 143
	IPSO	0	0	0	0	0.176 184
f_4	LinWPSO	5.859 102E-006	3.986 579	0.797 323	2.630 522	0.795 277
	NCSPSO	5.694 518E-006	9.956 955E-006	9.191 366E-006	9.111 346E-013	1.932 607
	CPSO	5.610 312E-006	3.986579	0.265 780	1.022 981	0.958 858
	IPSO	5.841 859E-006	9.991 565E-006	9.340 662E-006	7.380 784E-013	0.839 276
f_5	LinWPSO	-3.862 781	-3.514 660	-3.847 758	0.003 973	0.480 784
	NCSPSO	-3.862 780	-3.862 770	-3.862 774	7.802 219E-012	0.493 449
	CPSO	-3.862 781	-3.862 770	-3.862 774	8.184 724E-012	0.213 512
	IPSO	-3.862 782	-3.862 770	-3.862 774	1.004 657E-011	0.108 616
f_6	LinWPSO	3.000 000	3.000 009	3.000 005	1.002 263E-011	0.065 057
	NCSPSO	3.000 000	3.000 009	3.000 004	7.119 702E-012	0.372 724
	CPSO	3.000 000	3.000 009	3.000 004	1.090 695E-011	0.114 134
	IPSO	3.000 000	3.000 009	3.000 005	8.460 679E-012	0.054 177

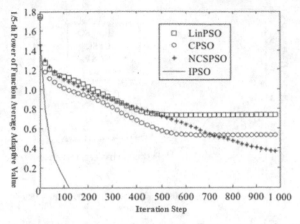

Fig. 1. Convergence Performance Comparison of Function f_1

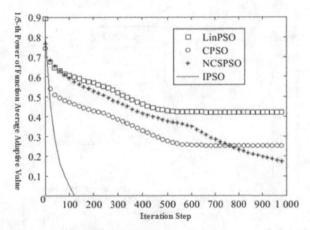

Fig. 2. Convergence Performance Comparison of Function f_2

The CPSO algorithm fitness value is 0.0001. The stopping principle is reaching the maximum iterative step of 2000 or achieving the expected accuracy of 0.000 01. In order to compare the performance of the algorithm further and reduce the influence of contingency, the tests of each function by each algorithm run 30 times independently. Then we can get the results of best and worst optimal value, average optimal value, variance and average time-consuming that are shown in Table2. Fig.1. to Fig.6. shows convergence curves of the four algorithms on six test functions, we set the iteration step to 1000 for high-dimensional function and fetch 500 step for low-dimensional function, evaluation criteria is used as 1/5-th power of average adaptive value in order to compare visually.

From table 2 it can be seen that for the three high-dimensional functions, the best value, worst value, average value, variance and computing time achieved by IPSO algorithm are better than the other three algorithms, especially the Quadirc function,

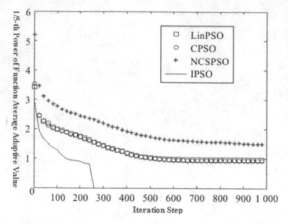

Fig. 3. Convergence Performance Comparison of Function f_3

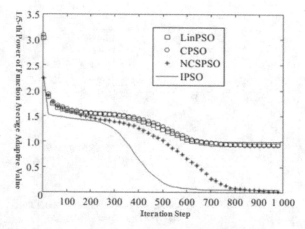

Fig. 4. Convergence Performance Comparison of Function f_4

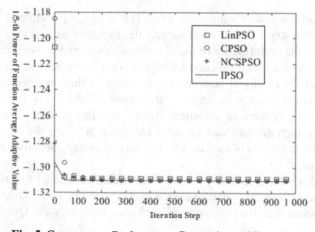

Fig. 5. Convergence Performance Comparison of Function f_5

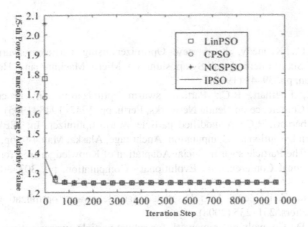

Fig. 6. Convergence Performance Comparison of Function f_6

and the optimization ability and computing time of IPSO algorithm is far better than other algorithms. For the medium-dimensional function Rosenbrock, the best value, worst value, average value and variance achieved by IPSO algorithm are roughly equal to NCSPSO algorithm, but the computing time is better than NCSPSO algorithm and the various indicators are also better than the other two algorithms. For low-dimensional function Hartman, the optimization capability of IPSO algorithm is basically the same as the NCSPSO、CPSO algorithms and better than LinWPSO algorithm. But the computing time of IPSO algorithm is better than other three algorithms. The variance indicator of IPSO algorithm is worst than NCSPSO and CPSO algorithms. For function Goldstein-Price, the optimization capability of IPSO algorithm is basically the same as the other three algorithms, but the computing time is better than other three algorithms. In summary, ISPO algorithm has a strong optimization ability and robustness, especially for the high-dimensional functions. As it is shown from Fig.1. to Fig.6., we can achieve the conclusion that the convergence speed, optimization ability and premature prevention of IPSO algorithm are superior to other algorithms for high-dimensional functions.

5   Conclusion

This paper improves particle swarm optimization algorithm, after the generation of the personal optimal solution in history and the global optimal solution by standard particle swarm algorithm, a quadratic interpolation function is created. Two positions from set of the current personal best position are closed at random. A new position is produced by the quadratic interpolation given through three positions, global best position and two other positions. The current personal best position and the global best position are updated by comparing with the new position. Simulation experimental results of six classic benchmark functions indicate that the new algorithm greatly improves the searching efficiency and the convergence rate of PSO.

References

1. Eberhart, R.C., Kennedy, J.: A new Optimizer using particles swarm theory. In: Proceedings Sixth International Symposium on Micro Machine and Human Science, Nagoya, Japan, pp. 39–43 (1995)
2. Kennedy, J., Eberhart, R.C.: Particle swarm optimization. In: Proceedings IEEE International Conference on Neural Networks, Perth, pp. 1942–1948 (1995)
3. Shi, Y.H., Eberhart, R.C.: A modified particle swarm optimizer. In: IEEE International Conference on Evolutionary Computation, Anchorage, Alaska, May 4-9, pp. 69–73 (1998)
4. Kennedy, J.: The Particle swarm: Social Adaptation of Knowledge. In: Proceedings of the 1997 International Conference on Evolutionary Computation, pp. 303–308. IEEE Press (1997)
5. Cai, X.J., Cui, Z.H., Zeng, J.C.: Dispersed particle swarm optimization. Information Processing Letters, 231–235 (2008)
6. Luo, Q., Yi, D.: Co-evolving framework for robust particle swarm optimization. Applied Mathematics and Computation, 611–622 (2008)
7. Shi, Y., Everhart, R.C.: Empirical study of particle swarm optimization. In: Proceedings of Congress on Computational Intelligence, Washington DC, USA, pp. 1945–1950 (1999)
8. Sheloka, P., Siarry, P., Jayaraman, V.: Particle swarm and ant colony algorithms hybridized for improved continuous optimization. Applied Mathematics and Computation, 129–142 (2007)
9. Gao, S., Yang, J.Y.: Swarm Intelligence Algorithm and Applications, pp. 112–117. China Water Power Press, Beijing (2006)
10. Gao, S., Tang, K.Z., Jiang, X.Z., Yang, J.Y.: Convergence Analysis of Particle Swarm Optimization Algorithm. Science Technology and Engineering 6(12), 1625–1627 (2006)
11. Wang, Y.S., Li, J.L.: Centroid Particle Swarm Optimization Algorithm. Computer Engineering and Application 47(3), 34–37 (2011)
12. Chi, Y.C., Fang, J.: Improved Particle Swarm Optimization Algorithm Based on Niche and Crossover Operator. Journal of System Simulation 22(1), 111–114 (2010)

A New Gene Selection Method for Microarray Data Based on PSO and Informativeness Metric

Jian Guan, Fei Han, and Shanxiu Yang

School of Computer Science and Telecommunication Engineering,
Jiangsu University, Zhenjiang, China
jianguan87@gmail.com, hanfei@ujs.edu.cn,
yangshanxiu_joy007@163.com

Abstract. In this paper, a new method encoding a priori information of informativeness metric of microarray data into particle swarm optimization (PSO) is proposed to select informative genes. The informativeness metric is an analysis of variance statistic that represents the regulation hide in the microarray data. In the new method, the informativeness metric is combined with the global searching algorithms PSO to perform gene selection. The genes selected by the new method reveal the data structure highly hided in the microarray data and therefore improve the classification accuracy rate. Experiment results on two microarray datasets achieved by the proposed method verify its effectiveness and efficiency.

Keywords: Gene selection, particle swarm optimization, informativeness metric, extreme learning machine.

1 Introduction

The increasingly complex gene expression data, characterized by huge number of genes and small group of samples (know as dimensional-disaster), pose major challenges to analysis. Due to the characteristics of gene expression data, majority of genes in the expression data are irrelevant to disease, which causes great difficulty for gene selection. Gene selection methods focus on how small genes sets influence the classification accuracy, and they can be divided into three categories: filter, wrapper and embedded methods [1]. Filter approaches, e.g., information gain [2], t-test [3], Relief-F [4], rely on general characteristics of training data. The advantage of filters is their computational efficiency and the disadvantage is easily selecting a large number of redundant genes. Wrapper approaches consider the classification accuracy as a factor in selecting gene subset [5]. Their advantage is taking the interaction among genes into account and the disadvantage is low computational efficiency [6]. Embedded methods combine the evaluation and selection processes together and accomplish the searching process guided by machine learning algorithms. They have both the advantages of filter and wrapper methods.

D.-S. Huang et al. (Eds.): ICIC 2013, LNAI 7996, pp. 145–154, 2013.

In recent years, particle swarm optimization (PSO) has been used increasingly as an effective technique for global minima searching. Similar to other evolutionary algorithms (EAs) [7], PSO is a population based optimization technique, which searches for optima by updating generations. However, different with other EAs, PSO has no evolution operators such as crossover and mutation, and it is easy to implement with few parameters to adjust [8]. PSO is also used to perform gene selection in this paper. To ensure the explication and high prediction performance of the selected genes, a priori information "informativeness metric" is combined with PSO to perform gene selection. Informativeness metric is based on simple analysis of covariance statistics that compares gene expression profiles between phenotypic groups and focuses on differences between groups rather than differences within groups. Compared with other priori information, the informativeness metric considers both the predictive and feature redundancy of gene set and it can reveal the data structure hidden in the microarray data effectively.

2   Particle Swarm Optimization

The Particle swarm optimization (PSO) algorithm was first introduced by Eberhart and Kennedy in 1995 [9]. In PSO algorithm, the position of i-th particle is represented as $X_i = (x_{i1}, x_{i2}, \cdots, x_{iD})$, which represents a point in D-dimensional solution space. The i-th particle holds the current best position $P_i = (p_{i1}, p_{i2}, \cdots, p_{iD})$ and a moving speed $V_i = (v_{i1}, v_{i2}, \cdots, v_{iD})$. The global best position of the particle swarm is denoted by $P_g = (p_{g1}, p_{g2}, \cdots, p_{gD})$. With the definition above, the basic evolution equation of particle swarm is:

$$v_{ij}(t+1) = v_{ij}(t) + c_1 r_{1j}(t)(p_{ij}(t) - x_{ij}(t)) + c_2 r_{2j}(t)(p_{gj}(t) - x_{ij}(t)) \tag{1}$$

$$x_{ij}(t+1) = x_{ij}(t) + v_{ij}(t+1) \tag{2}$$

c_1 and c_2 are both positive constants, called learning factors. r_{1j} and r_{2j} are two random number between 0 and 1. The subscript "i" represents i-th particle, the subscript "j" represents the j-th dimension of particles, and "t" represents the t-th generation.

The adaptive particle swarm optimization (APSO) algorithm is based on the original PSO algorithm, proposed by Shi & Eberhart [10]. The velocity update formula in APSO is as follows:

$$v_{ij}(t+1) = w v_{ij}(t) + c_1 r_{1j}(t)(p_{ij}(t) - x_{ij}(t)) + c_2 r_{2j}(t)(p_{gj}(t) - x_{ij}(t)) \tag{3}$$

where w is an inertial weight. The APSO is more effective than original PSO, because the searching space reduces step by step, not linearly.

3   The Proposed Gene Selection Method

3.1   The Informativeness Metric

The informativeness metric [11] is based on simple ANOVA statistics comparing the gene expression profiles between phenotypic groups. For simplicity, assume that a two class microarray data consists of s genes and n samples, m of n samples are from cancer patient, the other $n-m$ are normal, and the number of genes in a subset is denoted by g.

Y_{ijk} represents the expression value of the i-th gene of sample j for class k , where $i=1, \ldots , g$ genes; $j=1,\ldots, m$ or $n-m$ samples; k means the cancer or normal group. $\overline{Y}_{.jk}$ stands for the mean expression value for a sample j from class k averaged over the g genes, $\overline{Y}_{.jk} = \dfrac{1}{g}\sum_{i=1}^{g} Y_{ijk}$. $\overline{Y}_{..k}$ stands for average expression over the class k, such as $\overline{Y}_{..k} = \dfrac{1}{m}\sum_{j=1}^{m} \overline{Y}_{.j1}$ or $\dfrac{1}{n-m}\sum_{j=1}^{n-m}\overline{Y}_{.j2}$. The overall mean from the cancer and normal group is calculated as follows:

$$\overline{Y}_{...} = \frac{1}{2}(\overline{Y}_{..1}+\overline{Y}_{..2}) = \frac{1}{2}(\frac{1}{m}\sum_{j=1}^{m}\overline{Y}_{.j1}+\frac{1}{n-m}\sum_{j=1}^{n-m}\overline{Y}_{.j2}) \tag{4}$$

The triple dots indicate summations over genes, samples and then over the two classes. Once a gene subset selected, a statistic mean treatment sum of squares (MMS) which captures the amount of variation attribute to class-specific effects is denoted as follows [11]:

$$MMS = (n-m)(\overline{Y}_{..1}-\overline{Y}_{...})^2 + m(\overline{Y}_{..2}-\overline{Y}_{...})^2 \tag{5}$$

Another statistic mean error sum of squares (RSS) representing the residual variation remaining after class effect is denoted as follows [11]:

$$RSS = \frac{n-m}{n-2}\sum_{j=1}^{n-m}(\overline{Y}_{.j1}-\overline{Y}_{..1})^2 + \frac{m}{n-2}\sum_{j=1}^{m}(\overline{Y}_{.j2}-\overline{Y}_{..2})^2 \tag{6}$$

3.2   The Proposed Method

Based on the informativeness metric, if the MMS statistic is large and the RSS statistic shrinks to a relatively small value for a gene subset, it will result into high classification accuracy on the samples. That means an informative subset would yield a large MMS and relatively a small RSS. To select an effective informative gene subset, PSO combining with the informativeness metric information of microarray data is used to perform gene selection in the paper. The workflow of the

PSO-informativeness based gene selection approach is shown on figure 1, and the detailed steps of the approach are described as follows:

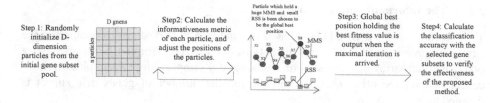

Fig. 1. The workflow of the PSO-informativeness based gene selection approach

Step 1: PSO is employed to select the best gene subset from the initial gene pool. The i-th particle $X_i = (x_{i1}, x_{i2}, \cdots, x_{iD})$ represents a candidate gene subset. The element $x_{ij}(1 \le j \le D)$ is the label of the gene in the initial pool. Each particle X_i contains a D matrix represent the D-dimensional gene subset.

Step 2: Fitness value is the critical factor for movement and evolution of particles, the informativeness metric MMS and RSS values for each particle are regarded as the corresponding fitness value. The current best position searched until now $P_i = (p_{i1}, p_{i2}, \cdots, p_{iD})$ of the i-th particle holds the biggest MMS and the global best position $P_g = (p_{g1}, p_{g2}, \cdots, p_{gD})$ holds the biggest MMS and meanwhile the comparatively smallest RSS of all particles. In each generation, the particle which holds the biggest MMS searched until now is selected as the current best position. If the MMS value of the current best position p_i is bigger and meanwhile the RSS is smaller than the values of global best position p_g, the current best position p_i will be selected to be the new global best position. Then all particles adjust their positions according to Eqs. (3) and (2).

Step 3: If the global best position which holds the best fitness value stays still for several generations or the maximal iterative generation is arrived, the particle with the best fitness value is output.

Step 4: With the selected gene subsets, extreme learning machine (ELM) [12] is used to perform sample classification. Conversely, the high prediction accuracy verify the effectiveness of the proposed gene selection method.

4   Experiment Results

The algorithm is tested on two widely used dataset, ALL-AML Leukemia and Colon datasets which are available on the website http: //www.broadinstitute.org/ cgi-bin/ cancer/ datasets .cgi.

ALL-AML Leukemia data contains 72 samples with 7129 gene expression levels, and the samples are belong to two classes, Acute Myeloid Leukemia (AML)

and Acute Lymphoblastic Leukemia (ALL). The data contains 25 AML and 47 ALL samples.

In Colon dataset, there are 62 microarray experiments with 2000 gene expression levels. Among them, 40 samples are from tumors and 22 samples are normal.

In the following experiments, half of the samples are used as the training subset and the rest are test subset. Among the train subset, the first third of samples are used to perform selection and the rest used to verify the performance of the selected genes.

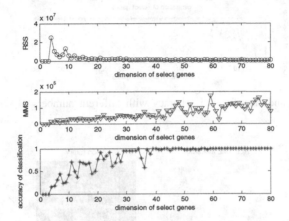

Fig. 2. RSS, MMS and classification accuracy with different numbers of selected genes on ALL-AML leukemia data

To focus on the issue that how many genes in microarray data are associated with different sub type, the first step is to find the relationship among evaluation functions MMS, RSS and classification rate. Figure 2 and 3 show the mean results of the algorithm on ALL-AML leukemia and Colon data with ten independent executions respectively. The X-axis is the dimension of select genes and the two above subfigures show the curves of RSS and MMS, the bottom subfigure shows the mean accuracy of classification. Our purpose is to find the compact gene subset when classification rate is high and simultaneously a large MSS and a small RSS value are yielded. It can be seen form Figure 2 that when dimension of selected genes is greater than 30, the new algorithm performs a 100% classification rate, meanwhile the maximum value of MMS reaches 1.8e10 and the RSS is relatively small. When the dimension of select genes above 59, the MMS curve reaches the biggest value, and the RRS stay at a small value level. 59 could be the appropriate dimension for correct data classification and meanwhile contains the whole data structure included in ALL-AML leukemia microarray data. From Figure 3, when dimension of selected genes is greater than 25, the new algorithm performs a 100% classification rate. When the dimension of select genes above 62 the MMS curve reaches the biggest value, the RRS cure stay at a small value. 62 could be the appropriate dimension which contains the whole data structure for colon tumor data.

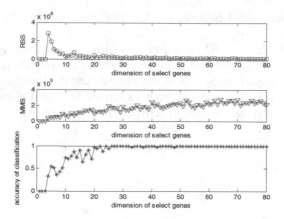

Fig. 3. RSS, MMS and classification accuracy with different numbers of selected genes on colon data

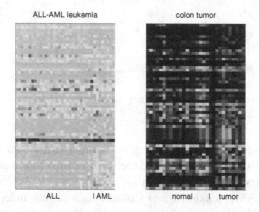

Fig. 4. Heatmap of expression levels based on the selected subset of genes on ALLAML Leukemia and Colon data

Figure 4 shows the heatmap of ALL-AML Leukemia and Colon gene expression levels based on the gene subsets selected in the experiment above. Significant differences can be easily identified among different tumor sub types. Hence, it is can be concluded that the selected genes contain the complete classify information. This experiment also shows that the informativeness based gene selection approach is effective. As for classification accuracy, gene subsets consisting of 59 and 62 genes is too large, so the selected gene subset may involve redundant information. The next step will choose smaller subset from these selected informative genes.

To reduce redundant genes selected with the informativeness metric information, PSO-informativeness based gene selection approach is used to perform further selection, which selects three to five genes and picks out the subsets obtaining the highest classification rate. The relationship between the informativeness metric MMS,

RSS and the classification rate on gene subsets is shown in Figure 5. For effectively display the result, the data has been normalized and only the results with classification accuracy higher than 70% are shown. From Figure 5, each point represents a gene subset and the corresponding MMS, RSS and the accuracy rate of classification. The depth of the red color represents the height of classification rate. According to the principle of the proposed algorithm, an informative subset would yield a large MMS statistic and relatively a small RSS, the points in deep red must concentrated in the area with a large MMS and a small RSS. It can be found from the two figures that the points in deep red are gathered in the space with a large MMS and a small RSS, which is match with our algorithm.

Table 1. Some gene subsets selected by the proposed method for ALL-AML leukemia and Colon data

	Gene subset	Classificatio
	ALL-AML leukemia	n accuracy %
1	{M27878_atU50327_s_atU05237_at}	100
2	{M84349_atU08006_s_atM11718_at M33653_at}	100
3	{X82224_atU08006_s_atM62994_at M33653_at}	100
4	{U08006_s_atM62994_atM11718_at M33653_at}	100
	Colon	
1	{J00231 T57619 T72175 H64489 }	96.2
2	{T92451 T57619 T58861 H64489 }	96.2
3	{M87789 T57619 T72175 H64489 }	96.2
4	{M63391T58861T57619H64489 M94132 }	93.6

Among all the gene subsets as shown in Figure 5, some gene subsets selected by the proposed method which obtain better 5-fold cross-validation classification accuracy by ELM for ALL-AML leukemia and Colon data are listed in Table 1. For ALL-AML leukemia data, the gene subset consisting of {M27878_at U50327_s_at U05237_at} reaches 100% classification accuracy. And for Colon data, the gene subset {J00231 T57619 T72175 H64489} reaches the highest 96.2% classification accuracy.

To show the relationship between gene expression and sample sub type, we select two genes form the gene subsets which perform the highest classification accuracy and draw distribution scatter plot based on these two genes. The gene subset {M27878_at X82224_at} for ALL-AML leukemia data and gene subset {M63391 T58861} for Colon tumor data are perform the best linear separability. Figure 6 shows the sample distribution scatter plot based on these two gene subsets. It can be found form the figures that there is a distinguishable boundary between the two types of samples. According to relevant literature, gene M27878_at is described as Zinc finger protein 84 (HPF2) and gene X82224_at is Glutamine transaminase K. For colon tumor data, gene M63391 is human desmin gene, complete cds and gene T58861 is ribosomal protein L30E.

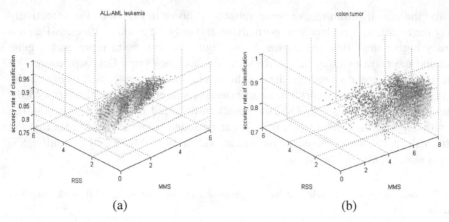

Fig. 5. The three-dimensional scatter plot of MMS, RSS and classification accuracy on two data (a)ALLAML Leukemia (b) Colon

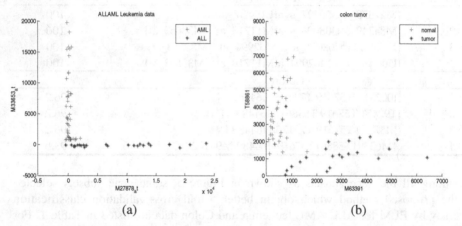

Fig. 6. (a) Distribution scatter plot based on gene subsets {M27878_at X82224_at} on ALL-AML leukemia data (b) Distribution scatter plot based on gene subsets {M63391 T58861} on Colon data

Table 2. The comparison between the proposed method and six classical methods

Methods	Leukemia	Colon
The proposed method	100(3)	96.2(4)
Two-Phase EA/K-NN	-	94.12(37)
SVM	100(3)	98(4)
GA-SVM	100(25)	99.41(10)
Evolutionary algorithm	100(4)	97(7)
Redundancy based	87.55(4)	93.55(4)
PSO-SVM	-	94(4)

To illustrate the performance of the PSO-informativeness based method, six widely-used methods, including Two-Phase EA/K-NN [13], SVM [14], GA-SVM [15], evolutionary algorithm [16], Redundancy based [17] and PSO-SVM [8], are compared with the proposed method. Results for all the datasets are shown in Table 2. The new approach obtains the highest classification accuracy with the smallest subset among all methods on ALLAML Leukemia data. As for Colon data, the proposed method obtain a relatively high classification accuracy with the smallest gene subset.

5   Conclusions

In this paper, a new method of gene selection for high dimensional DNA microarray data was proposed. This new method used particle swarm optimization (PSO) to perform gene selection. A priori information called informativeness metric containing the specific data structure of a dataset is regard as the fitness function of PSO to select the informative gene subsets. The experiment results on two open datasets verified the effectiveness of the proposed method. Future work will include applying the proposed method to multiclass microarray data.

Acknowledgments. This work was supported by the National Natural Science Foundation of China (Nos.61271385, 60702056) and the Initial Foundation of Science Research of Jiangsu University (No.07JDG033).

References

1. Saeys, Y., Inza, I., Larranaga, P.: A review of feature selection techniques in bioinformatics. Bioinformatics 23(19), 2507–2517 (2007)
2. Hobson, A., Cheng, B.: A comparison of the Shannon and Kullback information measures. Journal of Statistical Physics 7(4), 301–310 (1973)
3. Guyon, I., Elisseeff, A.: An introduction to variable and feature selection. The Journal of Machine Learning Research 3, 1157–1182 (2003)
4. Kononenko, Šimec, E., Robnik-Šikonja, M.: Overcoming the myopia of inductive learning algorithms with RELIEFF. Applied Intelligence 7(1), 39–55 (1997)
5. Blanco, R., Larranaga, P., Inza, I., Sierra, B.: Gene selection for cancer classification using wrapper approaches. International Journal of Pattern Recognition and Artificial Intelligence 18(8), 1373–1390 (2004)
6. Guyon, I., Gunn, S., Nikravesh, M., Zadeh, L.: Feature Extraction: Foundations and Applications. STUDFUZZ, vol. 207. Physica-Verlag, Springer (2006)
7. Eiben, E., Smith, J.E.: Introduction to Evolutionary Computing. Natural Computing Series. MIT Press, Springer, Berlin (2003)
8. She, Q., Shi, W.M., Kong, W., Ye, B.X.: A combination of modified particle swarm optimization algorithm and support vector machine for gene selection and tumor classification. Talanta 71, 1679–1683 (2007)
9. Kennedy, J., Eberhart, R.: Particle Swarm Optimization. In: IEEE International Conference on Neural Networks, vol. 4, pp. 1942–1948 (1995)

10. Shi, Y., Eberhart, R.C.: A modified particle swarm optimizer. In: Proceeding of IEEE World Conference on Computation Intelligence, pp. 69–73 (1998)
11. Mar, J.C., Wells, C.A., Quackenbush, J.: Defining an Informativeness Metric for Clustering Gene Expression Data. Bioinfromatics 27(8), 1094–1100 (2011)
12. Huang, G.-B., Zhu, Q.-Y., Siew, C.-K.: Extreme Learning Machine: Theory and Applications. Neurocomputing 70, 489–501 (2006)
13. Juliusdottir, T., Corne, D., Keedwell, E., Narayanan, A.: Two-Phase EA/k-NN for Feature Selection and Classification in Cancer Microarray Datasets, CIBCB, 1594891, pp. 1–8 (2005)
14. Guyon, J., Weston, S., Barnhill, Vapnik, V.: Gene selection for cancer classification using support vector machines. Machine Learning 46(1-3), 389–422 (2002)
15. Huerta, E.B., Duval, B., Hao, J.-K.: A Hybrid GA/SVM Approach for Gene Selection and Classification of Microarray Data. In: Rothlauf, F., Branke, J., Cagnoni, S., Costa, E., Cotta, C., Drechsler, R., Lutton, E., Machado, P., Moore, J.H., Romero, J., Smith, G.D., Squillero, G., Takagi, H. (eds.) EvoWorkshops 2006. LNCS, vol. 3907, pp. 34–44. Springer, Heidelberg (2006)
16. Deb, K., Reddy, A.R.: Classification of two-class cancer data reliably using evolutionary algorithms. Biosystems 72(1-2), 111–129 (2003)
17. Yu, L., Liu, H.: Redundancy based feature selection for microarray data. In: ACM SIGKDD International Conference on Knowledge Discovery and Data Mining, Seattle, Washington, August 22–25, pp. 737–742 (2004)

A Hybrid Attractive and Repulsive Particle Swarm Optimization Based on Gradient Search

Qing Liu and Fei Han

School of Computer Science and Telecommunication Engineering, Jiangsu University,
Zhenjiang, Jiangsu, 212013, China
qingjing_03@126.com, hanfei@ujs.edu.cn

Abstract. As an evolutionary computing technique, particle swarm optimization (PSO) has good global search ability, but its search performance is restricted because of stochastic search and premature convergence. In this paper, attractive and repulsive PSO (ARPSO) accompanied by gradient search is proposed to perform hybrid search. On one hand, ARPSO keeps the reasonable search space by controlling the swarm not to lose its diversity. On the other hand, gradient search makes the swarm converge to local minima quickly. In a proper solution space, gradient search certainly finds the optimal solution. In theory, The hybrid PSO converges to the global minima with higher probability than some stochastic PSO such as ARPSO. Finally, the experiment results show that the proposed hybrid algorithm has better convergence performance with better diversity than some classical PSOs.

Keywords: Particle swarm optimization, stochastic search, diversity, gradient search.

1 Introduction

As a global search algorithm, PSO is a population-based stochastic optimization technique developed by Kennedy and Eberhart [1,2]. Since PSO is easy to implement without complex evolutionary operations compared with genetic algorithm (GA) [3,4], it has been widely used in many fields such as power system optimization, process control, dynamic optimization, adaptive control and electromagnetic optimization [5,6]. Although PSO has shown good performance in solving many optimization problems, it suffers from the problem of premature convergence like most of the stochastic search techniques, particularly in multimodal optimization problems [6].

In order to improve the performance of PSO, many improved PSOs were proposed. Passive congregation PSO (PSOPC) introduced passive congregation for preserving swarm integrity to transfer information among individuals of the swarm [7]. PSO with a constriction (CPSO) defined a "no-hope" convergence criterion and a "rehope" method as well as one social/confidence parameter to re-initialize the swarm [8,9]. Attractive and repulsive PSO (ARPSO) alternates between phases of attraction and repulsion according to the diversity value of the swarm, which prevents premature

D.-S. Huang et al. (Eds.): ICIC 2013, LNAI 7996, pp. 155–162, 2013.

convergence to a high degree at a rapid convergence like adaptive PSO [10]. In fuzzy PSO, a fuzzy system was implemented to dynamically adapt the inertia weight of PSO, which is especially useful for optimization problems with a dynamic environment [11]. Gradient-based PSO (GPSO) combined the standard PSO with second derivative information to perform global exploration and accurate local exploration, respectively to avoid the PSO algorithm to converge to local minima [6].

Although random search such as adaptive PSO converges faster than GA, it is easy to converge to local minima because of losing the diversity of the swarm. On the other hand, some improved PSOs such as PSOPC, CPSO, fussy PSO and ARPSO, may improve their search ability, but they require more time to find the best solution. GPSO can find the global minima in most cases, but it is highly unstable.

In this paper, an improved hybrid particle swarm optimization called HARPSOGS is proposed to overcome the problems above. The proposed method combines random search with deterministic search in the optimization process. To avoid the swarm being trapped into local minima, firstly, ARPSO is used to perform stochastic and rough search. In the solution space obtained by ARPSO, gradient search is then used to perform further deterministic search. When gradient search converges to local minima, the proposed algorithm turns to ARPSO, which may establish a new solution space by improving the diversity of the swarm. Hence, the proposed hybrid PSO converges to the global minima with a high likelihood.

2   Particle Swarm Optimization

2.1   Basic Particle Swarm Optimization

PSO is an evolutionary computation technique in searching for the best solution by simulating the movement of birds in a flocking [1,2]. The population of the birds is called swarm, and the members of the population are particles. Each particle represents a possible solution to the optimizing problem. During each iteration, each particle flies independently in its own direction which guided by its own previous best position as well as the global best position of all the particles. Assume that the dimension of the search space is D, and the swarm is $S=(X_1, X_2, X_3, ..., X_{Np})$; each particle represents a position in the D dimension; the position of the i-th particle in the search space can be denoted as $X_i=(x_{i1}, x_{i2}, ..., x_{iD})$, $i=1, 2, ..., N_p$, where N_p is the number of all particles. The best position of the i-th particle being searched until now is called *pbest* which is expressed as $P_i=(p_{i1}, p_{i2}, ..., p_{iD})$. The best position of the all particles are called *gbest* which is denoted as $P_g=(p_{g1}, p_{g2}, ..., p_{gD})$. The velocity of the i-th particle is expressed as $V_i=(v_{i1}, v_{i2}, ..., v_{iD})$. According to the literature [1,2], the basic PSO was described as:

$$V_i(t+1) = V_i(t) + c1 * rand() * (P_i(t) - X_i(t)) + c2 * rand() * (P_g(t) - X_i(t)) \qquad (1)$$

$$X_i(t+1) = X_i(t) + V_i(t+1) \qquad (2)$$

where $c1$, $c2$ are the acceleration constants with positive values; *rand()* is a random number ranged from 0 to 1.

In order to obtain better performance, adaptive particle swarm optimization (APSO) algorithm was proposed [12], and the corresponding velocity update of particles was denoted as follows:

$$V_i(t+1) = W(t)*V_i(t) + c1*rand()*(P_i(t) - X_i(t)) + c2*rand()*(P_g(t) - X_i(t)) \tag{3}$$

where $W(t)$ can be computed by the following equation:

$$W(t) = W_{max} - t * (W_{max} - W_{min})/N_{pso} \tag{4}$$

In Eq.(4), W_{max}, W_{min} and N_{pso} are the initial inertial weight, the final inertial weight and the maximum iterations, respectively.

2.2 Attractive and Repulsive Particle Swarm Optimization (ARPSO)

In the literature [10], attractive and repulsive particle swarm optimization (ARPSO), a diversity-guided method, was proposed which was described as:

$$V_i(t+1) = w*V_i(t) + dir[c1*rand()*(P_i(t) - X_i(t)) + c2*rand()*(P_g(t) - X_i(t))] \tag{5}$$

where $dir = \begin{cases} -1 & diversity < d_{low} \\ 1 & diversity > d_{high} \end{cases}$.

In ARPSO, a function was proposed to calculate the diversity of the swarm as follows:

$$diversity() = \frac{1}{N_p * |L|} * \sum_{i=1}^{N_p} \sqrt{\sum_{j=1}^{D} (p_{ij} - \overline{p}_j)^2} \tag{6}$$

where $|L|$ is the length of the maximum radius of the search space; p_{ij} is the j-th component of the i-th particle and $\overline{p}_j$ is the j-th component of the average over all particles.

In the attraction phase ($dir=1$), the swarm is attracting, and consequently the diversity decreases. When the diversity drops below the lower bound, d_{low}, the swarm switches to the repulsion phase ($dir=-1$). When the diversity reaches the upper bound, d_{high}, the swarm switches back to the attraction phase. ARPSO alternates between phases of exploiting and exploring – attraction and repulsion – low diversity and high diversity and thus improve its search ability [10].

3   The Proposed Method

Compared with basic PSO and APSO, ARPSO keeps reasonable diversity effectively by adaptively attracting and repelling the particles each other, so ARPSO has better convergence performance. However, since ARPSO is still a stochastic evolutionary algorithm as basic PSO and APSO, it may converge to local minima or require more time to find the global minima. As a deterministic search method, gradient descent method is fast in convergence, but it is apt to converge to local minima. Obviously, ARPSO and gradient descent method are mutually complementary on improving the convergence accuracy and rate of the PSO.

Therefore, in this paper, ARPSO combining with gradient descent is proposed to improve the search ability in this paper. The detailed steps of the proposed algorithm called as HARPSOGS are as follows:

Step 1: Initialize all particles randomly, the maximum iteration number and optimization target.

Step 2: Calculate *pbest* of each particle, *gbest* of all particles and the normalized negative gradient of the fitness function to all particles.

Step 3: Each particle update its position according the following equations:

$$X' = X_i(t) + V_{arpso} \tag{7}$$

$$X'' = X_i(t) - \beta * \frac{\frac{\partial f(X_i(t))}{\partial X_i(t)}}{\left\| -\frac{\partial f(X_i(t))}{\partial X_i(t)} \right\|} \tag{8}$$

$$X_{i+1}(t) = \begin{cases} X' & if \ f(X') < f(X'') \\ X'' & else \end{cases} \tag{9}$$

In above equations, V_{arpso} is the velocity update of the i-th particle obtained by Eq.(5); $f()$ is the fitness function; The gain β represents the relative significance of the negative gradient.

Step 4: Once the new population is generated, return to Step 2 until the goal is met or the predetermined maximum learning epochs are completed.

4   Experiment Results

In this section, the performance of the proposed hybrid PSO algorithm is compared with the APSO, CPSO and ARPSO for some functions in the De Jong test suite of benchmark optimization problems. The problems provide common challenges that an evolutionary computation algorithm is expected to face, such as multiple local minima and flat regions surrounding global minimum. Table 1 shows some classic test functions used in the experiments. The Sphere functions is convex and unimodal (single local minimum). The Rosenbrock test function has a single global minimum located in a long narrow parabolic shaped flat valley and tests the ability of an optimization algorithm to navigate flat regions with small gradients. The Rastrigin and Griewangk functions are highly multimodal and test the ability of an optimization algorithm to escape from local minima. The Schaffer and LevyNo.5 both are two-dimensional functions with many local minima, which the former has infinite minima and the latter has 760 local minima.

Table 1. Test functions used for comparison of HARPSOGS with other PSOs

Test function	Equation	Search space	Global minima
Sphere (F1)	$\sum\limits_{i=1}^{n} x_i^2$	$(-100,100)^n$	0
Rosenbrock (F2)	$\sum\limits_{i=1}^{n-1}(100\left(x_{i+1}-x_i^2\right)^2+\left(1-x_i\right)^2)$	$(-100,100)^n$	0
Rastrigin (F3)	$10n+\sum\limits_{i=1}^{n}(x_i^2-10\cos(2\pi x_i))$	$(-100,100)^n$	0
Griewangk (F4)	$1+\dfrac{1}{4000}\sum\limits_{i=1}^{n}x_i^2-\prod\limits_{i-1}^{n}\cos(\dfrac{x_i}{\sqrt{i}})$	$(-100,100)^n$	0
Schaffer (F5)	$-0.5+\dfrac{\sin\sqrt{x_1^2+x_2^2}-0.5}{(1+0.001(x_1^2+x_2^2))^2}$	$(-100,100)^n$	-1
LevyNo.5 (F6)	$\sum\limits_{i=1}^{5}[i\cos((i-1)x_1+i)]\sum\limits_{j=1}^{5}[j\cos((j-1)x_2+j)]$ $(x_1+1.42513)^2+(x_2+0.80032)^2$	$(-100,100)^n$	-176.1376

The population size for all PSOs is 20 in all experiments. And the acceleration constants $c1$ and $c2$ for PSO and ARPSO both are set as 2.0. The constants $c1$ and $c2$ both are 2.05 in CPSO. In the proposed hybrid algorithms, $c1$ and $c2$ are 2.1.In ARPSO, the diversity threshold d_{low} and d_{high} are selected as 5e-6 and 0.25, respectively. The decaying inertia weight w starting at 0.9 and ending at 0.4 is set for all PSOs according to the literature [12]. All the results shown in this paper are the mean values of 20 trails.

Fig.1 shows the mean convergence curve of different PSOs on six functions. Without loss of the generality, the functions such as F1, F2, F3 and F4 are all in ten dimensions. The "Error" in the Y-Label in Fig. 1 is the absolute value of the difference between the real global minima and the global minima searched by PSOs. From Fig.1, the proposed algorithm has much better convergence accuracy than other PSOs on all six functions. Moreover, the convergence curve of CPSO, APSO and ARPSO are almost horizontal after some iterations, which indicates that these PSOs converge to local minima and could not jump out of the local minima. As for the proposed algorithm, its convergence curve has a downward trend with the increase of the iteration number, which shows that HARPSOGS has the ability of avoiding the premature convergence. Hence, the proposed algorithm has best searching ability in all PSOs.

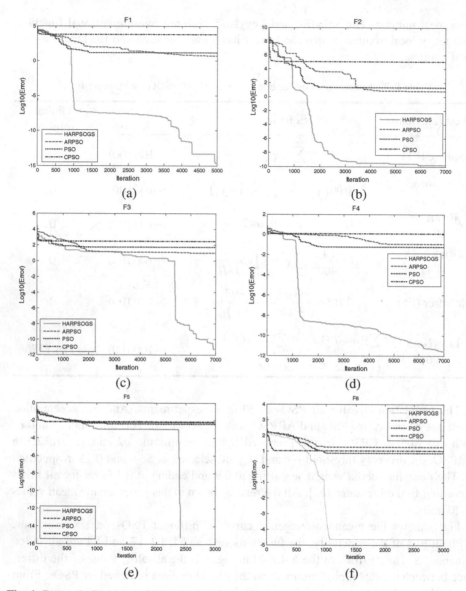

Fig. 1. Best solution versus iteration number for the four test functions by using four PSOs (a) Sphere (b) Rosenbrock (c) Rastrigin (d) Griewangk (e) Schaffer (f) LevyNo.5

Table 2 shows mean best solution for the six test functions by using four PSOs. It also can be found that the proposed algorithm has better convergence accuracy than other PSOs in all cases.

Fig. 2 shows the effect of the selection of the parameter, β, in the proposed algorithm on six test functions. From Fig. 2, the $\log_{10}$(error) has a upward trend as the parameter β increases for the functions F1, F2, F3 and F6. The $\log_{10}$(error) have ideal values while the $\text{Log}_{10}(\beta)$ is selected around -2 and -0.15 for the functions F4 and F5, respectively.

Table 2. Mean best solution for the six test functions by using four PSOs

Functions	Dimensions	APSO	CPSO	ARPSO	HARPSOGS
			Mean best solution		
F1	10	0.0182	103.6567	1.2219	5.3501e-015
	20	0.2634	490.0573	5.5009	7.4817e-13
	30	7.0057	6.5446e+3	15.4783	6.5369e-012
F2	10	4.0096	1.5347e+4	8.2513	5.4010e-010
	20	356.1296	4.6317e+5	59.8249	1.1348e-009
	30	5.6335e+3	1.2056e+7	1.8580e+3	5.6649e-007
F3	10	40.7362	173.8921	51.9232	2.2117e-012
	20	85.7184	422.2540	52.8668	6.8801e-011
	30	169.2328	1.3539e+3	76.0364	9.9437e-010
F4	10	0.0911	0.5917	0.108	2.5535e-14
	20	0.0816	1.4798	0.4276	1.0856e-13
	30	0.1706	1.5848	3.7071	2.2123e-9
F5	2	-0.9961	-0.9933	-0.9932	-1
F6	2	-168.8847	-166.0207	-158.7700	-176.1376

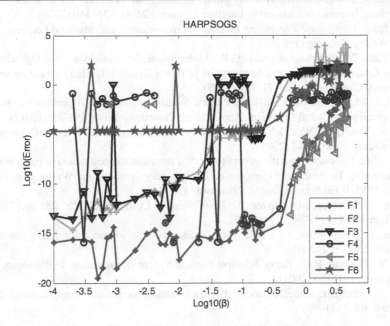

Fig. 2. The parameter β versus the best solutions for the six test functions

5 Conclusions

To enhance the search ability of ARPSO, the deterministic search method based on gradient descent was incorporated into ARPSO. The proposed algorithm could not only keep the effective search space but also perform local search with gradient descent method. The experiment results verified that the proposed method could effectively avoid premature convergence and had better convergence performance than other PSOs.

Acknowledgements. This work was supported by the National Natural Science Foundation of China (Nos.61271385, 60702056), Natural Science Foundation of Jiangsu Province (No.BK2009197) and the Initial Foundation of Science Research of Jiangsu University (No.07JDG033).

References

1. Kennedy, J., Eberhart, R.C.: Particle Swarm Optimization. In: IEEE International Conference on Neural Networks, vol. 4, pp. 1942–1948 (1995)
2. Eberhart, R.C., Kennedy, J.: A New Optimizer Using Particle Swarm Theory. In: Proceedings of the Sixth International Symposium on Micro Machines and Human Science, pp. 39–43 (1995)
3. Grosan, C., Abraham, A.: A novel global optimization technique for high dimensional functions. International Journal of Intelligent Systems 24(4), 421–440 (2009)
4. Goldberg, D.: Genetic Algorithms in Search, Optimization, and Machine Learning. Addison-Wesley, Reading (1989)
5. del Valle, Y., Venayagamoorthy, G.K., Mohagheghi, S.: Particle Swarm Optimization : Basic Concepts, Variants and Applications in Power Systems. IEEE Transactions on Evolutionary Computation 12, 171–195 (2008)
6. Noel, M.M.: A New Gradient Based Particle Swarm Optimization Algorithm for Accurate Computation of Global Minimum. Applied Soft Computing 12(1), 353–359 (2012)
7. He, S., Wu, Q.H., Wen, J.Y.: A Particle Swarm Optimizer with Passive Congregation. Biosystems 78, 135–147 (2004)
8. Clerc, M.: The swarm and the queen: towards a deterministic and adaptive particle swarm optimization. In: Proc. 1999 Congress on Evolutionary Computation, Washington, DC, pp. 1951–1957. IEEE Service Center, Piscataway (1999)
9. Corne, D., Dorigo, M., Glover, F.: New Ideas in Optimization, ch. 25, pp. 379–387. McGraw Hill (1999)
10. Riget, J., Vesterstrom, J.S.: A diversity-guided particle swarm optimizer - the arPSO, Technical report 2 (2002)
11. Shi, Y., Eberhart, R.C.: Fuzzy Adaptive Particle Swarm Optimization. Evolutionary Computation 1, 101–106 (2001)
12. Shi, Y., Eberhart, R.C.: A modified particle swarm optimizer. Computational Intelligence 6, 69–73 (1998)

Video Target Tracking Based on a New Adaptive Particle Swarm Optimization Particle Filter

Feng Liu[1], Shi-bin Xuan[1,2], and Xiang-pin Liu[1]

[1] College of Information Science and Engineering, Guangxi University for Nationalities,
Nanning 530006, China
[2] Guangxi Key Laboratory of Hybrid Computation and IC Design Analysis,
Nanning 530006, China
sbinxuan@gxun.cn

Abstract. To improve accuracy and robustness of video target tracking, a tracking algorithm based on a new adaptive particle swarm optimization particle filter (NAPSOPF) is proposed. A novel inertia weight generating strategy is proposed to balance adaptively the global and local searching ability of the algorithm. This strategy can adjust the particle search range to adapt to different motion levels. The possible position of moving target in the first frame image is predicted by particle filter. Then the proposed NAPSO is utilized to search the smallest Bhattacharyya distance which is most similar to the target template. As a result, the algorithm can reduce the search for matching and improve real-time performance. Experimental results show that the proposed algorithm has a good tracking accuracy and real-time in case of occlusions and fast moving target in video target tracking.

Keywords: Adaptive, Particle Swarm Optimization, Particle Filter, Video Target Tracking, Occlusions, Fast Moving Target, Bhattacharyya Distance.

1 Introduction

Video target tracking is a key problem in computer vision, which has a wide range of applications in human-computer interaction, visual surveillance, intelligent transportation and military guidance etc [1]. In spite of the substantial research effort expended to tackle this challenge, developing a robust and efficient tracking algorithm still remains unsolved due to the inherent difficulty of the tracking problem.

Recently, particle filters (PF) have been extensively used in video target tracking field [2-4], which have been proved to be a robust method of tracking due to the ability of solving non-Gaussian and non-linear problems [5]. Nevertheless, PF may confront with the problem of weight degradation which if solved by re-sampling method may result unavoidable particle impoverishment [6]. As a result, the accuracy and robustness of target tracking are influenced.

To address that problem, a tracking algorithm based on a new adaptive particle swarm optimization particle filter (NAPSOPF) is proposed. Bhattacharyya distance is

D.-S. Huang et al. (Eds.): ICIC 2013, LNAI 7996, pp. 163–171, 2013.

utilized as the measurement of similarity between models of target template and candidate region in the algorithm based on color histogram [7]. The possible position of moving target in the first frame image is predicted by particle filter, and matching the target template and candidate regions with the color histogram statistical characteristics in order to ensure the tracking accuracy. Then the proposed NAPSO is utilized to search the smallest Bhattacharyya distance which is most similar. As a result, the algorithm can estimate a more realistic state and reduce the re-sampling frequency of particle filter, so that the computational cost of particle filter is effectively reduced.

The experimental results prove that the proposed algorithm has a good tracking accuracy and real-time in case of occlusions and fast moving target in target tracking.

2   PSO-PF Algorithm

2.1   Basic PSO Algorithm

PSO is a stochastic, population-based optimization algorithm. It was originally proposed by Eberhart [8]. PSO algorithm can be expressed as follows: to randomly initialize a particle swarm whose number is m and dimension is d, in which the particle's position is $X_i = (x_{i1}, x_{i2}, \cdots, x_{id})$ and its speed is $V_i = (v_{i1}, v_{i2}, \cdots, v_{id})$. During each iteration, the particles can renew their own speed and position through partial extremum and global extremum so as to reach optimization. The update formula is:

$$V_{id} = \omega * V_{id} + c_1 * Rand() * (P_{id} - X_{id}) + c_2 * Rand() * (G - X_{id}) \tag{1}$$

$$X_{id} = X_{id} + V_{id} \tag{2}$$

Where $Rand$ is a random number within interval $(0,1)$, ω is the inertia coefficient, c_1 and c_2 are learning factors.

2.2   Principle of Standard PSO-PF Algorithm

In the re-sampling process of regular particle filtering, the particles with bigger weights will have more offspring, while the particles with smaller weights will have few or even on offspring. It inspires us to find more particles with small weights, which will make the proposal distribution closer to the posteriori distribution. And it is the aim of using particles swarm optimization in the particles filtering.

The most important issue of using particles swarm optimizer is the choice of fitness function. Giving the fitness function as follows:

$$z_k \sim fitness = \exp[-\frac{1}{2R_k}(z_k - \hat{z}_{k|k-1}^i)^2] \tag{3}$$

Where z_k is the latest observed, $\hat{z}_{k|k-1}^i$ is the predictive observed value.

Take N particles $\{x_{0:k}^i\}_{i=1}^N$ as samples from importance function at the initial time. Calculate the importance value: $\omega_k^i = \omega_{k-1}^i \exp[-\frac{1}{2R_k}(z_k - \hat{z}_{k|k-1}^i)^2]$

Then update the velocity and position of particle:

$$v_{k+1}^i = \omega * v_k^i + c_1 r_1 * (p_{pbest} - x_k^i) + c_2 r_2 * (p_{gbest} - x_k^i) \tag{4}$$

$$x_{k+1}^i = x_k^i + v_{k+1}^i \tag{5}$$

Where r_1, r_2 is a random number within interval $(0,1)$, ω is the inertia coefficient, c_1 and c_2 are learning factors.

Calculate the importance weight of the particle after optimization and perform normalization: $\omega_k^i = \omega_k^i / \sum_{i=1}^N \omega_k^i$

Note that particle weights after re-sampling step are all equal to $1/N$, finally state output: $\hat{x}_k = \sum_{i=1}^N \omega_k^i x_k^i$

3 A New Adaptive Particle Swarm Optimization (NAPSO)

To overcome the problems of PSO which is easy to fall into local optimum in the whole iterative process and has a low convergence rate in the late iterative process, NAPSO is proposed. The adaptive adjustment process: inertia weight is a key parameter to balance the ability of local search and global search, a larger inertia weight value facilitates global exploration which enables the algorithm to search new areas, while a smaller one tends to facilitate local exploitation to increase the search accuracy [9]. In view this; the inertia weight will be adjusted adaptively by the fitness of particles in this paper. Adaptive weight is defined as follows:

$$\omega = \begin{cases} \omega_{min} & \text{if } f_i \text{ is better than } f_{avg} \\ \omega_{min} + (\omega_{max} - \omega_{min}) \dfrac{|f_{best} - f_i|}{|f_{best} - f_{avg}|} & \text{if } f_i \text{ is better than } f_{avg}'' \text{ but worse than } f_{avg}' \\ \omega_{max} & \text{if } f_i \text{ is worse than } f_{avg}'' \end{cases} \tag{6}$$

Where f_i is the fitness of particle, $f_{avg} = \frac{1}{N}\sum_{i=1}^N f_i$ is the average fitness of particles, f_{avg}' is the average fitness of the particles which are better than f_{avg}, f_{avg}'' is the average fitness of the particles which are worse than f_{avg}, f_{best} is the fitness of the best particle, ω_{max} is the maximum inertia weight and ω_{min} is the minimum inertia weight. $0 < \dfrac{|f_{best} - f_i|}{|f_{best} - f_{avg}|} < 1$ Ensure that $\omega \in [\omega_{min}, \omega_{max}]$, and a better fitness of the particles tends to a smaller ω, which tends to facilitate local exploitation to increase the search accuracy.

The particles are divided into three group based on the fitness of the particles in order to adopt different inertia weight generating strategy. Combining with the degree of premature convergence and individual adaptive value adjust the inertia weight, this algorithm is effective in controlling the particle swarm diversity, and it also has good convergence rate and can avoid getting into the local optimum.

4   Video Target Tracking Based on NAPSOPF

4.1   The Dynamic Model

We represent the target regions by rectangles or ellipses, so that a sample is given as

$$S = [x, \hat{x}, y, \hat{y}, H_x, H_y]^T \tag{7}$$

Where (x, y) represent the location of the ellipse (or rectangle), $\hat{x}$, $\hat{y}$ are velocities in x and y directions, H_x, H_y are the length of the half axes.

The sample set is propagated through the application of a dynamic model [10]:

$$S_k = \begin{bmatrix} 1 & T & 0 & 0 & 0 & 0 \\ 0 & 1 & 0 & 0 & 0 & 0 \\ 0 & 0 & 1 & T & 0 & 0 \\ 0 & 0 & 0 & 1 & 0 & 0 \\ 0 & 0 & 0 & 0 & 1 & 0 \\ 0 & 0 & 0 & 0 & 0 & 1 \end{bmatrix} S_{k-1} + \begin{bmatrix} \frac{T^2}{2} & 0 & 0 & 0 & 0 & 0 \\ 0 & T & 0 & 0 & 0 & 0 \\ 0 & 0 & \frac{T^2}{2} & 0 & 0 & 0 \\ 0 & 0 & 0 & T & 0 & 0 \\ 0 & 0 & 0 & 0 & 1 & 0 \\ 0 & 0 & 0 & 0 & 0 & 1 \end{bmatrix} W_k \tag{7}$$

Where T is the sampling period, and W_k is a random vector drawn from the noise distribution of the system.

4.2   Observation Model and the PSO Fitness Function

The observation model is used to measure the observation likelihood of the samples, and is important issue for target tracking.

In our tracking algorithm we need a similarity measure which is based on color distributions. A popular measure between two distributions $p(u)$ and $q(u)$ is the Bhattacharyya coefficient,

$$\rho[p, q] = \int \sqrt{p(u)q(u)} du \tag{8}$$

Considering discrete densities such as our color histograms $p = \{p^{(u)}\}_{u=1...m}$ and $q = \{q^{(u)}\}_{u=1...m}$ the coefficient is defined as

$$\rho[p, q] = \sum_{u=1}^{m} \sqrt{p_u q_u} \tag{9}$$

As distance between two distributions we define the measure, which is called the Bhattacharyya distance: $d = \sqrt{1 - \rho[p, q]}$

The smaller d is, the more similar the distributions are, so we choose the Bhattacharyya distance d as the fitness function of the NAPSO algorithm to Force the particles to be closer to the real state.

4.3   NAPSOPF in the Tracking Algorithm Process

As we know, the standard PSO-PF can improve the particle degradation of PF and is easier for actualization. Unfortunately it is a process of iterative optimization which will prolong the calculation time because of the high iterative frequency, and it may be easily trapped into local optimization, influencing the accuracy of video target tracking. Moreover, the fitness function (3) can't fit the tracking. To solve those problems, NAPSOPF tracking algorithm is proposed.

The tracking algorithm based on NAPSOPF:

Firstly, we manually select the target template in the first frame and calculate the color histogram of the target region, generate a particle set of N particles, then predict each particle using the dynamic model and calculate the color histograms of each particle, the distance between the target region color histogram and the color histograms of each particle is calculated using the Bhattacharyya similarity coefficient.

We use d as the fitness function of NAPSO , the algorithm can drive the particles into regions of the smaller fitness value, the smaller the fitness value of d is, the better match the particle is, accordingly forcing the particles to be closer to the real state:

Based on the Bhattacharyya distance, a color likelihood model is defined as

$$p(z_k \mid S_k^i) = \frac{1}{\sqrt{2\pi}\sigma}\left(-\frac{d^2}{2\sigma^2}\right) \tag{10}$$

Finally state output: $\hat{S}_k = \sum_{i=1}^{N} \omega_k^i S_k^i = \sum_{i=1}^{N} \omega_{k-1}^i p(z_k \mid S_k^i) S_k^i$

The best matched candidates for tracking can be obtained by comparing the Bhattacharyya distance between the target region and each particle. The estimated state is closer to the real state and the tracking accuracy is improved, even in the case of occlusions. Moreover, NAPSOPF solve the problem of low precision and complicated calculation of the standard PSO-PF, which can improve improves the real-time capability in the target tracking.

As we know, inertia weight is a key parameter to balance the ability of local search and global search, a larger inertia weight value facilitates global exploration which enables the algorithm to search new areas, while a smaller one tends to facilitate local exploitation to increase the search accuracy. The NAPSOPF algorithm adapts to different motion levels by setting the value of the maximum inertia weight in NAPSO, which can solve the low accuracy problem of fast moving target in video target tracking, because the displacement of the target between two consecutive frames is large.

Based on a large amount of experiments, ω_{min} is set to 0.1. In case of fast motion object, ω_{max} is set to 0.6~0.8; normal speed object, ω_{max} is set to 0.4~0.6; slow motion object, ω_{max} is set to 0.2~0.4.

Process of NAPSOPF tracker algorithm:

(1) Initialize particles $\{S_{k-1}^i\}_{i=1}^{N}$ around the previous target and assign each particle an equal weight: $\omega_0^i = 1/N$, calculate the color histogram of the target region $H_0 = \{q^{(u)}\}_{u=1}^{\tau}$.

(2) Propagate each sample set $\{S_k^i\}_{i=1}^{N}$ from the set $\{S_{k-1}^i\}_{i=1}^{N}$ by equation (7).

(3) Observe the color distributions: calculate the color distribution for each sample of the set S_k^i ; calculate the Bhattacharyya distance between the target template and each sample of the set $S_k^i : d = \sqrt{1 - \rho[p,q]} = \sqrt{1 - \sum_{u=1}^{\tau} \sqrt{p_i^{(u)} q^{(u)}}}$

(4) The NAPSO algorithm is led in, and the particles are divided into three group based on the value of fitness function d in order to adopt different inertia weight generating strategy:

$$\omega = \begin{cases} \omega_{min} & d_i < d'_{avg} \\ \omega_{min} + (\omega_{max} - \omega_{min}) \dfrac{|d_{best} - d_i|}{|d_{best} - d_{avg}|} & d'_{avg} < d_i < d''_{avg} \\ \omega_{max} & d_i > d''_{avg} \end{cases} \qquad (11)$$

Where d_i is the fitness of particle, d_{avg} is the average fitness of particles, d'_{avg} is the average fitness of the particles which are smaller than d_{avg} , d''_{avg} is the average fitness of the particles which are bigger than d_{avg} , d_{best} is the fitness of the best particle, ω_{max} is the maximum inertia weight and ω_{min} is the minimum inertia weight. The velocity and position of the particles are updated according the following equations (4), (5):

(5) Calculate the observation likelihood: $p(z_k | S_k^i) = \dfrac{1}{\sqrt{2\pi}\sigma} \exp\left(-\dfrac{1 - \rho[H_i, H_0]}{2\sigma^2}\right)$

(6) Calculate the importance weight of the particles after optimization and perform normalization: $\omega_k^i = \omega_{k-1}^i p(z_k | S_k^i)$, $\omega_k^i = \omega_k^i / \sum_{i=1}^{N} \omega_k^i$

(7) Estimate the mean state: $\hat{s}_k = \sum_{i=1}^{N} \omega_k^i s_k^i$.

(8) Re-sampling step then $k = k + 1$ Go back to (2)

5   Simulation and Results Analysis

We run our algorithm on AMD Athlon (tm) II X2 B24 Processor 2.99GHz PC with 1.75GB memory. We implement all codes by MATLAB R2012a and visual studio.

5.1   Tracking Performance

Several real video sequences are used to compare the three implemented tracking algorithms. We illustrate our results with three test sequences A, B and C.

Sequence A is a football game video. As shown in Figure 1, the target is occluded by another athlete around frame 10; the figure shows that NAPSOPF algorithm is more accurate in case of occlusions than PF and PSOPF algorithm.

Sequence B is a racing video clip, the speed of the target is about 200km/h. To adjust the particle search range, we set the maximum inertia weight. Figure 2 shows the performance of NAPSOPF algorithm is more accurate than PF and PSOPF algorithm when tracking a very fast moving target.

Sequence C is a soccer match video clip, the speed of the ball is about 90-110km/h and the ball is occluded by the goal around frame 46, we set the maximum inertia weight $\omega_{max} = 0.5$. As shown in Figure 3, the NAPSOPF algorithm is more accurate and robust in case of fast moving target which is occluded.

Fig. 1. Trackers performance with Sequence A frame 5, 10, 15

PF   PSOPF   NAPSOPF

Fig. 2. Trackers performance with Sequence B frame 55, 65, 70, 75

PF   PSOPF   NAPSOPF

Fig. 3. Trackers performance with Sequence C frame 31,41,46,48

5.2   Results Analysis

Experimental results of Sequence A, B and C are shown in Table 1~Table 2.

Based on the results provided in these two tables, it is clear that, PF has the worst results, but has the best running time. While the NAPSOPF algorithm has the best results, it reduce the re-sampling frequency of particles filter, so that the computational cost of particle filter is effectively reduced, the algorithm can meet the requirement of real-time tracking by handling about 10 frames per second.

Table 1. comparison between PF, PSOPF and NAPSOPF simulation parameters

Algorithm	Sequence A		Sequence B		Sequence C	
N=300	Effective Number	Re-sampling	Effective Number	Re-sampling	Effective Number	Re-sampling
PF	155.969	13.3	167.087	14.4	141.391	15.9
PSOPF	173.124	10.7	178.369	12.6	147.827	14.3
NAPSOPF	189.317	7.7	187.570	9.9	150.945	13.0

Table 2. comparison of running time between PF, PSOPF and NAPSOPF

Particles Number	Algorithm	Sequence A (20 frames)	Sequence B (25 frames)	Sequence C (22 frames)	Average frames per second
	PF	1.954	2.249	2.076	10.649
N=300	PSOPF	2.174	2.632	2.372	9.324
	NAPSOPF	2.044	2.351	2.177	**10.175**
	PF	1.580	1.835	1.631	13.257
N=100	PSOPF	1.776	2.189	1.895	11.431
	NAPSOPF	1.689	1.941	1.747	**12.438**

6   Conclusions

The NAPSOPF tracking algorithm can enable the particles to fit the environment better and then estimate a more realistic state, thereby it improves the accuracy and robustness of occlusions and fast moving target in video target tracking. The experiment results indicate that the algorithm has a good tracking accuracy and real-time in case of occlusions and fast moving target in video target tracking.

Acknowledgements. This research was supported by the National Science Foundation Council of Guangxi (2012GXNSFAA053227).

References

1. Collns, R., Lipton, A., Kanadeandt.: A System for Video Surveillance and Monitoring VSAM Final report. Carnegie Mellon University (2000)
2. Belagiannis, V., Schubert, F., Navab, N., Ilic, S.: Segmentation based particle filtering for real-time 2D object tracking. In: Fitzgibbon, A., Lazebnik, S., Perona, P., Sato, Y., Schmid, C. (eds.) ECCV 2012, Part IV. LNCS, vol. 7575, pp. 842–855. Springer, Heidelberg (2012)

3. Loris, B., Marco, C., Vittorio, M.: Decentralized Particle Filters for Joint Individual-Group Tracking. In: CVPR (2012)
4. Chong, Y., Chen, R., Li, Q., Zheng, C.-H.: Particle filter based on multiple cues fusion for pedestrian tracking. In: Huang, D.-S., Gupta, P., Zhang, X., Premaratne, P. (eds.) ICIC 2012. CCIS, vol. 304, pp. 321–327. Springer, Heidelberg (2012)
5. Wang, A.X., Li, J.J.: Target Tracking Based on Multi-core Particle Filtering. Computer Science 39(8), 296–299 (2012)
6. Doucet, A., Godsill, S.: On Sequential Monte Carlo Sampling Methods for Bayesian Filtering. Statistics and Computing 10(1), 197–208 (2000)
7. Katja, N., Esther, K., Luc, V.G.: Object Tracking with and Adaptive Color-based Particle filter. In: Proceedings of the 24th DAGM Symposium on Pattern Recognition, Zurich, Switzerland, September 16-18, pp. 353–360 (2002)
8. Kennedy, J., Eberhart, R.C.: Particle Swarm Optimization. In: Proc of the IEEE Intl. Conf. on Neural Networks, Perth, Australia, pp. 1942–1948. IEEE Service Center, Piscataway (1995)
9. Shi, Y.H., Eberhart, R.C.: A Modified Particle Swarm Optimizer. In: Proceedings of The IEEE Congress on Evolutionary Computation, pp. 69–73. IEEE Service Center, Piscataway (1998)
10. Li, A.P.: Research on Tracking Algorithm for Visual Target under Complex Environments, pp. 20–31. Shanghai Jiao Tong University, Shanghai (2006)

Firefly Algorithm and Pattern Search Hybridized for Global Optimization

Mahdiyeh Eslami[1,*], Hussain Shareef[1], and Mohammad Khajehzadeh[2]

[1] Electrical, Electronic and Systems Engineering Department,
National University of Malaysia, Selangor, Malaysia
Mahdiyeh_eslami@yahoo.com
[2] Civil Engineering Department, Anar Branch, Islamic Azad University, Anar, Iran

Abstract. Firefly optimization algorithm is one of the latest swarm intelligence based optimization algorithm. A new hybrid optimization algorithm, which combines pattern search with firefly algorithm, namely FAPS, is proposed for numerical global optimization. There are two alternative phases of the proposed algorithm: the global exploration phase realized by firefly algorithm and the exploitation phase completed by pattern search. The performance of the proposed FAPS algorithm was tested on a comprehensive set of benchmark functions. The numerical experiments demonstrate that the new algorithm has high viability, accuracy and stability and the performance of firefly algorithm is much improved by introducing a pattern search method.

Keywords: global optimization, firefly algorithm, pattern search, hybridization.

1   Introduction

Global optimization could be a very challenging task because many objective functions and real world problems are multimodal, highly non-linear, with steep and flat regions and irregularities, thus better optimization algorithms are always needed. Generally, unconstrained optimization problems can be formulated as:

$$\min f(x), x = (x_1, x_2, \ldots, x_n) \tag{1}$$

where $f\colon \mathbf{R}^n \to \mathbf{R}$ is a real-valued objective function, $x \in \mathbf{R}^n$, and n is the number of the parameters to be optimized.

A variety of techniques may be used to solve unconstrained optimization problems; some are based on classical local optimization methods, such as linear or quadratic programming, while others use artificial intelligence or global heuristic algorithms. Classical techniques converge faster and they can obtain solutions with higher accuracy compared to global optimization methods. However, these approaches often rely deeply on the initial starting point and a good initial guess is vital for them to be executed effectively. On the other hand, global optimization methods, e.g., genetic

---

* Corresponding author.

D.-S. Huang et al. (Eds.): ICIC 2013, LNAI 7996, pp. 172–178, 2013.
© Springer-Verlag Berlin Heidelberg 2013

algorithm (GA) [1], particle swarm optimization (PSO) [2], artificial bee colony (ABC) [3], gravitational search algorithm (GSA) [4], etc, are good at finding the promising regions of search space and probability of getting trapped into local optima would be decreased. In addition, initial configuration has no effect on the solution of these methods. In recent years, these methods have been successfully applied for a wide variety of optimization tasks [5-10]. However, these methods have often exhibited unacceptable slow convergence rates due to their random search, especially near the area of the global optimum. Therefore, the hybrid algorithms that can benefit from the advantages of both methodologies and alleviate their inherent drawbacks are of interest.

The firefly algorithm (FA) is one of the latest global optimization algorithms, inspired by the flashing behavior of fireflies [11]. The FA is characterized as a simple concept that is both easy to implement and computationally efficient. The effectiveness and higher performance of FA as an excellent global optimization algorithm, compared with some well-know search methods (e.g. GA and PSO), infers from the results of experiment undertaken previously [10].

This paper develops a new hybrid algorithm combining the FA with pattern search (PS) methods, which is a subclass of direct search methods. During the initial optimization stages, the proposed algorithm starts with FA to find a near optimum solution and accelerate the convergence speed. The searching process is then switched to PS and the best solution found by FA will be taken as the initial starting point for the PS and will be fine-tuned. In this way, the hybrid FAPS algorithm may find an optimum solution more quickly and accurately.

2   Firefly Algorithm

Firefly algorithm (FA) is a population-based algorithm to find the global optima of objective inspired by the flashing behavior of fireflies. There are about two thousand firefly species, and most fireflies produce short and rhythmic flashes. The main part of a firefly's flash is to act as a signal system to attract other fireflies. By idealizing some of the flashing characteristics of fireflies, firefly-inspired algorithm is presented by Yang [11].

In the FA, physical entities (agents or fireflies) are randomly distributed in the search space. Agents are thought of as fireflies that carry a luminescence quality, called luciferin, that emit light proportional to this value. Each firefly is attracted by the brighter glow of other neighboring fireflies. The attractiveness decreases as their distance increases. If there is no brighter one than a particular firefly, it will move randomly.

The initial positions of agents are determined randomly in the search space, as:

$$x_{i,j} = x_{i,\min} + rand(0,1) \times (x_{i,\max} - x_{i,\min}), \qquad i = 1, 2, ..., N \qquad (2)$$

where. $x_{i,\min}$ and $x_{i,\max}$ are the lower and upper bounds for the ith variable, respectively, and N is the total number of agents.

In the Firefly Algorithm, there are two important issues: the variation of light intensity and formulation of the attractiveness. As mentioned before, the attractiveness of a firefly is determined by its brightness which is associated with the encoded objective function.

As light intensity and thus attractiveness decreases as the distance from the source increases, the variations of light intensity and attractiveness should be monotonically decreasing functions. Therefore, the main form of attractiveness function $\beta(r)$ can be formulated as the following generalized form:

$$\beta(r) = \beta_0 e^{-\gamma r^2} \tag{3}$$

where r is the distance between two fireflies, β_0 is the attractiveness at $r = 0$ and γ is the light adsorption coefficient. The distance affecting the attractiveness between any two fireflies i and j at positions x_i and x_j, can be defined as the Cartesian or Euclidean distance as follows:

$$r_{ij} = \left\| x_i - x_j \right\| \tag{4}$$

The movement of a firefly i is attracted at another more attractive (brighter) firefly j is determined by:

$$x_i^{t+1} = x_i^t + \beta_0 e^{-\gamma r^2} (x_j^t - x_i^t) + \alpha(rand - 0.5) \tag{5}$$

The second term in Eq. (22) is due to attraction. The third term introduces randomization with α being the randomization parameter and *rand* is a random number generated uniformly distributed between 0 and 1. Generally, the randomization parameter α, is considered as a decreasing function with a geometric progression reduction scheme starting from the initial α_0 according to:

$$\alpha = \alpha_0 \theta^t, \qquad 0 < \theta < 1 \tag{6}$$

The location of fireflies can be updated sequentially, by comparing and updating each pair of them in every iteration cycle. Iterations are performed until reaching maximum number of iterations or meeting the stop condition.

3   Pattern Search

The Pattern Search (PS) is a direct search method well capable of solving optimization problems of irregular, multimodal objective functions, without the need of calculating any gradient or curvature information, especially for addressing problems for which the objective functions are not differentiable, stochastic, or even discontinuous (Torczon, 1997). As opposed to more traditional local optimization methods that use information about the gradient or partial derivatives to search for an optimal solution, pattern search algorithms compute a sequence of points that get closer and closer to the globally optimal solution. The pattern search method can be briefly explained in a way that starts by establishing a set of points called a mesh around the given point, which could be computed from a previous step of the iteration or from the initial starting point provided by the user. The mesh is created by adding a scalar (called the mesh size) multiple set of vectors called a pattern to the current point, then it searches a set of points (mesh) around the current point to find a point where the objective function has a lower value. After a point with a lower objective function value is detected, the algorithm sets the point as its current point and the iteration can be considered successful. Then, the algorithm steps to the next iteration with extended mesh size which is induced by the expansion factor. If the algorithm does not find a point that improves the objective function, the iteration is called unsuccessful. The current points stay the same in the next iteration and the mesh size decreases due to the contraction factor (Lewis and Torczon, 2002). The pattern search optimization algorithm stops when the terminating conditions occurs.

4   Hybrid FAPS Algorithm

By combining the FA with PS, a new hybrid algorithm referred to as FAPS algorithm is formulated in this paper. In this approach the balance between exploration and exploitation is achieved using the FA as a global optimizer for global exploration and PS as a deterministic local search algorithm for local exploitation. This algorithm effectively uses the benefit of both FA and PS techniques and avoids their weaknesses. In FAPS, the searching process starts with the FA by initializing a group of random agents since the PS is sensitive to the starting point. The calculation continues with the FA for a specific number of iterations to search the global best position in the solution space. The best solution found by FA will be taken as the initial starting point for the PS and the searching process is then switched to PS to accelerate convergence to the global optimum. In this way, the hybrid algorithm may find an optimum more quickly and accurately. The procedure for the proposed FAPS algorithm is presented in Fig. 1.

5   Experimental Results

In this section, the efficiency and robustness of the proposed hybrid algorithm for numerical optimization will be investigated. In order to prove that an algorithm is able to perform sufficiently well over a wide range of feasible functions, the most commonly used strategy is the application of benchmark test comprising several functions.

In this study, a set of six well-known standard benchmark functions are employed. Although these functions may not necessarily give an accurate indication of the performance of an algorithm on real world problems, they can be used to investigate certain aspects of the algorithms under consideration. The functions, dimension, the admissible range of the variable and the optimum to be obtained are summarized

```
input: f(X), n, β₀, γ, α₀, MaxGeneration;
    for i =1 to n
        generate initial position of agent i, xᵢ, using Eq. (2);
    end
t = 1
while t < MaxGeneration
    for i =1 to n
        for j=1 to i
            if f(xⱼ) < f(xᵢ) then
                Calculate distance of fireflies i and j from Eq. (4)
                Evaluate attractiveness function β(r) using Eq. (3)
                Move firefly i toward j using Eq. (5)
            end
        end
    end
Rank the fireflies and find the current best
t = t +1
end
Consider the best solution obtained by FA as the initial guess for PS
Use PS algorithm to search around global best, which is found by FA
Output the solution obtained from PS
```

Fig. 1. The framework of the FAPS algorithm

Table 1. Standard benchmark functions

Test Function	Dimension (n)	Range	Optimum
$F_1(X) = \sum_{i=1}^{n} x_i^2$	30	$[-100,100]^n$	0
$F_2(X) = \sum_{i=1}^{n}|x_i| + \prod_{i=1}^{n}|x_i|$	30	$[-10,10]^n$	0
$F_3(X) = \sum_{i=1}^{n}(x_i^2 - 10\cos(2\pi x_i) + 10)$	30	$[-5.12,5.12]^n$	0
$F_4(x) = -20\exp\left(-0.2\sqrt{\frac{1}{n}\sum_{i=1}^{n}x_i^2}\right) -$ $\exp\left(\frac{1}{n}\sum_{i=1}^{n}\cos 2\pi x_i\right) + 20 + e$	30	$[-32,32]^n$	0
$F_5(X) = \frac{1}{4000}\sum_{i=1}^{n}x_i^2 - \prod_{i=1}^{n}\cos(\frac{x_i}{\sqrt{i}}) + 1$	30	$[-600,600]^n$	0

in Table 1. All the functions are to be minimized. The first three functions are unimodal functions whereas the last three functions are multimodal optimization problems with a considerable amount of local minima.

The presented benchmark functions are solved using both FA and FAPS algorithms. In the experiments, the parameters used are set as: population size (n) is

50, maximum iteration numbers (*MaxGeneration*) is 500; β_0, α_0 and γ are 0.2, 0.5 and 1, respectively. The algorithms are simulated 30 times independently and the results are recorded. Then the statistical analyses are carried out and for each method, the worst, mean, median, best and standard deviation are calculated. The performance comparison between two algorithms on five functions is presented in Table 2.

The results of Table 2 are encouraging. The findings in this table show that, for all test functions, the FAPS converged to a more significantly accurate final solution than the FA. In terms of mean and best fitness values, the new algorithm could provide a significantly better solution for all functions. At the same time, the standard deviation of the results obtained by the FAPS in 30 independent runs for all the functions are smaller than those computed by FA indicating the superior stability of the new method.

Table 2. Minimization result of benchmark functions

Function	Method	Worst	Mean	Median	Best	Standard deviation
F_1	FA	0.0046	0.0019	0.0016	7.54e-4	0.0011
	FAPS	2.61e-16	4.48 e-17	1.87 e-17	2.88 e-19	6.85 e-17
F_2	FA	0.6295	0.4145	0.3803	0.222	0.112
	FAPS	1.11 e-5	5.76 e-6	4.98 e-6	3.43 e-7	3.48 e-6
F_3	FA	73.8	47.85	48.205	27.72	10.78
	FAPS	2.18e-5	8.56e-6	6.45e-6	4.73e-8	8.09e-6
F_4	FA	0.082	0.043	0.041	0.0145	0.0181
	FAPS	2.29 e-7	8.46 e-8	7.06 e-8	5.36 e-9	6.76 e-8
F_5	FA	0.0126	0.0016	9.53 e-4	2.63 e-4	0.0026
	FAPS	0.00	0.00	0.00	0.00	0.00

6   Conclusion

A hybrid global optimization algorithm called FAPS has been proposed. It combines two search techniques: the firefly algorithm and pattern search method. Both FA and PS are derivative free algorithms for global optimization, so the proposed FAPS also does not require the derivatives of the optimized objective functions. In this approach, the FA successfully searches all space during the initial stages of a global search. Then, the algorithm switches to PS for accurate local exploitation around the best solution and to complement the global exploration provided by the FA. The performance of the proposed algorithm as a global optimization technique is investigated using a set of six well-known unimodal/multimodal benchmark functions. In comparison with the results obtained from the classical FA, the FAPS algorithm has been verified to possess excellent performance in terms of accuracy, convergence rate, stability and robustness.

References

1. Holland, J.: Adaptation in natural and artificial systems. University of Michigan Press (1975)
2. Kennedy, J., Eberhart, R.: Particle swarm optimization. In: IEEE International Conference on Neural Networks, vol. 4, pp. 1942–1948 (1995)
3. Karaboga, D., Basturk, B.: A powerful and efficient algorithm for numerical function optimization: artificial bee colony (ABC) algorithm. Journal of Global Optimization 39(3), 459–471 (2007)
4. Rashedi, E., Nezamabadi-pour, H., Saryazdi, S.: GSA: a gravitational search algorithm. Information Sciences 179(13), 2232–2248 (2009)
5. Khajehzadeh, M., Taha, M.R., El-Shafie, A., Eslami, M.: Modified particle swarm optimization for optimum design of spread footing and retaining wall. Journal of Zhejiang University-Science A 12(6), 415–427 (2011)
6. Khajehzadeh, M., Taha, M.R., El-Shafie, A., Eslami, M.: A modified gravitational search algorithm for slope stability analysis. Engineering Applications of Artificial Intelligence 25(8), 1589–1597 (2012)
7. Eslami, M., Shareef, H., Mohamed, A., Khajehzadeh, M.: An efficient particle swarm optimization technique with chaotic sequence for optimal tuning and placement of PSS in power systems. International Journal of Electrical Power & Energy Systems 43(1), 1467–1478 (2012)
8. Eslami, M., Shareef, H., Mohamed, A., Khajehzadeh, M.: Gravitational search algorithm for coordinated design of PSS and TCSC as damping controller. Journal of Central South University of Technology 19(4), 923–932 (2012)
9. Dong, Y., Tang, J., Xu, B., Wang, D.: An application of swarm optimization to nonlinear programming. Computers & Mathematics with Applications 49(11-12), 1655–1668 (2005)
10. Yang, X.-S.: Firefly algorithms for multimodal optimization. In: Watanabe, O., Zeugmann, T. (eds.) SAGA 2009. LNCS, vol. 5792, pp. 169–178. Springer, Heidelberg (2009)
11. Yang, X.S.: Nature-inspired metaheuristic algorithms. Luniver Press, Beckington (2008)

Differential Lévy-Flights Bat Algorithm for Minimization Makespan in Permutation Flow Shops

Jian Xie[1], Yongquan Zhou[1,2], and Zhonghua Tang[1]

[1] College of Information Science and Engineering,
Guangxi University for Nationalities, Nanning, Guangxi, 530006, China
[2] Guangxi Key Laboratory of Hybrid Computation and IC Design Analysis,
Nanning, Guangxi, 530006, China
yongquanzhou@126.com

Abstract. The permutation flow shop problem (PFSP) is an NP-hard problem with wide engineering and theoretical background. In this paper, a differential Lévy-flights bat algorithm (DLBA) is proposed to improve basic bat algorithm for PFSP. In DLBA, LOV rule is introduced to convert the continuous position in DLBA to the discrete job permutation, the combination of NEH heuristic and random initialization is used to initialize the population with certain quality and diversity, and a virtual population neighborhoods search is used to enhance the global optimal solution and help the algorithm to escape from local optimal. Experimental results and comparisons show the effectiveness of the proposed DLBA for PFSP.

Keywords: Bat Algorithm, Lévy-Flight, Minimization Makespan, Permutation Flow Shop Scheduling, Virtual Population Neighborhoods Search.

1   Introduction

Permutation Flow shop Scheduling Problem (PFSP) is one of best known production scheduling problem, which has been proved to be NP-hard problem [1]. Due to its significance in both academic and engineering applications, the permutation flow shop with the criterion of minimizing the makespan, maximum lateness of jobs, or minimizing total flow time are researched, a great diversity of methods have been proposed to solve PFSP and obtained some achievements.

Many methods have been introduced for solving PFSP with the objective of minimizing makespan. Generally, those methods can be broadly classified into three categories: exact methods [2], constructive heuristic methods (NEH can be viewed as the typical cases [3]) and meta-heuristic algorithms based on the constructive operation and neighborhood search. The meta-heuristics mainly include particle swarm optimization algorithm (PSO) [4], invasive weed optimization (IWO) [5] and genetic algorithm (GA) [6], etc. In [6], Wang used the well-known NEH combined with GA to generate the initial population, and applied multi-crossover operators to enhance the exploring potential. In [7], Li et al. applied a differential evolution (DE) based memetic algorithm, named ODDE, to solve PFSP by combining with NEH

D.-S. Huang et al. (Eds.): ICIC 2013, LNAI 7996, pp. 179–188, 2013.

heuristic initialization, opposition-based learning, pairwise local search and fast local search in ODDE.

Recently, a bat algorithm is proposed, which is inspired by the intelligent echolocation behavior of micro-bats when they foraging [8]. Up to now, most published works on BA mainly have focused on solving constrained optimization [9] and engineering optimization [10], etc. In this paper, we propose a differential Lévy-flights bat algorithm (DLBA) combining with some local search mechanisms for solving PFSP.

The rest of this paper is organized as follows. In Sections 2, PFSP, BA and relevant knowledge are described. In Section 3, the DLBA is proposed to improve BA for PFSP, and several mechanisms are integrated in DLBA. The experimental results of the DLBA and comparisons to other previous algorithms are shown in Section 4. In the last section, we conclude this paper and point out some future work.

2   Problem Descriptions and Bat Algorithm

2.1   Permutation Flow Shop Scheduling Problem

The PFSP in the paper consists of a set of jobs on a set of machines with the objective of minimizing makespan. In PFSP, n jobs are to be processed on a series of m machines, sequentially. All jobs are processed in the same permutation, meanwhile, every job is processed in one machine only once and each machine can only process one job at a time, all jobs are processed in an identical processing order on all machines. Let t_{ij} $(1 \leq i \leq n, 1 \leq j \leq m)$ is the times of job i processed on machine j, assuming preparation time for each job is zero or is included in the processing time t_{ij}, $\pi = (j_1, j_2, \cdots, j_n)$ is a scheduling permutation of all jobs. Π is set of all scheduling permutation. Completion time $C(j_i, k)$ of job j_i on machine k, and every job will be processed on machine 1 to machine m orderly. The completion time of the permutation flow shop scheduling problem according to the processing sequence $\pi = (j_1, j_2, \cdots, j_n)$ is shown as follows:

$$C(j_1, 1) = t_{j_1, 1}$$
$$C(j_i, 1) = C(j_{i-1}, 1) + t_{j_i, 1}, i = 2, 3, \cdots, n$$
$$C(j_1, k) = C(j_1, k-1) + t_{j_1, k}, k = 2, 3, \cdots, m$$
$$C(j_i, k) = \max\{C(j_{i-1}, k), C(j_i, k-1)\} + t_{j_i, k}, i = 2, 3, \cdots, n, k = 2, 3, \cdots, m$$
$$\pi_* = \arg\{C_{\max}(\pi) = C(j_n, m)\} \to \min, \forall \pi \in \Pi$$

π_* is the most suitable arrangement which is the goal of the permutation flow shop problem to finding, $C_{\max}(\pi_*)$ is the minimal makespan.

2.2   Basic Bat Algorithm and Lévy Flight

The BA is a meta-heuristic proposed by Yang in 2010 [8]. Under several ideal rules, firstly, the bat individual randomly selects a certain frequency of sonic pulse, and the position of bat individual is updated according to their selected frequency. Secondly, if a random number is greater than its pulse emission rate R , then a new position is generated around the current global best position for each individual, which is the equal of local search. At last, if the local search is effective and its loudness L is greater than a random number, then the new position is accepted, and its R and L is updated, where R is increased, L is decreased. In general, the bat algorithm has three procedures, position updating, local search and decreasing the probability of local search. The detail of BA could refer to [8].

Lévy flight is a random walk in which the step-lengths have a probability distribution that is heavy-tailed. Lévy flight has the several properties: statistical self-similarity, random fractal characteristics and infinite variance with an infinite mean value [11]. Due to the remarkable properties of stable Lévy distribution, Lévy flight has been applied to optimization and optimal search [12], and preliminary results show its promising capability.

3   Differential Lévy-Flights Bat Algorithm for PFSP

Inspired by Yang's method, the DLBA is proposed based on differential operator and Lévy-flights trajectory. The purpose of proposed DLBA is to improve the performance of the basic bat algorithm. The causes of improvement are that Lévy-flights can make the algorithm effectively jump out of local optimal, and differential operation can speed up the convergence rate. PFSP is a typical discrete problem. The key of discrete problem is sequence operation. Considering these sequences, the continuous algorithm also can be used to solve it. Many continuous algorithms have successfully solved the PFSP, such as GA [6], PSO [3] and IWO [5]. These algorithms construct a direct mapping relationship between the job sequence and the vector of individuals. The DLBA also adopts this kind of method. The proposed DLBA and BA are used to solving PFSP. In order to solve this specific discrete problem, several pretreatment and neighborhoods search are integrated into BA and DLBA. The pretreatment is NEH initialization. First of all, the DLBA is described in section 3.1.

3.1   Differential Lévy-Flights Bat Algorithm

The DLBA based on the basic structure of BA and re-estimate the characters used in the BA. In DLBA, not only the movement of the bat is quite different from the BA, but also the local search process is different from it.

In DLBA, the frequency fluctuates up and down, which can change self-adaptively, and the differential operator is introduced, which is similar to the mutation operation of DE. The frequency updating formula of a bat is defined by (1):

$$\begin{cases} f_{1i}^t = (t(f_{1,\min} - f_{1,\max})/n_t + f_{1,\max})\beta_1 \\ f_{2i}^t = (t(f_{2,\max} - f_{2,\min})/n_t + f_{2,\min})\beta_2 \end{cases}, \tag{1}$$

where n_t is a fixed parameters, $\beta_1, \beta_2 \in [0,1]$ is a random vector drawn from a uniform distribution, $f_{1,\max} = f_{2,\max}$, $f_{1,\min} = f_{2,\min}$. The position update by frequency and individual position of bats in the bats swarm, rather than get through the velocities in original BA.

$$x_i^{t+1} = x_{best}^t + f_{1i}^t(x_{r1}^t - x_{r2}^t) + f_{2i}^t(x_{r3}^t - x_{r4}^t), \tag{2}$$

where x_{best}^t is the current global best solution which is located after comparing all the solutions among all the n bats in t generation, x_{ri}^t is the random bat individual in the bat swarm, and this can be achieved by randomization. In additional, these bats perform the Lévy-flight by (3) before the position updating (2).

$$x_i^{t+1} = \hat{x}_i^t + \mu sign[rand - 0.5] \oplus Levy, \tag{3}$$

where $\hat{x}_i^t$ is the history best location of individual i, μ is a random parameter drawn from a uniform distribution, sign $\oplus$ means entry-wise multiplications, $rand \in [0,1]$, random step length $Levy$ obeys Lévy distribution.

On the other hand, each bat should have different values of loudness L and pulse emission rate R, while R is relatively low and L is relatively high. During the search process, the L usually decreases, while the R increases. The updated position is influenced by the pulse emission rate and loudness as well. In DLBA, the R_i^t cause fluctuation of position by (4), and the L_i^t cause fluctuation of position by (5).

$$x_{new} = x_{best} + \varepsilon L_t, \tag{4}$$

$$x_{new} = x_{best} + \eta R_t, \tag{5}$$

where $\eta, \varepsilon \in [-1,1]$ is a random number, $L_t = <L_i^t>$ is the average loudness of all the bats at current generation. x_{best} is the current global best location (solution) in whole bats swarm. $R_t = <R_i^t>$ is the average pulse emission rate of all the bats at this generation.

The new R_i^t and loudness L_i^t at t generation are updated by:

$$L_i^{t+1} = \alpha L_i^t, \ R_i^{t+1} = (1 + \exp(-10/t_{\max} \times (t - t_{\max}/2) + R_i^1)^{-1}, \tag{6}$$

where $\alpha \in [0,1]$ is a constant, t is iteration number, $t_{\max}$ denote maximum iteration number, R_i^1 denote initial pulse emission rate of each bat. Then, $R \in (0,1)$ is similar to sigmoid function.

Note that the loudness and emission rates will be updated only if the best solution in the current generation better than the best solution in last generation. The pseudo code of DLBA is depicted in Program1:

```
Program1•The pseudo code of DLBA:
  Input: Objective function and parameters in algorithm;
  Output: Optimal individual and optimal fitness;
  begin
    Initialize population and parameters;
    Calculating the initial fitness of each individual;
    repeat
      Update position by (1)~(3) for each individual;
        Evaluate each individual and find out x_best , x̂ ;
      if• rand > R_i •then
          Generate a x_new around the selected best solution by(4);
      end
      Evaluate x_new and find out x_best , x̂ ;
      if• rand < L_i •then
            Generate a x_new around the selected best solution
by(5);
      end
      Evaluate x_new and find out x_best , x̂ ;
      if ( f(x_best^{t+1}) < f(x_best^t) ) then
          Update R_i , L_i by(6) ;
      end
      t = t+1 ;
    until  t = t_max
  end.
```

3.2   Solution Representation and Initialization in DLBA

Since standard BA and DLBA is a continuous optimization algorithm, the standard continuous encoding scheme of BA and DLBA cannot be used to solve PFSP directly. In order to apply BA and DLBA to PFSP with minimal makespan, the largest-order-value (LOV) rule [13] is introduced to convert the individual to the job permutation.

In this paper, the DLBA is applied to explore the new search space. Initial swarm is generally generated randomly. However, an initial population with certain quality and diversity can speed up the convergence rate of algorithm. Meanwhile, recent studies have confirmed the superiority of NEH [3] over the most recent constructive heuristic. The initial strategy that using NEH heuristic to generate 1 sequence and the rest of the sequences are initialized randomly is adopted in DLBA.

3.3   Virtual Population Neighborhoods Search

In this paper, a virtual population neighborhoods search (VPNS) with same population size is easy embedded in DLBA for solving PFSP. The purpose of the virtual population neighborhoods search is to find a better solution from the neighborhood of the current global best solution. For PFSP, Four neighborhoods (Swap, Insert, Inverse and Crossover) are used to improve the diversity of population and enhance the quality of the solution. (1) *Swap*: choose two different positions from a job permutation randomly and swap them. (2) *Insert*: choose two different positions from a job permutation randomly and insert the back one before the front. (3) *Inverse*: inverse the subsequence between two different random positions of a job permutation. (4) *Crossover*: choose a subsequence in a random interval from another random job permutation and replace the corresponding part of subsequence.

In DLBA, the swap crossover probability p_i of bat individual i is relative to its own fitness, the better fitness, and the lower probability. The p_i is described by (7)

$$p_i = ((fit_i - fit_{\min})/(fit_{\max} - fit_{\min} + 1)), \qquad (7)$$

where fit_i is the fitness of individual i, $fit_{\min}$ and $fit_{\max}$ are the minimum and maximum fitness in current population, respectively. p_i reflects the quality of individual. Firstly, using the x_{best}^t to generate a virtual population, it is as well as bat population with same structure. Secondly, for each virtual individual, if a random number is greater than its p_i, the swap operation will be selected, then generate two swap position and execute swap operation; otherwise, the crossover operation will be selected, randomly choose a individual in original bat population and randomly generate a crossover interval, the virtual individual's position in crossover interval will be replaced by corresponding part of selected individual in original bat population. At last, an adjustment operation is used to guarantee the effectiveness of scheduling sequence.

In order to further enhance the local search ability and get a better solution, a new individual enhancement scheme is proposed to combine the inverse and insert operation. For the inverse operation, a step length s and inverse range rng is generated firstly, $s(t) = \lceil n/t \rceil$, $rng(t) = \max(1, \min(n - s, n))$, where n is the number of job, t is iteration. For each individual of the virtual population based on x_{best}^t, randomly select a starting point $k1$ of inverse in range $rng(t)$, and the ending point $k2 = \min(k1 + s, n)$. The subsequence (from $k1$ to $k2$) is inversed.

The insert operation is applied to generate a virtual population that is based on x_{best}^t. Firstly, a job position i is chosen randomly in x_{best}^t, the selected job i is inserted into the back of job i, orderly, until the population size ps is reach the ps of original bat population.

3.4   DLBA Framework for PFSP

In DLBA, all individual once update either in bat population or in virtual population, these individual will be evaluate and accept one solution as the current global best x_{best}^t if the objective fitness of it is better than the fitness of the last x_{best}^{t-1}. The continuous individual must convert to discrete job permutation by LOV rule before each individual is evaluated. The algorithm terminate until the stopping criterion is reached, the pseudo code of DLBA framework for PFSP is shown in Program 2:

```
Program2•The pseudo code of DLBA framework for PFSP:
  Input: Objective function and parameters in algorithm;
  Output: Optimal individual and optimal fitness;
  begin
    Initialize all parameters, use NEH heuristic to generate 1
    sequence and the rest of the sequences are initialized
    randomly;
    Evaluate initial individual and find out x^i_best ;
   repeat
      Update individual using DLBA
      Evaluate each individual and find out x^i_best ;
       Compute the probability p_i of each individual by(7);
      Carry out VPNS operation;
      Evaluate each individual and find out x^i_best ;
       t=t+1 ;
    until  t=t_max
  end.
```

4   Numerical Simulation Results and Comparisons

To test the performance of the proposed DLBA for PFSP, computational simulations are carried out with some well studied problems taken from the OR-Library. In this paper, eight Car instances are selected. The DLBA is coded in MATLAB 2012a, simulation experiments are performed on a PC with Pentium 2.0 GHz Processor and 1.0 GB memory. Each instance is independently run 15 times for every algorithm for comparison. The parameter setting for BA and DLBA are listed in Table 1, the parameters of BA are recommended in original article, "--" represents that there isn't this parameter in algorithm. A variant of DLBA, original BA and DLBA are tested and compared, whose abbreviations are as following:

- DLBA-VPNS: DLBA without virtual population neighborhoods search;
- BA-VPNS: The part of DLBA is replaced by original BA in DLBA framework, the BA with NEH initialization but without virtual population neighborhoods search.

Table 1. The parameter set of BA and DLBA

	ps	f_{min}	f_{max}	L^i	R^i	α	γ	n_t
BA	40	0	100	(1,2)	(0,0.1)	0.9	0.9	--
DLBA	40	0	10	(1,2)	(0,0.1)	0.9	--	5000

4.1   Comparisons of DLBA, DLBA-VPNS and BA-VPNS

The statistical performances of DLBA, DLBA-VPNS and BA-VPNS are shown in Table 2. In this table, C_* is the optimal makespan known so far, *BRE*, represents the best relative error to C_*, *ARE*, denotes the average relative error to C_*, and *WRE* represents the worst relative error to C_*, these measuring criteria reference [3]. *Tavg* is the average time of 15 times run in seconds, where the stopping criterion for all algorithm is set to a maximum iteration of 100. *Std* is the standard deviation of the makespan.

From Table 2, it can be observed that the DLBA obtains much better solutions than DLBA-VPNS and BA-VPNS. The purpose of comparison between DLBA-VPNS and BA-VPNS is to show the effectiveness of Lévy-flight and differential operation. From the columns 9-18 in Table 2, for *BRE*, there are 2 instances of DLBA-VPNS is superior to BA-VPNS; for *ARE*, there are 5 instances of DLBA-VPNS is superior to BA-VPNS, and for *WRE*, the DLBA-VPNS is also better than BA-VPNS. The computation time of DLBA-VPNS and BA-VPNS is nearly same. The comparison between DLBA and DLBA-VPNS is to show the effectiveness of virtual population neighborhoods search. From the columns 4-13 in Table 3.

Table 2. Experimental results of DLBA, DLBA-VPNS and BA-VPNS

Instance n|m	C*	DLBA					DLBA-VPNS					BA-VPNS				
		BRE	ARE	WRE	Std	Tavg	BRE	ARE	WRE	Std	Tavg	BRE	ARE	WRE	Std	Tavg
Car1	11|5 7038	0	0	0	0	1.88	0	0	0	0	0.86	0	0	0	0	0.75
Car2	13|4 7166	0	0	0	0	2.07	0	0	0	0	0.94	0	0.391	2.93	173.89	0.90
Car3	12|5 7312	0	0.555	1.190	42.45	2.13	1.190	1.190	1.190	0	0.91	1.190	1.190	1.190	0	0.91
Car4	14|4 8003	0	0	0	0	1.98	0	0	0	0	0.99	0	0	0	0	0.85
Car5	10|6 7720	0	0	0	0	2.06	0	0.144	1.321	28.13	0.83	1.308	1.393	1.490	7.230	0.82
Car6	8|9 8505	0	0.459	0.764	12.96	1.47	0	0.644	2.787	59.25	0.77	0	0.510	0.764	31.72	0.74
Car7	7|7 6590	0	0	0	0	1.39	0	0	0	0	0.73	0	0	0	0	0.71
Car8	8|8 8366	0	0	0	0	1.52	0	0	0	0	0.77	0	0.473	0.645	24.72	0.75

4.2   Comparisons of Experimental Results

In order to show the effective of DLBA, we carry out a simulation to compare DLBA with other state-of-art algorithms, for example, IWO [5] HGA [6] and PSOVNS, experimental results reference [3] in previous papers. The experimental results of Car instances are listed in Table 3. From Table 3, we can find that the DLBA, PSOVNS, IWO and HGA all can find the better solution for Car instances, and many instances can find optimal solution. The DLBA can find the optimal solution on *BRE*. However, 6 instances can find optimal solution on *ARE*. Fig.1 and Fig.2 show the Gantt chart of optimal schedule for Car5 and Car 7.

Table 3. Comparisons of results for Car instances

Instance n|m	C*	DLBA			PSOVNS[3]			IWO[5]			HGA[6]		
		BRE	ARE	WRE	BRE	ARE	WRE	BRE	ARE	WRE	BRE	ARE	WRE
Car1	11|5 7038	0	0	0	0	0	0	0	0	0	0	0	0
Car2	13|4 7166	0	0	0	0	0	0	0	0.004	0.029	0	0.88	2.93
Car3	12|5 7312	0	0.555	1.190	0	0.420	1.189	0	0.004	0.011	0	1.05	1.19
Car4	14|4 8003	0	0	0	0	0	0	0	0	0	0	0	0
Car5	10|6 7720	0	0	0	0	0.039	0.389	0	0.003	0.013	0	0.17	1.45
Car6	8|9 8505	0	0.459	0.764	0	0.076	0.764	0	0.003	0.007	0	0.50	0.76
Car7	7|7 6590	0	0	0	0	0	0	0	0	0	0	0	0
Car8	8|8 8366	0	0	0	0	0	0	0	0	0	0	0.48	1.96

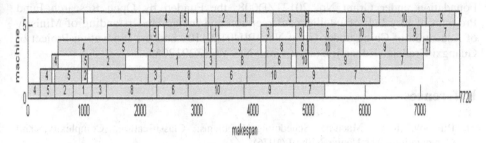

Fig. 1. Gantt chart of an optimal schedule for Car5, $\pi_* = [4,5,2,1,3,8,6,10,9,7]$

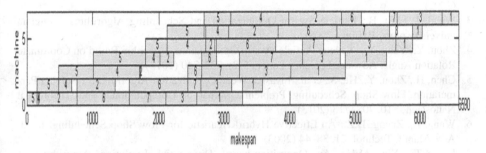

Fig. 2. Gantt chart of an optimal schedule for Car7, $\pi_* = [5,4,2,6,7,3,1]$

5   Conclusions

In this paper, a differential Lévy-flights bat algorithm (DLBA) is proposed to improve bat algorithm for solving PFSP. The differential operator is used to speed up the convergence rate, and Lévy-flights is used to increase the diversity of population, so that the algorithm can effectively jump out of local optimal, which can avoid algorithm is premature. PFSP is a typical discrete problem. However, the DLBA and BA is continuous algorithm. In order to solve this discrete problem, LOV rule is used to convert the continuous encoding to a discrete job permutation. In order to further improve the performance of the algorithm, the virtual population neighborhoods search is integrated into BA and DLBA. However, the virtual population neighborhoods search may expend more time to find a better solution. A good initial solution can decrease the computation time, thus an initial strategy with NEH heuristic is adopted. Experimental results and comparisons show the effectiveness of the proposed DLBA for PFSP. Moreover, due to the BA and DLBA is continuous algorithm, but the PFSP is a typical discrete problem, for these discrete problems and combinational optimization problem, such as vehicle routing problem, it is our further work to present a discrete bat algorithm.

Acknowledgment. This work is supported by National Science Foundation of China under Grant No. 61165015. Key Project of Guangxi Science Foundation under Grant No. 2012GXNSFDA053028, Key Project of Guangxi High School Science

Foundation under Grant No. 20121ZD008, the Funded by Open Research Fund Program of Key Lab of Intelligent Perception and Image Understanding of Ministry of Education of China under Grant No. IPIU01201100 and the Innovation Project of Guangxi Graduate Education under Grant No. YCSZ2012063.

References

1. Rinnooy, K.A.: Machine Scheduling Problems: Classification, Complexity, and Computations. The Hague, Nijhoff (1976)
2. Croce, F.D., Ghirardi, M., Tadei, R.: An Improved Branch-and-bound Algorithm for the Two Machine Total Completion Time Flow Shop Problem. Eur. J. Oper. Res. 139, 293–301 (2002)
3. Wand, L., Liu, B.: Particle Swarm Optimization and Scheduling Algorithms. Tsinghua university press, Beijing (2008)
4. Zhou, Y., Yang, Y.: A Mean Particle Swarm Optimization Algorithm Based on Coordinate Rotation Angle. Advanced Science Letters 5(2), 737–740 (2012)
5. Chen, H., Zhou, Y., He, S., et al.: Invasive Weed Optimization Algorithm for Solving Permutation Flow-Shop Scheduling Problem. Journal of Computational and Theoretical Nanoscience 10, 708–713 (2013)
6. Wang, L., Zheng, D.Z.: An Effective Hybrid Heuristic for Flow Shop Scheduling. Int. J. Adv. Manuf. Technol. 21, 38–44 (2003)
7. Li, X.T., Yin, M.H.: An Opposition-based Differential Evolution Algorithm for Permutation Flow Shop Scheduling Based on Diversity Measure. Advances in Engineering Software 55, 10–31 (2013)
8. Yang, X.S.: A New Metaheuristic Bat-Inspired Algorithm. Nature Inspired Cooperative Strategies for Optimization 284, 65–74 (2010)
9. Gandomi, A.H., Yang, X.S., Alavi, A.H., et al.: Bat Algorithm for Constrained Optimization Tasks. Neural Computing and Applications 22, 1239–1255 (2012)
10. Yang, X.S., Gandom, A.H.: Bat Algorithm: a Novel Approach for Global Engineering Optimization. Engineering Computations 29, 464–483 (2012)
11. Chechkin, A.V., Metzler, R., Klafter, J., et al.: Introduction to The Theory of Lévy Flights, Anomalous Transport: Foundations and Applications, pp. 129–162. Wiley-VCH (2008)
12. Yang, X.S., Deb, S.: Cuckoo Search via Lévy Flights. In: Proc. of World Congress on Nature & Biologically Inspired Computing, pp. 210–214 (2009)
13. Qian, B., Wang, L., Hu, R., et al.: A Hybrid Differential Evolution Method for Permutation Flow-shop Scheduling. The International Journal of Advanced Manufacturing Technology 38, 757–777 (2008)

Cloud Model Glowworm Swarm Optimization Algorithm for Functions Optimization

Qiang Zhou[1], Yongquan Zhou[2], and Xin Chen[1]

[1] College of Information Science and Engineering,
Guangxi University for Nationalities, Nanning, Guangxi, 530006, China
[2] Guangxi Key Laboratory of Hybrid Computation and IC Design Analysis,
Nanning, Guangxi, 530006, China
yongquanzhou@126.com

Abstract. For basic artificial glowworm swarm optimization algorithm has a slow convergence and easy to fall into local optimum, and the cloud model has excellent characteristics with uncertainty knowledge representation, an artificial glowworm swarm optimization algorithm based on cloud model is presented by utilizing these characteristics. The algorithm selects an optimal value of each generation as the center point of the cloud model, compares with cloud droplets and then achieves the better search value of groups which can avoid falling into the local optimum and can speed up the convergence rate of the algorithm. Finally, we use the standard function to test the algorithm. And the test results show that the convergence and the solution accuracy of our proposed algorithm have been greatly improved compared with the basic artificial glowworm swarm optimization algorithm.

Keywords: cloud model, glowworm swarm optimization algorithm, standard function, optimal value.

1 Introduction

In order to simulate the foraging or courtship behavior of glowworm swarm in nature, a novel bionic swarm intelligence optimization algorithm named Glowworm Swarm Optimization (GSO) [1] [2] had been designed. It was presented by the two Indian scholars Krishnanad K. N and D. Ghose. In recent years, GSO has been paid more and more attention and gradually becomes a new research focus in the field of computational intelligence. And it also has been successfully applied in the sensor noise test [3], the harmful gas leak positioning and simulation robot [4], etc.

Though GSO algorithm is more versatile and has a faster convergence rate, it is prone to reach a precocious state and lacks of precision. In order to make up for the defects of the basic GSO algorithm, a glowworm swarm optimization algorithm based on cloud model(CGSO) is presented by adopting the global and local search strategies of the GSO according to the tendentious stability and randomness characteristics of the cloud model. Compared to the basic GSO algorithm and the improved MMGSO [10], the overall optimization performance of the CGSO algorithm achieves a larger improvement and enhancement.

D.-S. Huang et al. (Eds.): ICIC 2013, LNAI 7996, pp. 189–197, 2013.

2   Glowworm Swarm Optimization Algorithm

In the basic GSO, each artificial glowworm is randomly distributed in the defined domain of objective function. They all carry the same luciferin, and have their local-decision range. Glowworms move and look for their neighbors, and then gather together. Finally, most of the glowworms will gather together in the position of the larger objective function value.

2.1   Deployment Stage

The aim in this phase is to let the glowworms evenly distribute within the defined domain of the objective function. At the same time, each glowworm carries the same initial brightness and sensing radius.

2.2   The Fluorescein Updating Stage

A glowworms' brightness represents whether the location is good or not. The more light brightness is, the better location is represented. And there are more possible to attract other glowworms to move towards this direction.

From the beginning of the next iteration, all glowworms' brightness will update. The formula for updating the glowworms' fluorescein value is described as follows:

$$l_i(t) = (1-\rho)l_i(t-1) + \gamma J(x_i(t)) \tag{1}$$

Among the formula, $l_i(t)$ is the i th glowworm's fluorescein value of the t th generation, $\rho \in (0,1)$ is a parameter to control the fluorescein value, γ is a parameter, and $J(x_i(t))$ is the value of fitness function.

2.3   Movement Stage

The glowworms will choose one direction of its neighbors at the movement stage. Each glowworm must consider two factors. The first one is the glowworms in its own radius area, and the second is the glowworms which are brighter than itself within its radius range.

We can use formula (2) to determine the number of glowworms within the decision range.

$$N_i(t) = \{ j : \|x_j(t) - x_i(t)\| < r_d^i ; l_i(t) < l_j(t) \} \tag{2}$$

where r_d^i is the decision radius of the i-th glowworm, $x_j(t)$ is the j-th glowworm's position of the t-th generation and $l_i(t)$ is the i-th glowworm's fluorescein value of the t-th generation.

The location updates according to the formula (3). Location updating formula:

$$x_i(t+1) = x_i(t) + s\left(\frac{x_j(t) - x_i(t)}{\|x_j(t) - x_i(t)\|}\right) \qquad (3)$$

Where s is a step size parameter of glowworm's movement.

2.4   Region Decision-Radius Updating

The purpose to update the radius is to make the regional decision radius be able to obtain adaptive adjustment with the change of the number of neighbors. If the neighbor density is small, the radius will increase, thereby the chance will increase to search the neighbors.

The decision radius updates according to the formula (4) after each iteration. Decision range updating formula is described as follows.

$$r_d^i(t+1) = \min\{r_s, \max\{0, r_d^i(t) + \beta(n_t - |N_i(t)|)\}\} \qquad (4)$$

Where $r_d^i(t+1)$ represents the decision radius of the i-th glowworm of t th generation after $t+1$ times iteration. r_s is sensing range. β is a constant to control adjacent glowworm variation range. n_t is a neighborhood threshold to control the number of adjacent glowworms. $N_i(t)$ is the number of glowworms with higher fluorescein value within the decision-making radius.

3   The Cloud-Model-Based Artificial Glowworm Swarm Optimization Algorithm

3.1   The Cloud Model

The cloud model introduced by Li De-yi in 1998 [6][7] is an uncertainty transformation model between a qualitative concept and its quantitative representation. The desired variable E_x, entropy E_n, and hyper entropy H are applied to represent the mathematical characteristics of the cloud. The fuzziness and randomness are integrated to constitute a mapping between the quantity and quality. Here, the center value of the cloud is the desired variable E_x. The entropy determines the width of the cloud distribution and the hyper entropy H determines cloud dispersion.

This paper uses MATLAB to generate an instance of cloud model C (18,2,0.2), which consists of 1000 cloud droplets. The cloud model converts the multiple information of the same concept to a single concept particle through three digital characteristics. It's time complexity is determined by the number of cloud droplets generated. Assume a cloud model generates n cloud droplets, and then its time complexity is $O(n)$.

The one-dimensional normal cloud algorithm steps are as follows:

(1). Use E_n as the desired value, H as Standard deviation and then generate a normal random number E_N.

(2). Use E_n as the desired value, the absolute value of E_N as the standard deviation to generate a normal random number x which is called the domain space cloud droplets.

(3). Repeat the steps (2) and (3), till produce N cloud droplets.

Due to the characteristics of stability in the uncertainty and varying in the stability of the cloud model [9]. Therefore, the cloud model is adopted.

3.2   The Normal-Cloud-Based Glowworm Swarm Optimal Search Strategy

The GSO algorithm looks for the glowworm which has a higher fluorescein value than itself. This process has a higher randomicity, and is not enough steady. The precision of search is difficult to increase when it searches a specific iteration times. Due to the steady and randomicity of the cloud model. This paper improves the glowworm's location updating so as to increase the local search ability of the algorithm and solution precision of GSO algorithm, and raise its convergence speed. Assume the location of current generation glowworm is X_i. There are three equations in the one dimension normal cloud $C(E_x, E_n, H)$, such as $E_x = X_i^j$, $E_n = et$, $H = E_n /10$ .Where, $j \in \{1,2,\cdots,m\}$ and et is a variable. Cloud droplets X_i^j generated through the normal cloud is the value of the j-th dimension. In order to guarantee the cloud droplets X_i^j in the definition domain, formula (5) is applied to process the cross-border of cloud droplets.

$$X_i^j = \begin{cases} a, & X_i^j < a \\ b, & X_i^j > b \end{cases} \tag{5}$$

Where, $X_i^j \in [a,b]$ a,b is the maximum and minimum of the definition domain respectively.

The lager the value E_n is, the more widely the cloud droplets distributes, otherwise, more narrowly. When the iteration reaches particular times, the value of the glowworm swarm is more close to the optimal value. So the nonlinear degression strategy is used to control the value of et to improved the precision of the solving [8].

$$et = -(E_{max} - E_{min})(t/T_{max})^2 + E_{max} \tag{6}$$

In the formula, we set the value of E_{max} and E_{min} according to the definition domain of the object function. And $t \in \{1,2,\cdots,T_{max}\}$, here t is current iteration times, and T_{max} is the maximum iteration times.

3.3   Implement Steps of the CGSO Algorithm

1). Initial the parameters $\rho, \gamma, \beta, s, l, m, n$ and the position, decision radius and the value of fluorescein of the glowworm.
2). Update the fluorescein value of each glowworm i referring to the formula (1).
3). Select the glowworm which satisfies the requirements according to the formula (2). Use the roulette way to choose the move direction of the glowworm j, and apply the formula (3) to update its location and process the glowworm's location which has been updated.
4). Compute the optimal value of the glowworms' position of the current generation. Use the optimal value as the center. Produce a particular amount of cloud droplets according to 2.2, and adopt the greedy algorithm to choose the best value of y cloud droplets. Then compare it with the optimal value of the glowworm of the current generation.
5). Update the decision radius of the glowworm referring to the formula (4).
6). Start the next iteration after one iteration, and judge whether it satisfy the condition or not, if it meets the needs, then quit the loop and record the result. Otherwise, turn to 2).

4   Simulation Experiment

In our simulation, MATLAB 7.0 is applied to compile the program codes. The experiment is operated on a PC with a Windows XP OS, a AMD Athlon 640 processor and a 4GB size memory.

4.1   Test Function and Initial Parameters [5]

In order to check the effectively of the algorithm, 8 base test functions are chose to compared with GSO algorithm and MMGSO. Which are as follows:

$$F_1 : f_1(x) = \sum_{i=1}^{m} x_i^2, -100 \le x_i \le 100 (i = 1.2, \cdots, m; m = 10),$$

The global minimum value is zero;

$$F_2 : f_2(x) = \sum_{i=1}^{m} (x_i^2 + 0.5)^2, -10 \le x_i \le 10 (i = 1, 2, \cdots, m; m = 10),$$

The global minimum value is zero;

$$F_3 : f_3(x) = -\exp(-0.5 \sum_{i=1}^{m} x_i^2), -1 \le x_i \le 1 (i = 1, 2, \cdots, m; m = 10),$$

The global minimum value is -1;

$$F_4 : f_4(x) = \sum_{i=1}^{m} ix_i^2, -5.12 < x < 5.12(i = 1, 2, \cdots, m; m = 10),$$

The global minimum value is zero;

$$F_5 : f_5(x) = [1 + (x_1 + x_2 + 1)^2 (19 - 14x_1 + 3x_1^2 - 14x_2 + 6x_1x_2 + 3x_2^2)]$$
$$[30 + (2x_1 - 3x_2)^2 (18 - 32x_1 + 12x_1^2 + 48x_2 - 36x_2 + 27x_2^2)],$$
$$-2 \le x_i \le 2, i = \{1, 2\}$$

The global minimum value in (0,0) is zero;

$$F_6 : f_6(x) = 0.5 - \frac{\sin\sqrt{x_1^2 + x_2^2} - 0.5}{[1 + 0.001(x_1^2 + x_2^2)]^2}, -100 \le x_i \le 100, i = \{1, 2\},$$

The global minimum value in (0,-1) is 3;

$$F_7 : f_7(x) = (x_1^2 + x_2^2)^{1/4}[\sin^2(50(x_1^2 + x_2^2)^{1/10}) + 1.0], -100 < x < 100,$$

The global minimum value in (0,0) is zero;

$$F_8 : f_8(x) = 10m + \sum_{i=1}^{m}[x_i^2 - 10\cos(2\pi x_i)], -5.12 < x_i < 5.12$$
$$(i = 1, 2, \cdots, m; m = 10),$$

The global minimum value is zero;
   Set the parameter of CGSO as follows:

$$\max t = 200, s = 0.03, n = 100, \beta = 0.08, \gamma = 0.6, l = 0.5, n_t = 5, y = 30.$$

4.2   Comparison of CGSO,GSO and MMGSO

In the experiment, functions $F_1 \sim F_8$ are tested for 20 times independently, the optimal value; the worst value and the average value are obtained. From the Table 1, we can see that the accuracy of the CGSO algorithm which is compared with the GSO and MMGSO algorithm has been significantly improved after optimizing the eight standard functions above. The CGSO algorithm is closer to the theoretically optimum value. In particularly, the algorithm has been greatly improved in accuracy on F_1, F_2, F_4 and F_7. F_1 reaches 10^{-2}, F_2 reaches 10^{-3}, F_4 reaches 10^{-4} and F_7 reaches 10^{-2}. From the test results, CGSO is More significantly superior to the basic GSO and MMGSO.

Table 1. Comparison of the results of the test functions

Function	Algorithm	Optimal value	Worst value	Average value
F_1	CGSO	4.3869369275E-02	3.5580026695E+00	1.0187481346E+00
	GSO	2.3588555795E+03	5.4519947600E+03	4.0954059282E+03
	MMGSO	2.4622496906E+ 03	3.5261024196E+ 03	3.0293045177E+ 03
F_2	CGSO	2.5625310318E-03	1.7153111934E-02	7.2781937529E-03
	GSO	2.5527377372E+01	5.8518932075E+01	4.2047405524E+01
	MMGSO	2.4511797054E+01	3.5948040784E+01	3.0402824004E+01
F_3	CGSO	-9.9999296438E-01	-9.9858049952E-01	-9.9979094498E-01
	GSO	-8.9889802232E-01	-7.5375454723E-01	-8.2389941922E-01
	MMGSO	-7.2261158393E-01	-6.9745197732E-01	-0.7108969951E-01
F_4	CGSO	2.0922935343E-04	1.9517179871E-02	1.6445397737E-03
	GSO	3.2499235773E+01	7.7112489815E+01	5.3377856031E+01
	MMGSO	2.7046904761E+01	2.8864106889E+01	2.8060252860E+01
F_5	CGSO	3.0000000435	3.0002932923	3.0000648932
	GSO	3.0039141969	5.4326001069	3.7716431499
	MMGSO	3.0000008381	3.0000066951	3.0000036463
F_6	CGSO	2.4558581811E-03	2.4561330150E-03	2.4559044092E-03
	GSO	2.4558595486E-03	1.5260097807E-01	5.4110901953E-02
	MMGSO	2.4560806E-03	3.2138358E-03	2.7391065E-03
F_7	CGSO	5.5270557680E-02	1.7868865702E-01	9.6506152596E-02
	GSO	1.0131463148	2.4575967511	1.7611353619
	MMGSO	0.5393870922	1.5793736647	1.0982276798
F_8	CGSO	2.8882505418E+01	5.9721144823E+01	4.2567094272E+01
	GSO	5.3984615100E+01	7.6788165292E+01	6.7553289291E+01
	MMGSO	5.4144331901E+01	6.8399067513E+01	6.3460667071E+01

Figure 3~10 is the convergence curves about the iteration times and the objective function value of CGSO, basic GSO and MMGSO[10].

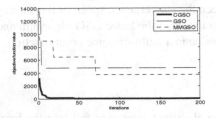 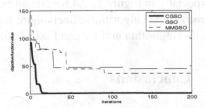

Fig. 1. The convergence of the function F_1 **Fig. 2.** The convergence of the function F_2

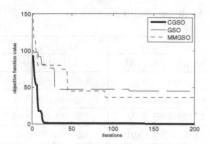

Fig. 3. The convergence of the function F_3

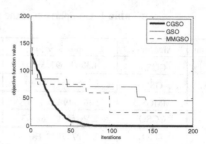

Fig. 4. The convergence of the function F_4

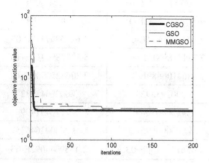

Fig. 5. The convergence of the function F_5

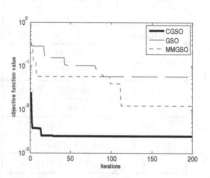

Fig. 6. The convergence of the function F_6

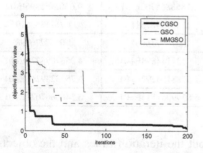

Fig. 7. The convergence of the function F_7

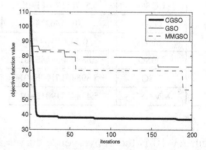

Fig. 8. The convergence of the function F_8

Especially in Figure 1, 2,4,6,8, it can be seen the convergence speed or the solving accuracy.CGSO algorithm effects more preferably than the basic GSO algorithm and MMGSO algorithm and we can find a better value in a multi-extreme point function.

5   Conclusions

For the basic glowworms swarm optimization algorithm is easy to fall into the local optimum, an artificial glowworms swarm optimization based on cloud model algorithm is presented in this paper. Experiments show that CGSO has a more rapid global search capability and can avoid premature convergence. In terms of solution precision and convergence rate, GSO algorithm has been improved.

Acknowledgment. This work is supported by National Science Foundation of China under Grant No. 61165015.Key Project of Guangxi Science Foundation under Grant No. 2012GXNSFDA053028, Key Project of Guangxi High School Science Foundation under Grant No. 20121ZD008.

References

1. Krishnan, K.N., Ghose, D.: Glowworm swarm optimization: a new method for optimizing multi-modal functions. Computational Intelligence Studies 1, 93–119 (2009)
2. Krishnanand, K.N.: Glowworm swarm optimization: a multimodal function optimization paradigm with applications to multiple signal source localization tasks. Indian Institute of Science, Indian (2007)
3. Krishnanand, K.N., Ghose, D.: A glowworm swarm optimization based multi-robot system for signal source localization. In: Liu, D., Wang, L., Tan, K.C. (eds.) Design and Control of Intelligent Robotic Systems. SCI, vol. 177, pp. 49–68. Springer, Heidelberg (2009)
4. Krishnanand, K.N., Ghose, D.: Multiple mobile signal sources:a gloworm swarm optimizationapproach. In: Proc. of the 3rd Indian International Conference on Artificial Intelligence (2007)
5. Liu, J.-K., Zhou, Y.-Q.: A parallel artificial glowworm algorithm With a master-slave structure. Computer Engineering and Applications 48, 33–38 (2012)
6. Li, D.-Y., Yang, Z.-H.: Two-dimensional cloud model and its application in predicting. Journal of Computers 21, 961–969 (1998)
7. Li, D.-Y., Meng, H.-J., Shi, X.-M.: Membership clouds and membership cloud generators. Computer Research and Development 32, 16–20 (1995)
8. Ye, D.-Y., Lin, X.-J.: The cloud variability artificial bee colony algorithm. Computer Application 32, 2538–2541 (2010)
9. Zhang, G.-W., He, R., Liu, Y.: Evolutionary algorithm based on cloud model. Journal of Computers 31, 1082–1091 (2008)
10. Liu, J.-K., Zhou, Y.-Q.: A artificial glowworm algorithm of a maximum and minimum fluorescein value. Application Research of Computers 28, 3662–3665 (2011)

An Improved Glowworm Swarm Optimization Algorithm Based on Parallel Hybrid Mutation

Zhonghua Tang[1], Yongquan Zhou[1,2], and Xin Chen[1]

[1] College of Information Science and Engineering,
Guangxi University for Nationalities, Nanning, Guangxi,530006, China
[2] Guangxi Key Laboratory of Hybrid Computation and IC Design Analysis,
Nanning, Guangxi, 530006, China
yongquanzhou@126.com

Abstract. Glowworm swarm optimization (GSO) algorithm is a novel algorithm based on swarm intelligence and inspired from light emission behavior of glowworms to attract a peer or prey in nature. The main application of this algorithm is to capture all local optima of multimodal function. GSO algorithm has shown some such weaknesses in global search as low accuracy computation and easy to fall into local optimum. In order to overcome above disadvantages of GSO, this paper presented an improved GSO algorithm, which called parallel hybrid mutation glowworm swarm optimization (PHMGSO) algorithm. Experimental results show that PHMGSO has higher calculation accuracy and convergence faster speed compared to standard GSO and PSO algorithms.

Keywords: GSO, PSO, Hybrid Mutation, Global Search.

1 Introduction

GSO algorithm was proposed by an Indian researcher in 2005[1]. GSO algorithm is of great ability in solving problems[6-7]. Nevertheless, this algorithm has some flaws, such as easy to fall into local optimum, low convergence rate and computational accuracy. In this paper, we presented PHMGSO algorithm to conquer above defects.

The following of this paper is organized as follows; section 2 introduces standard GSO. The proposed algorithm is introduced in section 3. In section 4, simulation results are presented and the final section is conclusions and future work.

2 The Standard GSO Algorithm

GSO algorithm consists of four phases: glowworms distribute phase, luciferin-update phase, glowworms movement and neighborhood range update phase[2].

2.1 Glowworms Distribute Phase

A set of n glowworms are randomly distributed in different location of the search space. All glowworms carry on an equal quantity luciferin value l_0.

D.-S. Huang et al. (Eds.): ICIC 2013, LNAI 7996, pp. 198–206, 2013.

2.2   Luciferin-Update Phase

The luciferin update depends on the objective function value and previous luciferin level. The luciferin update rule is given by:

$$l_i(t) = (1-\rho)l_i(t-1) + \gamma J(x_i(t)) \tag{1}$$

Where $l_i(t)$ denotes the luciferin value glowworm i at tth iteration; ρ and γ are the luciferin decay and enhancement constant,; $J(x_i(t))$ represents the value of the objective function at glowworm i's location.

2.3   Glowworms Movement Phase

During the movement step, each glowworm decides, using a probabilistic mechanism, to move toward a neighbor that has more luminosity than its own. For each glowworm i the probability of moving toward a neighbor j is given by

$$p_{ij}(t) = (l_j(t) - l_i(t)) / (\sum_{k \in N_i(t)} (l_k(t) - l_i(t))) \tag{2}$$

$$j \in N_i(t), N_i(t) = \{ j : d_{ij}(t) < r_d^i(t); l_i(t) < l_j(t) \} \tag{3}$$

where, $d_{ij}(t)$ denotes the Euclidean distance between glowworms i and j at time t , and $r_d^i(t)$ represents the variable neighborhood range associated with glowworm i at tth iteration. Then, the model of the glowworm movements can be stated as:

$$x_i(t+1) = x_i(t) + st * \left(\frac{x_j(t) - x_i(t)}{\|x_j(t) - x_i(t)\|} \right) \tag{4}$$

Where $x_i(t) \in R^m$ is the location of glowworm i at tth iteration, in the m-dimensional real space R^m, $\|\|$ represents the Euclidean norm operator, and s (> 0) is the step size.

2.4   Neighborhood Range Update Phase

With assuming r_0 as an initial neighborhood domain for each glowworm, at each of iterations of GSO algorithm, neighborhood domain of glowworm is given by:

$$r_d^i(t+1) = \min\{r_s, \max\{0, r_d^i(t) + \beta(n_t - N_i(t))\}\} \tag{5}$$

where, β is a constant, n_t is a parameter to control the number of neighborhoods , r_s denotes the sensory radius of glowworm, $N_i(t)$ denotes neighborhoods set.

3   Proposed Algorithm

GSO algorithm has shown some such weaknesses in global and high dimension search as slow convergence rate and low accuracy computation[3]. In PHMGSO, we take three improvement measures: Firstly, introduce the parallel hybrid mutation idea which combines the uniform distribution mutation with the Gaussian distribution mutation [4]. The former prompts a global search in a large range, while the latter searches in a small range with the high precision. Secondly, we adopt dynamic moving step length to every individual in proposed algorithm. Thirdly, we executive normal distribution variation to the glowworms if its' position unchanged in any generation.

3.1   Parallel Hybrid Mutation

The thought of parallel hybrid mutation derives form HSCPSO. In [4], the author has concluded that this model can be used to improve the diversity of the particle swarm. The process of parallel hybrid mutation is given as follows:

① Set the mutation capability value of each individual is given by:

$$mc_i = 0.05 + 0.45 \frac{(\exp(\ 5\,(i-1)\ /(\ ps\ -1)) - 1)}{\exp(\ 5) - 1} \tag{6}$$

Where, ps denotes the population size, i denotes the individual serial number, mc_i denotes the mutation capability value of ith individual.

② Choose the mode of the mutation via Algorithm 1.

ALGORITHM.1 The algorithm of choose mutation mode

```
for i=1:ps
    if ceil(mc_i+rand-1)==1
      If rand<=p_u
      X_i(t)=(1+rand)*x_i(t)
        Else
          X_i(t)=Gaussian( )*x_i(t)
      End if
End
```

In algorithm 1, p_u is called the mutation factor that denotes the ratio of the uniform distribution mutation and correspondingly, $1-p_u$ is the ratio of the Gaussian distribution mutation. The Gaussian(σ) returns a random number drawn from a Gaussian distribution with a standard deviation is σ. Ceil(p) rounds the elements of

p to the nearest integers greater than or equal to P. here we adopt linear mutation factor, the function is given as follow

$$p_u(t)=1-\frac{t}{gen} \tag{7}$$

Where, t denotes the current generation; gen denotes the maximum generation.

3.2   Dynamic Step Size

In GSO algorithm, step size is fixed value 0.03, which has been proved is very appropriate to capture all local optima of multimodal function [2]. However, in the global search process, move step size fixed 0.03 is unreasonable. The fixed value 0.03 can lead to the phenomenon that the glowworm at the global optimal stagnates in algorithm's initial phase. Because other glowworm need more steps to move to nearby the global optimal if the step size is too small. We have found this phenomenon in experiment. In order to conquer the above default, we adopted dynamic step size strategy, which let the initial moving step size value is 0.3 and it's value will decrease with the increase of the number of iterations via the formula(8). It is clear that the moving step value range is $[10^{-5}, 0.3]$. The minimum value of the moving step derives form the ASGSO [8]. Dynamic moving step size strategy can accelerate the convergence speed in the algorithm initial stage and improve the calculation accuracy in the algorithm later stage.

$$s(t) = st * (1 - \frac{t}{gen}) + 10^{-5} \tag{8}$$

Where, t denotes the current generation; gen denotes the maximum generation.

3.3   Normal Distribution Variation

In GSO algorithm, the glowworm that takes along the maximum luciferin in every generation is not change it's position. In proposed algorithm, glowworm will be forced to normal distribution variation according to the formula (9) if its position is not change in any generation.

$$x_i(t+1) = normrnd\ (0, st(t)) * x_i(t)) \tag{9}$$

Where, $normrnd(0, st(t))$ produce a normal distribution random number which mean is 0 standard deviation is $st(t)$. st is the step size.

3.4　PHMGSO Algorithm Pseudo

ALGORITHM.2 PHMGSO algorithm

```
Set number of dimensions=m
Set number of glowworms=n
Let s be the step size
Let x_i(t) be the location of glowworm i at tth iteration
```

%glowworms distribute phase;

```
for i=1 to n do  l_i(0) = l_0
```

$$r_i^d = r_0$$

```
set maximum iteration number=iter_max;
set t=1;
while (t<=iter_max) do:
{
    for each glowworm i do: % luciferin update phase
        luciferin update according to (1)
    for each glowworm i do: %glowworms movement phase
    {
        confirm the set of neighbors according to (3);
        compute the probability of movement according to
(2);
        select a neighbor j using probabilistic mechanism;
        glowworm i moves toward j according to (4);
        %glowworms movement phase
        update neighborhood range according to (5);
        %neighborhood range update phase
        if the position of glowworm i is not change
        normal distribution variation according to (9)
        % normal distribution variation
    }
    implementation algorithm 1; % parallel hybrid
mutation
    t=t+1;
}
```

4   Simulation Results and Comparison Analysis

4.1   Experimental Environment and Parameter Setting

Experimental platform: Processor(AMD II X4 640, 3.0 GHz), Memory(4.00 GB), Operating system(Windows7),Simulation language(Matlab 7.1). GSO algorithm parameters Setting: ρ =0.4, β =0.08, γ =0.6, n_t =5, l_0 =5, st =0.03. PHMGSO algorithm parameters: ρ =0.4, β =0.08, γ =0.6, n_t =5, l_0 =5, $st(0)$ =0.3. The standard deviation of Gaussian distribution σ=1. PSO parameters Setting[5]: accelerating factors are $c1$ =1.49445 and $c2$ =1.49445, speed range is $[-10,10]$. Population size and evolution generation is 100 and 500 respectively.

4.2   Test Function Expression and Domain

$$f_1(x) = \sum_{i=1}^{10} x_i^2 ; \qquad [-5.12, 5.12]$$

$$f_2(x) = \sum_{i=1}^{n-1} [100*(x_{i+1} - x_i^2) + (1 - x_i)^2]; \qquad [-2.048, 2.048]$$

$$f_3(x) = 10n + \sum_{i=1}^{10} [x_i^2 - 10*\cos(2\pi x_i)]; \qquad [-50, 50]$$

$$f_4(x) = 0.5 - ((\sin(\sqrt{x_1^2 + x_2^2}))^2 - 0.5)/(1 + 0.001*(x_1^2 + x_2^2))^2; \qquad [-1,1]$$

$$f_5(x) = -\exp(-0.5 \sum_{i=1}^{10} x_i^2); \qquad [-32.786, 32.786]$$

$$f_6(x) = -20\exp(-0.2\sqrt{\sum_{i=1}^{10} x_i^2} \cos(2\pi x_i)) + \exp(1) + 20; \qquad [-5.12, 5.12]$$

$$f_7(x) = \sum_{i=1}^{10} i*x_i^2; \qquad [-5.12, 5.12]$$

$$f_8(x) = \frac{1}{4000} \sum_{i=1}^{30} x_i^2 - \prod_{i=1}^{30,} \cos(\frac{x_i}{\sqrt{i}}) + 1; \qquad [-30, 30]$$

Above functions' optimal value are 0 except f5, which optimal value is -1.

4.3   The Results of Simulation and Analysis

For every test function, all algorithms were performed 30 times, the optimal value, worst value, mean and the standard deviation as shown in table 1 and 2. Convergence curves are shown in figure 2 to figure 9. From the table 2 and the convergence graphs, we can observe that PHMGSO be of higher calculation accuracy and convergence rate than GSO and PSO.

Table 1. The experimental data

ID	Algorithm	Optimal	Worst	Mean	Std.
f_1	PSO	6.90e-004	0.0562	0.0207	0.0174
	GSO	0.1673	2.1405	0.8709	0.48760
	PHMGSO	**0**	**2.6e-010**	**1.8e-011**	**6.1e-011**
f_2	PSO	7.81	12.06	9.81	1.099
	GSO	7.962	23.357	12.339	3.953
	PHMGSO	**8.777**	**8.998**	**8.953**	**0.066**
f_3	PSO	18.091	49.039	33.324	8.474
	GSO	17.931	45.595	33.484	6.936
	PHMGSO	**0**	**3.4e-008**	**2.7e-009**	**8.6e-009**
f_4	PSO	0.010	0.010	0.010	5.6e-007
	GSO	0.336	0.400	0.377	0.018
	PHMGSO	**0.001**	**0.029**	**0.004**	**0.006**

Table 2. The experimental data (Continued)

ID	Algorithm	Optimal	Worst	Mean	Std.
f_5	PSO	-0.999	-0.998	-0.999	4.9e-004
	GSO	-0.991	-0.999	-0.998	0.002
	PHMGSO	**-1**	**-1**	**-1**	**0**
f_6	PSO	0.265	2.222	0.953	0.616
	GSO	18.049	19.607	18.762	0.433
	PHMGSO	**8.8e-016**	**1.2e-005**	**1.1e-006**	**3.6e-006**
f_7	PSO	0.020	0.217	0.101	0.070
	GSO	0.749	12.933	4.2071	2.954
	PHMGSO	**0**	**1.3e-009**	**1.0e-010**	**3.3e-010**
f_8	PSO	7.5e-005	3.2e-004	2.2e-004	7.4e-005
	GSO	0.183	0.485	0.341	0.066
	PHMGSO	**0**	**1.0e-013**	**6.9e-015**	**2.4e-014**

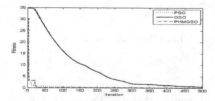

Fig. 1. Convergence Graph of f1

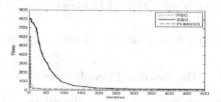

Fig. 2. Convergence Graph of f2

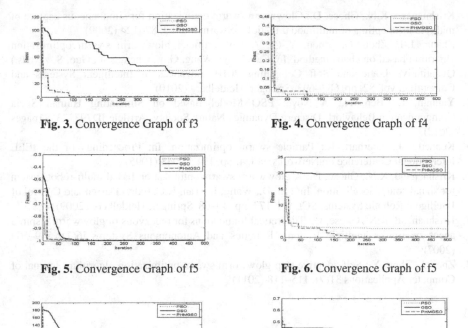

Fig. 3. Convergence Graph of f3

Fig. 4. Convergence Graph of f4

Fig. 5. Convergence Graph of f5

Fig. 6. Convergence Graph of f5

Fig. 7. Convergence Graph of f7

Fig. 8. Convergence Graph of f8

5 Conclusion

In this paper, a novel algorithm called PHMGSO was presented. Experimental results show that PHMGSO is more effective than GSO in global search. In future works, we will work to solve application model via PHMGSO.

Acknowledgment. This work is supported by National Science Foundation of China under Grant No. 61165015. Key Project of Guangxi Science Foundation under Grant No. 2012GXNSFDA053028, Key Project of Guangxi High School Science Foundation under Grant No. 20121ZD008, the Funded by Open Research Fund Program of Key Lab of Intelligent Perception and Image Understanding of Ministry of Education of China under Grant No. IPIU01201100 and the Innovation Project of Guangxi Graduate Education under Grant No. gxun-chx2012103.

References

1. Krishnanand, K.N., Ghose, D.: Detection of multiple source locations using a glowworm metaphor with applications to collective robotics. In: Proceedings of IEEE Swarm Intelligence Symposium, pp. 84–91 (2005)

2. Krishnanand, K.N., Ghose, D.: Glowworm swarm optimization for simultaneous capture of multiple local optima of multimodal fuctions. Swarm Intell 3, 87–124 (2009)
3. Zhang, J.-l., Zhou, G., Zhou, Y.-Q.: A new artificial glowworm swarm optimization algorithm based on chaos method. In: Cao, B.-y., Wang, G.-j., Chen, S.-l., Guo, S.-z. (eds.) Quantitative Logic and Soft Computing 2010. Advances in Intelligent Systems and Computing, vol. 82, pp. 683–693. Springer, Heidelberg (2010)
4. Yanmin, L., Ben, N.: A Novel PSO Model Based on Simulating Human Socia Communication Behaviorl. Discrere Dynamics Nature Society, Artilcle ID797373, 21pages (2012)
5. Kennedy, J., Eberhart, R.: Particle swarm optimization. In: Proceedings of the IEEE International Conference on Networks, vol. 4, pp. 1942–1948 (1995)
6. Krishnanand, K.N., Ghose, D.: A glowworm swarm optimization based multi-robot system for signal source localization. In: Liu, D., Wang, L., Tan, K.C. (eds.) Design and Control of Intelligent Robotic Systems. SCI, vol. 177, pp. 49–68. Springer, Heidelberg (2009)
7. Krishnanand, K.N., Ghose, D.: Theoretical foundations for redezvous of glowworm inspired agent swarms at multiple lcations. Robotics and Autonomous Systems 56(7), 549–569 (2007)
8. Zhe, O., Zhou, Y.: Self-adaptive step glowworm swarm optimization algorithm. Journal of Computer Applications 31(7), 115–118 (2011)

Bat Algorithm with Recollection

Wen Wang, Yong Wang, and Xiaowei Wang

College of Information Science and Engineering, Guangxi University for Nationalities,
Nanning, Guangxi, 530006, China
wangygxnn@sina.com

Abstract. Bat algorithm(BA) is a new swarm intelligence optimization algorithm. However, bat algorithm has the obvious phenomenon of the premature convergence problem and is easily trapped into local optimum. In order to overcome the shortcoming of the BA algorithm, we proposed an improved bat algorithm called bat algorithm with recollection(RBA). Experiment were conducted on some benchmark functions. The experimental results show that the RBA can effectively avoid the premature convergence problem and has a good performance of global convergence property.

Keywords: Bat algorithm(BA), Bat algorithm with recollection(RBA), disturbance factor.

1   Introduction

More and more scholars[1-12] have paid their attention to the swarm intelligent optimization based on the bionics, and have successfully put forward various meta heuristic swarm intelligent optimization inspired from the biological behavior characteristics (such as Particle Swarm Optimization(PSO)[1] presented by Kennedy and Eberhart, Ant Colony Algorithm[5] developed by Dorigo and Maniezzo, Artificial Immune System[6] proposed by Jiao Licheng et al, Artificial Fish Swarm Algorithm (AFSA)[7] presented by Li Xiaolei et al, GSO[9] presented by Krishnanand et al, Bat Algorithm(BA)[12] was presented by X.-S.Yang, etc). They are use some biological behavior as the search modes of optimization algorithm.

Bat Algorithm (BA) is based on the searching for food of bats. Although, BA has a characteristic of simplicity in design and easy to be implemented, it has the obvious phenomenon of the premature convergence problem, which is existing in most of the stochastic optimization algorithms, there are some deficiencies in BA: each component of bat's flight-velocity uses the same frequency increment, which makes most of bats' flight behavior lack of flexibility, and will make most of bats unable avoid the attack of their natural enemies and unable avoid itself being immersed into blind alley finally, it means that the performance of the algorithm will easily be trapped into local extremum. That the algorithm does not take it into consideration that the position of the prey is random changing with time. That is to say, each bat in BA flies directly to the position of its prey and does not consider the factor that the

D.-S. Huang et al. (Eds.): ICIC 2013, LNAI 7996, pp. 207–215, 2013.
© Springer-Verlag Berlin Heidelberg 2013

location of its prey has changed before it has arrived at. These mean that usually bats can not capture its prey in such manner, which implys the slow convergence of BA and difficulty of finding global optimum.

In order to overcome to the shortcomings of the BA, we put forward an improved bat algorithm called bat algorithm with recollection(RBA) in this paper.

2   Bat Algorithm

BA uses the following approximate or idealized rules: ① All bats use echolocation to sense distance, and they also 'know' the difference between food/prey and background barriers in some magical way; ② Bats fly randomly with velocity V_i at position X_i with a frequency f_{min}, varying wavelength λ and loudness A_0 to search for prey. They can automatically adjust the wavelength (or frequency) of their emitted pulses and adjust the rate of pulse emission $\gamma \in (0,1)$, depending on the proximity of their target; ③ Assume that the loudness varies from a large (positive) A_0 to a minimum constant value A_{min}.

In addition to above simplified assumptions, bat algorithm also assumes that the frequency f is in a range $[f_{min}, f_{max}]$.

Movement of Bats: In bat algorithm, as for the i bat, if its position and its velocity are X^{t-1} and V^{t-1} respectively at time step $t-1$, then at time step t, its new position X_i and new velocity V_i are updated according to the following formulas respectively

$$f_i = f_{min} + (f_{max} - f_{min})\beta \tag{1}$$

$$V_i^t = V_i^{t-1} + (X_i^{t-1} - X_*)f_i \tag{2}$$

$$X_i^t = X_i^{t-1} + V_i^t \tag{3}$$

Where $\beta \in [0,1]$ is a random vector drawn from a uniform distribution. Here X_* is the current global best location (solution) which is located after comparing all the solutions among the swarm of bats at each iteration t.

For the local search part, once a solution is selected among the current best solutions, a new solution for each bat is generated locally using random walk

$$X_{new} = X_{old} + \varepsilon A^t \tag{4}$$

where $\varepsilon \in [-1,1]$ is a random number, while $A^t =< A_i^t >$ is the average loudness of all the bats at this time step.

Loudness and Pulse Emission: In bat algorithm, as for the i bat, its loudness A_i and its rate γ_i of pulse emission are updated according to the following formulas

$$A_i^{t+1} = \alpha A_i^t \qquad r_i^{t+1} = r_i^0[1-\exp(-\gamma t)] \tag{5}$$

where α and γ are constants. For any $0 < \alpha < 1$ and $0 < \gamma$, we have

$$A_i^t \to 0, \quad \gamma_i^t \to \gamma_i^0, \quad \text{as } t \to \infty \tag{6}$$

3   Bat Algorithm with Recollection

Analyzing the main strategy which are used by BA, we found that there are some shortcomings existing in BA: ① That it does not be taken into consideration that the position of the prey is random changing with time continuously. ② That the formula (1) and formula (2) have restricted the bat's flight-abilities and flight-modes, and has leaded to search abilities being reduced finally.

3.1   Algorithm description

Through careful observation, we find: ① In real world, bats have superb flight-skills, they can use various flight modes to avoid them being immersed into blind alley. ② Bats usually have the function of conditioned reflex(we look it as a sort of a function of recollection) to their foraging history.

   According to the flight characteristics of bats described above, we put forward an improved bat algorithm called bat algorithm with recollection as follows.

   Let $X_i^{t-1} = (x_{i1}^{t-1}, \cdots, x_{in}^{t-1})$ and $V_i^{t-1} = (v_{i1}^{t-1}, \cdots, v_{in}^{t-1})$ be the position and velocity for bat i at time $t-1$, respectively, and the current best location is $X_* = (x_{*1}, \cdots, x_{*n})$.

   Firstly, according to the condition that bat i can use superb flight-skills to search for food in its search space, we design its flight mode and its emission frequency as follows:

$$f_{ij} = f_{min} + (f_{max} - f_{min})\beta_{ij} \tag{7}$$

$$v_{ij}^t = w(t)v_{ij}^{t-1} + (x_{ij}^{t-1} - x_{*j})f_{ij}, \ j = 1, \cdots, n. \tag{8}$$

In which, β_{ij} is a random number in the range $[0,1]$, f_{min} and f_{max} are the minimum frequency and maximum-frequency respectively, $w(t) = w_{min} + (w_{max} - w_{min})\exp(-\rho(\frac{t}{T_{max}})^2)$, w_{max} and w_{min} are the maximum value and the minimum value of $w(t)$ respectively, $1 \le \rho \le T_{max}$, T_{max} is the maximum iterations. $w(t)$ is the time-varying velocity inertia weight factor. It makes bat can smoothly close to the current best position X_*.

Secondly, we design the update of the position as follows

$$x_{ij}^t = x_{ij}^{t-1} + (1 - \frac{t}{T_{max}})^\theta v_{ij}^t , \quad j = 1, \cdots, n \tag{9}$$

In which, θ is a constant, $1 < \theta \leq M$, M is the population size. As for the bat i, the component $(1 - \frac{t}{T_{max}})^\theta$ control bats search range in the later period in order to make them close to the best position as soon as possible.

Thirdly, we will discuss the problem of disturbance caused by time-delay as follows.

Suppose a bat has found its prey at the location X_*^{t-1} at time $t-1$, then the prey is likely to move to the position $X_*^{t-1} + \gamma$ at time t, where $\gamma = (\gamma_1, \cdots, \gamma_n)$, γ_j is a random number in the range $[-\varepsilon, \varepsilon]$, ε is a small positive. If X_*^{t-1} is the current best location found by the swarm of bats till time $t-1$, then the prey which is at the point X_*^{t-1} at time $t-1$ is likely to move to the new position X_*^t at time t, where

$$X_*^t = X_*^{t-1} + (1 - g_i(t))^2 \cdot \mu \cdot X_*^{t-1} \tag{10}$$

$g_i(t) = (2\pi)^{-\frac{1}{2}} \exp(-\frac{|f(X_i^{t-1}) - f(X_*)|^2}{\sigma^2})$, $\mu = (\mu_1, \cdots, \mu_n)$ is a random vector, μ_j is a random number in the range $[-\varepsilon, \varepsilon]$, ε is a small positive. In this paper, we call the component $\mu(1 - g_i(t))^2$ to be the time-delay disturbance-factor.

3.2 The Pseudo Code of the RBA

```
Objective function f(x), x = (x_1,...,x_d)^T

Initialize the bat population X_i(i = 1,2,...,n) and V_i

Define pulse frequency f_i at X_i

Initialize pulse rates γ_i and the loudness A_i

While (t <Max number of iterations)

Generate new solutions by adjusting frequency,
and updating velocities and locations/solutions
[equations (7) to (9)]

if (rand > γ_i)

Select a solution among the best solutions
```

```
Generate a local solution around the selected best
solution use formula(4)
end if
Generate a new solution by flying randomly
if (rand < A_i & f(X_i)< f(X_*))
Accept the new solutions
Increase γ_i and reduce A_i use formula(5)
end if
disturbing the current best X_* [equations (10)]
Rank the bats and find the current best X_*
end while
```

4   Validation and Comparison

In order to test the performance of the RBA, we have tested it against the original BA[12]. For the ease of visualization, we have implemented our simulations using Matlab for various test functions.

4.1   Benchmark Functions

For the sake of having a fair and reasonable comparison between the RBA and the original BA, we have chosen five well-known high-dimensional functions and one 2 dimension function as our optimization simulation tests.

(a) Sphere function

$$f_1(x) = \sum_{i=1}^{50} x_i^2, -10 \le x_i \le 10 .$$

It is a unimodal function, and its global minimum $f_{min} = 0$ at $(0,\cdots,0)$.

(b) Rastrigin function

$$f_2(x) = \sum_{i=1}^{50} [x_i^2 - 10\cos(2\pi x_i) + 10], -5.12 \le x_i \le 5.12 .$$

The function is a complex multimodal function, When attempt to solve Rastrigin's function, algorithms may easily fall into a local optimum. It has a global minimum $f_{min} = 0$ at $(0,\cdots,0)$.

(c) Griewank function

$$f_3(x) = \frac{1}{4000} \sum_{i=1}^{50} x_i^2 - \prod_{i=1}^{50} \cos \frac{x_i}{\sqrt{i}} + 1, -600 \le x_i \le 600.$$

Griewank's function has a $\prod_{i=1}^{D} \cos(\frac{x_i}{\sqrt{i}})$ component causing linkages among variables, thereby making it difficult to reach the global optimum. The function has a global minimum $f_{min} = 0$ at $(0, \cdots, 0)$.

(d) Rosenbrock function

$$f_4(x) = \sum_{i=1}^{D-1} (100(x_{i+1} - x_i^2)^2 + (x_i - 1)^2), -30 \le x_i \le 30, D = 50.$$

Its global minimum is $f_{min} = 0$ at $x_i = 0, i = 1, \cdots, 50$.

(e) Ackley function

$$f_5(x) = 20 + e - 20 \exp(-0.2 \sqrt{\frac{1}{50} \sum_{i=1}^{50} x_i^2}) - \exp(\frac{1}{50} \sum_{i=1}^{50} \cos(2\pi x_i)), -30 \le x_i \le 30.$$

The function has one narrow global optimum basin and many minor local optima, and has a global optimum $f_{min} = 0$ at $(0, \cdots, 0)$.

(f) Eggcrate function

$$f_6(x) = x^2 + y^2 + 25(\sin^2 x + \sin^2 y), -2\pi \le x, y \le 2\pi.$$

The function has one global minimum $f_{min} = 0$ at the point $(0, 0)$.

4.2   Comparison of Experimental Results

In our simulations, We run each algorithms for 50 times so that we can do reasonable and meaningful analysis. In order to ensure the comparability of the experimental results, we let the parameter settings as consistent as possible for the RBA and the BA. For the two algorithms, the population size is set at 40 for all simulations. We set the maximum iterations at 100. Other coefficients are set as follows: the frequency f is in the range $[f_{min}, f_{max}]$, where $f_{min} = 0$, $f_{max} = 100$, and $\alpha = \gamma = 0.9$.

In our experiment, we select the best fitness value(BFV), the worst fitness value(WFV), the mean value(Mean), and the standard deviation(STDEV) as the evaluation indicators of optimization ability. For these evaluation indicators can not only reflect the optimization ability but also indicate the computing cost. We got the experimental results being listed in Table 1 as follows.

Table 1. Experimental results

function	algorithms	BFV	WFV	Mean	STDEV
f_1	BA	9.052151e+002	1.338033e+003	1.188659e+003	87.2966964
	RBA	8.857428e-012	8.250480e-004	2.300737e-005	1.178076e-004
f_2	BA	6.621783e+002	8.266144e+002	7.497180e+002	37.7913856
	RBA	1.776321e-010	4.784255e+002	1.007481e+002	1.681055e+002
f_3	BA	8.462324e+002	1.279952e+003	1.090307e+003	87.7748478
	RBA	6.727141e-011	1.0196547	0.09946002	0.2427918
f_4	BA	3.153734e+010	5.948311e+010	4.605235e+010	5.912543e+009
	RBA	48.9950846	49.0117029	48.9962432	0.0029465
f_5	BA	1.6980805	1.6981405	1.6980893	1.111032e-005
	RBA	1.6980804	1.6980806	1.6980805	2.455807e008
f_6	BA	8.052535e-004	21.0237849	9.0033055	6.5973327
	RBA	1.700144e-013	9.4881973	2.2771674	4.0933896

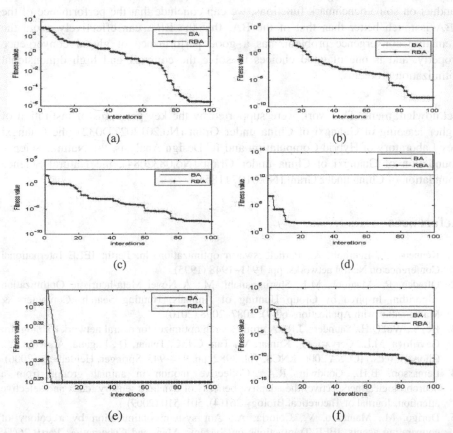

Fig. 1. The median convergence characteristics of test functions. (a) Sphere function. (b) Rastrigrin function. (c) Griewank function. (d) Rosenbrock function. (e) Ackley function. (f) Eggcrate function.

In order to more easily to contrast the convergence characteristics of the two algorithms, Fig. 1 presents the convergence characteristics in term of the best fitness value of the mean run of each algorithm for each test function.

Discussion: It can be see from Table1, every evaluation item of RBA is smaller than BA. By analyzing the results listed in the Table1, and the convergence curve simulation diagram Fig. 1, we can see that the RBA perform much better than the BA, that the RBA is much superior to the BA in terms of global convergence property, accuracy and efficiency. Therefore, we conclude that the performance of the RBA is much better than that of the BA.

5   Conclusions

In order to overcome the shortcoming of bat algorithm, we presents an improved bat algorithm called bat algorithm with recollection(RBA). From the experiment we conduct on some benchmark functions, we can conclude that the performance of the RBA is much better than that of the BA, that the RBA can effectively avoid the premature convergence problem, has a good performance of global convergence property, and is one of good choices to solve the complex and high dimensional optimization problems.

Acknowledgment. This work were supported by the key programs of institution of higher learning of Guangxi of China under Grant (No.201202ZD032), the Guangxi Key Laboratory of Hybrid Computation and IC Design Analysis, the Natural Science Foundation of Guangxi of China under Grant (No.0832084), the Natural Science Foundation of China under Grant (No.61074185).

References

1. Kennedy, J., Eberhort, R.: Particle swarm optimization. In: Perth: IEEE International Conference on Neural networks, pp. 1941–1948 (1995)
2. Oftadeh, R., Mahjoob, M.J., Shariatpanahi, M.: A Novel Meta-heuristic Optimization Algorithm Inspired by Group Hunting of Animals: Hunting Search. Computers & Mathematics with Applications 60(7), 2087–2098 (2010)
3. He, S., Wu, Q.H., Saunders, J.R.: A group search optimizer for neural network training. In: Gavrilova, M.L., Gervasi, O., Kumar, V., Tan, C.J.K., Taniar, D., Laganá, A., Mun, Y., Choo, H. (eds.) ICCSA 2006. LNCS, vol. 3982, pp. 934–943. Springer, Heidelberg (2006)
4. Lemasson, B.H., Goodwin, R.A.: Collective motion in animal groups from a neurobiological perspective: the adaptive benefits of dynamic sensory loads and selective attention. Journal of Theoretical Biology 261(4), 501–510 (2009)
5. Dorigo, M., Maniezzo, V., Coloria, A.: Ant system: optimization by a colony of cooperating agents. IEEE Transactions on Systems, Man, and Cybernetics: PartB 26(1), 29–41 (1996)
6. Jiao, L.C., Wang, L.: Anovel genetic algorithm based on immunity. IEEE Transaction on System, Man and Cybernetic 30(5), 552–561 (2000)

7. Li, X.-L., Qian, J.-X.: An optimizing method based on autonomous animals: fish-swarm algorithm. Systems engineering theory and Practice 22(11), 32–38 (2002)
8. Huang, D.-S., Zhang, X., Reyes García, C.A., Zhang, L. (eds.): ICIC 2010. LNCS, vol. 6216. Springer, Heidelberg (2010)
9. Krishnanand, K.N., Ghose, D.: Glowworm swarm based optimization algorithm for multimodel functions with collective robotics applications. Multiagent and Grid Systems 2(3), 209–222 (2006)
10. Chen, J.-R., Wang, Y.: Using fishing strategy optimization method. Computer engineering and Applications 45(9), 53–56 (2009)
11. Yang, X.-S., Deb, S.: Cuckoo search via Levy flights. In: Proc. of World Congress on Nature & Biologically Inspired Computing (NaBIC 2009), pp. 210–214. IEEE Publications, India (2009)
12. Yang, X.-S.: A new metaheuristic bat-inspired algorithm. In: González, J.R., Pelta, D.A., Cruz, C., Terrazas, G., Krasnogor, N. (eds.) NICSO 2010. SCI, vol. 284, pp. 65–74. Springer, Heidelberg (2010)

An Adaptive Bat Algorithm

Xiaowei Wang, Wen Wang, and Yong Wang

College of Information Science and Engineering, Guangxi University for Nationalities,
Nanning, Guangxi, 530006, China
wangygxnn@sina.com

Abstract. After analyzing the deficiencies of bat algorithm (BA), we proposed an improved bat algorithm called an adaptive bat algorithm(ABA). In the ABA, each bat can dynamic and adaptively adjust its flight speed and its flight direction while it is searching for food, and makes use of the hunting approach of combining random search with shrinking search. The experimental results show that the ABA not only has marked advantage of global convergence property but also can effectively avoid the premature convergence problem.

Keywords: Bat algorithm(BA), optimization, adaptive bat algorithm(ABA), premature convergence.

1 Introduction

More and more scholars[1-12] have paid their attention to the swarm intelligent optimization based on the bionics, and have successfully put forward various meta heuristic swarm intelligent optimization inspired from the biological behavior characteristics (such as Particle Swarm Optimization(PSO)[1] presented by Kennedy and Eberhart, Ant Colony Algorithm[5] developed by Dorigo and Maniezzo, Artificial Immune System[6] proposed by Jiao Licheng et al, Artificial Fish Swarm Algorithm (AFSA)[7] presented by Li Xiaolei et al, GSO[9] presented by Krishnanand et al, Bat Algorithm(BA)[12] was presented by X.-S.Yang, etc). They are use some biological behavior as the search modes of optimization algorithm.

Bat Algorithm (BA) is based on the searching for food of bats. Although, BA has a characteristic of simplicity in design and easy to be implemented, it has the obvious phenomenon of the premature convergence problem, which is existing in most of the stochastic optimization algorithms, there are some deficiencies in BA: each component of bat's flight-velocity uses the same frequency increment, which makes most of bats' flight behavior lack of flexibility, and will make most of bats unable avoid the attack of their natural enemies and unable avoid itself being immersed into blind alley finally, it means that the performance of the algorithm will easily be trapped into local extremum. That the algorithm does not take it into consideration that the position of the prey is random changing with time. That is to say, each bat in BA flies directly to the position of its prey and does not consider the factor that the

D.-S. Huang et al. (Eds.): ICIC 2013, LNAI 7996, pp. 216–223, 2013.
© Springer-Verlag Berlin Heidelberg 2013

location of its prey has changed before it has arrived at. These mean that usually bats can not capture its prey in such manner, which implys the slow convergence of BA and difficulty of finding global optimum.

In order to overcome to the shortcomings of the BA, we put forward an improved bat algorithm called bat algorithm with recollection(RBA) in this paper.

2   Bat Algorithm

BA uses the following approximate or idealized rules: ①All bats use echolocation to sense distance, and they also 'know' the difference between food/prey and background barriers in some magical way; ②Bats fly randomly with velocity V_i at position X_i with a frequency f_{min}, varying wavelength λ and loudness A_0 to search for prey. They can automatically adjust the wavelength (or frequency) of their emitted pulses and adjust the rate of pulse emission $\gamma \in (0,1)$, depending on the proximity of their target; ③Assume that the loudness varies from a large (positive) A_0 to a minimum constant value A_{min}.

In addition to above simplified assumptions, bat algorithm also assumes that the frequency f is in a range $[f_{min}, f_{max}]$.

Movement of Bats: In bat algorithm, as for the i bat, if its position and its velocity are X^{t-1} and V^{t-1} respectively at time step $t-1$, then at time step t, its new position X_i and new velocity V_i are updated according to the following formulas respectively

$$f_i = f_{min} + (f_{max} - f_{min})\beta \tag{1}$$

$$V_i^t = V_i^{t-1} + (X_i^{t-1} - X_*)f_i \tag{2}$$

$$X_i^t = X_i^{t-1} + V_i^t \tag{3}$$

Where $\beta \in [0,1]$ is a random vector drawn from a uniform distribution. Here X_* is the current global best location (solution) which is located after comparing all the solutions among the swarm of bats at each iteration t.

For the local search part, once a solution is selected among the current best solutions, a new solution for each bat is generated locally using random walk

$$X_{new} = X_{old} + \varepsilon A^t \tag{4}$$

where $\varepsilon \in [-1,1]$ is a random number, while $A^t = <A_i^t>$ is the average loudness of all the bats at this time step.

Loudness and Pulse Emission: In bat algorithm, as for the i bat, its loudness A_i and its rate γ_i of pulse emission are updated according to the following formulas

$$A_i^{t+1} = \alpha A_i^t \qquad r_i^{t+1} = r_i^0[1-\exp(-\gamma t)] \tag{5}$$

where α and γ are constants. For any $0 < \alpha < 1$ and $0 < \gamma$, we have

$$A_i^t \to 0, \quad \gamma_i^t \to \gamma_i^0 \ , \quad \text{as } t \to \infty \tag{6}$$

3   Adaptive Bat Algorithm

By analyzing the main strategy used by the original bat algorithm (BA), we found that there are some shortcomings existing in BA: That it does not be taken into consideration that the position of the prey is continuously changing with time. That the formula (1) and formula (2) have restricted the bat's flight-modes and flight-skills. So we will improve the original bat algorithm (BA) from the following aspects: We will improve each bat's flight-skills and flight-modes, let each bat dynamic and adaptively adjust its flight speed and its flight direction. Considering that the location of the prey is continuous changing with time, we will make each bat use the hunting approach of combining random search with shrinking search.

3.1   Algorithm Description

According to the flight characteristics of bats described above, we will present an improved bat algorithm called an adaptive bat algorithm as follows.

Let $X_i^{t-1} = (x_{i1}^{t-1}, \cdots, x_{in}^{t-1})$ and $V_i^{t-1} = (v_{i1}^{t-1}, \cdots, v_{in}^{t-1})$ be the position and velocity for bat i at time $t-1$, respectively, and $X_*^{t-1} = (x_{*1}^{t-1}, \cdots, x_{*n}^{t-1})$ be the current best location found by the swarm till time instant $t-1$.

Firstly, we designed the ABA as follows: That the flight speed of a bat has something to do with the distance between the bat and its prey. That is to say, the farther the distance between the bat and its prey, the faster the speed flying to its prey. According to flight characteristics described above, we designed bat's flight-modes as follows:

$$f_{ij} = f_{\min} + (f_{\max} - f_{\min})\beta_{ij} \tag{7}$$

$$\omega_{ij}^t = \omega_0(1 - \exp(-\mu \,|\, x_{ij}^{t-1} - x_{*j}^{t-1} \,|)) \tag{8}$$

$$v_{ij}^t = \omega_{ij}^t \cdot v_{ij}^{t-1} + (x_{ij}^{t-1} - x_{*j}^{t-1})f_{ij} \tag{9}$$

$$x_{ij}^t = x_{ij}^{t-1} + v_{ij}^t, \quad j = 1, \cdots, n \tag{10}$$

In which, μ is a positive constant, ω_0 is a positive constant and $1 \le \omega_0$. In our simulations, we set $\mu \in [1,10], \omega_0 \in [4,50]$.

Explanation: From the formula (8), (9) and (10), we can see that the jth component of V^t has something to do with the distance between the jth component of X_i^{t-1} and the jth component of X_*, or we say, if the distance between the jth component of X_i^{t-1} and the jth component of X_* is more farther(that is $|x_{ij}^{t-1} - x_{*j}^{t-1}|$ is more bigger accordingly), then the jth speed-component of the bat i will relatively become bigger.

Secondly, considering that the position of the prey is continuous changing with time, so we have introduced the search method that each bat uses the hunting approach of combining random search with shrinking search. Suppose the current best objective (prey) found by the swarm is at the position $X_*^{t-1} = (x_{*1}^{t-1}, \cdots, x_{*n}^{t-1})$ at time instant $t-1$, then at the time instant t, the prey will move to a new location according to the following formula (11):

$$x_{*j}^t = x_{*j}^{t-1} + r_j^t, \quad j = 1, \cdots, n, \tag{11}$$

Where r_j^t is a random number in the range $[-\zeta^t, \zeta^t]$, $\zeta^t = \zeta^0(1 - \dfrac{t}{T_{max}})^\theta$, $0 < \zeta^0 < 1$, θ is a positive integer and $\theta > 1$, T_{max} is the maximum iterations. In our simulation, we set $\zeta^0 = 0.6, \theta = 4$.

3.2   The Procedure of Adaptive Bat Algorithm

Step1. Specify the optimization problem and parameters of the ABA.

Step2. Initialize every bat's location X_i in the search space, and initialize every bat's pulse frequency f_i, pulse rate r_i and loudness A_i ($i = 1, \cdots, M$), and get the current best solution $Sol(i)$.

Step3. If $t < T_{max}$, turn to   Step 3. Otherwise, turn to Step8.

Step4. Each bat gets its frequency f_i according to the formula (7), and updates its position and velocity according to the formula (9) and the formula (10).

Step5.   If rand$> r_i$, then according to the formula (11) to search another new optimal solution. If a new better solutions $S(i)$ has been found, then let $X_* \leftarrow S(i)$.

Step6. If rand< A_i and $f(X_i) < f(X_*)$, then accept the new solution $S(i)$, and update Sol(i) with $S(i)$. According to the formula (5) to update the A_i and r_i.

Step7. Rank the swarm according to the ascending sort of the fitness value and get the current optimal solution X_*.

Step8. $t \leftarrow t+1$, turn to Step 2.

Step9. Output the optimal solution.

4   Experimental Results and Discussion

In order to test the performance of the ABA, we have tested it against the BA [1], and the PSO [4]. For the ease of visualization, we have implemented our simulations using Matlab for all test functions.

4.1   Test Functions

We have chosen six well-known benchmark functions as our optimization simulation tests. The properties and the formula of these functions are presented below.

(a) Circles function

$$f_1(X) = (x_1^2 + x_2^2)^{0.25}[\sin^2(50(x_1^2 + x_2^2)^{0.1}) + 1], -100 \leq x_i \leq 100$$

The function has a global optimum $f_{min} = 0$ at the point $(0,0)$.

(b) Griewank function

$$f_3(x) = \frac{1}{4000}\sum_{i=1}^{50} x_i^2 - \prod_{i=1}^{50} \cos\frac{x_i}{\sqrt{i}} + 1, -600 \leq x_i \leq 600 \cdot$$

Griewank's function has a $\prod_{i=1}^{D}\cos(\frac{x_i}{\sqrt{i}})$ component causing linkages among variables, thereby making it difficult to reach the global optimum. The function has a global minimum $f_{min} = 0$ at $(0,\cdots,0)$.

(c) Sphere function

$$f_1(x) = \sum_{i=1}^{50} x_i^2, -10 \leq x_i \leq 10 \cdot$$

It is a unimodal function, and its global minimum $f_{min} = 0$ at $(0,\cdots,0)$.

(d) Rosenbrock function

$$f_4(x) = \sum_{i=1}^{D-1}(100(x_{i+1} - x_i^2)^2 + (x_i - 1)^2), -30 \leq x_i \leq 30, D = 50.$$

Its global minimum is $f_{min} = 0$ at $x_i = 0, i = 1,\cdots,50$.

(e) Ackley function

$$f_5(x) = 20 + e - 20\exp(-0.2\sqrt{\frac{1}{50}\sum_{i=1}^{50} x_i^2}) - \exp(\frac{1}{50}\sum_{i=1}^{50}\cos(2\pi x_i)), -30 \le x_i \le 30 \cdot$$

Its global minimum is $f_{\min} = 0$ at $(0,\cdots,0)$.

(f) Rastrigrin function

$$f_2(x) = \sum_{i=1}^{50}[x_i^2 - 10\cos(2\pi x_i) + 10], -5.12 \le x_i \le 5.12.$$

Its global minimum is $f_{\min} = 0$ at $(0,\cdots,0)$.

4.2   Comparison of Experimental Results

For the three algorithms, the population size is set at 40 for all simulations. As for the PSO, the acceleration coefficients are set $c_1 = c_2 = 2$, and the inertia weight is set $w = 0.628$. Other coefficients are set as follows: $\alpha = \gamma = 0.5$, the frequency f is in the range $[f_{\min}, f_{\max}]$, where $f_{\min} = 0$, $f_{\max} = 2$, and $A_0 = 0.5$.

In our simulations, we select the best fitness value(BFV), the worst fitness value(WFV), the mean value(Mean), the standard deviation(STDEV), and the average of iterations(AVI) as the evaluation indicators of optimization ability for the BA, the RBA, and the PSO. For these evaluation indicators can not only reflect the optimization ability but also indicate the computing cost. We run each algorithm for 50 times so that we can do a reasonable analysis and got the experimental results being listed in Table 1 as follows.

Table 1. Experimental results

function	algorithms	BFV	WFV	Mean	STDEV	AVI
f_1	ABA	1.7532e-007	3.5067e-004	1.4255e-004	1.2258e-004	76.06
	BA	0.0114	0.7251	0.3862	0.1583	100
	PSO	0.0889	7.7057	2.5810	2.2330	100
f_2	ABA	7.6605e-015	0.0613	0.0129	0.0161	333.92
	BA	1.5302	3.5300	2.3832	0.4154	500
	PSO	0.0272	0.4632	0.1538	0.0988	500
f_3	ABA	4.1693e-015	3.1540e-014	1.2417e-014	5.6010e-015	17.98
	BA	1.2781e-005	2.4617e-005	1.8091e-005	2.6871e-006	255.24
	PSO	1.4591e-005	8.0167e-004	2.0646e-004	2.0320e-004	496.02
f_4	ABA	7.3515e-008	4.0325	1.2006	1.8526	474.36
	BA	2.4591e+004	1.7460e+006	2.8298e+005	2.9526e+005	500
	PSO	36.4962	9.0064e+006	1.8699e+005	1.2727e+006	500
f_5	ABA	0.0228	1.8295	0.2847	0.4050	1000
	BA	5.2503	11.9699	8.1288	1.8181	1000
	PSO	0.9008	2.8495	1.8191	0.4036	1000
f_6	IBA	35.8185	110.9306	70.7998	18.4798	1000
	BA	66.6705	162.1979	99.4527	21.1198	1000
	PSO	116.3957	293.6785	185.0561	33.5121	1000

In order to more easily to contrast the convergence characteristics of the two algorithms, Fig. 1 presents the convergence characteristics in term of the best fitness value of the mean run of each algorithm for each test function.

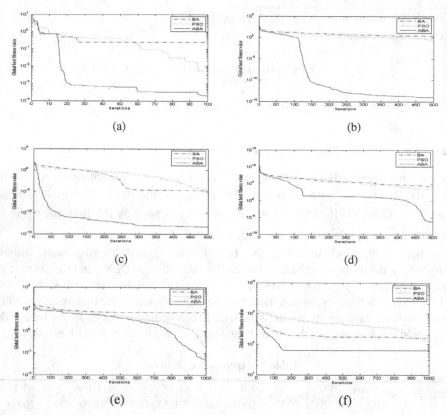

(a)   (b)

(c)   (d)

(e)   (f)

Fig. 1. The median convergence characteristics of 50-D test functions. (a) Circles function. (b) Griewank function. (c) Sphere function. (d) Rosenbrock function. (e) Ackley function. (f) Rastrigrin function.

Discussion: By analyzing the results listed in the Table1, and the convergence curve simulation diagram Fig. 1, we can see that the ABA perform better both than the original BA and the original PSO, that the ABA is superior to the original BA and the original PSO in terms of global convergence property, accuracy and efficiency.

Therefore, we conclude that the performance of the ABA is better than that of the original BA and the original PSO.

5   Conclusions

In order to overcome the premature convergence problem of bat algorithm, we present an improved bat algorithm called adaptive bat algorithm (ABA). From the experimental simulation results we conduct on some benchmark functions such as Circles function,

Griewank function, Sphere function, Rosenbrock function, Ackley function, and Rastrigin function, we can conclude that the performance of the ABA is better than that of the BA and that of PSO, that the ABA has a good performance of global convergence property, and is one of good choices to solve the complex and high dimensional optimization problems.

Acknowledgment. This work were supported by the key programs of institution of higher learning of Guangxi of China under Grant (No.201202ZD032), the Guangxi Key Laboratory of Hybrid Computation and IC Design Analysis, the Natural Science Foundation of Guangxi of China under Grant (No.0832084), the Natural Science Foundation of China under Grant (No.61074185).

References

1. Kennedy, J., Eberhort, R.: Particle swarm optimization. In: Perth: IEEE International Conference on Neural networks, pp. 1941–1948 (1995)
2. Oftadeh, R., Mahjoob, M.J., Shariatpanahi, M.: A Novel Meta-heuristic Optimization Algorithm Inspired by Group Hunting of Animals: Hunting Search. Computers & Mathematics with Applications 60(7), 2087–2098 (2010)
3. He, S., Wu, Q.H., Saunders, J.R.: A group search optimizer for neural network training. In: Gavrilova, M.L., Gervasi, O., Kumar, V., Tan, C.J.K., Taniar, D., Laganá, A., Mun, Y., Choo, H. (eds.) ICCSA 2006. LNCS, vol. 3982, pp. 934–943. Springer, Heidelberg (2006)
4. Lemasson, B.H., Anderson, J.J., Goodwin, R.A.: Collective motion in animal groups from a neurobiological perspective: the adaptive benefits of dynamic sensory loads and selective attention. Journal of Theoretical Biology 261(4), 501–510 (2009)
5. Dorigo, M., Maniezzo, V., Coloria, A.: Ant system: optimization by a colony of cooperating agents. IEEE Transactions on Systems, Man, and Cybernetics: PartB 26(1), 29–41 (1996)
6. Jiao, L.C., Wang, L.: Anovel genetic algorithm based on immunity. IEEE Transaction on System, Man and Cybernetic 30(5), 552–561 (2000)
7. Li, X.-L., Shao, Z.-J., Qian, J.-X.: An optimizing method based on autonomous animals: fish-swarm algorithm. Systems Engineering Theory and Practice 22(11), 32–38 (2002)
8. Huang, D.-S., Zhang, X., Reyes García, C.A., Zhang, L. (eds.): ICIC 2010. LNCS, vol. 6216. Springer, Heidelberg (2010)
9. Krishnanand, K.N., Ghose, D.: Glowworm swarm based optimization algorithm for multimodel functions with collective robotics applications. Multiagent and Grid Systems 2(3), 209–222 (2006)
10. Chen, J.-R., Wang, Y.: Using fishing strategy optimization method. Computer engineering and Applications 45(9), 53–56 (2009)
11. Yang, X.-S., Deb, S.: Cuckoo search via Levy flights. In: Proc. of World Congress on Nature & Biologically Inspired Computing (NaBIC 2009), pp. 210–214. IEEE Publications, India (2009)
12. Yang, X.-S.: A new metaheuristic bat-inspired algorithm. In: González, J.R., Pelta, D.A., Cruz, C., Terrazas, G., Krasnogor, N. (eds.) NICSO 2010. SCI, vol. 284, pp. 65–74. Springer, Heidelberg (2010)

Comparative Study of Artificial Bee Colony Algorithms with Heuristic Swap Operators for Traveling Salesman Problem

Zhonghua Li[1], Zijing Zhou[1,2], Xuedong Sun[3], and Dongliang Guo[1]

[1] School of Information Science and Technology, Sun Yat-sen University, Guangzhou, China
lizhongh@mail.sysu.edu.cn
[2] Department of Information Technology, China Guangfa Bank, Guangzhou, China
zijingsysu@foxmail.com
[3] School of Software, Sun Yat-sen University, Guangzhou, China
xuedong_sun2004@163.com

Abstract. Because the traveling salesman problem (TSP) is one type of classical NP-hard problems, it is not easy to find the optimal tour in polynomial time. Some conventional deterministic methods and exhaustive algorithms are applied to small-scale TSP; whereas, heuristic algorithms are more advantageous for the large-scale TSP. Inspired by the behavior of honey bee swarm, Artificial Bee Colony (ABC) algorithms have been developed as potential optimization approaches and performed well in solving scientific researches and engineering applications. This paper proposes two efficient ABC algorithms with heuristic swap operators (i.e., ABC-HS1 and ABC-HS2) for TSP, which are used to search its better tour solutions. A series of numerical experiments are arranged between the proposed two ABC algorithms and the other three ABC algorithms for TSP. Experimental results demonstrate that ABC-HS1 and ABC-HS2 are both effective and efficient optimization methods.

Keywords: Artificial Bee Colony Algorithm, Heuristic Swap Operator, Optimization, Traveling Salesman Problem.

1 Introduction

The Traveling Salesman Problem (TSP) is one class of combinatorial optimization problems, and is also an NP-hard problem. Usually, the optimal solution of TSP is represented by a Hamiltonian path with minimum cost. Hence, many real problems can be modeled as TSP and/or its variants. Up to date, TSP has been extended to such industrial applications as intelligent transportation, flow shop scheduling, logistics, circuit layout, robot control and wireless router.

To solve TSP, many algorithms/methods have been proposed focusing on both solution accuracy and computational complexity. These algorithms/methods are usually classified into two types: deterministic algorithms and heuristic algorithms. Deterministic algorithms include dynamic programming [1], branch and bound [2]

D.-S. Huang et al. (Eds.): ICIC 2013, LNAI 7996, pp. 224–233, 2013.

and minimal spanning tree method [3], while heuristic algorithms usually take advantage of such intelligent computational approaches as genetic algorithm, artificial neural network, simulated annealing, and artificial immune network.

When the number of cities to be visited is small, deterministic algorithms are effective for solving TSP. For any deterministic algorithm, the optimal solution is expected to obtain within finite steps. In most cases, deterministic methods are also become one part of some complex algorithms for TSP. When the number of cities to be visited becomes large, time consumption of these complex methods is very costly. Besides, some exhaustive methods were used to find the shortest closed tour. In reality, it is verified that exhaustive methods are improper for solving large-scale TSP.

Further, a number of heuristic algorithms and their variants are proposed to capture an acceptable approximate solution for large-scale TSP and even some of these algorithms are able to get the global optimal solution. In the literature [4], a greedy sub tour mutation (GSTM) operator is designed to increase the performance of genetic algorithm for TSP. Meanwhile, a simulated annealing algorithm with greedy search is used to solve TSP and some improvements are gained in both solution accuracy and time consumption [5]. The document [6] proposes a parallel immune algorithm for TSP to improve the computing efficiency. In the article [7], a novel artificial neural network is designed to solve TSP. Later, some hybrid algorithms, e.g., combination of genetic algorithm and immune algorithm [8], integration of genetic algorithm and artificial neural network [9], are proposed to improve the searching efficiency of optimal tour solution. Compared to deterministic algorithms, heuristic algorithms are capable of searching satisfactory solutions for TSP in a relatively short time.

In the previous two decades, heuristic algorithms along with swarm intelligence became a research focus in the field of intelligent computing. The major heuristic algorithms are composed of particle swarm optimization, ant colony optimization, artificial bee colony (ABC) algorithm, etc. ABC algorithm is an emerging evolutionary algorithm in the families of swarm intelligence inspired by bee's behavior. The basic version of ABC algorithm was firstly proposed by Karaboga for numerical optimization in 2005. Because the evolving mechanism of ABC is simple and effective, many versions of ABC are developed to apply to different applications. The document [10] extends the ABC algorithm with greedy sub tour crossover to solve the TSP, and provides people with a new viewpoint to solve TSP. Because TSP is very complicated and neighborhoods of its feasible solution are not continuous, the operator in the basic ABC which is used to produce new solutions for numerical optimization is not at all available for TSP. In the article [11], the proposed neighborhood operators are looked as solution updating operators to solve TSP. However, these neighborhood operators are random, as a result, the update of new solutions become blind. Consequently, it is hard to find satisfactory solutions near to the optimal solution.

To improve the computational efficiency and solution accuracy, this paper proposes a couple of revised ABC algorithms with heuristic swap operators for solving TSP and focuses on the solution updating strategies within the neighborhood. On the one hand, a 2-opt neighborhood heuristic swap operator is proposed as the first updating operator, denoted by ABC-HS1. On the other hand, a combinational framework of 2-opt neighborhood heuristic swap operator and Or-opt neighborhood heuristic swap operator is proposed as the second updating operator, denoted by ABC-HS2. ABC-HS1 and ABC-HS2 are used to update solutions in both the employed bee phase and the onlooker bee phase of the ABC algorithm for TSP.

The remainder of this paper is organized as below. Section 2 reviews the problem description of TSP. Section 3 describes the proposed ABC-HS1 and ABC-HS2 algorithms for TSP in detail. In section 4, Numerical experiments are arranged, and theirs results and the corresponding discussions are given. Finally, some conclusions are drawn in Section 5.

2   Review of Traveling Salesman Problem

A typical TSP means that a salesman is required to visit all cities defined by the problem in a round-trip tour, and each city can be visited only once. Measured by the traveling distance between any two cities, the salesman is likely expected to find the minimum length of the closed tour. It is assumed that C_i represents the ith city and C_j represents the jth city of the tour. And $\pi = \{C_1, C_2, ..., C_i, ..., C_j, ..., C_N\}$ ($1 \le i \le j \le N$) is a feasible tour permutation which consists of the sequence number of cities to be visited. Hence, a typical TSP can be formulated mathematically,

$$
\begin{cases}
\min f(\pi) = \sum_{i=1}^{N-1} d_{C_i C_{i+1}} + d_{C_1 C_N} \\
s.t. \quad
\begin{aligned}
&C_i \ne C_j (i \ne j) \\
&C_i \in \{1, 2, ..., N\}
\end{aligned}
\end{cases}
\tag{1}
$$

where N is the total number of cities to be visited. $f(\pi)$ is the path length of a feasible tour π, and our task is to find the optimal value of π to minimize $f(\pi)$. If $d_{C_i C_j} = d_{C_j C_i}$, the problem is called a *symmetric* TSP (STSP). Otherwise, the problem is called an *asymmetric* TSP (ATSP). For the sake of research interests, we focus on only STSP in this paper.

3   Proposed ABC Algorithms with Heuristic Swap Operators

3.1   Principles of ABC Algorithms

The ABC algorithms simulate bees in the bee colony how to cooperate for foraging in nature. In terms of the distribution of work, these bees are classified into three roles: scout bees, employed bees and onlooker bees. And the roles of bees will change at some phase. The scout bees are responsible for searching new food sources at the whole search space. The employed bees exploit the food sources and memorize their positions. After exploiting, they will return to hive to share the searched information about the food sources with dancing. And the onlooker bees choose the employed bees to follow. For the sake of understanding, the pseudocode of the ABC algorithm is given in Fig.1.

```
PROCEDURE ABC ( )
//Initialization phase
Initialize parameters and generate candidate solutions;
REPEAT
    //Employed bee phase
    Update candidate solutions;
    Evaluate candidate solutions;
    //Onlooker bee phase
    Select employed bees;
    Update candidate solutions;
    Evaluate candidate solutions;
    //Scout bee phase
    IF (exceeding the trial limit)
        Abandon solution and recruit candidate solution randomly;
    END IF
    Memorize the best obtained solution;
UNTIL (the maximum number of iterations)
END PROCEDURE
```

Fig. 1. Pseudocode of ABC

During the initialization phase of ABC algorithm, some parameters are required to be set, including the colony size CS, the maximum number MNI of iterations and the trial limit TL of searching the same food source. The number of employed bees is always equal to the number of onlooker bees, i.e., half of the colony size. And then ABC generates candidate feasible solutions randomly.

During the employed bee phase, an employed bee (food source) denotes a feasible solution. We need to find the best food source which owns the largest amount of nectar, i.e., the optimal solution of TSP. Because the optimal solution is the shortest tour and the tour length is non-negative, the following fitness function can be defined to evaluate solutions, that is

$$F(\pi_i) = 1 / (1 + f(\pi_i)),\qquad(2)$$

where $f(\pi_i)$ is the objective function of the ith solution. It is easy to find out that the shorter tour length means the greater fitness. The fitness function can help us evaluate which one is better between two food sources. Each employed bee will try to find another new food source with local search, to choose the better food source and to memorize it.

During the onlooker bee phase, every onlooker bee will select an employed bee to follow with a certain probability. The selection probability is defined in terms of selection methods, which ensures that the better food source has more chance to be chosen. When an onlooker bee has decided which employed bee to follow, it will follow the desired employed bee and search another food source with local search. If the searched food source is better than the food source memorized by the employed bee, their roles will be exchanged.

During the scout bee phase, the food sources which have been exploited exceeding TL, will be abandoned. And the scout bee will randomly search food sources in the global space. And then, ABC algorithm will go back to the employed bee phase again. The local search process of employed bees and onlooker bees will be executed repeatedly until the algorithm iterates for MNI times.

3.2   The Proposed ABC with Heuristic Swap for TSP

In this paper, a couple of heuristic swap operators are proposed as the candidate solutions updating operators. Observed from the best tour of some TSP instances, cities far from a specific city will not become the next city to be visited. The next city is chosen within the neighborhood. And the neighborhood is constructed from some cities closer to the city. In this paper, we use a parameter N_{max} to determine the scale of neighborhood. Inspired from 2-opt swap method and Or-opt swap method, we proposed heuristic swap operator 1 (HS1) and heuristic swap operator 2 (HS2) as the updating operator for ABC algorithm to solve TSP, respectively.

3.2.1   Heuristic Swap Operator 1 (HS1)

Take a feasible solution π as an example, and $edge(C_i, C_{i+1})$ denotes the arc from C_i to C_{i+1}, the pseudocode of 2-opt neighborhood heuristic swap operator (i.e., HS1) for employed bees and onlooker bees is shown in Fig.2.

```
PROCEDURE HS1( )
FOR each employed bee and onlooker bee
      Select a starting city  C_i  randomly;
      Select a city  C_j  randomly within the  N_max -city neighborhood of  C_i ;
      Generate  new  solution  by  performing  2-opt  swap  with  edge(C_i, C_{i+1})  and
edge(C_j, C_{j+1});
      Select the value of searching direction parameter  Φ  randomly;
      // Φ  obey  [0,1]  uniform distribution.
      IF ( Φ > P_N )      // P_N  is the selection probability of starting city.
            The  C_{j+1}  will be the starting city for next-time neighborhood updating;
      ELSE
            The  C_{i+1}  will be the starting city for next-time neighborhood updating;
      END IF
END FOR
END PROCEDURE
```

Fig. 2. Pseudocode of HS1

HS1 for onlooker bees needs to select the city determined at previous updating for the same position of the employed bees. And the rest of the procedure is the same as those of employed bees. For example, we construct five cities which are nearest to the city selected as the neighborhood. Thus, the procedure of HS1 is shown in Fig. 3.

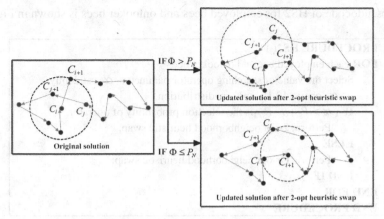

Fig. 3. Operational Procedure of HS1

3.2.2   Heuristic Swap Operator 2 (HS2)

The pseudocode of the Or-opt neighborhood heuristic swap operator is the same as the 2-opt neighborhood heuristic swap operator except for new solution generation mechanism. And new feasible solutions are produced by performing Or-opt exchange with $\{edge(C_i, C_{i+1}), edge(C_{i+1}, C_{i+2}), edge(C_j, C_{j+1})\}$ and $\{edge(C_i, C_{i+2}), edge(C_j, C_{i+1}),$ $edge(C_{i+1}, C_{j+1})\}$. Heuristic swap operator 2 (HS2) uses a parameter α to determine performing either 2-opt neighborhood heuristic swap or Or-opt neighborhood heuristic swap. And the procedure of HS2 is shown in Fig. 4.

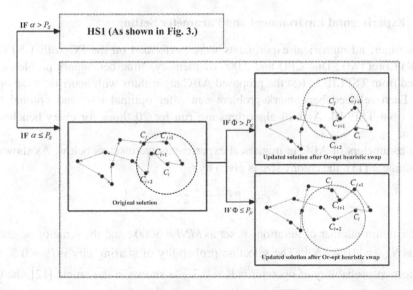

Fig. 4. Operational Procedure of HS2

The pseudocode of HS2 for employed bees and onlooker bees is shown in Fig.5.

```
PROCEDURE HS2( )
FOR each employed bee and onlooker bee
    Select the value of searching operator parameter  α  randomly;
    // α obeys  [0,1]  uniform distribution.
    IF ( α > P_S ) // P_S  is the selection probability of operator.
        Perform 2-opt neighborhood heuristic swap;
    ELSE
        Perform Or-opt neighborhood heuristic swap;
    END IF
END FOR
END PROCEDURE
```

Fig. 5. Pseudocode of HS2

4   Numerical Experiments and Results

This section will arrange a series of numerical experiments to examine the performance of the proposed ABC-HS1 algorithm and ABC-HS2 algorithm for TSP, and to make some comparisons with other algorithms (e.g., ABC-RS, ABC-RI and ABC-RR). Besides, some technical details, simulated experimental results and corresponding discussions are given.

4.1   Experimental Environment and Parameter Setting

In this paper, all numerical experiments were performed on the PC with 1.50 GHz Intel(R) Core(TM)2 Duo CPU and 2.00 GB memory. Nine benchmark problems are selected from TSPLIB to test the proposed ABC algorithms with heuristic swap operators. Each selected benchmark problem can offer optimal tour and optimal tour length from TSPLIB. And all algorithms are run for 20 times for every benchmark problem.

The parameters of ABC in numerical experiments were set as below. As shown in the document [11], the colony size is given by

$$CS = \lceil N / 2 \rceil \times 2 .$$ (3)

The maximum number of iterations is set as $MNI = 3000$, and the scale of neighborhood is $N_{max} = \lceil 10\% \times N \rceil$. The selection probability of starting city is $P_N = 0.5$, and the selection probability of operator is $P_S = 0.3$. As shown in the article [12], the trial limit TL of neighborhood updating operation for the same food source is given by

$$TL = CS \times N / 3 .$$ (4)

4.2   Experimental Results

For the sake of clear evaluation, some performance metrics should be defined. Assume that BE denotes the relative error of the best solution, which can be given by

$$BE = (Best - Opt) / Opt \times 100\%, \qquad (5)$$

where $Best$ is the tour length of the best solution gained from 20 experiments, and Opt is the tour length of the known optimal solution from TSPLIB. It is assumed that AE denotes the average relative error, which can be determined by

$$AE = (Avg - Opt) / Opt \times 100\%, \qquad (6)$$

where Avg is the average tour length of the solutions gained from 20 experiments.

The experimental results of the proposed ABC algorithms for 9 benchmark problems are shown in Table 1. Observed from Table 1, ABC-HS1 and ABC-HS2 can both obtain a final solution close to the optimal solution for all benchmark problems. As the number of cities increases, the average error also increases for ABC-HS1 and ABC-HS2. When the number of cities to be visited is less than 100, AE of ABC-HS1 is less than 3%, and AE of ABC-HS2 is less than 2%. When the number of cities to be visited is less than 200, AE of ABC-HS1 is less than 4%, and AE of ABC-HS2 is less than 3% except for the KroB150 problem. It is clear that ABC-HS2 outperforms ABC-HS1 in solution accuracy.

Table 1. Performance of the Proposed ABC Algorithms for TSP

Problem	N	Opt	HS1				HS2			
			Best	BE(%)	Avg	AE(%)	Best	BE(%)	Avg	AE(%)
eil51	51	428.87	432.12	0.76	435.95	1.65	429.53	0.15	433.92	**1.18**
birlin52	52	7544.4	7544.4	0	7640.6	1.28	7544.4	0	7612.3	**0.90**
st70	70	677.11	680.5	0.50	691.89	2.18	677.19	0.01	684.42	**1.08**
eil76	76	545.39	552.15	1.24	559.79	2.64	550.41	0.91	554.45	**1.66**
KroC100	100	20751	20941	0.91	21461	3.42	20798	0.22	21076	**1.57**
eil101	101	642.31	649.58	0.52	658.24	2.48	645.64	1.13	652.14	**1.53**
lin105	105	14383	14593	1.46	14890	3.52	14569	1.27	14617	**1.63**
KroB150	150	26130	26544	1.58	27133	3.84	26799	2.52	26988	**3.28**
KroA200	200	29368	30140	2.63	30484	3.80	30161	2.63	30229	**2.93**

Table 2. Results of Comparisons between 5 ABC Algorithms for TSP

Problem	Index	ABC-RS	ABC-RI	ABC-RR	ABC-HS1	ABC-HS2
eil51	BE(%)	11.49	5.44	0.86	0.76	**0.15**
	AE(%)	18.06	9.04	2.59	1.65	**1.18**
st70	BE(%)	27.11	9.18	1.10	0.50	**0.01**
	AE(%)	37.01	15.95	2.67	2.18	**1.08**
Eil101	BE(%)	23.42	10.66	3.59	1.13	**0.52**
	AE(%)	30.37	13.66	5.30	2.48	**1.53**

Table 2 shows experimental results of two proposed ABC algorithms (ABC-HS1 and ABC-HS2) and three ABC algorithms selected from literature [11], which are ABC with random swap operator (ABC-RS), ABC with random insertion operator (ABC-RI) and ABC with random reversing operator (ABC-RR). Compared to ABC-RS, ABC-RI and ABC-RR, both of ABC-HS1 and ABC-HS2 perform better in both BE and AE. The results indicate that the proposed ABC algorithms with heuristic swap operators are efficient and competitive for solving TSP.

4.3   Tecshnical Discussions

The proposed ABC-HS1 and ABC-HS2 for travel salesman problems can obtain final solutions approximate to the optimal solutions of the benchmark problems. The neighborhood heuristic swap mechanism is simple, and there is much enrichment and development space in the future researches. (1) Some useful and efficient mechanisms are employed to make ABC-HS1 and ABC-HS2 more accurate. Using some heuristic swap operators, e.g., a neighborhood updating operator, we can further improve the method of neighborhood definition in the future. For example, by combining with crossover operators of GA for TSP, ABC can utilize the corresponding excellent information of candidate solutions, and thus guide the evolution of the whole candidate solution population. (2) Some new mechanisms are used to avoid being trapped into the local optimum. Known form Table 1, the proposed ABC algorithms may be possibly faced with local optimum. It is because that the mechanism of abandoning solutions in ABC-HS1 and ABC-HS2 has not strong enough capability of finding global optimum. Therefore, to add some novel mechanisms may be alternative options.

5   Conclusions

In this paper, two ABC algorithms integrating with heuristic swap operators, i.e., ABC-HS1 and ABC-HS2, are proposed for TSP. Considering that updating candidate solutions near neighborhood will be more effective, ABC-HS1 updates solutions by using 2-opt heuristic swap method, while ABC-HS2 updates solutions by using 2-opt heuristic swap and Or-opt heuristic swap method. In order to examine the performance of the proposed ABC algorithms, a series of numerical experiments are arranged and the performance of the proposed ABC algorithms is testified on the nine benchmark TSP problems. In the numerical experiments, ABC-HS1 and ABC-HS2 show some promising results. Compared to other ABC algorithms with neighborhood operator, the proposed ABC algorithms perform better in solution accuracy.

Acknowledgement. This work is supported by the National Science Foundation of China (No.61201087 and No.61203060) and the Guangdong Natural Science Foundation under Grant (No.S2011010001492).

References

1. Aicardi, M., Giglio, D., Minciardi, R.: Determination of Optimal Control Strategies for TSP by Dynamic Programming. In: 47th IEEE Conference on Decision and Control, pp. 2160–2167. IEEE Press, Piscataway (2008)
2. Lopez, M.R., Tunon, M.I.C.: Design and Use of the CPAN Branch & Bound for the Solution of the Travelling Salesman Problem (TSP). In: 15th International Conference on Electronics, Communications and Computers, pp. 262–267. IEEE Press, United States (2005)
3. Kahng, A.B., Reda, S.: Match Twice and Stitch: A New TSP Tour Construction Heuristic. Operations Research Letters 32(6), 499–509 (2004)
4. Albayrak, M., Allahverdi, N.: Development A New Mutation Operator to Solve the Traveling Salesman Problem by Aid of Genetic Algorithms. Expert Systems with Applications 38(3), 1313–1320 (2011)
5. Geng, X., Chen, Z., Yang, W., Shi, D., Zhao, K.: Solving the Traveling Salesman Problem Based on an Adaptive Simulated Annealing Algorithm with Greedy Search. Applied Soft Computing 11(4), 3680–3689 (2011)
6. Zhao, J., Liu, Q., Wang, W., Wei, Z., Shi, P.: A Parallel Immune Algorithm for Traveling Salesman Problem and Its Application on Cold Rolling Scheduling. Information Sciences 181(7), 1212–1223 (2011)
7. Saadatmand-Tarzjan, M., Khademi, M., Akbarzadeh-T, M., Abrishami, M.H.: A Novel Constructive-Optimizer Neural Network for the Traveling Salesman Problem. IEEE Transactions on Systems, Man, and Cybernetics, Part B: Cybernetics 37(4), 754–770 (2007)
8. Lu, J., Fang, N., Shao, D., Liu, C.: An Improved Immune-Genetic Algorithm for the Traveling Salesman Problem. In: 3rd International Conference on Natural Computation, pp. 297–301. IEEE Press, Piscataway (2007)
9. Vahdati, G., Ghouchani, S.Y., Yaghoobi, M.: A Hybrid Search Algorithm with Hopfield Neural Network and Genetic algorithm for Solving Traveling Salesman Problem. In: 2nd International Conference on Computer and Automation Engineering, pp. 435–439. IEEE Press, Piscataway (2010)
10. Banharnsakun, A., Achalakul, T., Sirinaovakul, B.: ABC-GSX: A Hybrid Method for Solving the Traveling Salesman Problem. In: 2nd World Congress on Nature and Biologically Inspired Computing, pp. 7–12. IEEE Press, Piscataway (2010)
11. Kıran, M.S., İşcan, H., Gündüz, M.: The Analysis of Discrete Artificial Bee Colony Algorithm with Neighborhood Operator on Traveling Salesman Problem. In: Neural Computing & Applications, pp. 1–13. Springer, London (2012) (in press)
12. Karaboga, D., Gorkemli, B.: A Combinatorial Artificial Bee Colony Algorithm for Traveling Salesman Problem. In: 2011 International Symposium on Innovations in Intelligent Systems and Applications, pp. 50–53. IEEE Press, Piscataway (2011)
13. Li, G., Niu, P., Xiao, X.: Development and Investigation of Efficient Artificial Bee Colony Algorithm for Numerical Function Optimization. Applied Soft Computing 12(1), 320–332 (2012)
14. Su, Z., Wang, P., Shen, J., Li, Y., Zhang, Y., Hu, E.: Automatic Fuzzy Partitioning Approach Using Variable String Length Artificial Bee Colony (VABC) Algorithm. Applied Soft Computing 12(11), 3421–3441 (2012)
15. Kashan, M.H., Nahavandi, N., Kashan, A.H.: DisABC: A New Artificial Bee Colony Algorithm for Binary Optimization. Applied Soft Computing 12(1), 342–352 (2012)

Kernel *k'*-means Algorithm for Clustering Analysis

Yue Zhao, Shuyi Zhang, and Jinwen Ma[*]

Department of Information Science, School of Mathematical Sciences
And LMAM, Peking University, Beijing, 100871, China
jwma@math.pku.edu.cn

Abstract. *k'*-means algorithm is a new improvement of *k*-means algorithm. It implements a rewarding and penalizing competitive learning mechanism into the *k*-means paradigm such that the number of clusters can be automatically determined for a given dataset. This paper further proposes the kernelized versions of *k'*-means algorithms with four different discrepancy metrics. It is demonstrated by the experiments on both synthetic and real-world datasets that these kernel *k'*-means algorithms can automatically detect the number of actual clusters in a dataset, with a classification accuracy rate being considerably better than those of the corresponding *k'*-means algorithms.

Keywords: Clustering analysis, *k*-means algorithm, Mercer kernels, Kernel method.

1 Introduction

Clustering analysis is a powerful technique applied in many areas of data analysis and information processing, such as data mining and compression, pattern recognition and vector quantization. The aim of clustering analysis is to discover the hidden data structure of a dataset according to a certain similarity criterion such that all the data points are assigned into a number of distinctive clusters where points in the same cluster are similar to each other, while points from different clusters are dissimilar [1].

k-means algorithm is a classical algorithm for clustering analysis. It is widely used for its simplicity, speediness and effectiveness [2], [3]. In the *k*-means paradigm, each cluster is represented by an adaptively-changing cluster center starting from *k* initial points. Essentially, *k*-means algorithm tries to minimize the sum of the mean squared distances between the data points and their nearest centers.

However, *k*-means algorithm needs a strong assumption that the number k' of actual clusters in the dataset must be previously known and $k = k'$. Actually, if *k* is not set correctly, *k*-means algorithm will lead to a wrong clustering result [4], [5]. In the general situation, we cannot know this critical information in advance. In order to solve this problem, many approaches have been suggested, such as the Bayesian Ying-Yang harmony learning algorithm [6], [7] and the rival penalized competitive learning algorithm [8], [9]. But those clustering methods generally involve in complicated mathematical models and heavy computations.

---

[*] Corresponding author.

D.-S. Huang et al. (Eds.): ICIC 2013, LNAI 7996, pp. 234–243, 2013.

Following the k-means paradigm, Zalik [10] proposed a new kind of k-means algorithm called k'-means algorithm. It sets the number k of clusters to be larger than the true number k' of actual clusters in the dataset and then implements a rewarding and penalizing competitive learning mechanism via just using a new kind of data discrepancy metric between an data point and a cluster center instead of the conventional Euclidean distance. As a result, the correct number of clusters can be detected, with the extra clusters becoming empty at last. In order to improve the performance of the original k'-means algorithm, Fang et al. [11], [12] have recently proposed some new k'-means algorithms with frequency sensitive discrepancy metrics.

On the other hand, k-means algorithm can be also improved by using kernel functions or kernelization trick. The idea of the kernel method is to map the observed data to a higher dimensional space in a nonlinear manner [13], [14]. If the nonlinear mapping is smooth and continuous, the topographic ordering of the data will be preserved in the feature space so that points clustered in the data space can be also clustered in the feature space [15]. Moreover, the overlapped clusters may be separated in the higher dimensional feature space. The kernel function behaved as the inner product of the projected points can avoid the explicit expression of the map function [16].

In this paper, we propose the kernelized versions of four k'-means algorithms for clustering analysis. We test the four kernel k'-means algorithms on three synthetic datasets and three real-world datasets, being compared with the general k'-means algorithms. It is demonstrated by experimental results that the kernel k'-means algorithms not only detect the number of actual clusters correctly, but also outperform the general k'-means algorithms on classification accuracy rate. Moreover, the kernel k'-means algorithms are successfully applied to unsupervised color image segmentation.

The rest of this paper is organized as follows. In Section 2, we give a brief review of k'-means algorithm. In Section 3, we present the kernelized versions of k'-means algorithms with four frequency sensitive discrepancy metrics. The experimental results and comparisons are conducted in Section 4. Finally, we give a brief conclusion in Section 5.

2 k'-means Algorithms

Mathematically, the clustering problem can be described as follows: given a dataset containing N points in the d-dimensional real space, as well as its cluster number k ($< N$), we need to select k cluster centers or means for k clusters in the data space under certain data discrepancy metric or criterion according to which, the data points are assigned into one of the k clusters. As for the classical k-means algorithm [2], each x_t is assigned to a cluster via the classification membership function given by

$$I(x_t, i) = \begin{cases} 1, & \text{if } i = \arg\min_j \| x_t - c_j \|^2; \\ 0, & \text{otherwise.} \end{cases}$$

That is, x_t belongs to C_i if and only if $I(x_t, i) = 1$. In each iteration, the i-th cluster center c_i is updated as

$$c_i = \frac{1}{|C_i|} \sum_{x_t \in C_i} x_t,$$

where $|C_i|$ denotes the number of the data points in Cluster i, C_i.

k'-means algorithm works in the same clustering paradigm as the k-means algorithm, but differs in the expression of the cluster membership function $I(x_t, i)$. It implements a rewarding and penalizing competitive learning mechanism for selecting the correct number of clusters for a dataset [10]. In this situation, k is assumed to be larger than the true number k' of actual clusters in the dataset. For eliminating those extra clusters, the discrepancy metric in [10] is defined as

$$dm(x, c_i) = \|x - c_i\|^2 - E\log_2 p_i, \tag{1}$$

where $E > 0$ is a constant serving as a penalty factor and $p_i = P(C_i)$ is the frequency that an input data points is in the C_i cluster. If $p_i = 0$, $-E\log_2 p_i$ is considered as the positive infinity.

With such a special discrepancy metric, the clusters with few data points will be penalized and become empty in the sequential iterations. Then the correct number of clusters will be finally detected. Besides the cluster centers would be located in the places where the data points are densely accumulated. However, the original k'-means algorithm with this kind of data discrepancy metric is unstable and thus the clustering result is not so good on the large-scale real-world dataset. In order to improve the performance of k'-means algorithm, Fang et al. [12] proposed three new discrepancy metrics, which were respectively defined as

$$d_1(x, c_i) = \|x - c_i\|^2 + \lambda / p_i; \tag{2}$$

$$d_2(x, c_i) = p_i \|x - c_i\|^2 + \lambda / p_i; \tag{3}$$

$$d_3(x, c_i) = p_i \|x - c_i\|^2 - \lambda \log p_i; \tag{4}$$

where λ serves as a penalty factor, being a positive constant. When some p_i becomes zero, λ / p_i and $-\lambda \log p_i$ are also considered as the positive infinity for any input data point x and this cluster becomes empty.

With each of the three discrepancy metrics, say $d(x_t, c_j)$, we can compute the corresponding cluster membership function by

$$I(x_t, i) = \begin{cases} 1, & \text{if } i = \arg\min_j d(x_t, c_j), j = 1, ..., k; \\ 0, & \text{otherwise.} \end{cases}$$

In the same way, the new k'-means algorithm can be implemented for clustering analysis.

3   Kernel k'-means Algorithms

We further propose the kernelized versions of these k'-means algorithm. Let $\phi: z \in R^d \rightarrow \phi(z) \in F$ be a nonlinear mapping from the input space to a higher dimensional feature space F. The idea of the kernel method is to make a nonlinear classification problem be linear in a higher dimensional feature space so that the classification problem can be effectively and efficiently solved [17]. Generally, the kernelization of a conventional method makes use of a kernel function $\kappa(.)$ which serves as the inner product in the higher dimensional space, i.e., $\kappa(x, y) = < \phi(x), \phi(y) >$, and actually allows us to avoid the explicit expression of the mapping ϕ [18].

Here, we try to use the kernel method to make the projected clusters linearly separated in the feature space. We then conduct the k'-means algorithm in the higher dimensional feature space. In the feature space, we need to cluster the projected samples (i.e., data points) $\phi(x_1), \phi(x_2), \cdots, \phi(x_N)$. So, the discrepancy metrics becomes

$$kdm(\phi(x), \phi(c_i)) = \| \phi(x) - \phi(c_i) \|^2 - E\log_2 p_i; \tag{5}$$

$$kd_1(\phi(x), \phi(c_i)) = \| \phi(x) - \phi(c_i) \|^2 + \lambda / p_i; \tag{6}$$

$$kd_2(\phi(x), \phi(c_i)) = p_i \| \phi(x) - \phi(c_i) \|^2 + \lambda / p_i; \tag{7}$$

$$kd_3(\phi(x), \phi(c_i)) = p_i \| \phi(x) - \phi(c_i) \|^2 - \lambda \log p_i. \tag{8}$$

Here, $\phi(c_i)$ is the center of cluster C_i in the feature space, set as

$$\phi(c_i) = \frac{1}{|C_i|} \sum_{x_j \in C_i} \phi(x_j).$$

Since $\kappa(x, y) = < \phi(x), \phi(y) >$, the distance between two projected samples $\phi(x_i), \phi(x_j)$ becomes $\| \phi(x_i) - \phi(x_j) \|^2 = \kappa(x_i, x_i) - 2\kappa(x_i, x_j) + \kappa(x_j, x_j)$.

Then, the distance between projected sample $\phi(x)$ and center $\phi(c_i)$ becomes

$$\| \phi(x) - \phi(c_i) \|^2 = \kappa(x, x) - \frac{2}{|C_i|} \sum_{x_j \in C_i} \kappa(x, x_i) + \frac{1}{|C_i|^2} \sum_{x_j \in C_i} \sum_{x_k \in C_i} \kappa(x_j, x_k).$$

In this way, it can be seen that the explicit expression of the mapping ϕ can be avoided by using a kernel function.

Similarly, for each discrepancy metrics $kd(x_t, c_j)$, the corresponding cluster membership function can be computed by

$$I(x_t, i) = \begin{cases} 1, & \text{if } i = \arg\min_j kd(x_t, c_j), j = 1, ..., k; \\ 0, & \text{otherwise.} \end{cases}$$

4   Experimental Results

In this section, the kernel k'-means algorithms of four frequency sensitive discrepancy metrics is tested and compared with the corresponding k'-means algorithms on three typical synthetic datasets and three real-world datasets on both cluster number detection and classification accuracy rate. Besides, we also apply the kernel k'-means algorithms to unsupervised image segmentation. Actually, the Gaussian kernel function is used in the experiments, being expressed by

$$\kappa(x_i, x_j) = \exp\left(-\frac{\|x_i - x_j\|^2}{2\sigma^2}\right)$$

where σ is the parameter depending on the scale of sample data. In the experiments, k is set to be larger than k', i.e., $k \geq k'$. However, if k is too large, k'-means algorithms may not converge effectively. Generally, it is reasonable to set $k \in [k', 3k']$. For the penalty factor λ or E, its feasible intervals for the four discrepancy metrics from Eq.(1) to Eq.(4) or corresponding k'-means algorithms are [0.1, 0.3], [0.03, 0.1], [0.01, 0.15], and [0.03, 0.5], respectively. Although the feasible value of the penalty factor can lead the k'-means algorithm to the correct detection of cluster number, it affects the performance of classification on a dataset. We will select the optimal value of the penalty factor by experience.

Table 1. The parameter values for the three synthetic datasets.

Dataset	Gaussian	μ	σ_{11}	$\sigma_{12}(\sigma_{21})$	σ_{22}	α
S1	G1	(-1, 0)	0.04	0	0.04	0.25
(N=1200)	G2	(1, 0)	0.04	0	0.04	0.25
	G3	(0, 1)	0.04	0	0.04	0.25
	G4	(0,-1)	0.04	0	0.04	0.25
S2	G1	(-1, 0)	0.09	0	0.09	0.25
(N=1200)	G2	(1, 0)	0.09	0	0.09	0.25
	G3	(0, 1)	0.09	0	0.09	0.25
	G4	(0,-1)	0.09	0	0.09	0.25
S3	G1	(-1, 0)	0.16	0	0.16	0.25
(N=1200)	G2	(1, 0)	0.16	0	0.16	0.25
	G3	(0, 1)	0.16	0	0.16	0.25
	G4	(0,-1)	0.16	0	0.16	0.25

We begin to introduce the three synthetic datasets S1, S2, S3 used in the experiments. Actually, they are generated from a mixture of four bivariate Gaussian distributions on the plane coordinate system (i.e., $d = 2$). Thus, a cluster or class takes the form of a Gaussian distribution. Particularly, all the Gaussian distributions are cap-shaped, that is, their covariance matrices have the form of $\sigma^2 I$, where I is the d-dimensional identity matrix. The values of the parameters for the three synthetic datasets are listed in Table 1. Figure 1 also shows the sketches of the three synthetic datasets. The degree of overlap among the actual clusters or Gaussian distributions in the dataset increases considerably from S1 to S3 and thus the corresponding classification problem becomes more complicated. For these three synthetic datasets, we typically set $k = 8$ for the four kernel k'-means algorithms.

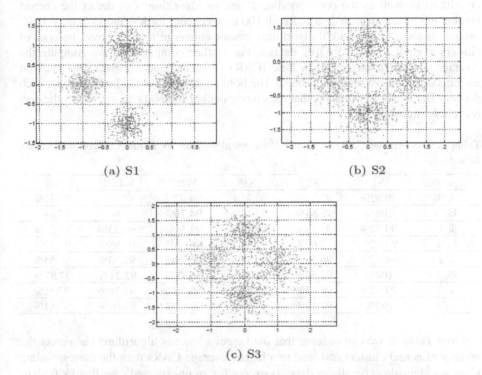

(a) S1   (b) S2

(c) S3

Fig. 1. The sketches of three typical synthetic datasets used in the experiments

We further introduce three real-world datasets used in the experiments. Actually, we select Wisconsin Breast Cancer (WBC) dataset, Wisconsin Diagnostic Breast Cancer (WDBC) dataset and Landsat Satellite (LS) dataset from UCI Machine Learning Repository [19], which are also used in [12]. WBC dataset contains 699 9-dimensional sample points belong to two classes. There are 458 sample points and 241 sample points in each class, respectively. In fact, there are 16 missing values in some sample points and we just set them as zeros. WDBC dataset contains 569 30-dimensional sample points also belong to two classes. There are 357 sample points and 212 sample points in each class, respectively. LS dataset is also a high dimensional and complicated real-world dataset. For simplicity, we only consider the first and second classes in the original dataset. So, the LS dataset used here contains 1551 37-dimensional sample points again belong to two classes. Particularly, there are 1072 sample points and 479 sample points in two classes, respectively. For these three real-world datasets, we typically set $k = 5$ for the four kernel k'-means algorithms.

To evaluate these kernel k'-means algorithm, we implement them on each of the above six datasets for 100 times with randomly selected initial clusters centers or means. For comparison, we also implement the corresponding four k'-means algorithms in the same way. For clarity, we refer to the k'-means algorithms with the discrepancy metrics given by Eq.(1) to Eq.(4) as the original, first, second, and third ones (denoted as k'0 to k'3). In the same way, we address their kernel versions and denote them as kk'0 to kk'3. It is found by the experiments that the four kernel k'-means algorithms as well as the corresponding k'-means algorithms can detect the correct number of clusters in each dataset for all 100 times except for k'0 and kk'0 on WDBC dataset. In fact, k'0 and kk'0 algorithms cannot converge to the correct number of clusters at each time on WDBC dataset. For further comparisons, we compute the Average Classification Accuracy Rate (CAR) of each algorithm over 100 trials on each datasets and list them in Table 2. The bold characters are the best CAR for each dataset. The symbol '--' means that the corresponding method can not obtain the correct cluster number.

Table 2. The average CARs of kernel k'-means algorithms vs k'-means algorithms on six datasets

Algorithm	S1	S2	S3	WBC	WDBC	LS
k'0	99.92%	97.49%	90.62%	94.02%	--	93.16%
kk'0	**100%**	**97.86%**	93.15%	94.70%	--	92.84%
k'1	99.92%	96.56%	89.52%	94.51%	90.33%	93.23%
kk'1	99.92%	97.62%	92.85%	96.35%	90.98%	93.32%
k'2	99.92%	96.55%	89.57%	96.70%	91.73%	97.55%
kk'2	**100%**	97.58%	**93.23%**	**96.85%**	**92.21%**	**97.87%**
k'3	99.92%	96.63%	89.68%	96.70%	91.38%	97.35%
kk'3	**100%**	97.42%	93.15%	**96.85%**	92.09%	97.81%

From Table 2, we can observe that the kernel k'-means algorithms can detect the number of actual clusters and lead to a higher average CARs than the corresponding k'-means algorithms on all the datasets except for an unexpected case that kk'0 algorithm has a slightly lower CAR than k'0 algorithm on LS dataset. Moreover, kk'2 behaves better than the other algorithms in most cases. However, kk'0 does not work well on WDBC and LS datasets, which may be related to the used discrepancy metric, kernel function as well as the particular structures of these two real-world datasets.

At last, we apply our kernel k'-means algorithms to unsupervised color image segmentation, which is a fundamental problem in image processing and can be treated as a clustering problem. Usually, the number of objects in an image is not pre-known. The image segmentation is in an unsupervised mode to automatically determine the number of objects and background in the image [20]. Kernel k'-means algorithms provide a new tool for unsupervised image segmentation. Here, we try to apply these

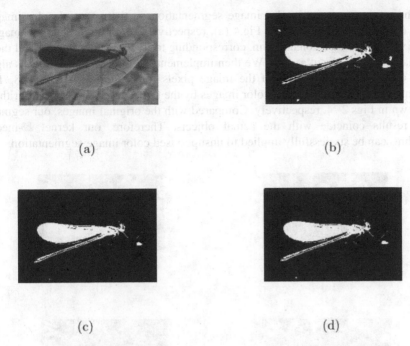

(a)

(b)

(c)

(d)

Fig. 2. The segmentation results on the color image of an insect. (a) The original image. (b)-(d) The segmentation results of the kernel k'-means algorithms kk'1, kk'2, and kk'3, respectively.

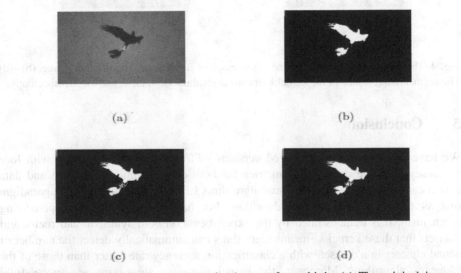

(a)

(b)

(c)

(d)

Fig. 3. The segmentation results on the color image of two birds. (a) The original image. (b)--(d) The segmentation results of the kernel k'-means algorithms kk'1, kk'2, and kk'3, respectively.

algorithms to unsupervised color image segmentation on three typical color images shown in Fig.2 (a), Fig. 3 (a) and Fig.4 (a), respectively. In these three color images, each pixel is a 3-dimensional datum, corresponding to its RGB coordinates, and these pixel data are normalized at first. We then implement the three kernel k'-means algorithms on the normalized data of the image pixels for clustering with $k = 8$. The segmentation results of the three color images by the three kernel k'-means algorithms are shown in Figs 2--4, respectively. Compared with the original images, our segmentation results coincide with the actual objects. Therefore, our kernel k'-means algorithms can be successfully applied to unsupervised color image segmentation.

(a)   (b)

(c)   (d)

Fig. 4. The segmentation results on the color image of flowers. (a) The original image. (b)--(d) The segmentation results of the kernel k'-means algorithms kk'1, kk'2, and kk'3, respectively.

5   Conclusion

We have established the kernelized versions of four k'-means algorithms with four frequency sensitive discrepancy metrics for both cluster number detection and data classification. These kernel k'-means algorithms still keep a simple learning paradigm just as the classical k-means algorithm, but have a rewarding and penalizing mechanism. It is demonstrated by the experiments on both synthetic and real-world datasets that these kernel k'-means algorithms can automatically detect the number of actual clusters in a dataset, with a classification accuracy rate better than those of the k'-means algorithms. Moreover, the kernel k'-means algorithms are successfully applied to unsupervised color image segmentation.

Acknowledgments. This work was supported by the Natural Science Foundation of China for Grant 61171138 and BGP Inc., China National Petroleum Corporation.

References

1. Ma, J., Wang, T., Xu, L.: A Gradient BYY Harmony Learning Rule on Gaussian Mixture with Automated Model Selection. Neurocomputing 56, 481–487 (2004)
2. MacQueen, J.: Some Methods for Classification and Analysis of Multivariate Observations. In: Proceedings of the fifth Berkeley Symposium on Mathematical Statistics and Probability, vol. 1, pp. 281–297 (1967)
3. Kanungo, T., Mount, D., Netanyahu, N., et al.: An Efficient k-means Clustering Algorithm: Analysis and Implementation. IEEE Transactions on Pattern Analysis and Machine Intelligence 24(7), 881–892 (2002)
4. Ma, J., Wang, T.: A Cost-function Approach to Rival Penalized Competitive Learning (RPCL). IEEE Transactions on Systems, Man, and Cybernetics, Part B: Cybernetics 36(4), 722–737 (2006)
5. Ma, J., He, X.: A Fast Fixed-point BYY Harmony Learning Algorithm on Gaussian Mixture with Automated Model Selection. Pattern Recognition Letters 29, 701–711 (2008)
6. Ma, J., Liu, J., Ren, Z.: Parameter Estimation of Poisson Mixture with Automated Model Selection through BYY Harmony Learning. Pattern Recognition 42, 2659–2670 (2009)
7. Ma, J., Liu, J.: The BYY Annealing Learning Algorithm for Gaussian Mixture with Automated Model Selection. Pattern Recognition 40, 2029–2037 (2007)
8. Xu, L., Krzyzak, A., Oja, E.: Rival Penalized Competitive Learning for Clustering Analysis. RBF Net, and Curve Detection. IEEE Transactions on Neural Networks 4(4), 636–649 (1993)
9. Ma, J., Cao, B.: The Mahalanobis Distance Based Rival Penalized Competitive Learning Algorithm. In: Wang, J., Yi, Z., Żurada, J.M., Lu, B.-L., Yin, H. (eds.) ISNN 2006. LNCS, vol. 3971, pp. 442–447. Springer, Heidelberg (2006)
10. Zalik, K.R.: An Efficient k'-means Clustering Algorithm. Pattern Recognition Letters 29, 1385–1391 (2008)
11. Fang, C., Ma, J.: A Novel k'-means Algorithm for Clustering Analysis. In: Proceedings of the 2nd International Conference on Biomedical Engineering and Informatics (2009)
12. Fang, C., Jin, W., Ma, J.: k'-means Algorithms for Clustering Analysis with Frequency Sensitive Discrepancy Metrics. Pattern Recognition Letters 34, 580–586 (2013)
13. Girolami, M.: Mercer Kernel-Based Clustering in Feature Space. IEEE Transactions on Neural Networks 13(3), 780–784 (2002)
14. Taylor, J.S., Cristianini, N.: Kernel Methods for Pattern Analysis. Cambridge University Press, London (2004)
15. Filippone, M., Camastra, F., Masulli, F., Rovetta, S.: A Survey of Kernel and Spectral Methods for Clustering. Pattern Recognition 41(1), 176–190 (2008)
16. Muller, K., Mika, S., Ratsch, G., Tsuda, K., Scholkopf, B.: An Introduction to Kernel-Based Learning Algorithms. IEEE Transactions on Neural Networks 12(2), 181–201 (2001)
17. Dhillon, I.S., Guan, Y., Kulis, B.: Kernel K-means, Spectral Clustering and Normalized Cuts. In: Proceedings of the tenth ACM SIGKDD International Conference on Knowledge Discovery and Data Mining, pp. 551–556 (2004)
18. Baudat, G., Anouar, F.: Generalized Discriminant Analysis Using a Kernel Approach. Neural Computation 12, 2385–2404 (2000)
19. UCI Machine Learning Repository, http://mlearn.ics.uci.edu/databases
20. Shi, J., Malik, J.: Normalized Cuts and Image Segmentation. IEEE Transactions on Pattern Analysis and Machine Intelligence 22(8), 888–905 (2000)

Recursive Feature Elimination Based on Linear Discriminant Analysis for Molecular Selection and Classification of Diseases

Edmundo Bonilla Huerta, Roberto Morales Caporal, Marco Antonio Arjona, and José Crispín Hernández Hernández

Laboratorio de Investigación en Tecnologías Inteligentes
Instituto Tecnológico de Apizaco Av. Instituto Tecnológico s/n C.P 90300
Apizaco, Tlaxcala, México
edbonn@hotmail.com, josechh@yahoo.com

Abstract. We propose an effective Recursive Feature Elimination based on Linear Discriminant Analysis (RFELDA) method for gene selection and classification of diseases obtained from DNA microarray technology. LDA is proposed not only as an LDA classifier, but also as an LDA's discriminant coefficients to obtain ranks for each gene. The performance of the proposed algorithm was tested against four well-known datasets from the literature and compared with recent state of the art algorithms. The experiment results on these datasets show that RFELDA outperforms similar methods reported in the literature, and obtains high classification accuracies with a relatively small number of genes.

Keywords: Gene Selection, Classification, LDA, RFE, Microarray, Filter.

1   Introduction

The DNA microarray is a recent and powerful technology that allows to be monitored and measured gene expression levels for tens of thousands of genes in parallel for a single tissue sample. This technology enables considering cancer diagnosis based on gene expressions [1,2,3]. Microarray gene expression data presents five important characteristics: small-sample size, high-throughput data, noise, irrelevant and redundant data. This causes a lot of difficulties for classifiers. Given the very high number of genes, it is important to apply an effective gene selection strategy to select a small number of relevant genes for classifying tissue samples.

This paper presents a RFELDA method to select the most informative genes by removing irrelevant or redundant genes for classification of diseases. We claim that the hyperplane obtained by SVM is as effective as the eigen vector obtained by LDA. For this study, four gene selection filters are used on the embedded approach and the Fisher's Linear Discriminant Analysis (LDA) is used to provide useful information to eliminate recursively irrelevant and redundant genes. LDA is a well-known method of dimension reduction and classification, where the data vectors are transformed into a

D.-S. Huang et al. (Eds.): ICIC 2013, LNAI 7996, pp. 244–251, 2013.

low-dimensional subspace so that the class centroids are spread out as much as possible. Recently, LDA has been used for several classification problems especially for microarray gene expression data [4,5,6].

2 State of the Art

One recent well-known embedded method for feature/gene selection is Support Vector Machine Recursive Feature Elimination (SVMRFE), proposed by Guyon et al. [7], The main idea of SVMRFE is that the orientation of the separating hyper-plane found by the SVM can be used to select informative genes. A similar idea is presented by Tang et al. [8] they propose a RFE algorithm named FCM-SVM-RFE for the gene selection of microarray gene expression data. In a first stage Fuzzy C-Means is proposed to build groups genes with similar functions into clusters. In a second stage RFE-SVM is used to select the most informative genes from each cluster. Recently, an improved method for cluster elimination based on SVM has been proposed by Luo et al. [9] called ISVM-RCE. This algorithm is composed of three steps: clustering, scoring gene with SVM and recursive elimination by removing small clusters of non relevant genes at each iteration step. Another recent work is Lagging Prediction Peephole Optimization (LPPO), proposed by Liu et al [10], which combines supervised learning and statistical similarity measures to choose the final optimal feature/gene set. This method deals with redundancy issues and improves classification for microarray data. A new idea of RFE is presented by Yang et al. [11], they developed a multicriterion fusion-based recursive feature elimination (MCF-RFE) to improve the robustness and stability of feature selection results.

3 Methodology Used

We propose a method to reduce the initial dimension of microarray datasets and to select a relevant gene subset for classification. Our model involves three basic steps: First, four statistical filters are proposed to filter relevant genes. Then the problem of gene selection and classification is treated by the RFE-SVM that guides the gene elimination process step by step.

3.1 Filter Methods

Microarray data generally contain less than one hundred samples described by at least several thousands of genes. In this first-stage, we limit this high dimensionality by a first pre-selection step (filter). In this paper, we use four well-known filters reported in the literature: BSS/WSS(BW)[4], t-statistics(TT)[12], Wilcoxon-test (WT)[13] and SNR (SN)[14]. In this work, genes were selected according to these filters to identify differentially expressed genes in two tissue types. In our study we retain $p=35$ informative genes for each disease obtained from the ADN microarray technology.

3.2   Linear Discriminant Analysis

LDA is one of the most commonly used technique for data classification and dimension reduction. Linear Discriminant Analysis projects the data into a low dimension space according to Fisher's criterion, that is to maximize the ratio of between-class scatter matrix variance S_B to the within-class scatter matrix variance S_W with the maximal separability. LDA considers to maximize the following criterion (J):

$$J(w)\frac{w^T S_B w}{w^T S_W w} \qquad (1)$$

The scatter matrices S_B and S_W are defined as follows:

$$S_B = \sum_c (\mu_c - \mu)(\mu_c - \mu)^T \qquad (2)$$

$$S_w = \sum_c \sum_{i \in c} (x_i - \mu_c)(x_i - \mu_c)^T \qquad (3)$$

where c are the classes, μ is the overall mean of the data-classes. Thus w is the eigen vector associated to the sole eigen value of $S_W^{-} S_B$ when S_W^{-1} exists. Once axis w is obtained, LDA provides a eigen vector and the absolute value of this vector indicates the relevancy of the q initial variables (genes) for the class discrimination.

In this process, LDA is used as a classification method to evaluate the classification accuracy that can be achieved on a selected gene subset. Moreover the coefficients of the eigen vector calculated by LDA are used to evaluate the relevancy of each gene for class discrimination. For a selected gene subset of size p, if $p << n$ we rely on the classical LDA to obtain the projection vector w, otherwise we apply the generalized LDA to obtain this vector.

3.3   RFE-LDA

LDA is proposed to train and to obtain the weight (eigen value) of each gene by removing the first-one with the smallest weight iteratively, until the last gene is obtained. In each iteration we use a validation method to train LDA classifier and to calculate the eigen value of each gene. Thus, we find a final gene subset that contains the most relevant genes with the highest performance. Our method RFELDA can be described as follows:

1. Given the reduced gene set obtained from a filter, set $G=\{g_1,g_2,...,g_p\}$.
2. Train LDA classifier.
3. Obtain discriminant coefficients of each gene from LDA classifier by using a validation method.
4. Eliminate the gene less relevant g_i from G and updates G, set $G=\{G-g_i\}$, set $p=p-1$.
5. Go to step 2 until $p=1$.

This recursive procedure is based on the SVMRFE strategy proposed by Guyon et al. [7], where each gene from initial data is evaluated in terms of its corresponding coefficient in the SVM classifier. In our case, LDA is proposed for this gene elimination and thus to find a small gene subset with the highest performance. We only remove a gene in each iteration. This process leads to an increase or decrease in the accuracy of the gene subset until a reduced gene subset is obtained.

4   Experimental Results

4.1   DNA Datasets

To evaluate the performance of our model, we have done extensive experiments on four DNA microarray gene expression datasets (see table 1 for a detailed description).

Table 1. DNA dataset characteristics

Dataset	Genes	Samples	# class 1	# class 2	Author
Leukemia	7129	72	25	47	[3]
Colon	2000	62	22	40	[1]
CNS	7129	60	21	39	[15]
Prostate	12600	109	59	77	[16]

For a leukemia dataset we use the following preprocessing process [4]:

1. Thresholding : Floor of 100 and ceiling of 16,000.
2. Filtering : Exclusion of genes with max/min<=5 or (max-min) <=500, where min and max refers to minimum and maximum intensities for a particular gene across the samples.
3. Natural logarithm: The natural logarithm $(\ln(x))$ was taken for each value.
4. Standardizing: Each sample was standardized to have a mean of 0 and a standard deviation of 1.

For the rest of the datasets, we normalized the gene expression levels of each dataset into intervals [0, 1] using the minimum and maximum expression values of each gene.

4.2   Experiments on Leukemia Dataset

In this section the performance of our model is evaluated by two validation methods: 10-FOLD Cross Validation and Bootstrap .632. We apply both methods 100 times. The final performance for leukemia is shown in Figure 1(a). We note that

RFELDA$_{BW}$+100FCV gives a perfect classification (100%) with only 3 genes. In contrast, RFELDA$_{BW}$+B.632 (Figure 1(b)) offers a very high performance by using a subset of 10 genes. Two similar results are generated by RFELDA$_{TT}$+100FCV and RFELDA$_{TT}$+B.632 that offer a minimal gene subset of 3 genes with an accuracy rate of 100% respectively. However, the best overall performance (100%) is obtained by RFELDA$_{SN}$+100FCV and RFELDA$_{SN}$+B.632, from a subset of 15 genes to a subset of 3 genes (see Figure 2(b)). Table 2 shows the small gene subset with only 3 genes obtained by this model.

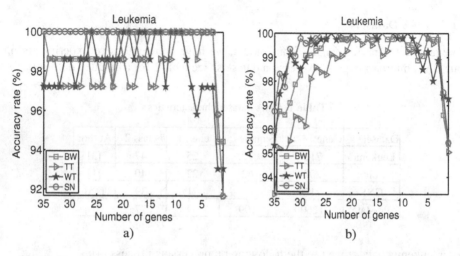

Fig. 1. Simulation results on leukemia dataset with validation method a) FCV b) B.632

Table 2. A perfect gene subset for leukemia dataset with the highest accuracy

Id-gene	Description
3847	U82759
2121	M63138
4377	X62654

4.3   Experiments on Colon Tumor Dataset

The performance vs. gene subset obtained by our model by using the four filters is shown in Figure 2. We observe that RFELDA$_{BW}$+100FCV and RFELDA$_{TT}$+100FCV give a subset of 8 genes with a performance of 95.1% (Figure 2(a)). However the best performance (96.2%) with a small gene subset (4) is obtained by RFELDA$_{BW}$+B.632 (Figure 2(b)). Nevertheless, RFELDA$_{SN}$+B.632 and RFELDA$_{BW}$+B.632 offer an accuracy rate of 98.5% and 98.8% respectively, the first-one by using 13 genes and the second-one 20 genes (see Table 3).

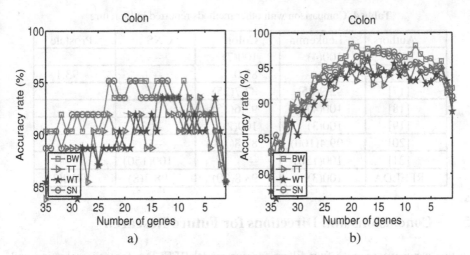

Fig. 2. Simulation results on colon tumor dataset with validation method a) FCV b) B.632

Table 3. Best gene subset for colon tumor dataset

Id-gene	Description	Id-gene	Description
267	M76378	365	X14958
765	M76378	918	X16356
377	Z50753	391	D31885
739	X12369	822	T92451
1494	X86693	419	R44418
1058	M80815	1873	L07648
625	X12671	882	R33367
138	M26697	187	T51023
581	T51571	802	X70326
1974	M64110	1280	X16354

4.4   Performance Comparisons

Our model is coded in matlab using a laptop intel® core ™ i7 CPU Q480 1.87 GHz and 4.00 GB RAM.

RFELDA is compared with several results from different works reported in the literature. Table 4 summarizes the best accuracies obtained by our model. The first column indicates a work reported in the literature. Second, third, four and fifth columns shown the accuracy obtained for each dataset. Each cell contains the classification accuracy and the minimal gene subset.

It can be seen that RFELDA reaches higher performances with a smaller number of genes for leukemia and prostate dataset. These results are obtained by two exhaustive validation methods. For the other works the validation methods are obtained by LOOCV or 10-Fold only in the training set. We apply 100 times each validation method to assess the stability and robustness of our results.

Table 4. Comparison with other methods reported in literature

Authors	Leukemia	Colon	CNS	Prostate
[7]	100(4)	100(7)	--	--
[9]	98.1	82.0	--	93.1
[17]	95.9(25)	87.7(25)	--	--
[18]	100(4)	93.6(15)	--	--
[19]	100(3)	100(2)	--	--
[20]	99.4(100)	87.8(100)	--	--
[21]	100(150)	--	100(150)	--
RFELDA	100(3)	98.8(20)	98.3(8)	98.0(3)

5   Conclusions and Directions for Future Research

In this paper we propose four filters for our model RFELDA for gene selection and classification task. In comparison with the original SVM-RFE our model gives much better performance in four microarray datasets. This approach can be further improved on several aspects. One way, involves finding gene subsets with higher classification and a small size with another classifier (SVM, GLDA). Another way is to include a multiple criteria or multi-objective criteria. In the future, it will be possible to incorporate more gene selection filters such as: Mutual Information or Minimal Redundancy-Maximal Relevancy method.

Acknowledgments. We would like to thank the reviewers for their comments. This work is carried out within the DGEST project PROIFOPEP: 4505.12-P.

References

1. Alon, U., Barkai, N., et al.: Broad patterns of gene expression revealed by clustering analysis of tumor and normal colon tissues probed by oligonucleotide arrays. Proc. Nat. Acad. Sci. USA (1999)
2. Alizadeh, A., Eisen, M.B., et al.: Distinct types of diffuse large (b)–cell lymphoma identified by gene expression profiling. Nature, 503–511 (2000)
3. Golub, T., Slonim, D., et al.: Molecular classification of cancer: Class discovery and class prediction by gene expression monitoring. Science 537, 286 (1999)
4. Dudoit, S., Fridlyand, J., Speed, T.: Comparison of discrimination methods for the classification of tumors using gene expression data. Journal of the American Statistical Association 97, 77–87 (2002)
5. Ye, J., Li, T., Xiong, T., Janardan, R.: Using uncorrelated discriminant analysis for tissue classification with gene expression data. IEEE/ACM Trans. Comput. 1(4), 181–190 (2004)
6. Yue, F., Wang, K., Zuo, W.: Informative gene selection and tumor classification by null space LDA for microarray data. In: Chen, B., Paterson, M., Zhang, G. (eds.) ESCAPE 2007. LNCS, vol. 4614, pp. 435–446. Springer, Heidelberg (2007)

7. Guyon, I., Weston, J., Barnhill, S., Vapnik, V.: Gene selection for cancer classification using support vector machines. Machine Learning 46(1-3), 389–422 (2002)
8. Tang, Y., Zhang, Y.-Q., Huang, Z.: Fcmsv- rfe gene feature selection algorithm for leukemia classification from microarray gene expression data. In: IEEE International Conference on Fuzzy Systems, pp. 97–10 (2005)
9. Luo, L.-K., Feng, D., Ye, L.-J., Zhou, Q.-F., Shao, G.-F., Peng, H.: Improving the computational efficiency of recursive cluster elimination for gene selection. IEEE/ACMTransactions on Computational Biology and Bioinformatics 8(1), 122–129 (2011)
10. Liu, Q., Sung, H.: Gene selection and classification for cancer microarray data based on machine learning and similarity measures. BMC Genomics 12(5), 1–12 (2011)
11. Yang, F., Mao, K.: Robust feature selection for microarray based on multicreterion fusion. IEEE/ACMTrans. Comput. Biology 8(4), 1080–1092 (2011)
12. Li, Z., Zeng, X.-Q., Yang, J.-Y., Yang, M.-Q.: Partial Least Squares based dimension reduction with gene selection for tumor classification. In: BIBE 2007, pp. 1439–1444 (2007)
13. Deng, L., Pei, J., Ma, J., Lee, D.L.: Rank sum test method for informative gene discovery. In: 10th ACM SIGKDD International Conference on Knowledge Discovery and Data Mining (KDD 2004), pp. 410–419 (2004)
14. Mishra, D., Sahu, B.: Feature selection for cancer classification: A signal-to-noise ratio approach. International Journal of Scientific & Engineering Research 2(4), 1–7 (2011)
15. Pomeroy, S.-L., Tamayo, P., et al.: Prediction of central nervous system embryonal tumour outcome based on gene expression. Nature 415, 436–442 (2002)
16. Singh, D., Febbo, P., Ross, K., Jackson, D., Manola, J., Ladd, C., Tamayo, P., Renshaw, A., D'Amico, A., Richie, J.: Gene expression correlates of clinical prostate cancer behavior. Cancer Cell 1, 203–209 (2002)
17. Cho, S.-B., Won, H.-H.: Cancer classification using ensemble of neural networks with multiple significant gene subsets. Applied Intelligence 26(3), 243–250 (2007)
18. Li, S., Wu, X., Hu, X.: Gene selection using genetic algorithm and support vectors machines. Soft Computing 12(7), 693–698 (2008)
19. Alba, E., García-Nieto, J., Jourdan, L., Talbi, E.-G.: Gene selection in cancer classification using PSO/SVM and GA/SVM hybrid algorithms. In: Congress on Evolutionary Computation, pages, pp. 284–290 (2007)
20. Satoshi, N., Okuno, Y.: Lapalacian linear discriminant analysis to unsupervised feature selection. IEEE/Transactions on Biology and Bioinformatics 6(4), 605–614 (2009)
21. Li, X., Peng, S., Zhan, X., Zhang, J., Xu, Y.: Comparison of feature selection methods for multiclass cancer classification based on microarray data. In: 4th International Conference on Biomedical Engineering and Informatics (BMEI), pp. 1692–1696 (2011)

Inferring Transcriptional Modules from Microarray and ChIP-Chip Data Using Penalized Matrix Decomposition

Chun-Hou Zheng, Wen Sha, Zhan-Li Sun, and Jun Zhang

College of Electrical Engineering and Automation, Anhui University, Hefei, Anhui, China
zhengch99@126.com

Abstract. Inferring transcriptional regulatory modules is a useful work for elucidating molecular mechanism. In this paper, we propose a new method for transcriptional regulatory module discovering. The algorithm uses penalized matrix decomposition to model microarray data. Which takes into account the sparse a prior information of transcription factors--gene (TFs--gene) interactions. At the same time, the ChIP-chip data are used as constraints for penalized matrix decomposition of gene expression data. Finally the regulatory modules can be inferred based on the factor matrix. Experiment on yeast dataset shows that our method can identifies more meaningful transcriptional modules relating to specific TFs.

Keywords: Transcriptional modules, penalized matrix decomposition, gene expression data, ChIP-chip data.

1 Introduction

Genes are co-expressed under regulation by transcriptional factors to carry out complex functions in living cells. Reconstruction the regulatory networks from the high throughput "omics" data is one of the foremost challenges of current bioinformatics research [1]. In despite of the development of high throughput technologies such as macroarray, next generation sequencing[2], etc., the problem of network reconstruction remains highly underdetermined since the experiment data is not sufficient to estimate all parameters of the network models. Fortunately, many studies have unveiled that regulatory networks are modular and hierarchically organized [3]. So, we can reconstruct the whole regulatory network start with indentifying the modules. Moreover, inferring modules instead of full networks drastically reduces the complexity of the inference problem and shows great promise for systems biology research.

Traditionally, identification methods inferring modules from different 'omics' data sources separately, e.g., solely based on microarrays. Until now, microarray gene expression data have been extensively used for inferring transcriptional regulatory modules. Based on microarray data, many computational methods have been proposed for transcriptional modules identification. These methods can be roughly

D.-S. Huang et al. (Eds.): ICIC 2013, LNAI 7996, pp. 252–259, 2013.

divided into three categories: clustering methods [4,5], model-based methods [6] and projection methods [7] .

Clustering methods identify transcriptional modules by assuming that genes with similar expression profiles share similar functions or the same pathway. However, experimental results have shown that genes involved in the same biological process or pathway can have different expression patterns [8]. Model-based approaches usually model gene expression data as a linear mixture of latent variables corresponding to specific biological sources. One challenge of these approaches is the lack of sufficient data to estimate the parameters[9]. Projection methods decompose the dataset matrix into desired components that are constrained to be mutually either uncorrelated or independent. Since matrix decomposition methods do not cluster genes based on pairwise similarity measurement, functionally related genes showing different expression patterns can be clustered[14].

Whereas many achievements about modules inferring have been obtained using gene expression data solely, however, simultaneous analysis of distinct data sources has a major advantage over their separate analysis: their integration allows gaining holistic insight into the network and a more refined definition of transcriptional modules can be derived [10]. Therefore, several recent approaches, including GRAM [11], COGRIM [12] and ReMoDiscovery [13], have been developed to infer transcriptional regulatory modules by combine several data sources such as gene expression data and transcription factor (TF) binding information.

Gene expression is believed to be regulated by a small (compared to the total number of genes) number of transcription factors (TFs) which act together to maintain the steady-state abundance of specific mRNAs. It can be assumed that the expression of each gene is influenced by only a small subset of the possible TFs and that these TFs influence their targets to various degrees.

From another view point, each TF also only regulate a small number of genes. According to this fact, we developed a PMD based method to discover the transcriptional modules from microarray data[14]. In this paper, we extend that PMD based method to incorporating TF binding data for inferring TF-mediated regulatory modules. The contribution of the paper is that a new transcriptional modules discovery method based on PMD is proposed.

2   Penalized Matrix Decomposition

Supposing that there is a a matrix X with size $p \times n$. Assuming that the column and row means of X are zero, the singular value decomposition (SVD) of X can be defined as follows:

$$X = UDV^T, \quad U^TU = I_p, \quad V^TV = I_n \quad (1)$$

By imposing additional sparse constraints on U and V, Witten et al. proposed the PMD method [15]. Which is a new framework for computing a rank K approximation for the matrix X.

For simplicity, let's first considering the following rank-1 optimization problem:

$$\underset{d,\mathbf{u},\mathbf{v}}{\text{minimize}}\frac{1}{2}\left\|X - d\mathbf{u}\mathbf{v}^T\right\|_F^2 \tag{2}$$

$$\text{s.t. } \left\|\mathbf{u}\right\|_2^2 = 1, \left\|\mathbf{v}\right\|_2^2 = 1, P_1(\mathbf{u}) \le \alpha_1, P_2(\mathbf{v}) \le \alpha_2, d \ge 0.$$

where u is a column of U, v is a column of V, d is a diagonal element of D, $\left\|\bullet\right\|_F$ is the Frobenius norm and P_1 and P_2 are convex penalty functions. It can be proven that only certain ranges of α_1 and α_2 can lead to feasible solutions, and the value of d fit for Eq.(2) is $d = \mathbf{u}^T X\mathbf{v}$ [15].

It has been proved in literature [15] that:

$$\frac{1}{2}\left\|X - UDV^T\right\|_F^2 = \frac{1}{2}\left\|X\right\|_F^2 - \sum_{k=1}^{K}\mathbf{u}_k^T X\mathbf{v}_k d_k + \frac{1}{2}\sum_{k=1}^{K}d_k^2 \tag{3}$$

Here U and V are orthogonal matrices with size $p\times K$ and $n\times K$ respectively, D is a diagonal matrix with diagonal elements d_k. According to Eq.(3) it can be seen that, when K=1, the u and v which satisfy Eq.(2) can also solve the following optimization problem:

$$\underset{\mathbf{u},\mathbf{v}}{\text{maxmize}}\ \mathbf{u}^T X\mathbf{v} \tag{4}$$

$$\text{s.t.} \left\|\mathbf{u}\right\|_2^2 = 1, \left\|\mathbf{v}\right\|_2^2 = 1, P_1(\mathbf{u}) \le \alpha_1, P_2(\mathbf{v}) \le \alpha_2$$

To solve this optimization problem, we first finesse Eq.(4) to the following biconvex optimization:

$$\underset{\mathbf{u},\mathbf{v}}{\text{maxmize}}\ \mathbf{u}^T X\mathbf{v} \tag{5}$$

$$\text{s.t.} \left\|\mathbf{u}\right\|_2^2 \le 1, \left\|\mathbf{v}\right\|_2^2 \le 1, P_1(\mathbf{u}) \le \alpha_1, P_2(\mathbf{v}) \le \alpha_2$$

It can be proven that the solution to Eq.(5) also satisfies Eq.(4) provided that α is chosen appropriately [15]. Eq.(5) is named rank-1 PMD. Since it is biconvex, it can be optimized by an iterative algorithm. The detailed iterative algorithm can be found in [15].

In order to obtain multiple factors of U and V, i.e., $\mathbf{u}_1\cdots\mathbf{u}_k$ and $\mathbf{v}_1\cdots\mathbf{v}_k$, using PMD, we can maximize the single factor criterion in Eq.(5) repeatedly, each time replacing the matrix X using the residuals matrix, i.e., $X^{k+1} \leftarrow X^k - d_k\mathbf{u}_k\mathbf{v}_k^T$.

The detailed algorithm to obtain multiple factors of the PMD and simulated examples to show the efficacy of PMD for discovering the latent factors can be found in [15]. In this paper, we take the l_1-norm of u and v as the penalty function, i.e. $\left\|\mathbf{u}\right\|_1 \le \alpha_1, \left\|\mathbf{v}\right\|_1 \le \alpha_2$. By choosing appropriate parameters α_1 and α_2, PMD can result in sparse factors u and v. Generally speaking, α_1 and α_2 should be restricted to the ranges $1 \le \alpha_1 \le \sqrt{p}$ and $1 \le \alpha_2 \le \sqrt{n}$ [15].

3   Inferring Transcriptional Modules Based on PMD

Considering the gene expression data matrix X with size $p \times n$, each row of X containing the expression levels of a gene in all the n samples, and each column of X containing the expression levels of all the p genes in one sample. Typically, the number of genes, i.e., p, is in the thousands, while the number of samples, i.e., n, is typically less than hundreds.

Mathematically, we factorize matrix X into two matrices

$$X \sim UH \tag{6}$$

Where H is a $k \times n$ matrix. Each row of H is a latent factor, i.e. metagene. U is a $p \times k$ matrix with each entry u_{ij} representing the coefficient of metagene j in gene i.

Suppose there is a configuration matrix C with size $p \times l$ obtained from ChIP-chip data with l TFs. Where $c_{ij} = 1$ represents gene i is regulated by TF j. Combined the matrix C, we develop a constrained PMD approach to model the gene expression data, which take into account of the combinatorial and sparse nature of gene regulation by TFs. Detailedly, during inferring U in Equation (6), we done the factorization $X \sim UDV^T = UH$ by using PMD with constraints that

$$U = U \cdot C \tag{7}$$

Where ' $\cdot$ ' denotes an element-by-element product, which means that the element u_{ij} is non-zero only when $c_{ij} = 1$. Also, sparse constraint is imposed on U. Since D is a diagonal matrix, H is only a linear resize of V^T. By choosing appropriate α_1, we can achieve a sparse matrix U, i.e. many entries of U are zero. By incorporating TF-target gene relationships obtained from ChIP-chip data into PMD, we can infers the regulate network structure more accurate and robust.

In PMD algorithm, how to choose k is an open question. Generally speaking, when combined with ChIP-chip data, k should take 1 since C with size $p \times l$. In PMD, the factors u and v are calculated one by one. So in this paper, we can take $k=1$, and run the PMD l times. Each time using c_j (the j-th column of C, $j = 1, \cdots, l$) as constraint. Since c_j represents the probability of all the genes regulated by TF j, we can select the genes regulated by a special TF one time.

The proposed constrained rank-one PMD algorithm is summarized as follows:

Step 1. Initialize v to have L_2 norm 1.
Step 2. Iterate until convergence:

    (a) $u \leftarrow \arg\max_u u^T X v$ subject to $P_1(u) \leq \alpha_1$ and $\|u\|_2^2 \leq 1$.

    (b) $v \leftarrow \arg\max_v u^T X v$ subject to $P_2(v) \leq \alpha_2$ and $\|v\|_2^2 \leq 1$.

    (c) $u \leftarrow u \cdot c$.

Step 3. $d \leftarrow u^T X v$.

4   Experimental Results

4.1   Datasets

Two publicly available yeast gene expression datasets were used for assessing the algorithm proposed in this paper. The first one is yeast cell cycle dataset, which was detected using DNA microarray under the normal growth condition[16]. This dataset contains 73 samples. The second is a stress response dataset with 173 samples, determined under different experimental conditions such as temperature shocks, amino acid starvation, nitrogen source deletion and progression into stationary phase [17]. The missing data in the two datasets were filled using the KNNimpute approach [18].

The yeast ChIP-chip dataset used for the studies was obtained from a published work [19]. The dataset contains 203 TFs, with all profiled in a rich medium and 84 profiled under multiple stress conditions.We used the TF-gene pairs that have P-values <0.1 to construct the configuration matrix C in our analysis.

There are 6089 genes contained both in the two yeast gene expression datasets and the ChIP-chip dataset. So in this study, the microarray data matrix X with size 6089×246, and the configuration matrix C with size 6089×203.

4.2   Biological Assessment of the Results

To assess biological relevance of the inferred regulatory modules, we examined whether the identified TF-regulated transcriptional modules accumulate in certain Gene Ontology (GO) categories by conducting over-representation analysis. Which was used to detect whether a GO term is enriched in a transcriptional network. In this study, we take advantage of GO:TermFinder (http://search.cpan.org/dist/GO-TermFinder) to conduct over-representation analysis[20]. Since the over-representation analysis was applied to many GO categories, we further conducted the false discovery rate (FDR) correction, which provides strong control to have less false negatives at the cost of a few more false positives.

4.3   Over-Representation Analysis of Transcriptional Modules

We identified TF-mediated transcriptional modules using our method based on the microarray data and ChIP-chip data derived from the yeast. 19 TFs that are involved in the cell cycle and stress response, such as ABF1, ACE2, FKH1, etc. (see Table 1), are chosen to identify their regulated genes. We then examined the functional relevance of the target gene clusters using GO:TermFinder. The results were listed in table 1. Which shows the statistical enrichment of functional GO terms in the target gene clusters. The enrichment level was calculated by transforming the enrichment P-values after FDR correction to the negative log values and averaged over all functional modules for corrected P < 0.05. If no functional modules are found for corrected P < 0.05, the smallest value of corrected P is taken for calculating the enrichment level.

To show the effective of our algorithm, we also identified the modules using other three similar methods, i.e. GRAM [11], COGRIM [12] and ReMoDiscovery [13]. These three methods can be used to identify transcriptional modules that are coregulated by TFs through integrated analysis of microarray, ChIP-chip and TF motif data. The results are also listed in Table 1.

Table 1. The over-representation analysis of transcriptional modules identified by different methods

Clusters	Method			
	PMD	GRAM	CORRIM	ReMoDis
ABF1	8.7852	6.12	5.74	4.55
ACE2	7.7655	1.60	3.59	4.47
FKH1	6.9020	4.79	1.76	4.05
FKH2	5.1541	1.28	5.82	6.09
GCN4	3.6021	6.64	7.40	7.61
LEU3	3.6540	7.29	6.44	5.20
MBP1	5.2753	6.15	4.93	6.17
MCM1	3.6617	6.87	5.97	6.74
NDD1	3.7601	1.82	1.66	4.49
RAP1	2.7273	7.37	8.92	6.60
REB1	4.5098	5.03	5.53	5.10
STB1	17.4937	4.43	3.91	6.51
SWI4	4.4123	5.61	5.76	6.42
SWI5	3.5236	2.36	5.70	4.78
SWI6	5.1594	4.58	4.79	4.90
HSF1	6.7823	4.42	4.40	4.38
MSN4	6.8485	5.51	5.67	5.72
SKN7	5.6919	6.75	4.55	2.91
YAP1	2.9443	5.69	6.64	5.92
Average	5.7186	4.96	5.22	5.40

From Table 1 it can be seen that, our algorithm out-performs the other methods on the functional enrichment in the target gene clusters. The averaged enrichment level over all the 19 clusters in over-representation analysis was the highest by our algorithm (5.72), followed by ReMoDiscovery (5.40), COGRIM (5.22) and GRAM (4.96). This implies that our PMD based method can identify more functionally coherent genes regulated by specific TFs.

5   Conclusions

In this paper, we present a novel method for discovering transcriptional regulatory modules by integrated analysis of microarray and ChIP-chip data. Compared published methods, our method take full advantage of the sparse a prior information of regulation between genes and TFs. Experimental results on yeast dataset demonstrated the usefulness of our method. Novel TF-target interactions were predicted and new insights into the regulatory mechanisms of the cell can be inferring from these interactions. In the future, more efficient sparse model suit for gene expression data should be studied to discover more regulate modules.

Acknowledgements. This work was supported by the National Science Foundation of China under Grant Nos. 61272339&61271098, the Natural Science Foundation of Anhui Province under Grant No. 1308085MF85, and the Key Project of Anhui Educational Committee under Grant No. KJ2012A005.

References

1. Cavalieri, D., De Filippo, C.: Bioinformatic Methods for Integrating Whole-Genome Expression Results Into Cellular Networks. Drug Discov. Today 10, 727–734 (2005)
2. The Cancer Genome Atlas Network.: Comprehensive molecular characterization of human colon and rectal cancer. Nature 487, 330–337 (2012)
3. Segal, E., et al.: A Module Map Showing Conditional Activity of Expression Modules in Cancer. Nat. Genet. 36, 1090–1098 (2004)
4. Mukhopadhyay, A., Maulik, U.: Towards Improving Fuzzy Clustering Using Support Vector Machine: Application to Gene Expression Data. Pattern Recognition 42(11), 2744–2763 (2009)
5. Fernandez, E.A., Balzarini, M.: Improving Cluster Visualization in Self-Organizing Maps: Application in Gene Expression Data Analysis. Computers in Biology and Medicine 37(12), 1677–1689 (2007)
6. Dueck, D., Morris, Q.D., Frey, B.J.: Multi-way Clustering of Microarray Data Using Probabilistic Sparse Matrix Factorization. Bioinformatics 21(suppl. 1), i144–i151 (2005)
7. Huang, D.S., Zheng, C.H.: Independent Component Analysis Based Penalized Discriminant Method for Tumor Classification using Gene Expression Data. Bioinformatics 22(15), 1855–1862 (2006)
8. Zhou, X.J., et al.: Functional Annotation and Network Reconstruction Through Cross-Platform Integration of Microarray Data. Nat. Biotechnol. 23, 238–243 (2005)
9. Liao, J.C., Boscolo, R., Yang, Y.L., Tran, L.M., Sabatti, C., Roychowdhury, V.P.: Network Component Analysis: Reconstruction of Regulatory Signals In Biological Systems. Proc. Natl. Acad. Sci. USA 100, 15522–15527 (2003)
10. Van den Bulcke, T., Lemmens, K., Van de Peer, Y., Marchal, K.: Inferring Transcriptional Networks by Mining 'Omics' Data. Current Bioinformatics 1(3), 301–313 (2006)
11. Bar-Joseph, Z., et al.: Computational Discovery of Gene Modules And Regulatory Networks. Nat. Biotechnol. 21, 1337–1342 (2003)
12. Chen, G., et al.: Clustering of Genes Into Regulons using Integrated Modeling-COGRIM. Genome Biol. 8, R4 (2007)

13. Lemmens, K., et al.: Inferring Transcriptional Modules from Chip-Chip, Motif and Microarray Data. Genome Biol. 7, R37 (2006)
14. Zhang, J., Zheng, C.H., Liu, J.X., Wang, H.: Discovering the Transcriptional Modules using Microarray Data by Penalized Matrix Decomposition. Computers in Biology and Medicine 41(11), 1041–1050 (2011)
15. Witten, D.M., Tibshirani, R., Hastie, T.: A Penalized Matrix Decomposition, with Applications to Sparse Principal Components and Canonical Correlation Analysis. Biostatistics 10(3), 515–534 (2009)
16. Spellman, P.T., Sherlock, G., Zhang, M.Q., Iyer, V.R., et al.: Comprehensive Identification of Cell Cycle-Regulated Genes of the Yeast Saccharomyces Cerevisiae by Microarray Hybridization. Mol. Biol. Cell 9, 3273–3297 (1998)
17. Gasch, A.P., Spellman, P.T., Kao, C.M., Carmel-Harel, O., Eisen, M.B., Storz, G., Botstein, D., Brown, P.O.: Genomic Expression Programs in the Response of Yeast Cells to Environmental Changes. Mol. Biol. Cell 11, 4241–4257 (2000)
18. Troyanskaya, O., Canto'r, M.: Missing Value Estimation Methods for DNA Microarrays. Bioinformatics 17, 520–525 (2001)
19. Harbison, C.T., et al.: Transcriptional Regulatory Code of a Eukaryotic Genome. Nature 431, 99–104 (2004)
20. Boyle, E.I., Weng, S., Gollub, J., Jin, H., Botstein, D., Cherry, J.M., Sherlock, G.: GO:TermFinder–Open Source Software for Accessing Gene Ontology Information and Finding Significantly Enriched Gene Ontology Terms Associated With A List Of Genes. Bioinformatics 20(18), 3710–3715 (2004)

Mathematical Inference and Application of Expectation-Maximization Algorithm in the Construction of Phylogenetic Tree

Kai Yang and Deshuang Huang

College of Electronic and Information Engineering, Tongji University,
Shanghai 201804, China
yangzhenhaoyu@163.com

Abstract. The central task in the research of molecular evolution may be the reconstruction of a phylogenetic tree from sequences of taxa. So, phylogenetic tree is also known as the evolutionary tree, which uses a kind of graph similar to the tree branch to represent the genetic relationship among the organisms and infers the evolutionary history of the species based on the study of biological sequences. The result of phylogenetic analysis is usually expressed in the form of phylogenetic tree. This paper makes estimations on the parameters of the phylogenetic tree by EM algorithm to estimate the optimum length of the branch under parameter model to pave the way for constructing a better phylogenetic tree.

Keywords: Mathematical inference, EM algorithm, Phylogenetic analysis.

1 Introduction

Molecular phylogenetic analysis is considered to be one of the most important fields of Bio-informatics which aims to reconstruct a phylogenetic tree from a group of homologous DNA, protein sequences or other biological sequences, by calculating the evolutionary distance among them to show their evolutionary relationship [1]. Generally speaking, phylogenetic analysis includes the following steps: Sequence alignment, the establishment of nucleotide substitution model, the construction of the phylogenetic tree and the estimation of the phylogenetic tree.

The traditional probability model of nucleotide substitution is a useful preliminary approximation, such as dealing with the sequences without insertions or deletions. This paper extends the study to this assumption, we mainly focus on the evolution of the DNA sequences containing defect data to pave the way of the establishment of phylogenetic tree with good performance, then we establish the parameters estimation theorem based on the defect data under the parameter model.

The remainder of this paper is organized as follows: Section 2 introduces the basic knowledge about molecular phylogenetic analysis. Section 3 shows the mathematical inference procedure and the parameter estimation of two models: the Jukes-Cantor single parameter model and the Kimura double parameter model, in Section 4, we

D.-S. Huang et al. (Eds.): ICIC 2013, LNAI 7996, pp. 260–266, 2013.

present the experiment results and make a comparison with the other related models, and then analyse the results in Section 5. Conclusion and future directions are depicted in Section 6.

2   Phylogenetic Analysis and Estimation Models

The general understanding that many biological sequences share a single common origin is fundamental to biology. Such a set of contemporary sequences that diverged from their ancestral sequence in a tree-like fashion. Inferring this phylogenetic tree has been a major research problem since the dawn of computational molecular biology.

2.1   Phylogenetic Analysis

The central task in the study of molecular evolution is the reconstruction of a phylogenetic tree from the sequences of taxa. In theory, a phylogenetic tree is usually a binary tree, whose leaf nodes stand for the species or the organisms [2]. The tree topology indicates the phylogenetic relationship, and the length of branches figures out the evolutionary distance. Considering a phylogenetic tree of N leaves, we call the observed d sequences external nodes, Each site nucleotide expectations for the algebra of two sequences is defined as the evolutionary distance which is equal to the branch length d . Nucleotide substitution may have transition or transversion. Translation means a purine is replaced by another different purine, or a pyrimidine by another different pyrimidine; Other nucleotide substitution is called transversion [3].

2.2   Single Parameter Model

In Jukes-Cantor single parameter model, d is the evolutionary distance between two sequences, $\tilde{P}_{xx}(d)$ represents the invariant probability of nucleotide, $\tilde{P}_{xy}(d)$ represents the probability when nucleotide changes. the nucleotide substitution probability of each site between two sequences is

$$P_{xy}(d) = \begin{cases} \frac{1}{4}\left[1+3\exp\left(-\frac{4}{3}d\right)\right] \equiv \tilde{P}_{xx}(d), \text{ when } x = y; \\ \frac{1}{4}\left[1-\exp\left(-\frac{4}{3}d\right)\right] \equiv \tilde{P}_{xy}(d), \text{ when } x \neq y, \end{cases} \tag{1}$$

2.3   Double Parameter Model

In Kimura double parameter model, s represents the probability of nucleotide transversion, u represents the probability of nucleotide transition, r represents nucleotide invariant probability.the nucleotide substitution probability of each site between two sequences can be expressed as

$$P_{xy}(\gamma,\phi) = \begin{cases} \dfrac{1}{4}[1 - \exp(-\phi)] \cong s \\[2mm] \dfrac{1}{4}[1 + \exp(-\phi) - 2\exp(-\gamma)] \cong u \\[2mm] \dfrac{1}{4}[1 + \exp(-\phi) + 2\exp(-\gamma)] \cong r \end{cases} \tag{2}$$

3   Parameter Estimation

The root node is the common ancestor of the observation sequence, which represents the development direction of the test species. According to the stationary distribution definition and the ergodic property of Markov chain, the probability of the root node obeys the stable distribution π_j, assuming r_j represents the site number where the root nodes is j, then the probability likelihood of the root node is $\pi_j^{r_j}$, the evolution probability likelihood of the topology tree is the production of those nodes that have evolutionary relationship [9].

$$L(d) = \prod_{j \in D} \pi_j^{r_j} \prod_{i=1}^{n} \prod_{(k,h) \in E_{T_0}} P_{s_k v_h}(d_i) \prod_{j=1}^{n-2} \prod_{(h,q) \in E_{T_1}} P_{v_h v_q}(d_j) \tag{3}$$

$$L(d) = C + \sum_{i=1}^{2n-2} \sum_{(k,h) \in E_T} \sum_{(x,y) \in D} \left\{ l_0^i \log \tilde{P}_{xx}(d_i) + (m_1^i + m_2^i) \log \tilde{P}_{xy}(d_i) \right\} \tag{4}$$

Then we use EM algorithm. In the E-Step, we use the current tree topology and edge lengths to compute expected sufficient statistics, which summarize the data. In the M-Step, we search for a topology that maximizes the likelihood with respect to these expected sufficient statistics:

E-step:

$$Q(d, d^{t-1}) = E[l(d) | l_0, m_1, d^{t-1}] \tag{5}$$

M-step:

$$\frac{\partial Q}{\partial d_1} = 0, \frac{\partial Q}{\partial d_2} = 0, ..., \frac{\partial Q}{\partial d_{2n-2}} = 0 \tag{6}$$

$$d_i(\max) = -\frac{3}{4}\log\left(1 - \frac{4}{3}\frac{m_1^i + E(m_2^i)}{l_0^i + m_1^i + m_2^i}\right) \qquad (i = 1, 2, ..., 2n-2) \tag{7}$$

4   Experiments

To construct the phylogenetic tree, firstly we should obtain the evolutionary distance among all taxonomical groups [10]. The construction of phylogenetic tree is based on the relationships among these distance values, the phylogenetic tree is constructed by limited data [11], so we also need to measure the reliability and accuracy of the tree in order to make comparisons of the performance among various methods. The EM algorithm to estimate the branch length has better statistical performance. However when the number of the species is too large, it will require considerable computation time, the accuracy will also has a certain decline, while it still performs better than phylogenetic tree constructed in general conditions [12]. By simulation, we consider that when the sequence divergence is larger, the method of parameter estimation has superiority in constructing the phylogenetic trees.

In order to prove the accuracy of parameter estimation and the superiority of the statistics, the following experiment is be done: By computer simulation, we use a model tree, and let a group of DNA sequences evolve in accordance with the nucleotide substitution model to reconstruct a tree by the DNA sequences, and then compare it with the actual tree.

Tree A and Tree B respectively represent the model tree and the actual tree constructed by these sequences used in the simulation. Assuming that the replacement of the nucleotide proceeds according to certain proportion, by using the pseudo random number, we can obtain the number of substitution in the actual tree branches [13], in the simulation, the number of nucleotides in each sequence is 500, tree A and tree B are identified by the replacement number of each nucleotide sequence, tree C shows the phylogenetic tree constructed according to the best branch length obtained by parameter estimation, at the same time we obtain the evolutionary tree consisted of

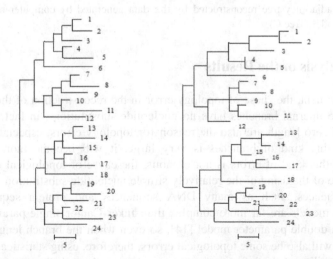

Fig. 1. The model tree composed of 24 nucleotide sequences (Tree A, Left); The actual tree obtained by computer simulation (Tree B, Right)

these sequences, when comparing to the actual tree, we found that tree C has an error, the trichotomy sites of sequence 8,9,10 in the actual tree is decomposed into two consecutive and different sites in tree C, although the branch length of the two sites is close to 0. The cause of this error is that we use the NJ method to construct the phylogenetic tree, except the small difference, the tree is basically the same as the actual tree, the branch length is also close to the tree B, what is worth noticing is that here we use the Jukes-Cantor distance instead of the Kimura distance, because the former is more appropriate in current case, and the difference is not large as well.

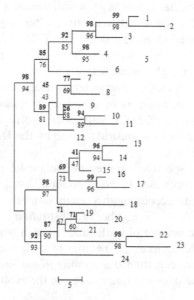

Fig. 2. The adjacency tree reconstructed by the data generated by computer and parameter estimation model (Tree C)

5　Analysis of the Results

In the experiment, the cause of topology error in the reconstruction of the tree is that one or more internal branches have no nucleotide substitutions, in fact, the internal branches of zero length are also the reason for topology errors, especially when the amount of this kind of branches is very large, it will become more prominent. Fortunately this kind of errors is not obvious, the cause of topological errors might also because of the using of the relatively simple nucleotide substitution model, most of the sequences, such as many DNA sequences and protein sequences, real substitution models are far more complex than Jukes-Cantor single parameter model and Kimura double parameter model [14], so even when the branch length d is not great, there will also be some topological errors, therefore, using statistical methods to test the reliability of the inference trees is necessary, through the way of estimating the optimum branch length, we basically accurately reconstruct the true tree.

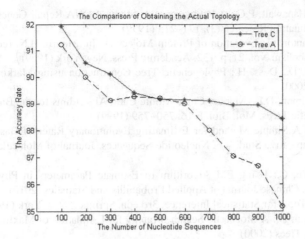

Fig. 3. Comparison of the probability of obtaining the topology of the actual tree

6   Conclusion

This paper, we use DNA sequences with defect data to construct phylogenetic tree, which has the best branch length, and we form the parameter estimation theorem in both Jukes-Cantor single parameter model and Kimura two parameter model, then we prove its correctness by using the EM algorithm. the results obtained by EM algorithm and the experiments are proved to have a better performance than the other related methods. Generally speaking, the accuracy of a reconstructed tree must satisfy at least two factors: 1.the linear relationship between the distance; 2.the standard error or coefficient of variation of the distance estimation. So far, some people have described the evolution of a single character along a single lineage. However, phylogeny deals with lots of species. They are assumed to be the descendants of a single ancestral species. The pattern or topology of this divergence process is usually unknown, and its inference is still the main goal of this study. Also there is no generally statistical methods accepted by the public that can be used to choose the appropriate distance to build the tree topology, and we are glad to pin our hope and efforts on finding a more useful method to construct a better phylogenetic tree.

Acknowledgments. This work was supported by the grant of the National Science Foundation of China, Nos. 61133010 & 31071168.

References

1. Barbara, E.E., Michael, I.J., Kathryn, E.M.: Protein Molecular Function Prediction by Bayesian Phylogenomics. PLoS Computational Biology 1(5), e45 (2005)
2. Felsentein, J.: Phy logenies from mo lecular sequences: Inference and Reliability. Annual Review of Genetic 22, 521–565 (1988)

3. Eisen, J.A., Hanawalt, P.C.: A Phylogenomics Study of DNA Repair Genes, Proteins, and Processes. Mutation Research (3), 171–213 (1999)
4. Jukes, T., Cantor, C.: Evolution of Protein Molecules. In: Munro, H.N. (ed.) Mammalian Protein Metabolism, vol. 21, p. 132. Academic Press, New York (1969)
5. Li, S., Pearl, D., Doss, H.: Phylogenetic Tree Construction using Markov chain Monte Carlo, 451 (2000)
6. Larget, B., Simon, D.L.: Markov Chain Monte Carlo Algorithms for the Bayesian analysis of Phylogenetic Trees. Mol. Biol. E 16, 750–759 (1999)
7. Kimura, M.: A Simple Method for Estimating Evolutionary Rates of Base Substitutions through Comparative Studies of Nucleotide Sequences. Journal of Molecular Evolution 16 (1980)
8. Tang, X., Wu, C.: Using EM Algorithm to Estimate Parameters in Phylogenetic Tree Construction. Chinese Journal of Applied Probability and Statistics (2010)
9. Tanner, M.: Tools for Statistical Inference, 3rd edn. Springer, New York (1996)
10. Salter, L., Peal, D.: A Stochastic Search Strategy for Estimation of Maximum Likelihood Phylogenetic Trees (2000)
11. Hostand, O., Bjorklund, M.: Nucloetide Substitution Models and Estimation of Phylogeny. Mol. Biol. E 15, 1381–1389 (1998)
12. Nei, M., Tajima, F., Tateno, Y.: Accuracy of Estimated Phylogenetic Trees from Molecular Data.II.Gene frequency data. J. Mol. Evol. 19, 153–170 (1983)
13. Goldman, N.: The Statistical Tests of the Models of the DNA Substitution. J. Mol. E 36, 182–198 (1993)
14. Li, W.-H.: A Statistical Test of the Phylogenies Estimated from Sequence Data. Mol. Biol. E 6, 424–435 (1989)

Dimensionality Reduction for Microarray Data Using Local Mean Based Discriminant Analysis

Yan Cui[1], Chun-Hou Zheng[2], and Jian Yang[1]

[1] School of Computer Science and Engineering, Nanjing University of Science and Technology,
Nanjing, Jiangsu, China
yancui128@gmail.com, csjyang@mail.njust.edu.cn
[2] College of Electrical Engineering and Automation, Anhui University, Hefei, Anhui, China
zhengch99@126.com

Abstract. In this paper we propose a new method for finding a low dimensional subspace of high dimensional microarray data. We developed a new criterion for constructing the weight matrix by using local neighborhood information to discover the intrinsic discriminant structure in the data. Also this approach applies regularized least square technique to extract relevant features. We assess the performance of the proposed methodology by applying it to four publicly available tumor datasets. In a low dimensional subspace, the proposed method classified these tumors accurately and reliably. Also, through a comparison study, the reliability of the dimensionality reduction and discrimination results is verified.

Keywords: Dimensionality reduction, Discriminant analysis, Gene expression, Local mean.

1 Introduction

Microarray data analysis is characterized by extremely high data dimensionality due to a large number of gene expression values measured for each tissue sample on an array. At the same time, the sample size is typically far smaller than the data dimension. This situation necessitates dimensionality reduction through extracting a small number of relevant features to avoid data over-fitting and improve generalization of discriminant analysis. Which demands that the reduced dimension can discover the intrinsic structure from the high dimensional data and preserve this structure information in the low-dimensional subspace.

Dimensionality reduction techniques mainly includes discriminant analysis and gene selection. Until now various gene selection methods have been reported[1,2]. In this paper, we introduce discriminant analysis method. Approaches of discriminant analysis range from traditional techniques, such as Principal Component Analysis(PCA)[3] and Linear Discriminant Analysis(LDA)[4] to various manifold learning techniques, e.g. Locally Linear Embedding(LLE)[5] and Locality Preserving Projection(LPP)[6]. These methods do yield impressive results on some benchmark artificial datasets. However, there exit two main problems. First, when the data is

D.-S. Huang et al. (Eds.): ICIC 2013, LNAI 7996, pp. 267–276, 2013.

sampled from a nonlinear manifold, PCA and LDA fail to discover the underlying structure, due to taking the global Euclidean structure into account instead of the local geometry structure of the original data. Second, manifold learning methods can preserve local structures in small neighborhoods and successfully derive the intrinsic features of nonlinear manifolds, nevertheless, these methods as well as PCA and LDA involve eigen-decomposition of dense matrices. Moreover, for gene expression data, the scatter matrices are singular when the number of genes is larger than the number of samples. To overcome this problem, some additional preprocessing steps such as PCA and Singular Value Decomposition (SVD) are required to guarantee the non-singularity of scatter matrices[7], which would be time-consuming for microarray data.

In this study, we propose a discriminant analysis method for dimensionality reduction named local mean based discriminant analysis (LM-DA). Its character is mainly in following two terms. First, We construct the weight matrix using local mean of K-nearest neighbors for each sample to discover the intrinsic discriminant structure in the data. Then, we adopt regularized least square technique without extra preprocessing step to extract a set of relevant features. Based on this character, our approach can provide an efficient and effective way for dimensionality reduction. Finally, several publicly available microarray datasets are used to evaluate the performance of our proposed method.

2   Materials and Methods

2.1   Microarray Gene Expression Data

The leukemia data consist of 72 tissue samples, each with 7129 gene expression values[8]. The samples include 47 acute lymphoblastic leukemia (ALL) and 25 acute myeloid leukemia (AML). The original data have been divided into a training set of 38 samples and a test set of 34 samples. The data were produced from Affymetrix gene chips.

The prostate gene expression profiles were derived from 52 prostate tumors and 50 nontumor prostate samples from patients undergoing surgery[9]. Oligonucleotide microarrays containing probes for approximately 12600 genes and ESTs. The training set are 102 samples, and he test ones are 34 samples.

The SRBCT dataset has 2308 genes (including expressed sequence tags) that were preliminarily chosen by[10] and a total of 83 samples (we removed five non-SRBCT samples). It is classified into four classes, Burkitt's lymphoma (BL), Ewing's sarcoma (EWS), rhabdomyosarcoma (RMS), and neuroblastoma (NB).

The lung cancer data consists of 203 samples, each with 12600 gene expression values[11]. The samples include histologically defined lung adenocarcinomas, squamous cell lung carcinomas, pulmonary carcinoids, SCLC cases, and normal lung specimens.

2.2   Mathematical Formulation of LM-DA

The basic idea of LM-DA method is to extract a set of relevant features which can reveal the intrinsic discriminant structure of microarray data, when the high dimensional data is embedded into a low dimensional space. Moreover, the samples of different classes are far from each other while samples of the same class are close to each other in the low dimensional space.

Suppose matrix $X = [\mathbf{x}_1, \mathbf{x}_2, ..., \mathbf{x}_n]$ be the gene expression data matrix including all the samples $\{\mathbf{x}_i\}_{i=1}^n \in R^m$ in its columns, while each row represents the profile of a gene. In order to introduce LM-DA, we first reformulate LDA in a pairwise manner.

$$S_w = \frac{1}{2} \sum_{i=1}^n \sum_{j=1}^n W_{ij}^{(w)} (\mathbf{x}_i - \mathbf{x}_j)(\mathbf{x}_i - \mathbf{x}_j)^T \tag{1}$$

$$S_b = \frac{1}{2} \sum_{i=1}^n \sum_{j=1}^n W_{ij}^{(b)} (\mathbf{x}_i - \mathbf{x}_j)(\mathbf{x}_i - \mathbf{x}_j)^T \tag{2}$$

where

$$W_{ij}^{(w)} = \begin{cases} \dfrac{1}{n_r} & \text{if } \mathbf{x}_i \text{ and } \mathbf{x}_j \text{ are in the same class} \\ 0 & \text{otherwise} \end{cases}$$

$$W_{ij}^{(b)} = \begin{cases} \dfrac{1}{n} - \dfrac{1}{n_r} & \text{if } \mathbf{x}_i \text{ and } \mathbf{x}_j \text{ are in the same class} \\ \dfrac{1}{n} & \text{otherwise} \end{cases} \tag{3}$$

The objective function of LDA is:

$$\mathbf{a}^* = \arg\max_{\mathbf{a}} \frac{\mathbf{a}^T S_b \mathbf{a}}{\mathbf{a}^T S_w \mathbf{a}} \tag{4}$$

Similar to LDA, our approach LM-DA also seeks directions on which the sample pairs in the same class are close and the samples pairs in different classes are apart.

In order to preserve the local structure of the gene expression data when high dimensional data is embedded to a low dimensional space, we define the weight matrix as follows:

$$\tilde{W}_{ij}^{(w)} = \begin{cases} \dfrac{A_{ij}}{n_r} & \text{if } \mathbf{x}_i \text{ and } \mathbf{x}_j \text{ are in the same class} \\ 0 & \text{otherwise} \end{cases}$$

$$\tilde{W}_{ij}^{(b)} = \begin{cases} A_{ij}\left(\dfrac{1}{n} - \dfrac{1}{n_r}\right) & \text{if } \mathbf{x}_i \text{ and } \mathbf{x}_j \text{ are in the same class} \\ \dfrac{1}{n} & \text{otherwise} \end{cases} \tag{5}$$

$$A_{ij} = \exp\left(-\frac{\|\mathbf{x}_i - \mathbf{x}_j\|^2}{\sigma_i \sigma_j} \right) \tag{6}$$

$$\sigma_i = \|\mathbf{x}_i - \mathbf{m}_r(\mathbf{x}_i)\| \tag{7}$$

In our approach, the weight matrix definition is computed locally. Based on the definition (i.e., Eq.(5)) in[12], we construct σ_i using local mean of K-nearest neighbors for each sample. In Eq.(7), the K-nearest neighbors of $\mathbf{x}_i$ in Class r are $\mathbf{x}_{rk}$, where $k = 1,2,...,K$., $\mathbf{m}_r(\mathbf{x}_i)$ is called the local mean of the sample $\mathbf{x}_i$ in Class r, i.e., $\mathbf{m}_r(\mathbf{x}_i) = \frac{1}{K}\sum_{k=1}^{K}\mathbf{x}_{rk}$.

σ_i is a local scaling parameter which represents the local scaling of the data samples around $\mathbf{x}_i$. A specific scaling parameter for each sample allowing self-tuning of the point-to-point distances according to the local statistics of the neighborhoods surrounding samples $\mathbf{x}_i$ and $\mathbf{x}_j$.

Based on the above weight matrix, we define the local between-class scatter matrix $\tilde{S}_b$ and the local within-class scatter matrix $\tilde{S}_w$ as follows:

$$\tilde{S}_b = \frac{1}{2}\sum_{i=1}^{n}\sum_{j=1}^{n}\tilde{W}_{ij}^{(b)}(\mathbf{x}_i - \mathbf{x}_j)(\mathbf{x}_i - \mathbf{x}_j)^T \tag{8}$$

$$\tilde{S}_w = \frac{1}{2}\sum_{i=1}^{n}\sum_{j=1}^{n}\tilde{W}_{ij}^{(w)}(\mathbf{x}_i - \mathbf{x}_j)(\mathbf{x}_i - \mathbf{x}_j)^T \tag{9}$$

In Eqs.(8) and (9), we weight the values for the sample pairs in the same class. This means that far apart sample pairs in the same class have less influence on $\tilde{S}_b$ and $\tilde{S}_w$. Therefore, our method can discover the intrinsic discriminant structure in the data well.

Then, the objective function of LM-DA can be written as:

$$\mathbf{a}^* = \arg\max_{\mathbf{a}} \frac{\mathbf{a}^T \tilde{S}_b \mathbf{a}}{\mathbf{a}^T \tilde{S}_w \mathbf{a}} \tag{10}$$

The general solution to Eq.(10) is eigen-decomposition. Moreover, to get a stable solution, the matrix $\tilde{S}_w$ is required to be non-singular[13], which is not available for microarray data. In this case, some additional preprocessing steps(e.g., SVD, PCA) are needed. However, these are time-consuming. In our method, we use regularized least square to extract relevant features.

The objective function can be rewritten as:

$$\mathbf{a}^* = \arg\max_{\mathbf{a}} \frac{\mathbf{a}^T X\left(\widetilde{D}^{(b)} - \widetilde{W}^{(b)}\right)X^T\mathbf{a}}{\mathbf{a}^T X\left(\widetilde{D}^{(w)} - \widetilde{W}^{(w)}\right)X^T\mathbf{a}} \tag{11}$$

where $\widetilde{D}$ is a diagonal matrix whose entries are column (or row, since $\widetilde{W}$ is symmetric) sum of $\widetilde{W}$.

Generally, for classification purpose, a mapping for all samples, including new test samples, is required. If we choose a linear function, i.e., $y_i = \mathbf{a}^T \mathbf{x}_i$, Eq.(11) can be rewritten as:

$$\mathbf{y}^* = \arg\max \frac{\mathbf{y}^T\left(\widetilde{D}^{(b)} - \widetilde{W}^{(b)}\right)\mathbf{y}}{\mathbf{y}^T\left(\widetilde{D}^{(w)} - \widetilde{W}^{(w)}\right)\mathbf{y}} \tag{12}$$

In order to avoid SVD step in solving the eigen-problem in Eq.(10), we can first solve the eigen-problem in Eq.(12) to get $\mathbf{y}$, and then find $\mathbf{a}$ which satisfies $X\mathbf{a}^T = \mathbf{y}$, the proof can be found in[14]. In our approach, we find $\mathbf{a}$ via regularized least squares.

$$\mathbf{a} = \arg\min_{\mathbf{a}}\left(\sum_{i=1}^{n}\left(\mathbf{a}^T\mathbf{x}_i - y_i\right)^2 + \alpha\|\mathbf{a}\|^2\right) \tag{13}$$

where $\alpha \geq 0$ is a parameter to control the amounts of shrinkage, y_i is the $i-th$ element of $\mathbf{y}$.

The advantages of using regularized least squares are as follows: Firstly, The matrix $\widetilde{D}^{(w)} - \widetilde{W}^{(w)}$ is guaranteed to be nonsingular and therefore the eigen-problem in Eq.(12) can be stably solved. Secondly, the technique to solve the least square problem is already matured[13] and there exist many efficient algorithms (e.g., LSQR[15]) to handle large scale least square problems. Finally, regularized least square techniques can produce more stable and meaningful solutions, especially for high-dimensional data sets.

$$\mathbf{z} = A^T\mathbf{x} \tag{14}$$

At the end of LM-DA, the original gene expression data can be embedded into low dimensional subspace using Eq. (14), A is a $m \times d$ transformation matrix, $A = [\mathbf{a}_1, \mathbf{a}_2, ..., \mathbf{a}_d]$. Then, we can use certain classifiers to classify the data. Different from LDA, LM-DA can well capture the intrinsic discriminant structure in the data and without additional preprocessing steps.

3   Results

We used the proposed method to analyze the leukemia, prostate, SRBCT and lung cancer datasets. During our experiments, we adopt the nearest neighbor(NN) classifier to classify the data. The training and test datasets were obtained by splitting the total dataset into two parts containing 80% and 20% of the total samples (similar to fivefold cross-validation), respectively, while ensuring that the proportion of classes was balanced in the two sets. Considering the arbitrariness of this partitioning, we repeated the above cross-validation procedure 100 times. By this approach, we obtained the point that represents the minimum of the mean test error (mean value of 100 test errors) and indicates the optimal dimension of subspace.

3.1   Dimensionality Reduction and Cross-Validation Results

The leukemia dataset consists of training and test sets comprising 38 and 34 samples, respectively. In this study, we combined these two datasets into a single set of 72 samples; thus, about 80% randomly selected samples were used for training in each cross-validation step. The cross-validation result is illustrated in Fig. 1(a). From Fig. 1(a), we recognize that the optimal dimension of subspace, which produces the minimum test error (overall mean cross-validation error, 0.83%), is 6. In other word, we can obtain an optimal dimension of subspace by extracting 6 relevant features that are most frequently used over 100 cross-validations at the minimum error point. For prostate dataset, the cross-validation result for the dimensionality reduction is depicted in Fig. 1(b). Following an approach similar to that used for the leukemia dataset, we reduce dimension to 7 that achieves the minimum CV test error (5.00%). The cross-validation results for the SRBCT and lung cancer datasets are illustrated in Fig. 1(c) and Fig. 1(d), respectively. For SRBCT dataset, the optimal dimension of subspace is 4 which produces no misclassification. The optimal dimension of sunspace for lung cancer dataset is 9 and the minimum CV test error is 3.65%. Fig. 2 shows the precision-recall results of four datasets.

3.2   Comparison Study

To evaluate the performance of the proposed dimensionality reduction method, we compared our method with Locality Preserving Projection(LPP) and Linear Discriminant Analysis(LDA).

Table 1. The minimum mean cross-validation error (%) and the standard deviations (stds) of LPP, LDA and LM-DA with NN classifiers on four datasets and the corresponding dimension of subspace, computational time

Methods		Datasets			
		Leukemia	Prostate	SRBCT	Lung cancer
LPP	Error	7.35±3.95	20.00±7.65	3.12±6.88	9.66±2.32
	Dimension	13	12	9	14
	Computational time	3.559	12.008	1.519	21.095
LDA	Error	4.12±2.77	7.72±3.80	0	5.00±4.74
	Dimension	1	1	2	4
	Computational time	3.474	10.919	1.726	18.581
LM-DA	Error	0.83±0.86	5.00±1.59	0	3.65±1.45
	Dimension	6	7	4	9
	Computational time	1.780	5.889	0.792	10.061

Table 1 shows the comparison results. For these datasets, the dimensions of subspace using our method LM-DA are smaller than that using LPP, and the minimum cross-validation results are also better than LPP and LDA. In our method, we construct the weight matrix using local mean of K-nearest neighbors($K=5$) for each sample to discover the intrinsic discriminant structure in the data. When the high dimensional data is embedded into low dimensional space, the samples of different classes are far from each other while samples of the same class are close to each other in subspace. Therefore our method can extract useful features to achieve good classification results.

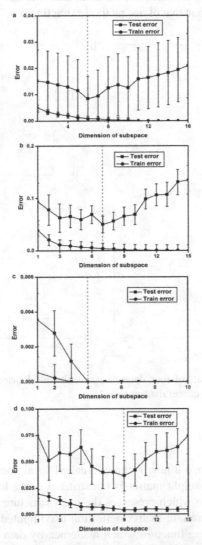

Fig. 1. Cross-validation results of four datasets. (a) leukemia dataset: at the minimum test error point (indicated by dashed line), the dimension of subspace is 6, (b) prostate dataset: the dimension is 7, (c) SRBCT dataset: 4 and (d) lung cancer dataset: 9.

LDA can only extract limited feature due to the intrinsic nature of algorithm (*i.e.*, it only extract one feature for two classes dataset). Although LDA with NN classifier achieve a good classification result, this small number of feature is obviously not enough to represent numeral pattern for classification purpose.

In addition, the computational time of LDA and LPP is more than our method, since LDA and LPP need SVD step to guarantee solution procedure nonsingular for high dimensional data. However, since our method using regularized least square technique, which need not extra preprocessing step, to extract the relevant features so that it using less time than LPP and LDA. Moreover, the computational time of LDA which extracts only one feature is more than that of our method extracting 6 or 7 features.

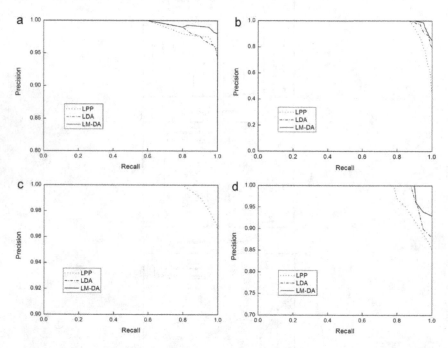

Fig. 2. Precision-recall results of four datasets. (a) leukemia dataset, (b) prostate dataset, (c) SRBCT dataset and (d) lung cancer dataset.

4   Discussion

In this paper, we developed a new technique called local mean based discrimanant analysis (LM-DA). The weight matrix is constructed using local mean of K-nearest neighbors for each sample which preserves the local structure of the gene expression data. In this method, the weight matrix definition is computed locally so that we can find the optimal embedding function which made nearby data pairs in the same class close and the data pairs in different classes far from each other. Moreover, by using the local neighborhood information, LM-DA can well capture the intrinsic discriminant structure in the data. Differing from other dimensionality reduction methods such as LDA and LPP, LM-DA avoids using SVD step to solve singular problem. It introduces regularized least square technique to extract relevant features which facilitates efficient computation.

The ultimate goal of this paper is to identify the optimal dimension of subspace that high dimensional data are projected to low dimensional space. Which can facilitate the works such as constructing a diagnosis system and predicting a disease based on gene expression data. The benefits of employing the present method to study a disease would be enhanced through associating with biological experiments related to the disease.

Acknowledgments. This work was partially supported by the National Science Fund for Distinguished Young Scholars under Grant No. 61125305, the National Natural Science Foundation of China under Grant No. 61272339, and the Key Project of Anhui Educational Committee, under Grant No. KJ2012A005.

References

1. Wang, H.Q., Huang, D.S.: A Gene Selection Algorithm based on the Gene Regulation Probability using Maximal Likelihood Estimation. Bitotechnol. Lett. 27, 597–603 (2005)
2. Payton, P., Kottapalli, K.R., Kebede, H., Mahan, J.R., Wright, R.J., Allen, R.D.: Examining the Drought Stress Transcriptome in Cotton Leaf And Root Tissue. Bitotechnol. Lett. 33, 821–828 (2011)
3. Joliffe, I.: Principal Component Analysis. Springer (1986)
4. Fukunnaga, K.: Introduction to Statistical Pattern Recognition, 2nd edn. Academic Press (1991)
5. Roweis, S.T., Saul, L.K.: Nonlinear Dimensionality Reduction by Locally Linear Embedding. Science 290, 2323–2326 (2000)
6. He, X., Niyogi, P.: Locality Preserving Projections. In: Adv. Neural Inf. Process Syst. MIT Press (2004)
7. Ye, J.: Characterization of a Family of Algorithms for Generalized Discriminant Analysis on Undersampled Problems. J. Mach. Learn. Res. 6, 483–502 (2005)
8. Golub, T.R., Slonim, D.K., Tamayo, P., Huard, C., Gaasenbeek, M., Mesirov, J.P., Coller, H., Loh, M.L., Downing, J.R., Caligiuri, M.A., Bloomfield, C.D., Lander, E.S.: Molecular Classification of Cancer: Class Discovery and Class Prediction by Gene Expression Monitoring. Science 286, 531–537 (1999)
9. Singh, D., Febbo, P.G., Ross, K., Jackson, D.G., Manola, J., Ladd, C., Tamayo, P., Renshaw, A.A., D'Amico, A.V., Richie, J.P., Lander, E.S., Loda, M., Kantoff, P.W., Golub, T.R., Sellers, W.R.: Gene Expression Correlates of Clinical Prostate Cancer Behavior. Cancer Cell 1, 203–209 (2002)
10. Khan, J., Wei, J.S., Ringner, M., Saal, L.H., Ladanyi, M., Westermann, F., Berthold, F., Schwab, M., Antonescu, C.R., Peterson, C., Meltzer, P.S.: Classification and Diagnostic Prediction of Cancers using Gene Expression Profiling and Artificial Neural Networks. Nat. Med. 7, 673–679 (2001)
11. Bhattacharjee, A., Richards, W.G., Staunton, J., Li, C., Monti, S., Vasa, P., Ladd, C., Beheshti, J., Bueno, R., Gillette, M.: Classification of Human Lung Carcinomas by mRNA Expression Profiling Reveals Distinct Adenocarcinoma Subclasses. Proc. Natl. Acad. Sci. USA 98, 13790–13795 (2001)

12. Zelnik-Manor, L., Perona, P.: Self-tuning Spectral Clustering. In: Adv. Neural. Inf. Process Syst., vol. 17, pp. 1601–1608. MIT Press, Cambridge (2005)
13. Golub, G.H., Loan, C.F.V.: Matrix Computations, 3rd edn. Johns Hopkins University Press (1996)
14. Cai D., He X., Han J.: Spectral Regression for Dimensionality Reduction. UIUCDCS-R-2007-2856 (2007)
15. Paige, C.C., Saunders, M.A.: LSQR: An Algorithm for Sparse Linear Equations and Sparse Least Squares. ACM Trans. Math. Softw. 8, 43–71 (1982)

Machine Learning-Based Approaches Identify a Key Physicochemical Property for Accurately Predicting Polyadenlylation Signals in Genomic Sequences

HaiBo Cui[1,*] and Jia Wang[2]

[1] Faculty of Mathematics and Computer Science, Hubei University, Wuhan, 430062, China
child@hubu.edu.cn
[2] College of Science, Huazhong Agricultural University, Wuhan, 430070, China
wang.jia@mail.hzau.edu.cn

Abstract. Accurately predicting poly(A) signals (PASs) is one of important topics in bioinformatics for high-quality genome annotation and transcription regulation mechanism investigation. In this study, we identified a powerful physicochemical property of DNA sequence for computationally predicting PASs using machine learning technologies. On the basis of this feature, we built a PAS prediction model by capturing the position-specific information from the region surrounding PASs. The cross-validation results demonstrated that the prediction accuracies of our constructed model on 12 categories of human PASs are comparable to those of recently published PAS predictor Dragon PolyA Spotter. Further analysis revealed that the region 25 nucleotides downstream of PASs is the most important region for the accurate prediction of PASs.

Keywords: polyadenlylation site, poly(A), genomic sequence, physicochemical property, dinucleotide, random forest, machine learning, bioinformatics.

1 Introduction

The polyadenylation (poly(A)) tail is a series of adenine bases added to the 3'end of RNA molecules during RNA processing as an essential element for the nuclear export, translation, and stability of mRNAs [1-2]. In eukaryotes, the poly(A) tail is characterized with three cis-elements including the cleavage site where the pre-mRNA is cut off from the genomic sequence, and the poly(A) signal (PAS) where the RNA polymerase binds to the genomic sequence for the cleavage. The PAS is usually a conserved hexanucleotide (e.g., AATAAA and AUUAAA) located 10-30 nucleotides upstream of the cleavage site [3]. Thus accurately predicting PAS in genomic sequences is important for the high-quality genome annotation and the investigation of transcription regulation mechanisms.

Currently, several approaches have been developed to predict PASs [4-8], most of them were focused on the sequence-based features, such as the position weight

---

* Corresponding author.

D.-S. Huang et al. (Eds.): ICIC 2013, LNAI 7996, pp. 277–285, 2013.

matrices around PASs, and k-mer compositions in both upstream and downstream regions. Although the performance of PAS predictor has been steadily improved, the prediction accuracy is still unsatisfied with both low sensitivity and low specificity, caused by the utilization of partial function information in DNA molecules [9-10]. Therefore, mining other features from nucleotide sequences is of great importance for accurately predicting PASs. The conformational and physicochemical properties, such as the DNA curvature and bendability, provide a novel view of DNA sequence at the structure level. The power of some DNA structure-based features have been demonstrated in the prediction of promoters and transcription factor binding sites [11-12]. Recently, Friedel et al. generated more than 100 structural-based features from the pairs of two bases [13]. By integrating these dinucleotide properties with 163 sequence-based features, Kalkatawi and colleagues developed a new PAS predictor, named Dragon PolyA Spotter, to predict 12 categories of PASs with the prediction accuracy higher than 88% [9].

In this study, we aimed at (1) identifying key dinucleotide properties in the prediction of PASs using machine learning approaches; (2) constructing a new PAS prediction model using the random forest algorithm; (3) investigating the contribution of different regions surrounding PASs to the prediction. Using 12 categories of human PASs, we found that the B-DNA mobility to bend towards major groove is the most powerful dinucleotide property in accurately predicting PASs. On the basis of this feature, we constructed a new PAS prediction model, named polyA-Predict, which makes use of the position-specific information and is able to more accurately predict PASs than Dragon PolyA Spotter for most tested cases. The further analysis revealed that the region downstream of PASs contributes the most to the prediction.

2 Material and Methods

2.1 Dataset

The positive and negative samples used in this study were originally collected by Kalkatawi and colleagues [9], and can be downloaded from the website at http://cbrc.kaust.edu.sa/dps. The positive samples were real PASs in human mRNA sequences obtained from the UCSC Genome Brower (http://www.genome.ucse.edu). To obtain the genomic sequences around these PASs, the poly(A) tails and their upstream 96 nucleotides of mRNA sequences were aligned to human genome reference sequence (human genome assembly 19, hg19) using BLASTN. After filtering the alignments with stringent mapping criteria (E-value $\leq$ 1.0E-20, Identify $\geq$ 96%), 14799 positive samples were obtained, in which there were 12 poly(A) motifs including AAAAAG (1402), AAGAAA (1391), AATAAA (5164), AATACA (724), AATAGA (332), AATATA (392), ACTAAA (553). AGTAAA (697), ATTAAA (2433), CATAAA (474), GATAAA (481) and TATAAA (755). The 200 nucleotides upstream and downstream of these motifs were extracted for the dinucleotide property analysis.

The negative samples were false PASs randomly selected from human genomic sequences. For each category of PAS motifs, the number of negative samples was equal to that of positive samples. More detailed information about the construction of this database can be referred in [9].

2.2   Encoding Positive and Negative Samples

The DNA sequences of positive and negative samples were respectively encoded with 110 features including 30 physicochemical (e.g., free energy), 73 conformational (e.g., twist) and 7 letter-based (e.g., GC content) properties. The dinucleotide values of these features were collected from literatures and obtained from the DiProGB (Dinucleotide Properties Genome Browser) database [13]. Using the sliding window approach [14-15], we transferred each genomic sequence to a numeric vector by translating the adjacent two nucleotides with the corresponding dinucleotides value of the feature of interest. The values in this numeric vector were further smoothed with a running average method. For instance, the value in position j was averaged with values from positions j-1 to j+1.   At last, each sequence has 110 numeric vectors corresponding to all these dinucleotide properties.

2.3   Constructing Poly(A) Predictor with Random Forest

Random forest (RF) is a widely used machine learning algorithm for the prediction and classification problems in areas of bioinformatics and computational biology [16-18], due to the advantages of high prediction performance and fast run speed [19]. RF constructs prediction models using an ensemble of many decision trees generated with two advanced ML technologies (i.e., bootstrapping and random selection of subset of features and samples) and performs the prediction based on the major vote of these decision trees. The RF algorithm in this study was implemented with the randomForest package in R programming language (http://cran.r-project.org).

Due to the limit number of positive and negative samples, the PolyA predictor was constructed with only one dinucleotide property through implementing RF algorithm. Thus, for each sample, the input is one numeric vector characterized by the dinucleotide property of interest, and the output is the predicted score to be a true sample.

2.4   Performance Evaluation

In order to fairly evaluate the performance of PAS predictors, we utilized the cross-validation method to construct training and testing databases for machine learning. For N-fold cross-validation, the positive and negative samples were firstly separated into N parts, and then N-1 parts were selected to form the training dataset for training prediction model and the left one part was regards as the testing dataset for testing prediction model. This process was repeated N times until all samples have been used as testing data. The results of these N fold experiments were finally averaged as the performance of prediction model.

Based on the prediction results on testing dataset in each fold experiment, the performance of prediction model was assessed through the analysis of Receiver

Operating Characteristic (ROC) curve, which reveals the relationships between true positive rate (TPR) and false positive rate (FPR) at different thresholds. The area under the ROC curve, termed as AUC, was used to quantify the overall performance of prediction model with a numeric value ranging from 0 to 1.0. Thus five AUC values can be generated from five-fold cross validation experiment. The prediction model achieves a better performance when the mean of these five AUC values is more closer to 1.0.

Besides the ROC curve analysis, three other measures including sensitivity $Se = 100 \times TP/(TP+FN)$, specificity $Sp = 100 \times TN/(TN+FP)$ and accuracy $Acc = 100 \times (TP+FN)/(TP+TN+FP+FN)$ were also calculated for the comparison of the performance of Dragon polyA Spoter reported in [9]. Here TP, TN, FP and FN are respectively the number of true positives, true negatives, false positives and false negatives.

3   Results and Discussions

3.1   Identification of Important Dinucleotide Properties for Accurate PAS Prediction Using Machine Learning-Based Approach

Selecting informative features is one of the most important steps toward the construction of accurate prediction models in machine learning. Here the importance of each dinucleotide property was analyzed with the performance of RF-based prediction models in identifying 12 classes of human PASs. In brief, we respectively built 110 RF-based PAS predictors for each class of PAS motifs with the corresponding dinucleotide properties, and evaluated the performance of constructed PAS predictors using 5-fold cross validation experiment. The evaluation results were then recorded in a matrix, where each row represents one dinucleotide property and each column indicates the AUC value of prediction model tested on one type of PASs. As we were interested in the property which is powerful to identify most classes of PAS motifs, the rank of AUC (i.e., RAUC) instead of the absolute value was used to represented the importance of one dinucleotide property. Thus RAUC is an integer ranged from 1 to 110. The lower the RAUC, the more powerful the feature in the poly(A) prediction. For an ideal feature, the RAUC should be 1 for all classes of PAS motifs.

For each dinucleotide property, the distribution of 12 RAUC values for all tested classes of Poly(A) motifs was visualized with a boxplot. Thus a lower boxplot indicates a more important feature in predicting PASs. The boxplots of all 110 features were shown in Figure 1. Although all 110 dinucleotide properties were derived from the same genomic sequences, the distributions of RAUC values were remarkably different (Figure 1), indicating that these dinucleotide properties contribute differently to the prediction of PASs.

This feature importance analysis indicates that the B-DNA mobility to bend towards major groove is the most powerful feature in PAS prediction. This feature is a DNA physicochemical property that measures the mobility of protein-DNA complex to major groove for more favorable interactions [20]. Figure 2 shows the average profiles of this identified feature in positive and negative samples for different categories of PASs. All these profiles have a peak at the position where PASs locate in genomic sequences, consisted with the fact that the PAS is bound with the RNA polymerase for efficient cleavage of pre-mRNA from genomic sequences. The curve was significantly decreased after the peak in negative samples, whereas in positive samples, the curve was decreased much slower. The WebLogo analysis results also indicate the conservation of nucleotide composition in this region (Figure 3). These results are consistent with the fact that the DNA sequences downstream PASs need to be bound with proteins for transcription termination [21]. Interestingly, we found that there was another peak in the profile of some categories of PASs including AAAAAG, AATAAA, AATAGA, AATATA, AGTAAA, CATAAA and GATAAA. This phenomenon was not observed for the PAS categories such as AAGAAA and TATAAA. This result indicates that alternative mechanisms of termination may exist in eukaryotic cells [2].

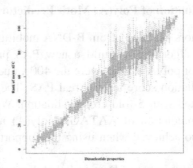

Fig. 1. The importance of dinucleotide properties estimated with the rank of AUC values

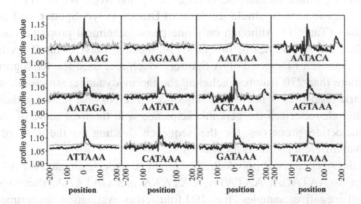

Fig. 2. Profiles of B-DNA mobility to bend towards major groove in the vicinity of PASs in positive and negative samples. "0" indicates the position of the first nucleotide in PASs. The black and gray curves respectively represent the profiles from positive and negative samples.

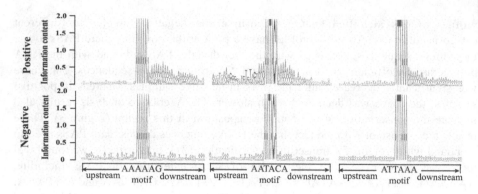

Fig. 3. The sequence logos of the genomic sequences around PASs in positive and negative samples

3.2   Performance Evaluation of Poly(A) Motif Predictors

Based on the numeric vectors generated from B-DNA mobility to bend towards major groove, we used the RF algorithm to build a new PAS prediction model, named polyA-Predict. The input of polyA-Predict were the 400 values in the numeric vector, and the output is the prediction score to be a real PAS. The performance of polyA-Predict was firstly evaluated with 5-fold cross validation. We found the AUC value was about 0.98 on the   prediction of AATAGA poly(A) motifs. A slightly lower AUC value (0.94) was also achieved when using the support vector machine instead of RF algorithm.

To compare with the performance of Dragon PolyA Spotter reported in [9], we also performed the 100-fold cross validation experiment and evaluated the performance of polyA-Predict with three measurements (i.e., Sn, Sp and Acc). We fixed the Sn values of polyA-Predict so as to be similar to those of Dragon PolyA Spotter, and compared the Acc values (Table 1). Although only one physicochemical property was used in polyA-Predict, the Acc values of this newly introduced poly(A) motif predictor are comparable to those of Dragon PolyA Spotter, which also used the RF-algorithm but integrated more than 270 features including the thermodynamic, structural, statistical and bioelectric properties of nucleotide sequences. The possible reason is that Dragon PolyA Spotter characterized the genomic sequence with the total score of the profile of the dinucloetide properties for the sequence, leading to the loss of feature information at different positions around PASs.

Of note, the Acc values of polyA-Predict might be under-estimated for the poly(A) motifs, such as AATAGA, AATATA, CATAAA and GATAAA, due to 332~481 positive and negative samples for 100-fold cross validation experiment. The experimental results from 5-fold cross validation indicates that the Acc values of polyA-Predict on these motifs are higher than 0.90 (Table 1).

Table 1. Comparison of PolyA-Predict performance with Dragon PolyA Spotter. "CV" denotes cross validation.

Poly(A) motifs	Dragon PolyA Spotter (100-fold CV)			polyA-Predict (100-fold CV)			polyA-Predict (5-fold CV)		
	Sn	Sp	Acc	Sn	Sp	Acc	Sn	Sp	Acc
AAAAAG	93.2	95.6	94.4	92.9	97.9	**95.4**	93.1	98.6	**95.9**
AAGAAA	88.7	94.1	91.4	84.8	98.6	**91.7**	88.7	98.7	**93.7**
AATAAA	86.1	91.6	88.9	84.5	93.8	**89.2**	86.5	89.9	88.2
AATACA	87.3	92.5	89.9	86.4	96.2	**91.3**	87.0	94.6	**90.8**
AATAGA	96.7	91.3	89.0	68.2	99.3	83.8	87.0	97.6	**92.3**
AATATA	87.2	93.6	90.4	68.2	100.0	84.1	87.2	98.7	**93.0**
ACTAAA	85.0	91.1	88.1	81.0	96.8	**89.0**	84.6	94.9	**89.8**
AGTAAA	83.1	94.5	88.8	67.0	99.7	83.3	82.9	98.6	**90.7**
ATTAAA	85.2	92.6	88.9	83.6	97.0	**90.3**	85.5	96.0	**90.8**
CATAAA	83.5	92.4	88.0	76.6	97.2	86.9	83.8	97.9	**90.8**
GATAAA	87.9	92.5	90.2	75.6	99.3	87.4	88.4	97.1	**92.7**
TATAAA	86.1	94.2	90.1	86.0	96.7	**91.4**	86.4	97.5	**91.9**

3.3 Detection of Important Regions for Accurately Predicting PASs

Although the differences between positive and negative samples have been observed in upstream and downstream regions of PASs (Figure 2), the importance of different regions for the PAS prediction remains unknown quantitatively. Using genomic sequences around PASs with different lengths, we assessed the performance of polyA-Predict with the ROC curve analysis. As shown in Figure 4, the main region contributes to the AATAGA and AATATA poly(A) motif prediction is the genomic sequence downstream of PASs. The similar results were also observed for other poly(A) motifs. Future investigating these regions would be helpful for more thoroughly understanding the transcription termination mechanisms.

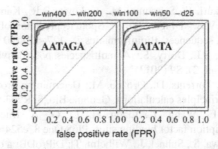

Fig. 4. ROC curve analysis of the performance of polyA-Predict with different window lengths. "win400", "win200", "win100" and "win50" respectively indicated the 200, 100, 50 and 25 nucleotides upstream and downstream of poly(A) motifs were used to encode poly(A) motifs. "d25" denotes only the genomic sequences with 25 nucleotides downstream poly(A) motifs were used to characterize poly(A) motifs.

4   Conclusion

In this study, we identified a key physicochemical property of genomic sequences, which has the capability of accurately predicting PASs when considering the position-specific information around PASs. We also found that region downstream of PASs is important for the prediction of PASs. The conformational and physicochemical properties of genomic sequences provide novel insights of DNA molecules different from those are characterized by the commonly used sequence-based features. Using these structural-based features would be helpful for accurate prediction of functional elements in genomic sequences.

References

1. Fuke, H., Ohno, M.: Role of poly (A) tail as an identity element for mRNA nuclear export. Nucleic Acids Res. 36, 1037–1049 (2008)
2. Kuehner, J.N., Pearson, E.L., Moore, C.: Unravelling the means to an end: RNA polymerase II transcription termination. Nature reviews. Mol. Cell Biol. 12, 283–294 (2011)
3. Beaudoing, E., Freier, S., Wyatt, J.R., Claverie, J.M., Gautheret, D.: Patterns of variant polyadenylation signal usage in human genes. Genome Res. 10, 1001–1010 (2000)
4. Ji, G., Wu, X., Shen, Y., Huang, J., Quinn Li, Q.: A classification-based prediction model of messenger RNA polyadenylation sites. J. Theor. Biol. 265, 287–296 (2010)
5. Goni, J., Zheng, J., Shen, Y., Wu, X., Jiang, R., Lin, Y., Loke, J.C., Davis, K.M., Reese, G.J., Li, Q.Q.: Predictive modeling of plant messenger RNA polyadenylation sites. BMC Bioinformatics 8, 43 (2007)
6. Chang, T.H., Wu, L.C., Chen, Y.T., Huang, H.D., Liu, B.J., Cheng, K.F., Horng, J.T.: Characterization and prediction of mRNA polyadenylation sites in human genes. Med. Biol. Eng. Comput. 49, 463–472 (2011)
7. Cheng, Y., Miura, R.M., Tian, B.: Prediction of mRNA polyadenylation sites by support vector machine. Bioinformatics 22, 2320–2325 (2006)
8. Wu, X., Ji, G., Zeng, Y.: In silico prediction of mRNA poly(A) sites in Chlamydomonas reinhardtii. Mol. Genet. Genomics 287, 895–907 (2012)
9. Kalkatawi, M., Rangkuti, F., Schramm, M., Jankovic, B.R., Kamau, A., Chowdhary, R., Archer, J.A., Bajic, V.B.: Dragon PolyA Spotter: predictor of poly(A) motifs within human genomic DNA sequences. Bioinformatics 28, 127–129 (2012)
10. Ho, E.S., Gunderson, S.I., Duffy, S.: A multispecies polyadenylation site model. BMC Bioinformatics 14(suppl. 2), S9 (2013)
11. Goni, J.R., Perez, A., Torrents, D., Orozco, M.: Determining promoter location based on DNA structure first-principles calculations. Genome Bio. 8, R263 (2007)
12. Xu, B., Schones, D.E., Wang, Y., Liang, H., Li, G.: A structural-based strategy for recognition of transcription factor binding sites. PloS One 8, e52460 (2013)
13. Friedel, M., Nikolajewa, S., Suhnel, J., Wilhelm, T.: DiProDB: a database for dinucleotide properties. Nucleic Acids Res. 37, D37–D40 (2009)
14. Ma, C., Chen, H., Xin, M., Yang, R., Wang, X.: KGBassembler: a karyotype-based genome assembler for Brassicaceae species. Bioinformatics 28, 3141–3143 (2012)
15. Gan, Y., Guan, J., Zhou, S.: A comparison study on feature selection of DNA structural properties for promoter prediction. BMC Bioinformatics 13, 4 (2012)

16. Rajagopal, N., Xie, W., Li, Y., Wagner, U., Wang, W., Stamatoyannopoulos, J., Ernst, J., Kellis, M., Ren, B.: RFECS: A Random-Forest based algorithm for enhancer identification from chromatin state. PLoS Comput. Biol. 9, e1002968 (2013)
17. Li, Z.C., Lai, Y.H., Chen, L.L., Chen, C., Xie, Y., Dai, Z., Zou, X.Y.: Identifying subcellular localizations of mammalian protein complexes based on graph theory with a random forest algorithm. Mol. Biosyst. 9, 658–667 (2013)
18. Wang, J., Kou, Z., Duan, M., Ma, C., Zhou, Y.: Using amino acid factor scores to predict avian-to-human transmission of avian influenza viruses: A machine learning study. Protein and Peptide Letters (2013)
19. Touw, W.G., Bayjanov, J.R., Overmars, L., Backus, L., Boekhorst, J., Wels, M., van Hijum, S.A.: Data mining in the life sciences with random forest: a walk in the park or lost in the jungle? Brief. Bioinform (2012)
20. Gartenberg, M.R., Crothers, D.M.: DNA sequence determinants of CAP-induced bending and protein binding affinity. Nature 333, 824–829 (1988)
21. Rosonina, E., Kaneko, S., Manley, J.L.: Terminating the transcript: breaking up is hard to do. Genes Dev. 20, 1050–1056 (2006)

Disease-Related Gene Expression Analysis Using an Ensemble Statistical Test Method

Bing Wang[1,2,3] and Zhiwei Ji[3]

[1] The Advanced Research Institute of Intelligent Sensing Network, Tongji University,
shanghai 201804, China
[2] The Key Laboratory of Embedded System and Service Computing, Ministry of Education,
Tongji University, Shanghai 201804, China
[3] School of Electronics and Information Engineering, Tongji University,
Shanghai 201804, China
wangbing@ustc.edu

Abstract. The development of novel high-throughput experimental techniques makes it possible to comprehensively analyze biological data in health and disease. However, a large amount of data generated results in dramatic data-analytic challenges in discovery of 'signature' molecules, which are specific to different biological conditions (e.g. normal vs. disease, treated vs. untreated). Current statistical methods are effective only in the case their hypothesis can be matched. In this paper, we apply an ensemble statistical method to infer significant molecules. In our approach, four well-done and well-understanding statistical techniques had been used for the analysis to the experimental data, and then the results will be collected into an ensemble framework to find the high confident "significant" molecules which can distinguish the different experimental conditions. We evaluate the performance of our approach on a test dataset which deposited on GEO database with an access number of GSE45114.

Keywords: Statistical Tests, Ensemble framework, signature molecules, gene expression profile.

1 Introduction

Significant molecules discovery plays a vital important role in treatment as well as in detection and prevention of diseases. Advanced in omic profiling technologies allow the systemic analysis and characterization of alternations in genes, RNA, proteins and metabolites, and offer the possibility of discovering novel signal molecules and pathways activated in disease or associated with disease conditions [1]. Meanwhile, current high-throughput experiments also generated vast amounts of data from genomics, proteomics to metabolomics, by which data analysis is becoming the bottleneck of scientific investigation, especially for signal molecule discovery[2-6].

In recent years, various powerful data mining methods have been propagated for identifying, prioritizing and classifying signal molecules in different omics level [7]. Kwon et al. analyzed multiple single nucleotide polymorphisms (SNPs) in genome-wide

D.-S. Huang et al. (Eds.): ICIC 2013, LNAI 7996, pp. 286–291, 2013.

association studies (GWAS) using Bayesian classification with a singular value decomposition method [8]. Deng et al. presented a novel Bayesian network model to incorporate mass spectrometry and microarray data for prostate cancer signal molecule discovery [9]. Decision tree classification method also has been used to diagnose gastric cancer based on mass spectral data [10]. Neural network-based approaches had been used in analyzing omic data, e.g. in cancer for their capability of learning from examples [11-13]. These methods tried to identify candidate signal molecules in data sets gathered from well-phenotyped cohort studies. However, most of them are performed on small populations, which may result in potential false signal molecules finding.

Statistical test is another more popular strategy in signal molecule discovery for some advantages, such as well-developed and easily implement. Typically, a statistical assumption has been made that the object features (genes, proteins, or metabolites) are expressed in the same level in the different phenotypes, then statistical test methods can calculate the probability (the p value) based on their statistical parameters for each individual feature to reject or accept the assumption. In general, the appropriate test statistic will depend on the experimental design and data distribution. For example, T-test or Mann-Whitney test can be used for binary experiments, and F-test might be employed for polytomous labels. However, no specific test method is ideal or applicable to all study design. T-test assumes normality and constant variance for each feature across all samples; Wilcoxon rank sum test (WRST) can not work for multi-model distribution; Kolmogorov-Smirnov test (K-S test) is non-parametric and has no assumption about distribution, but it only applies to continuous distributions and is too sensitive near the center of the distribution. Usually the distribution of experimental data is not clear, which make the selection of test method very difficult.

To address the above problem, this work proposed an ensemble framework to identify signal molecules. Our approach can combine the advantages of current data mining and statistical test method, in which different powerful statistical test approaches, i.e., T-test, KS-test, Brunner-Munzel test (BM-test) [14] and Baumagartner-Weiss-Schindler test (BWS-test) [15] have been used to find potential signal molecules in the original data, then a simple vote will be implemented under a framework to decide the final list of significant molecules. Our proposed platform was applied into gene expression profile analysis in which the data was downloaded from Gene Expression Omnius (GEO) database. The results shows our approach is effective to infer differentially expressed genes.

2   Experiments and Algorithm

2.1   Dataset Preparation

The gene expression data in this work was obtained from GEO database with an access ID of GSE45114, which consists of mRNA transcriptome extract form three tissues: HCC, peri-cancerous liver and normal liver. This microarray has 49 samples: 24 HCC samples (HCC vs. Normal) and 25 Peri samples (Pericancer vs.

Normal). The microarray of GSE45114 is 22K Human Genome Array (http://bioservices.capitalbio.com/xzzq/wj/3879.shtml), which contains 23232 probes. The microarray data could be obtained using matlab bioinformatics toolbox from GEO platform, which had been preprocessed. Probe set is mapped to NCBI entrez genes, and for genes with multiple probe set, the average expression value of all corresponding probe sets is used. Finally, 15,857 genes had been used in the analysis.

2.2   Statistical Tests

Currently, there are many different statistical test methods had been developed and applied to discover gene differentially expressed between two or more conditions from microarray data. In this work, four statistical test approaches have been selected for differential expressions analysis. One parametric test, i.e., T test, one non-parametric test, i.e., K-S test, two rank-based tests, i.e., BM test and BWS test were selected for the data analysis in this work.

2.3   FDR Control

Typically, signal molecule discovery analysis based on statistical significance test approaches is a multiple testing procedure. For each compound expressed in different groups, statistical test methods generally judge its significance by p-value. The p-value is a measure of how likely the expression of a compound will appear if no real difference existed. Therefore, p-value, usually 0.05, means there is a 5% chance that the decision is wrong. If there are 1000 compounds in the data and 0.05 of p-value is the cutoff, 50 false positives by chance alone will be expected to get. To overcoming this problem, false discovery rate (FDR) control has been applied in this work to find an adjusted p-value using an automatic optimized FDR approach.

3   Results and Discussion

In this work, we inferred differentially expressed genes which are related to HCC from microarray dataset. Four statistical test methods, i.e., t test, K-S test, BM test and BWS test, had been adopted to find the significant genes based on the gene expression data, where the significance threshold of p-value for all four test methods was set up to 0.05. Because the high gene number which is larger than 15,000, to decrease the false discovery rate in each statistical test, the FDR control had been used, and the proportion of false positives desired was also set up to 0.05.

Table 1 shows the number of significant genes after FDR control for all four different types of statistical test methods. It can be found that the number difference among these four test methods is small. T test can find 4812 significant genes, 5114 genes for K-S test, 5494 for BM test and 5230 for BWS test. The mean ratio of 23.6% (30.4% to 34.7%) of all original genes can be inferred as significant genes. And among these, over- and under-expressed genes are almost half and half.

Table 1. The number of over- and under-expressed genes

	t test	K-S test	BM test	BWS test
Number of Over-expressed genes	2322	2463	2565	2655
Number of Under-expressed genes	2490	2651	2839	2665
Number of significant genes	4812	5114	5494	5230

Although the significant genes are got after FDR control, about 5,000 is still a big amount for biologists to extract feasible information to conduct experiments or diagnosis. To get high confident target genes, we analyze the results for the statistical test methods further within a vote framework. Obviously, if one gene can be inferred as significant by two or more statistical methods, the probability that it is a true biomarker will be bigger than that detected just by one test. Figure 1 and 2 shows the number of over- and under- expressed genes cross the different statistical test methods.

Taking *t* test as example, 2322 genes can be detected as over-expressed genes, and 1918 of them can be detected by BM test and K-S test (denoted as T-BM-KS gene set), 107 of them can not be inferred by either of BM test and BWS test (denoted as T-Not-BM-KS gene set). Obviously, the 1,918 genes in T-BM-KS set will have high confidences which are true significant molecules in comparison to the 107 genes in T-Not-BM-KS gene set.

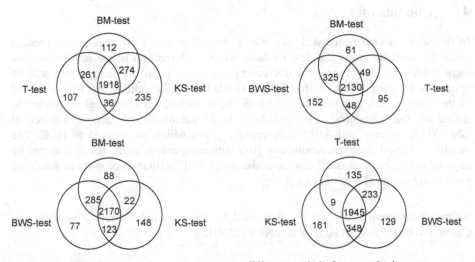

Fig. 1. Over-expressed genes cross different statistical test methods

To make the analysis results more confident, we adopted a vote framework to decide which gene can be seen as significant gene. Within this framework, only genes detected by all the members (statistical methods here) can be seen as the final significant genes. Therefore, 1,913 over-expressed and 2,122 under-expressed genes were selected from the original gene sets. 16.2% genes in the result of *t* test, 21.1% in K-S test, 26.6% in BM test, and 22.9% in BWS test were remove from the final significant gene set, which will drop the complexity very much for the biologists to select genes which will be verified in experiments for only four fifth genes they should pay attention to.

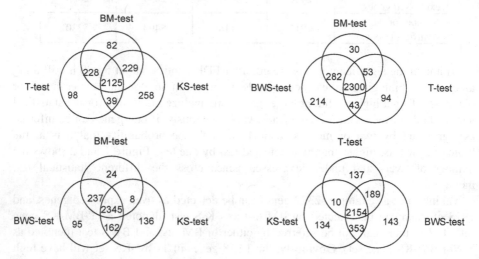

Fig. 2. Under-expressed genes cross different statistical test methods

4   Conclusions

In this study, a vote framework has been proposed to select differentially expressed genes by multiple statistical test method, where different statistical test techniques were implement on the original microarray dataset respectively, and a vote will be applied to the gene selection after FDR control. The final candidate list was consisted of the genes which can be detected by all the statistical tests we used in this work. Based on the microarray data got from GEO database with access number of GSE45114, we can find 4,015 differentially genes which are related to HCC. The number is almost decrease around one fifth with comparison to the result detected by any one of the four statistical test methods, which will facilitate biologists to select the gene with biological significance.

Acknowledgement. This work was funded by the National Science Foundation of China (No.61272269, 61102119 and No.61133010).

References

1. Baumgartner, C., Osl, M., Netzer, M., Baumgartner, D.: Bioinformatic-driven search for metabolic biomarkers in disease. J. Clin Bioinformatics 1(2) (2011), doi:10.1186/2043-9113-1181-1182

2. Wang, B., Chen, P., Wang, P., Zhao, G., Zhang, X.: Radial basis function neural network ensemble for predicting protein-protein interaction sites in heterocomplexes. Protein Pept. Lett. 17(9), 1111–1116 (2010)
3. Huang, D.S., Zheng, C.H.: Independent component analysis-based penalized discriminant method for tumor classification using gene expression data. Bioinformatics 22(15), 1855–1862 (2006)
4. Wang, B., Chen, P., Zhang, J., Zhao, G., Zhang, X.: Inferring protein-protein interactions using a hybrid genetic algorithm/support vector machine method. Protein Pept. Lett. 17(9), 1079–1084 (2010)
5. Zheng, C.H., Huang, D.S., Zhang, L., Kong, X.Z.: Tumor clustering using nonnegative matrix factorization with gene selection. IEEE Trans. Inf. Technol. Biomed. 13(4), 599–607 (2009)
6. Wang, B., Wong, H.S., Huang, D.S.: Inferring protein-protein interacting sites using residue conservation and evolutionary information. Protein Pept. Lett. 13(10), 999–1005 (2006)
7. Zhang, F., Chen, J.Y.: Data mining methods in Omics-based biomarker discovery. Methods in Molecular Biology 719, 511–526 (2011)
8. Kwon, S., Cui, J., Rhodes, S.L., Tsiang, D., Rotter, J.I., Guo, X.: Application of Bayesian classification with singular value decomposition method in genome-wide association studies. BMC Proceedings 3(suppl. 7), S9 (2009)
9. Deng, X., Geng, H., Ali, H.H.: Cross-platform analysis of cancer biomarkers: a Bayesian network approach to incorporating mass spectrometry and microarray data. Cancer Informatics 3, 183–202 (2007)
10. Su, Y.H., Shen, J., Qian, H.G., Ma, H.C., Ji, J.F., Ma, H., Ma, L.H., Zhang, W.H., Meng, L., Li, Z.F., Wu, J., Jin, G.L., Zhang, J.Z., Shou, C.C.: Diagnosis of gastric cancer using decision tree classification of mass spectral data. Cancer Sci. 98(1), 37–43 (2007)
11. Wang, H.Q., Wong, H.S., Zhu, H., Yip, T.T.: A neural network-based biomarker association information extraction approach for cancer classification. Journal of Biomedical Informatics 42(4), 654–666 (2009)
12. Chi, C.L., Street, W.N., Wolberg, W.H.: Application of artificial neural network-based survival analysis on two breast cancer datasets. In: AMIA.. Annual Symposium Proceedings/AMIA Symposium, pp. 130–134 (2007)
13. Amiri, Z., Mohammad, K., Mahmoudi, M., Zeraati, H., Fotouhi, A.: Assessment of gastric cancer survival: using an artificial hierarchical neural network. Pak J. Biol. Sci. 11(8), 1076–1084 (2008)
14. Brunner, E., Munzel, U.: The nonparametric Behrens-Fisher problem: Asymptotic theory and a small-sample approximation. Biometrical Journal 42(1), 17–25 (2000)
15. Baumgartner, W., Weiss, P., Schindler, H.: A nonparametric test for the general two-sample problem. Biometrics 54(3), 1129–1135 (1998)

A Novel Method for Palmprint Feature Extraction Based on Modified Pulse-Coupled Neural Network

Wen-Jun Huai and Li Shang

College of Electronic & Information Engineering, Suzhou Vocational University, Suzhou
215104, Jiangsu, China
{hwj,sl0930}@jssvc.edu.cn

Abstract. Most of these methods succeeded to achieve the invariance against object translation, rotation and scaling, but that could not neutralize the bright background effect and non-uniform light on the quality of the generated features. To eliminate the limit that the recent subspace learning methods for facial feature extraction are sensitive to the variations of orientation, position and illumination in capturing palmprint images, a novel palmprint feature extraction approach is proposed. Palmprint images are decomposed into a sequence of binary images using a Modified Pulse-Coupled Neural Network (M-PCNN), and then the information entropy of each binary image are calculated and regarded as features. A classifier based Support Vector Machine (SVM) is employed to implement recognition and classification. Simultaneously, it overcomes the disadvantage of standard PCNN model with high number of parameters. Theoretical and experimental results show that the proposed approach is robust to the variations of orientation, position and illumination conditions in comparison with other methods based subspace.

Keywords: Pulse-Coupled Neural Network (PCNN), Palmprint recognition, Feature extraction, Information entropy, Support vector machine (SVM).

1 Introduction

For years, palmprint verification is one of the emerging technologies, and a variety of recognition algorithms have been developed. Among the works that appear in the literature are phase-difference information [1], palmprint principal lines [2], online palmprint identification [3] and ordinal palmprint representation [4]. These systems depend mainly on the segmentation step, which is very critical. A new trend for pattern recognition mainly depends on image understanding without segmentation. PCNN, which originated from Eckhorn [5] who focused on the synchronous oscillation phenomenon in the visual cortex of cats, has been extensively used in image processing, including image recognition [6–9]. It has been acquired from the study of synchronous pulse bursts in cat visual cortex. Then different PCNN algorithm has been developed and has all the assumptions for dimension decreasing of image recognition. Examples of these modifications are modified PCNN, feedback

D.-S. Huang et al. (Eds.): ICIC 2013, LNAI 7996, pp. 292–298, 2013.

PCNN, and fast linking PCNN and optimized PCNN [10, 11]. PCNN is applied to image filtering, smoothing, segmentation and fusion [12].

2   Basic Principle

2.1   Modified Pulse-Coupled Neural Network

Over the last two decades, PCNN model is developed for a variety of applications in the engineering field. Typical network topology for PCNN is a two-dimensional planar lattice with a one-to-one pixel to neuron correspondence. The model can be generally divided into three parts: input field, modulation field, and pulse generator [13]. In the input field, each neuron receives a stimulus from two channels between feeding input and linking input, which communicate with the eight-neighboring neurons Nx through synaptic weights M and W with amplification V_F and V_L, respectively. Because of very strict conditions, the application is very complicated. There are several simplify for the exact physiological PCNN neuron, while still keeping the main features of the general theory. Each neuron in the M-PCNN could be described by the following set of equations [14, 15]:

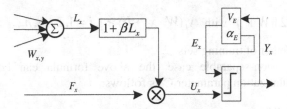

Fig. 1. The Structure of Modified PCNN

Parameters αL, αF and αE are decay coefficients, Yy is the firing information that indicates whether the surrounding neurons have fired or not, and Yx indicates whether this neuron fires or not. * is convolution operator. A schematic block diagram of the modified PCNN neuron is shows in Fig. 1. The formula is described as follows:

$$L_x(n) = L_x(n-1) \cdot e^{-\alpha L} + V_L \cdot (R * Y_y(n-1)) \qquad (1)$$

$$F_x(n) = S + F_x(n-1) \cdot e^{-\alpha F} + V_F \cdot (R * Y_y(n-1)) \qquad (2)$$

Where: index x or y denotes neuron position corresponding to a single pixel in the image space. Y_y is the pulse output of neuron y. Lx presents the intensity of image pixel corresponding to neuron x stimulus input; and αF and αL are the decay time constants of inputs F and L, respectively. Internal activity is calculated by:

$$U_x(n) = F_x(n) \cdot [1 + \beta \cdot L_x(n)] \qquad (3)$$

Where: linking coefficient β controls combination strength. Similar to sub-threshold method, pulse output of a neuron is generated when internal activity exceeds corresponding neural threshold as follows:

$$Y_x(n) = \begin{cases} 1 & U_x(n) > E_x(n-1) \\ 0 & others \end{cases} \tag{4}$$

In Eq. (4), Ex is a threshold of neuron x which is given by:

$$E_x(n) = e^{-\alpha E}E_x(n-1) + V_E Y_x(n) \tag{5}$$

M-PCNN model is a simplified on standard PCNN model, simply set three parameters αE, V_E, and β, greatly reducing the amount of calculation.

2.2   SVM Classification Algorithm

SVM has been used successfully in many object recognition applications. For the two-class case, this optimal hyperplane bisects the shortest line between the convex hulls of the two classes. An optimal hyperplane is required to satisfy the following constrained minimization as fellow:

$$\min\{1/2 \parallel W \parallel^2\} \text{ with } y_i(W \cdot x_i + b) \ge 1, i = 1, 2, \cdots, n . \tag{7}$$

Where n is the number of training sets.

For a linearly non-separable case, the above formula can be extended by introducing a regularization parameter C as follows:

$$\max\{\sum_{i=1}^{n} \lambda_i - \frac{1}{2}\sum_{i,j=1}^{n} \lambda_i \lambda_j y_i y_j (x_i \cdot x_j)\} \text{ with } \sum_{i=1}^{n} y_i \lambda_i = 0, 0 \le \lambda_i \le C, i = 1, 2, \cdots, n . \tag{8}$$

Where, the λ_i are the Lagrangian multipliers and nonzero.

The decision function is obtained as follows:

$$f(x) = \text{sgn}\{\sum_{i=1}^{n} y_i \lambda_i (x_i \cdot x) + b\} . \tag{9}$$

A kernel can be used to perform this transformation and the dot product in a single step which provided the transformation can be replaced by an equivalent kernel function. There are some used kernels, such as polynomial, radial basis and sigmoid.

3   New Feature Generation Method

In this algorithm, palmprint image input in M-PCNN network to simulate biological visual perception of the process, and break it down into a sequence consisting of several binary image cognition, computing cognitive sequences of each binary image entropy, as identification features of palmprint images. Then, SVM is used for classification. Figure 2 shows the basic flow of the algorithm.

Fig. 2. The Pulse Image Sequence From Original Image by the M-PCNN

After many experiments, the parameters of M-PCNN is used in this paper as shown in table 1, the connection weighted coefficient matrix $W_{x,y}$ is determined as fellow.

$$W_{x,y} = \begin{bmatrix} 0.5 & 1 & 0.5 \\ 1 & 0 & 1 \\ 0.5 & 1 & 0.5 \end{bmatrix}$$

Table 1. The parameters of PCNN in the experiment

Parameters	V_E	αE	β
Value	127	0.4	0.5

Table 2. The Pulse Image Sequence From Original Image by the M-PCNN

N=1	N=2	N=3	N=4	N=5	N=6	N=7
N=8	N=9	N=10	N=11	N=12	N=13	N=14
N=15	N=16	N=17	N=18	N=19	N=20	N=21
N=22	N=23	N=24	N=25	N=26	N=27	N=28
N=29	N=30	N=31	N=32	N=33	N=34	N=35
N=36	N=37	N=38	N=39	N=40	N=41	Original Image

The inherent potential V_E to take half of the original image grey value biggest, some value is determine which are the decay time constant αE of dynamic threshold and the coefficient β of connection strength after many experiments

The palmprint image generated by M-PCNN composed of multiple binary image recognition sequence as table 2, which are 41 of all, to calculation of information entropy each binary image as the recognition feature.

Information entropy is defined as:

$$H(P) = -P_1 \log_2(P_1) - P_0 \log_2(P_0). \tag{10}$$

Among them, $H(P)$ is information entropy of each binary image, P_1 and P_0 mean the probability of binary image pixel values for 1 and 0 respectively. The generated cognition sequence has obvious difference which composed by multiple binary images because of the distribution differences from different image pixel gray value.

4   Experimental Results

4.1   Data Preprocessing

Simulation experiments programming to realize using software Matlab7.1, hardware environment for the Intel core i5 3.4 GHz CPU, 4GB memory, the operating system is Windows XP professional edition. This work is carried on the PolyU Palmprint Database in order to verify the effectiveness of the proposed method.

4.2   Palm Classification Performances

In finishing the classification mission, the features of palmprint were extracted by using M-PCNN. The SVM was used to testing feature extraction method based on the kernel of RBF and polynomial function. For comparison with based on characteristic subspace method, the features was obtained by using PCA, Wavelet, and FRIT have also been tested. The experiment results are shown in Table 3.

Table 3. Comparison of Different Recognition Algorithm

Methods	PCA	Wavelet	FRIT	M-PCNN
Recognition rate (%)	89.1	93.34	95.83	96.55

Regarding to Table 2, Classification results of the proposed method are slightly higher recognition accuracy than the former, its best recognition rate can reach up to 96.55% t, and only around to 93.34% using Wavelet which is almost close to 93.75% in literature [15]. Along with the increase of sample size and training, recognition rate will be more stable. Both polynomial and radial basis kernel were chosen as the discriminate function to train and classify in the experiment.

5   Conclusion and Future Work

This paper proposes a novel palmprint feature extraction algorithm which combines PCNN and SVM. This algorithm simulating biological visual perception of the process, the palmprint image is decomposed into cognition sequence which composed of several binary image and extraction entropy from each binary image as recognition features of palmprint image. Theory and experiments show that palmprint features extracted by this algorithm have stronger robustness to location, illumination changes. The experimental results show that the feature information has a strong discriminate ability and still have potential to be improved. In the future work, we will investigate the M-PCNN method based on more reasonable mechanism to extract characteristic information more efficiently.

Acknowledgments. The authors would like to express their sincere thanks to BRC at the Hong Kong Polytechnic University for providing us the PolyU Palmprint Database. This work was supported by NNSF of Jiangsu Province (No. BK2009131), Startup Foundation Research for Young Reachers (No. 20102SZDQ05), Practice and Innovation Training Project of Jiangsu Province (No. 2012JSSPITP4089).

References

1. Badrinath, G.S., Gupta, P.: Palmprint based recognition system using phase-difference information. Future Generation Computer Systems 28, 287–305 (2012)
2. Yuan, W.Q., Gu, Z.H.: Research on feature selection method based on five classes of palmprint principal lines on the whole palm. Chinese Journal of Scientific Instrument 33, 942–948 (2012) (in Chinese)
3. Zhang, D., Kong, W.K., You, J., Wong, M.: Online palmprint identification. Pattern Analysis and Machine Intelligence 25, 1041–1050 (2003)
4. Sun, Z., Tan, T., Wang, Y., Li, S.Z.: Ordinal palmprint represention for personal identification. In: CVPR 2005, pp. 279–284. IEEE Press, New York (2005)
5. Eckhorn, R., Reitboeck, H.J., Arndt, M., Dicke, P.: Feature linking via synchronization among distributed assemblies: simulations of results from cat visual cortex. Neural Comput. 2, 293–307 (1990)
6. Rava, T.H., Bettaiah, V., Ranganath, H.S.: Adaptive pulse coupled neural network parameters for image segmentation. World Acad. Sci. Eng. Technol. 73, 1046–1052 (2011)
7. Wei, S., Hong, Q., Hou, M.: Automatic image segmentation based on PCNN with adaptive threshold time constant. Neurocomputing, 1485–1491 (2011)
8. Chen, Y., Park, S.K., Ma, Y., Ala, R.: A new automatic parameters setting method of a simplified PCNN for image segmentation. IEEE Trans. Neural Network, 880–892 (2011)
9. Fu, J.C., Chen, C.C., Chai, J.W., Wong, S.T.C., Li, I.: Image segmentation by EM-based adaptive pulse coupled neural networks in brain magnetic resonance imaging. In: Wong, S. (ed.) Computerized Medical Imaging and Graphics, pp. 308–320. Elsevier (2010)
10. Forgac, R., Mokris, I.: Feature generation improving by optimized PCNN. In: 6th International Symposium on SAMI 2008, pp. 203–207. IEEE Press, New York (2008)

11. Tolba, M.F., Abdellwahab, M.S., Aboul-Ela, M., Samir, A.: Image signature improving by PCNN for Arabic sign language recognition. Machine Learning & Pattern Recognition, 1–6 (2010)
12. Weili, S., Yu, M., Zhanfang, C., Hongbiao, Z.: Research of automatic medical image segmentation based on Tsallis entropy and improved PCNN. In: Fukuda, T., et al. (eds.) ICMA 2009, pp. 1004–1008. IEEE Press, New York (2009)
13. Yide, M., Lian, L., Kun, D.: Pulse coupled neural network with digital image processing, pp. 225–237. Science Press (2006)
14. Huai, W.-J., Shang, L.: A Palmprint Classification Method Based on Finite Ridgelet Transformation and SVM. In: Huang, D.-S., Gan, Y., Bevilacqua, V., Figueroa, J.C. (eds.) ICIC 2011. LNCS, vol. 6838, pp. 398–404. Springer, Heidelberg (2011)

An Improved Squeaky Wheel Optimization Approach to the Airport Gates Assignment Problem

Cuiling Yu and Xueyan Song

School of Computer Science and Technology, Tianjin University, Tianjin, China

Abstract. The optimization of the airport gates assignment problem aims to improve the efficiency of gates allocation as well as the satisfaction degree of passengers. In this paper, an SWO algorithm combined with Tabu Search strategy was proposed to optimize the allocation of airport gates. Experimental results show that the improved SWO approach is more efficient than the conventional SWO approach.

Keywords: Airport Gates Assignment Problem, SWO algorithm, Tabu Search.

1 Introduction

The airport gates assignment problem (AGAP) attempts to assign suitable gates to flights in order to optimize a number of objectives. The walking length of passengers is one of the important optimization objectives of AGAP. A number of heuristic algorithms have been proposed in literature. Yan Chang formulated the AGAP as a multi-commodity network problem and a Lagrangian heuristic algorithm was applied to obtain the solutions [1]. Ahmet Bolat proposed a mixed-binary mathematical model with a quadratic function for minimizing the variance of idle times at the gates to make the initial assignments insensitive to variations in flight schedules [3]. H.Ding et al. designed a greedy algorithm and used a Tabu Search meta-heuristic to solve the problem [4]. In this paper, a Squeaky Wheel Optimization (SWO) algorithm is proposed to optimize the allocation of airport gates. In SWO, a greedy algorithm is used to construct a solution; then the solution is analyzed to find the trouble spots which might have brought negative effects to the overall performance. Through the analysis, a new priorities order isestabished to construct a solution in the next solution. This Construct/Analyze/Prioritize cycle continues until a stop criterion is reached [6]. SWO can be viewed as searching in two search spaces: solutions and prioritizations, as shown in figure 1. Successive solutions are only indirectly related through the re-prioritization that results from analyzing the prior solution. Similarly, successive prioritizations are generated by constructing and analyzing solutions.

Although SWO has the ability to make large and coherent moves, it is poor at making small "tuning" moves in the solution space. Therefore, an SWO algorithm combined with tabu search method is proposed in this paper to improve the search ability.

D.-S. Huang et al. (Eds.): ICIC 2013, LNAI 7996, pp. 299–306, 2013.

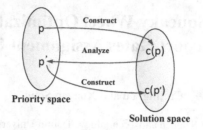

Fig. 1. The coupling space diagram of SWO

The paper is organized as follows. The formulation of the AGAP is given in Section 2. Section 3 present the improved SWO algorithm. Experimental results are presented in Section 4. Conclusion remarks is given in Section 5.

2   Formulation of AGAP

2.1   Objective Function

In this paper, we take the shortest average walking distance as optimization objective , which is shown as follows:

$$\min f = \sum_{i=1}^{N}\sum_{j=1}^{M} R_i D_j y_{i,j} \Big/ \sum_{i=1}^{N} R_i \tag{1}$$

Where:

R_i : the number of passengers on flight i ;

D_j : the distance from the airport gate j to the departure lounge;

$y_{i,j}$: whether the flight i is assigned to the airport gate j . If it is so then the $y_{i,j}=1$, or $y_{i,j}=0$.

2.2   Constraint Conditions

Take into account the security factors in actual airport gates assignment, the constraints considered in this paper are expressed as follows.

$$\sum_{j=1}^{M} y_{i,j}=1 \tag{2}$$

Equation (2) ensures that each flight can only be assigned to one airport gate.

$$E_{j,k}\text{-}L_{i,k} \geq ST \quad \text{if } x_{ijk}=1 \tag{3}$$

Formula (3) ensures that the adjacent flights which are in the same airport gate have a safe and reasonable time interval so that the flights could go in and out of the airport gate smoothly.

Where:

$E_{j,k}$: the time that the flight j go into the airport gate k ;

$L_{i,k}$: the time that the flight i leave the airport gate k ;

ST : the safety interval between the adjacent flight;

x_{ijk} : the flight i and the flight j are the adjacent flights in the same airport gate k . If the flight i is in front of the flight j , then the $x_{ijk}=1$, or $x_{ijk}=0$.

$$G_i \geq min \tag{4}$$

Formula (4) ensures that the stop time of the flights must be longer than the shortest ground service time.

Where:

G_i : the stop time of flight i ;

min : the shortest ground service time of the flight.

$$R_i, D_j, L_{i,k}, E_{j,k}, G_i \geq 0 \tag{5}$$

$i=1,2....N$, $j=1,2,....M$;

Formula (5) is a non-negative restriction to ensure that all of the variables could make sense.

3   Algorithm

3.1   Traditional SWO Algorithm

The core of SWO is a "Construct/Analyze/Prioritize" cycle. A greedy algorithm is applied to construct a solution by the order of priority of each element. That solution is then analyzed to find the elements of "Trouble Makers". The priorities of the trouble makers are then increased, which cause the greedy constructor to deal with them sooner on the next iteration. This cycle repeats until a termination condition occurs.

3.2   An Improved SWO Algorithm

It is observed that a small change in the sequence generated by the Prioritizer will cause a large change to the corresponding solution. In order to make SWO algorithm achieve small move, we combine SWO algorithm with Tabu Search algorithm in the solution space. Through the application of the Tabu Search, the improved SWO algorithm will search in the neighborhood of good solution provided by SWO Constructor, with the aim to obtain improved solution and avoid trapping in a small cycle.

The improved SWO algorithm incorporates Tabu search algorithm after Constructor procedure. As shown in Figure 2, the SWO constructor generate good solution as the input (i.e. initial solution)of the Tabu Search algorithm, then the Tabu Search algorithm try to find better solution in its neighborhood. Instead of the constructing a solution, the output of the Tabu Search goes into the "Constructor/Analyzer/Prioritizer" iteration. Considering the running time, Constructor invokes the Tabu Search part by probability P.

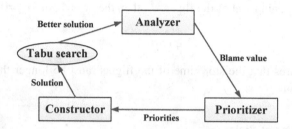

Fig. 2. The frame diagram of improved SWO algorithm

The solution structure of Tabu Search part is consistent with SWO part. Both are based on integer coding. The Tabu Search algorithm use the swap neighborhood strategy, namely, it swaps the airport gates assigned to each of a pair of flights.

When Tabu Search algorithm obtains a better solution than the optimal solution in history, it checks the tabu list. If the solution is already in the list, it set the solution as the current solution.

4   Simulation Studies

The flight data adopted in the experiments is based on the BCIA (Beijing Capital International Airport) [7] . The passengers' information was also added to the data, which is shown in table 1. The unit of the flight's arrival and departure time is minutes in the table.

The information of airport gates is based on the data of the table 2. The unit of the distances from the airport gates to the waiting hall is meters in the table 2.

It took 2.9712 seconds to run the original SWO algorithm for 500 times and got the optimal solution at the generation 188 firstly. The shortest average passenger walking distance is 571.4635 meters according to the results, which was shown in figure 4. G11 in figure 4 is the parking apron. Since the parking apron has no restriction, it can accommodate many flights at the same time and doesn't have to consider the safe interval between flights or uniqueness constraints.

Table 1. The flights information

Flight No.	Arrival Time	Leave Time	Passengers	Flight No.	Arrival Time	Leave Time	Passengers	Flight No.	Arrival Time	Leave Time	Passengers
1	2	45	189	40	196	266	138	79	395	438	183
2	6	47	134	41	204	247	126	80	396	438	227
3	14	57	136	42	208	250	198	81	402	469	162
4	16	60	253	43	214	256	215	82	409	451	284
5	21	91	124	44	217	261	98	83	412	454	274
6	27	95	342	45	224	265	312	84	417	461	186
7	31	72	68	46	227	268	147	85	425	493	173
8	37	107	147	47	235	302	338	86	430	496	228
9	43	87	233	48	236	279	128	87	434	478	255
10	49	92	155	49	243	285	126	88	437	482	173
11	54	98	167	50	248	315	190	89	442	485	228
12	59	102	215	51	252	296	152	90	447	489	295
14	70	111	345	53	262	306	93	92	459	500	173
15	71	116	178	54	266	308	226	93	462	504	183
16	76	118	167	55	271	315	289	94	466	535	228
17	85	152	192	56	280	325	246	95	475	543	295
18	87	132	175	57	285	353	236	96	479	524	284
19	94	135	283	58	287	330	214	97	484	553	268
20	100	141	235	59	291	361	216	98	490	533	275
21	101	142	290	60	296	340	187	99	495	565	183
22	106	151	267	61	301	371	263	100	498	539	195
23	113	157	263	62	306	350	265	101	503	544	255
24	119	185	136	63	312	355	274	102	507	549	276
25	125	193	157	64	320	365	263	103	513	557	248
26	130	174	136	65	323	366	153	104	517	560	369
27	133	175	158	66	328	394	182	105	521	563	288
28	137	182	247	67	331	400	284	106	526	569	156
29	144	185	268	68	337	378	257	107	532	573	367
30	148	218	246	69	342	387	266	108	540	581	462
31	152	194	89	70	347	389	238	109	544	613	783
32	160	226	251	71	355	422	267	110	547	613	157
33	162	205	122	72	356	400	284	111	555	599	179
34	166	211	86	73	362	429	295	112	556	625	246
35	171	214	119	74	369	436	261	113	562	607	348
36	177	244	118	75	373	414	273	114	568	613	126
37	183	252	253	76	377	421	283	115	573	639	157
38	189	233	361	77	382	423	166	116	577	620	248
39	191	261	139	78	390	434	241	117	582	651	190

Table 2. The airport gates data

Airport Gate No.	Distance to Waiting Hall(m)
G1	100
G2	100
G3	200
G4	200
G5	300
G6	300
G7	400
G8	400
G9	500
G10	500
G11 (Apron)	1500

Airport Gate No.	Flight No.
G1	7 18 30 47 64 76 87 98 109
G2	11 29 41 52 68 79 91 105 116
G3	5 23 37 54 65 78 95 111
G4	6 26 38 51 63 74 90 101 112
G5	4 19 33 49 60 72 83 94 110
G6	2 12 24 45 59 75 86 102 113
G7	8 28 40 61 80 92 103 114
G8	9 21 35 48 62 73 89 100 115
G9	1 13 25 43 55 67 85 104 117
G10	3 20 36 53 69 82 93 107
G11	10 14 15 16 17 22 27 31 32 34 39 42 44 46 50 56 57 58 66 70 71 77 81 84 88 96 97 99 106 108

Fig. 3. The schematic diagram of airport gates assignment

Since the Constructor calls Tabu Search procedure by probability P, the improved SWO algorithm has certain randomness. We got the average operation data from ten experiments. It took 13.2745 seconds to run the improved SWO algorithm 500 times and got the optimal solution at the generation 164. The shortest average passenger walking distance is 560.2347 meters according to the optimal solution. The comparison of experimental results between the original SWO algorithm and the improved SWO algorithm was shown in figure 5.

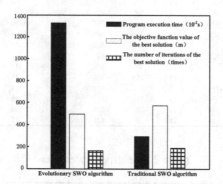

Fig. 4. Experimental results contrast diagram

According to the results of experiments, the quality of the solution has been improved although the improved SWO algorithm spent more time than the original SWO algorithm.

The coupling of solution space and priority space is an important characteristic of SWO algorithm. When a small change occurs in the sequence generated by the Prioritizer, a very big change may happen to the solution from the Constructor. It enables the SWO algorithm to have a large continuous move in the solution space. Thus, SWO can jump out of local optimal more easily than general local search algorithms. But this advantage can also become the weakness of SWO algorithm at the same time.

The SWO algorithm can't make the small move in the solution space, which may lose the opportunity to get a better solution and get poor convergence. With the Tabu Search algorithm, we use the good solution from the Constructor as the initial solution of the Tabu Search, and search in its neighborhood for more optimal solution. Once we obtain better solution, it will replace the solution from the Constructor as the input of Analyzer in next iteration. This makes the improved SWO algorithm have the ability of achieving small move in the solution space. Consequently, the improved SWO algorithm can improve the quality of solution and obtain better convergence.

Due to the coupling spaces of SWO algorithm, Solution space and sort space have interrelationship between each other. With Tabu Search algorithm, we can change the solution from Constructor, which means that corresponding change will happen to the current priority sequence.

The original SWO algorithm always pay attention to "trouble makers". However, there also exist some elements easy to deal with in the large-scale problems. In the whole SWO iteration process, we found that the elements which have low blame values didn't update their positions in the priority sequence very often. The improved SWO algorithm makes the elements which have low blame values reach a better level, so that the optimal solution may be found earlier and better.

We introduced the probability P for Tabu Search procedure in the improved SWO algorithm. Namely, the solution generated from Tabu Search part will replace the solution generated by Constructor in next iteration with probability P. On the one hand, it will ensure that the improved SWO algorithm can obtain high running speed; on the other hand, it will protect the inherent integrity of SWO algorithm, which is helpful to maintain its own advantage.

5   Conclusions

In this paper, we establish a mathematical model for the airport gates assignment problem model and use the SWO algorithm to search for solution. Due to the compact structure and the oriented search way of SWO, the experiments results explain that the SWO algorithm is efficient. Considering the weakness of the SWO algorithm, we proposed an improved SWO algorithm in this paper. Compared with the conventional SWO algorithm, the improved SWO algorithm bring improved solution, although the running time is little bit increased. Compared with other complicated algorithms, it still has advantage on the speed. Thus, the improved SWO algorithm is feasible and effective.

In the future, we will use the SWO algorithm for more complicated airport gate assignment problem.

Acknowledgements. This study is supported by grants from National Natural Science Foundation of China (61039001) and Tianjin Municipal Science and Technology Commission(11ZCKFGX04200).

References

1. Chang, Y.: A Network Model for Gate Assignment. Journal of Advanced Transportation 32(2), 176–189 (1998)
2. Yan, S.Y., Shieh, C.Y., Chen, M.: A Simulation Framework for Evaluating Airport Gate Assignments. Transportation Research 36(10), 885–898 (2002)
3. Bolat, A.: Procedures for Providing Robust Gate Assignments for Arriving Aircrafts. European Journal of Operational Research 120(1), 63–80 (2000)
4. Ding, H., Lim, A., Rodrigues, B., Zhu, Y.: Aircraft and Gate Scheduling Optimization at Airports. In: Proeeedings of the 37th Hawaii International Conference on System Sciences (2004)
5. Cheng, Y.: A Rule-based Reactive Model for The Simulation of Aircraft on Airport Gate. Knowledge-based Systems 10, 225–236 (1998)
6. Joslin, D.E., Clements, D.P.: "Squeaky Wheel" Optimization. Journal of Artificial Intelligence Research 10, 353–373 (1999)
7. Wei, D.X., Liu, C.Y.: Optimizing Gate Assignment at Airport Based on Genetic-Tabu Algorithm. In: International Conference on Automation and Logistics, pp. 1135–1140 (2007)

Study on Rumor Spreading Simulation Model in Collective Violent Events

Bu Fanliang and Dang Huisen

Department of Engineering of Security&Protection System,
Chinese People's Public Security University, Beijing, China
bufanliang@sina.com, danghs@sohu.com

Abstract. Rumor spreading plays an important role in the development of collective violent events. Using Agent-based modeling and simulation, this paper presents a new rumor spreading simulation model to represent the specific process of rumor spreading by extracting characters of rumor and individual factors. Simulation results have shown that the more inflammatory the rumor is, the more quickly it spreads and the wider the influence scope is; the higher the proportion of individuals who know the truth in the crowd is, the more slowly rumor spreads and the smaller the influence scope is; however the raise of individual's grievance could accelerate the spreading of rumor easily. The conclusions could provide meaningful proposals for related departments to deal with similar problems.

Keywords: Collective violent events, rumor spreading, Agent, inflammatory.

1   Introduction

In recent years, the frequent collective violent events have caused great losses to the public which has extremely bad influence to the construction of harmonious society, such as Shaoguan events, Shishou events and so on. Scholars have been performing their researches about collective violent events one after another and making notable achievements [1-3]. Throughout these events, we could find that rumor spreading has great influence to the expanding and worsens the events. Previous studies on rumor spreading in collective violent events mostly discuss and analyze in theory, which belong to sensory perceptions and qualitative analysis. With the amalgamation of social science and natural science, there are more and more computer simulations in the study of social science. Agent-based modeling and simulation depicts the whole system by using simple descriptions from top to down, it suits well to the study of rumor spreading [4]. This paper uses Agent-based modeling and simulation to model the individuals and rumors in the context of collective violent events, explores the influences of rumor properties and individual characters to rumor spreading and event development.

D.-S. Huang et al. (Eds.): ICIC 2013, LNAI 7996, pp. 307–313, 2013.

2   Collective Violent Events and Rumor Spreading

Conflict of interest is the source of collective violent events. With the deepening of reform and open, China is undergoing tremendous changes. Some problems that concerns people's interest would turn out to be social problems and collective violent events would erupt suddenly as the social conflicts accumulate. Nowadays, collective violent events usually occur between the government and the mass, there are more and more non-direct-interest collective violent events [5]. The participators usually have no direct interest appeal to the conflict events, while the accumulated grievance mood drives them take part in the events to throw their criticism. The rumor has a good environment to propagate and spread when concerned departments could not make the information available to the public. Meanwhile, rumor spreading would worsen grievances among the people which promote the expanding of collective violent events.

Since Knapp analyzed rumor scientifically in 1942 [6], numerous scholars have begun their studies of rumor from the perspective of mathematic computation, analog simulation and so on. Rumor, some kind of unconfirmed message or information, has two basic key elements: importance and fuzziness, individual characters and environment factors also have great influence to the spread of rumor [7]. Different from other conditions, in the context of collective violent events the rumors often have stronger inflammatory and the emotions of individuals are much higher, so we have to study the macroscopic phenomenon of the emergence of rumor spreading from individual level.

3   Agent-Based Model of Rumor Spreading

Agent-based modeling and simulation depicts the whole system by describing Agent [8]. We present rumor spreading model using this method [9]. By abstracting related factors to represent the process of rumor spreading, we focus on the characters of rumor and individual to observe the different influences of each factor.

3.1   Rumor Properties

Importance and fuzziness are two basic elements of rumor. The more people care about the event and the less the authority information is available, the more the number of rumors is and the wider they expand. In collective violent events, some agitators mislead people who don't know the truth by starting and spreading rumors recklessly which may lead to a bad result. We abstract parameters L and S to indicate the importance and fuzziness of rumors respectively. L refers to inflammatory, the bigger L is the more people pay close attention to it. S refers to the proportion of people who know the truth in the crowd, the higher the proportion is the less S is. L is distributed in [0, 1] while S values range from 0 to 100%.

3.2   Individual Characters

According to the content above, there are two types of individuals: people who know the truth and who don't. The people who don't know the truth, moreover, contains some rumormongers which could be designated randomly during the simulation. Our study is largely concerned with the spreading among the people.

As we know, a considerable number of participators in collective violent events have no direct interest appeal to the conflict; it is the grievances that drive them join the events. We abstract G to indicate individual's grievance to the society. $G \in [0, 1]$, individuals have greater chance to spread rumors as G increases.

Rationality R refers to individuals' trust level in rumors. The probability of rumor spreading decreases as people trust the rumors less. $R \in [0, 1]$.

F refers individual's influence force to others. In the real world, leaders, friends often have more influence to individual's attitude towards rumors. $F \in [0, 1]$.

B refers the trust level of individual to the rumors. $B \in [0, 1]$. When the value of B exceed the threshold b_0, individuals would develop into infectors. During the simulation, the value of B changes when individual encounters rumormongers or infectors, as shown in equation (1).

$$B(t + 1) = B(t) + m_k * F * G * L \qquad (1)$$

m_k is the influence coefficient when heard some rumor k times, F refers the influence force of rumormongers or infectors, G and L refers to individual's grievance and rumor's inflammatory.

Individual spreads the rumor with certain probability after turning into infectors, the probability is expressed in equation (2).

$$P_1 = G * (1 - R) \qquad (2)$$

The higher individual's grievance is, the more nonrational the individual is and the bigger the spread probability is.

For those individuals who know the truth, they also spread the truth during the communications with others and the probability is shown is equation (3).

$$P_2 = R * (1 - G) \qquad (3)$$

At the same time, the trust level B of individual heard the truth is changing to

$$B(t + 1) = B(t) - n_k * F * R \qquad (4)$$

Where n_k is the influence coefficient when heard the truth k times and F is influence force of the people who spread the truth.

3.3   Simulation Environment and Settings

There are N participators divided into people who know the truth and who don't. At the beginning of simulation, we designate some individuals as rumormonger randomly and set properties for each individual.

Individuals start to move randomly during the simulation. Individuals who don't know the truth could perceive the environment and communicate with rumormongers, infectors or people who know the truth. Meanwhile they update their trust level B to the rumors. When B exceeds some threshold b_0, the individual develop into infectors.

We change concerned parameters to observe their influence to rumor spreading. The interface diagram of simulation is shown in fig. 1.

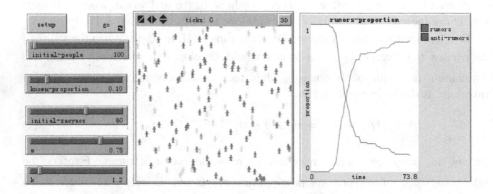

Fig. 1. Interface diagram of simulation

4   Simulation Results

Rumor spreading is closely related with the properties of rumor itself and individual. Next we will change the parameters of our interest to observe their influence to the spreading.

4.1   Influence of Rumor's Inflammatory

Set the inflammatory of rumor L as 0.2, 0.4 and 0.6 respectively and the proportion of people who know the truth is 10%, the grievance of individual G obeys normal distribution, $\mu=0.4$. Run the simulation respectively and we obtain the time-varying curves of rumor spreading as shown in fig. 2 (red curves refer to the proportion of infectors while blue curves refer to the proportion of people who don't believe the rumors).

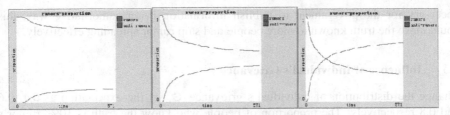

Fig. 2. Conditions of rumor spreading with different inflammatory of rumors (L is 0.2, 0.4 and 0.6 respectively from left to right)

From fig. 2 we can see that, with the increase of rumor's inflammatory, the proportion of infectors raise all the time and the rate of change increases too. This indicates that the speed of rumor spreading is accelerating and the spread scope expands wider as well. In the case of low inflammatory, the rumor spreads only on a small scale, we could clearly observe the latent period and outbreak period. As the inflammatory increases, the speed of rumor spreading quickens rapidly and it breaks out after a short time. Therefore, it may be just temporary from the birth of rumors with high inflammatory to the consequence of collective violent events in real world. Related departments should gear up in advance, monitor public opinions and cut off the diffuse channels before the breakout of rumor spreading.

4.2   Influence of the Proportion of People Who Know the Truth

Set the proportion of people who know the truth as 5%, 10% and 15%, rumors' inflammatory is 0.4, other parameters keep the same. Run the simulations and we obtain the time-varying curves of rumor spreading as shown in fig. 3.

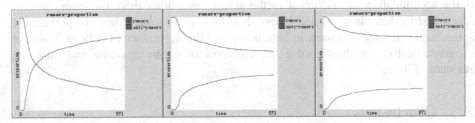

Fig. 3. Conditions of rumor spreading with different proportions of people who know the truth (the proportion is 5%, 10% and 15% respectively from left to right)

It's obvious that with the expansion of people who know the truth, the probability that individual discerns the truth is increasing and therefore the probability that individual believes rumor reduces. It makes the infectors less and less, the spreading speed slows down and rumor spreads only among a small minority finally. In real world, when some hot-button issue happens, certain departments could not adequately disclose information in time for various reasons, this strengthen the fuzziness of the event and result in many rumors to eliminate information's uncertainty in turn. It's necessary for the government to publish information in an authoritative way, this could make the truth known to more people and stop rumor spreading effectively.

4.3   Influence of Individual's Grievance

Change the distribution of individual's grievance G, set the expectation as 0.4, 0.6 and 0.8 respectively. The proportion of people who know the truth is 10%, rumor's inflammatory is 0.4, and other parameters keep the same. Run the simulations and obtain the time-varying curves of rumor spreading as shown in fig. 4.

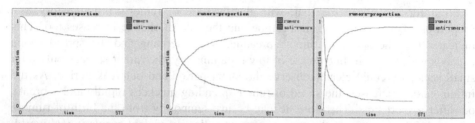

Fig. 4. Conditions of rumor spreading with different distributions of individual's grievance (the expectation of the normal distribution is 0.4, 0.6and 0.8 respectively from left to right)

We could notice that, the most noticeable thing is the slope of curves' change which refers to the proportion of people who believe the rumor. The rise of individual's grievance increases the slope rapidly; it's almost a vertical line at last. This indicates that the rumor spreading is in a situation of breakout, even a rumor with low inflammatory could also spread widely. From another point of view, it validates the development of non-direct-interest collective violent events. Grievance is the accumulation of discontent to the society in a long time, the emergence of rumor just fit their emotions and relieve their feelings to some extent. To prevent the spreading of rumors with some inflammatory and the occurrence of collective violent events radically, we should solve the problems of public attention and raise the standard of living.

5   Conclusions

The emergence of behaviors at the group level could be observed by applying simple individual behavior rules [10]. By abstracting key characters of rumor and individual we have built a rumor spreading model to represent the rumor spreading process in collective violent events. Simulations have shown that, the more inflammatory the rumor is, the more quickly it spreads and the wider it expands; the higher the proportion of people who know the truth, the more slowly the rumor spreads and the smaller it expands; the raise of individual's grievance could accelerate the spreading of rumor easily. So concerned departments should public information by all means to make more people to know the truth at the beginning of the conflict and pay more attention to the inflammatory rumors, cut off the spreading medium in time. To set up harmonious society and handle problems people really care about, improve people's lives and reduce individual's grievance to the society could prevent the occurrences of collective violent events radically.

The model reflects the process of rumor spreading; it also has some shortcomings, such as the different influences of individual in other social relations. We would make further efforts to study the rumors in collective violent events in detail.

References

1. Bu, F.L., Feng, P.Y.: Analysis of Agent-based Non-organization and Direct Interest Collective Event. In: Automatic Control and Artificial Intelligence, pp. 1762–1767 (2012)
2. Bu, F.L., Sun, J.Z.: Particle Swarm Optimization-based Simulation and Modeling System for Mass Violence Events. In: Automatic Control and Artificial Intelligence, pp. 2093–2097 (2012)
3. Bu, F.L., Zhao, Y.N.: Modeling and Warning Analysis of Mass Violence Events. In: Automatic Control and Artificial Intelligence, pp. 2048–2052 (2012)
4. Erlc, B.: Agent-based Modeling: Methods and Techniques for Simulating Human Systems. PNAS 99(3), 7280–7287 (2002)
5. Zhang, J.L.: Summary of Collective Events. Beijing. Intellectual Property Right Press (2011)
6. Knapp, R.H.: A Psychology of Rumor. Public Opinion Quarterly 8(1), 22–37 (1944)
7. Gordon, W.A., Leo, P.: The Psychology of Rumor. Henry Holt and Company (1947)
8. Liao, S.Y., Dai, J.Y.: Study On Complex Adaptive System and Agent-Based Modeling&Simulation. Journal of System Simulation 16(1), 113–117 (2004)
9. Galam, S.: Modeling Rumors: The No Plane Pentagon French Hoax Case. Physical A: Statistical Mechanics and its Applications 320, 571–580 (2003)
10. Nanda, W.: Understanding Crowd Behavior: Simulating Situated Individuals. University of Groningen Press, Groningen (2011)

Particle Swarm Optimization-Neural Network Algorithm and Its Application in the Genericarameter of Microstrip Line

Guangbo Wang, Jichou Huang, Pengwei Chen, Xuelian Gao, and Ya Wang

School of Electrical and Electronic Engineering,
North China Electric Power University, Beijing, China, 102206
wang_guang_bo@163.com

Abstract. To solve the general model for S-parameter of microstrip line quickly, this paper proposes Particle Swarm Optimization-Neural Network (PSO-NN) algorithm, which is based on the research of Particle Swarm Optimization (PSO) algorithm and neural network algorithm. By testing and analyzing PSO-NN, PSO and BP neural network algorithm respectively with the performance check function, we find PSO-NN the best performance. Finally, PSO-NN algorithm is applied to the general model for S-parameter of microstrip line which has made use of CST software to get the training data and validation data of the S-parameter of microstrip line. By training and validating PSO-NN, PSO and BP neural network algorithm, we prove that PSO-NN algorithm has the minimum average error and standard deviation in acceptable time. Compared with CST software, the PSO-NN algorithm has shorter simulation time at the same precision level .Therefore, this paper is of great value to the research of PCB board.

Keywords: Particle Swarm Optimization-Neural Network (PSO-NN) algorithm, S-parameter of microstrip line, Algorithms Performance Check Function

1   Introduction

Printed Circuit Board (referred to as PCB) is an integral part of the electronic equipment, and research on electromagnetic compatibility characteristics is the key to the design of advanced and complex electronic systems, as well as the bottleneck to improve the system. The PCB microstrip lines, an important part of the structure of the PCB, its EMC characteristics is mainly concentrated in the crosstalk and radiation [1-2]. These problems directly determine the transmission performance of the PCB. Thus, the study on PCB board micro-with a line is of great importance.

Currently, some commercial software can insure accurate calculation of the loss distribution and field distribution under different operating conditions of the microstrip line. CST electromagnetic compatibility simulation software can provide

D.-S. Huang et al. (Eds.): ICIC 2013, LNAI 7996, pp. 314–323, 2013.

users with a complete system-level and component-level numerical simulation analysis. The software covers the entire electromagnetic band providing a complete time-domain and frequency-domain full-wave algorithm. The simulation model can accurately simulate the physical environment [3-4]. But as for the actual wiring of the PCB, representing the PCB and the actual wiring microstrip line with generic model and simulation with CST software cost a lot of time and money, moreover, the hardware requirements are relatively high.

The article mainly solves the S-parameters of microstrip line, it's also a further study of the problem of crosstalk and radiation of the microstrip line, according to the S-parameters. S-parameters of the microstrip line has been calculated by now. Article [5] uses combination of generalized transmission line equation and Pade approximation method. [6-7] use the finite-difference time-domain method. [8] uses The moments method. But they are all the calculation of the S parameter in some particular cases. This paper establishes a generic model of the S-parameters of a PCB board microstrip line, where the S-parameters can be obtained by the structural parameters of the microstrip line.

In this paper, we use Particle Swarm Optimization (PSO) algorithm optimize artificial Neural Networks (referred to as NN) weights, NN proposes PSO - neural network (referred to as PSO-NN) algorithm using BP algorithm, and apply it in General microstrip line S-parameter model. We use parameters obtained by CST software and S parameters of the microstrip line as the input data of the S-parameter model of generic microstrip line, the training data and the verification data. After the model training, the error of the S-parameters of a microstrip line obtained is very small, which indicates that the application of this algorithm S-parameter model of the microstrip line is feasible. In this paper, the S-parameters of the microstrip line generic model can get precise S-parameter based on the structural parameters of the microstrip line. And it costs less time compared to CST software.

2   Particle Swarm Optimization - Neural Network Algorithm

PSO and NN, the two algorithms, are both inspired from biological treatment mode, and both methods established to solve practical complex problems. With the in-depth study of these two algorithms, there are also mangy problems found. PSO has high speed, high efficiency and simple algorithm suitable for the real value of the type of treatment, but doesn't have the adaptive learning ability [9-10]; BP neural network has a strong adaptive learning ability and nonlinear simulation capabilities, but easy to fall into local minimum. And the design of the network only depends on the designer's experience and trial sample space with no theoretical guidance, so the possibility of getting the global optimal solution is small. Therefore, the combination of PSO and NN algorithms can not only take advantage of the NN nonlinear mapping ability, but also make NN have a strong ability to learn and get global optimization ability.

PSO-NN algorithm flow chart is shown in Figure 1, where the degree of adaptation of the particle is calculated by the average function.

3   Algorithm Verification

In this paper, we use the performance of the algorithm check function to verify the PSO-NN algorithm performance, and compared it with the PSO, NN algorithm. The performance of the algorithm is shown in Table 1.

3.1   Simulation Parameters Set

In order to facilitate the algorithm validation, set PSO-NN algorithm PSO, NN population sizes are all 10, the maximum number of iterations are 2000 generation, and make the program terminate run when it reaches the maximum number of iterations. PSO-NN, PSO algorithm inertia weight value is 0.72, and learning rate is 1.5. The PSO-NN NN algorithm parameter settings are three input node, an output node and a 3-layer BP network with 7hidden layer nodes.

3.2   Simulation Data Extraction

According to the function of the variable range in Table 1, extract 500 data from small to large as the input data of the simulation, and get output data calculated via function, then use PSO-NN, PSO and NN to fit the output data, and compare the error after fitting. Because the input data is randomly obtained with a test function corresponding PSO-NN NN algorithm, PSO-NN and PSO algorithm, the input data is not necessarily the same, so the results are credible. Since the number of the input data is the same, the program running time is not affected.

3.3   Simulation Results and Analysis

The simulation results are shown in Table 2, Table 3, Figure 2 and Figure 3. Table 2 gives PSO-NN and PSO algorithm, PSO-NN and NN algorithm's average error, standard deviation of the running time and number of iterations of the three test functions.

The error curves of the two PSO-NN and PSO algorithms corresponding to DeJeng function, Rosenbrock function and Rastrigrin function are shown in Figure 2 (a), (b) and (c) as below.

The error curves of the two PSO-NN and NN algorithms corresponding to DeJeng function, Rosenbrock function and Rastrigrin function are shown in Figure 3 (a), (b) and (c) as below.

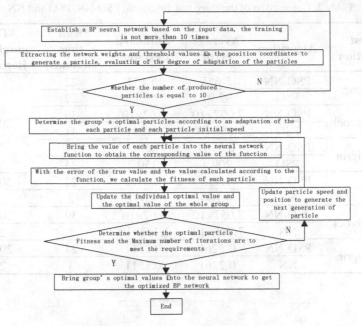

Fig. 1. The flow chart of PSO-NN algorithm

Table 1. The of characteristics of three test function

Function name	Formula	Feature	Variable scope
DeJong	$$f_0(x) = \sum_{i=1}^{n} x_i^2$$	Square summation form continuous	[-100•100]
Rosen-brock	$$f_1(x) = \sum_{i=1}^{n-1} (100(x_{i+1} - x_i^2)^2 + (x_i - 1)^2)$$	The single peak morbid function	[-100•100]
Rastrigrin	$$f_2(x) = \sum_{i=1}^{n} (x_i^2 - 10\cos(2\pi x_i) + 10)$$	The non-linear multi-modal function	[-10•10]

According to Table 2, Figure 2 and Figure 3, the running time of PSO is relatively close to PSO-NN and outweighs NN, and the mean error and standard deviation of PSO is greater than that of PSO-NN; NN's running time is the least, and the number of iterations is superior to PSO-NN, but the error and standard deviation are worse

Table 2. Comparison of the three test functions: PSO-NN, PSO and NN

Test function	Algorithm	Average error	Standard deviation	Run time (seconds)	The number of iterations
DeJeng	PSO-NN	5.655	5.132	22	
	PSO	92.97	68.11	22	
Rosenbrock	PSO-NN	7.057×10^6	2.62×10^6	39	2000
	PSO	1.382×10^7	3.412×10^6	41	
Rastrigrin	PSO-NN	0.5509	0.389	24	
	PSO	209.1	150.2	23	
DeJeng	PSO-NN	1.556	0.8975	22	2000
	NN	4.286	1.592	2	20
Rosenbrock	PSO-NN	7.057×10^6	2.62×10^6	39	2000
	NN	7.745×10^7	1.143×10^7	2	20
Rastrigrin	PSO-NN	0.2261	0.1511	24	2000
	NN	0.2261	0.1511	3	20

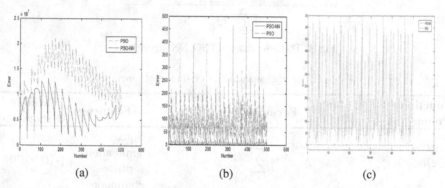

(a)   (b)   (c)

Fig. 2. Error curves of PSO-NN and PSO in the three test functions (a) DeJeng function (b) Rosenbrock function (c) Rastrigrin function

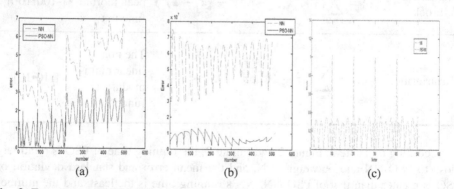

(a)   (b)   (c)

Fig. 3. Error curves of PSO-NN and NN in the three test functions (a) DeJeng function; (b) Rosenbrock function; (c) Rastrigrin function

than PSO-NN. Visibly, the accuracy and stability of the PSO-NN are superior to that of PSO and NN. Because the PSO-NN algorithm proposed in our paper is to carry out the training of NN firstly, then extract weight and use PSO to optimize the two processes, last substitute them into the NN algorithm, so the program lasts a long time.

4   Algorithm Apply

Since the data needed to be learned, trained and validated of S-parameter model of the PCB microstrip lines is very large, we take the PSO-NN algorithm into account to solve this complex nonlinear problem.

4.1   The Physical Structure of the Microstrip Line

We use the existing general microstrip line model in this paper. The model uses certain key parameters to describe the physical structure of the microstrip line, with the segmentation parameters to describe the microstrip line. Figure 4 and Figure 5 show the details.

Figure 4 is a cross section of the microstrip line. ε is the microstrip line material conductivity, with a value of 5.76×10^7 S / m; ω is the microstrip line width, with a value of 0.2 mm; σ is the substrate material permittivity, with a value of 4.5; d is the substrate thickness with a value of 0.1 mm; t is the microstrip line thickness with a value of 0.035 mm.

Figure 5 is a plan configuration diagram of the microstrip line. The microstrip line connection is composed of three segments, where $V(n)$ is the microstrip line segment corner direction, the n-dimensional vector, with the value of 0°, 45°, - 45°; $L(n)$ is the microstrip line segment length of the n-dimensional vector, with the value of 0-20 mm.

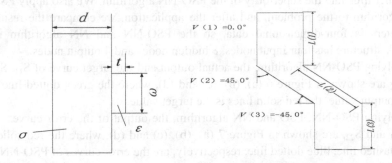

Fig. 4. Cross-sectional configuration diagram   **Fig. 5.** Microstrip line overlooking diagram of the microstrip line

Numerical Calculation Results

Because of the high accuracy of the mature commercial simulation software, we use the CST software to simulate the common microstrip line model. In the calculation process, we set the parameters of the microstrip line structure model for the segment total number n as 3, the corner direction V (1) as $0°$, V (2) as $45°$ and V (2) as $-45°$; L (1) changes from 9mm to 14mm, with the step length of 1mm; L (2) changes from 1.5mm to 3mm, with the step length of 0.5mm; L (3) changes from 9mm to 14mm, with the step length of 1mm.

By CST software simulation, the S-parameters of the different structure of microstrip lines can be obtained within the range of 0 to 10GHz, since the microstrip line is a two-terminal element, the S parameter has four components, namely S_{11}, S_{12}, S_{21} and S_{22}.

4.2   The Application of PSO-NN in Microstrip Line S-Parameter Model

Training Data Acquisition

Using PSO-NN algorithm, with the structure parameters of the microstrip lines L (1), L (2), L (3) and the frequency f as the input data, S_{11},S_{12}, S_{21}, and S_{22} as the desired output data, we establish the microstrip line S parameter model.

The CST software takes about 7 hours and obtains 100,000 data in total, with each four components of the S-parameter S_{11}, S_{12}, S_{21}, and S_{22} as a data. In order to improve the randomness of the sampling, the data is divided into 20 000 groups, and every group has 5 data. Take the first four as training data, the last data as the validation data. Then it can fully reflect the input and output characteristics of the modeling problem required.

Comparison and Analysis

To further illustrate the superiority of the PSO-NN algorithm. We also apply PSO and NN algorithm to the problem, and after the application, we compare the results. S-parameters is four-dimensional data, so the PSO-NN and NN algorithm in BP network structure has four input nodes, 8 hidden nodes and 4 output nodes.

Applying PSO-NN algorithm, the actual output and the target curve of S_{11}, S_{12}, S_{21} and S_{22} are shown in Figure 6 (a), (b), (c) and (d), where the green dotted line is the actual output value, the red solid lines is the target value.

Applying PSO-NN, PSO and NN algorithm, the output of the error curves of S_{11}, S_{12}, S_{21} and S_{22}, are shown in Figure 7 (a), (b), (c) and (d), where the red solid line, green dashed line, blue dotted line, respectively, are the error curves of PSO-NN, PSO and NN.

Table 3 shows the average error of the three algorithms used in the S-parameters of the microstrip line model, as well as the standard deviation, the running time and the number of iterations.

Table 3. The comparison of PSO-NN, PSO and NN algorithms

Algorithm	Parameters	Average error	Standard deviation	The number of iterations	Run time (seconds)
PSO-NN	S_{11}	0.8792	0.7808	10000	52
	S_{12}	0.01935	0.01291		
	S_{21}	0.01432	0.01141		
	S_{22}	0.8608	0.07775		
PSO	S_{11}	1.361	1.132	10000	53
	S_{12}	0.02269	0.01197		
	S_{21}	0.03281	0.01397		
	S_{22}	1.305	1.066		
NN	S_{11}	1.127	0.9533	5	1
	S_{12}	0.02593	0.01667		
	S_{21}	0.02475	0.01657		
	S_{22}	1.056	0.9591		

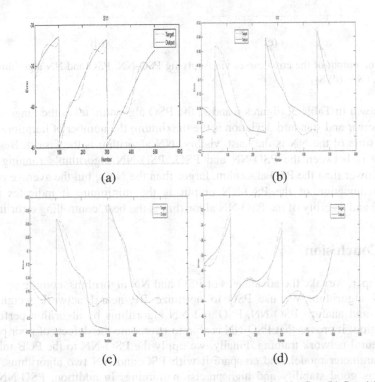

(a)   (b)

(c)   (d)

Fig. 6. The actual output and the target curve via PSO-NN algorithm (a) S_{11}; (b) S_{12}; (c) S_{21}; (d) S_{22}

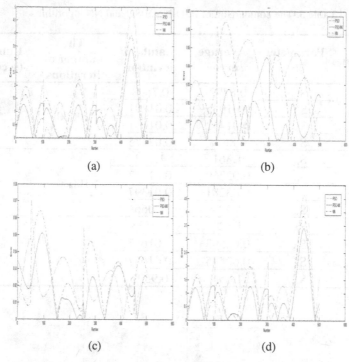

(a)   (b)

(c)   (d)

Fig. 7. The output of the error curves via Applying PSO-NN, PSO and NN algorithm (a) S_{11}; (b) S_{12}; (c) S_{21}; (d) S_{22}

As shown in Table 3, Figures 6 and 7, the PSO algorithm takes the longest, and its average error and standard deviation is the maximum; the number of iterations and the running time of the NN is the least, visibly, NN algorithm is the fastest. Besides, its accuracy is between the PSO-NN and PSO. PSO-NN algorithm's running time is slightly lower than the PSO algorithm, larger than the NN's, but the average error and standard deviation of the PSO-NN output is the minimum. It indicates that the accuracy and stability of the PSO-NN algorithm is the best, controlling error in 5%.

5   Conclusion

In this paper, we take the advantages of PSO and NN algorithms, coming up with the PSO-NN algorithm. We use PSO to optimize BP neural network weights, then compare and analyze PSO-NN, PSO and NN algorithms by algorithm performance test function. It proves that PSO-NN is able to overcome the defects of poor precision of BP neural network training. Finally, we apply the PSO-NN to the PCB microstrip line S-parameter model, and compare it with PSO and NN two algorithms, finding that it has good stability and high precision training. In addition, PSO-NN's total running time is far less than the CST software simulation's, and the degree of similarity of the two results is high. Therefore, the proposal of PSO-NN for the PCB microstrip line S-parameter modeling problem is feasible.

References

1. Bai, X., Xu, L.J.: Research on the Effects of Substrates on the Crosstalk in Microstrips. Journal of Electronics & Information Technology 32(11), 2768 (2010) (in Chinese)
2. An, J., Wu, J.F., Wu, Y.H.: Research of Suppressing Crosstalk of the Microstrip Lines by Using Stripe Protection. Transactions of Beijing Institute of Technology 31(3), 344 (2011) (in Chinese)
3. Liu, Q.: Interface Between C++Builder and CST Microwave Studio Andits Application Based on OLE. Journal of Computer Applications 32(z1), 77 (2012) (in Chinese)
4. Zhang, F., Du, X.Y., Zhou, D.F.: Application of Software CST to the Design of LPDA. Computer Applications and Software 24(11), 16 (2007) (in Chinese)
5. Liu, L.L., Wang, Y.Q.: Generalized Transmission Lline Equations Combined with Pade Approximation for the Analysis of Signal Iintegrity. Journal of Nanjing University of Posts and Telecommunications(Natural Science) 31(6), 17 (2011) (in Chinese)
6. Wang, H.Y., Zhang, X.Z.: Study on Coupling Between Microstrip and Patch Antenna in FDTD Method. Journal of Harbin University of Commerce(Natural Sciences Edition) 27(2), 219 (2011) (in Chinese)
7. Sun, R.H., Gao, W.D., Liu, H.: Analysis on the Characteristics of CSRR Microstrip Based on FDTD. Journal of Anhui University (Natural Sciences) 36(1), 76 (2012) (in Chinese)
8. Wang, K.W., Wang, J.H.: Study on the Transmission and Renection Properties 0f the Microstrip Line Bend. Journal of Microwaves 22(3), 32 (2006) (in Chinese)
9. Yuan, Z.H., Geng, J.P., Jin, R.H.: Pattern Synthesis of 2-D Arrays Based on a Modified Particle Swarm Optimization Algorithm. Jounal of Electronics&Information Technology 29(5), 1236–1237 (2007) (in Chinese)
10. Hu, L.L., Guo, Y.C.: An Orthogonal Wavelet Transform Blind Equalization Algorithmbased on the Optimization of Particle Swarm. Jounal of Electronics&Information Technology 33(5), 1254 (2011) (in Chinese)

Fair Virtual Network Embedding Algorithm with Repeatable Pre-configuration Mechanism

Cong Wang[1], Ying Yuan[2], and Ying Yang[3]

[1] School of Computer and Communication Engineering,
Northeastern University at Qinhuangdao, 066004, China
congw1981@gmail.com
[2] School of Information Science and Engineering, Northeastern University,
Shenyang, 11004, China
[3] Liren College of Yanshan University, Qinhuangdao, 066004, China

Abstract. Virtual network embedding (VNE) is crucial mechanism for network virtualization. In addition to keep efficient mapping of virtual networks onto substrate network, the fairness of virtual networks during embedding must be guaranteed. This paper proposes a fair VNE algorithm to ensure fairness during embedding. Leveraging repeatable features in virtual machine deployment, the algorithm allows virtual nodes from same VN mapped onto single physical nodes to reduce virtual links mapping so as to save physical link bandwidth. To increases mapping possibility of big virtual networks and achieve fairness the algorithm contains a pre-configuration step to reduce differences in number of virtual nodes among virtual networks before embedding. Simulation results show that our algorithm achieves higher profit and more fairness on same substrate network than un-repeatable approach.

Keywords: virtual network embedding, resource allocation, repeatable node mapping, pre-configuration.

1   Introduction

The core technique for network virtualization is how to efficiently map virtual networks with constraints on both nodes and links onto substrate network, which known as the Virtual Network Embedding (VNE) problem. Such problem is known to be NP-hard even part of the constraints are relaxed [1]. Therefore, its solutions mainly rely on heuristic algorithms.

Particularly, in virtual machine deployment, there are some techniques about ram data switch between VMs co-resident on the same single physical machine [2], such repeatable feature can provide a fully transparent and high performance data switch through ram channel instead of the traditional link switch. This provides a progressive way in VNE, the major advantage is that such repeatable techniques can save much more bandwidth between VMs of the same virtual network by mapping them on a same physical machine.

D.-S. Huang et al. (Eds.): ICIC 2013, LNAI 7996, pp. 324–331, 2013.

To make high utility of the physical resource, VNE algorithms must calculate many redundant requests and map feasible requests based free real-time physical resource and the resource needed by the virtual networks. For this reason, virtual networks with lighter resource requirement are easily to be mapped and bigger ones are hardly to be implement even they are in front of the queue than others.

We argue that in addition to keep efficient use of substrate resource, the fairness of the virtual network requests should also be guaranteed. So this paper presents a VNE algorithm with pre-configuration mechanism based repeatable features to guarantee fairness for the queuing virtual networks, which also can increase utility of the substrate network at the same time.

2   Related Work

Many algorithms have been proposed for the VNE problem. They can be classified to one-stage VNE algorithm and two stage VNE algorithm. The main different of the two styles is whether the node mapping and the link mapping for a VN is completed at the same time.

In one-stage VNE solution, Lischka et al. [3] proposed a backtracking-based VN embedding algorithm using subgraph isomorphism detection that extensively searches the solution space in a single stage. I. Houidi [4] proposed a distributed algorithm for mapping the VN. However, they only consider an offline version of the problem and assume the substrate network has enough resources to handle all the VN requests. In two-stage VNE solutions, Minlan Y et al. [5] have provided a two stage algorithm for embedding the VNs. Firstly, they embedding the virtual nodes. Secondly they proceed to map the virtual links using shortest paths and multi-commodity flow (MCF) algorithms in order to increase the acceptance ratio and the revenue. Y. Zhu and M. Ammar [6] proposed an algorithm greedily chooses the substrate nodes that are lightly loaded to map the virtual nodes and uses the shortest path between the selected nodes to map the virtual links. N. Chowdhury et al. [7] introduce co-ordination among the node mapping and link mapping phases to improve the performance of the mapping algorithm. They solve the problem by formulating a mixed integer program and further relaxing the constraints to obtain the linear programming formulation. Mosharaf C et al. [1] present an algorithm leverages better coordination between the two mapping phases and relax the integer constraints of VNE to obtain a linear program and devise two online VN embedding algorithms. Xiang, C et al. [8] present a Particle Swarm Optimization based algorithms named VEN-R-PSO, as a heuristic algorithm they did not consider the repeatable node mapping and so there is make no sense to improve the efficiency by adjust the position.

Our solution is also Particle Swarm Optimization based like [8]. However the difference from previous studies are: Firstly, on design concept, our algorithm bring into play the repeatable features so as to save more physical bandwidth, thus can accept more VNR in the same time; Secondly, to guarantee fairness, we introduce a pre-configuration step before embedding step. Virtual topologies with big number of nodes will be prune to new topologies with small number of nodes but equal to the original ones.

3   VNE Problem Description

We model the problem of cost efficient reconfiguration and embedding of a virtual network as a mathematical optimization problem using integer linear programming (ILP). The goal is to minimize the usage of the substrate resources. Because node resources cannot be reduced through certain mechanism, however leveraging the advantage of ram switch between VMs host on same physical machine, we can map virtual nodes of same VN onto same machine as far as possible instead of allocation them a physical link capacity for their communication, i.e. try to embed each VNR to the least number of physical machines to save physical bandwidth. Thus the object of our optimization problem just needs to calculate link cost. It is defined as follows:

Object:

$$\text{Minimize} \sum_{(i,j) \in P^s} \varphi_{ij}^w \times bw(e^w)$$

$$\text{s.t.} \quad \forall n \in N^V, \forall j \in N^S \quad Cpu(j) - \sum_{n^v \to j} Cpu(n^v) \geq Rcpu(n) \tag{1}$$

$$\forall w \in E^v, \quad \forall (i,j) \in p^s, \quad w \to p^{ij} \quad \min_{e^s \in p^{ij}} Cbw(e^s) \geq Rbw(w)$$

Where φ_{ij}^w is a binary variable:

$$\forall w \in E^v, w \to p^{ij}, \forall i, j \in N^s \quad \varphi_{ij}^w \begin{cases} 0 & i = j \\ 1 & i \neq j \end{cases} \tag{2}$$

The first qualification is node resource constraints, where $\sum Cpu(n^v)$ is the total amount of CPU capacity which has already been allocated; $Cpu(j)$ is the total amount of CPU capacity of the substrate node j.

The second qualification is link resource constraints, where $\min Cbw(e^s)$ is the minimum bandwidth of links in the path p^{ij}, that means if a virtual link w is embedded onto a substrate path p^{ij}, the capacity of each link in this path must be higher than the request bandwidth of virtual link w.

Then we can use long-term revenue to cost ratio shown in equation (3) to measure a VNE algorithm. In the equation, R is resource request of virtual network V; C is real physical resource provided by substrate network to implement the virtual network V. Because virtual node resources request cannot be reduced, so the main impact to equation (3) is links mapping results. For repeatable features, if a virtual link is mapped onto several physical links, C will be bigger than R. In contrast, if a virtual link is mapped into ram, i.e. to end virtual nodes are mapped onto same physical, so there is no cost to sustain such link. So the revenue to cost ratio is bigger, the better.

$$R/C = \lim_{T \to \infty} \frac{\sum_{t=0}^{T} R(V(t))}{\sum_{t=0}^{T} C(V(t))} \tag{3}$$

4   Fair Virtual Network Embedding Algorithm

In this section, we will firstly discuss the pre-configuration mechanism for virtual topology pruning, and then describe the discrete PSO algorithm in details to solve fair VNE problem.

4.1   Pre-configuration Mechanism

To reduce topological differences among virtual requests in the waiting queue, we introduce a pre-configuration step in VNE solution. We use repeatable mechanism before embedding step. With repeatable features, when a virtual network request is added into waiting queue, its topology will be convert to a smaller one but logically equivalent to the original one by merging virtual nodes, as shown in Fig. 1.

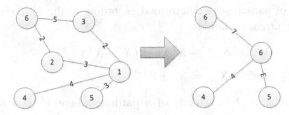

Fig. 1. Demonstration of topology pruning

Such mechanism also brings out another advantage, because some virtual nodes are merged into one big virtual node, the virtual links between them can be ignored, thus can save some calculate works in embedding step and is also helpful to improve the mapping access ratio. The detail of such mechanism is shown as follows:

Algorithm 1. The pre-configuration algorithm

Input: a virtual network topology T_{vn}
Output: a pruned topology T_p
1.   If number of nodes of T_{vn} is bigger than *Nodes_Threshold*, it is need to be pruned, otherwise return;
2.   Sort virtual nodes of T_{vn} in descending order of CPU capacity requirement and add them to a FIFO queue;
3.   Get an node from queue, add it to a set *Temp*;
4.   If the sum of CPU capacity requirement of all nodes in *Temp* is less than *Capacity_Threshold*, go to 3;
5.   If *Temp* is NULL then return, else merge all nodes in *Temp* to a new node with summed CPU capacity requirement.
6.   For all virtual links in T_{vn}, if both destination and source node are in *Temp*, delete it; if only destination node in *Temp*, change its destination to the merged node; for only source node in *Temp* is similar; combine links with same destination and source node.
7.   If the number of nodes of current topology still bigger than *Nodes_Threshold*, go to 2;

The above algorithm can reduce topological differences of virtual networks by try to make big virtual networks have similar numbers of virtual nodes. In algorithm, we use *Nodes_Threshold* to determine whether the topology of a virtual network is needed to be pruned, i.e. to filter big ones; *Capacity_Threshold* is used to control single virtual node resource requirement, because more bigger the request CPU capacity more harder to embed.

4.2   DPSO Solution for Fair VNE Problem

We use Disperse Particle Swarm Optimization (DPSO) to solve the problem discussed in section 3. For a VNR, the search space is N-dimensional, where N is the number of node of the VN. Then a particle swarm is used to search the optimal position $X^i = [x_1^i, x_2^i, ..., x_N^i]$ to map the virtual nodes of a VN. To VNE problem the position and velocity of particles are determined according to the following velocity and position update recurrence relations:

$$V^{k+1} = \varphi_1(X_p^k - X^k) + \varphi_2(X_g^k - X^k)$$
$$X^{k+1} = X^k \oplus V^{k+1}$$

(4)

Where $V^{k+1} = [v_1^i, v_2^i, ..., v_N^i]$ is velocity of a particle, where v_i^k is a binary variable. For each v_i^k, if $v_i^k = 1$, the corresponding virtual node's position in the current VNE solution should be preserved; otherwise, should be adjusted by selecting another node.

Because we use DPSO to calculate the optimal position, we must give the relevant discrete quantity operation definitions:

Definition 1: Subtraction of Position $X_* - X$ If X_* and X have the same values at the same dimension, the resulted value of the corresponding dimension is 1, otherwise, the resulted value of the corresponding dimension is 0.

Definition 2: Addition of Multiple $\varphi_1 X' + \varphi_2 X''$ a new velocity that corresponds to a new virtual network embedding solution, where $\varphi_1 + \varphi_2 = 1$. If X' and X'' have the same values at the same dimension, the resulted value of the corresponding dimension will be kept; otherwise, keep X' with probability φ_1 and keep X'' with probability φ_2.

Definition 3: Addition of Position and Velocity $X^k \oplus V^k$ a new position that corresponds to a new virtual network embedding solution. If the value of v_i^k equals to 1, the value of x_i^k will be kept; otherwise, the value of x_i^k should be adjust by selecting another substrate node.

We use equation (4) to update position and velocity in our DPSO algorithm. After every update, particles calculate their fitness according to equation (1). In each round, if the position cannot match the two qualifications, the fitness will be set to $+\infty$. In our solution we use FloydWarshall algorithm to calculate shortest path between every virtual node pairs, the fitness is gain by the object function in (1); otherwise, the fitness is set to be $+\infty$. The detail of our algorithm is shown as follows:

```
program Fair Mapping (Gv,Gs){
Pre_reconfiguration();
  Generate particles for Gv
  remove nodes and links which capacity is less than
  minimal request of Gv
  for(int i=0;i<ParticleCount;i++){
    particle[i].initialposition()}
  for(int i=0;i<MaxItCount;i++){
  gbestpre=particles.getgBest()
    for(int j=0;j<particle numbers;j++){
      particle[j].calculateFitness()
      particle[j].updateSpeed()
      particle[j].updatePosition()}
    if(gbestpre==particles.getgBest()){
  numfoit++}else{numfoit=0}
    if(numfoit==10){break}}
  if(particles.getgBest()!=+∞){
    this.solution=Particle.getGbsolution()}
  return this. Solution
}
```

In the algorithm, function *Pre_reconfiguration()* is implementation of algorithm 1. Removing nodes and links which capacity is less than minimal node and link capacity request of VNR is to reduce the search space for particles.

5   Performance Evaluation

We implemented the algorithm present in this paper using CloudSim. The main parameters of our simulation are listed in Table 1.

Table 1. Parameters in simulation

Topology:	Substrate Network	Virtual Network
Number of Nodes:	60	2-10
Connectivity:	0.2	0.5
Node Capacity:	100 unit	3-30 unit uniform distribution
Bandwidth Capacity	100 unit	3-30 unit uniform distribution

In the experiment, *Nodes_Threshold* and *Capacity_Threshold* in algorithm 1 are set to 5 and 15 respectively, $\varphi_1 = \varphi_2 = 0.5$ in equation (4). We confirmed the pre-configuration mechanism and analyzed the performance of the whole Fair VNE algorithm present in this paper with and without pre-configuration.

We simulated 1000 VNRs in one test. Each test run 10 times. When the substrate network accepts about 18 VNRs it reaches full load condition, then the other VNRs should wait to be implement. Because of the connectivity of VN is bigger than substrate network, which may lead VNE solution is hard to be calculate, thus we can measure the extreme effect of the algorithm. Firstly, we plot the average original number of nodes of virtual networks which are accepted during every 500 time unit.

From Fig. 2 we can see that the pre-configuration mechanism can gain a more smooth average number of nodes than without it. That is to say there is little difference in embedding difficulty between big and small topology of virtual network. Through this mechanism we can improve the fairness during VNE. The reason is for bigger virtual topology we pruned it to a logically equivalent but smaller one. Note that though the two topologies ate logically equivalent, but by repeatable features the requirement of virtual link capacity is reduced. Thus some virtual link between merged virtual nodes is ignored. This can reduce mapping difficulty of big virtual networks.

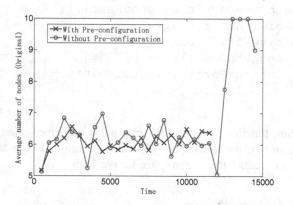

Fig. 2. The effect of pre-configuration mechanism

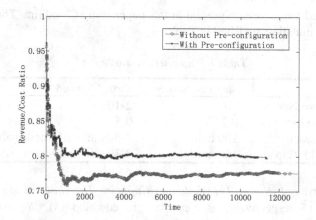

Fig. 3. The Revenue/Cost Ratio with and without pre-configuration

Fig. 3 plots the Revenue/Cost Ratio according to equation (3) of the algorithm with and without pre-configuration mechanism. Note that if virtual links are mapped onto substrate links we calculate cost; if they are mapped as a ram switch link the cost is 0. From the figure we can see that the pre-configuration mechanism get higher ratio and with such mechanism same 1000 virtual network request are accomplished earlier. The reason why the pre-configuration mechanism can improve the Revenue/Cost Ratio is it reduces some virtual link mapping work. Thus it can gain higher virtual network accept ratio than without it, so the same 1000 virtual networks are accomplished earlier. Actually, both repeatable pre-configuration and mapping in subsection 4.2 can save substrate bandwidth, many virtual nodes from same VN are mapped on to single substrate node and the links between them are just instead by ram switch.

6   Conclusions

This paper introduced a fair VNE algorithm, the major characteristics of our work is that we introduce a pre-configuration mechanism by leveraging the advantage of inter ram switch techniques. Such mechanism can achieve higher revenue/cost ratio of substrate network and more fairness of virtual networks during embedding. We also present a whole repeatable DPSO based VNE solution and test the effect of our algorithm. In our future work, we will continually study the pre-configuration mechanism and try to design a distribute algorithm for searching the mapping solution for VNE problems.

Acknowledgments. This work was supported by The Central University Fundamental Research Foundation, under Grant. N110323009.

References

1. Mosharaf, C., Muntasir, R.R., Raouf, B.: ViNEYard: Virtual Network Embedding Algorithms With Coordinated Node and Link Mapping. IEEE/ACM Transactions on Networking 20(1), 206–219 (2011)
2. Jian, W., Kwame, L.W., Kartik, G.: XenLoop: a ransparent high performance inter-vm network loopback. In: HPDC 2008: Proceedings of the 17th International Symposium on High Performance Distributed Computing, pp. 109–118. ACM, New York (2008)
3. Jens, L., Holger, K.: A virtual network mapping algorithm based on subgraph isomorphism detection. In: VISA 2009, pp. 81–88. ACM, New York (2009)
4. Ines, H., Wajdi, L., Djamal, Z.: A distributed virtual network mapping algorithm. In: IEEE ICC 2008, pp. 5634–5640. IEEE Press, Beijing (2008)
5. Minlan, Y., Mung, R., Rethinking, C.: virtual network embedding: substrate support for path splitting and migration. ACM SIGCOMM CCR 38(2), 17–29 (2008)
6. Yong, Z., Mostafa, A.: Algorithms for assigning substrate network resources to virtual network components. In: INFOCOM 2006. 25th IEEE International Conference on Computer Communications, pp. 1–12. IEEE Press, Barcelona (2006)
7. Chowdhury, N.M.M.K., Boutaba, R.: Network virtualization: state of the art and research challenges. IEEE Communications Magazine 47(7), 20–26 (2009)
8. Xiang, C., Zhong, B.Z., Sen, S., Kai, S., Fang, C.Y.: Virtual Network Embedding Based on Particle Swarm Optimization. Acta Electronica Sinica 39(10), 2240–2244 (2011)

Dynamic Obstacle-Avoiding Path Planning for Robots Based on Modified Potential Field Method

Qi Zhang, Shi-guang Yue, Quan-jun Yin, and Ya-bing Zha

Simulation Engineering Institute, School of Information System and Management,
National University of Defense Technology, Hunan Province, Changsha, 410073, China
zhangqiy123@126.com

Abstract. The potential field method is widely used for autonomous robots due to its simplicity and high efficiency in dynamic motion planning. However, there is still drawback of unnecessary obstacle avoidance of former methods in dynamic obstacle avoidance planning. This paper proposes a new potential field method to solve the problem, whose new virtual force is deduced through introducing the restriction of collision angle with exponential form and both the information of angle and magnitude of relative velocity. The simulation results prove that the robot can not only avoid their obstacles and move to the target safely and quickly in dynamic environments, but remove largely the unnecessary obstacle avoidance by using the proposed method.

Keywords: robot motion planning, artificial potential field, dynamic obstacle avoidance, the local minimum.

1   Introduction

Autonomous robot path planning is searching out a path from starting point to goal point in static and moving obstacles environment, and the path must be no collision and optimization. As the environment is complex, real-time and uncertain, obstacle avoidance problem has become a key of mobile robot in obstacle avoidance path planning. There have been many methods to solve the problem in the past, and the most commonly used are artificial potential field(APF) method, grid method, genetic algorithm, fuzzy methods, neural network and so on[1][4]. In which, the potential field method is commonly used for robot path planning and collision avoidance because of its simple mathematical analysis, little computation and well real-time control.

As most of the previous potential field methods are just adjusted to deal with path planning in stationary environments, researchers have paid much attention to improving the APF method in dynamic obstacle avoidance path planning. Fujimura[2][3] regarded time as a parameter in planning model, proposed a dynamic artificial potential field method, Han Yong[5] introduced the velocity of the target into the potential field, Ge[6][7] added relative velocity of the goal with respect to the robot into attractive potential function and relative velocity into repulsive potential function, Lu Yin[4][8] added relative acceleration into potential function on the basis, Li Guanghui[9] introduced velocity vector field into the repulsive potential field to improve

D.-S. Huang et al. (Eds.): ICIC 2013, LNAI 7996, pp. 332–342, 2013.

the agility in obstacle avoidance. All of those improved robot motion planning in dynamic environments well. However, because of the complexity of environment and lack of beforehand planning making use of the relative velocity, there is still drawback of unnecessary obstacle avoidance. It is a fatal bug in situations of soccer robot games with a strong requirement to speediness. The unnecessary obstacle avoidance means that robot is driven by unnecessary or excess repulsive force in need of less change to avoid obstacles. To solve the problem, Zhang Feng[10] proposed a method to adjust the robot velocity value and direction to escaping collision based on relative coordinates. Duan Hua[11] uses a collision time parameter as the substitute for the relative position to remove meaningless collision motions, Xie Hongjian[12] analyzes the angle between the relative speed and the relative position of the robot to change the obstacle avoidance direction speed to avoid the dynamic obstacles.

In this paper, we propose a new potential field method for motion planning for mobile robots to overcome the drawback in dynamic environments. The new repulsive potential is defined as not just the function of relative position of the obstacle and robot, but also relied on angle and magnitude of relative velocity. Besides, the collision angle with exponential form is introduced into the repulsive potential as a restriction to remove obvious unnecessary avoidance motions. The new virtual force not only keeps the strongpoint of elegant mathematical analysis and simplicity of traditional potential field method, but also removes largely unnecessary obstacle-avoiding motions, improving celerity and availability to avoid static and moving obstacles in dynamic environments.

2   The APF Method and Its Limitations in Dynamic Environments

The APF method is first proposed by Khatib in 1986. Its basic concept[2][4] is that mobile robot is moving in a virtual force field in which the goal is filled with attractive potential U_{att} and the obstacles are filled with repulsive potential U_{rep}. The attractive potential U_{att} decreases as the relative distance between the robot and the goal decreases and the repulsive potential U_{rep} increases as the relative distance between the robot and the goal decreases. Then the virtual force is equal to the negative gradient of all the attractive and repulsive potential, and the robot is driven by which to get a continuous path from the initial position to the goal position.

Generally, the traditional APF method is effective by adjusting the parameters of attractive and repulsive force in obstacle avoidance planning, while its drawback is also obvious in local planning. In former methods, the robot is driven by repulsive force as soon as the distance between the robot and obstacle is in the given range repulsive potential works, leading to the unnecessary obstacle avoidance in the following two situations[11][13]. In one situation, the robot has no possibility of collision with the obstacle although the distance between them is shorter, and the robot is driven by repulsive force. In another situation, the robot gets excess repulsive force rather than little correct, deviating from the optimal path. Obviously, the obstacle avoidance planning does not appear optimal and takes more time. The root cause is lack of forecast planning making full use of relative velocity and obstacle information.

3   The Modified APF Method and Its Analysis

The motion planning problem of mobile robots in a dynamic environment is to plan and control the robot motion from an initial position to track the target's state in a desired manner while avoiding moving and static obstacles. To simplify the analysis, we make the following assumptions. Assumption 1: The shape of the robot and obstacles are expanded to round for simplicity. Assumption 2: The relative position and velocity of the obstacle with respect to the robot are measured by robot sensor.

3.1   The Modified Repulsive Potential Field and Repulsive Force Function

In view of repulsive force working least in avoidance motions, the repulsive potential should just work in some collision angle, and its magnitude not only changes with the relative position and velocity of the robot and the obstacle, but also decrease as the angle of the relative velocity and relative position of the robot and the obstacle increases in collision angle. The collision angle is the minimum angle ensuring that the robot cannot collide with the obstacle along the current advancing direction[11][13].

As is shown in Fig.1, y label is the current direction of robot moving, $\mathbf{V}$ is the relative velocity of the obstacle $\mathbf{O}$ with respect to the robot $\mathbf{R}$, $\mathbf{n}_{RO}$ is the unit vector pointing from the robot to the obstacle center, d_m is the shortest allowed distance between the robot and the obstacle. θ_m is the defined collision angle, which increases as the distance between the robot and the obstacle decreases. From Fig.1, we can see the relation $\sin \theta_m = d_m/d$, Where d is equal to $\|\mathbf{p}_{obs}(t) - \mathbf{p}(t)\|$, denotes the distance between the obstacle and the robot. θ is the angle between the relative velocity of the robot with respect to the obstacle and the relative position of the robot with respect to the target, it meets the relation $\theta = \theta_v - \theta_o$.

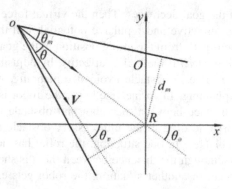

Fig. 1. Relation of the position and velocity of the robot and obstacle

Assuming the robot moves along the current advancing direction in the next period of time, then when meeting the condition $|\theta| = \theta_m$, the robot **R** will pass the obstacle **O** by without collision; meeting the condition $|\theta| > \theta_m$, there is no possibility to have a collision; meeting the condition $|\theta| < \theta_m$, the robot **R** will collide with the obstacle **O** in some time. Besides, the bigger θ is, the smaller the tangential force normal to the relative position direction of the robot and the obstacle is needed. Therefore, the repulsive potential should meet the condition that it increases as the relative position of the robot and the obstacle decreases, decreases as θ increases with the absolute value θ smaller than the collision angle θ_m. Considering the method to solve the problem of goals unreachable with obstacles nearby when using potential field methods in Ge and Cui(2002)[6], the repulsive potential function in this paper is presented as follows:

$$
U_{rep}(\mathbf{P},\mathbf{V}) = \begin{cases} not \quad defined \quad , & \rho_s(P,P_{obs}) - d_m < 0 \\ \varepsilon(e^{(\theta_m - \theta)} - 1)(\dfrac{1}{\rho_s(P,P_{obs}) - d_m} - \dfrac{1}{\rho_0})\rho^2{}_s(P,P_{goal})/2, \\ \qquad |\theta| < \theta_m \wedge 0 < \rho_s(P,P_{obs}) - d_m < \rho_0 \\ 0 \quad , & other wise \end{cases}
$$

(1)

where $U_{rep}(\mathbf{P},\mathbf{V})$ denotes the repulsive potential generated by the obstacle; ρ_0 is a positive constant describing the influence range of the obstacle; ε is a positive constant adjusting magnitude of potential; d_m is equal to the summation of R_{ob}, R_{ro} and R_{so}, denoting the smallest radius of round enveloping the obstacle, the radius of robot, the reserved safe distance between the robot and the obstacle respectively. $d_g = \rho_s(P,P_{goal}) = \|\mathbf{p}_{goal}(t) - \mathbf{p}(t)\|$, denotes the distance from the robot to the target. In the Fig.3, we can see that θ_m is located in the scale $\begin{bmatrix} 0 & 90^0 \end{bmatrix}$, and meets the condition $\sin \theta_m = \dfrac{d_m}{d}$. To simply the operation, we substitute θ_m for $\dfrac{d_m}{d - R_{so}}$, which is feasible in actual data scale.

Note that when $\rho_s(P,P_{obs}) < d_m$, the repulsive potential is not defined, since there is no possible solution to avoid collision with the obstacle in the aforementioned case where the robot moves toward the robot. When $|\theta| < \theta_m \wedge 0 < \rho_s(P,P_{obs}) - d_m < \rho_0$,it means the robot moves into the influence range of the repulsive potential and has possibility to collide with the obstacle, then should be driven by the repulsive force.

The corresponding new repulsive force is defined as the negative gradient of the repulsive potential in terms of both position and velocity:

$$\mathbf{F}_{rep}(\mathbf{P},\mathbf{V}) = -\nabla U_{rep}(\mathbf{P},\mathbf{V})$$
$$= -\nabla_p U_{rep}(\mathbf{P},\mathbf{V}) - \nabla_v U_{rep}(\mathbf{P},\mathbf{V}) \quad . \tag{2}$$

Note that
$$U_{rep1}(\mathbf{P},\mathbf{V}) = \varepsilon(e^{\theta_m - \theta} - 1)(\frac{1}{d - d_m} - \frac{1}{\rho_0}) ,$$

So
$$-\nabla_p U_{rep}(\mathbf{P},\mathbf{V}) = -\nabla_p U_{rep1}(\mathbf{P},\mathbf{V})\frac{d_g^2}{2}\mathbf{n}_{RO} - U_{rep1}(\mathbf{P},\mathbf{V})d_g\mathbf{n}_{Rg}$$

$$-\nabla_v U_{rep}(\mathbf{P},\mathbf{V}) = -\nabla_v U_{rep1}(\mathbf{P},\mathbf{V})\frac{d_g^2}{2}$$

$$\nabla_p U_{rep1}(\mathbf{P},\mathbf{V}) = \varepsilon(\frac{1}{d - d_m} - \frac{1}{\rho_0})e^{\theta_m - \theta}\frac{d_m}{(d - R_{so})^2}\mathbf{n}_{RO} + \varepsilon(e^{\theta_m - \theta} - 1)\frac{1}{(d - d_m)^2}\mathbf{n}_{RO} \tag{3}$$

We have the following mathematic relation:

$$V_{RO\perp} = \|\mathbf{V}\|\sin\theta, V_{RO} = \|\mathbf{V}\|\cos\theta .$$

$$\frac{\partial\theta}{\partial\mathbf{V}} = \frac{\partial\theta_v}{\partial\mathbf{V}} = \left[\frac{\partial\theta_v}{\partial V_x} \quad \frac{\partial\theta_v}{\partial V_y}\right]^T = \frac{1}{\|\mathbf{V}\|^2}\left[\begin{matrix}-V_y \\ V_x\end{matrix}\right] = \frac{1}{\|\mathbf{V}\|^2}\mathbf{V}\perp , \tag{4}$$

Where $\mathbf{V}\perp$ is the vector counterclockwise rotating $\mathbf{V}$ by $90°$, $\mathbf{n}_{RO\perp}$ is the unit vector vertical with the direction pointing from the robot to the target.

$$\mathbf{V}\perp = -V_{RO\perp}\mathbf{n}_{RO} + V_{RO}\mathbf{n}_{RO\perp}$$

$$\nabla_v U_{rep1}(\mathbf{P},\mathbf{V}) = \varepsilon(\frac{1}{d - d_m} - \frac{1}{\rho_0})\frac{e^{\theta_m - \theta}}{\|\mathbf{V}\|^2}(-\mathbf{V}\perp) \tag{5}$$

$$-\nabla_v U_{rep}(\mathbf{P},\mathbf{V}) = -\varepsilon\frac{d_g^2}{2}(\frac{1}{d - d_m} - \frac{1}{\rho_0})\frac{e^{\theta_m - \theta}}{\|\mathbf{V}\|}(\sin\theta\mathbf{n}_{RO} - \cos\theta\mathbf{n}_{RO\perp}) \tag{6}$$

The virtual repulsive force generated by the obstacle is then given by

$$\mathbf{F}_{rep}(\mathbf{P},\mathbf{V}) = \begin{cases} not \quad defined \quad , & \rho_s(P,P_{obs}) - d_m < 0 \\ \mathbf{F}_{rep1} + \mathbf{F}_{rep2} + \mathbf{F}_{rep3} \quad , & \\ \quad |\theta| < \theta_m \wedge 0 < \rho_s(P,P_{obs}) - d_m < \rho_0 \\ 0 \quad , & other\ wise \end{cases} \tag{7}$$

Where

$$\mathbf{F}_{rep1} = -\varepsilon((\frac{1}{d-d_m} - \frac{1}{\rho_0})e^{\theta_m - \theta}(\frac{d_m}{(d-R_{so})^2} + \frac{\sin\theta}{\|\mathbf{V}\|}) + \frac{1}{(d-d_m)^2}(e^{\theta_m-\theta}-1) - F_{ro0})\frac{d_g^2}{2}\mathbf{n}_{RO}$$

$$\mathbf{F}_{rep2} = -\varepsilon((\frac{1}{d-d_m} - \frac{1}{\rho_0})e^{\theta_m-\theta}\frac{\cos\theta}{\|\mathbf{V}\|} - F_{roT0})\frac{d_g^2}{2}\mathbf{n}_{RO\perp}$$

$$\mathbf{F}_{rep3} = \varepsilon(e^{(\theta_m-\theta)}-1)(\frac{1}{d-d_m} - \frac{1}{\rho_0})d_g\mathbf{n}_{Rg}$$

F_{ro0} and $F_{ro\perp}$ are introduced to ensure the continuity of repulsive force, which is equal to the repulsive force with the same magnitude of velocity and distance when $|\theta| = \theta_m$.

$$F_{ro0} = (\frac{1}{d-d_m} - \frac{1}{\rho_0})(\frac{d_m}{(d-R_{so})^2} + \frac{\sin\theta_m}{\|\mathbf{V}\|})$$

$$F_{ro\perp} = (\frac{1}{d-d_m} - \frac{1}{\rho_0})\frac{\cos\theta_m}{\|\mathbf{V}\|}$$

3.2   The Analysis of the New Repulsive Force

Analyzing (7), we can see that the repulsive force component $\mathbf{F}_{rep1}$ is in the opposite direction of $\mathbf{n}_{RO}$ which will keep the robot away from the obstacle, whose magnitude increases as the relative distance of the robot and the obstacle decreases, increases as the angle θ of relative velocity and the relative position of the robot and the obstacle. It means that the more obvious the trend of collision is, the bigger the repulsive force $\mathbf{F}_{rep1}$ is. From Fig.2 we can see that the repulsive force component $\mathbf{F}_{rep2}$ is in the same direction of $\mathbf{n}_{RO\perp}$, and it will act as a steering force for detouring the obstacle, whose magnitude keeps proportional to the projection magnitude of the relative velocity along the direction of heading from the robot to the obstacle and keeps inversely proportional to θ. It means that the easier the robot detours the obstacle, the smaller the repulsive force $\mathbf{F}_{rep2}$ is, avoiding excess force to deviate the optimizing path more. Obviously, when the robot, the obstacle and the target move in the same direction along the same line and the robot is trapped in the so-called local minimum, the $\mathbf{F}_{rep2}$ supplies the robot with the biggest force to detour, which can remove the local minimum in dynamic obstacle-avoiding planning. The repulsive force component $\mathbf{F}_{rep3}$ is in the same direction of $\mathbf{n}_{Rg}$, which can be added in the attractive force. All of the three force components have the common factor d_g, which ensures that the repulsive force approaches to zero even if in the situation the obstacle is near the target.

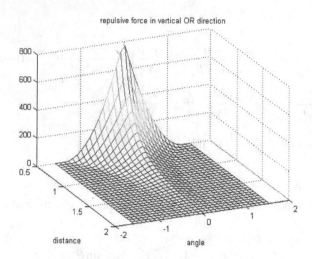

Fig. 2. The repulsive force F_{rep2} in vertical OR direction

4   Simulation Results

Comprehensive simulation studies are carried out to validate the effectiveness and efficiency of the new potential field conformation in visual C++ environment.

In the simulation, the robot is round mobile robot, whose radius R_{ro} is 0.3 m. The shape of the obstacle is expanded to round, its radius R_{ob} is set as 0.3 m. The reserved safe distance between the robot and the obstacle R_{so} is 0.15 m, the influence range of obstacle ρ_0 is 2.0 m, the decision cycle of robot is 100 ms. The mass of the robot is 20 kg, its max velocity is 1.5 m/s, max acceleration is 0.9 m/s^2, $\alpha_p = 10$, $\alpha_v = 8$, $\varepsilon = 5$. The unit of the coordinate in simulation environment is mm, which is 1:30 with the actual size.

Because the model of Ge is a classic and effective method of the obstacle avoidance path planning in dynamic environment, so the simulation experiment is organized as follows: Section 1 validates the basic ability of the robot to avoid the moving obstacles in dynamic environment. Section 2 presents the improvement for the unnecessary obstacle avoidance.

4.1   Avoiding the Static and Moving Obstacles in Dynamic Environment

As is shown in Fig.3, the robot **R** is $[100,200]^T$, the target is moving from point $[250,200]^T$ at constant velocity $[10,-15]^T$. The obstacle O_1 is static at position $[150,200]^T$, O_2 is moving from point $[180,100]^T$ at constant velocity $[0,15]^T$, O_3 is moving from point $[300, 250]^T$ at velocity $[-13,-18]^T$, O_4 is static and its position is $[280,80]^T$. The robot is attracted by the moving target. When detecting the position and velocity of the robot meeting the condition of collision angle, the robot starts to avoid the obstacles by the repulsive force, finally it avoids all the obstacles in the path and tracks the state of the target at 117th period. Fig.3 shows the well performance of the robot in modified method to avoid the static and moving obstacles effectively. The result demonstrates the effectiveness of the proposed method in dynamic environment.

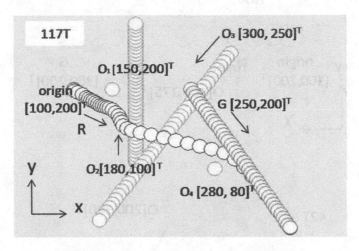

Fig. 3. Avoiding the static and moving obstacles in complex environment

4.2   The Improvement for Unnecessary Obstacle Avoidance

The proposed method makes improvements for unnecessary obstacle avoidance in two kinds of situations. On one hand, the robot needs no avoidance for the situation that robot has no possibility of collision with obstacle by restriction of the collision angle. On the other hand, the repulsive force in $\mathbf{n}_{RO\perp}$ direction decreases as θ increases to avoid deviating from the optimal path excessively when the absolute value θ is smaller than the collision angle.

In Fig. 4, the initial position of the robot $\mathbf{R}$ is $[100,200]^T$, the target $\mathbf{G}$ is   static and at position $[300,200]^T$, (a) and (b) present the unnecessary obstacle avoidance and their improvement with static and moving obstacle. The black track denotes the path planned in Ge's method and the blue one denotes the modified in this paper. The simulation exhibits the performances of Ge's method with parameters of 1500 and the new method with $\varepsilon = 5$. The robot avoids the obstacle driven by the repulsive force and covers a curving path in the former method. By comparison, the situation does not meet the condition of collision angle in the proposed method, and the robot makes no avoiding motion. In (b), we can see that the robot in the modified method has reached the target while there is a few distance about 20 to cover of former method. The robot takes 48 and 52 periods respectively to reach target completely in the two methods.

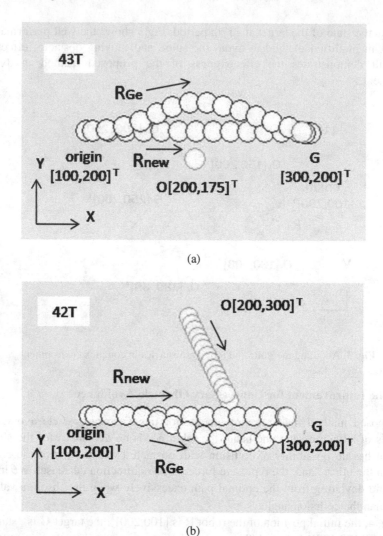

Fig. 4. A kind of obvious unnecessary obstacle-avoidance of former method and improvement

In addition, the change of repulsive force in $\mathbf{n}_{RO\perp}$ direction helps robot to avoid deviating from the optimal path excessively in some occasions. The initial position of the robot R is $[100,200]^T$, the target G is static and its initial position is $[250,200]^T$, The obstacle is moving from point $[200,300]^T$ at constant velocity $[0,-36]^T$ and $[0,-50]^T$ respectively. The simulation exhibits the performances of Ge's method with different parameters of 2000, 1500, 500 and the proposed method with $\varepsilon = 5$.

In Table.1, the modified method can plan a well path which costs the least periods of 32 and 37 and covers the shortest length 155.28 and 150.05 when the velocity of the moving obstacle is -36 and -50 respectively. The former method can plan a better path costing less time with little parameter such as 500 in Ge's method when the velocity of the obstacle is small like -36. However, there is collision when the velocity of the obstacle increases to -50 because the repulsive force is too small to avoid the moving obstacle. With the parameter of Ge's method increasing to 2000, the collision above is removed, but the excess avoidance is emerged with low velocity of obstacle, whose costing time reaches 39 periods and path length reaches 185.18. By comparison, the modified model plans effective and safe path in situations with different velocity of moving obstacle, make a little change to avoid obstacle by the change of the repulsive force and reach to the target.

Table 1. The performance in different methods with different velocities of obstacle

Method	parameter	Vo	collision	Length of path	Costing time
Ge method	500	-36	no	156.27	33
		-50	yes	—	—
Ge method	1500	-36	no	174.57	37
		-50	yes	—	—
Ge method	2000	-36	no	185.18	39
		-50	no	200.73	42
Modified method	5	-36	no	155.28	32
		-50	no	150.05	37

Simulation results show that the former model takes actions to avoid collision in some occasions without possibility to collide or make excess avoidance, while the modified model has no need or make a little change to avoid obstacle by restriction of collision angle or control the change of repulsive force in $\mathbf{n}_{RO\perp}$ direction, takes less time and less distance to achieve task than before. Obviously, the modified method removed largely the unnecessary obstacle avoidance.

5   Conclusion

In this paper, a modified potential field method has been proposed for mobile robot motion planning in dynamic environments. By defining the collision angle with exponential form as the restriction having the repulsive potential field, making full use of the information of both angle and magnitude of relative velocity of the robot and the obstacle, we deduce the new potential function and virtual force function meeting our anticipant trend. The robot can avoid the static and moving obstacles effectively with an excellent path, and remove largely the unnecessary obstacle avoidance. Computer simulations have demonstrated the effectiveness of the mobile robot motion planning schemes based on the new potential field method. To simply the research, we expand the obstacle shape the robot sensed to round, so further research should be made to demonstrate the universality of the arithmetic applying in the obstacle-avoiding planning in actual complex environments.

References

1. Zhu, D.Q., Yan, C.M.: Survey on technology of mobile robot path planning. Control and Decision 25(7), 961–967 (2010)
2. Fujimura, K., Samet, H.: A hierarchical strategy for path planning among moving obstacles. IEEE Trans. on Robotic Automation 5(1), 61–69 (1989)
3. Conn, R.A., Kam, M.: Robot motion planning on N dimensional star worlds among moving obstacles. IEEE Trans. on Robotic Automation 14(2), 320–325 (1998)
4. Khatib, O.: Real time obstacle avoidance for manipulators and mobile robots. The International of Robotics Research 5(1), 90–98 (1986)
5. Han, Y., Liu, G.D.: Dynamic motion planning for mobile robots based on potential field method. Robot (1), 45–49 (2006)
6. Ge, S.S., Cui, Y.J.: Dynamic motion planning for mobile robots using potential field method. Autonomous Robots 13(2), 207–222 (2002)
7. Ge, S.S., Cui, Y.J.: New potential functions for mobile robot path planning. IEEE Trans. on Robotic Automation 16(5), 615–620 (2000)
8. Yin, L., Yin, Y.X., Lin, C.J.: A new potential field method for mobile robot path planning in the dynamic environments. Asian Journal of Control 11(2), 214–225 (2009)
9. Li, G.H., Wei, T.: Robot Soccer Obstacle Avoidance Path Plan Based on Velocity Vector for Dynamic Environment. Journal of Wuhan University of Technology 31(13), 133–136 (2009)
10. Zhang, F., Tan, D.L.: A new real-time and dynamic collision avoidance method of mobile robots based on relative coordinates. Robot 25(1), 31–35 (2003)
11. Duan, H., Zhao, D.B.: Potential field-based obstacle avoidance algorithm for dynamic environment. Huazhong University of Science and Technology (Nature Science Edition) 34(9), 39–42 (2006)
12. Xie, H.J., Wang, H.Z.: Dynamic Collision Avoidance Planning of Mobile Robot Based on Velocity Resolution. Journal of East China University of Science and Technology (Natural Science Edition) 37(2), 234–238 (2011)
13. Ko, N.Y., Lee, B.H.: Avoidability Measure in Moving Obstacle Avoidance Problem and Its Use for Robot Motion Planning. In: IEEE/RSJ Int. Conf. on Intelligent Robots and Systems, vol. (2), pp. 1296–1303 (1996)

On Incremental Adaptive Strategies

Mingxuan Sun

College of Information Engineering
Zhejiang University of Technology
Hangzhou 310023, China
mxsun@zjut.edu.cn

Abstract. In this paper, we propose in a continuous-time framework incremental adaptation strategies for parametric uncertain systems. The issue of the bounded estimation is posed by considering merely the case of unsaturated adaptation. The partially- or fully-saturated adaptation laws are characterized analytically, to illustrate the flexibility in choice of adaptation laws. The presented analysis shows the asymptotic convergence of the system undertaken.

Keywords: Incremental adaptation, saturated adaptive algorithms, adaptive control, continuous-time systems.

1   Introduction

Integral adaptation is a common method of estimation involved in the conventional continuous-time adaptive systems. The adaptation mechanism tunes controller parameter online to achieve desirable control performance. Lyapunov synthesis is used to derive such adaptive laws, and establish the closed-loop stability and convergence properties. We refer to [1, 2] for fundamental theoretical results of adaptive systems in continuous-time domain.

The results reported in the published literature indicate that integral adaptive algorithms are effective in handling parametric uncertainties, where the unknown parameters remain constant over time. Implementing such an adaptive law, one needs to find a discrete approximation to the integral equation (or the differential equation) for estimate calculation. In this paper, we focus our attention on developing incremental adaptive systems. The integral form of mechanisms for adaptation, used in the conventional adaptive systems, are avoided, while the incremental adaptive strategies are developed and evaluated.

The presented work in this paper is closely related to repetitive control, which has been shown suitable to track/reject periodic references/disturbances asymptotically [3-6]. We refer to a special case of repetitive control as adaptive repetitive control (ARC), which is applied to periodic systems, and a periodic adaptive mechanism is involved for estimating the periodically time-varying parameters. ARC is applicable to deal with periodic disturbances, as the disturbances can be considered as parameters of the system undertaken [7-8]. One important issue in ARC is the saturated adaptation, by which the bounded estimation is ensured [9, 10].

D.-S. Huang et al. (Eds.): ICIC 2013, LNAI 7996, pp. 343–352, 2013.

This paper presents a control method for uncertain systems with unknown constant parameters, which lies within the framework of adaptive control. It can be considered an extension of the theory of ARC [7-10] to its adaptive control counterpart. The unsaturated   and saturated adaptive mechanisms are detailed analyzed and compared.

2   Problem Formulation

The following is a typical system with parametric uncertainty:

$$\dot{x} = \theta^T \varphi(x) + u \tag{1}$$

where x and u are the scalar output and input of the system, respectively, $\varphi(\cdot)$ is a continuous nonlinear function, and θ is the unknown n_θ-dimensional parameter vector.

The objective of this paper is to develop adaptive mechanisms for system (1) such that the system output converges to zero, as time increases, i.e., $x(t) \to 0$, as $t \to \infty$, while all of the signals in the closed-loop remain bounded. It is seen that θ appears linearly in (1), which indicates that similar to most of the conventional adaptive systems, the main point of this paper is to deal with the linear-in-the-parameters uncertainty. To achieve the objective, in this paper we shall focus our attention on the development of the incremental adaptive mechanisms, instead of the integral adaptive ones.

Let us start with an investigation of an integral adaptive mechanism. We use the certainty equivalent control of the form

$$u = -\hat{\theta}^T \varphi(x) - \beta x, \beta > 0 \tag{2}$$

with the adaptation law being applied, to give the estimate $\hat{\theta}$ for θ,

$$\dot{\hat{\theta}} = \Gamma \varphi(x) x \tag{3}$$

where $\Gamma > 0$ is a diagonal matrix.

With the use of (2), the closed-loop system can be described by

$$\dot{x} = \tilde{\theta}^T \varphi(x) - \beta x \tag{4}$$

where $\tilde{\theta} = \theta - \hat{\theta}$ denotes the estimation error.

Consider the following candidate Lyapunov-like function

$$V(x, \hat{\theta}) = \frac{1}{2} \tilde{\theta}^T \Gamma^{-1} \tilde{\theta} + \frac{1}{2} x^2 \tag{5}$$

The notation of $V(t)$ for $V\left(x(t),\hat{\theta}(t)\right)$ is used in the sequel, for simplicity of presentation. The time derivative of $V(t)$ along (4) can be calculated as

$$\dot{V} = \tilde{\theta}^T \Gamma^{-1} \dot{\hat{\theta}} + x\dot{x}$$
$$= -\tilde{\theta}^T \Gamma^{-1} \dot{\hat{\theta}} + \tilde{\theta}^T \varphi(x)x - \beta x^2 \tag{6}$$

It follows by applying (3) that

$$\dot{V} = -\beta x^2 \tag{7}$$

Hence, $\dot{V}(t)$ is semi-definite, implying that $V(t) \le V(0)$. By the definition of $V(t)$, $x(t) \in L_\infty$ and $\theta(t) \in L_\infty$, as both $x(0)$ and $\hat{\theta}(0)$ are bounded. Consequently, we obtain that from (2), $u(t) \in L_\infty$, and $\dot{x}(t) \in L_\infty$ by using (4).

Integrating both sides of (7), we obtain

$$\int_0^t x^2(s)\,ds = \frac{1}{\beta}\left(V(0) - V(t)\right) \le \frac{1}{\beta}V(0) \tag{8}$$

implying that $x(t) \in L_2$. We can now invoke Barbalat's lemma to show that the $x(t)$ converges to zero, as $t \to \infty$, i.e., $\lim_{t\to\infty} x(t) = 0$.

The adaptive law (3) is referred to as an integral law, as it can be written as the following form of

$$\hat{\theta}(t) = \hat{\theta}(0) + \int_0^t \Gamma\varphi(x(s))x(s)\,ds \tag{9}$$

As for $t > T$,

$$\hat{\theta}(t-T) = \hat{\theta}(0) + \int_0^{t-T} \Gamma\varphi(x(s))x(s)\,ds \tag{10}$$

and an incremental form of (3) is obtained as follows:

$$\hat{\theta}(t) = \hat{\theta}(t-T) + \int_{t-T}^t \Gamma\varphi(x(s))x(s)\,ds, t > T \tag{11}$$

Then, appealing to the integral mean-value theorem, the incremental algorithm (11) becomes

$$\hat{\theta}(t) = \hat{\theta}(t-T) + T\Gamma\varphi(x(\xi))x(\xi), t > T \tag{12}$$

where ξ lies between $t-T$ and t. It depends on t, and takes different values at different instants of time.

Our concern focuses on the second term of the right-hand side of (12), where T appears. As T is set to be small, the update becomes weak, while the parameter estimation is not in time updated with the measured data, when setting it too large. In order to void the situation to occur, the efficient adaptation mechanisms are given in this paper. We call them incremental adaptation mechanisms.

3   An Incremental Adaptive System

The development of adaptive systems in closed-loop configuration is carried out in this section and the next section. We shall apply the approach of incremental adaptation, which is in contrast to the conventional designs where the integral adaptive mechanisms are used. We will clarify that the estimates may not be bounded, as an unsaturated adaptation mechanism is used, and further illustrate that boundedness of the estimates could be guaranteed when using a saturated-adaptation mechanism.

At first, let us consider the incremental adaptation law of the form

$$\hat{\theta}(t) = \begin{cases} \hat{\theta}(t-T) + \Gamma \varphi(x(t))x(t), t \geq 0 \\ \hat{\theta}_0, \qquad\qquad\qquad t \in [-T, 0) \end{cases} \tag{13}$$

where $\Gamma > 0$ is a diagonal and positive definite matrix, and $\hat{\theta}_0$ is to be chosen by designer, which may not be non-zero, in order to ensure the continuity of $\hat{\theta}(t)$.

Theorem 1. *For the incremental adaptive system, composed of the control law (2) and the adaptation law (13), (i) $x \in L_\infty$ and $\hat{\theta} \in L_{2T}$; and (ii) the actual trajectory converges to zero asymptotically as time increases, i.e., $\lim_{t \to \infty} x(t) = 0$.*

Proof: Let us choose the following candidate Lyapunov-Krasovski functional:

$$V(t) = \frac{1}{2} \int_{t-T}^{t} \tilde{\theta}^T(s)\Gamma^{-1}\tilde{\theta}(s)\,ds + \frac{1}{2}x^2(t) \tag{14}$$

By using the following algebraic relationship

$$\left[\tilde{\theta}^T(t)\Gamma^{-1}\tilde{\theta}(t) - \tilde{\theta}^T(t-T)\Gamma^{-1}\tilde{\theta}(t-T)\right] + \left[\hat{\theta}(t) - \hat{\theta}(t-T)\right]^T \Gamma^{-1}\left[\hat{\theta}(t) - \hat{\theta}(t-T)\right]$$
$$= -2\tilde{\theta}^T(t)\Gamma^{-1}\left[\hat{\theta}(t) - \hat{\theta}(t-T)\right]$$

the time-derivative of V along the solution of (4) can be calculated as

$$\begin{aligned} \dot{V}(t) &= \tilde{\theta}^T(t)\Gamma^{-1}\tilde{\theta}(t) - \tilde{\theta}^T(t-T)\Gamma^{-1}\tilde{\theta}(t-T) + x(t)\dot{x}(t) \\ &= -\tilde{\theta}^T(t)\Gamma^{-1}\left[\hat{\theta}(t) - \hat{\theta}(t-T)\right] \\ &\quad - \frac{1}{2}\left[\hat{\theta}(t) - \hat{\theta}(t-T)\right]^T \Gamma^{-1}\left[\hat{\theta}(t) - \hat{\theta}(t-T)\right] \\ &\quad + x(t)\dot{x}(t) \end{aligned} \tag{15}$$

With the adaptation law (13), (15) becomes

$$\dot{V} = -\frac{1}{2}\varphi^T(x)\Gamma\varphi(x)x^2 - \beta x^2 \leq -\beta x^2 \tag{16}$$

Hence, $\dot{V} \leq 0$, resulting in $V(t) \leq V(t-T)$. $V(t)$ is bounded for $t \in [0, \infty)$, due to the boundedness of $V(t)$, $t \in [-T, 0)$. This in turn implies that $x \in L_\infty$. Using (16) again, we obtain that

$$\int_{t-T}^{t} x^2(s)ds \leq V(t-T) - V(T) \leq V(t-T)$$

implying that $x \in L_{2T}$, and

$$\lim_{t \to \infty} \int_{t-T}^{t} x^2(s)ds = 0 \tag{17}$$

Also from the boundedness of V, we know that the term $\int_{t-T}^{t} \tilde{\theta}^T(s)\tilde{\theta}(s)ds$ is bounded, which means that $\hat{\theta} \in L_{2T}$.

Using that $(a-b)^2 \leq 2a^2 + 2b^2$ and from (4), we obtain

$$\dot{x}^2 \leq 2\left(\tilde{\theta}^T\varphi(x)\right)^2 + 2\beta^2 x^2 \tag{18}$$

Due to the boundedness of x and the continuity of $\varphi(x)$, there exist constants c_1 and c_2 such that

$$\int_{t-T}^{t} \dot{x}^2(s)ds \leq c_1 + c_2 \int_{t-T}^{t} \left\|\tilde{\theta}(s)\right\|^2 ds \tag{19}$$

By Lemma 1 in [10], the conclusion (ii) follows.   ∎

It is seen from Theorem 1 that the boundedness of $\hat{\theta}$ itself, however, is not guaranteed. One can rewrite the adaptation law (13) as, for $t_i = t_0 + iT$, $t_0 \in [0, T]$, $\hat{\theta}(t_i) = \hat{\theta}(t_0) + \Gamma \sum_{j=1}^{i} \varphi(x(t_j))x(t_j)$. The boundedness of $\hat{\theta}$ requires that the sum of $\sum_{j=1}^{i} \varphi(x(t_j))x(t_j)$ should be finite. Unfortunately, there is no direct way to fix that with the current version of adaptation. This is the main motivation for us to consider saturated-adaptation mechanisms.

In comparison with (12), the adaptation law described by (13) updates the parameter estimate with the term where no T appears. This may lead fast convergence rate, due to the use of the strong update mechanism.

4   The Saturated Mechanisms

The incremental adaptive law is modified by introducing the partial-saturation in the form of

$$\hat{\theta}(t) = \text{sat}_{\bar{\theta}}\left(\hat{\theta}(t-T)\right) + \Gamma\varphi(x(t))x(t) \tag{20}$$

where $\bar{\theta}$ are the lower and upper bounds on $\theta(\cdot)$, assumed to be known *a priori* .

The boundedness and convergence properties of the adaptive system, with the same control law as given by (2), are summarized in the following theorem.

Theorem 2. *The incremental adaptive system, composed of the control law (2) and the partially-saturated adaptation law (20), is ensured that (i) all the variables in the closed-loop are bounded; and (ii) the actual trajectory converges to zero asymptotically as $t \to \infty$, i. e., $\lim_{t\to\infty} x(t) = 0$.*

Proof: The time derivative of V along the solution of (4) is calculated as

$$\dot{V}(t) = \frac{1}{2}\Big\{\tilde{\theta}^T(t)\Gamma^{-1}\tilde{\theta}(t)$$
$$-\left[\theta(t-T)-\text{sat}_{\bar{\theta}}\left(\hat{\theta}(t-T)\right)\right]^T \Gamma^{-1}\left[\theta(t-T)-\text{sat}_{\bar{\theta}}\left(\hat{\theta}(t-T)\right)\right]$$
$$+\left[2\theta(t)-\text{sat}_{\bar{\theta}}\left(\hat{\theta}(t-T)\right)-\hat{\theta}(t-T)\right]^T \Gamma^{-1}\left[\hat{\theta}(t-T)-\text{sat}_{\bar{\theta}}\left(\hat{\theta}(t-T)\right)\right]\Big\}$$
$$+x(t)\dot{x}(t) \tag{21}$$

Appealing to Lemma 1 in [8], for $q_1 = 1$, and $q_2 = 1$, to tackle the third term in the bracket of the right hand side of (21), we obtain

$$\left[2\theta(t)-\text{sat}_{\bar{\theta}}\left(\hat{\theta}(t-T)\right)-\hat{\theta}(t-T)\right]^T \Gamma^{-1}\left[\hat{\theta}(t-T)-\text{sat}_{\bar{\theta}}\left(\hat{\theta}(t-T)\right)\right] \leq 0$$

which leads to

$$\dot{V}(t) \leq -\tilde{\theta}^T(t)\Gamma^{-1}\left[\hat{\theta}(t)-\text{sat}_{\bar{\theta}}\left(\hat{\theta}(t-T)\right)\right]$$
$$-\frac{1}{2}\left[\hat{\theta}(t)-\text{sat}_{\bar{\theta}}\left(\hat{\theta}(t-T)\right)\right]^T \Gamma^{-1}\left[\hat{\theta}(t)-\text{sat}_{\bar{\theta}}\left(\hat{\theta}(t-T)\right)\right]$$
$$+x(t)\dot{x}(t) \tag{22}$$

It follows by applying (20) that

$$\dot{V} = -\frac{1}{2}\varphi^T(x)\Gamma\varphi(x)x^2 - \beta x^2 \leq -\beta x^2 \tag{23}$$

Hence, $\dot{V} \leq 0$, and we conclude the boundedness of V, which implies that $x \in L_{\infty}$. From (20), it in turn follows that $\hat{\theta}(t) \in L_{\infty}$, and from (2) $u \in L_{\infty}$. All the variables in the closed-loop system are thus bounded. Invoking (23) again, $\dot{V} \leq -\beta x^2$, and we integrate both sides to obtain

$$\int_0^t x^2(s)ds \leq \frac{1}{\beta}(V(0) - V(t))$$

i.e., $x \in L_2$. Thus, by Barbalat's lemma, $\lim_{t \to \infty} x(t) = 0$. ■

It is observed from Theorem 2 that the asymptotic tracking of the actual trajectory is achieved, while the bounded estimation is ensured. With the introduction of saturation in the adaptation law, the problem arising from the unsaturated one has been solved, which only ensure $\hat{\theta} \in L_{2T}$.

The incremental adaptive control works well, although the saturated adaptation mechanism is introduced. Adaptation law (20), however, cannot ensure the estimate $\hat{\theta}(t)$ to be within a pre-specified region, due to the unsaturated update term of the adaptation law.

We now consider the fully-saturated adaptation law of the form

$$\hat{\theta}(t) = \mathrm{sat}_{\bar{\theta}}\left(\hat{\theta}^*(t)\right) \tag{24}$$

$$\hat{\theta}^*(t) = \mathrm{sat}_{\bar{\theta}}\left(\hat{\theta}^*(t-T)\right) + \Gamma\varphi(x(t))x(t) \tag{25}$$

where the entire right-hand side of the adaptation law is saturated, and the estimate $\hat{\theta}$ satisfies that $\hat{\theta} = \mathrm{sat}_{\bar{\theta}}\left(\hat{\theta}\right)$. Namely, $\hat{\theta}$ lies within the saturation bounds, and $\bar{\theta}$ represent the lower and upper bounds for $\theta(\cdot)$.

Going through the same derivations as in the proof for Theorem 2, we obtain

$$\dot{V}(t) = -\tilde{\theta}^T(t)\Gamma^{-1}\left[\hat{\theta}(t) - \hat{\theta}(t-T)\right] - \frac{1}{2}\left[\hat{\theta}(t) - \hat{\theta}(t-T)\right]^T \Gamma^{-1}\left[\hat{\theta}(t) - \hat{\theta}(t-T)\right]$$

$$+ x(t)\dot{x}(t)$$

$$= -\tilde{\theta}^T(t)\Gamma^{-1}\left[\hat{\theta}(t) - \hat{\theta}^*(t)\right] - \tilde{\theta}^T(t)\Gamma^{-1}\left[\hat{\theta}^*(t) - \hat{\theta}(t-T)\right] \tag{26}$$

$$- \frac{1}{2}\left[\hat{\theta}(t) - \hat{\theta}(t-T)\right]^T \Gamma^{-1}\left[\hat{\theta}(t) - \hat{\theta}(t-T)\right]$$

$$+ x(t)\dot{x}(t)$$

Setting $q_1 = 1$ and $q_2 = 0$ in Lemma 1 of [8], and noting that $\hat{\theta} = \mathrm{sat}_{\bar{\theta}}\left(\hat{\theta}^*\right)$,

$$\left[\theta - \hat{\theta}\right]^T \Gamma^{-1}\left[\hat{\theta}^* - \hat{\theta}\right] \leq 0$$

implying that

$$\dot{V}(t) \le -\tilde{\theta}^T(t)\Gamma^{-1}\left[\hat{\theta}^*(t)-\hat{\theta}(t-T)\right]$$
$$-\frac{1}{2}\left[\hat{\theta}(t)-\hat{\theta}(t-T)\right]^T \Gamma^{-1}\left[\hat{\theta}(t)-\hat{\theta}(t-T)\right]$$
$$+x(t)\dot{x}(t)$$

Using the adaptive law (24)-(25), the derivative of V with respect to time can be calculated as

$$\dot{V}(t) \le -\frac{1}{2}\left[\hat{\theta}(t)-\hat{\theta}(t-T)\right]^T \Gamma^{-1}\left[\hat{\theta}(t)-\hat{\theta}(t-T)\right]-\beta x^2(t) \le -\beta x^2(t) \qquad (27)$$

from which the stability and convergence result of the adaptive system can be established, which is summarized in the following theorem.

Theorem 3. *For the incremental adaptive system, composed of the control law (2) and the fully-saturated adaptation law (24),(i) all the variables in the closed-loop remain uniformly bounded, and (ii) the actual trajectory converges to zero asymptotically as time increases, i.e., $\lim_{t\to\infty} x(t) = 0$.*

The above-mentioned incremental adaptive systems are in closed-loop configuration. The alternatives in open-loop configuration are now presented, for illustrating flexible ways to develop incremental adaptive systems.

For applying the open-loop adaptation, one more term is to be introduced needed in the control law (2), which offers compensation due to the use of the open-loop adaptation laws. For the partially-saturated adaptation case,

$$u_c = \varphi^T(x)\left[\frac{1}{2}\Gamma\varphi(x)x+\text{sat}(\hat{\theta})-\hat{\theta}\right] \qquad (28)$$

and for the fully-saturated adaptation case,

$$u_c = \frac{1}{2}\varphi^T(x(t))\Gamma\varphi(x(t))x(t) \qquad (29)$$

and the control law (2) is modified as

$$u = -\hat{\theta}^T\varphi(x)-\beta x-u_c, \beta > 0 \qquad (30)$$

Then, with the control law (30), the system dynamics (4) becomes,

$$\dot{x} = \tilde{\theta}^T\varphi(x)-\beta x-u_c \qquad (31)$$

The partially-saturated incremental adaptive law is given as

$$\hat{\theta}(t+T) = \text{sat}_{\bar{a}}(\hat{\theta}(t))+\Gamma\varphi(x(t))x(t) \qquad (32)$$

and the fully-saturated one is in the form of

$$\hat{\theta}^*(t+T) = \hat{\theta}(t) + \Gamma\varphi(x(t))x(t)$$
$$\hat{\theta}(t+T) = \text{sat}_{\bar{\theta}}\left(\hat{\theta}^*(t+T)\right)$$

(33)

Both Theorem 2 and Theorem 3 still hold even when the above adaptive designs are applied, respectively. The following candidate Lyapunov-Krasovski functional is chosen for the analysis purpose:

$$V(t) = \frac{1}{2}\int_t^{t+T} \tilde{\theta}^T(s)\Gamma^{-1}\tilde{\theta}(s)\,ds + \frac{1}{2}x^2(t)$$

5   Conclusion

In this paper, the incremental adaptive strategies are presented for developing adaptive systems, which are different from the integral adaptive algorithms adopted in the conventional adaptive systems. With the unsaturated adaptive mechanism, it is shown that the asymptotic convergence of the system is guaranteed, while the estimates are bounded in the sense of L_{2T}. To introduce saturation in an adaptation mechanism is shown to be a simple way to obtain bounded estimates. Both partially- or fully-saturated adaptation laws have been characterized analytically.

Acknowledgements. This research is supported by the National Natural Science Foundation of China (No. 61174034).

References

1. Krstic, M., Kanellakopoulos, I., Kokotovic, P.V.: Nonlinear and Adaptive Control Design. Wiley, NY (1995)
2. Ioannou, P.A., Sun, J.: Robust Adaptive Control. Prentice-Hall, Upper Saddle River (1996)
3. Sadegh, N., Horowitz, R., Kao, W.W., Tomizuka, M.: A unified approach to design of adaptive and repetitive controllers for robotic manipulators. ASME Journal of Dynamic Systems Measurement Control 112, 618–629 (1990)
4. Moore, J.B., Horowitz, R., Messner, W.: Functional persistence of excitation and observability for learning control systems. ASME Journal of Dynamic Systems Measurement Control 114, 500–507 (1992)
5. Jiang, Y.A., Clements, D.J., Hesketh, T.: Adaptive repetitive control of nonlinear systems. In: Proceedings of the 34th Conference on Decision & Control, New Orleans, LA, pp. 1708–1713 (1995)
6. Dixon, W.E., Zergeroglu, E., Dawson, D.M., Costic, B.T.: Repetitive learning control: A Lyapunov-based approach. IEEE Transactions on Systems, Man, and Cybernetics 32, 538–545 (2002)

7. Sun, M., Ge, S.S.: Adaptive rejection of periodic & non-periodic disturbances for a class of nonlinearly parametrized systems. In: Proceedings of the 2004 American Control Conference, Boston, Massachusetts, June 30-July 2, pp. 1229–1234 (2004)
8. Sun, M., Ge, S.S.: Adaptive repetitive control for a class of nonlinearly parametrized systems. IEEE Transactions on Automatic Control 51, 1684–1688 (2006)
9. Sun, M.: Adaptive controller designs for nonlinear periodic systems. In: Proceedings of 2009 IEEE International Conference on Intelligent Computing and Intelligent Systems, Shanghai, China, November 20-22, pp. 798–802 (2009)
10. Sun, M.: Partial-period adaptive repetitive control by symmetry. Automatica 48, 2137–2144 (2012)

Quantitative Evaluation across Software Development Life Cycle Based on Evidence Theory

Weixiang Zhang, Wenhong Liu, and Xin Wu

Beijing Institute of Tracking and Telecommunications Technology, Beijing, China
wxchung@msn.com

Abstract. The paper brings out a method on quantitative software trustworthy evaluation across software development life cycle. First, build hierarchical assessment model with decomposition of software trustworthy on every stages and design appropriate quantitative or qualitative metrics; then, take advantage of knowledge discovery in database techniques to obtain the weights of all software trustworthy characteristics; finally, make use of evidence theory to pretreatment and reason a huge number of multi-type measurement data. The example from engineering shows that it can effectively improve objectiveness and accuracy of the assessment results.

Keywords: Software Quantitative Evaluation, Evidence Theory, Software Development Life Cycle (SDLC), Data Fusion, Trustworthy.

1   Introduction

With the rapid development of information technology, computer software has been used widely, and the consequences of software failure has become even more serious, especially in the aerospace, finance, telecommunications and other important areas of people's livelihood. Although the software quality is a growing concern, there is lack of effective methods to evaluate software quality accurately and objectively.

Software trustworthy in software engineering was brought into existence in the 1990s, the research on which wins widespread attention. In short, software trustworthy refers to the capabilities of the system can be trusted to deliver services in the time and the environment. It includes usability, reliability, confidentiality, integrity and other features at first and then is expanded continuously [1]. Compared to other concepts, it is more suitable for aerospace and other field with high demands to safety and reliability.

Elements of software quality evaluation in general consist of quality model, software metrics and assessment method. In quality model, the more common approach is to decompose the software quality into several levels and every level including some factors, and then to break down every factor in the lowest level into some quantitative indicators [2-3]. Zhang [4] propose a hierarchic trustworthy decomposition model, which can take advantage of software testing cases to get pretty good results; however, due to the measurement data for the model is limited to

D.-S. Huang et al. (Eds.): ICIC 2013, LNAI 7996, pp. 353–362, 2013.

software testing stage, the acquisition and utilization of the trustworthy information is not comprehensive enough.

In software metrics and assessment method, fuzzy comprehensive evaluation algorithm with its simple operation has come to attention recently [5-6]. But due to its defects mainly on fuzzy calculation and weight distribution, there is likely a large deviation in the course of assessing a complex system, especially when comparing similar things. Yang [7] proposes a trustworthy software evaluation based evidence theory; however, it is too much reliance on expert scoring and does not involve full life cycle assessment.

Data fusion can be broadly summarized as the process that in accordance with established rules to combine data from multiple sources into a comprehensive intelligence and then to provide users upon its demand with information [8]. Benefiting from our practice on aerospace software, the paper presents a trustworthy assessment method across full software life cycle, which based on hierarchical trustworthy model, acquainting measurement data from each stage of the life cycle, and using of data fusion theory and KDD for effective analysis and integration of multi-type data, to finally obtain more objective and accurate quantitative assessment result.

2   Trustworthy Assessment Model across Software Development Life Cycle

2.1   Structure of the Model

The trustworthy assessment model (called QUEST II model) is a phased hierarchy model, as shown in Figure 1.

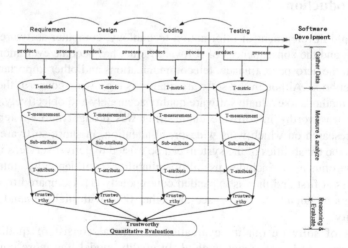

Fig. 1. structure of the QUEST II model

QUEST II model is made up of five-layer hierarchy. Firstly, software trustworthy (denoted by Trustworthy) is considered as the image of the software quality, and then Trustworthy is decomposed into several trustworthy attributes (denoted by T-attribute) including availability, performance, reliability, security, real-time, maintainability and survivability; then, every T-attribute is decomposed into a number of trustworthy sub-attributes (denoted by Sub-attribute), every Sub-attribute is decomposed into a number of trustworthy measurement (denoted by T-measurement); finally, every T-measurement is decomposed into several trustworthy metrics (denoted by T-metric) to the implementation of the trustworthy assessment data acquisition. T-attribute, Sub-attribute, T-measurement and T-metric are collectively referred to trustworthy characteristics.

The model assembles trustworthy assessment data in every stage of software development life cycle (SDLC) such as software requirements, design, coding, and testing. T-attribute and Sub-attribute are designed consistently in every stage, but T-measurement and T-metric are not completely consistent. Due to space limitations, do not give the detailed description of the model here.

2.2   Design of T-Metrics

According to the QUEST II model, T-metric is the lowest level of trustworthy characteristics and so it can affect directly the quality of evaluation. Trying to use quantitative metrics with objective and intuitive features will improve the accuracy of the assessment results. However, some metrics cannot be evaluated in a quantitative manner and requires the use of expert judgment for qualitative assessment.

In the QUEST II model, both quantitative and qualitative metrics are defined, as shown in Table 1. It is strongly recommended to define appropriately quantitative or qualitative metrics according to their respective features.

After obtaining original measurement data, it needs to do data pretreatment either for quantitative or qualitative T-metrics, which will be described in the next section.

Table 1. count of trustworthy characteristics in standard QUEST II model

	T-attribute	Sub-attribute	T-measurement	T-metric	
				Qualitative	Quantitative
Requirement	7	22	67	36	51
Design	7	22	71	35	64
Coding	7	22	67	21	61
Testing	7	22	95	7	112

3   Quantitative Evaluation across Software Development Life Cycle

3.1   Trustworthy Characteristic Tree

For every stages of SDLC, establish the software trustworthy characteristic tree (TcTree), as shown in Figure 2, whose nodes are various trustworthy characteristics. T-attribute denotes by e_i, Sub-attribute denotes by $e_{i,j}$, T-measurement denotes by

$e_{i,j,p}$, T-metric denotes by $e_{i,j,p,q}$, and w represents weights which mean relative importance of the same level characteristics and satisfy the normalization. TcTree of different stages is not exactly the same because the metrics of different stages are not entirely consistent.

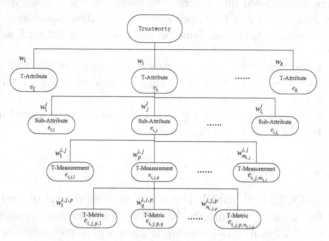

Fig. 2. software trustworthy characteristic tree

3.2   Obtaining Weight Base on KDD

To obtain the weight, make use of empirical databases and Knowledge Discovery in Databases (KDD) technique [4]:

Firstly, classify the software according to its technical complexity and management complexity (denoted by STC or SMC respectively). Both STC and SMC are divided into five grades using of the software nature and the ability of software developers.

Secondly, based on the empirical databases, get the weights of all trustworthy characteristics of the software with known STC and SMC. Fine-tuning can be done to the weights according to the feature of the software and the evaluation requirement.

Finally, determine all the weights in every stages of SDLC (in different TcTree).

3.3   Pretreatment of Measurement Data Base on Unified Utility

For qualitative metrics, assessment experts would have a different preference, it may lead to that the same degree of good or bad corresponds to a different score. For quantitative metrics, due to the different difficulty to realize, there are similar things [7]. Therefore, it needs to transform the original measurement data of qualitative or quantitative metrics into unified utility, which can resolves the problem of multi-type and multi-dimensional information fusion in addition.

For convenience, assume utility grade is $G = \{G_i, i = 1, 2, \cdots, n\}$, e is any T-metric and its measurement value is v_e and its utility value after transformation is μ_e.

3.3.1   Pretreatment of Qualitative Metrics

Make use of Formula (1) to transform the original measurement data of qualitative metric to unified utility. Where, $u(G_i)$ denotes the utility of grade G_i to the assessment expert, λ_i denotes the confidence of the expert.

$$\mu_e = \sum \lambda_i u(G_i)$$ (1)

For example, if the evaluation information given by an expert to e is $v_e = \{(C,0.3),(B,0.6),(A,0.1)\}$, and the utility of $average, good, excellent$ to the expert is $0.6, 0.8, 0.9$ respectively. Then:

$$\mu_e = 0.3 \bullet u(C) + 0.6 \bullet u(B) + 0.1 \bullet u(A) = 0.3 \times 0.6 + 0.6 \times 0.8 + 0.1 \times 0.9 = 0.85.$$

3.3.2   Pretreatment of Quantitative Metrics

For every quantitative metric, define its utility curve $u(v)$ according to its difficulty degree on implementation. Simply, it can be achieved by configuring utility value in the sampling point $S = \{S_i\}$.

After obtained $u(v)$ of the metric e, the transformation from its original measurement data v_e to utility value $u(v_e)$ can be done by Formula (2):

$$u(v_e) = u(S_{i+1}) - \frac{S_{i+1} - v_e}{S_{i+1} - S_i}(u(S_{i+1}) - u(S_i)), if\ v_e \in [S_i, S_{i+1}]$$ (2)

3.3.3   Evidence Theory and Dempster's Combination Rule

There must be many "uncertain" or "do not know" information during evaluation, so uncertainty reasoning methods need to be taken. Evidence theory is one of the most useful methods and widely used in multi-attribute decision problems [9]. Evidence theory is known as Dempster-Shafer theory or D-S theory, where; the mass function is introduced which satisfy axiom weaker than the probability theory and able to distinguish between "uncertainty" and "don't know". In the evidence theory, accurate data and subjective judgments with uncertainty characteristics can be compatible modeled in a unified framework.

Suppose Θ as the frame of discernment, $P(\Theta)$ as the power set of Θ constituted by all subsets of Θ. Subsets $A \subseteq \Theta$ called proposition. The mass function, also known as BPA (Basic Probability Assignment), is defined by a mapping $M : P(\Theta) \rightarrow [0,1]$, with $m(\varnothing) = 0$ and $\sum_{A \subseteq \Theta} m(A) = 1$.

According to Dempster's combination rule [10], the orthogonal sum is defined as:

$$m_{1,2,\cdots,n}(Z) = \begin{cases} m(\varnothing) = 0, \ Z = \varnothing; \\ \dfrac{\displaystyle\sum_{\cap Z_i = Z}\left(\displaystyle\prod_{1 \le i \le n} m_i(Z_i)\right)}{1-K}, \ \varnothing \subset Z \subseteq \Theta. \end{cases} \tag{3}$$

Where $K = \displaystyle\sum_{\cap Z_i = \varnothing} m_i(Z_i)$ named confliction factor, equivalent to the confidence

degree of the empty set in the process. In order to ensure the validity of Dempster's combination rule (the denominator of Formula (3) is not zero), here assumes that the assessment information taken as evidence is not complete conflicted.

3.3.4   Quantitative Assessment Algorithm Based on Data Fusion

After pretreatment of qualitative or quantitative T-metrics, evidence reasoning can be done. There are two ways , in general, distributed computing or center computing. Taking into account the actual situation of software evaluation, use the distributed computing way. The specific steps of the algorithms are:

STEP1. Suppose $H = \{H_s, s = 0,1,2,\cdots,5\} = \{0, 0.2, 0.4, 0.6, 0.8, 1.0\}$ as a unified software trustworthy assessment grades. For every stage C_x in SDLC, get the degree of confidence of the utility μ_e of any T-metric $e_{i,j,p,*}$ (if no confusion, abbreviated as e) on H using Formula (4):

$$\beta_s = \frac{H_{s+1} - \mu_e}{H_{s+1} - H_s}, \beta_{s+1} = 1 - \beta_s, \ if \ (H_s \le \mu_e \le H_{s+1}). \tag{4}$$

STEP2. For any T-metric, calculate its mass function given by every experts. Where, $\alpha \in (0,1]$ is discount factor which equals to 0.95 here, L denotes as L-th expert.

For the key T-metric e_k (with maximum weight w_k), use Formula (5) to get its mass function:

$$\begin{aligned} m^L(H_s \mid e_k) &= \alpha\beta_s, \\ m^L(H_\Theta \mid e_k) &= 1 - \sum_s m^L(H_s \mid e_k). \end{aligned} \tag{5}$$

For non-key T-metric e_t with weight w_t, use Formula (5) to get its mass function:

$$\begin{aligned} m^L(H_s \mid e_t) &= \frac{w_t}{w_k}\alpha\beta_s, \\ m^L(H_\Theta \mid e_t) &= 1 - \sum_s m^L(H_s \mid e_t). \end{aligned} \tag{6}$$

STEP3. For any T-metric e (that is, leaf node of TcTree), using Dempster's rule (Formula(3)), combines its multiple mass function gived by the experts to obtain its synthetic mass function $\overline{m}(H_s \mid e)$ and $\overline{m}(H_\Theta \mid e)$. That is, data fusion of multi-expert assessment data.

STEP4. For any T-metric $e_{i,j,p,*}$, get its confidence degree on H using Formula (4), wherein, the utility of $e_{i,j,p,*}$ can be achieved by below formula:

$$\mu_{i,j,p,*} \triangleq \mu_{e_{i,j,p,*}} = \sum_s \left(H_s \times \overline{m}\left(H_s \mid e_{i,j,p,*} \right) \right) \tag{7}$$

STEP5. To obtain the mass function $m\left(H_s \mid e_{i,j,p,*} \right)$ and $m\left(H_\Theta \mid e_{i,j,p,*} \right)$ of T-metric $e_{i,j,p,*}$ using Formula (5) and Formula(6).

STEP6. For any parent node $e_{i,j,*}$, to obtain its mass function $m\left(H_s \mid e_{i,j,*} \right)$ and $m\left(H_\Theta \mid e_{i,j,*} \right)$ using of all mass function of its children $e_{i,j,p,*}$ by Dempster's rule.

STEP7. Loop STEP 4 ~ STEP 6 to combine upwards until the trustworthy characteristics to be dealt with is Trustworty (thatis, the root node of TcTree).

STEP8. Repeat the above steps to iterate through all stages C_* of the SDLC. Then, go to the next step.

STEP9. For any T-attribute a_*, combine all its mass function of every SDLC stage to obtain its synthetic mass function $M(H_s \mid a_*)$ and $M(H_\Theta \mid a_*)$ using of Dempster's rule again. That is, data fusion of multi-cycle assessment data.

STEP10. To obtain the mass function $M(H_s)$ and $M(H_\Theta)$ of Trustworty across SDLC using of $M(H_s \mid a_*)$ and $M(H_\Theta \mid a_*)$ by Dempster's rule.

STEP11. At last, use Formula(8) to achieve the final quantitative result of trustworty evaluation:

$$T = \sum_s \left(H_s \times M\left(H_s \right) \right) \tag{8}$$

Without loss of generality, the above algorithm using of Dempster's rule. Other evidence synthesis rules (such as [11], [12], etc.) can be used with simple substitution.

4   Example

DPS (Data Processing Software) is a key engineering software, whose main function is to gather and analyze multi-source and multi-format measurement data in order to generate real-time processing results. The method proposed in this paper has been applied to DPS and parts of the evaluation data are shown below.

According to the aforementioned weight acquisition method based on KDD, the technical complexity of DPS is 4 and the management complexity is 3. Based on the trustworthy assessment model, weights distribution of DPS's various level trustworthy characteristics is obtained, as shown in Table 2.

Using of the quantitative assessment algorithm, mass functions of all trustworthy characteristics can be obtained, as shown in Table 3. EP1, EP2 and EP3 are assessment experts. After circular reasoning, quantitative evaluation results of DPS are achieved, as shown in Table 4.

As a key software, DPS has gone through many engineering tasks and been received by users. The result shown in Table 4 is consistent with the actual situation, and proves the effectiveness of the method proposed in this paper.

Table 2. DPS's trustworthy characteristics and weight distribution in requirement stage (partial)

T-attribute	Sub-attribute	T-measurement	T-metric
Availability 0.195	Suitability 0.30	Function defined adequacy 0.25	Rate of function defined adequacy
		Function defined integrity 0.25	Rate of function defined integrity
		Function defined correctness 0.25	Rate of function defined correctness
		Data element definition 0.25	Rate of data element definition
	Accuracy 0.25	Accuracy of data processing 0.5	Rate of accuracy of data processing
		Accuracy of calculation 0.5	Rate of accuracy of calculation
	Interoperability 0.20	Interface protocol definition 0.2	Rate of interface protocol definition
		Interface data to identify 0.2	Rate of interface data to identify
		Interface mode definition 0.2	Rate of interface mode definition
		Interface requirements documentation 0.2	Rate of requirements documentation
		Interface requirements scalability 0.2	Interface requirements scalability
	Ease of operation 0.15	Operating consistency 0.4	Operating consistency 0.5
			Manner consistency 0.5
		Correct to wrong operation 0.2	Definition of correcting operation
		Default value definition 0.2	Rate of default value definition
		Easy to monitor operating status 0.2	Easy to monitor operating status
	Compliance 0.10	Compliance requirements definition 1.0	Rate of compliance definition
Real-time 0.155	Processing timeliness 0.60	Processing timeliness definition 0.5	Rate of processing timeliness definition
		Worst processing time 0.5	Rate of worst processing time
	Real-time stability 0.40	Processing time jitter definition 1.0	Rate of processing time jitter definition
Reliability 0.120	Maturity 0.40	Maturity requirements definition 0.6	Maturity requirements definition
		Measures after failure 0.4	Measures after failure
	Fault tolerance 0.30	Error handling rules identify 1.0	Rate of error handling rules identify
	Persistent 0.30	Persistence requirements definition 0.6	Persistence requirements definition
		Contingency plans definition 0.4	Contingency plans definition
Security 0.175	Safety 0.40	Permissions control definition 0.4	Rate of permissions control definition
		Critical transaction identification 0.3	Rate of critical transaction identification
		Resource security definition 0.3	Rate of resource security definition
	Integrity 0.30	Critical data identification 0.4	Rate of critical data identification
		Integrity requirements definition 0.6	Integrity requirements definition
	Anti-risk 0.30	Safety requirements definition 1.0	Safety requirements definition

Table 3. DPS's mass functions of trustworthy characteristics in requirement stage (partial)

E	EP1		EP2		EP3	
e(1,1,1,1)	H4:0.05, H5:0.95					
e(1,1,2,1)	H4:0.05, H5:0.95					
e(1,1,3,1)	H4:0.43, H5:0.52					
e(1,1,4,1)	H4:0.36, H5:0.59					
e(1,2,1,1)	H3:0.08, H4:0.87					
e(1,2,2,1)	H4:0.12, H5:0.83					
e(1,3,1,1)	H4:0.12, H5:0.83					
e(1,3,2,1)	H4:0.00, H5:0.95					
e(1,3,3,1)	H4:0.19, H5:0.76					
e(1,3,4,1)	H4:0.00, H5:0.95					
e(1,3,5,1)	H4:0.95	H5:0.00	H3:0.24	H4:0.71	H4:0.95	H5:0.00
e(1,4,1,1)	H2:0.19	H3:0.76	H4:0.33	H5:0.62	H4:0.95	H5:0.00
e(1,4,1,2)	H3:0.38	H4:0.57	H4:0.52	H5:0.43	H3:0.19	H4:0.76
e(1,4,2,1)	H4:0.57	H5:0.38	H4:0.21	H5:0.74	H4:0.57	H5:0.38
e(1,4,3,1)	H4:0.00, H5:0.95					
e(1,4,4,1)	H3:0.10	H4:0.85	H4:0.33	H5:0.62	H3:0.19	H4:0.76
e(1,5,1,1)	H4:0.00, H5:0.95					
e(2,1,1,1)	H4:0.24, H5:0.71					
e(2,1,2,1)	H2:0.14, H3:0.81					
e(2,2,1,1)	H3:0.10, H4:0.86					
e(3,1,1,1)	H3:0.19	H4:0.76	H4:0.48	H5:0.47	H3:0.76	H4:0.19
e(3,1,2,1)	H3:0.38	H4:0.57	H3:0.24	H4:0.71	H4:0.38	H5:0.57
e(3,2,1,1)	H3:0.34, H4:0.61					
e(3,3,1,1)	H3:0.57	H4:0.38	H4:0.67	H5:0.29	H4:0.38	H5:0.57
e(3,3,2,1)	H4:0.48	H5:0.47	H4:0.62	H5:0.33	H2:0.47	H3:0.48
e(4,1,1,1)	H4:0.57, H5:0.38					
e(4,1,2,1)	H4:0.67, H5:0.29					
e(4,1,3,1)	H3:0.55, H4:0.40					
e(4,2,1,1)	H3:0.01, H4:0.94					
e(4,2,2,1)	H3:0.47	H4:0.48	H4:0.52	H5:0.43	H3:0.76	H4:0.19
e(4,3,1,1)	H3:0.95	H4:0.00	H3:0.43	H4:0.52	H3:0.95	H4:0.00

Table 4. Quantitative evaluation results of DPS

	Requirement	Design	Coding	Testing	Combination
Availability	0.909	0.933	0.901	0.839	**0.905**
Real-time	0.885	0.876	0.888	0.839	**0.839**
Reliability	0.787	0.797	0.731	0.611	**0.792**
Security	0.768	0.766	0.784	0.786	**0.798**
Survivability	0.679	0.732	0.760	0.644	**0.749**
Performance	0.758	0.744	0.766	0.795	**0.795**
Maintainability	0.716	0.625	0.752	0.853	**0.714**
Trustworthy					**0.799**

5   Conclusion

Software quality evaluation is an important and difficult subject in software engineering. In this paper, the proposed trustworthy assessment model can comprehensive gather various data from every stages of SDLC, the designed quantitative evaluation algorithm based on evidence theory can effectively deal with multi-stage multi-type multi-dimensional data, the taken methods for rights acquisition based on KDD can effectively reduce the subjectivity in the evaluation process. The example from engineering has proved that the method is able to give a quite accurate quantitative evaluation.

Next, more in-depth study would be conducted on how to improve the suitability of T-metrics and other trustworthy characteristics in order to gather more useful and comprehensive measurement data of every SDLC stages.

References

1. Liu, K., Shan, Z., Wang, J.: Overview on major research plan of trustworthy software. Bulletin of National Natural Science Foundation of China 22(3), 145–151 (2008)
2. McCall, J.: The Automated Meaz of Software Quality. In: 5th COMMPSAC (1981)
3. ISO/IEC 9126 Information Technology—Software Product Evaluation—Quality Characteristics and Guidelines for Their Use, 1st edn. (1991)
4. Zhang, W., Liu, W., Du, H.: A software quantitative assessment method based on software testing. In: Huang, D.-S., Ma, J., Jo, K.-H., Gromiha, M.M. (eds.) ICIC 2012. LNCS (LNAI), vol. 7390, pp. 300–307. Springer, Heidelberg (2012)
5. Wang, S.Z., Xian, M., Wang, X.S.: Study on synthetic evaluation method of software quality. Computer Engineering and Design 23(4), 16–18 (2002) (in Chinese)
6. Dong, J.L., Shi, N.G.: Research and improvement of the fuzzy synthesis evaluation algorithm based on software quality. Computer Engineering and Science 29(1), 66–68 (2007) (in Chinese)
7. Yang, S.L., Ding, S., Chu, W.: Trustworthy Software Evaluation Using Utility Based Evidence Theory. Journal of Computer Research and Development 46(7), 1152–1159 (2009) (in Chinese)
8. Kang, Y.H.: Data Fusion Theory and Application. Electronic Technology University Press, Xi'an (1997) (in Chinese)
9. Li, Y., Cai, Y., Yin, R.P.: Supprot vector machine ensemble based on evidence theory for multi-class classification. Journal of Computer Research and Development 45(4), 571–578 (2008) (in Chinese)
10. Shafer, G.: A Mathematical Theory of Evidence. Princeton U P, Princetion (1976)
11. Yager, R.R.: On the D-S framework and new combination rules. Information Sciences 41(2), 93–138 (1987)
12. Sun, Q., Ye, X.Q., Gu, W.K.: A New Combination Rules of Evidence Theory. ACTA Electronicia Sinica 28(8), 117–119 (2000)

B1 Signal Acquisition Method for BDS Software Receiver

Jiang Liu[1,2,*], Baigen Cai[1], and Jian Wang[1]

[1] School of Electronic and Information Engineering,
Beijing Jiaotong University, Beijing, China
[2] School of Transportation Science and Engineering, Beihang University, Beijing, China
jiangliu.lj@gmail.com

Abstract. Development and application of Global Navigation Satellite Systems (GNSSs) are of great importance for the increasing demands of location-based services in both the civil and military fields. As a novel GNSS system, BeiDou Navigation Satellite System (BDS) is experiencing a rapid developing period in recent years. Software-based receiver is proved an inevitable trend of the GNSS techniques, where BDS software receiver has been concentrated in both the engineering and industry field. As the initial step in signal processing of the BDS computation, the acquisition module detects the presence of BDS satellite signal from the antenna and provides coarse estimation of Doppler frequency and code phase. The FFT-based acquisition method for BDS B1 signal is investigated in this paper, where the cross-correlation is simplified according to selected frequency points. From results of experiments with a BDS intermediate frequency signal collector and a reference receiver, it is illustrated that the presented FFT-based scheme is capable of realizing effective BDS B1 signal acquisition and is with great potential for development of complete BDS software receiver.

Keywords: BeiDou Navigation Satellite System, satellite navigation, signal acquisition, software receiver.

1   Introduction

Satellite navigation systems have such characteristics of real-time and high precision which can make the target object continuously and accurately locate itself [1]. In recent years, GNSSs (Global Navigation Satellite Systems) are experiencing a rapid development in both system development and technique application in many civil and military branches [2~4]. The GNSS receiver techniques pursuing a low cost and high performance solution will promote more application fields for GNSS industry and that will be beneficial for much of the social and economic progress.

In recent years, a new trend in designing GNSS receivers has emerged that implements digitization closer to the receiver antenna front-end to create a system that works at increasingly higher frequencies and a wider bandwidth, and this development draws on an software receiver (SR) or software defined radio (SDR) approach originating from signal processing technologies available in many applications [5]. A

---

* Corresponding author.

D.-S. Huang et al. (Eds.): ICIC 2013, LNAI 7996, pp. 363–372, 2013.
© Springer-Verlag Berlin Heidelberg 2013

software receiver, which is with architecture as shown in Fig. 1 [6], can use different signals without the requirement for a new correlator chip, and new frequencies and new pseudo-random number (PRN) codes can be used simply by making software changes. Thus, software receiver will lessen the risks involved for designers during the period of transition to the new signals [7]. Thus, it is of great necessity to develop practical real-time software GNSS receivers.

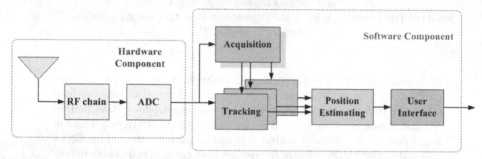

Fig. 1. Architecture of a GNSS software receiver

The BeiDou Navigation Satellite System (BDS, also named COMPASS) in China, is an important member of GNSS community [8]. In recent years, BDS has been developed as a successful GNSS direction and now is capable of providing local services to the Asian-Pacific areas with 16 satellites in orbit from late 2012. In addition, it is with a key feature to provide a short messaging communication service which distinguishes from other satellite navigation systems. In order to achieve more adaptive ability to multi-mode calculation and system services, a software receiver is with great potential to be reprogrammed to use BDS navigation satellite signal and even multiple constellations with high level integrity and availability, which provides benefits from the software radio architecture. It can also enable tightly-coupled BDS/INS positioning and navigation solutions for several critical fields and scenarios. Signal acquisition is an important foundation to achieve a software architecture receiver. As one of the key issues in software receiver development, there have been many results in GPS signals and multi-mode positioning calculation [9,10]. For development of software receiver for BDS-based applications, it is necessary to solve the signal acquisition issues for current B1 signals.

Based on the architecture of a software-based BDS receiver, this paper collects the real BDS IF (Intermediate Frequency) signal by specific IF device to carry out B1I code acquisition. With the analysis of BDS B1I ranging code generation, the FFT (Fast Fourier Transform) based signal acquisition solution is presented and discussed. By comparing the results with the actual BDS receiver, effectiveness and availability of this trail for BDS are identified for application.

2   BDS B1 Signal Generation

The BDS B1 signal is the sum of channel I and Q which are in phase quadrature of each other. The ranging code and NAV message are modulated on carrier. The signal is composed of the carrier frequency, ranging code and NAV message. The B1 signal is expressed as follows [11]:

$$S^j(t) = A_I C_I^j(t) D_I^j(t) \cos\left(2\pi f_0 t + \varphi^j\right) + A_Q C_Q^j(t) D_Q^j(t) \sin\left(2\pi f_0 t + \varphi^j\right) \quad (1)$$

where j denotes the satellite number, I and Q illustrate the channel I and Q, A is the signal amplitude, $C(t)$ is the ranging code, $D(t)$ is the data modulated on ranging code, f_0 and φ are the carrier frequency and initial phase respectively.

The B1I ranging code is a balanced Gold code truncated with the last one chip, where the Gold code has been proved with desirable correlation properties. With a similar structure strategy with GPS signal, for BDS system the Gold code is generated by means of Modulo-2 addition of G1 and G2 sequences which are respectively derived from two 11-bit linear shift registers. The generator polynomials for G1 and G2 are as follows [11].

$$G1(X) = 1 + X + X^7 + X^8 + X^9 + X^{10} + X^{11} \quad (2)$$

$$G2(X) = 1 + X + X^2 + X^3 + X^4 + X^5 + X^8 + X^9 + X^{11} \quad (3)$$

The initial phases of the target G1 and G2 sequences are G1:01010101010 and G2:01010101010. Principle of the B1I signal generator is shown in Fig. 2.

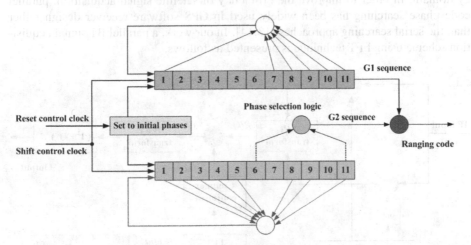

Fig. 2. Principle of the BDS B1I signal generator

To make different B1 codes for the BDS satellites, output of the two shift registers are combined as specific rules. The G1 register supplies its output, and the G2 register supplies two of its states to a Modulo-2 adder to generate its output. Thus, by means of Modulo-2 addition of G2 with different phase shift and G1, a ranging code is generated for different satellites. Ref [11] provides the phase assignment of G2 sequence for all the 37 BDS satellites.

3   B1 Signal Acquisition Method

Signal acquisition is the first step to track the specific visible BDS satellites. It realizes a rough estimation to the carrier frequency and the code phase [12]. Considering the downconversion corresponds to the BDS IF, since the frequency may deviate from the primary value caused by the relative movement of the visible BDS satellites, the Doppler shift of the carrier frequency could not be neglected when we extract the frequency from B1I signal nominal frequency (1561.098 MHz) and the mixers in the downconverter. In order to generate suitable local carrier signal, the frequency deviation should be estimated as accurately to enhance the acquisition performance. Besides, as for the code phase, it is crucial to associate the PRN code to current data, thus the generated local PRN code may be effectively time aligned with the incoming code. Different with the conventional satellite receiver with a hardware-based structure, the software functions are developed to achieve the acquisition efficiently.

Signal acquisition is actually a process to identify the aligned carrier frequency and code phase, by searching the parameter space with specific ranges. Assuming an expected maximum Doppler shift f_B with a searching step δ_f, there are $\left(\dfrac{2f_B}{\delta_f}+1\right)$ possible frequency results in frequency axis, and the code sweep will be performed within all 2046 possible code phases, which is with a larger size than that in frequency domain. In order to improve the efficiency of satellite signal acquisition, parallel code phase searching has been widely used in GPS software receiver design rather than the serial searching approaches [13,14]. In our work, a parallel B1 signal acquisition scheme using FFT technique is presented as follows.

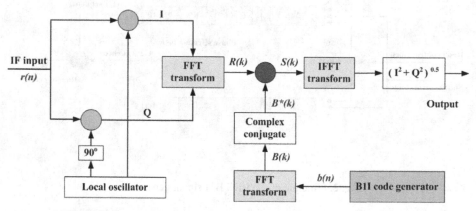

Fig. 3. Structure of BDS B1I signal acquisition based on the FFT scheme

Different from the searching space with $\left(\dfrac{2f_B}{\delta_f}+1\right)\times 2046$ iteration steps in the serial scheme, circular cross correlation between the input and PRN code without the code-phase-shifting promotes a time-efficient solution to the acquisition when the

Fourier transformation is involved. Thus, there are only $\left(\dfrac{2f_B}{\delta_f}+1\right)$ steps in calcula-

tion process. Fig. 3 gives the structure of the parallel searching algorithm for BDS B1I signal based on the FFT scheme.

The discrete Fourier transform of the sequence $x(n)$ with a finite length L can be performed as

$$x(n)\overset{F}{\rightarrow}X(k):X(k)=\sum_{n=0}^{L-1}x(n)e^{-\frac{j2\pi kn}{L}} \tag{4}$$

The cross-correlation between two finite length sequence $b(n)$ and $r(n)$, which are local generated B1I signal and the input BDS signal respectively, is performed to realize the parallel acquisition logic. That can be written as

$$s(n)=\frac{1}{L}\sum_{m=0}^{L-1}b(m)r(n+m) \tag{5}$$

According to the definition of Fourier transform and cross correlation, the discrete Fourier transforms of cross correlation between the sequence $b(n)$ and $r(n)$ could be computed as

$$s(n)\overset{F}{\rightarrow}S(k):S(k)=\frac{1}{L}\sum_{n=0}^{L-1}\sum_{m=0}^{L-1}b(m)r(n+m)e^{-\frac{j2\pi kn}{L}} \tag{6}$$

$$=\frac{1}{L}\sum_{m=0}^{L-1}b(m)e^{\frac{j2\pi km}{L}}\sum_{n=0}^{L-1}r(n+m)e^{-\frac{j2\pi k(n+m)}{L}}=\frac{1}{L}B^*(k)R(k)$$

where $B^*(k)=\sum_{m=0}^{L-1}b(m)e^{\frac{j2\pi km}{L}}$ is the complex conjugate of $B(k)$.

When using the FFT method in the stated process, both the local generated signal and the input signal are transformed into the frequency domain, and the original calculation of cross-correlation as equation (5) can be simplified with multiply operation, and the IFFT operation can directly provide the result sequence $s(n)$ in time domain. The absolute value of $s(n)$ reflects the correlation level between the local PRN code and the input signal. Identification of satellite acquisition for a satellite is determined by of peak value. Here a criterion considering both the maximum peak value and the second large one is presented, which can be written as

$$\sigma_j=\frac{peak(s)_1}{peak(s)_2}>\sigma_T,j=1,2,\cdots,37 \tag{7}$$

where σ_T denotes the detection threshold, and j is the PRN number.

In practical application, the computation efficiency and frequency resolution correspond to the length of the involved data, and the parameters f_B, δ_f and σ_T should be selected properly for an effective acquisition performance.

4   Test Results Analysis

The experiment was performed with an actual BDS IF signal acquisition platform in 2013. A hardware structure-based BDS receiver is taken as the reference to validate the satellite identification capability of the presented FFT-based acquisition scheme. Fig. 4 shows the devices used in experiment.

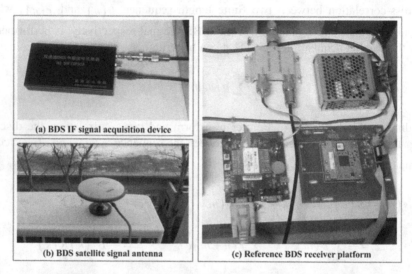

(a) BDS IF signal acquisition device

(b) BDS satellite signal antenna

(c) Reference BDS receiver platform

Fig. 4. BDS B1 signal collection devices and reference BDS receiver

Fig. 5 illustrates the raw collection signal that is used for acquisition calculation in frequency domain. By using the presented FFT-based approach, acquisition for BDS B1 signal is calculated in the developed software platform. There were seven satellites identified by the acquisition logic. Fig. 6 and Fig. 7 show the calculation results, and Table 1 shows the parameters and settings in the experiment.

Table 1. Parameters and settings in the experiment

Settings	Value	Parameters	Value
Intermediate Frequency	4.092MHz	Maximum Doppler shift f_B	14 kHz
Sampling Frequency	16.368MHz	Searching step δ_f	0.5kHz
Code Length	2046	Detection criterion threshold σ_T	2.5

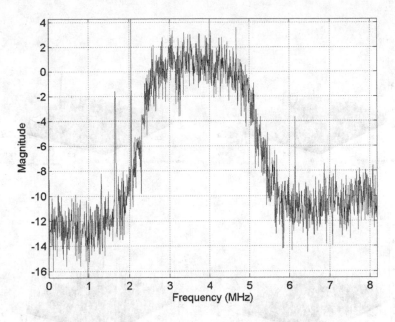

Fig. 5. Raw signal in the frequency domain

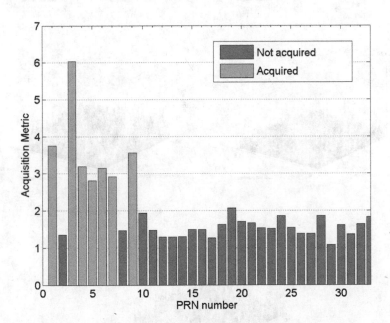

Fig. 6. Acquisition results with the presented FFT-based scheme

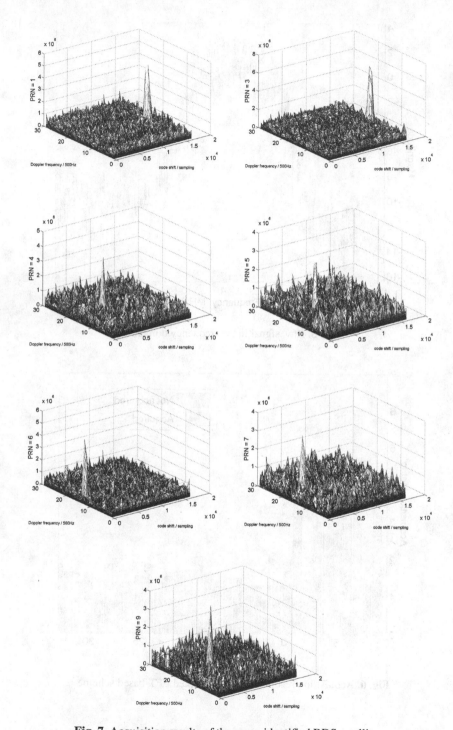

Fig. 7. Acquisition results of the seven identified BDS satellites

From the figures it is can be seen that the presented B1 signal acquisition method is capable to identify the BDS satellites in the undesirable visibility conditions, since the BDS antenna was set on the window, where half of the satellites in space could not be directly received. In order to investigate the acquisition capability, signal strength of the seven captured satellites is listed as shown in Fig. 8. From the signal-noise-ratios (SNRs) of the corresponding satellites, it can be confirmed that the adopted solution is with expected function and effectiveness.

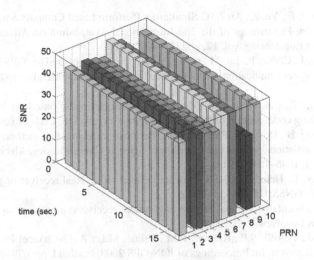

Fig. 8. Signal-noise-rate of the identified BDS satellites from reference receiver

As for the time efficiency of the presented method, due to utilization of FFT-based scheme, the acquisition is parallelized in the code phase dimension which means only 57 steps were performed. Totally 24.16s is expensed for software acquisition module processing for all the probable 37 BDS satellites, and the operation time with single frequency point is 0.42s. Although it has not been practically applied in specific platform, it provides certain reflection of performance level for the presented method.

5   Conclusions

This paper focuses on the key issue of signal acquisition in a software-based BDS receiver design. The software algorithm, which takes BDS IF samples as the input, performs a fundamental signal processing for the whole position calculation process. The presented BDS B1 signal acquisition scheme employs the FFT technique to realize the simplification of computing, where the cross-correlation can be carried out in the frequency domain and the performance of BDS acquisition is guaranteed. Current progress concentrates on certain modular function of the BDS software receiver. As the development of BDS capability, more issues of the whole system will be investigated and engineering application with specific platforms may be highly concerned.

Acknowledgement. This research was supported by China Postdoctoral Science Foundation funded project (2011M500219, 2012T50037), National Natural Science Foundation of China (61104162, 61273897), and the Specialized Research Fund for the Doctoral Program of Higher Education (20120009110029).

References

1. Geng, Z., Zhao, F., Ye, Z.: An ATC Simulation Platform based Compass Satellite Navigation System. In: Proceedings of the 2nd International Symposium on Aircraft Airworthiness, Procedia Engineering, vol. 17, pp. 422–427 (2011)
2. Gao, X., Guo, J., Cheng, P., Lu, M., et al.: Fusion positioning of BeiDou/GPS based on spatio temporal system unification. Acta Geodaetica et Cartographica Sinica 41(5), 743–748 (2012)
3. Dai, L., Tang, T., Cai, B., Liu, J.: Method of train positioning based on Beidou carrier phase smoothing code. Journal of the China Railway Society 34(8), 64–69 (2012)
4. Yang, Y., Peng, B., Dong, X., Fan, L., et al.: Performance evaluation method of hybrid navigation constellations with invalid satellites. Science China Physics, Mechanics & Astronomy 54(6), 1046–1050 (2011)
5. Won, J., Pany, T., Hein, G.: GNSS software defined radio: Real receiver or just a tool for experts. Inside GNSS 1(5), 48–56 (2006)
6. Tsui, J.: Fundamentals of global positioning system receivers: a software approach. John Wiley & Sons Inc., New York (2005)
7. Ledvina, B.M., Powell, S.P., Kintner, P.M., Psiaki, M.L.: A 12-Channel Real-Time GPS L1 Software Receiver. In: Proceedings of ION GPS 2003, Portland, pp. 679–688 (2003)
8. Wu, X., Hu, X., Wang, G., Zhong, H., Tang, C.: Evaluation of COMPASS ionospheric model in GNSS positioning. Advances in Space Research 51(6), 959–968 (2013)
9. Lin, W.H., Mao, W.L., Tsao, H.W., Chang, F.R., Huang, W.H.: Acquisition of GPS software receiver using Split-Radix FFT. In: Proceedings of IEEE Conference on Systems, Man, and Cybernetics, Taipei, pp. 4607–4613 (2006)
10. Principe, F., Bacci, G., Giannetti, F., Luise, M.: Software-defined radio technologies for GNSS receivers: A tutorial approach to a simple design and implementation. International Journal of Navigation and Observation 2011, 1–27 (2011)
11. Dou, B.: Navigation Satellite System Signa. In: Space Interface Control Document. Open Service Signal B1I (Version 1.0). China Satellite Navigation Office, 2012, pp.3–8 (2012)
12. Borre, K., Akos, D.M., Bertelsen, N., Rinder, P., Jensen, S.H.: A Software-defined GPS and Galileo receiver - A single-frequency approach. Springer (2006)
13. Cui, S., Yao, X., Qin, X., Fang, J.: An optimized method based on DSP in embedded software GPS receiver. In: Proceedings of the 19th International Conference on Geoinformatics, pp. 1–5. IEEE (2011)
14. Thompson, E., Clem, N., Renninger, I.: Software-defined GPS receiver on USRP-platform. Journal of Network and Computer Applications 35(4), 1352–1360 (2012)

Isomerism Multiple-Path Routing Algorithm
of Intelligent Terminals

Liu Di

Department of Physics and Electronic Engineering Hechi University Yizhou, Guangxi, China
liudi611@126.com

Abstract. The intelligent information terminal of the Internet of Things are usually connected by isomerism multiple-path network, So its link switching need to be down on the application layer which cause the transmit interrupt. By using the two-way ants between the destination and source node to monitor the link status, an isomerization multi-path routing algorithm basing on ant algorithm is proposed, the algorithm is able to complete fast routing and re-package of packet header on the network layer and it can achieve the seamless of isomerization multi-link and then to avoid transmit interrupt. Compared with the existing multi-path routing algorithm the simulation results show that the algorithm has improved the performance and stability of packet loss rate, transmission delay and the ability to adapt to the busy link.

Keywords: Multi-Path Routing, Two-way ants, intelligent information terminal, Isomerization network.

1 Introduction

As a data terminal node, the intelligent information terminal of IoT application system, bears the important task of data acquisition, processing, transmission. And how to ensure the network real-time communication is especially important, this is the premise of all subsequent operating smoothly. In existing information terminal communication scheme, independent multilink[1] is one of the high reliability solutions. But due to the multiple link is heterogeneous, and in the middle of the transmission network is not controlled, the plan currently can only be done in the application layer link switching, in terms of response time, portability and reliability more obvious flaws still exist. Therefore, how to reliable multipath routing algorithm is introduced into the Internet of things terminal information, heterogeneous link switching in the underlying network is complete, in order to avoid the switch link business interruption caused by the system, has become an urgent problem in the Internet of things application system, current research is one of the hot issues[2-4].

In the existing transmission network, OSPF (Open Shortest Path First algorithm) is one of the widely used routing protocol, which uses the link state routing algorithm (link state routing, LS) to implement the routing. LS determines its innate characteristics of multipath and congestion avoidance function, when the link failures, it is simply

D.-S. Huang et al. (Eds.): ICIC 2013, LNAI 7996, pp. 373–381, 2013.

discarded packets instead of initiative to address the other available even free link. Therefore, OSPF can't let the Internet of things terminal to realize heterogeneous information more efficient management and use of a link [5].

Compared with other routing algorithm, the ant algorithm [6] don't need to establish a mathematical model of the large and complex probability calculation, and is a kind of high efficiency, low consumption and routing algorithm is relatively easy to implement. As a result, the algorithm is widely used in the multipath communication network and decision optimization problems. In routing research, literature [7] to improve the ant algorithm, puts forward a routing algorithm with congestion avoidance properties. The algorithm can quickly select the optimal path between two nodes, but it does not have active congestion avoidance mechanism, response characteristic is not ideal. Literature [8] proposed a QoS multipath routing algorithm based on Ant algorithm, AMP (Ant Multiple-Path). The algorithm can quickly will congestion nodes data using the pathfinding scattered ants has explore the new path to send. But real-time response characteristics of the proposed algorithm is not stable, nor in the heterogeneous multipath network deployment, cannot satisfy the IoT information terminal, especially the safety class application system for network transmission requirement of real-time.

2   Ant Routing Algorithm

Network established by the directed graph $G(V,E)$ [9], among them: $V = \{v_1, v_2, \cdots v_n\}$ for the set of nodes (routing); $E = \{e_1, e_2, \cdots e_m\}$ is a collection of edges (link). AMP algorithm with $PT(Vs,t)$ from the source node to destination node, the time delay Δ, bandwidth B constraint condition respectively, the following type of (1) (2):

$$de(PT(Vs,t)) = de(Vs) + \cdots + de(t) + de(Vs,Vm) + \cdots + de(e(Vn,t)) \leq \Delta \qquad (1)$$

$$bw(PT(Vs,t)) = \min\{bw(e(Vs,Vm), \cdots, e(Vn,t))\} \geq B \qquad (2)$$

Algorithm's aim is to explore into an optimal path from v to t, the path P conform to the following type (3) type as shown in (4) characteristics:

1. Path P link bandwidth is greater than B

$$\min\{bd(J)\} \geq B, (J \in P) \qquad (3)$$

2. The sum of the node delay of path P is less than $\triangle$

$$\sum_{J \in P, V \in P} de(J) + de(V) \leq \Delta \qquad (4)$$

In under the premise of meet the above two requirements, is the "optimal" measure of the cost smallest, namely:

$$\cos t(P) = \min \left\{ \sum_{J \in P} \cos t(J) \right\}$$ (5)

Simulation results show that the AMP algorithm under different network load conditions, the residual bandwidth, delay, packet loss rate on the performance, LB than mentioned in the literature [7] algorithm is improved substantially, the search path of many conditions, and should has the first type of congestion avoidance strategy. But requires multiple link performance close to the algorithm can guarantee the transmission delay, and the Internet of things terminal information link is often heterogeneous, transmission delay difference is bigger. So, need to be targeted for improvement.

3   Heterogeneous Multipath Routing Algorithm Based on Ant Algorithm

In a multipath links of the Internet of things application system, the intermediate transmission network often belong to different managers, its technology has very strong heterogeneous characteristics and realization of the user is not completely controllable. Therefore, AIMR algorithm design for information in Internet of things terminal as the center of the lightweight multipath routing algorithm, complete multipath routing in the network layer, to improve the efficiency of routing nodes as far as possible, at the same time also can improve the information terminal of the ability to adapt.

3.1   Network Model

Simplified network model is shown in figure 1. Nodes 1, 2, for the IoT information terminals, through three separate links (LAN, 3G, WIFI) to transmit data, define it as access node; Node 3 for telecom operators in 3G base station; Node 4 abstraction for local area network (LAN), its size in the actual deployment can freely adjustable; Node 5 for the wireless AP, and its concrete realization technology also has nothing to do with this algorithm. Defines node 3, 4, 5 as transport nodes; Node for the data nodes in June and July. When carries on the simulation and actual project deployment, nodes can access through the switches to connect multiple data node; Each transport nodes can connect to multiple access nodes.

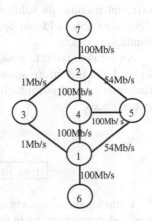

Fig. 1. Network model

3.2   Algorithm Thought

Algorithm aims at real-time, active monitoring under the premise of link-state, reduces the complexity of the optimal path in the access node selection, thus in the underlying implementation on-demand network link fast switching. Real-time control of link-state depend on improved packet structure detection and state ants to achieve, this is the foundation for later target was achieved. By simplifying the pheromone table structure and multipath link number matching to reduce the complexity of the access node pathfinding, accelerate the multipath routing convergence speed [10-12].

3.3   Algorithm Used to Describe

In access node, with a pheromone table instead of a routing table as a basis for the link to choose, in the table records every link of real-time state, and that the table records the information is reliable, its structure as shown in table 1, the article records in the table number link and node number is equal, i.e. each link in the pheromone table always keep only a routing record, was in a query operation, to link status and QoS values as the basis.

Table 1. Pheromone table structure

N-ID	L-ID	LDN-IP	LSN-IP	NN-IP	QoS	ST

N-ID to access the serial number of nodes, as a result of the existence of multipath, IP, using it as the basis of identification of node; L - the serial number of the ID for the current link; LDN-IP for the link to the destination node corresponding to the current interface IP address; LSN - IP for the link corresponding to the source node to the current interface IP address; NN-IP is the IP for the link to the next node; QoS is used to record the current link transmission delay value, that value as the key link to choose parameters; ST field is a Boolean, used to mark this record for efficient routing, its value by type (4) in the preset threshold values of delay Δ to determine, this field can be used to block failure records of routing, to speed up the routing operation.

An ant is to point to a packet, the packet structure as shown in table 2, can be used to send link test information and feedback link state information, the link status of the real-time control is through the mechanism of ants of two-way communication between nodes.

Table 2. The ant packet structure

L-ID	LDN-IP	LSN-IP	LSN-ID	TTL	QoS

Table top three fields in conformity with the pheromone table; LSN-ID for the current link corresponding to the source node code; TTL for the ant's life cycle, to avoid the ants unlimited cycle; QoS is used to evaluate the link status.

All access nodes while they are in electrical initialize it to destination node No1 and No2 link available, and on this basis to establish the original pheromone table. During application initialization, randomly select a from the node of the available IP, but it must give a destination node Numbers (N-ID), N-ID is used to query information table, complete the message encapsulation, so as to realize compatible with the existing transmission network, strengthen information terminal compatibility. Algorithm steps are as follows:

Step 1: In accordance with table 1 initialization pheromone table structure;

Step 2: In a separate thread to start the timer, the use of ant inspection of each link status and QoS values, regularly updating pheromone table;

Step 3: According to the application layer delivery down N-ID, choose a pheromone ST field is really the first record in the table, complete link option. If there is no link available to choose, to perform this step starts the timer and giving prompt information;

Step 4: According to the format of the TCP header, use the link corresponding IPSx related fields and replacement IPDx packet in the packet header, complete encapsulation, and forwarding a message through IPNx interface.

Step 5: Into the next communication cycle, turn to step3.

4   Simulation Test

General routing algorithm performance can be from the average end-to-end delay and network packet loss rate two aspects to evaluation. AIMR algorithm as a kind of lightweight routing algorithm applied to Internet information terminal, its performance is mainly affected by the following two factors: one is from the link fails to be detected; 2 it is link to switch the data encapsulation time again. Intermediate transmission network delay, packet loss rate, is not the controllable factors of this algorithm.

Use OPNET simulation platform for performance simulation test, the topology of the network as shown in figure 1, each link transmission delay is set as the 10 ms, at 6 nodes have a data source, the destination address is 7 nodes. For data source of traffic changes during the process of simulation, the more common approach is to use negative exponent data source, that is a given initial flow rate, and then press a certain step length increment, intermittent pauses between send, until the upper and lower latency than preset threshold value to determine the link congestion. Using this test method can objectively reflect the general performance of the routing algorithm. But IoT information terminals used public transport link of the actual status is often random changes, has a strong uncertainty, which should be paid more attention to in random case, the effectiveness of the algorithm. So we will node in the data source is set to 6 in 1 Mb/s to 10 Mb/s between random changes, while transport nodes 3, 4, and 5 can be controlled by the timer intermittent failure simulation nodes. Considering the existence of the wireless link, the algorithm for determining each link congestion conditions upper and lower threshold value were 0.6 and 0.3, and at the same time for more than three ants status is the same. Adopting these measures, in order to the simulation results are close to the real application scenario, and has carried on the simulation and analysis for the following key performance.

4.1   Network Packet Loss Rate

In this algorithm to establish network model, the normal load of Ethernet environment packet loss rate stable at about 0.2 to 0.3 levels. Introducing node intermittent failure

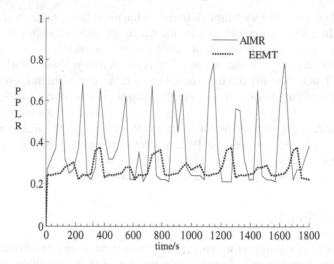

Fig. 2. Network packet loss rate charts

factors, using this algorithm simulation run half an hour after the network packet loss rate statistics as shown in figure 2.

By statistical curve in the figure, the frequent switching between multiple heterogeneous link did not increase overall network packet loss rate. This indicates that algorithm can link frequent failure to timely detection of the state, and complete the switch in the underlying on-demand. Because of link state control and switch can't do absolute in the sense of "real time", in carries on the statistics link switching point, network packet loss rate curve appeared some peak, but overall still control within 0.8, which means link switching action will not affect the application layer of the business, the users feel link switching action, show that algorithm is feasible in the index.

4.2 Network Time Delay

Test data shows that in reference [1] if done in the application layer link switch, according to the different links, time needed for switching between 15 seconds and 37 seconds. During this period, interruption of business needs and is easy to cause application error. AIMR algorithm, the switch link performed by the application layer to network layer. Intermittent failure control data random changes, link, the simulation run half an hour, the statistical results as shown in figure 3.

Can be found through the data analysis , the statistics link after the start of the first node can achieve stable transmission state. Timer to control the current link fails, then because of two-way ants between the nodes, the algorithm can quickly perceived link state changes, and immediately began to multipath routing process, a message header to packaging, the packet is forwarded to the optimal link available at this time. During this period, the network delay there will be a small peak, but all within 10 ms,

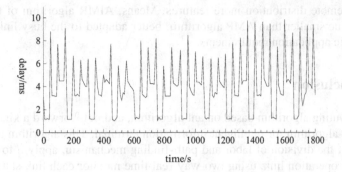

Fig. 3. Network latency Statistics

still have low delay characteristics. Considering the actual working condition, again will link failure frequency is much lower than the simulation environment, can think logically algorithm to realize the seamless switching node of the link between 6 and 7.

IoT data needs to be in the public network and public network can appear in addition to the foregoing test simulation of direct failure situation, may also continue to busy at peak hours. Adaptive testing algorithm, therefore, to continue to busy link is very necessary. Data source of the data sending rate fixed set as network bandwidth 0.8 and intermittent control link failure, simulation run half an hour later, the packet delay value carries on the statistical analysis and compared with algorithm based on the AMP, the result is shown in figure 4.

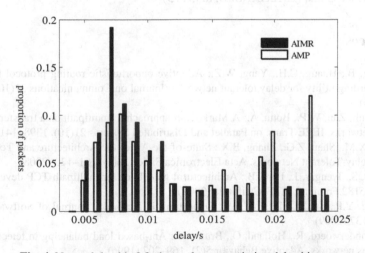

Fig. 4. Network load is 0.8, the packet transmission delay histogram

The statistical result shows that using the algorithm of AMP, an average delay of all packets is 11.32 milliseconds. By AIMR algorithm, an average delay of all packets is 8.87 milliseconds. As can be seen from the map, AIMR algorithm more packets delay distribution range, near the average delay and packets of AMP algorithm presents a remote distribution more features. Means, AIMR algorithm of time delay variance value smaller than AMP algorithm, better adapted to the busy link, also is a more realistic application requirements.

5   Conclusion

Multipath routing algorithm based on ant algorithm, and puts forward a kind of multipath routing algorithm for intelligent information terminal. This algorithm to simulate the nature of the division of labor and path-finding mechanism, apply it to the multipath routing operation link, using two-way real-time monitor each link-state, ant time delay as the main basis to update the routing table and packaging again according to the complete packet header, link switching process of application layer is not visible. AIMR simulation results indicate that the algorithm in the network, transmission delay and packet loss rate to adapt to the continuous busy link capacity than the existing in such aspects as general multipath routing algorithm has certain superiority, also more close to the Internet of things terminal connected to the characteristics of the information. But, as a result of the existence of two-way ants and need to encapsulate on packet header, in link in the process of switching algorithm takes up a certain node processor time, resulting in packet loss rate is not stable, it is also the key and focus of further research in the future.

Acknowledgment. This project was supported by Natural Science foundation of Guangxi Province China (2012GXNSFBA053175)

References

[1] Wang, B., Huang, C.H., Yang, W.Z.: Adaptive opportunistic routing protocol based on forwarding-utility for delay tolerant networks. Journal on Communications 31(10), 36–47 (2010)

[2] Buivinh, Zhu, W.P., Botta, A.: A Markovian approach to multipath data transfer in overlay networks. IEEE Trans. on Parallel and Distributed Systems 21(10), 1398–1411 (2010)

[3] Fan, X.M., Shan, Z.G., Zhang, B.X.: State-of-the-Art of the Architecture and Techniques for Delay-Tolerant Networks. Acta Electronica Sinica 36(1), 161–170 (2008)

[4] Barre, S., Iyengar, J., Ford, B.: Architectural guidelines for multipath TCP development, RFC 6182.Fremont. IETF, California (2009)

[5] Xiong, Y.P., Sun, L.M., Niu, J.W.: Opportunistic Networks. Journal of Software 20(1), 124–137 (2009)

[6] Schoonderwoerd, R., Holland, O., Bruten, J.: Ant-based load balancing in telecommunications networks. Adaptive Behavior 5(2), 169–207 (1996)

[7] Zhang, L.Z., Xian, W., Wang, J.P.: Routing Protocols for Delay and Disruption Tolerant Networks. Journal of Software 21(10), 2554–2572 (2010)

[8] Su, J.S., Hu, Q.L., Zhao, B.K.: Routing Techniques on Delay/Disruption Tolerant Networks. Journal of Software 21(1), 119–132 (2010)

[9] Li, X.Y., Wan, P.J.: Fault "tolerant deployment and topology control in wireless networks. In: Proceedings of The ACM Symposium on Mobile Ad Hoc Networking and Computing (MobiHoc), pp. 17–128 (2003)

[10] Hajiaghayi, M., Immorlica, N., Mirrokni, V.S.: Power optimization in fault-tolerant topology control algorithms for wireless multihop networks. In: Proc. ACM International Conference on Mobile Computing and Networking (MOBICOM), pp. 300–312 (2003)

[11] Perotto, F., Casetti, C., Galante, G.: SCTP-based transport protocols for concurrent multipath transfer. In: Proc. of IEEE Wireless Communications and Networking Conference (WCNC), pp. 2969–2974 (2007)

[12] Hui, P., Crowcroft, J., Yoneki, E.: Bubble rap: Social-based forwarding in delay-tolerant networks. In: ACM MobiHoc, pp. 241–250 (2008)

Prediction Based Quantile Filter for Top-*k* Query Processing in Wireless Sensor Networks

Hui Zhang[1], Jiping Zheng[1,2,*], Qiuting Han[1], Baoli Song[1], and Haixiang Wang[1]

[1] Department of Computer Science and Technology,
Nanjing University of Aeronautics and Astronautics, P.R. China
{zhanghui,jzh,hanqiuting,songbaoli,wanghaixiang}@nuaa.edu.cn
[2] State Key Laboratory for Novel Software Technology, Nanjing University, P.R.China

Abstract. Processing top-*k* queries in energy-efficient manner is an important topic in wireless sensor networks. It can keep sensor nodes from transmitting redundant data to base station by filtering methods utilizing thresholds on sensor nodes, which decreases the communication cost between the base station and sensor nodes. Quantiles installed on sensor nodes as thresholds can filter many unlikely top-*k* results from transmission for saving energy. However, existing quantile filter methods consume much energy when getting the thresholds. In this paper, we develop a new top-*k* query algorithm named QFBP which is to get thresholds by prediction. That is, QFBP algorithm predicts the next threshold on a sensor node based on historical information by <u>A</u>utoreg<u>R</u>essive <u>I</u>ntegrated <u>M</u>oving <u>A</u>verage models. By predicting using ARIMA time series models, QFBF can decrease the communication cost of maintaining thresholds. Experimental results show that our QFBP algorithm is more energy-efficient than existing quantile filter algorithms.

Keywords: Wireless Sensor Networks, top-*k*, Quantile Filter, Time Series, ARIMA.

1 Introduction

With the improved recognition to the physical world and the rapid development on technologies such as electron and wireless communication, wireless sensor networks have been applied to many areas such as military, medical treatment, environment surveillance, and so on. The bright future of wireless sensor networks has attracted the attention of so many scholars[3]. In wireless sensor networks, sensor nodes are energy-limited which are powered by batteries. In addition, various kinds of sensors generate a large amount of data, and the data are dense. However, users are always

* This research is supported by the Ph.D. Programs Foundation of Ministry of Education of China under Grant No. 20103218110017, NUAA Research Fundings of China under Grand No. NN2012102, NS2013089 and KFJJ120222, the Priority Academic Program Development of Jiangsu Higher Education Institutions of China, the Fundamental Research Funds for the Central Universities of China.

D.-S. Huang et al. (Eds.): ICIC 2013, LNAI 7996, pp. 382–392, 2013.

interested in max or min k objects among them. So, top-k query processing is crucial in many applications, which is an aggregate query technology and ripe in uncertain databases and relational databases[4,7,13,12]. But, the top-k query in wireless sensor networks is different from general queries such as SUM, AVG etc., because the sensor node could not determine whether its data would be included in the final results because this will be decided by the base station after collecting all data from sensor nodes. The centralized query processing manner produces a large of communication cost and wastes lots of energy. So how to process top-k query in an energy-efficient manner is an important topic in wireless sensor networks, which can supply top-k results to users by minimizing the energy consumption.

Nowadays, filter-based monitoring approach to process top-k queries is a mainstream in wireless sensor networks. QF (Quantile Filter, QF)[5] is an energy-efficient approach among existing algorithms, which is based on quantiles of sensor values. The basic idea is to choose one quantile from all children nodes as threshold and installs the threshold on children sensor nodes. Then, each child node sends related data to parent node which the data are larger than the quantile which avoids redundant data from transmitting to parent nodes. However, the obtaining of thresholds in QF depends on the interaction between child and parent nodes which consumes more unnecessary communication energy.

In this paper, we propose a novel top-k monitoring approach named QFBP (Quantile Filter Based on Predicted Values, QFBP), which gets thresholds by prediction according to time series model. We adopt ARIMA (Autoregressive Integrated Moving Average Model) time series model for prediction because of the convenience and more suitable for sensor data. The proposed algorithm decreases energy consumption on communication and transmission by predicting the thresholds. We evaluate our proposed methods by extensive experiments to compare with QF algorithm by simulations. Experimental results show that our QFBP outperforms the existing QF algorithm in terms of both energy-efficiency and correctness.

2   Related Work

A naive implementation of monitoring top-k query is the centralized approach in which the base station computes the top-k result set after collecting all sensor readings from sensor nodes periodically. However, a wireless sensor network can be viewed as a distributed network which consists of lots of energy-limited sensor nodes, and communication cost is the main energy consumption. The transmission of massive sensed data and the interaction among these sensor nodes will consume extra energy. In order to reduce the communication cost in the period of data collection, Madden et al.[10] introduced an in-network aggregation technique known as TAG(Tiny AGregation, TAG). In TAG algorithm, the data are sent by nodes from lower level to higher level over the network. If the length of points set is less than k, the node forwards all the points to parent node. Otherwise, forwards k points with highest values. In the end, base station computes the final results according to these points collected from all sensor nodes. The algorithm keeps unavailable data from transmission, but it still produce unnecessary energy consumption and is not really energy-efficient.

Wu et al.[16,17] proposed filtering based on monitoring approach called FILA(Filter-based Monitoring Approach, FILA). Its basic idea is to install a filter on each sensor node to filter out unnecessary data which are not contributed to final results. Reevaluation and filter settings are two critical aspects that ensure the correctness and effectiveness of the algorithm. But, when sensing values on nodes vary widely, base station needs updating filters for related nodes frequently which leads to large amount of updating cost and makes the performance of algorithm worse. Based on FILA, Mai et al.[11,14] proposed DAFM(Distributed Adaptive Filter based Monitoring, DAFM) algorithm which aims to reduce the communication cost of sending probe messages in reevaluation process and lower down the transmission cost when updating filter in FILA. With different query semantics and system model, Chen et al.[5] proposed quantile filter which views sensing values and its sensor as a point. A top-k query is to return k points with highest sensing values. Its basic idea is to filter out the redundant data based on a threshold which is a value of the point sets from all child nodes. But, the frequent interaction between parent and child nodes will produce more energy consumption on obtaining the thresholds.

Filter-based and aggregation are two main strategies at present that they cope with each other to process top-k query in wireless sensor networks. In [9], Liu et al. develop a new cross pruning (XP) aggregation framework for top-k query in wireless sensor network. There is a cluster-tree routing structure to aggregate more objects locally and a broadcast-then-filter approach in the framework. In addition, it provides an in-network aggregation technique to filter out redundant values which enhances in-network filtering effectiveness. Cho et al.[6] proposed POT algorithm (Partial Ordered Tree, POT) which considers the space correlation to maintain k sensor nodes with highest sensing values. The effectiveness and correctness of algorithms in [9] and [6] depend on the interaction of cluster nodes in time. Abbasi et al.[2] proposed MOTE (Model-based Optimization Technique, MOTE) approach based on assigning filters on nodes by model-based optimization. Nevertheless, it is an NP-hard problem that how to get optimal filter setting for top-k set.

Energy-efficient is a critical issue in wireless sensor networks, and also an important indicator to evaluate the effectiveness and practicality of algorithms. In recent years, some researchers introduce time series models to wireless sensor networks. For example, Tulone et al.[15] and Liu et al.[8] apply time series predicting models to the data collection in wireless sensor networks. The basic idea is to build a same model on the base station and each node. The base station predicts the values of nodes until outlier readings produced, the nodes send sensing readings to base station. When a reading is not properly predicted by the model, model is relearned to adapt changes. The main difference is that the previous one [15] is based on AR(Autoregression model) while the other [8] adopts ARIMA models. Although time series have been applied to wireless sensor networks, they are just utilized to minimize energy consumption for data collection, not applied to top-k query processing. In this paper, based on quantile filter method, we propose a new top-k monitoring algorithm with time series forecasting models, especially ARIMA models, called

QFBP algorithms, which reduces the communication cost when getting thresholds and minimizes the energy consumption in an efficient manner.

3 Top-k and Quantile Filter Based Algorithms

3.1 Preliminaries

Fig.1 is a system architecture of wireless sensor networks which consists of many sensor nodes. Each node in the network transmits the sensed value to the base station by cooperating with others nodes. Base station has enough energy while sensor nodes are powered by batteries and energy-limited. When the base station is beyond a sensor node's radio coverage, data are transmitted to base station by other nodes through multiple jumps, otherwise, data are sent to base station directly. We think that the path of the data transmission is viewed as a routing tree [17].

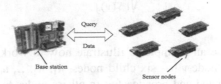

Fig. 1. The system architecture

In the wireless sensor network given as Fig.1, we view the *id* of a node and its sensed value as a point, assume that $P(v_i)$ is the set of points generated at sensor v_i, then $P=\cup_{i=1}^{N}P(v_i)$ is the point set of the whole sensor networks. Top-k query is to return k ($1 \leq k \leq N$) points with highest generated readings in P, if the length of results is larger than k, k of them are selected as the results according to the sensing attributes.

3.2 Overview of QF

1. QF Algorithm

The QF algorithm proposed by Chen et al.[5] for top-k query mainly has three phases: first, each sensor node sends its quantile filter value to parent node after sorting its points in decreasing order of sensed values; second, parent node selects one of the received quantile values as filter, the chosen filter is called the threshold Q_{filter}, then broadcasts the filter to all children; third, the children send those points whose values are no less than Q_{filter} to their parents.

In the following, we will illustrate the determination of threshold Q_{filter}. Consider a sensor v in the routing tree with d_v children u_i, $1 \leq i \leq d_v$. Assume that $L(u_i)=\{p(i)_1,\ldots, p(i)_{l(i)}\}$ is a points set at u_i, and, if $1 \leq j_1 \leq j_2 \leq l(i)$, then $p(i)_{j_1}.val \geq p(i)_{j_2}.val$. When $l(i)<k$, $l(i)$ is the number of the points at u_i, otherwise, $l(i)=k$. That is, there are at most k points in points set $L(u_i)$. In given $L(u_i)$, the α-quantile is referred to the point $p(i)_{\lceil \alpha|L(u_i)|\rceil}$ whose rank is $\lceil \alpha|L(u_i)|\rceil$. We view $p(i)_{\lceil \alpha|L(u_i)|\rceil}.val$ as the α-quantile of $L(u_i)$,

$0<\alpha<1$. Then, $S(v)=\cup_{i=1}^{d_v}L(u_i)$ is the potential top-k points set from children of v_i, and the length of $S(v)$ is satisfied with: $|S(v)| \leq k*d_v$. Each chileren u_i sends values to its parent: $p(i)\lceil_{\alpha*l(i)}\rceil.val$ and $l(i)$, $1\leq i\leq d_v$. parent v receives d_v quantile filter and sorts them in decreasing order. Assume that the sequence is $q_{i_1}, q_{i_2},..., q_{i_{d_v}}$ after sorting. $q_{i_j}(=p(i_j)$ $\lceil_{\alpha*l(i_j)}\rceil.val)$ is sent by u_{i_j}. If m is found such that $\sum_{t=1}^{m-1} \alpha*l(i_t)<k$, $\sum_{t=1}^{m}\alpha*l(i_t)\geq k$, parent node chooses m_{th} quantile filter as threshold Q_{filter} which ensures that there are at least k point in $S(v)$.

As shown from above, because there are at least k points whose values are larger than Q_{filter}, other points in children nodes are impossible to be in result set and can be filtered out. Compared with Naive-k, quantile filter approach can suppress some points and save energy. However, performance is affected by the value of α, QF algorithm [5] introduced how to get optimal α. Limited by space of article, we do not discuss how to get optimal value of α and give the formula of α directly:

$$\alpha = \sqrt{\frac{k-1}{|S(v)|}} \tag{1}$$

2. Example
In the following, we use an example to illustrate how to work of the algorithm. As shown in Fig.2, sensor node v has six child nodes, $u_1, u_2,..., u_6$, each u_i has a points set. Top-9 query is initiated and broadcasted to others nodes by the base station. Because $k=9$ and $|S(v)|=30$, we can get that $\alpha = \sqrt{\dfrac{k-1}{|S(v)|}}$ =0.516 according to formula(1).

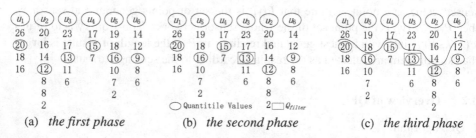

(a) *the first phase*   (b) *the second phase*   (c) *the third phase*

Fig. 2. An example of algorithm Quantile-Filter

In the first phase, The nodes sort their values and send their α-quantile and $l(i)$ to v. In the second phase, v sorts the received quantile values in decreasing order, the sequence after sorting is that $q_{i1=1}(=20)$, $q_{i2=5}(=16)$, $q_{i3=4}(=15)$, $q_{i4=3}(=13)$, $q_{i5=2}(=12)$, $q_{i6=6}(=9)$. We choose $q_{i4=3}(=13)$ as threshold Q_{filter}, because that $0.516*l(1)$ + $0.516*l(5)$ + $0.516*l(4)$+ $0.516*l(3)$=9.28$\geq k$. v broadcasts q_{i4} to other nodes. In the third phase, each u_i sends points whose values are larger than Q_{filter} to v.

3.3   The Shortcoming of QF

Seen from QF algorithm, the threshold was determined after sorting all quantile filters from children, that is, it is one of the quantile filter values which are from their children as shown in Fig.2, the determination of threshold $q_{i4}(=13)$ relies on six messages from all children. We can see that the amount of saved energy relies on the interaction between children and parent nodes. However, the frequent interaction will incur extra communication cost. So, we propose a top-k query processing algorithm in which threshold is predicted by historical thresholds and not relies on quantile values from children. When there is an error between actual value and predicted value which is not satisfied by users, predicted model is relearned by noticing children to send quantile values.

4   The Proposed Method

In this paper, since threshold is actually the characteristic values such as temperature, humidity and light which are sensed by sensor nodes and the sequence of thresholds is satisfied with the time series. In addition, considering the limitation in computational and storage requirements of sensor nodes, we adopt time series predicted model called ARIMA. Besides above advantages, ARIMA is flexible and suitable to many applications in which data satisfies different distributions.

4.1   ARIMA Model

ARIMA is a time series predicted model which forecasts the next value according to historical data. Nowadays, the ARIMA model is used for univariate time series widely which consists of three components: AR (Auto Regressive), the integrated and MA (Moving Average). AR and MR can be combined into ARMA, and ARIMA model is actually ARMA model after d differencing. If there is a time series Z_t, then the value at time t can be predicted by ARMA (p, q), p represents the number of historical data, q represents the number of latest random shocks on time series. ARMA is showed below:

$$Z_t = \varphi_0 + \sum_{i=1}^{p} \varphi_i Z_{t-i} + a_t - \sum_{i=1}^{q} \theta_i a_{t-i} \qquad (2)$$

In general, ARMA model usually supposed that data are stationary, that is, the statistical properties of data do not change over time. In the equation of (2), the parameters of the model are $\varphi_i(1 \le i \le p)$ in autoregressive and $\theta_i(1 \le i \le q)$ are parameters in moving average.

4.2   Model-Based Prediction

Prediction by ARIMA model usually consists of several aspects:
1. Model identification and parameter estimation
Each parent node builds an ARIMA model to predict the next threshold Q_{filter}. In order to predict, parent v node maintains enough thresholds, $L_i = \{<t_1, z_{i,1}>, <t_2, z_{i,2}>, \ldots, <t_T, z_{i,T}>\}$, where t_j $(1 \le j \le T)$ is an epoch number and $z_{i,j}$ is the sensor reading of v_i at epoch t_j. ARIMA model on node v_i is built based on L_i. To simplify the process, suppose $d=0$, that is, the time series of L_i is stationary and is satisfied with ARMA (p, q)

model. In order to obtain p and q, we introduce functions of autocorrelation (ACF) and partial autocorrelation (PACF). Autocorrelation describes the simple correlation between values in time series. The correlation coefficient between Z_t and Z_{t-1} is called the lag-l autocorrelation of Z_t and is commonly denoted by γ_l. The computation is shown as below:

$$\gamma_l = \frac{\sum_{t=1}^{T-l}(Z_t - \overline{Z})(Z_{t+l} - \overline{Z})}{\sum_{t=1}^{T}(Z_t - \overline{Z})^2} \tag{3}$$

T is the number of samples; l is the distance of interval; Z is used to estimate the expectation of time series which describes the average value in arithmetic.

PACF describes the conditional correlation between Z_t and Z_{t-1} when given time series of that $Z_{t-1}, Z_{t-2},\ldots, Z_{t-l+1}$. The degree of correlation is measured by φ_{ll} and is estimated by partial autocorrelation parameters.

$$\varphi = \begin{cases} \gamma_1 & l=1 \\ \dfrac{\gamma_l - \sum_{j=1}^{l-1}\varphi_{l-1,j}\gamma_{l-j}}{1 - \sum_{j=1}^{l-1}\varphi_{l-1,j}\gamma_j} & l=2,3,\cdots \end{cases} \tag{4}$$

We can get the estimated values of p and q according to PACF and ACF. That is, the sample PACF cuts off at lag p; as for q, we can get it by ACF. For a time series Z_t with γ_l, if $\gamma_q \neq 0$ and $\gamma_l = 0$ when $l > q$, then, the order of MA is q. However, we just get the estimated p and q, they can be decided by information criteria called AIC (Akaike information criterion). AIC criteria finds the minimum p^* and q^* to minimize the value of AIC. We view as p^* and q^* as the optimal estimated values of p and q. The computation of AIC is shown as equation (5). T is the sample size. $\hat{\sigma}^2$ is the maximum-likelihood estimate of σ^2.

$$AIC = \ln \hat{\sigma}^2 + \frac{2(p+q)}{T} \tag{5}$$

$$\hat{\sigma}^2 = \frac{\sum_{i=c+1}^{T}(Z_t - \sum_{j=1}^{c}\varphi_{jj} \times Z_{i-j})^2}{T-c} \tag{6}$$

In the equation (6), c represents the number of parameters, $c = p+q$. After order determination of ARMA model, we estimate parameters $(\varphi_0, \varphi_1,\ldots, \varphi_k)$ by the conditional least-squares method. Now, we can predict the next value of threshold by ARIMA model. In this procedure, the order of model would not be changed, but the parameters of model is self-learning while processing query which ensures that prediction error is acceptable.

2. Forecasting

As illustrated by (1) in Section 4.2, suppose that we are at time index h, Z is a time series formed by previous index of h, 1-step-ahead forecasted value of time series Z, is v^{pre}.

$$v^{pre} = E(Z_{h+1} \mid Z_h, Z_{h-1},\ldots,Z_1) = \varphi_0 + \sum_{m=1}^{p}\varphi_m Z_{h+1-m} - \sum_{m=1}^{q}\theta_m a_{h+1-m} \tag{7}$$

4.3   Relearning of Model

Our QFBP algorithm includes two parts: Algorithm 1, function Child-sensor() and Algorithm 2, function Parent-sensor. Once acquiring threshold Q_{filter} of the next time, if a top-k query is needed, QFBP algorithm does not depend on the interaction of child and parent nodes which reduces the energy consumption and saves energy. Due to that there exists accumulated predication errors, when the amount of received points on parents is larger than k obviously, we think that the model is unsuitable to predict and need to be relearned. Then, parent notifies the children to send quantlile filters which are used to relearn model parameters. In this algorithm, we measure the performance of prediction according to the number of received points on parent nodes, once num is larger than k obviously, we think that prediction error is beyond the limitation we can accept. After receiving the quantile filters sent by children, ARIMA model adjusts its parameters to adapt to the change of data distribution.

Algorithm 1 Algorithm QFBP: *Child-sensor()*

1: *receive(Query(k))*;
2: *compute* $|S(v)|$ and α;
3: **if** $\alpha \leq 0$ or $\alpha \geq 1$ or $E_{filter}(S(v)) < E_{Naive}(S(v))$ or $E_{filter}(L(u_i)) < E_{naive}(L(u_i))$ **then**
4:     sends top-$l(i)$ to parent node v
5: **else**
6:     **if** receive *Request_Quantile* from v **then**
7:                  **send**$(p(i)_{\lceil \alpha * l(i) \rceil}.val, l(i))$;
8:     **else**
9:        **if** receive Q_{filter} from v **then**
10:        send the points whose values are no less than Q_{filter}
11:        **end if**
12:     **end if**
13: **end if**

Algorithm 2 Algorithm QFBP: *Parent-sensor()*

1: *receive(Query(k))*;
2: *compute* $|S(v)| = \cup_{i=1}^{d_i} L(u_i)$ and α;
3: receive the values from its children
4: **if** $num > 4*k$ **then**
5:     sends *RequestQuantile* to *children*;
6: **else**
7:     predict Q_{filter} by model named ARIMA
8:     broadcast the Q_{filter} to *children*
9: **end if**
10: **if** receive all the points from *children* **then**
11:     compute top-k points from the received points and v's own point;
12: **end if**

In our QFBP algorithm, $E_{naive}(S(v))$ and $E_{filter}(S(v))$ represent the total energy consumption in different situations whether installing threshold; $E_{naive}(L(u_i))$ and $E_{filter}(L(u_i))$ represent the energy consumption on each node that installed the threshold or not; In order to save more energy after installing the threshold, there is $E_{filter}(S(v)) < E_{naive}(S(v))$, $E_{filter}(L(u_i)) < E_{naive}(L(u_i))$. At the same time, the optimal quantile of α should be larger than 0, so, $|S(v)| \geq d_v/(1-\alpha)+(k-1)/\alpha$.

5   Experiments

5.1   Experiment Setting

We evaluate the performance of the proposed algorithm QFBP based on MATLAB simulation. The experimental data is from the Intel Lab [1]. We use *temperature*, *humidity* and *light* data on March 1*st*, 2004 as experimental data. The data were collected from 54 sensor nodes every 31 *seconds*. In the paper, we treat them as data collected about every 1 *min.*, that is we view the average of two sampling values as one.

We demonstrate the effectiveness of the proposed algorithm QFBP compared with QF [5]. The number of communication in a query is the measurement under different specified k. Considering the difference in energy consumption between sending and receiving, we simulate Mica2Dot sensor node which the message energy consumption of sending is 0.37 times of receiving.

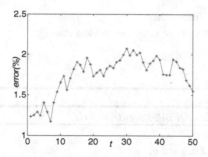

Fig. 3. Prediction error

From Fig.3, the relative error of 50-steps-head prediction is below 2%. In addition, we adopt 1-step-head prediction in this paper, that is, max distance between true value and predicted value is below 0.015. So we think that prediction using ARIMA models is suitable for our proposed method.

5.2   Experiment Results

Fig.4 and Fig.5 are about the performance of the two algorithms on different datasets. Fig.4 is about the comparison of total communications in different k. Fig.5 is about the times of sending data under different *k*. As shown from the comparison of two algorithms for different datasets: no matter what distribution the data is, the performance of proposed QFBP algorithm is better than existing QF algorithms.

Take Fig.4 (b) as an example, when *k*=19, QFBP algorithm is 3000 communications less than QF algorithm. It is mainly caused by that QF algorithm needs children sensor nodes sending many quantile filters to their parents which incurs extra communication cost. However, the proposed algorithm just acquires thresholds by predicting until the prediction error is not satisfied with algorithm's demand. Also, from Fig.5 (b), we can see that when *k*=19, QFBP is 2000 communications less than QF algorithm.

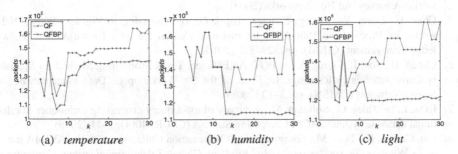

(a) *temperature*   (b) *humidity*   (c) *light*

Fig. 4. Performance of different algorithm on different data

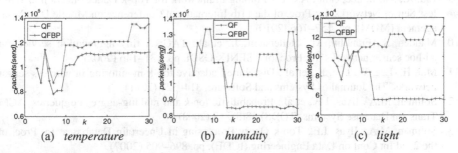

(a) *temperature*   (b) *humidity*   (c) *light*

Fig. 5. The number of packets of sending

6   Conclusion

In this paper, we proposed a new top-*k* query algorithm QFBP based on prediction which predicts the next threshold based on ARIMA models according to historical data. Though QFBP algorithm needs much computing resources when learning ARIMA models in energy-enough base station, it can avoid the interaction when getting the threshold between children and parent which decreases the energy consumption and prolong the lifetime of sensor networks. Comparison analysis demonstrates the correctness and effectiveness of our proposed QFBF algorithm.

References

1. Intel Berkeley Research Lab,
   http://www.select.cs.cmu.edu/data/labapp3/index.html
2. Abbasi, A., Khonsari, A., Farri, N.: MOTE: Efficient Monitoring of Top-k Set in Sensor Networks. In: IEEE Symposium on Computers and Communications (ISCC), pp. 957–962 (2008)
3. Akyildiz, I., Su, W., Sankarasubramaniam, Y., et al.: Wireless sensor networks: a survey. The International Joural of Computer and Telecommunications Networking 38(4), 393–422 (2002)
4. Anastasi, G., Conti, M., Francesco, M., et al.: Energy conservation in wireless sensor networks: A survey. Ad Hoc Networks (2009)
5. Chen, B., Liang, W.: Energy-Efficient Top-k Query Processing in Wireless Sensor Networks. In: Proc. of the 19th ACM International Conference on Information and Knowledge Management (CIKM), pp. 329–338 (2010)
6. Cho, Y.H., Son, J., Chung, Y.D.: POT: An Efficient Top-k Monitoring Method for Spatially Correlated Sensor Readings. In: Proc. of the 5th Workshop on Data Management for Sensor Networks (DMSN), pp. 8–13 (2008)
7. Iiyas, I., Beskales, G., Soliman, M.: A survey of top-k query processing techniques in relational database systems. ACM Computing Surveys (CSUR) 40(4), 1–11 (2008)
8. Liu, C., Wu, K., Tsao, M.: Energy Efficient Information Collection with the ARIMA model in Wireless Sensor Networks. In: Proc. of Global Telecommunications Conference, pp. 2470–2474. IEEE (2005)
9. Liu, X., Xu, J., Lee, et al.: A Cross Pruning Framework for Top-k Data Collection in Wireless Sensor Networks. In: Proc. of the 11th International Conference on Mobile Data Management (MDM), pp. 157–166 (2010)
10. Madden, S., Franklin, M., Hellerstein, J., et al.: TAG: A tiny aggregation service for ad-hoc sensor networks. In: Proc. of USENIX OSDI, pp. 131–146 (2002)
11. Mai, H., Lee, Y., Lee, K., et al.: Distributed adaptive top-k monitoring in wireless sensor networks. The Journal of Systems and Software, 314–327 (2011)
12. Soliman, M.A., Ilyas, I.F., et al.: Probabilistic top-k and ranking-aggregate queries. ACM Trans. on Database Systems (TODS) 33(3), 13 (2008)
13. Soliman, M.A., Ilyas, I.F.: Top-k Query Processing in Uncertain Databases. In: Proc. of the 23nd Int Conf on Data Engineering (ICDE), pp. 896–905 (2007)
14. Thanh, M., Lee, K., Lee, Y., et al.: Processing Top-k Monitoring Queries in Wireless Sensor Networks. In: Proc. of Third International Conference on Sensor Technologies and Applications, pp. 545–552. 545-552 (2009)
15. Tulone, D., Madden, S.: PAQ: Time series forecasting for approximate query answering in sensor networks. In: Römer, K., Karl, H., Mattern, F. (eds.) EWSN 2006. LNCS, vol. 3868, pp. 21–37. Springer, Heidelberg (2006)
16. Wu, M., Xu, J., Tang, X., et al.: Top-k Monitoring Top-k Query in Wireless Sensor Networks. IEEE Trans. on Knowledge and Data Engineering, 962–976 (2006)
17. Wu, M., Xu, J., Tang, X., et al.: Monitoring Top-k Query in Wireless Sensor Networks. In: Proc. of the 22nd International Conference on Data Engineering, ICDE (2006)

An Automatic Generation Strategy
for Test Cases Based on Constraints

Dandan He, Lijuan Wang, and Ruijie Liu

Dalian Institute of Science and Technology, Dalian, China
dlmu_ddh@163.com, znhy.wang@163.com, lrj951190@163.com

Abstract. We propose an automatic generation strategy for test cases. We use the Unified Modeling Language (UML) use cases to describe system requirements, as it can serve as a good basis of the test case generation. Meanwhile, in order to ensure that the test cases can automatically be generated, sequence diagram is used to describe the scene because of its superiority on message transferring between systems. A formal description of the sequence diagram is given, in order to achieve the automated generation of test cases. However, due to the characteristics of the use case description, the internal demands of the system are ignored, leading to inaccurate and rough test cases directly generated by use cases. In order to effectively control such adverse effects, we adopt contraction method that formal constraints demands described by the use cases and keep the constraints through the sequence diagram. This contraction method can not only enhance the test scene accuracy, but also ensure the final generated test cases more accurate with more detailed coverage.

Keywords: Test case, Test Scenario, Sequence Diagram.

1 Introduction

Generating and executing test cases is the core task for software testing which is a crucial step to ensure the software in good quality. During the case generation process, it is required that each test case is designed for one function and should be easy to read, and the fine execute granularity of the test case is preferred. The testing steps should be clear, and ambiguous words should be avoided in the comments of the test cases.

The Unified Modeling Language (UML) use cases are often used to describe the system requirements, making them more concise and more easily understood by the users. The use cases can drive the generation of test cases, and reduce the cost of test, even used for retrospective validation of the requirements. However, as the demands are frequently changed and the demand descriptions by use cases tend to ignore the internal demand, test cases will be inaccurate and poorly traceable. Therefore, how to automatically generate test cases to make them accurate with high coverage is essential for the study.

The contribution of this paper is a test case generation strategy based on the use case contract, using light weight formal idea in use case analysis phase, i.e. formal contract constraints are given after transforming the demand into use cases. Based on our

D.-S. Huang et al. (Eds.): ICIC 2013, LNAI 7996, pp. 393–400, 2013.

method, the consistency and correctness of the use cases can be checked and all valid sequences of the use cases are explicitly constructed as a model. As a result, the vague semantics is overcome which may arise in the process of automatically generated test cases using natural language to describe demands, and the traceability of .test cases from use cases is enhanced.

2   Relative Works

The use case is a good basis for system testing. In software development and systems engineering, a use case is a behavioral sequence, which describe the interaction between the system and the actor to obtain a goal for the actor with an observable and valuable result. The relationship between all the use cases or use case sets and the actors is denoted by a graphical way in UML, called use case diagram.

Although direct generating test cases from use cases have many advantages, the drawbacks should not be underestimated such as mechanical, rough, imprecise. This paper introduces the idea of contract designed by Meyer to constrain the demand.

The basic idea of contract designed by Meyer is known as contract programming, the programming of the contract or design by contract programming, which is a software design method [1, 2]. Its essence is the standardization of the software component interfaces for each developer in the software development process, i.e. formalized, precise and verifiable interface constraints must be defined. This idea can be further extended to the abstract datatype standardization, by adding pre-conditions, post-conditions and invariants. These constraints are called"contract".

Inspired by the ideas of Meyer's contract design, we add the "contracts" (i.e. constraints) before the use cases are executed [1]. The mathematical logic is used in the constraints of the use cases, which is combined with first-order logic expressions and the predicate with logical operators, as it is the theoretical basis of Meyer's contract. Therefore, the contract will be involved in the FORALL (universal quantifier), EXISTS (existential quantifier), Boolean logic AND (conjunctive), OR (disjunction), NOT (negation) and→ (logical implication) [1].

Besides using use cases to describe demands as the basis of our proposed strategy, we need to get more accurate test objectives (test scenarios) in order to generate test cases. The UML sequence diagram is adopted as the basis for the test scenarios in our paper, which is one of the scene representation methods widely used in the software development practice. It is mainly to describe the communication between the system components and the environment, and process the test case generation and execution [3-7]. The basic idea is to find the expected behaviors from the sequence diagram firstly, and then get the conditions that the input information must be met in order to achieve these behaviors, which in turn to generate test cases to test the system for the actual behavior of the system. Meanwhile, sequence diagram contains more information than use cases, because it contains more model elements with more accurate description than use cases. Therefore, using the sequence diagram to describe the scene just reflects both the characteristics of the sequence diagram and the scenes, meaning that with the timing of a series of messages, the use case behaviors of the sequence are shown and the functional requirements of the system can be completely indicated.

3   Test Case Generation

Our test case generation strategy is first to use cases contract, then get the test scenarios from the sequence diagram, and finally generate test cases.

3.1   Use Case Contract

3.1.1   Use Case Parameterization

Use case parameterization is a prelude to the formal use cases, in which the parameters can be actors, or some key concepts that may be embodied in the design process in the application. The parameters from a use case are often used in the scenarios they affiliated, and the actor involved in the use cases is a specific parameter [3]. The input and output of the use case are defined as parameters which will be used as the part of the use case contract, and all types of parameters are set as the enumerated type.

The parameters of a use case can be defined using UC as follows:

```
UC Usecase_name(use case parameterlist)
```

3.1.2   Use Case Contract Construction

When use case parameters are determined, the pre-conditions and post-conditions and invariants can be used to add contracts in use cases. The use case contracts are denoted in the form of predicate logic expressions combining predicates and logical operators, attached to each use case as UML comments in order to accurately depict the semantics of use cases. The use case contract can be used as the basis to construct test cases and determine test adequacy. The use case with contracts is defined as the following format:

```
UC Usecase_Name(Parameter List)
PRE PreCondition Logical Expression
POST PostCondition Logical Expression
```

3.2   Determine the Test Scenario

When the use cases are formalized, we need to analyze the use cases to obtain the corresponding situations, i.e. the sequence of events (also called stream of events), and use the sequence diagram to describe in order to determine the test scenario. The test scenario is actually a sequence of messages which are important references for test case generation [7-9]. Therefore, the generation of the test scenario is also called the test message sequence generation. The order of each message in the test scenario is determined by the sequential order of the sending events, and a valid events-sending queue can determine the corresponding test scenario. In order to get the test scenarios, we need to formally describe the sequence diagram which is defined as a quintuple group(O, M,E, $\rightarrow$,<), where O ={O1, O2,···, Om} is the object set in the sequence diagram, M is the collection of messages, E = M {s, r} is an event collection, in which an event is the sending or receiving of a message, $\rightarrow$ is a total ordering relation on the message set M, and < shows the orders of the messages on the longitudinal timeline in the sequence diagram[10-13]. The test scenarios generation using the use case is the core task to complete, and then the corresponding algorithm is essentially traversal on a directed graph [8]. After traversing all the events by breadth-first search or depth-first search method, we can export all of the scenarios in the sequence diagram.

3.3   Test Case Generation

Through the above prepared work, the test case generation algorithm can be given as follows:

(i)Traverse every object in the sequence diagram, extract the state information and add them to the sequence diagram.

Check each object in the sequence diagram, if it owns a state diagram, then find out all of the initial states of the object; for each initial state, use the format "object name. in (state name) "to place it to the appropriate position on the object's life line before this object receives a message; multiple initial states are connected by‖(logical or) operator .If an object stays in two states, the two states will be joined in the sequence diagram;

(ii)Traversing the sequence of messages in the sequence diagram in order to identify all the scenarios [14];

(iii)Select a scenario to traverse in the sequence diagram based on the sequence of messages, and AND operator is used to connect various constraints. The generated constraint expressions is simplified, in which if OR operators exist in the constraints, the constraint conditions will be reduced into multiple constraints without OR operator. The constraint condition of the scenario, the input and final output are recorded.

(iv)Determine the environmental conditions of each scenario. First, all of the test units are determined from the sequence diagram, wherein each interactive object is a test unit. Then each category is divided into a series of choices according to some constraint information in the sequence diagram, and the choices associated with each scenario are be combined together to construct the environmental conditions of the scenario.

As a result, the test case includes four parts: environmental conditions, input, function calling sequence and expected output, all of which can be generated by the above four steps. Combining them together, a test case is generated.

4   Explored Instances

We use a simple ATM system to show a detailed description of our test case generation process. The simplified ATM machine only has withdrawal, deposit and other services functions which will not be in detailed description. Actors include the bank, the client and the ATM machine. The use case diagram is shown in Fig.1.

Using the contract, we give the use cases withdrawal and PIN_Authentication in a formalized description as follows:

```
UC withdrawal (anum:accountNum, am:amount)
PRE insertCard (anum) and validateCard(anum) and
PIN_Authentication (anum) and Balance(anum, am)
POST withdrawaled (anum)
```

```
UC PIN_Authentication (anum:accountNum,p:password)
PRE insertCard (anum) and validateCard(anum) and NOT
passwordGiven=true
POST passwordGiven=true and password=p
```

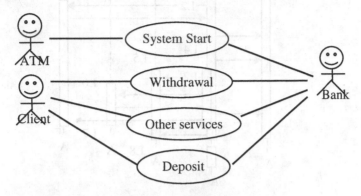

Fig. 1. ATM Test case

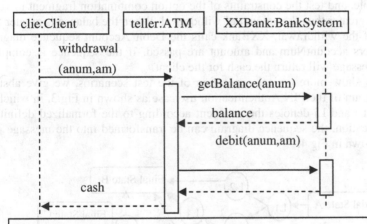

noConflicts<<PRECONDITION>>:
{list_clie→forALL(clie→include(anum,am)⇒debit
(anum,am))∧getBalance (anum,am)

withdrawal<<POSTCONDITION>>:
{let mtg:getPIN(aum,p) and mtg.updateBalance(anum,am) and
list_clie→forALL(clie→include(anum,am)⇒debit
(anum,am))∧getBalance (anum,am) withdrawal(anum)=true

Fig. 2. The withdrawal sequence

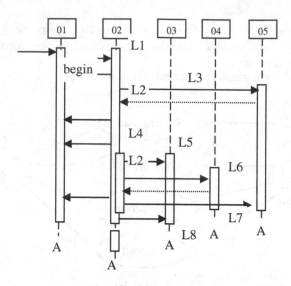

Fig. 3. The abstract sequence diagram of PIN_Authentication

Fig. 2 shows the withdrawal sequence, which is starting from the upper left corner, and the sequence of messages must meet certain constraints before passing. The client sends a message to the teller object, who subsequently pass the message to XXBank object. Then, XXBank calls the Balance Lookup sequence diagram, in which accountNum is passed as a parameter. Balance Lookup sequence diagram returns the balance variable, and test the constraints of the option combination fragment to verify the balance is greater than the amount of withdrawal. When the balance is greater than the amount of the withdrawal, XXBank calls the Debit Account sequence diagram, with parameters accountNum and amount are passed. If the sequence is complete, withdrawal message will return the cash for the client.

In order to show more typical generation of the test scenarios, we give abstract sequence diagram of the PIN Authentication use case as shown in Fig.3, in which Oi denotes object i and Lj denotes the j-thevent according to the formalized definition. After simplification, the sequence diagram can be transformed into the message path diagram as shown in Fig.4.

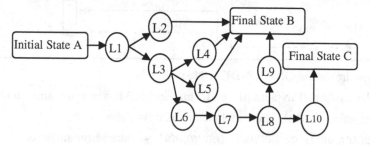

Fig. 4. Message path diagram of Fig.3

Fig. 4 Contains the initial state A, final state B and C, and a series of message events. All the message sequences in the sequence diagram are described in a very straight forward pattern; each node in the diagram are mapped to the nteraction between two objects through a sequence of messages, and the node stores the necessary information for the interaction, which in turn constructs the foundation test case generation. Finally, according to the traversal algorithm, we can obtain the test scene table as shown in Table 1.

Table 1. Test scenario tables

Test Scenario	Description
TSC1	A→L1→L2→B
TSC2	A→L1→L3→L4→B
TSC3	A→L1→L3→L5→B
TSC4	A→L1→L3→L6→L7→L8→L9→B
TSC5	A→L1→L3→L6→L7→L8→L10→C

Finally, we use the test case generation algorithm in section 3 and traverse the formalized sequence diagram to generate test cases.

5 Conclusion

The final target in software testing process is to generate test cases automatically and reduce the errors generated by manual testing. Based on predicate logic and set theory, we use the contract of the use case to make the formal specification of the generation process from use cases to test cases, and overcome the ambiguous semantics that may arise in the process of automatically generated test cases using natural language demands description. Combined with the formalized sequence diagram, test cases cannot only be automatically generated, but also with more enhanced traceability from use cases to test cases, making the final generate test cases more accurate and more detailed coverage.

In the future, we should study further formal test case generation method to explore the development of automatic test cases generated technology, and set up a test case automatic generation system for fully automated software testing.

References

1. Meyer, B.: Applying Design by Contract. IEEE Computer (1992)
2. Klaeren, H., Pulvermüller, E., Rashid, A., Speck, A.: Aspect composition applying the design by contract principle. In: Butler, G., Jarzabek, S. (eds.) GCSE 2000. LNCS, vol. 2177, pp. 57–69. Springer, Heidelberg (2001)

3. Hanna, S., Munro, M.: An Approach for Specification-based Test Case Generation for Web Services. In: 2007. IEEE/ACS International Conference on Computer Systems and Applications, pp. 16–23. IEEE Press, Amman (2007)
4. Fraiki, F., Leonhardt, T.: SeDiTeC-testing based on sequence diagrams. In: 17th IEEE International Conference on Automated Software Engineering, pp. 261–266. IEEE Press (2002)
5. Tsiolakis, A.: Semantic Analysis and Consistency Checking of UML Sequence Diagrams, Diplomm'beit, TU-Berlin (2001)
6. Monalisa, S., Debasish, K., Rajib, M.: Automatic Test Case Generation from UML Sequence Diagram. In: ADCOM 2007, International Conference on Advanced Computing and Communications, pp. 60–67. IEEE Press, Guwahati (2007)
7. Gnesi, S., Latella, D., Massink, M.: Formal Test-case Generation for UML Statecharts. In: Ninth IEEE International Conference on Engineering Complex Computer Systems, pp. 75–84. IEEE Press (2004)
8. Cartaxo, E.G., Neto, F.G.O., Machado, P.L.: Test Case Generation by means of UML Sequence Diagrams and Labeled Transition Systems. In: IEEE International Conference on Systems, Man and Cybernetics, pp. 1292–1297. IEEE Press, Montreal (2007)
9. Kuball, S., Hughes, G., Gilchrist, I.: Scenario-Based Unit Testing For Reliability. In: Annual on Reliability and Maintainability Symposium, pp. 222–227. IEEE Press, Seattle (2002)
10. Baker, P., Bristow, P., Jervis, C., King, D., Mitchell, B.: Automatic generation of conformance tests from message sequence charts. In: Sherratt, E. (ed.) SAM 2002. LNCS, vol. 2599, pp. 170–198. Springer, Heidelberg (2003)
11. Jorgensen, P.: Software Testing: A Craftsman's Approach. CRC Press (2002)
12. Kaschner, K., Lohmann, N.: Automatic Test Case Generation for Interacting Services. In: Feuerlicht, G., Lamersdorf, W. (eds.) ICSOC 2008. LNCS, vol. 5472, pp. 66–78. Springer, Heidelberg (2009)
13. Bai, X.Y., Dong, W.L., Tsai, W.T., Chen, Y.N.: WSDL-based automatic test case generation for Web services testing. In: IEEE International Conference on Service-Oriented System Engineering, pp. 207–212. IEEE Press (2005)
14. Beyer, M., Dulz, W., Zhen, F.: Automated TTCN-3 test case generation by means of UML sequence diagrams and Markov chains. In: 12th Asian Conference on Test Symposium, pp. 102–105 (2003)

Nonlinear Dynamic Analysis of Pathological Voices

Fang Chunying[1,2], Li Haifeng[1], Ma Lin[1], and Zhang Xiaopeng[1]

[1] School of Computer Science and Technology, Harbin Institute of Technology, Harbin, China
{Lihaifeng,malin_li}@hit.edu.cn, zhangxiaopeng6@126.com
[2] School of Computer and Information Engineering,
Heilongjiang Institute of Science and Technology, Harbin, China
fcy3333@163.com

Abstract. Research on the human health evaluation through sound analysis is now attracting more and more researchers in the world. Acoustic analysis could be a useful tool to diagnose the disease. Therefore, pathological voices can be used to evaluate the health status as a complementary technique, such as bronchitis. In this article, we proposed a nonlinear dynamic method to analysis pathological voices. Firstly, pathological voices were preprocessed and numerous features were extracted. Secondly, a binary coded chromosome genetic algorithm (GA) was applied as feature selection method to optimize feature descriptor set. The experimental results show that GA, PCA along with support vector machine (SVM) has the best performance in the pathology voices diagnosis.

Keywords: GA, SVM, PCA, Pathology voice.

1 Introduction

The sound is mainly generated through lung, larynx, epiglottis, tongue, teeth, lips, nose and other organs [1-2]. Researches have shown that voice can be used as a objective diagnosis method to diagnose diseases [3-5], which mainly focused on detecting different pathological voice by speech signal processing from feature extraction to the classifiers, Godino-Lorente used the MFCC features, F ratio, Fisher's discriminated ratio, and Gaussian mixture model, to assess pathological voices. The classification accuracy of 94.07% is achieved [6]. Arjmandi identified voice disorders using long-time features and SVM with different feature reduction methods, the experimental results denoted that linear discriminate analysis (LDA) along with SVM got the best performance [7]. Nayak identified voice disorder by applying DTW to extract efficient features and neural network as classifier [8]. Fonseca identified the voice disorders by discrete wavelet transform, linear prediction coefficient, and least-square SVM [9]. MFCC features of the heart sound are extracted and dynamic time warping is used to identify heart sound [10]. Cohen extracted the linear prediction coefficients and energy envelope features to classify seven types of breath sounds [11]. Kenneth Anderson and N. Gavriely studied the sounds generated by breathing in asthma, or distinguished between breath sound from subjects suffering from various diseases or conditions [12-13].

D.-S. Huang et al. (Eds.): ICIC 2013, LNAI 7996, pp. 401–409, 2013.

At present, experiments demonstrate that pathological speech identification can be improved considerably by using suitable features set with efficient feature reduction method and an appropriate classifier. We attempt to find an optimum algorithm to deal with the extracted feature set. In the following parts of this article, three questions are attentioned. Firstly, we introduce our methods used in this paper. Secondly, the implementation of methods will be introduced. Thirdly, results are reported in the two corpuses. Finally, we summarize this paper.

2   Methodology of Pathological Voices Analysis

The proposed methods consist of five phases (see Fig.1), preprocessing, feature extraction, feature selection, feature reduction, with a classifier that voice finally is classified into health or pathology voice. In many pattern recognition applications, feature extraction method is critical to improve the recognition accuracy [14]. This research attempts to find an optimum algorithm of dealing with the pathology voices in increasing accuracy.

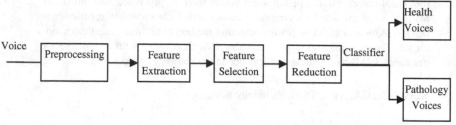

Fig. 1. The method's working flow

2.1   Preprocessing

In order to avoid high-frequency interval dropping, voice signal need to be pre-emphasized for further analysis. Pathological voice signal is z. The digital filter is shown as follow.

$$H(z) = 1 - uz^{-1}$$

(1)

Where, u=0.97. In this paper, the window function is Hamming function. Frame length is 32ms. Frame shift is 16ms.

2.2   Feature Extraction

We utilize TUM's open-source software openSMILE extract pathological features [19]. The whole feature set contains 6125-dimensional features including energy, spectral and voicing related low-level descriptors; a few LLDs are added including algorithmic harmonic-to-noise ratio(HNR),spectral harmonicity, and psychoacoustic spectral sharpness, a variety of functional are used as in [20].

2.3   Feature Selection

Feature selection algorithms divide into two categories based on whether or not they perform feature selection independently of the learning algorithm. If the technique performs feature selection independently of the learning algorithm, it follows a filter approach. Otherwise, it follows a wrapper approach [15]. GA can find the most efficient features of the whole space, Genetic algorithm (GA) has been demonstrated the most efficient feature selection method for learning areas and has less chance to get local optimal solution than other algorithms [16] [17] [18]. GA simulates Darwin's biological evolution to search the optimal solution. It is originally proposed by Professor J. Holland of the University of Michigan. Genetic algorithm means that a population may be the potential solution set of a problem. A population is composed of a certain number of gene encoding individuals. GA belongs to the wrapper approach. Therefore, classifier is very important and we use support vector machine (SVM) classifiers in this paper. The algorithm flow is shown in Figure 2.

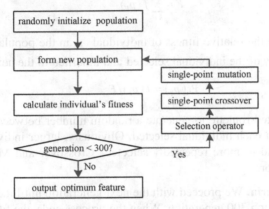

Fig. 2. The genetic algorithm flow

Encode Problem. Firstly, we must encode the problem. Here, we use a binary coded chromosome GA to select optimum feature descriptor subset for pathology voice. Each code, namely a chromosome, corresponds to a solution of the problem. Each gene on chromosome represents an input of the independent feature and its value can only be "0" or "1". If a gene's value is "1", it indicates that the corresponding feature is involved in the selected feature subset. On the contrary, the "0" indicates that the corresponding feature is not involved in the selected feature subset.

Initial Population. Let m as the number of feature descriptors, N is the size of population. Commonly, population size N is $20 < N < 100$ and is 60 in this paper. Chromosome of m genes is used to represent whether the corresponding feature is involved in the selected feature descriptor subset. In initial population $P = \{p_1, p_2, \cdots, p_N\}$, the genes of all individuals are randomly generated. Namely, each gene in a chromosome has value "0" or "1" randomly. **Fitness Function.** Classification accuracy of SVM classifier is used to evaluate the fitness of individuals. In detail, we use the reciprocal of sum of the test set's squared errors as the fitness function, it is quite straightforward to see that.

$$f(p_k) = 1 \bigg/ \sum_{i=1}^{n} (\hat{t}_i - t_i)^2 \tag{2}$$

Where $\hat{T} = \{\hat{t}_1, \hat{t}_2, \cdots, \hat{t}_n\}$ represent the predictive value of the test set classified by SVM using feature descriptor subset which p_k represents, $T = \{t_1, t_1, \cdots, t_n\}$ represent the true value of the test set, and n is the test data size.

Selection Operator. The selection operator determines an individual's genetic probability to the next generation population based on the individual's fitness. The processing is as follows:

Firstly, sum the fitness of all individual in the population:

$$F = \sum_{k=1}^{N} f(p_k) \tag{3}$$

Secondly, calculate the relative fitness of individual p_k in the population, which indicates the probability of the individual selected and inherited to the next generation:

$$\Pr(p_k) = f(p_k)/F \tag{4}$$

Finally, the simulated roulette to generate a random number between (0, 1), to determine the number of each individual selected. Obviously, larger individual's selection probability will lead to more repeatedly selected. Crossover and Mutation use both single-point operator.

Termination Criteria. We proceed with the next generation until the process reaches the maximum iteration 300 generation. When the process ends, the fittest individual is output as the optimum feature selection result.

2.4   Feature Reduction

In this paper, Principal Component Analysis(PCA) is experimented as feature reduction methods to achieve better optimum feature set. PCA is a technique that is widely used for applications such as dimensionality reduction, glossy data compression, feature extraction, and data visualization. PCA can be defined as the orthogonal projection of the data onto a lower dimensional linear space, known as the principal subspace, such that the variance of the projected data is maximized. The PCA algorithm flow is shown as follow.

1. Compute covariance matrix of the whole sample data X.

2. Compute eigenvalue λ_i and eigenvector e_i, according to the equation: $Se_i = \lambda_i e_i$.

3. Rank λ_i is in descending order as $\lambda_1 \geq \lambda_2 \geq \cdots \geq \lambda_n$,select d' eigenvectors $\Phi = (e_1, e_2, \cdots, e_{d'})$ corresponding to top d' eigenvalues according to the contri-

bution rate: $p = \dfrac{\sum\limits_{k=1}^{d'} \lambda_k}{\sum\limits_{k=1}^{d} \lambda_k} \geq 97.9\%$.

4. Project initial sample data into low dimension PCA subspace using matrix Φ as $Y = \Phi^T \cdot X$.

3   Experiments

SVM follows a procedure to find the separating hyperplane with the largest margin between two classes. It is based on statistical learning theory. The open source LibSVM is used to realize the SVM classifier.

3.1   Corpus

PVC.
PVC(pathology voice corpus) contains recordings in First Affiliated Hospital of Heilongjiang University of Chinese Medicine, which is pathology voice from two to six years old children including cold cough, cry ,breath, etc. In order to evaluate the proposed methods, normal breath sound is captured after three weeks when they are cure.

NCSC.
The "NKI CCRT Speech Corpus" (NCSC) recorded at the Department of Head and Neck Oncology and Surgery of the Netherlands Cancer Institute as described in [8]. The corpus contains recordings and perceptual evaluations of 55 speakers (10 female, 45 male) who underwent concomitant chemo-radiation treatment (CCRT) for inoperable tumors of the head and neck. Recordings and evaluations in the corpus were made before and after CCRT. The original samples were segmented at the sentence boundaries. Samples are discretized into binary class labels (I-intelligible, NI-nonintelligible), as shown in Table1.

Table 1. Partitioning of NCSC

NCSC	Train	Devel
I	384	341
NI	517	405

3.2   Performance Measures

In this paper, several measures are used to evaluate performance of different methods. Confusion matrix is a table with two rows and two columns that reports the number of True Negatives, False Positives. These measures are described as follows in Table2.

Table 2. The confusion matrix of classification evaluation criteria

		Predicted Positive	Predicted Negative
Actual	Positive	True Positive(TP)	False Negative(FN)
Actual	Negative	False Positive(FP)	True Negative(TN)

The unweight average recall is used in the given case of two classes('I' and 'NI'), it is calculated as (Recall(I)+Recall(NI))/2. Where, ACC and Recall is as shown in formulas 5, 6:

$$ACC = \frac{(TP + TN)}{(TP + FN + FP + TN)} \tag{5}$$

$$Recall = \frac{(TP)}{(TP + FP)} \tag{6}$$

3.3   Feature Select Process

The 6125 features are selected to form the new feature subset by GA. Fig3 shows the classified results of different feature dimension. The horizontal axis stands for the logarithm of feature dimension; the vertical axis shows the accuracy rate of the different feature set. The curves represent the highest accuracy rate is got in one hundreds dimensional features. However, it is very clear that the accuracy rate can be further improved; we combined the every feature subset of ten times GA feature selection which is 756-dimensional features.

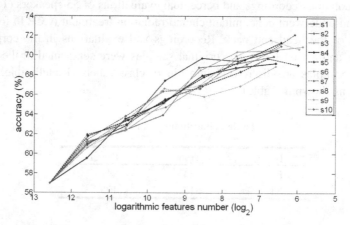

Fig. 3. The accuracy rates of the GA feature selection ten times

For GA, the accuracy rate constantly increases and then converges at some point. On the contrary, the accuracy rate declines for smaller feature numbers and greater feature numbers, indicating that more features lead to worse classification.

4   Results

4.1   The First Experiment Result

Feature selection is finished for ten times by GA in PVC. The ten optimized feature subset is consisted of the new feature set which is 756-dimension, Samples are discredited into binary class labels(H-Health, B-Bronchitis),the 30 samples are trained and tested, Results are got on the test set in Table 3.

Table 3. The confusion matrix of GA and SVM

	H	B	Results
H	12	3	80%
B	0	15	100%

Furthermore, feature reduction is finished by PCA, the feature subset is from 756 to 400 dimensions. Result is shown inTable 4.

Table 4. The classifier results after PCA

	H	B	Results
H	15	0	100%
B	0	15	100%

4.2   The Second Experiment

NCSC corpus is tested on the basis of the INTERSPEECH 2012.We train the models using the training set and test on the development set and results are as shown in Table 5; finally, unweight average recall (UA) is 75.5%.

Table 5. The classifier results after feature selection and reduction

	I	NI	Results
I	260	81	76.2%
NI	101	304	75.1%

Table 6. The comparision with baseline

	UA%
Our method	**75.5%**
Baseline(SVM)	61.4%
Baseline(Random Tree)	65.1%

Our results are compared with baseline results as shown in Table 6.

5   Conclusions

In this article, the strength of a kind of method to do with the high-dimensional feature set of pathology voices is extensively investigated by applying it to the NKI CCRT Speech Corpus. The optimized feature set is consisted of the ten subsets by GA. The results demonstrated that GA feature selection, PCA feature reduction and SVM classifier have established an adequate result of identified pathological voices. We got exclusive results obtained by application of a broad large of reveal algorithms. But many of the results are not satisfactory. i.e., LDA is used to feature reduction. The Unweight Average Recall is only 65.9%. Regarding further future research, this scheme has to be tested using pathology voices. Only a small tuning would be required in the features in the reduced space which physical meaning is not defined.

Acknowledgements. Our thanks to supports from the National Natural Science Foundation of China (61171186, 61271345), Key Laboratory Opening Funding of MOE-Microsoft Key Laboratory of Natural Language Processing and Speech (HIT.KLOF.20110xx), and the Fundamental Research Funds for the Central Universities (HIT.NSRIF.2012047). The authors are grateful for the anonymous reviewers who made constructive comments.

References

1. Fang, C.Y., Li, H.F., Ma, L., Hong, W.X.: Status and Development of Human Health Evaluation Based on Sound Analysis. In: First International Conference on Cellar Molecular Biology Biophysics and Bioengineering, p.66 (2010)
2. Fang, C.Y., Li, H.F.: Sound Analysis for Diagnosis of Children Health Based on MFCCE and GMM. In: International Review on Computers and Software, pp. 1153–1156 (2011)
3. Parsa, V., Jamieson, D.G.: Interactions Between Speech Coders and Disordered Speech. Speech Commun. 40, 365–385 (2003)
4. Hadjitodorov, S., Mitev, P.: A Computer System for Acoustic Analysis of Pathological Voices And Laryngeal Diseases Screening. Med. Eng. Phys. 24, 419–429 (2002)
5. Godino-Llorente, J.I., Gomez-Vilda, P.: Automatic Detection of Voice Impairments by Means of Short-term Cepstral Parameters and Neural Network Based Detectors. IEEE Trans. Biomed. Eng. 51, 380–384 (2004)
6. Godino-Lorente, J.I., Gomez-Vilda, P., Blanco-Velasco, M.: Dimensionality Reduction of A Pathological Voice Quality Assessment System Based on Gaussian Mixture Models And Short-term Cepstral Parameters. IEEE Trans. Biomed. Eng. 53, 1943–1953 (2006)
7. Arjmandi, M.K.: Identifies Voice Disorders Using Long-time Features And Support Vector Machine With Different Feature Reduction Methods. Journal of Voice (2011)
8. Nayak, J., Bhat, P.S.: Identification of voice disorders using speech samples. IEEE Trans. 37, 951–953 (2003)
9. Fonseca, E.S., Gudio, R.C., Scalassara, P.R.: Wavelet Time-frequency Analysis And Least Squares Support Vector Machines Forthe Identification of Voice Disorders. Comput. Biol. Med. 37, 571–578 (2007)

10. Fu, W.J., Yang, X.H., Wang, Y.T.: Heart Sound Diagnosis Based on DTW and MFCC. In: 2010 3rd International Congress on Image and Signal Processing, p. 2920 (2010)
11. Cohen, A., Landsberg, D.: Analysis And Automatic Classification of Breath Sounds. IEEE Transactions on Biomedical Engineering 31, 585–590 (1984)
12. Anderson, K., Qiu, Y.H., Arthur, R.: Whittaker:Breath Sounds Asthma And The Mobile Phone. The Lancet 358(9290), 1343–1344 (2001)
13. Gavriely, N., Airflow, D.W.: Effects on Amplitude And Spectral Content of Normal Breath Sounds. Journal of Applied Physiology 80(1), 5–13 (1996)
14. Peng, H.C., Long, F.H., Ding, C.: Feature Selection Based on Mutual Information: Criteria of Max-Dependency, Max-Relevance,and Min-Redundancy. IEEE Trans. Pattern Analysis and Machine Intelligence 27(8), 1226–1238 (2005)
15. Yang, J., Ames, I.A., Honavar, V.: Feature Subset Selection Using A Genetic Algorithm. Intelligent Systems and Their Applications, 44–49 (1998)
16. Ren, J., Qiu, Z., Fan, W., Cheng, H., Yu, P.S.: Forward semi-supervised feature selection. In: Washio, T., Suzuki, E., Ting, K.M., Inokuchi, A. (eds.) PAKDD 2008. LNCS (LNAI), vol. 5012, pp. 970–976. Springer, Heidelberg (2008)
17. Bu, H.L., Zheng, S.Z., Xia, J.: Genetic Algorithm Based Semi-feature Selection Method. In: 2009 International Joint Conference on Bioinformatics Systems Biology and Intelligent Computing, pp. 521–524 (2009)
18. Oh, I.S., Lee, J.S., Moon, B.R.: Hybrid Genetic Algorithms for Feature Selection. IEEE Transactions on Pattern Analysis and Machine Intelligence 26(11), 1424–1437 (2004)
19. Eyben, F., Wöllmer, M., Schuller, B.: OpenSMILE - The Munich Versatile and Fast Open-Source Audio Feature Extractor. In: Proc. ACM Multimedia, pp. 1459–1462. ACM, Florence (2010)
20. The INTERSPEECH 2012 Speaker Trait Challenge, http://emotion-research.net/sigs/speech-sig/is12-speaker-trait-challenge
21. Guo, D.M., Zhang, D., Li, N.M., Zhang, L., Yang, J.H.: A Novel Breath Analysis System Based on Electronic Olfaction. IEEE Trans. Biomedical Engineering 57(11), 2753–2760 (2010)

A New Algorithm of Frequency Estimation for Frequency-Hopping Signal

Jun Lv[*], Weitao Sun, and Tong Li

Department of Information Engineering,
Academy of Armored Forces Engineering Beijing, China
344243278@qq.com

Abstract. The performance of ESPRIT (Estimated Signal Parameters via Rotational Invariance Technique) algorithm about frequency estimation is obviously declined, when the SNR of signal is relatively low. Aiming at this problem, an improved ESPRIT algorithm is presented. The algorithm is based on spectrum interception of noise suppression, and can narrow the selection area of β in the improved ESPRIT algorithm based on rotational transformation of autocorrelation matrix, using r local maximum point of DFT spectrum. The algorithm not only reduces the computational complexity, but also weakens the impact of noise on frequency estimation of signal. Theoretical analysis and simulation results verify the effectiveness and feasibility of the proposed algorithm.

Keywords: Frequency estimation, an ESPRIT algorithm, noise suppression.

1   Introduction

The algorithm of ESPRIT (Estimated Signal Parameters via Rotational Invariance Technique) was proposed by Roy in 1986 [1]. The ESPRIT algorithm requires that the data is decomposed two or more of the same sub-matrices. And that the algorithm requires the spectral decomposition of the covariance matrix of the signal data [2-4]. Since the ESPRIT algorithm don't need to search spectral peak, so the computation of the algorithm is small. This ESPRIT algorithm has become an important method of modern signal processing, and has been quite a wide range of applications. But when the SNR (Signal-to-Noise Ratio) of signal is low, the effect of noise of Covariance matrix is larger, and leading to performance of ESPRIT algorithm reduce. So the estimation error of ESPRIT algorithm has some difference to the CRLB (Cramer-Rao Lower Bound) [6][7]. Therefore the development of the ESPRIT algorithm is restricted in the field of frequency estimation.

This paper presents an improved ESPRIT algorithm, and this algorithm are based on the NS-ESPRIT (Noise Suppressed algorithm based on ESPRIT) algorithm and modulus curve of correlation matrix's trace of ESPRIT. The algorithm retains the advantage of ESPRIT algorithm, and weakens the impact of noise on frequency estimation of signal.

---

[*] Corresponding author.

D.-S. Huang et al. (Eds.): ICIC 2013, LNAI 7996, pp. 410–417, 2013.

2   Basic ESPRIT Algorithm

Assuming the signal is composed by r multiplexed sine wave

$$x(n) = \sum_{i=1}^{r} A_i e^{j(\omega_i n + \varphi_i)} + w(n) \tag{1}$$

Where A_i and W_i are the Complex amplitude and frequency of the i th harmonic. $w(n)$ is complex-valued Gaussian white noise ,its mean is   zero and its variance is σ_w^2.

Define $y(n) = x(n+1)$, so

$$X_1(n) = (x(n), x(n+1),\dots,x(n+M-1))^T = AS(n) + W(n) \tag{2}$$

$$Y_1(n) = (y(n), y(n+1),\cdots, y(n+M-1))^T$$

$$= (x(n+1), x(n+2),\cdots, x(n+M))^T$$

$$= A\boldsymbol{\varphi}_1 S(n) + W(n+1) \tag{3}$$

Where $\boldsymbol{\varphi}_1 = \mathrm{diag}(e^{jw_1}, e^{jw_2},\cdots, e^{jw_r})^T$. Assume

$$\boldsymbol{a}(w_i) = (1, e^{jw_i},\cdots, e^{j(M-1)w_i}) \tag{4}$$

$$A(w) = (a(w_1), a(w_2),\cdots, a(w_r)) \tag{5}$$

$$S(n) = (A_1 e^{j(w_1 n + \theta_1)}, A_2 e^{j(w_2 n + \theta_2)},\cdots, A_n e^{j(w_r n + \theta_r)}) \tag{6}$$

$$W(n) = (w(n), w(n+1),\cdots, w(n+M-1))^T \tag{7}$$

For the observation vector $X_1(n)$, its autocorrelation matrix

$$R_{X_1 X_1} = E\{X_1(n) X^*_1(n)\} = APA^H + \sigma_w^2 I \tag{8}$$

Cross-correlation matrix of $X_1(n)$ and $Y_1(n)$ is

$$R_{X_1 Y_1} = E\{X_1(n) Y_1(n)^H\} = AP\boldsymbol{\varphi}^H A^H + \sigma_w^2 Z_1 \tag{9}$$

Where $\sigma_w^2 Z_1 = E\{W(n) W(n+1)^H\}$

3   The Improved Algorithm Based on ESPRIT

3.1   Solving Self-covariance Matrix Based on NS-ESPRIT Algorithm

The performance of ESPRIT algorithm about frequency estimation is obviously de-clined, when the SNR of signal is relatively low. Aiming at this problem, Literature [8] proposes a frequency estimation algorithm of noise suppression based on ESPRIT (NS-ESPRIT algorithm). Adding rectangular window to the original signal, then we do DFT to it. Selecting the frequency spectrum around narrowband, we can get the value of auto covariance function of signal by doing IDFT to intercepted frequency spectrum.

$$\hat{r}'(m) = \sum_{i=1}^{r} \sum_{k=k_i-\Delta}^{k_i+\Delta} \frac{1}{N} |X_{2N}(k)|^2 e^{jw_i m} \qquad (10)$$

Where $m = 1, 2, \cdots, N-1$. In order to get the best estimation of auto-covariance func-tion $\hat{r}(m)$, the range of sub-band of Frequency domain Δ is set to width of the main lobe of the rectangular window $B_0 = 4\pi/N$. The reason for the success of the algorithm is that it only estimates a particular frequency point. If we add a window to the signal, the Spectrum of outside the main lobe is discarded. Useful harmonic in-formation of signal is lost, but the noise is filtered out, so significant feature of the algorithm is the anti-noise ability.

Assuming the normalized frequency of signal is 0.323, and the Sampling points are 500. The algorithm estimates the signal's frequency after interception of spectrum by different length. SNR range from-16dB to 10dB, each SNR repeat test 200 times, the MSE (Mean-Square Error) of estimated frequency is shown in Figure 1.

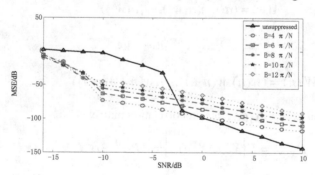

Fig. 1. The MSE of estimated frequency after interception of spectrum by different length

From the figure, we can see that the estimated accuracy of frequency is high in low SNR when we use algorithm of noise suppression based on ESPRIT. And the MSE is minimum when the interception of width B is equal to the width of main lobe of rectangular window $4\pi/N$. But an obvious defect of the algorithm is the high degree of complexity of the algorithm; the hardware implementation is more difficult.

3.2   ESPRIT Algorithm Based on Modulus Value of Curve of Cross-Correlation Matrix Trace

Literature [9] proposed method is exploiting received signals from two translating identical arrays like ESPRIT algorithm, but it need not have generalized eigenvalue decomposition to resolve the problem of high-resolution DOA estimation. Simplifying the complexity of calculation of algorithm, it can easy to achieve in engineering. Minor adjustments for this algorithm, it can also be applied to the frequency estimation. And it can reduce the complexity of the ESPRIT algorithm when estimating the signal frequency.

Doing rotation transformation to $X_1(n)$, the result is noted as $X_2(n)$

$$X_2(n) = -e^{j\beta}(AS(n) + W(n))$$

$$= A\begin{bmatrix} -e^{j\beta} & 0 & \cdots & 0 \\ 0 & -e^{j\beta} & \cdots & 0 \\ \vdots & \vdots & & \ddots \\ 0 & 0 & \cdots & -e^{j\beta} \end{bmatrix} S - e^{j\beta} \cdot W(n) \tag{11}$$

Now using $Y_1(n)$ and $X_2(n)$ construct $Y_2(n)$

$$Y_2(n) = Y_1(n) + X_2(n) = A\varphi_1 S + W(n+1) - e^{j\beta}(AS + W(n))$$

$$= A\begin{bmatrix} e^{jw_1} & -e^{j\beta} & 0 & \cdots & 0 \\ 0 & e^{jw_2} & -e^{j\beta} & \cdots & 0 \\ \vdots & \vdots & \vdots & \ddots & \vdots \\ 0 & 0 & \cdots & e^{jw_r} & -e^{j\beta} \end{bmatrix} S + W(n+1) - e^{j\beta} \cdot W(n)$$

$$= A(\varphi_1 - e^{j\beta} \cdot I)S + W(n+1) - e^{j\beta}W \cdot (n) \tag{12}$$

Let $\varphi_2 = \varphi_1 - e^{j\beta} \cdot I$, and $Y_2(n) = A\varphi_2 S + W(n+1) - e^{j\beta} \cdot W(n)$.

The cross-correlation matrix of $X_1(n)$ and $Y_2(n)$ Can be obtained.

$$R_{X_1, Y_2} = E\{X_1(n) \cdot Y_2(n)^H\} = E\{[AS + W(n)] \cdot [A\varphi_2 S + W(n+1) - e^{j\beta}W(n)]\}$$

$$= AE(SS^H)\varphi_2^H A^H + E(W(n)W(n+1)^H) - e^{-j\beta}E(W(n)W(n)^H)$$

$$= AP\varphi_2^H A^H + \sigma_w^2(Z_1 - e^{j\beta} \cdot I) \tag{13}$$

Let $Z_2 = Z_1 - e^{j\beta} \cdot I$, then $R_{X_1Y_2} = AP\varphi_2^H A^H + \sigma_w^2 Z_2$

From the principle of basic ESPRIT algorithm, we can know the frequency of the harmonic signal can be estimated by generalized eigenvalue of Matrix Pencil $(R_{X_1X_1}, R_{X_1Y_2})$ when β has an initial value. Solving steps of generalized eigenvalue of Matrix Pencil $(R_{X_1X_1}, R_{X_1Y_2})$ are as follow:

Step 1: Autocorrelation matrix $R_{X_1X_1}$ is constructed according to the data sequence

Step 2: making $\beta = -\pi$, data matrix $X_2(n)$ can be get by making rotation transformation to matrix $X_1(n)$. Constructing $Y_2(n) = Y_1(n) + X_2(n)$, we can calculate the cross-correlation matrix $R_{X_1Y_2}$ of $X_1(n)$ and $Y_2(n)$.

Step 3: we can calculate trace's the modulus value $\left|\sum \lambda_i\right| = |\text{tr}C|$ of matrix $C = R_{X_1Y_2}^{-1} R_{X_1X_1}$

Step 4: making $\beta = \beta + \Delta$, we can repeat steps 2 -4, until $\beta = \pi$.

Step 5: Searching the peak of $\left|\sum \lambda_i\right|$, we can get r times β, such as $w_i(i = 1,2,\cdots,r)$

The frequency resolution is limited by β, when β is very small, needing to search a lot points, it will lead to the large amount of computation. And when the β is big, the deviation of frequency estimation will be large. We can see those from Fig 2.

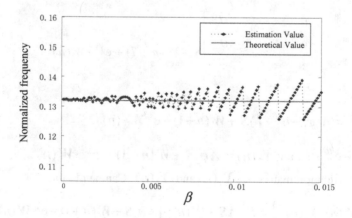

Fig. 2. Estimated values of normalized frequencies with the changing of β

3.3 Improved Algorithm of ESPRIT

Combing the NS-ESPRIT algorithm and the ESPRIT algorithm based on modulus value of curve of the cross-correlation matrix trace, this paper proposes an improved algorithm. The algorithm use crude estimation of each frequency point to narrow the range of β, and its ranges is $[\hat{w}_i' - B_0/2, \hat{w}_i' + B_0/2]$, $1 \leq i \leq r$. It can reduce the calculation when it searches the peak point of curve of the mode value. Figure 3 is shown the flowchart of the improved algorithm, and the dashed part will continuous cycle with the changing of β until the end.

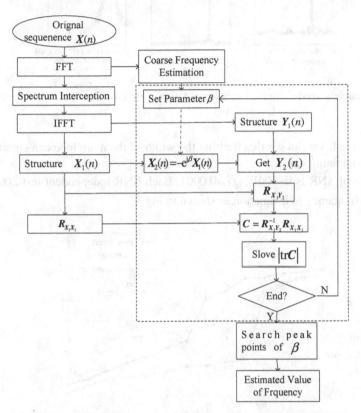

Fig. 3. The flowchart of improved algorithm based on ESPRIT

4   Simulation and Result Analysis

In order to verify the effectiveness and feasibility of the improved algorithm, the paper using the improved algorithm to estimate the frequency of harmonic signal. Setting the signal has one frequency; its normalized frequency is 0.132. The order of covariance matrix is set to 20. The signal' sampling points is $N = 1000$. Using the basic ESPRIT algorithm, its trace's mode value of curve of cross-correlation matrix is shows as Fig.4 (a) with changing of β; The curve of using NS-ESPRIT is show as Fig4(b); The improved algorithm limit the width of the main lobe, then the searching range of β is shown as Fig.4(c).

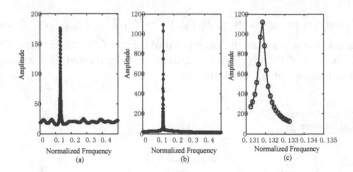

Fig. 4. The curve's mode value of cross-correlation matrix of the trace with the changing of β

From the Fig4, we can see that limiting the width of the main lobe can significantly reduce the computational.

The range of SNR is 0~10dB, β=0.0001. Each SNR independent test 200 times, the MSE of frequency of estimation as shown in Fig 5.

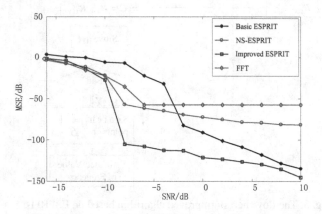

Fig. 5. The MSE of the improved frequency estimation algorithm

Table 1. Time consuming of each algorithm   Unit s

Algorithm	Estimated Time Consuming of 200 Times	Estimated Time Consuming of 1 Times (10^{-3})
FFT	0. 0790	0. 395
ESPRIT	0. 2242	1. 120
NS-ESPRIT	0. 3592	1. 800
Improved	0. 2273	1. 140

From the Fig 5, we can see that the MSE of frequency estimation of the improved algorithm is less than the FFT algorithm, the basic ESPRIT algorithm and NS-ESPRIT algorithm, and the Noise immunity of improved algorithm is better than the FFT algorithm, the basic ESPRIT algorithm and NS-ESPRIT algorithm. In order to compare the amount of computation of each algorithm, we count the consuming time of each algorithm and it shows as Table 1. From the Table we can see that compared with the basic ERPIST, the computation of the improved algorithm is basically unchanged.

5   Summary

The performance of ESPRIT algorithm about frequency estimation is obviously declined, when the SNR of signal is relatively low. Aiming at this problem, an improved ESPRIT algorithm based on modulus curve of correlation matrix's trace is presented, and independent variable of the curve is restricted using the frequency's pre-estimated value by NS-ESPRIT algorithm, which can reduce the computational cost. The improved algorithm not only has the advantages, such as fast calculation and high estimated accuracy, but also restricts effect of frequency estimation by noise.

References

1. Roy, R., Kailath, T.: ESPRIT:Estimation of Signal Parameter via Rotational Invariance Technique. IEEE Trans. Acoust, Speech Signal Processing 37(7), 984–995 (1989)
2. Bao, Z.Q., Wu, S.J., Zhang, L.R.: A Novel and Low Complexity ESPRIT Method. Journal of Electronics & Information Technology 29(9), 2042–2046 (2007)
3. Li, Y.B., Wang, T., Wang, J.L.: The Echo Frequency Estimation of Anti-torpedo Torpedo Acoustic Fuse Based on Fast ESPRIT. Command Control & Simulation 33(6), 99–101 (2011)
4. Wang, Q., Chen, W.P.: Rearch on the Frequency Estimation Based on ESPRIT Algorithm. Electronics R & D, 9–11 (2010)
5. Al-Ardi, E.M., Shubair, R.M., AL-Mualla, M.E.: Performance Evaluation of Direction Finding Algorithms for Adaptive Antenna Arrays. In: Proceedings of the 2003 10th IEEE International Conference on Electronics, Circuits and Systems, pp. 735–738. IEEE, Sharha (2003)
6. El-Shafey, M.H.: ESPRIT Condition in Signal Parameter Estimation. In: Proceedings of 32nd International Conference on Sarnoff Symposium, vol. 3, pp. 1–5 (2009)
7. Cui, H., Qian, Y.Y., Li, D.H.: Direction of arrival research on extracting signal subspace of ESPRIT algorithm. Computer Engineering and Design 30(13), 3507–3509 (2009)
8. Cui, Y.: Noise Suppression Frequency Estimation Algorithm Based on ESPRIT. Computer Engineering 36(14), 245–248 (2010)
9. Li, Y., Yang, J.P.: An ESPRIT Based Improved Algorithm. Signal Processing 25(3), 508–511 (2009)

Seizure Detection in Clinical EEG Based on Multi-feature Integration and SVM

Shanshan Chen[1,2], Qingfang Meng[1,2,*], Weidong Zhou[3], and Xinghai Yang[1,2]

[1] School of Information Science and Engineering, University of Jinan, Jinan 250022, China
[2] Shandong Provincial Key laboratory of Network Based Intelligent Computing,
Jinan 250022, China
[3] School of Information Science and Engineering, Shandong University, Jinan 250100, China
ise_mengqf@ujn.edu.cn

Abstract. Recurrence Quantification Analysis (RQA) was a nonlinear analysis method and widely used to analyze EEG signals. In this work, a feature extraction method based on the RQA measures was proposed to detect the epileptic EEG from EEG recordings. To combine the time-frequency characteristic of epileptic EEG, variation coefficient and fluctuation index were used to analyze epileptic EEG. The multi-feature combination of RQA and linear parameters had better performance in analyzing the nonlinear dynamic characteristics and time-frequency characteristic of epileptic EEG. For features selection and improving the classification accuracy, a support vector machine (SVM) classifier was used. The experimental results showed that the proposed method could classify the ictal EEG and interictal EEG with accuracy of 97.98%.

Keywords: Epileptic EEG, Recurrence quantification analysis (RQA), Multi-feature integration, Support vector machine (SVM).

1 Introduction

Epilepsy is one of most common neurological disorder characterized by the presence of recurring seizures. Brain activity during seizure differs greatly from that of normal state with respect to patterns of neuronal firing. The EEG contains important information about the conditions and functions of the brain, epilepsy can be assessed by the EEG. However, visual inspection of the long-term EEG signals for the presence of seizure activities is tedious and time-consuming. So the automatic seizure detection in EEG is significant for diagnosing epilepsy in clinical practice.

Considering the scalp EEG signals are complex, nonlinear and non-stationary, many techniques have developed for epileptic activity detection used nonlinear features extracted from the EEG signals in classification models. The nonlinear dynamical methods have been widely applied to analyze EEG signals, the nonlinear parameters [1-5,17] like the correlation dimension[3], the largest Lyapunov exponent[2] and entropy[4,5] are usually used as the extracted feature value. In Reference [2] the short-term largest Lyapunov exponent of EEG waveforms was used for

---

* Corresponding author.

D.-S. Huang et al. (Eds.): ICIC 2013, LNAI 7996, pp. 418–426, 2013.

the detection and prediction of the epileptic seizure. In Reference [5], four entropy features were extracted from the collected EEG signals and these entropy values were fed into seven classifiers to discriminate normal, pre-ictal and ictal states.

Recurrence Plot (RP) method developed by Eckmann [6], is also a nonlinear analysis method based on complexity [7, 9]. It provides a simple graphical method initially designed to display recurring patterns. In order to do more research the dynamical system, Zbilut and Webber Jr. have introduced recurrence quantification analysis (RQA) to quantify the RP [11, 16]. It has the advantage of analyzing linear and non-linear time signals without making any initial assumptions about stationarity, length or noise [8]. In recent years, recurrence plot has been successfully applied to the analysis of physiological data, such that epileptic EEG signals [8, 10, 13], Heart Rate Variability Data [9, 15], EMG signals [12] and blood pressure signals [14].

Support vector machine (SVM) is a very popular tool for solving supervised classification problems [17, 18]. SVM has a strong advantage in solving supervised classification problems of small sample data. The performance of the classifier depends on the features selection [10]. At the same time, feature extraction is also an efficient approach to analysis EEG signals. In this paper, we propose and evaluate the feature extraction method using recurrence quantification analysis (RQA), based on nonlinear analysis method, to automatically detect the epileptic EEG signals from the EEG signals. Usually, to combine the time-frequency characteristic of epileptic EEG signals, variation coefficient and fluctuation index were widely applied to analyze EEG signals. RQA combine with linear parameters to get fusion feature vectors as the input features using a support vector machine. The combination of linear parameters and nonlinear parameters has better performance in the nonlinear dynamic characteristics and time-frequency characteristic of epileptic EEG. The fusion features and SVM improve the performance of classification of epileptic EEG automatically.

2   Feature Extraction Based on Multi-feature Integration Method and SVM

2.1   Recurrence Plot

Recurrence pattern is an important feature of the deterministic dynamical systems. Recurrence plots describe the recurrence property of a deterministic dynamical system, visualizing the recurrence of states $x(i)$ in phase space. The methodology of RP is as follows.

For a given time series $\{x_1, x_2, \cdots, x_n\}$, the phase space vector $X(i)$ can be reconstructed by using the Taken's time delay method:

$$X(n) = \left[x_n, x_{n-\tau}, \cdots, x_{n-(m-1)\tau} \right]^T, n = (m-1)\tau+1, (m-1)\tau+2, \cdots, N \quad (1)$$

Recurrence plot is expressed by calculating the distance between any two points in the phase space:

$$R_{i,j} = \Theta\left(\varepsilon - \|X(i) - X(j)\|\right), i, j = 1, 2, \cdots, N - (m-1)\tau \tag{2}$$

Where ε is a cutoff distance, $\|\cdot\|$ is the Euclidean norm, $\Theta(x)$ is a Heaviside function.

$$\Theta(x) = \begin{cases} 1, x > 0 \\ 0, x \leq 0 \end{cases} \tag{3}$$

The cutoff distance ε defines a sphere centered at $X(j)$, and if $X(i)$ falls within this sphere, the state is close to $X(j)$, then $R_{i,j} = 1$; otherwise $R_{i,j} = 0$. Therefore, the Recurrence Plot (RP) is a $N \times N$ matrix of points (i, j) where each point is said to be recurrent and marked with a dot if the distance between the delayed vectors $X(i)$ and $X(j)$ is less than a given threshold ε.

2.2   Recurrence Quantification Analysis (RQA)

In order to further investigate the properties of RP, recurrence quantification analysis (RQA) is used to quantify the RP. RQA gives some insight in the detection of deterministic structure among the EEGs, and in the characterization of transients prior to the seizure that might predict its onset. The extracted RQA parameters could reflect the nonlinear dynamic characteristics of epilepsy EEG. Selected the following nonlinear features:

Recurrence Rate (RR): the density of recurrence point in the recurrence plot.

$$RR = \frac{1}{N^2} \sum_{i,j=1}^{N} R_{i,j} \tag{4}$$

Determinism (DET): DET is the ratio of recurrence points forming the diagonal structures to all recurrence points. It reveals the existence of a deterministic structure and indicates the predictability of the system. It is given by (l_{min} is the length of minimal diagonal line, $P(l)$ is the frequency distribution of diagonal structures)

$$DET = \frac{\sum_{l=l_{min}}^{N} lP(l)}{\sum_{i,j}^{N} R_{i,j}} \tag{5}$$

Entropy (ENTR): the Shannon entropy of the distribution of the length of line segments parallel to the main diagonal line. The entropy measure the complexity of the recurrence plot.

$$ENTR = -\sum_{l=l_{min}}^{N} p(l)\ln p(l) \tag{6}$$

Laminarity (LAM): the ratio between the recurrence points forming the vertical structures and the entire set of recurrence points can be computed by

$$LAM = \frac{\sum_{v=v_{min}}^{N} vp(v)}{\sum_{i,j} R_{i,j}} \tag{7}$$

Trapping (TT): The average length of vertical structures is given by

$$TT - ave = \frac{\sum_{v=v_{min}}^{N} vP(v)}{\sum_{v=v_{min}}^{N} P(v)} \tag{8}$$

Then, the maximal length of the vertical lines in the RP is given by

$$TT - \max = \max\left(\{v_l\}_{l=1}^{N_v}\right) \tag{9}$$

Here N_v is the absolute number of vertical lines.

2.3   Linear Features

In order to visualize the time-frequency characteristic of epileptic EEG, variation coefficient and fluctuation index were widely applied to analyze the signals.

Variation coefficient: the feature expresses the change of the signal's amplitude. It defined by:

$$V_c = \sqrt{\delta^2 / u^2} \tag{10}$$

Where, δ is the standard deviation and u is the mean value.

Fluctuation index: the index measures the intensity fluctuations of signals; the value of signals during seizure is higher than that of signals during seizure free interval.

$$F = \frac{1}{n}\sum_{i=1}^{n} |x(i+1) - x(i)| \tag{11}$$

2.4   SVM

SVM is a supervised learning technique used for classification. The main concept of SVM is to implicitly map the data into the feature space where a hyperplane (decision boundary) separating the classes may exist.

For a set of training samples $G = \{(x_i, y_i)\}_i^N$ (x_i is the input vector, y_i is the desired value and N is the total number of data patterns), the training data (x_i, y_i) and $y_i \in \{-1, 1\}$, must satisfy the constraints:

$$y_i[(w \cdot x_i) + b] - 1 \geq 0, i = 1, 2, \cdots N \tag{12}$$

The points for which the equalities in the above equations hold have the smallest distance to the decision boundary and they are called the support vectors. Then, the hyperplane is solved using Lagrange optimization framework. The general solution is given by

$$\sum_{i=1}^{N} a_i y_i = 0, a_i \in [0, C], i = 1, 2, \cdots N \tag{13}$$

In the case of non-linear classification, Kernels (functions of varying shapes) are used to map the data into a higher dimensional feature space in which a linear separating hyperplane could be found. The general solution is then of the form:

$$f(x) = \text{sgn}(\sum_{i=1}^{N} a_i^* y_i K(x, x_i) + b^*) \tag{14}$$

Depending on the choice of the Kernel function, SVM can provide both linear and non-linear classification.

2.5   Seizure Detection Process in CLINICAL EEG

Recurrence plot could describe the recurrence property of a determinism dynamical system. In the work, the key is extract feature values to distinguish the ictal EEG from the epileptic EEG signals.

Firstly, we propose and evaluate the feature extraction method based on RQA to quantify the structures in recurrence plot and extract six nonlinear dynamic features. These feature values could reflect the nonlinear dynamic characteristics of EEG data. Secondly, selected an appropriate threshold value as the discrimination criteria, which is cable to detect the epilepsy seizure signals from EEG recordings. In addition, EEG signals during seizure differ greatly from that of normal state with disordered neuronal firing activity. Considering the time-frequency characteristics of epileptic EEG, then calculated the variation coefficient and fluctuation index that reflect the change of amplitude and waveform. Finally, combined two linear parameters and six nonlinear dynamic parameters to get fusion feature vectors as the input features for the automatic detection of epilepsy signals using a support vector machine. SVM was used for feature selection and classification. The flow chart is showed in Fig.1.

The combination of linear parameters and nonlinear dynamic parameters, which have better performance in the nonlinear dynamic characteristics and the time-frequency characteristic of epileptic EEG, improve the performance of classification of epileptic EEG automatically.

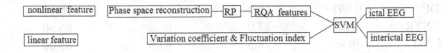

Fig. 1. Flow chart of EEG classification algorithm

3   Experiment Results and Analysis

The data used in this research were obtained from the six clinical diagnosed epilepsy cases in the Qilu Hospital of Shandong University. All EEG signals were sampled at a rate of 128 Hz. For the purpose of comparison, we select 200 episodes of epileptic data during a seizure free interval and 200 episodes of data during seizure, each episode of 1024 points.

The RQA measures are calculated for each time series. The embedding parameters used are m=7, t=1, the recurrence threshold ε is 0.3 times of the maximum distance. Fig.2 shows the recurrence rate and determinism analysis results of ictal and interictal EEG episodes. For recurrence rate (RR), select 0.68 as the threshold can make a distinction between ictal EEG and interictal EEG. Similarity, the best threshold of determinism (DET) was determined as 0.9975 which is shown in Fig. 2(b).

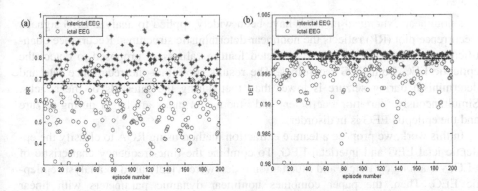

Fig. 2. The RR (a) and DET (b) analysis of epileptic EEG episodes

Here, taken the ictal EEG as the positive class, and the intermitted period EEG as the negative class. Then, Select appropriate threshold for each of the RQA parameters to distinguish ictal EEG from epileptic EEG. The classification results of ictal and interictal EEG are shown in Table 1.

Table 1. The classification of ictal EEG and interictal EEG using RQA features

	threshold	Sensitivity	Specificity	Accuracy (%)
RR	0.68	86.5	88.0	87.00
DET	0.9975	86.0	94.5	90.25
ENTR	5.80	87.0	86.0	86.50
LAM	99.96	78.5	87.5	83.00
TT-ave	25.00	85.0	86.5	85.75
TT-max	200.00	88.5	89.5	89.00

The six RQA values and linear parameters-fluctuation index and variation coefficient are used to form 8 dimensional vectors as the input feature for automated seizure detection using a Support Vector Machine (SVM). Select 150 episodes of ictal EEG and intertal EEG respectively, as the training sample to optimize the parameters of SVM. The rest episodes are used to test the classification. The result is showed in Table 2.

Table 2. The classification results based on multi-features fusion and SVM

P1	C	Train (%)			Test (%)		
		Sensitivity	Specificity	Accuracy	Sensitivity	Specificity	Accuracy
300	1000	97.33	98.66	98.00	95.98	99.98	97.98

4   Conclusions

The nonlinear dynamical methods have been widely applied to analyze EEG signals. Recurrence plot (RP) reflects the nonlinear deterministic structures. Recurrence quantification analysis (RQA) as the extracted feature values are proposed to detect the epileptic EEG from EEG recordings. The results show that the recurrence rate and determinism during seizure is lower than that during the intervals of the attacks. Simultaneously, Variation coefficient and Fluctuation index are larger during seizure and the epileptic EEG is in disorder.

In this work, we propose a feature extraction method using RQA to classify the epileptic ictal EEG and interictal EEG. To combine the time-frequency charteristic of EEG, variation coefficient and Fluctuation index were widely used to analysis epileptic EEG. Then, the paper combines nonlinear dynamic parameters with linear parameters, which have better performance in the nonlinear dynamic characteristics and the time-frequency characteristic of epileptic EEG. Finally, a SVM classifier was used to improve the performance of the automatic classification of epileptic EEG.

Acknowledgement. Project supported by the National Natural Science Foundation of China (Grant No. 61201428, 61070130), the Natural Science Foundation of Shandong Province, China (Grant No. ZR2010FQ020), the Shandong Distinguished Middle-aged and Young Scientist Encourage and Reward Foundation, China (Grant No. BS2009SW003), the China Postdoctoral Science Foundation (Grant No. 20100470081), and the Shandong Province Higher Educational Science and Technology Program (Grant NO.J11LF02).

References

1. Acharya, U.R., Chua, K.C., Lim, T.C., Dorithy, Suri, J.S.: Automatic Identification of Epileptic EEG Signals Using Nonlinear Parameters. Journal of Mechanics in Medicine and Biology 9, 539–553 (2009)
2. Swiderski, B., Osowski, S., Rysz, A.: Lyapunov Exponent of EEG Signal for Epileptic Seizure Characterization. Chaos 5, 82–87 (1995)
3. Wang, X.Y., Meng, J., Qiu, T.S.: Research on Chaotic Behavior of Epilepsy Electroencephalogram of Ehildren Based on Independent Component Analysis Algorithm. J. Biomed. Engin. 24, 835–841 (2007)
4. Ocak, H.: Automatic Detection of Epileptic Seizures in EEG Using Discrete Wavelet Transform and Approximate Entropy. Expert Systems with Applications 36, 2027–2036 (2009)
5. Achary, U.R., Molinari, F., Sree, S.V., Chattopadhyay, S., Ng, K.H., Suri, J.S.: Automated Diagnosis of Epileptic EEG Using Entropies. Biomedical Signal Processing and Control 7, 401–408 (2012)
6. Eckmann, J.P., Kamphorst, S.O., Ruelle, D.: Recurrence Plots of Dynamical Systems. Europhysics Letters 5, 973–977 (1987)
7. Marwan, N., Romano, M.C., Thiel, M., Kurths, J.: Recurrence Plots for the Analysis of Complex Systems. Physics Reports 438, 237–329 (2007)
8. Thomasson, N., Hoeppner, T.J., Webber, C.L., Zbilut, J.P.: Recurrence Quantification in Epileptic EEGs. Physics Letters. A 279, 94–101 (2001)
9. Marwan, N., Wessel, N., Kurths, J.: Recurrence Plot Based Measures of Complexity and its Application to Heart Rate Variability Data. Phys. Rev. E 66, 26702 (2002)
10. Ouyang, G., Xie, L., Chen, H., Li, X., Guan, X., Wu, H.: Automated Prediction of Epileptic Seizures in Rats with Recurrence Quantification Analysis. In: Conf. Proc. IEEE Eng. Med. Biol. Soc., pp. 153–156 (2006)
11. Ouyang, G.X., Li, X.L., Dang, C.Y., Richards, D.A.: Using Recurrence Plot for Determinism Analysis of EEG Recordings in Genetic Absence Epilepsy Rats. Clinical Neurophysiology 119, 1747–1755 (2008)
12. Zhong, J.K., Song, Z.H., Hao, W.Q.: Application of Recurrence Qualification Analysis to EMG. Acta Phys. Sin. 18, 241–245 (2002)
13. Acharya, U.R., Sree, V., Chattopadhyay, S., Yu, W.W., Alvin, P.C.: Application of Recurrence Quantification Analysis for the Automatic EEG Signals. International Journal of Neural Systems 21, 199–211 (2011)
14. Liu, X., Cheng, J.H., Lu, H.B., Zhang, L.F., Ma, J., Dong, X.Z.: Recurrence Quantification Analysis of Blood Pressure Signal in Rats after Simulated Weightlessness. Space Medicine& Medica l Engineering 19, 394–398 (2006)

15. Chen, X.M., Qiu, Y.H., Zhu, Y.S.: Recurrence Plot Analysis of HRV for Brain Ischemia and Asphyxia. Journal of Biomedical Engineering 25, 39–43 (2008)
16. Yan, R.Q., Zhu, Y.S.: Voiced/unvoiced Decision Based on Recurrence Quantification Analysis. Journal of Electronics & Information Technology 29, 1703–1706 (2007)
17. Nicolaou, N., Georgiou, J.: Detection of Epileptic Electroencephalogram Based on Permutation Entropy and Support Vector Machines. Expert Systems with Applications 39, 202–209 (2012)
18. Zhao, J.L., Zhou, W.D., Liu, K., Cai, D.M.: EEG Signal Classification Based on SVM and Wavelet Analysis. Computer Applications and Software 28, 114–116 (2011)

PEEC Modeling for Linear and Platy Structures with Efficient Capacitance Calculations

Yanchao Sun, Junjun Wang, Xinwei Song, and Wen Li

EMC Technology Institute, Beihang University
Beijing 100191, P.R. China
sycspecial2011@163.com

Abstract. Electromagnetic solvers based on the partial element equivalent circuit (PEEC) approach have been proven to be well suited for the solution of combined circuit and EM problems. In this paper, we focused on structures with small cross-section such as linear and platy conductors which are quite common in modeling various issues. According to their special geometries, only one-dimensional or two-dimensional subdivision rather than three-dimensional mesh is required, to which case basic PEEC method is applied inefficiently. Depending on this kind of situation, a novel equivalent capacitance calculation method for PEEC partial elements is proposed for modeling linear and platy structures in order to simplify the solution and improve the efficiency. Accordingly, the solving time and other resources can be reduced greatly.

Keywords: PEEC method, linear and platy Structures, equivalent capacitance calculation.

1 Introduction

The partial element equivalent circuit (PEEC) method is oriented towards the solution of very large practical EMC (mainly EMI) and Electrical Interconnect and Package (EIP) problems [1-2]. This method is a full wave technique for the solution of hetero-geneous, mixed circuit and fields problems in both the time as well as the frequency domains. The flexibility of this approach is illustrated by the different problems which have been solved using PEEC [3-4].

The meshing of geometrical objects is a key first step in the electromagnetic modeling of EMC problems using numerical techniques, and the overall solution efficiency strongly depends on the geometrical mesh algorithms [5]. When modeling various issues, structures with small cross-section such as linear and platy conductors are quite common. For instance, to accurately predict the interaction between signals on cables arranged closely, signal integrity analysis has become an essential part of the whole electromagnetic compatibility design, so PEEC method for modeling the cables as a high efficient and accurate approach is required. When the cross-section is smaller than the maximum dimension of discrete cells, subdivision is only necessary in the direction which is perpendicular to it, which means that for special structures such

D.-S. Huang et al. (Eds.): ICIC 2013, LNAI 7996, pp. 427–434, 2013.
© Springer-Verlag Berlin Heidelberg 2013

as linear and platy conductor, less than three-dimensional partition is enough. In the case for cables, we only need a one-dimensional partition along the length direction. Given this situation, the partial elements calculation in the basic PEEC method is inefficient [6].

In this paper we introduce a novel way to calculate the partial capacitance in PEEC model, which is especially applied to the structure with small cross-section contrasted to its length. It provides a concise mesh without the reduction of accuracy, the efficiency consequently improving.

The paper is organized as follows: Section 2 briefly recalls the derivation of PEEC models. In section 3, the specific partial capacitance calculation method is discussed; Section 4 reports simple example based on the novel calculation approach; Finally, Section 5 draws the conclusions.

2   Basic Formulation for PEEC Model

Substituting the closed forms of magnetic vector potential $\vec{A}(\vec{r},t)$ and electric scalar potential $\Phi(\vec{r},t)$ to the electric field integral equation [7-9], we can obtain:

$$\overline{E}^i(\vec{r},t) = \frac{\overline{J}(\vec{r},t)}{\sigma} + \frac{\partial}{\partial t} \frac{\mu}{4\pi} \int_{v'} \frac{\overline{J}(\vec{r'},t)}{|\vec{r}-\vec{r'}|} dV' + \overline{\nabla} \frac{1}{4\pi\varepsilon} \int_{S'} \frac{\rho(\vec{r'},t)}{|\vec{r}-\vec{r'}|} dS' \tag{1}$$

Achieving the discretization of equation (1) through the method of moments, and then using the so-called Galerkin's weighting process to generate a system of equations for the unknowns weights $I_n(\omega)$ and $Q_m(\omega)$. Result in the evaluation of the average value of $\Phi(\vec{r},\omega)$ over the surface of each patch is (in the frequency domain):

$$\Phi_l(\vec{r_l},\omega) = \frac{1}{S_l} \int_{S_l} \Phi(\vec{r_l},\omega) dS_l = \sum_{m=1}^{N_s} [\frac{1}{4\pi\varepsilon} \frac{1}{S_l} \frac{1}{S_m} \int_{S_l} \int_{S_m} \frac{e^{-j\omega\tau}}{|\vec{r_l}-\vec{r_m}|} dS_m dS_l] Q_m(\omega) \tag{2}$$

$$= P_{lm}(\omega) Q_m(\omega) \qquad for \quad l=1...N_s$$

Thus, the potential of the N_s patches can be related to the charges located on the same patches by the partial potential, at the angular frequency ω, by:

$$\Phi(\omega) = P(\omega)Q(\omega) \tag{3}$$

Equation (2), after the Galerkin scheme is applied, can be rewritten as:

$$E_0(\vec{r_i},\omega)l_i = \frac{l_i I_i(\omega)}{\sigma a_i} + \frac{j\omega\mu}{4\pi} \sum_{n=1}^{N_v} \frac{1}{a_i} \frac{1}{a_n} \int_{V_i} \int_{V_n} \vec{\mu_i} \cdot \vec{\mu_n} I_n(\omega) \frac{e^{-j\omega\tau}}{|\vec{r_i}-\vec{r_m}|} dV_n dV_i + \Phi_{2i}(\omega) - \Phi_{1i}(\omega) \quad for \ i=1...N_v \tag{4}$$

Each term of equation (4) represents a voltage drop across volume V_i along the direction and, thus, it can be rewritten as:

$$\Phi_{1i}(\omega) - \Phi_{2i}(\omega) = V_{0i}(\omega) + R_i I_i + j\omega \sum_{n=1}^{N_v} L_{p,in} I_n(\omega) \tag{5}$$

Where:

$$V_{0i}(\omega) = -E_0(r_i,\omega)l_i \;, R_i = \frac{l_i}{\sigma a_i}, L_{p,in}(\omega) = \frac{\mu}{4\pi}\frac{1}{a_i}\frac{1}{a_n}\int_{V_i}\int_{V_n}\vec{\mu_i}\cdot\vec{\mu_n}\frac{e^{-j\omega\tau}}{|\vec{r_i}-\vec{r_n}|}dV_n dV_i \tag{6}$$

$L_{p,in}$ is so called partial inductance between volume cells i and n.

In a more compact matrix form equation (5) can be written as:

$$-A\Phi(\omega) - RI_L(\omega) - j\omega L_P(\omega)I_L(\omega) - V_0(\omega) = 0 \tag{7}$$

where vectors $\Phi(\omega)$ and $I_L(\omega)$ collect the potentials to infinity and the currents flowing through the longitudinal branches, respectively, and the matrix A is the connectivity matrix[8-9].Circuit Equations (3) and (7) represent electric and magnetic field couplings, respectively, thus leading to partial element equivalent circuit.

3   An Efficient Capacitance Calculation Approach

Meshing is an important issue in accurate and effective PEEC modeling since the nodes determine the complexity of the networks. The number of unknowns directly impacts the solution time and the size of the problems which can be solved. We restrict the meshes to hexahedral volume and quadrilateral surface cells and use a fixed number of cells per shortest wavelength to assure sufficient accuracy. It is defined to 20 cells for a good performance in PEEC models.

The nodes in meshing coincide with the nodes in the circuit model. For reducing the total nodes (unknowns) number to get an efficient PEEC model, nodes of different directions meshing should be identical. To ensure this rule, the edge volume cells are divided half of the inner volume cells. Fig.1 (a) shows the volume discretization and nodes placement.

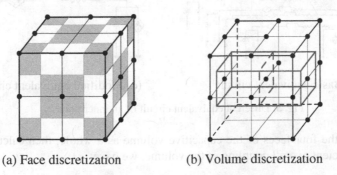

(a) Face discretization   (b) Volume discretization

Fig. 1. Simple structure conductor meshing

It should be noticed that the volume cell is shifted from the face cell half unit on the outside surfaces of conductors.

In this paper, we concentrate on modeling special structure like linear and platy conductor using PEEC method. Taking a microstrip line for example, the meshing can be carried out only in the length direction as shown in Fig.2.

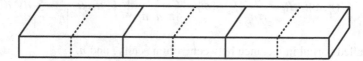

Fig. 2. Microstrip line meshing (Solid lines are inductive-resistive partitions, and dotted lines are capacitive partitions)

According to the basic PEEC method [8-9], the cross-couplings are represented by controlled source. The value can be calculated by:

$$U_i^{(L)} = \sum_{\substack{j=1 \\ j \neq i}}^{N_v} \frac{L_{ij}}{L_{jj}} V_j^{(L)} \quad , \quad I_k^{(C)} = \sum_{\substack{n=1 \\ n \neq k}}^{N_s} \frac{P_{kn}}{P_{kk}} I_{cn} \tag{8}$$

According to basic PEEC method, a node in a face cell decides a capacitive branch. Taking an inner volume cell in Fig.3 (a), it includes fours nodes. To make sure the inductive and capacitive circuits compatible with each other, four volume cells are required and four inductive branches are necessary, as Fig.3 (b), resulting in complicated circuits and long computing time. By introducing the equivalent potential coefficient, the number of nodes can be greatly reduced. The equivalent circuit will be like Fig.3 (c).

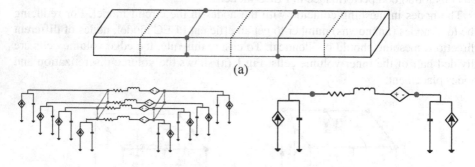

(a)

(b) Basic equivalent circuit   (c) Modified equivalent circuit

Fig. 3. Different equivalent circuit of an inner volume

Consider the four faces of the capacitive volume as a whole, then calculating potential coefficient overall coupling of the volume, we have:

$$\Phi = \Phi_1 = \Phi_2 = \Phi_3 = \Phi_4 \tag{9}$$

$$Q_t = Q_1 + Q_2 + Q_3 + Q_4 \tag{10}$$

Φ_1, Φ_2, Φ_3, Φ_4 represents the vector of face potential separately. Φ is the vector of cell potential. Q_1, Q_2, Q_3, Q_4 represents the face charge of S_1, S_2, S_3, S_4 (as Fig.4 shows) respectively, and Q_t is the total cell charge.

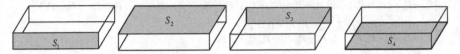

Fig. 4. Four faces of a capacitive volume

Taking any two capacitive volumes for example, the calculation steps are as follows.

Original equation can be written as:

$$
\begin{bmatrix} \Phi_1 \\ \Phi_2 \\ \Phi_3 \\ \Phi_4 \\ \Phi_5 \\ \Phi_6 \\ \Phi_7 \\ \Phi_8 \end{bmatrix}
=
\begin{bmatrix}
P_{11} & P_{12} & P_{13} & P_{14} & P_{15} & P_{16} & P_{17} & P_{18} \\
P_{21} & P_{22} & P_{23} & P_{24} & P_{25} & P_{26} & P_{27} & P_{28} \\
P_{31} & P_{32} & P_{33} & P_{34} & P_{35} & P_{36} & P_{37} & P_{38} \\
P_{41} & P_{42} & P_{43} & P_{44} & P_{45} & P_{46} & P_{47} & P_{48} \\
P_{51} & P_{52} & P_{53} & P_{54} & P_{55} & P_{56} & P_{57} & P_{58} \\
P_{61} & P_{62} & P_{63} & P_{64} & P_{65} & P_{66} & P_{67} & P_{68} \\
P_{71} & P_{72} & P_{73} & P_{74} & P_{75} & P_{76} & P_{77} & P_{78} \\
P_{81} & P_{82} & P_{83} & P_{84} & P_{85} & P_{86} & P_{87} & P_{88}
\end{bmatrix}
\begin{bmatrix} Q_1 \\ Q_2 \\ Q_3 \\ Q_4 \\ Q_5 \\ Q_6 \\ Q_7 \\ Q_8 \end{bmatrix}
\tag{11}
$$

Φ_1, Φ_2, Φ_3, Φ_4 are potentials of the first volume; Φ_5, Φ_6, Φ_7, Φ_8 are potentials of the second volume.

First, making other columns of each capacitive volume subtract its first column.

$$
\begin{bmatrix} \Phi_1 \\ \Phi_2 \\ \Phi_3 \\ \Phi_4 \\ \Phi_5 \\ \Phi_6 \\ \Phi_7 \\ \Phi_8 \end{bmatrix}
=
\begin{bmatrix}
P_{11} & P_{12}-P_{11} & P_{13}-P_{11} & P_{14}-P_{11} & P_{15} & P_{16}-P_{15} & P_{17}-P_{15} & P_{18}-P_{15} \\
P_{21} & P_{22}-P_{21} & P_{23}-P_{21} & P_{24}-P_{21} & P_{25} & P_{26}-P_{25} & P_{27}-P_{25} & P_{28}-P_{25} \\
P_{31} & P_{32}-P_{31} & P_{33}-P_{31} & P_{34}-P_{31} & P_{35} & P_{36}-P_{35} & P_{37}-P_{35} & P_{38}-P_{35} \\
P_{41} & P_{42}-P_{41} & P_{43}-P_{41} & P_{44}-P_{41} & P_{45} & P_{46}-P_{45} & P_{47}-P_{45} & P_{48}-P_{45} \\
P_{51} & P_{52}-P_{51} & P_{53}-P_{51} & P_{54}-P_{51} & P_{55} & P_{56}-P_{55} & P_{57}-P_{55} & P_{58}-P_{55} \\
P_{61} & P_{62}-P_{61} & P_{63}-P_{61} & P_{64}-P_{61} & P_{65} & P_{66}-P_{65} & P_{67}-P_{65} & P_{68}-P_{65} \\
P_{71} & P_{72}-P_{71} & P_{73}-P_{71} & P_{74}-P_{71} & P_{75} & P_{76}-P_{75} & P_{77}-P_{75} & P_{78}-P_{75} \\
P_{81} & P_{82}-P_{81} & P_{83}-P_{81} & P_{84}-P_{81} & P_{85} & P_{86}-P_{85} & P_{87}-P_{85} & P_{88}-P_{85}
\end{bmatrix}
\begin{bmatrix} Q_1+Q_2+Q_3+Q_4 \\ Q_2 \\ Q_3 \\ Q_4 \\ Q_5+Q_6+Q_7+Q_8 \\ Q_6 \\ Q_7 \\ Q_8 \end{bmatrix}
\tag{12}
$$

Then, making other rows of the capacitive volume subtract its first row.

$$
\begin{bmatrix} \Phi_1 \\ 0 \\ 0 \\ 0 \\ \Phi_5 \\ 0 \\ 0 \\ 0 \end{bmatrix}
=
\begin{bmatrix}
P_{11} & \cdots & P_{14}-P_{11} & P_{15} & \cdots & P_{18}-P_{15} \\
P_{21}-P_{11} & \cdots & P_{24}-P_{21}-(P_{14}-P_{11}) & P_{25}-P_{15} & \cdots & P_{28}-P_{25}-(P_{18}-P_{15}) \\
P_{31}-P_{11} & \cdots & P_{34}-P_{31}-(P_{14}-P_{11}) & P_{35}-P_{15} & \cdots & P_{38}-P_{35}-(P_{18}-P_{15}) \\
P_{41}-P_{11} & \cdots & P_{44}-P_{41}-(P_{14}-P_{11}) & P_{45}-P_{15} & \cdots & P_{48}-P_{45}-(P_{18}-P_{15}) \\
P_{51} & \cdots & P_{54}-P_{51} & P_{55} & \cdots & P_{58}-P_{55} \\
P_{61}-P_{51} & \cdots & P_{64}-P_{61}-(P_{54}-P_{51}) & P_{65}-P_{55} & \cdots & P_{68}-P_{65}-(P_{58}-P_{55}) \\
P_{71}-P_{51} & \cdots & P_{74}-P_{71}-(P_{54}-P_{51}) & P_{75}-P_{55} & \cdots & P_{78}-P_{75}-(P_{58}-P_{55}) \\
P_{81}-P_{51} & \cdots & P_{84}-P_{81}-(P_{54}-P_{51}) & P_{85}-P_{55} & \cdots & P_{88}-P_{85}-(P_{58}-P_{55})
\end{bmatrix}
\begin{bmatrix} Q_1+Q_2+Q_3+Q_4 \\ Q_2 \\ Q_3 \\ Q_4 \\ Q_5+Q_6+Q_7+Q_8 \\ Q_6 \\ Q_7 \\ Q_8 \end{bmatrix}
\tag{13}
$$

From the transformation matrix, four relation matrices yield as follows:

$$
\begin{aligned}
M_1 &= \begin{bmatrix}
P_{22}-P_{21}-(P_{12}-P_{11}) & \cdots & P_{24}-P_{21}-(P_{14}-P_{11}) & P_{26}-P_{25}-(P_{16}-P_{15}) & \cdots & P_{28}-P_{25}-(P_{18}-P_{15}) \\
P_{32}-P_{31}-(P_{12}-P_{11}) & \cdots & P_{34}-P_{31}-(P_{14}-P_{11}) & P_{36}-P_{35}-(P_{16}-P_{15}) & \cdots & P_{38}-P_{35}-(P_{18}-P_{15}) \\
P_{42}-P_{41}-(P_{12}-P_{11}) & \cdots & P_{44}-P_{41}-(P_{14}-P_{11}) & P_{46}-P_{45}-(P_{16}-P_{15}) & \cdots & P_{48}-P_{45}-(P_{18}-P_{15}) \\
P_{62}-P_{61}-(P_{52}-P_{51}) & \cdots & P_{64}-P_{61}-(P_{54}-P_{51}) & P_{66}-P_{65}-(P_{56}-P_{55}) & \cdots & P_{68}-P_{65}-(P_{58}-P_{55}) \\
P_{72}-P_{71}-(P_{52}-P_{51}) & \cdots & P_{74}-P_{71}-(P_{54}-P_{51}) & P_{76}-P_{75}-(P_{56}-P_{55}) & \cdots & P_{78}-P_{75}-(P_{58}-P_{55}) \\
P_{82}-P_{81}-(P_{52}-P_{51}) & \cdots & P_{84}-P_{81}-(P_{54}-P_{51}) & P_{86}-P_{85}-(P_{56}-P_{55}) & \cdots & P_{88}-P_{85}-(P_{58}-P_{55})
\end{bmatrix} \\[4pt]
M_2 &= \begin{bmatrix}
P_{12}-P_{11} & P_{13}-P_{11} & P_{14}-P_{11} & P_{16}-P_{15} & P_{17}-P_{15} & P_{18}-P_{15} \\
P_{52}-P_{51} & P_{53}-P_{51} & P_{54}-P_{51} & P_{56}-P_{55} & P_{57}-P_{55} & P_{58}-P_{55}
\end{bmatrix} \\[4pt]
M_3 &= -\begin{bmatrix}
P_{21}-P_{11} & P_{25}-P_{15} \\
P_{31}-P_{11} & P_{35}-P_{15} \\
P_{41}-P_{11} & P_{45}-P_{15} \\
P_{61}-P_{51} & P_{65}-P_{55} \\
P_{71}-P_{51} & P_{75}-P_{55} \\
P_{81}-P_{51} & P_{85}-P_{55}
\end{bmatrix} \\[4pt]
M_4 &= \begin{bmatrix}
P_{11} & P_{15} \\
P_{51} & P_{55}
\end{bmatrix}
\end{aligned}
\tag{14}
$$

With the above steps, the equivalent potential coefficient matrix can be obtained:

$$
\begin{bmatrix} P_{eq1} \\ P_{eq2} \end{bmatrix} = M_2 * (M_1^{-1} * M_3) + M_4
\tag{15}
$$

P_{eq1}, P_{eq2} represents the equivalent potential coefficient of capacitive volume 1 and 2, respectively.

In this way, simplified equivalent circuit is obtained without reducing the accuracy of partial elements. The approach in this paper is applicable to the structures with no need for three-dimensional meshing.

4   Examples

The following example considers the propagation of a signal on a microstrip with ground structure in Fig.5.

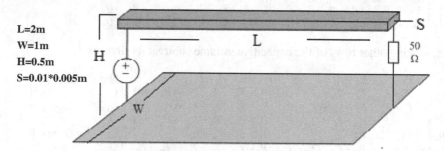

L=2m
W=1m
H=0.5m
S=0.01*0.005m

Fig. 5. Microstrip-ground structure

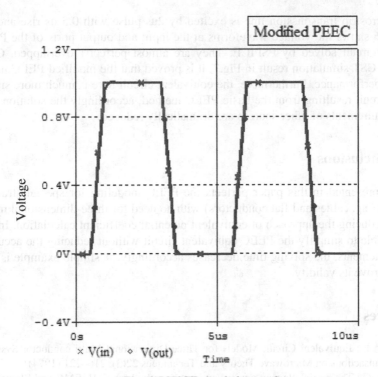

Fig. 6. PSPICE simulation result

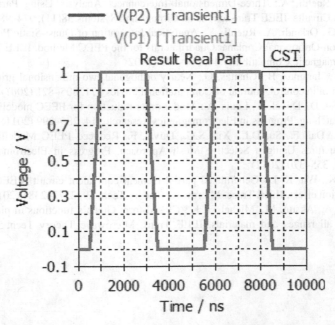

Fig. 7. CST simulation result (V(P1) is the input voltage, V(P2) is the output voltage.)

The microstrip transmission line is excited by 2us pulse with 0.5 us rise and fall time. Fig. 6 shows the voltage waveforms at the input and output ports of the PEEC equivalent circuit solved by PSPICE. They are almost perfectly overlapped. Compared with CST simulation result in Fig.7, it is proved that the modified PEEC model gets good performance. Furthermore, the equivalent circuit here is much more simple than the circuit resulting from the basic PEEC method, accordingly the solution time and the requirements of other sources are dramatically reduced.

5   Conclusions

The work presented in this paper perfects the PEEC modeling for special structure conductors (e.g. cables and flat conductors) with no need for three-dimensional meshing by introducing the approach of equivalent potential coefficient calculation. In this way, it is able to simplify the PEEC equivalent circuit without reducing the accuracy of partial elements, the solving time decreases accordingly. A simple example is illustrated to prove its validity.

References

1. Ruehli, A.E.: Equivalent Circuit Models for Three Dimensional Multi-conductor Systems. IEEE Transactions on Microwave Theory and Techniques 22(3), 216–221 (1974)
2. Antonini, G.: The partial element equivalent circuit method for EMI, EMC and SI analysis. ACES Newsletter 21(1), 8–32 (2006)
3. Heeb, H., Ruehli, A.: Three-Dimensional Interconnect Analysis Using Partial Element Equivalent Circuits. IEEE Transactions on Circuits and Systems 38(11), 974–981 (1992)
4. Antonini, G., Orlandi, A., Ruehli, A.: Analytical Integration of Quasi-Static Potential Integrals on Non-Orthogonal Coplanar Quadrilaterals for the PEEC Method. IEEE Transactions on Electromagnetic Compatibility 44, 399–403 (2002)
5. Akcelik, V., Jaramaz, B., Ghattas, O.: Nearly orthogonal two-dimensional grid generation with aspect ratio control. Journal of Computational Phyics 171, 805–821 (2001)
6. Song, Z., Su, D., Duval, F., Louis, A.: Model order reduction for PEEC modeling based on moment matching. Progress in Electromagnetics Research 114, 285–299 (2011)
7. Song, Z.F., Dai, F., Su, D.L., Xie, S.G., Duval, F.: Reduced PEEC Modeling of Wire-ground Sructures Using a Selective Mesh Approach. Progress in Electromagnetics Research 123, 355–370 (2012)
8. Yeung, L.K., Wu, K.-L.: Generalized partial element equivalent circuit (PEEC) modeling with radiation effect. IEEE Trans. Microwave Theory Tech. 59, 2377–2384 (2011)
9. Alparslan, A., Aksun, M., Michalski, K.: Closed-form Green's functions in planar layered media for all ranges and materials. IEEE Trans. Microwave Theory Tech. 58, 602–613 (2010)

A Novel Image Retrieval Method Based on Mutual Information Descriptors

Gang Hou[1,2], Ke Zhang[3], Xiaoxue Zhang[3], Jun Kong[3], and Ming Zhang[3,4,*]

[1] College of Communication Engineering of Jilin University, Changchun, China
[2] College of Humanities & Sciences of Northeast Normal University, Changchun, China
[3] School of Computer Science and Information Technology,
Northeast Normal University, Changchun, China
[4] Key Laboratory of Symbolic Computation and Knowledge Engineering of Ministry of
Education, Jilin University, Changchun, China
zhangm545@nenu.edu.cn

Abstract. In this paper, we propose a novel image retrieval method based on mutual information descriptors (MIDs). Under the physiological property of human eyes and human visual perception theory, MIDs are extracted to encode the internal correlation relationship among multiple image feature spaces, characterizing image contents with mutual information features based on the low-level image features, such as color, shape etc., then, the mutual information features fusion strategy is used to imitate the information transfer process in nervous system. When using the MIDs proposed to image retrieval, we can get many advantages such as low dimensionality, a certain robustness of geometric distortions and noise, and describing the human visual retrieval mechanism effectively. Experimental results show that MIDs have high indexing and retrieving performance compared with existing methods for content-based image retrieval (CBIR).

Keywords: Mutual Information Descriptors (MIDs), Content-based Image Retrieval (CBIR), Human Visual Perception, Feature Fusion.

1 Introduction

Image retrieval driven by human visual perception strategy can valid description image content, improve retrieval performance and efficiency [1]. To date, it has become one of the research trends and hot subjects in content-based image retrieval technology domain. Hence, how to make use of human visual perception mechanism for image feature description is still the key part.

In this paper, we propose a novel method called mutual information descriptors (MIDs) for content-based image retrieval task. MIDs can implement the image retrieval efficiently by imitating the human visual perception mechanism. This paper

---

* Corresponding author.

D.-S. Huang et al. (Eds.): ICIC 2013, LNAI 7996, pp. 435–442, 2013.

is organized as follows: Section 2 introduces the related theories and methods of human visual perception mechanism and content-based image retrieval. We detail MIDs method including feature extraction and fusion strategy in Section 3. In Section 4, the retrieval experiments based on MIDs and performance evaluation are reported. Conclusions are given in Section 5.

2   Related Theoretical Basis

- Human visual perception mechanism

The source of the visual attention is the difference of visual information feature. Various color, shape and texture, as well as velocity change of a mobile can cause visual difference, results in visual attention. Visual attention and perception are similar to feature extraction and fusion processes in pattern recognition fields.

Psychological and neurological studies indicate, most researchers agree with two-stage model hypothesis on visual perception mechanism [2-7]. Treisman proposed human visual process includes two stages: a pre-attentive stage for visual attention and an attentive stage for visual perception. At the pre-attentive stage, all low-level visual information features are concurrently extracted and roughly expressed by retina cells, they serve as the bases of feature fusion process, at the attentive stage, balance the importance factor according to the visual feature difference, and re-combine all of the bases to reconstructed objects.

- Micro-structure descriptor

Micro-structure descriptor is proposed by Liu [8] based on statistical texture, which considered natural image texture is composed of many micro-structures, and the output take the form of color layout information. MSD method is a novel and effective approach for CBIR integrating color feature and shape feature, but MSD pays more attentions to color than the other features for retrieval image content. However, in practice, we can act up to the visual perception rule for the image retrieval, taking into account of the internal correlation relationships among multiple image feature spaces, and make use of color and shape feature representation in parallel.

3   Mutual Information Descriptors (MIDs)

Application in reality, color features and shape features can effectively characterize the image structure. Therefore, color and shape feature extraction acts an important role in content-based image retrieval task, and mine the feature internal correlation relationships when extract features, be interdependent and blend well. Taking into account the shape information when perceiving color features, as well as the color information based on shape features, this is the main idea of the proposed Mutual information descriptors (MIDs) for content-based image retrieval in this paper.

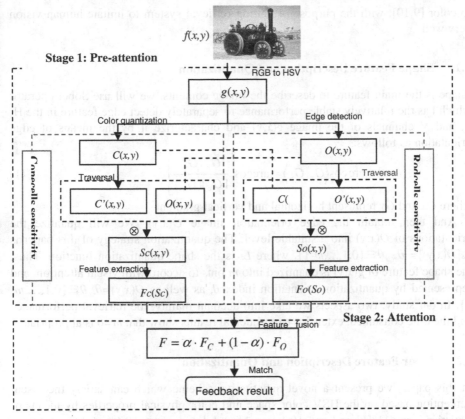

Fig. 1. The flow chart of MIDs-based image retrieval

Mutual information descriptors are based on human visual perception mechanism, utilizing the internal correlation relationships among multiple image feature spaces, and extracting features to describe the images for retrieval. The flow chart of MIDs-based image retrieval is as shown in Fig.1.

Firstly, at the pre-attentive stage, construct the color (shape) basal structure map, and construct mutual structure map utilizing the feature internal correlation relationships based on the above basal structures, then, extract mutual information features, including: shape-based color structure feature (SCSF) and color-based shape structure feature (CSSF), they separately correspond to the color sensitive map (CSM) and shape sensitive map (SSM) charged by the cones and cods.

Secondly, at the attentive stage, fuse the features of SCSF and CSSF to propose the mutual information descriptors for describing the image, in order to imitate the visual perception process that the nervous system deals with visual information.

At last, extract the query image features and the dataset image features using the mutual information descriptors with reference to the above steps, respectively, match feature, and implement retrieval.

For a full color nature image $f(x,y)$, we express it as $g(x,y)$ in the HSV color space and, which is perceptually uniform color space and in line with the perception of eyes to color [9,10], with the purpose of getting retrieval system to imitate human vision very well.

3.1   Shape Feature Description and Quantization

Shape is the main feature to describe the image contents, we will use Sobel operator which has the relatively stable performance to separately detect edge feature in the H, S, and V channels of the image $g(x,y)$ and characterize it by the means of edge orientation as follows:

$$\theta = \arccos\langle G_x, G_y \rangle = \arccos\left[\frac{G_x \cdot G_y}{|G_x \| G_y|}\right]$$ (1)

where Gx and Gy represent horizontal and vertical gradient.

And then, obtain the edge orientation image $O(x,y)$. We will quantize the orientations of $O(x,y)$ into a suitable level. The quantization strategy of this paper is $Ls[O(x,y)] = \theta$, $\theta \in \{0,1,...,m-1\}$, where Ls is the shape quantization function. Thus, the shape feature of $g(x,y)$ is quantized into m bins to accord with visual attention, and represented by quantization orientation index θ, as well as $O(x,y)=\theta$, $\theta \in \{0,1,...,m-1\}$. In our study, two aspects we take into account mainly, the retrieval performance and the time consumed, extensive experimental results show that $m=6$ is appropriate.

3.2   Color Feature Description and Quantization

In this paper, we present a novel quantitative scheme which can satisfy the visual perception based on the HSV color space and color physical properties by means of statistics, as $Lc[h]=a$, $a \in \{0,1,...,n-1\}$, $Lc[s]=b$, $b \in \{0,1,...,k-1\}$, $Lc[v]=c$, $c \in \{0,1,...,l-1\}$, where Lc is the color quantization function. In our scheme H component is quantized into 8 non-equal interval bins and the others are 3 bins with equal interval according to the scheme.

3.3   Mutual Information for Image Retrieval

- Mutual information extraction and roughly expression at the pre-attentive stage

In this paper, we will adopt the strategy of two-level structure maps to extract mutual information features. Two-level structure maps include basic structure map and mutual structure map, as following:

Level 1. For color and shape index images, build the basic color structure map $C'(x,y)$ and the basic shape structure map $O'(x,y)$ in parallel, respectively: the index images are manipulated using template by keeping the point with the same value of center, and deleting the ones who have jumpy values (Section 4 in details), shown in Fig.2.

Level 2 (Color/shape). According to basic structure map, build mutual structure map, and extract mutual information features.

As shown in biology, color and shape features are obtained through the cones and cods without order, respectively, and they are interdependent and blend well. Therefore, we can use MIDs to build the mutual structure map $S(x,y)$ (shown in Fig. 3) by mutual fashion, as follows:

$$S(x,y) = \begin{cases} S_c(x,y) = O(x,y) \otimes C'(x,y) \\ S_o(x,y) = O'(x,y) \otimes C(x,y) \end{cases} \tag{2}$$

where $S_c(x,y)$ is shape-based color structure map(SCSM), $So(x,y)$ is color-based shape structure map(CSSM). Here $\otimes$ indicates the filtering operation as follows:

Characterize shape-based structure map using color feature to preserve the color within the shape-based structure map and obtain SCSM, $S_c(x,y)$.

Characterize color-based structure map using shape feature, to preserve the orientation within the color-based structure map and obtain CSSM, $S_O(x,y)$.

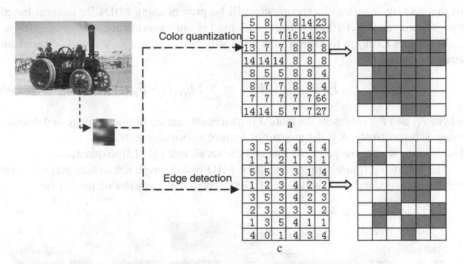

Fig. 2. Basic structure extraction process. a. color index image, b. basic color structure map, c. orientation index image, d. basic shape structure map.

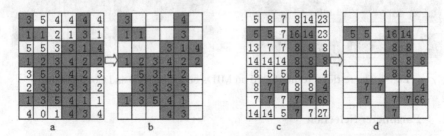

Fig. 3. Features extraction strategy a. SCSM, b.Sc(x,y), c.CSSM, d.So (x,y)

Based on the mutual structure maps, extract mutual structure features the $F_C(Sc)$ and the $F_O(So)$, as follows:

$$F_C(Sc)=[q(0),q(1),...,q(\omega),...,q(nkl-1)]^T \qquad (3)$$
$$F_O(So)=[p(0),p(1),...,p(\theta),...,p(m-1)]^T \qquad (4)$$

- Mutual information feature fusion and optimization at the attentive stage

We present the mutual information feature fusion strategy as follows:

$$F = \alpha \cdot F_C + (1-\alpha) \cdot F_O \qquad (5)$$

where α is balance parameter.

4   Experiments and Evaluation

In this section, a series of experiments will be present using MIDs for content-based image retrieval. Corel-10000 dataset is adopted in our experiments.

We select *L1*-norm to measure the similarity between images through comparison analysis, as follows:

$$L(F_Q, \ F_D)=\|F_Q-F_D\|_1 = \sum_{i=1}^{K} |F_Q(i) - F_D(i)| \qquad (6)$$

where F_Q and F_D, represent the mutual information feature of query image and dataset images, respectively, K is the number of feature vector elements.
In addition, we use the precision rate and the recall rate [11, 12] to quantitatively.

The following figures are the results of MID-based image retrieval experiments on the datasets of Corel-5000 and Corel-10000, where the query lies on the left top.

Fig. 4. The retrieval results based on MIDs on Corel-5000 and Corel-10000

4.1   Parameter Selection

The basic structure maps are built at the pre-attentive stage. For color (shape) index image, we select 3*3 templates as in [8].

To reduce computational complexity, when the balance parameter α is 0.15, we can achieve a better retrieval result. Thus, for Corel datasets, we uniformly choose α=0.15 in our experiments.

4.2 Experiment Comparison

To further demonstrate the performance of MIDs-based retrieval, we compare MIDs with other existing effective methods on the Corel datasets. They are MPEG-7 edge histogram descriptor (EHD) [13], and micro-structure descriptor (MSD) [8]. Experiment results are shown in Table 1.

From Table 1, we can see that the precision rate of MIDs is much higher than that of EHD, even higher 4 percent than that of MSD, recall rate has the same results on the Corel-5000. And we are glad to see that the precision rate and recall rate of MIDs aslo keep ahead compared to others, and be more stable and robust.

Table 1. Precision rate and recall rate of EHD, MSD, and MIDs (%)

	Corel-5000		Corel-10000	
	Precision rate	Recall rate	Precision rate	Recall rate
EHD	21.45	2.57	17.25	2.07
MSD	55.92	6.71	45.62	5.48
MIDs	59.37	7.12	48.89	5.97

5   Conclusions

In this paper, we propose a novel method for content-based image retrieval, mutual information descriptors (MIDs). The method depends on the visual perception mechanism. For cones and cods are sensitive to color and shape, respectively, we extract mutual information features which include color and shape information based on the internal correlation relationships among multiple image feature spaces at the pre-attentive stage, at the attentive stage, imitate the information transfer process of nervous system to fuse mutual information features, and retrieve image finally. The mutual information extraction and presentation conform to the nature of information obtained by retina cells. MIDs effectively characterize human visual retrieval mechanism. Experiments results indicated, MIDs compared with existing methods can precisely and comprehensively retrieve images we want, and have a high performance.

Acknowledgements. This work is supported by the Young Scientific Research Foundation of Jilin Province Science and Technology Development Project (No. 201201063), Key Laboratory of Symbolic Engineering of Ministry of Education (No. 93K172012K13), the Jilin Provincial Natural Science Foundation (No. 201115003), the Fund of Jilin Provincial Science & Technology Department (No. 20111804), the Education Department of Jilin province "Twelfth Five-Year" science and technology research projects (No. 2012413).

References

1. Fua, H., Chia, Z., Feng, D.: Attention-driven image interpretation with application to image retrieval. Pattern Recognition 39, 1604–1621 (2006)
2. Treisman, A.: A feature in integration theory of attention. Cognitive Psychology 12(1), 97–136 (1980)
3. Theeuwes, J.: Visual selective attention: a theoretical analysis. Acta Psychologica 83, 93–154 (1993)
4. Steinman, S., Steinman, B.: Computational models of visual attention. In: Hung, G.K., Ciuffreda, K.J. (eds.) Models of the Visual System, ch. 14, pp. 521–563 (2002)
5. Wolfe, J.: The level of attention: mediating between the stimulus and perception. In: Harris, L., Jenkin, M. (eds.) Levels of Perception, ch. 9, pp. 169–191 (2002)
6. Joseph, J., Chun, M., Nakayama, K.: Attentional requirements in a 'preattentive' feature search task. Nature 387, 805–807 (1997)
7. Niebur, E., Koch, C.: Computational architectures for attention. In: Parasuraman, R. (ed.) The Attention Brain. MIT Press, Cambridge (1998)
8. Liu, G.H., Li, Z.Y., Zhang, L., Xu, Y.: Image retrieval based on micro-structure descriptor. Pattern Recognition 44, 2123–2133 (2011)
9. Julesz, B.: Textons, the elements of texture perception and their interactions. Nature 290(5802), 91–97 (1981)
10. Burger, W., Burge, M.J.: Principles of Digital image processing: Core Algorithms. Springer (2009)
11. Müller, H., Müller, W., Squire, D.G., Maillet, S.M., Pun, T.: Performance evaluation in content-based image retrieval: over view and proposals. Pattern Recognition Letter 22(5), 593–601 (2001)
12. Yang, Y.: An evaluation of statistical approaches to text categorization. Information Retrieval 1(1-2), 69–90 (1999)
13. Yoon, S.J.: Image retrieval using a novel relevance feedback for edge histogram descriptor of MPEG-7

Tumor Gene Expressive Data Classification Based on Locally Linear Representation Fisher Criterion

Bo Li[1,2], Bei-Bei Tian[1,2], and Jin Liu[3]

[1] School of Computer Science and Technology, Wuhan University of Science
and Technology, 430081, Wuhan, China
[2] Hubei Province Key Laboratory of Intelligent Information Processing and
Real-time Industrial System, Wuhan China
[3] State Key Lab. of Software Engineering, Wuhan, China
liberol@126.com

Abstract. In this paper, a discriminant manifold learning method based on Locally Linear Embedding (LLE), which is named Locally Linear Representation Fisher Criterion (LLRFC), is proposed for the classification of tumor gene expressive data. In the proposed LLRFC, an inter-class graph and intra-class graph is constructed based on the class information of tumor gene expressive data, where the weights between nodes in both graph are optimized using locally linear representation trick. Moreover, a Fisher criterion is modeled to maximize the inter-class scatter and minimize the intra-class scatter simultaneously. Experiments on some benchmark tumor gene expressive data validate its efficiency.

Keywords: LLE, Fisher criterion, Tumor gene expressive data.

1   Introduction

With the emergence of DNA microarrys, it occurs to be highlight to simultaneously monitor expression of all genes in the genome. Increasingly, it is still a challenge how to interpret such data and make insight into biological processes and the mechanisms of human disease. Up to now, many studies have been reported on microarray gene expression data analysis for molecular classification of cancer [1, 2, 3]. Among all the methods, principal component analysis (PCA) [4] is a frequently used approach which can obtain an up-front characterization of the structure of the data. However, the resulting components by performing PCA are global interpretations and lack intuitive meanings. Independent component analysis (ICA) [5, 6] is a useful modification of PCA, where context has been extended with blind separation of independent sources from their linear mixtures instead of the mutually independent coefficients [6]. Thus it will result in that higher-order statistics are indispensable in determining the ICA expansion. ICA aims to find the maxima of a target function in a large-dimensional configuration space, which had also been resolved by Chiappetta [7]. There exists a disadvantage in the two methods mentioned above that they characterize the expression data with the linear tricks and most nonlinear information hidden in the

D.-S. Huang et al. (Eds.): ICIC 2013, LNAI 7996, pp. 443–449, 2013.

original data will be destroyed. To overcome the problem described above, a new technique to extract relevant biological correlations or "molecular logic" in gene expression data is proposed, where manifold method based on Locally Linear Embedding (LLE) with discriminant learning is involved. LLE is firstly presented by Roweis and Saul, which constitutes local coordinates with the least constructed cost and then maps them to a global one [8,9]. In the proposed method, the local linear reconstruction trick is also introduced to approach the weights between nodes in the inter-class graph and the intra-class graph, both of which can be constructed according to the labels. At last a Fisher crieterion based on inter-class scatter and the intra-class scatter can be reasoned to find the optimal projections.

The paper is organized as follows. Section 2 describes classical LLE algorithm. Section 3 presents the proposed algorithm. Some experimental results and simulations are offered in Section 4. Then the whole paper is finished with some conclusions in Section 5.

2   Locally Linear Embedding

Let $X = [X_1, X_2, ..., X_n] \in R^{D \times n}$ be n points in a high dimensional space. The data points are well sampled from a nonlinear manifold, of which the intrinsic dimensionality is d ($d \ll D$). The goal of LLE is to map the high dimensional data into a low dimensional manifold space. Let us denote the corresponding set of n points in the embedding space as $Y = [Y_1, Y_2, ..., Y_n] \in R^{d \times n}$. The outline of LLE can be summarized as follows:

Step1: For each data point X_i, identify its k nearest neighbors by kNN criterion or ε – ball criterion;

Step2: Compute the optimal reconstruction weights which can minimize the error of linearly reconstructing X_i by its k nearest neighbors;

Step3: Compute the low-dimensional embedding Y for X that best preserves the local geometry represented by the reconstruction weights

Step1 is typically done by using Euclidean distance to define neighborhood, although more sophisticated criteria may also be used, such as Euclidean distance in kernel space or cosine distance. *Step 2* seeks the best reconstruction weights. Optimality is achieved by minimizing the local reconstruction error of X_i,

$$\varepsilon_i(W) = \arg\min \left\| X_i - \sum_{j=1}^{k} W_{ij} X_j \right\|^2 \tag{1}$$

Step 3 computes the optimal low dimensional embedding Y based on the weight matrix W obtained from *Step 2*.

$$\varepsilon(Y) = tr\left\{ \sum_{ij} M_{ij} Y_i^T Y_j \right\} = tr\left\{ YMY^T \right\} \tag{2}$$

So based on the weighted matrix W , a sparse, symmetric and positive semi-definite matrix M can be defined as follows.

$$M = (I - W)^T (I - W) \tag{3}$$

3   Method

3.1   Motivation

The original LLE is often applied for data visualization because it can probe the intrinsic low dimensional manifold structure embedded in high dimensional data space. However, LLE can not efficiently extract the features for classification. The reason probably lies in that LLE nonlinearly mines the high dimensional data by preserving the local manifold structure without considering the class information, i.e. LLE explores manifold locality by KNN graph, which is composed of data with the same label as well as data with different labels according to the sorted Euclidean distances. Due to the prosperity of locality preserving, the neighborhood for any point will be mapped into a low dimensional space without any alteration. Firstly, for any point, its k nearest neighbors will not change. Secondly, the reconstruction weights among a point and its k nearest neighbors will be preserved. At last, the same conclusion can be drawn to the labels of any point and its k nearest neighbors. However, in most cases, any points and its k nearest neighbors are labeled with various classes, which results in great difficulties in data classifying.

Therefore, both manifold locality preserving and class information should be combined to improve the recognition performance of LLE. On the one hand, the locality should be preserved, which helps to mine manifold distributed data; on the other hand, the class information can be introduced to supervise the construction of neighborhood to improve the classification performance. Based on the above motivations, a novel supervised method is proposed, where an inter-class graph and an intra-class graph characterizing the corresponding data are constructed on the basis of data labels. In the inter-class graph, the nearest neighbors of any point are sampled from points with different labels. While in the intra-class graph, any neighborhood contains points with the same class. Moreover, either the inter-class graph or the intra-class graph is constructed using KNN criterion, which can approximately approach the locality of manifold. Thus, both class information and manifold locality preserving can be well integrated for dimensionality reduction.

3.2   The Intra-class Graph and The Inter-class Graph

On the basis of the data labels, an intra-class graph can be constructed. For any sample, its k nearest neighbors should be of the same label as the sample; meanwhile, these neighbors are selected with the first k bottom Euclidean distances to the sample. In the intra-class graph, due to its local linearity, the sample can be well reconstructed by its k nearest neighbors with the optimal weights, which is stated as follows.

$$\varepsilon\left(W_{int\,ra}\right) = \min \left\| X_i - \sum_{j=1}^{k} W_{ij} X_j \right\|^2 \tag{4}$$

Suffered the sum-to-one constraint for the reconstruction weights, Eq (4) can also be rewritten as follows.

$$\varepsilon\left(W_{int\,ra}\right) = \min \left\| \sum_{j=1}^{k} W_{ij} X_i - \sum_{j=1}^{k} W_{ij} X_j \right\|^2 = \min \left\| \sum_{j=1}^{k} W_{ij} (X_i - X_j) \right\|^2 \tag{5}$$

where $IntraN(X_i)$ denotes the intra-class neighborhood of point X_i.

Meanwhile, the inter-class graph can also be established as follows. Firstly, for any point X_i, super-ball or KNN criterion is introduced to determine its pre-defined neighborhood; secondly, the points located in the pre-defined neighborhood is sorted according to their ascending Euclidean distances to point X_i and then the bottom k points with different labels to that of point X_i are selected as its inter-class neighbors, which comprise the inter-class neighborhood of point X_i. The same process as in the intra-class graph is repeated to achieve the optimal reconstruction weights in the inter-class graph.

$$\varepsilon\left(W_{int\,er}\right) = \min \left\| \sum_{j=1}^{k} W_{ij} X_i - \sum_{j=1}^{k} W_{ij} X_j \right\|^2 = \min \left\| \sum_{j=1}^{k} W_{ij} (X_i - X_j) \right\|^2 \tag{6}$$

where $InterN(X_i)$ stands for the inter-class neighborhood of point X_i.

3.3   Justfication

As mentioned above, both the intra-class graph and the inter-class graph can be constructed with the corresponding optimal weights between nodes. Moreover, the proposed LLRFC aims to explore a low dimensional space with the best classification accuracy, where the data with the same label should be more clustered and the points with different labels should be projected farther. So based on the intra-class graph and the inter-class graph, a Fisher criterion can be modeled to maximize the inter-class graph scatter and minimize the intra-class graph scatter simultaneously.

$$S(Y) = \max \frac{tr\left\{YM_{int\,er}Y^T\right\}}{tr\left\{YM_{int\,ra}Y^T\right\}} \tag{7}$$

where

$$M_{int\,ra} = (I - W_{int\,ra})^T (I - W_{int\,ra}) \tag{8}$$

$$M_{int\,er} = (I - W_{int\,er})^T (I - W_{int\,er}) \tag{9}$$

For Eq. (7), a linear transformation $Y = A^T X$ is introduced and Lagrange multiplier method can also be adopted to find the linear transformation A as expected with the following Eq (10).

$$XM_{\text{int}\,er}X^T A = \lambda XM_{\text{int}\,ra}X^T A \qquad (10)$$

It can be concluded that A is spanned by the eigenvectors corresponding to top d eigenvalues of the above generalized eigen-equation.

4   Experiments

In this Section, the proposed LLRFC is compared with several related dimensionality reduction methods including LDA, LLDE [10], and neighborhood preserving embedding (NPE) [11]. And then, LDA, NPE, LLDE and LLRFC are exploited. At last, a classifier with the nearest neighbor criterion is adopted to identify the features extracted by LDA, NPE, LLDE and the proposed LLRFC, respectively.

4.1   Experiment on DLBCL Data

Diffuse large B-cell lymphomas (DLBCL) and follicular lymphomas (FL) are two B-cell lineage malignancies that have very different clinical presentations, natural histories and response to therapy, where DLBCL and FL contain 58 and 19 examples with 7070 genes, respectively. However, FLs frequently evolve over time and acquire the morphologic and clinical features of DLBCLs and some subsets of DLBCLs have chromosomal translocations characteristic of FLs. The gene-expression based classification model was built to distinguish DLBCL data. Shown in Table.1 is the optimal accuracy. It is found that the proposed method outperforms the other techniques such as LDA, NPE and LLDE.

Table 1. Performances comparison on DLBCL data

Approaches	Recognition rate	Dimensions
LLRFC	95.56%	6
LDA	93.33%	2
LLDE	93.33%	2
NPE	93.33%	2

4.2   Experiments on High-Grade Glioma Data

High-grade glioma samples were carefully selected including 28 glioblastomas and 22 anaplastic oligodendrogliomas, all of which were primary tumors sampled before therapy. The classic subset of tumors was cases diagnosed similarly by all examining pathologists, and each case resembled typical depictions in standard textbooks. A total of 21 classic tumors were selected, and the remaining 29 samples were considered nonclassic tumors, lesions for which diagnosis might be controversial. The goal here is to separate the glioblastomas from the anaplastic oligodendrogliomas, which allows appropriate therapeutic decisions and prognostic estimation. Shown in Table 2 is the separation result by using LLRFC, LDA, LLDE and NPE, respectively.

Table 2. Performances comparision on high-grade glioma data

Approaches	Recognition rate	Dimensions
LLRFC	84.17%	18
LDA	75.79%	2
LLDE	80.72%	20
NPE	80.72%	20

5   Conclusions

In this paper, a discriminant manifold learning method, namely Locally Linear Representation Fisher Criterion (LLRFC), is proposed for tumor gene expressive data classification. The proposed algorithm uses the label information to construct the inter-class graph and the intra-class graph respectively and then local linear reconstruction is introduced in both graphs. At last a Fisher criterion is constructed to explore the discriminate subspace for classification based on the inter-class scatter and the intra-class scatter. So the proposed algorithm becomes more suitable for the tasks of classification. This result is validated by experiments on real-world data set.

Acknowledgments. This work was partly supported by the grants of the National Natural Science Foundation of China (6100507, 61100106, 61273303, 61273225 &U1135005), China Postdoctoral Science Foundation (20100470613 & 201104173), Natural Science Foundation of Hubei Province (2010CDB03302 &2011CDC076), the Research Foundation of Education Bureau of Hubei Province (Q20121115), the Program of Wuhan Subject Chief Scientist (201150530152), Hong Kong Scholars Program (XJ2012012) and the Open Project Program of the National Laboratory of Pattern Recognition (201104212).

References

1. Alon, A.: Broad Patterns of Gene Expression Revealed by Clustering Analysis of Tumorand Normal Colon Tissues Probed by Clustering Analysis of Tumor and Normal Colon Tissues Probed by Oligonucleotide Arrays. Proc. Natl Acad. Sci. 96, 6745–6750 (1999)
2. Bittner, M.: Molecular Classification of Cetaceous Malignant Melanoma by Gene Expression Profiling. Nature 406, 536–540 (2000)
3. Furey, T.S., Cristianini, N., Duffy, N., Bednarski, D.W., Schummer, M., Haussler, D.: Support Vector Machines Classification and Validation of Cancer Tissue Samples Using Microarray Expression Data. Bioinformatics 16, 906–914 (2000)
4. Hoyer, P.O.: Non-negative Matrix Factorization with Sparseness Constraints. J. Mach.Learn. Res. 5, 1457–1469 (2004)

5. Gao, Y., Church, G.: Improving Molecular Cancer Class Discovery Through Sparse Nonnegative Matrix Factorization. Bioinformatics 21, 3970–3975 (2005)
6. Huang, D.S., Zheng, C.H.: Independent Component Analysis Based Penalized Discriminate Method for Tumor Classification Using Gene Expression Data. Bioinformatics 22, 1855–1862 (2006)
7. Comon, P.: Independent Component Analysis— A New Concept. Signal Processing 36, 287–314 (1994)
8. Chiappetta, P., Roubaud, M.C., Torresani, B.: Blind Source Separation and the Analysis of Microarray Data. Journal of Computational Biology 11, 1090–1109 (2004)
9. Roweis, S.T., Saul, L.K.: Nonlinear dimensionality reduction by locally linear embedding. Science 290, 2323–2326 (2000)
10. Saul, L.K., Roweis, S.T.: Think Globally, Fit Locally: Unsupervised Learning of Low Dimensional Manifolds. J. Mach. Learning Res. 4, 119–155 (2003)
11. Li, B., Zheng, C., Huang, D.: Locally linear discriminant embedding: an efficient method for face recognition. Pattern Recognition 41(12), 3813–3821 (2008)
12. He, X., Cai, D., Yan, S.: Neighborhood preserving embedding. In: Proceedings of the 10th IEEE International Conference on Computer Vision, pp. 1208–1213 (2005)

Sparse Signal Analysis Using Ramanujan Sums

Guangyi Chen[1], Sridhar Krishnan[2], Weihua Liu[3], and Wenfang Xie[4]

[1] Department of Computer Science and Software Engineering,
Concordia University, Montreal, Quebec, Canada H3G 1M8
guang_c@cse.concordia.ca

[2] Department of Computer and Electrical Engineering, Ryerson University, Toronto,
Ontario, Canada
Krishnan@ee.ryerson.ca

[3] State Key Lab. of Virtual Reality Technology and Systems, Beihang University,
ZipCode 100191, No 37, Xueyuan Rd., Haidian District, Beijing, P.R. China
liuwh_99@hotmail.com

[4] Department of Mechanical & Industrial Engineering, Concordia University,
Montreal, Quebec, Canada H3G 1M8
wfxie@encs.concordia.ca

Abstract. In this paper, we perform sparse signal analysis by using the Ramanujan Sums (RS). The RS are orthogonal in nature and therefore offer excellent energy conservation. Our analysis shows that the RS can compress the energy of a periodic impulse chain signal into fewer number of RS coefficients than the Fourier transform (FT). In addition, the RS are faster than the FT in computation time because we can calculate the RS basis functions only once and save them to a file. We can retrieve these RS basis functions for our calculation instead of computing them online. To process a signal of 128 samples, we spend 1.0 millisecond for the RS and 5.82 milliseconds for the FT by using our unoptimized Matlab code.

Keywords: Ramanujan Sums (RS), Fourier transform (FT), sparse representation, Gaussian white noise.

1 Introduction

Sparse and overcomplete representations have been studied by researchers for many years. Given an input signal, we want to project it onto some basis functions so that they can generate sparse coefficients, which are very important in a number of applications. The RS were invented by R. Ramanujan in 1918 [1], and were introduced to the signal processing community recently ([2]-[6]). The RS are orthogonal in nature and therefore offer excellent energy conservation, similar to the Fourier transform (FT). The RS are operated on integers and hence can obtain a reduced quantization error implementation. In addition, the RS are much faster than the FT in computing time because we can retrieve the RS basis functions omputed

D.-S. Huang et al. (Eds.): ICIC 2013, LNAI 7996, pp. 450–456, 2013.

before, instead of computing them online. Even though the RS have so many important properties, they have not been studied for sparse signal analysis. In this paper, we find that the RS can compress the energy of a periodic impulse chain signal into fewer number of RS coefficients than the FT. In addition, the RS are faster than the FT in computation time. It is an interesting job to find other types of signals that the RS can serve as a sparse basis. We believe that the RS can only be a sparse basis for a few classes of signals instead of for all signal classes.

2   Ramanujan Sums (RS) for Sparse Signal Analysis

The RS are the nth powers of qth primitive roots of unity, defined as

$$c_q(n) = \sum_{p=1;(p,q)=1}^{q} \exp(2i\pi \frac{p}{q} n) \tag{1}$$

where $(p,q) = 1$ means that the greatest common divisor (GCD) is unity, i.e., p and q are co-primes. An alternate computation of RS can be given as

$$c_q(n) = \mu(\frac{q}{(q,n)}) \frac{\phi(q)}{\phi(\frac{q}{(q,n)})} \tag{2}$$

Let $q = \prod_i q_i^{\alpha_i}$ (q_i prime). Then, we have $\phi(q) = q \prod_i (1 - \frac{1}{q_i})$. The mobius function $\mu(n)$ is equal to 0 if n contain a square number; 1 if $n=1$; and $(-1)^k$ if n is a product of k distinct prime numbers. The first few values of $c_q(n)$ are given as follows:

$$c_1 = <1>, \ c_2 = <-1,1>, \ c_3 = <-1,-1,2>, \ c_4 = <0,-2,0,2>, \ ...$$

where $<>$ indicates the period. For example, $c_4(1) = 0$, $c_4(2) = -2$, $c_4(3) = 0$, $c_4(4) = 2$, $c_4(5) = 0$, $c_4(6) = -2$, etc. We give Table 1 for $c_q(n)$ with $q \in [1,15]$ here.

The RS has the following multiplicative property:

$$c_{qq'}(n) = c_q(n)c_{q'}(n) \ \text{if} \ (q,q')=1. \tag{3}$$

and also the orthogonal property:

$$\sum_{n=1}^{qq'} c_q(n)c_{q'}(n) = 0 \ \text{if} \ q \neq q'. \tag{4}$$

$$\sum_{n=1}^{q} c_q^2(n) = q\phi(q) \ \text{otherwise.} \tag{5}$$

The 1D RS transform of a signal $x(n)$ is defined as

$$r_q = \frac{1}{\phi(q)} \lim_{M \to \infty} \frac{1}{M} \sum_{m=1}^{M} x(m)c_q(m) \tag{6}$$

Table 1. The RS basis $c_q(n)$ for $q \in [1,15]$

q															
1	1														
2	-1	1													
3	-1	-1	2												
4	0	-2	0	2											
5	-1	-1	-1	-1	4										
6	1	-1	-2	-1	1	2									
7	-1	-1	-1	-1	-1	-1	6								
8	0	0	0	-4	0	0	0	4							
9	0	0	-3	0	0	-3	0	0	6						
10	1	-1	1	-1	-4	-1	1	-1	1	4					
11	-1	-1	-1	-1	-1	-1	-1	-1	-1	-1	10				
12	0	2	0	-2	0	-4	0	-2	0	2	0	4			
13	-1	-1	-1	-1	-1	-1	-1	-1	-1	-1	-1	-1	12		
14	1	-1	1	-1	1	-1	-6	-1	1	-1	1	-1	1	6	
15	1	1	-2	1	-4	-2	1	1	-2	-4	1	-2	1	1	8

In this paper, we study the applications of RS to signal analysis, especially their sparse representation. We find out that the RS coefficients are sparser than the FT coefficients for representing periodic impulse chain signals. In order to speed up the calculation, we precompute $c_q(n)$ only once and save them into a file for later retrieval. This can save a huge amount of computation time.

We now present a few examples to demonstrate that the RS coefficients are sparser than the FT coefficients for periodic impulse chain signals.

Example 1. Define the following simulated 1D periodic impulse chain signal:

$$x(n) = 4\sum_k \delta(n - kT_1) \tag{7}$$

where $T_1=7$ is the distance between two adjacent impulses. Fig.1 shows the RS coefficients and the FT coefficients of this 1D function. It can be seen that the RS have captured the period $T_1=7$, and the RS coefficients are sparser than the FT coefficients.

Example 2. Let us define the following simulated 1D periodic signal:

$$x(n) = \sin(2\pi \frac{n}{T_1}) + 4\sum_k \delta(n - kT_2) \tag{8}$$

where $T_1=5$, $T_2=7$, $n \in [1,N]$ and $N=128$. This signal contains a periodic component (first term) and an impulse chain (second term). The sine waves repeat themselves every five samples and the distance between two impulses is seven samples. Fig.2 shows the RS coefficients and the FT coefficients of this signal. It can be seen that RS coefficients are sparser than the FT coefficients.

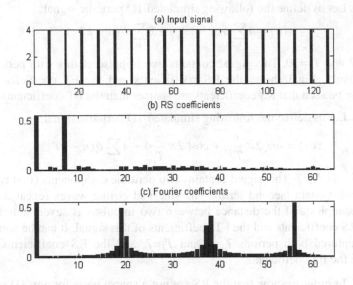

Fig. 1. The RS coefficients and the FT coefficients of a 1D periodic impulse chain signal

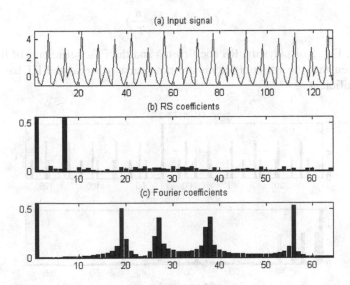

Fig. 2. The RS coefficients and the FT coefficients of a 1D periodic signal consisting of a sinusoidal function and an impulse chain

This example is troublesome for the FT because the period of the sinusoid is not a multiple of 1/N and the FT of the impulse chain is still an impulse chain in the frequency domain. Therefore, several frequency peaks will appear in the FT coefficient spectrum.

Example 3. Let us define the following simulated 1D periodic signal:

$$x(n) = 4\sum_k \delta(n - kT_1) + 8\sum_k \delta(n - kT_2) \qquad (9)$$

where $T_1=7$ and $T_2=10$. This signal contains two impulse chains with periods 7 and 10, respectively. Fig.3 shows the RS coefficients and the FT coefficients of this signal. It can be seen that RS coefficients are sparser than the FT coefficients.

Example 4. Let us define the following simulated 1D periodic signal:

$$x(n) = \sin(2\pi\frac{n}{T_1}) + \cos(2\pi\frac{n}{T_1}) + 4\sum_k \delta(n - kT_2) \qquad (10)$$

where $T_1=5$ and $T_2=7$. This signal contains two periodic components (first two terms) and an impulse chain (second term). The sine and cosine waves repeat themselves every five samples and the distance between two impulses is seven samples. Fig.4 shows the RS coefficients and the FT coefficients of this signal. It can be seen that the RS have captured both periods $T_1=5$ and $T_2=7$, and the RS coefficients are also sparser than the FT coefficients.

Example 5. In order to show that the RS are not a sparse basis for any 1D signal, we test the following periodic signal:

$$x(n) = \sin(2\pi\frac{n}{T_1}) \qquad (11)$$

where $T_1=5$. Fig. 5 displays the RS coefficients and the FT coefficients of this signal. It can be seen that the RS coefficients are not sparse whereas the FT has only one nonzero coefficient.

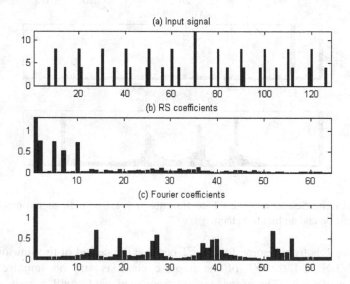

Fig. 3. The RS coefficients and the FT coefficients of a 1D periodic signal with two impulse chains

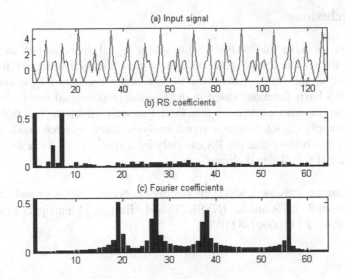

Fig. 4. The RS coefficients and the FT coefficients of a 1D periodic signal consisting of two sinusoidal functions and an impulse chain

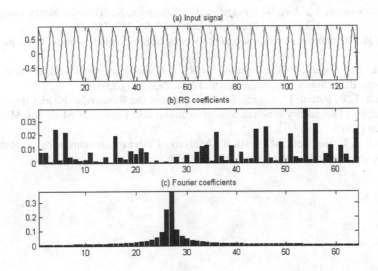

Fig. 5. The RS coefficients and the FT coefficients of a 1D periodic signal

We also count the time used to process a signal of 128 samples for 20 times and then calculate the average time for the RS and the FT. We use a personal computer (PC) with 4 GB memory and a 1.6 GHz CPU. We spend 1.0 millisecond for the RS and 5.82 milliseconds for the FT by using our unoptimized Matlab code. This indicates that the RS transform is faster than the FT. We did not use the built-in *fft* function provided in Matlab because it is highly optimized, which will make our comparison unfair.

3   Conclusions

In this paper, we have studied the RS for sparse signal analysis. Our experimental results have shown that the RS generate sparser coefficients than the FT for periodic impulse chain signals. Furthermore, the RS are faster than the FT because we can retrieve the RS basis functions saved in the file that was created before. The RS are more robust to Gaussian white noise than the FT as well. Given the fact that this is the first paper to apply the RS to sparse signal analysis, more research needs to be done on this topic. We believe that the RS can only be a sparse basis for a few classes of signals instead of for all signal classes.

Acknowledgment. This work was supported by the Natural Sciences and Engineering Research Council of Canada (NSERC) and Beijing Municipal Science and Technology Plan: Z111100074811001.

References

1. Ramanujan, R.: On Certain Trigonometric Sums And Their Applications. Trans. Cambridge Philos. Soc. 22, 259–276 (1918)
2. Sugavaneswaran, L., Xie, S., Umapathy, K., Krishnan, S.: Time-frequency analysis via Ramanujan sums. IEEE Signal Processing Letters 19(6), 352–355 (2012)
3. Planat, M.: Ramanujan sums for signal processing of low frequency noise. Phys. Rev. E. 66 (2002)
4. Samadi, S., Ahmad, M.O., Swamy, M.N.S.: Ramanujan sums and discrete fourier transform. IEEE Signal Processing Letters 12(4), 293–296 (2005)
5. Mainardi, L.T., Pattini, L., Cerutti, S.: Application of the Ramanujan Fourier transform for the analysis of secondary structure content in amino acid sequences. Meth. Inf. Med. 46(2), 126–129 (2007)
6. Mainardi, L.T., Bertinelli, M., Sassi, R.: Analysis of T-wave alternans using the Ramanujan Sums. Computer in Cardiology 35, 605–608 (2008)

Eigenface-Based Sparse Representation
for Face Recognition

Yi-Fu Hou[1], Wen-Juan Pei[1], Yan-Wen Chong[2], and Chun-Hou Zheng[1,*]

[1] College of Electrical Engineering and Automation, Anhui University, Hefei, China
zhengch99@126.com
[2] State Key Laboratory for Information Engineering in Surveying,
Mapping and Remote Sensing, Wuhan University, Wuhan, China
apollobest@126.com

Abstract. Face recognition has been a challenging task in computer vision. In this paper, we propose a new method for face recognition. Firstly, we extract HOG (Histogram of Orientated Gradient) features of each class face images in used Face databases. Then, we select the so-called eigenfaces from HOG features corresponding to each class face images and finally use them to build a overcomplete dictionary for ESRC (the Eigenface-based Sparse Representation Classification). Experiments show that our method receives better results by comparison.

Keywords: face recognition, histogram of orientated gradient, sparse representation, ESRC.

1   Introduction

Face recognition has attracted significant attention owing to the accessibility of inexpensive digital products, such as cameras and computers etc., and its applications in various domains. Recent years, many literatures [1][2][3][4][5][6][17][18]have been published about face recognition including its applications in many domains and how to extract feature and select classifier. Actually, the central of success of face recognition are feature extraction and classification method. How to extract effective and discriminant feature and select suitable classifier will directly associate with the classification accuracy.

The first step of recognition of face images is to extract effective and discriminant feature. Until now, a large number of features have been proposed in literatures and display good performance, such as Eigenfaces[1], Gabor wavelets[2], LBP(Local Binary Patterns)[3] etc. Different from these features, Dalal et al.[5]proposed locally normalized histogram of orientated gradient descriptor and provided excellent performance in human detection. Recently, Albiol et al.[7] successfully applied HOG descriptor to the domain of face recognition. O.deniz et al.[8] further explored the

---

* Corresponding author.

D.-S. Huang et al. (Eds.): ICIC 2013, LNAI 7996, pp. 457–465, 2013.

representational capability of HOG feature for face recognition and proposed a simple but powerful approach to build robust HOG descriptors, which lead to a great forward step for the application of HOG feature in face recognition.

For classifier, recently, inspired by successes in classical signal processing applications [9][10], Wright et al.[6] applied sparse representation for face recognition with the recent program of $l1\_$norm minimization method. This method has been used in many applications such as image super-resolution [20], compressive sensing [19], supervised denoising and inpainting [22]. For SR technique, an input testing face image can be well presented by the training samples as a coefficient vector. We can easily identify the class of a testing sample by analyzing its sparse representation coefficient vector. The standard SRC[6] (Sparse Representation Classification) method has been successfully applied in face recognition because it lessen overfitting problem, which is unlike conventional supervised learning methods, whose classification model for testing obtained by a training procedure.

However, in SRC the original training samples may not be as efficient as the eigenfaces[13], which can be extracted from the original training samples and capture the inherent structural information of training data. Actually, the eigenfaces can be viewed as few faces selected from training samples and contain intrinsic structural information of the same class samples. By extracting these eigenfaces of different classes samples and using them to build a overcomplete dictionary for ESRC, we can receive a much better recognition rate.

In this paper, we firstly extract HOG features of each class face images in Face databases, which are invariant to varying illumination and even shadowing etc. Then, we select the eigenfaces from these HOG features corresponding to each class face images and finally use them to build a overcomplete dictionary for ESRC improved from SRC.

2   Method

2.1   HOG Features of Face Images

Dalal and Triggs detailedly introduced the process of HOG feature extraction in literature [5]. The HOG descriptor of face images can be simply described as follows: face image window can be divided into small rectangular spatial regions, i.e., alleged "cells". To keep invariance to complex background environment, such as various illumination and even shadowing, each cell is analyzed in a somewhat lager spatial region so-called "block", which is beneficial to eliminate influence of illumination or other noise. We calculate histogram of gradient directions or edge orientations over the pixels of blocks and then combine them to obtain HOG feature representation of the total image window. In addition, the overlap between adjacent blocks significantly improves the performance, although this may seem redundant. More details of how to extract HOG feature from image can be seen from literatures [5][8].

2.2   Eigenfaces of Face Images

Generally speaking, eigenfaces [13] of each class face images can be defined as a linear combination of several special faces which may capture intrinsic structural information of the same class training samples. It is similar to a group of basis to linear space in matrix theory. In SR, the number of training samples, which compose the overcomplete directory, may be considerable. This probably results in more time expending during practical applications. Actually, by applying mathematical method to extract eigenfaces of the same class samples to reduce the number of atoms of dictionary in sparse representation, we still can obtain a sparse solution using $l1$_regularized least square [21] problem for a new testing sample. Mathematically, the face images data set matrix A can be rewritten as product of two matrices:

$$A = W \times B \tag{1}$$

where matrix A is of size $m \times n$ with m dimensions and n samples, matrix W is of size $m \times p$ with each of the p columns defining a Eigenface, matrix B is of size $p \times n$ with each of the n columns representing the eigenface expression pattern of the corresponding sample.

There are many literatures have been published about how to extract eigenfaces (eigengenes or metasamples) [11][12][13]. Alter et al.[11] extracted eigengenes by using SVD (Singular Value Decomposition) to transform "genes × samples" space to diagonalized " eigengenes × eigenarrays " space. Zheng et al.[14][15]used ICA (Independent Component Analysis) to model genes expression data regarding samples as a linear mixture of independent basis snapshots(i.e.metasamples). Brunet et al.[13] similarly used a small number of metasamples to represent samples by utilizing NMF(Non-negative Matrix Factorization) to gene expression data. So we can extract eigenfaces from the original face images data using several methods such as SVD, ICA, NMF and so on.

2.3   Eigenface-Based Sparse Representation Classification

Owing to eigenfaces contain the intrinsic structural information of faces. In this paper we will extract eigenfaces of each class face images, respectively, and then use them instead of the original training data to design a sparse representation classifier. In practice, each sub-dataset (i.e., including only one class face images) A_i will be factorized into two matrices:

$$A_i = W_i \times B_i \tag{2}$$

where i represents the ith class samples. Matrix W_i is of size $m \times p_i$ and B_i is of size $p_i \times n$, p_i is the number of eigenfaces of the ith class.

After computing the eigenfaces matrices W_i of all classes of face images, we combine them to build a overcomplete dictionary W for sparse representation, i.e.

$W = [W_1 \, W_2 \, \ldots \ldots W_k]$ where k presents the number of classes. Now, given a testing sample y, we can compute its sparse representational solution in terms of the eigenfaces of all training samples i.e. $y = W \, x$. Here, ideally, $x = [0 \, ; \cdots \cdots 0 \, , \, a_{i,1}, \cdots \cdots, a_{i,n_i}, 0, \cdots \cdots, 0]$, is a sparse coefficient vector whose entries are zero except for those associate with the ith class. We can identify the class to which a testing sample belongs by analyzing its nonzero entries of coefficient representation [6][16].

Obviously, for a $m \times n$ matrix W, if $m > n$,the system of equation $y = Wx$ is overdetermined and its correct solution is unique. But in our face recognition system, $y = Wx$ typically presents underdetermined system so that its solution is not unique. But this difficulty can be resolved by choosing the minimum $l2\_norm$ solution:

$$\hat{x}_2 = \arg \min \|x\|_2 \quad s.t. \quad y = Wx \tag{3}$$

Actually, $\hat{x}_2$ is generally not sparse with many nonzero entries corresponding to eigenfaces from different classes. In addition, we seek the sparsest solution to $y = Wx$ by solving the following optimization problem:

$$x_0 = \arg \min \|x\|_0 \quad s.t. \quad y = Wx \tag{4}$$

where $\|\cdot\|_0$ denotes the $l0\_norm$, which counts the number of nonzero entries in a vector. But in our situation of underdetermined system finding the sparsest solution is NP_hard.

Recent development in emerging theory of sparse representation and compressed sensing reveals that if the x_0 solution is sparse enough, the solution of the $l0\_minimization$ problem is equal to the solution to the following $l1\_minimization$ problem i.e.

$$x_1 = \arg \min \|x\|_1 \quad s.t. \quad y = Wx \tag{5}$$

This problem can be solved in polynomial time by standard linear programming methods. However, to certain degree of small and dense noise, we use the following objective function to solve x ,i.e.

$$J(x, \lambda) = \min \left\{ \ \|y - Wx\|_2 + \lambda \|x\|_1 \ \right\} \tag{6}$$

where the positive parameter λ is a scalar regularization.

In practice, modeling error and certain degree of noise will inevitably cause some nonzero entries in the zone of different classes so that testing samples cannot be classified successfully. For a more robust classifier, we use reconstruct residual as the basis of the judge .i.e.

$$\min r_i(y) = \left\| y - W\theta_i(x) \right\|_2 \tag{7}$$

where $\theta_i(x)$ is a vector whose nonzero entries are the ones from class i in x.

The classification algorithm is summarized as following:

Input: matrix of training samples $A = \left[A_1 \ A_2 \ \cdots\cdots \ A_k \right]$ for k classes; testing sample y.

Setp1: Extract the eigenfaces matrix $W = \left[W_1 \ W_2 \ \cdots\cdots \ W_k \right]$ for k classes using SVD or NMF.

Setp2: Solve the optimization problem defined in (6).

Setp3: Compute the residuals through formula (7).

Output: $identify\ (y) = \arg \min_i r_i(y)$.

2.4   Our Method

This section gives an overview of our method flow chain which is summarized in Fig.1.

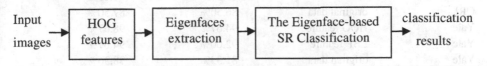

Fig. 1. The overview of our method flow chain

We firstly extract HOG features of each class face images in Face databases, which are invariant to varying illumination and even shadowing etc. Then, we select the eigenfaces from HOG features corresponding to each class face images by SVD in our work and subsequently use them to build a overcomplete dictionary for SRC, that is also why we name it "ESRC". Finally we can identify the class to which a testing sample belongs by comparing reconstruct residual i.e.the above formula (7), the class to the minimum reconstruct residual corresponds to the class of testing sample.

3   Experimental Results

3.1   Classification on the ORL Face Database and Yale Face Database

The ORL database of faces contains ten different images of each of 40 distinct subjects. For some subjects, the images were taken at different times, varying the lighting, facial expression including open/closed eyes, smiling/not smiling and facial details (glasses/not glasses). All the images are upright, frontal position (with toler-ance for some side movement). The Yale face database contains 165 grayscale images of 15individuals. There are 11 images per subjects with different facial expression or configuration, such as light-variation, glasses or not.

In this section we will extract HOG features of each class face images in two Face databases, i.e. the ORL Dataset and the Yale Database, and apply them to the proposed ESRC. In ORL Database, the former five face images of each person are selected as the training set and the remaining as testing set. While in Yale Database, we choose the former six face images of each person as training set and the others as testing set. In ESRC, SVD is used to extract eigenfaces of each class. The number of eigenfaces in each class is set to 3 because the number of samples of each person is small. Actually, its optimal number can be determined through the following experiment in Section 3.2.

To illustrate HOG feature's discrimination and effectiveness to face recognition, we compare HOG feature with the popular LBP feature and the original data of face images. For classifier, we show performances of the proposed ESRC in contract to the performances of the standard SRC.

Table 1. The accuracy rate of SRC and ESRC using HOG feature、 LBP feature and the original data on the ORL face database and the Yale face database, respectively

Database	Feature	SRC	ESRC
ORL	HOG feature	93.00%	94.00%
ORL	LBP feature	87.00%	87.00%
ORL	original data	86.50%	87.50%
Yale	HOG feature	94.67%	98.67%
Yale	LBP feature	92.00%	94.67%
Yale	original data	81.33%	78.36%

The results of classification are showed in Table 1. It can be seen from Table 1 that the application of HOG feature improves the accuracy of classification results comparing with LBP feature and original data. For classifier, the proposed ESRC can obtain relatively good classification accuracy compared with the standard SRC except for classification result in Yale Dataset, where we do not apply HOG features to classifier and the accuracy of SRC is higher than ESRC. Because we determine 3 eigenfaces in our experiment and they may not enough to express integrally intrinsic structural information of each class face images, so the accuracy of the standard SRC is slightly higher than the proposed ESRC. But we can conclude from the above Table 1 that the performance of combination of HOG feature and ESRC is superior to the performance produced by any one of them alone.

3.2 The Number of Eigenfaces

One key point we care more about is that the number of eigenfaces of each class face images. Different number of eigenfaces will present different performance. For example, too large number of eigenfaces will result in time consuming. However, if the number of eigenfaces is reduced too serious to integrally express inherent structural information of face images, classification accuracy will not be guaranteed. So how to determine adaptive number of eigenfaces is significant.

It can be seen clearly from Table 1 that HOG feature and ESRC make a positive effect on the classification results. In the experiment we select 3 eigenfaces to represent the samples of each class. Actually we can decide the optimal number of eigenfaces through experiments. Zheng et al.[16] mentioned that the value of P can be determined using the nested stratified 10-fold cross validation. In paper [16], it was pointed out that MSRC (Metasample-based Sparse Representation Classification) will not have clear advantages over SRC when the number of training samples is less than ten. If there are 10 or more than 10 training samples, MSRC will be a good choice for tumor classification. But in our experiment we can see from the Table 1 that ESRC can also obtain better classification results than SRC when the number of samples belows ten except for classification result in Yale Dataset when we do not apply HOG feature to classifier so that the result of SRC is higher than ESRC.

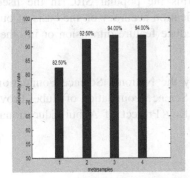 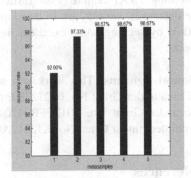

Fig. 2. Accuracy rate in ORL Database using different number of eigenfaces

Fig. 3. Accuracy rate in Yale Database using different number of eigenfaces

We choose different number of eigenfaces to determine the optimal number of eigenfaces in our work. The results are presented in Fig.2 and Fig.3. The two figures show that when the number of eigenfaces reaches a certain point, classification accuracy rate will keep stable while the relatively low accuracy rate is achieved when the number of eigenfaces belows the optimal number. This shows that three is the best choice to the number of eigenfaces of each class face images in our experiment because three eigenfaces can already integrally express all intrinsic structural information. Certainly, to other face datasets, the optimal number of eigenfaces still conducts experiments to determine.

4   Discussion

Extracting effective and discriminant feature and selecting suitable classification method are the two significant aspects of classification because they directly associate with the classification results. In this paper we extract HOG features which are invariant to varying illumination and shadow and then apply them to the Eigenface-based Sparse Representation Classification, which expresses each testing sample as a linear combination of a set of eigenfaces extracted from the training face images using

SVD[11] or ICA[14] or NMF[13]. The proposed ESRC is compared with the standard SRC on the ORL face database and the Yale face database. Experiments results show that HOG feature can efficient enhance classification performance and outperforms the popular LBP feature and the original data of face images and the combination of HOG feature and ESRC is a good choice for face recognition. But on Yale face database the standard SRC is slightly superior to ESRC when the HOG feature is not used since the inherent structural information contained in few eigenfaces is not enough to represent all of the same class samples. We also conduct experiments to seek the optimal number of eigenfaces of each class. The results show that 3 is the best choice for classification in our experiments because accuracy rate keeps stable when the number of eigenfaces beyonds three. Certainly, different face datasets have different optimal number of eigenfaces and we should determine it though experiments. Although combining HOG feature with EMRC outperforms the popular SRC in the used face databases, we may try to develop the classification accuracy on the basis of our method by seeking some excellent methods to reduce feature dimension or in other aspects.

Acknowledgments. This work was supported by the National Science Foundation of China under Grant No. 61272339, the Natural Science Foundation of Hubei Province under Grant No. 2012FFB04204, and the Key Project of Anhui Educational Committee, under Grant No. KJ2012A005.

References

1. Matthew, T., Alex, P.: Eigenfaces for Recognition. In: IEEE International Conference on Computer Vision and Pattern Recognition (1991)
2. Amin, M.A., Yan, H.: An Empirical Study on the Characteristics of Gabor Representations for Face Recognition. IJPRAI 23(3), 401–431 (2009)
3. Ahonen, T., Hadid, A., Pietikäinen, M.: Face recognition with local binary patterns. In: Pajdla, T., Matas, J(G.) (eds.) ECCV 2004. LNCS, vol. 3021, pp. 469–481. Springer, Heidelberg (2004)
4. Bartlett, M.S., Movellan, J.R., Sejnowski, T.J.: Face Recognition by Independent Component Analysis. IEEE Transactions on Neural Networks 13(6), 1450–1464 (2002)
5. Lowe, D.G.: Distinctive Image Features from Scale-Invariant Keypoints. International Journal of Computer Vision 60(2), 91 (2004)
6. John, W., Allen, Y.Y., Arvind, G., Shankar, S., Yi, M.: Robust Face Recognition via Sparse Representation. IEEE Transactions on Pattern Analysis and Machine Intelligence 31, 210–227 (2009)
7. Albiol, A., Monzo, D., Martin, A., Sastre, J., Albiol, A.: Face Recognition using HOG-EBGM. Pattern Recognition Letters 29(10), 1537–1543 (2008)
8. Déniz, O., Bueno, G., Salido, J.: Face Recognition using Histograms of Oriented Gradients. Pattern Recognition Letters 32, 1598–1603 (2011)
9. Emmanuel, J., Justin, R., Terence, T.: Robust Uncertainty Principles: Exact Signal Reconstruction from Highly Incomplete Frequency Information. IEEE Transactions on Information Theory 52, 489–509 (2006)

10. Emmanuel, J., Terence, T.: Near-Optimal Signal Recovery from Random Projections: Universal Encoding Strategies? IEEE Transactions on Information Theory 52(12), 5406–5425 (2006)
11. Orly, A., Patrick, O.B., David, B.: Singular Value Decomposition for Genome-Wide Expression Data Processing and Modeling. Proceedings of the National Academy of Sciences 97, 10101–10106 (2000)
12. Wolfram, L.: Linear Modes of Gene Expression Determined by Independent Component Analysis. Bioinformatics 18, 51–60 (2002)
13. Jean, P.B., Pablo, T., Todd, R.G., Jill, P.M.: Metagenes and Molecular Pattern Discovery Using Matrix Factorization. Proceedings of the National Academy of Sciences 101, 4164–4416 (2004)
14. Wang, H.Q., Huang, D.S.: Regulation Probability Method for Gene Selection. Pattern Recognition Letter 27(2), 116–122 (2006)
15. Huang, D.S., Zheng, C.H.: Independent Component Analysis-Based Penalized Discriminant Method for Tumor Classification Using Gene Expression Data. Bioinformatics 22, 1855–1862 (2006)
16. Zheng, C.H., Zhang, L., Ng, T.-Y., Simon, C.K.S., Huang, D.S.: Metasample-Based Sparse Representation for Tumor Classification. IEEE/ACM Transactions on Computational Biology and Bioinformatics 8(5) (2011)
17. Sandrine, D., Jane, F., Terence, P.S.: Comparison of Discrimination Methods for the Classification of Tumor Using Gene Expression Data. Journal of the American Statistical Association 97, 77–87 (2002)
18. Wang, S.L., Li, X., Zhang, S., Gui, J., Huang, D.S.: Tumor Classification by Combining PNN Classifier Ensemble with Neighborhood Rough Set Based Gene Reduction. Computers in Biology and Medicine 40(2), 179–189 (2010)
19. David, L.D.: Compressed Sensing. IEEE Transactions on Information Theory 52(4), 1289–1306 (2006)
20. Yang, J.C., John, W., Thomas, H., Yi, M.: Image Super-resolution as Sparse Representation of Raw Patches. In: IEEE International Conference on Computer Vision and Pattern Recognition (2008)
21. Seung, K., Kwangmoo, K., Stephen, B.: An Interior-Point Method for Large-Scale l1-Regularized Least Squares. IEEE Journal of Selected Topics in Signal Processing 1(4), 606–617 (2007)
22. Julien, M., Guillermo, S., Michael, E.: Learning Multiscale Sparse Representations for Image and Video Restoration 7(1), 214–241 (2008)

Facial Expression Recognition Based on Adaptive Weighted Fusion Histograms

Min Hu[1,2], Yanxia Xu[1,2], Liangfeng Xu[1], and Xiaohua Wang[1,2]

[1] College of Computer and Information, Hefei University of Technology;
[2] Affective Computing and Advanced Intelligent Machines AnHui Key Laboratory,
Hefei 230009
uhnim@163.com

Abstract. In order to improve the performance of expression recognition, this paper proposes a facial expression recognition method based on adaptive weighted fusion histograms. Firstly, the method obtains expression sub-regions by pretreatment, and calculates the contribution maps (CM) of each expression sub-region. Secondly, this method extracts Histograms of Oriented Gradient by the Kirsch operator and extracts histograms of intensity by centralized binary pattern (CBP), then the paper fuses the Histograms by parallel manner and the fused histograms are weighted by CMs. At last, the weighted fused histograms are used to classify by the Euclidean Distance and the nearest neighbor method. Experimental results which are obtained by applying the proposed algorithm and the Gabor wavelet, LBP, LBP+LPP, Local Gabor and AAM on JAFFE face expression dataset show that the proposed method achieves better performance for the face expression recognition.

Keywords: Facial Expression Recognition, Contribution Map, Histograms of Oriented Gradient, Histograms of Intensity, Centralized Binary Pattern.

1 Introduction

Facial expression shows wealth emotions of human and plays a vital role in people's day-to-day communication. Facial expression recognition (FER) [1-2] involves image processing, artificial intelligence, pattern recognition, and social psychology, etc. FER has three parts: face detection and pretreatment, feature extraction, and facial expression classification. Feature extraction is the key step as the quality of the features extracted directly affects the efficiency of the recognition.

At present, the main methods used for FER includes active appearance model (AAM) [3], principal component analysis (PCA) [4], Gabor wavelet transform [5, 6], scale invariant feature transform (SIFT) [7] and local binary pattern (LBP) [8, 9]. The disadvantages of AAM algorithm are that the calculation is complex and the initial parameters are difficult to obtain. PCA method extracts global features, but FER focus on local features. The method based on Gabor wavelet is time-consuming and memory-intensive so that it is not suitable for the establishment of a quick FER system; SIFT

D.-S. Huang et al. (Eds.): ICIC 2013, LNAI 7996, pp. 466–474, 2013.
© Springer-Verlag Berlin Heidelberg 2013

extracts the local features, but it has high computational complexity and brings about mismatching easily, dimension of the features extracted by SIFT is higher. LBP is an effective texture description operator, it has a small amount of calculation, resistance to light interference and a lower dimensional features. However, the features extracted by LBP contain only the texture information of images, but do not contain the shape information, moreover, LBP does not consider the influence of center pixel.

Based on the above analyses, this paper puts forward the method that facial expression recognition based on adaptive weighted fusion histograms. The method includes three parts: Firstly, the face images are cut into five expression sub-regions of eyebrows, eyes, nose, mouth and contribution maps (CM) of these sub-regions are calculated according to the amount of information they contained. Secondly, histograms of oriented gradients (HOG) [10,11] of expression sub-regions are extracted by the Kirsch operator [12] and histograms of intensity are extracted by centralized binary pattern (CBP)[13]. Then, the joint histograms are derived by connecting the HOG and histograms of intensity of every sub-region respectively, and the weighted fusion histograms are conducted by using CM to weight the sub-region. At last, the fusion histograms are used to classify by using Euclidean distance and the nearest neighbor method. The features extracted by this method not only contain the face shape information but also contain the texture information, what's more, CBP not only consider the impact of the center pixel and the dimensions of the histograms are lower.

2   Adaptive Contribution Map

According to the theorem of Shannon [14], a discrete random variable $V(v_1, v_2, ..., v_t)$ with a probability $P(V)(p(v_1), p(v_2), ..., p(v_n))$ can be described by entropy $H(V)$ as follows:

$$H(v) = \sum_{i=1}^{n} p(v_i) \log\left(\frac{1}{p(v_i)}\right) = -\sum_{i=1}^{n} p(v_i) \log(p(v_i)) \tag{1}$$

For a digital image $f(x, y)$, entropy can be defined as:

$$H[f(x, y)] = -\sum_{i=1}^{n} p_i \log(p_i) \tag{2}$$

In the above formula, p_i is the probability of the i th gray level value in the image $f(x, y)$; t is the total number of gray level values. For a digital image of facial expression, the entropy of the overall image expresses the amount of information of the image. But local features are more important in the process of facial expression recognition. Therefore, the paper calculates the entropy of expression sub-regions of eyebrows, eyes, nose, and mouth. The information of the sub-regions not only expresses the amount of information but also describe the richness of the sub-region texture. Nanni in [15] puts forward the improved methods of the calculation of the local information entropy. In the proposed method, the entropy map (EM) of a pixel is computed as follows:

$$EM \ (i, j) = H \ (F \ (i, j)_{sw}) \tag{3}$$

In formula (3), sw is the size of the sliding window; $H(F(i, j)_{sw})$ is the function of entropy, and $F(i, j)_{sw}$ is a sub-region in the sliding window centered at (i, j), which can be defined as:

$$F \ (i, j)_{sw} \left\{ f(x, y) \mid x \in \left[i - \frac{sw}{2}, i + \frac{sw}{2} - 1 \right], y \in \left[j - \frac{sw}{2}, j + \frac{sw}{2} - 1 \right] \right\} \tag{4}$$

Since the entropy of each pixel in the local information entropy can be derived by the distribution of the pixels around it, the contribution map (CM) of each sub-region can be expressed as follows:

$$CM \ = \frac{1}{M \times N} \sum_{x=1, y=1}^{M, N} EM \ (x, y) \tag{5}$$

Where $M \times N$ is the size of the sub-region, and $EM \ (x, y)$ is the entropy map of the (x, y).

The CM of different sizes of sliding windows are shown in Fig.1. In this paper, we set the sw to 7.

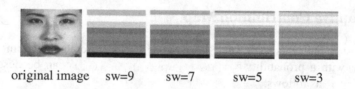

original image   sw=9   sw=7   sw=5   sw=3

Fig. 1. Contribution maps of different sizes of sliding window

3   The Generation of Histogram of Oriented Gradient and Histogram of Intensity

3.1   Histogram of Oriented Gradient (HOG)

Dalal proposed the HOG which is used for pedestrian recognition. HOG is the third part of the algorithm SIFT. The features described by HOG react the shape information of the image. Gradient corresponds the first derivative of the image. For a continuous image function $f(x, y)$, gradient in an arbitrary pixel point is a vector which can be defined as follows:

$$\nabla f(x, y) = [G_x, G_y]^T = [\frac{\partial f}{\partial x}, \frac{\partial f}{\partial y}] \tag{6}$$

Where G_x is the gradient of the X direction and G_y is the gradient of the Y direction. The magnitude and direction angles of the gradient are as follows:

$$\left|\nabla f(x,y)\right| = (G_x^2 + G_y^2)^{1/2} \tag{7}$$

$$\phi(x,y) = \arctan \frac{G_y}{G_x} \tag{8}$$

For a digital image $f(x,y)$, the gradient can be defined as:

$$\left|\nabla f(x,y)\right| = \left\{\left[f(x,y)-f(x+1,y)\right]^2 + \left[f(x,y)-f(x,y+1)\right]^2\right\}^{1/2} \tag{9}$$

In this paper, we use Kirsch operator to compute histogram of oriented gradient of a digital image. Because Kirsch operator is more accurate to locate the edge and has the effect to smooth the noise. Kirsch operator concludes eight window masks of 3*3. Each mask represents a direction. So Kirsch operator has eight directions which are shown as Fig.2.

Fig. 2. The eight directions of Kirsch operator

In a digital image $f(x,y)$, a pixel point whose coordinates are (i,j) has neighborhood window of 3*3 which is named p, $Q_k(k = 0,1,...7)$ are the eight edges response values, which can be obtained as follow:

$$Q_k = M_k * P(k = 0,1,...,7) \tag{10}$$

The response values are not equally important in all directions, so we select the direction of the max Q_k as the gradient direction of the pixel.

Following are the major stages to get the histogram of oriented gradient of a pixel:

1) Kirsch operator is applied to each pixel and then gets the gradient direction of each pixel;

2) We select 7*7 windows to compute the gradient direction of the center pixel; the gradient direction is shown in Fig.4;

3) Cycle2), until the gradient directions of each pixel are obtained.

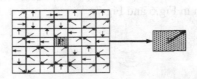

Fig. 3. The gradient direction of pixel P

3.2   Histogram of Intensity

The well known LBP operator labels each pixel of an image by threshold its P-neighbor values with the center value. The calculation of LBP is simple, but it does not consider the effect of the center pixel. In addition, the dimension of LBP histograms of gray is 256. Therefore, we select the CBP to make the histograms of gray. CBP takes the center pixel into consideration and the dimension of the histograms is 32. CBP can restrain the noise.

The CBP (8,1) operator is shown in Fig.4:

Fig. 4. CBP (8,1) Operator

$$CBP\ (8,1) = \sum_{i=0}^{3} u\left(g_i - g_{i+4}\right)2^i + u\left(g_c i - g_T\right)2^4 \tag{11}$$

$$g_T = \left(g_0 + g_1 + \ldots + g_7 + g_c\right)/9 \tag{12}$$

The texture image transformed by CBP is shown in Fig.5:

Fig. 5. Original image and texture image

4   Experiments and Results Analysis

4.1   The Process of Experiment

The experimental procedures of the algorithm are shown as follows:

1) Pretreatment: we cut out the expressions sub-region from the expression image by using the theory of three atriums and five eyes rule. [16] and calculate the CM of the sub-regions, which are denoted as CMb, CMe, CMn, and CMm. The process of cutting the sub-regions are shown in Fig.6 and Fig.7:

Fig. 6. Three atriums and five eyes rule

Fig. 7. The results of segmentation

2) Features Extraction: Firstly, we extract HOG of each expression sub-region by using the Kirsch Operator; Secondly, we make the histograms of intensity of expression sub-regions by using CBP. Thirdly, we fuse HOG and histograms of intensity by giving two weight values to the two parts histograms. In this paper, we set the two weights to 0.5 and 0.5. And then fused histograms are weighted by CMb, CMe, CMn, and CMm.

3) Classification: The weighted fusion histograms are classified by euclidean distance and the nearest neighbor method.

4.2   Experimental Results and Analysis

In order to verify the effectiveness of the method proposed, we do five circular experiences on JAFFE expression database. Each experiment has 105 images as training samples and 105 images as testing samples. Results are shown in Table 1. On the other hand, we compare the method LBP, Gabor, AAM ,LBP+LPP ,Local Gabor, which are shown in Fig.8 and table2.

Table 1. The results of the five experiments

(%)	Angry	Disgust	Fear	Happy	Neutral	Sad	Surprise	Average
1	94,29	93.33	94.29	96.19	95.24	93.33	100.0	95.24
2	93.33	93,33	95.24	94.29	94.29	94.29	99.06	94.83
3	94.29	92.38	95.24	95.24	94.29	95.24	100.0	95.24
4	94.29	93.33	93.33	94.29	95.24	94.29	98.10	94.70
5	93.33	91.43	94.29	95.24	96.19	93.33	99.06	94.57
Average	93.91	92.76	94.48	95.05	95.05	94.10	99.24	94.93

We can see that there are error recognitions, the reasons are as follows: 1) some expressions are very similar;2) The images have slight posture changes; 3) Some expression images are micro-expression.

Compared with other algorithms, the proposed algorithm has the highest recognition rate and lower recognition time. The reasons are:1)the features extracted by the proposed method contain the shape information and texture information; 2) the method extracts the features of sub-regions weighted by CMs which represents the importance of the sub-region; 3) the feature vector dimension is reduced by using CBP histograms and Kirsch operator.

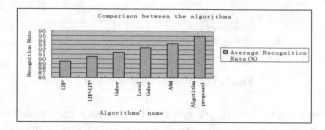

Fig. 8. The comparison of recognition rate with other algorithms

Table 2. The comparison of recognition time with other algorithms

Algorithm	Recognition Time(ms)
LBP	25.6
LBP+LPP	43.3
Gabor	2400
Local Gabor	2000
AAM	9500
Algorithm proposed	133

5   Conclusions

The paper proposes facial expression recognition based on adaptively weighted fusion histograms. Compared with LBP, Gabor wavelet, AAM, the proposed algorithm has the following advantages:

1)The features extracted by LBP are local features which are only texture information, But the features extracted by the proposed method not only contain shape information which is global features, but also contain texture information which is local features, what's more, we have significantly reduced the dimension of the LBP feature vectors.

2)Compared with the methods based AAM and Gabor, the proposed method is low computational complexity and has good real time ability .The features extracted by AAM not only take the shape information into consideration but also consider the texture information. However, AAM pparameters are difficult to determine, the fitting procedure is an iterative process so that it does not have real-time. Gabor can extract rich texture information as it extracts features from different directions and different scales. However, the dimension of the features vector is much higher, so it is easy to fall into the "curse of dimensionality", Gabor is also not interactive and real-time.

3)At last, the algorithm weights the fusion histograms according to the information that the sub-regions included. The weight values are adaptive.

Our future works aim to design further methods to obtain more precisely the importance of each area of the face during the recognition process and improve the results and finding a solution for the problems of drastic changes in the expression or great variations in pose.

Acknowledgment. This work was supported in part by the National High-Tech Research & Development Program of China under Grant No.2012AA011103), Key Scientific and Technological Project of Anhui Province under Grant No 1206c0805039 and by the NSFC-Guangdong Joint Foundation Key Project under Grant No. U1135003.The authors thank Dr. M. Lyons for providing the JAFFE database.

References

1. Jain, S., Hu, C.B., Aggarwal, J.K.: Facial expression recognition with temporal modeling of shapes. In: IEEE International Conference on Computer Vision Workshops, pp. 1642–1649 (2011)
2. Song, K., Shuo, C.: Facial expression recognition based on mixture of basic expression and intensities. In: IEEE International Conference on System, Man and Cybernetics, pp. 3123–3128 (2012)
3. Hommel, S., Handmann, U.: AAM based continuous expression recognition for face image sequence. In: IEEE 12th International Symposium on Computational Intelligence and Informatics, pp. 189–194 (2011)
4. Zhao, Y., Shen, X., Georganas, N.D., Petriu, E.M.: Part-based PCA for facial feature extraction and classification. In: IEEE International Workshop on Haptic Audio Visual Environments and Games, pp. 99–104 (2009)
5. Sun, X., Xu, H., Zhao, C.X., Yang, J.Y.: Facial expression recognition based on histogram sequence of local Gabor binary patterns. In: IEEE Conference on Cybernetics and Intelligent Systems, pp. 158–163 (2008)
6. Chen, T., Su, F.: Facial expression recognition via Gabor wavelet and structured sparse representation. In: 3rd IEEE International Conference on Network Infrastructure and Digital Content, pp. 420–424 (2012)
7. Zhao, W., Ngo, C.: Flip-Invariant SIFT for copy and object detection. IEEE Transactions on Image Processing, 980–991 (2013)
8. Kim, T., Kim, D.: MMI-based optimal LBP code selection for facial expression recognition. In: IEEE International Symposium on Signal Processing and Information Technology, pp. 384–391 (2009)
9. Jabid, T., Kabir, M.H., Chae, O.: Facial expression recognition using local directional pattern. In: Proc. of the 17th IEEE International Conference on Image Processing, pp. 1605–1608. IEEE Press, S.I (2010)
10. Dalal, N., Triggs, B.: Histograms of oriented gradients for human detection. In: Proceedings of Conference on Computer Vision and Pattern Recognition, pp. 886–893. IEEE Compute Society Press, Los Alamitos (2005)
11. Park, W., Kim, D.H., Suryanto, L.C.G., Roh, T.M., Ko, S.: Fast human detection using selective block-based HOG-LBP. In: 19th IEEE International Conference on Image Processing, pp. 601–604 (2012)

12. Shao, P., Yang, L.: Fast Kirsch edge detection based on templates decomposition and integral image. ACTA Automatic Sinica 33(8) (2007)
13. Fu, X., Wei, W.: Facial expression recognition based on multi-scale centralized binary pattern. Control Theory & Applications 26(6), 629–633 (2009)
14. Shannon, C.E.: A mathemalthtical theory of communication. Bell Syst. Tech. J. 27, 379–423, 623–656 (1948)
15. Loris, N., Brahnam, S., Alessandra, L.: A local approach based on a Local Binary Patterns variant texture descriptor for classifying pain states. Expert Systems with Applications 37(12), 7888–7894 (2010)

An Efficient Indexing Scheme for Iris Biometric Using K-d-b Trees

Hunny Mehrotra and Banshidhar Majhi

Department of Computer Science and Engineering
National Institute of Technology Rourkela
Odisha 769008, India
{hunny,bmajhi}@nitrkl.ac.in

Abstract. In this paper, an indexing approach is proposed for clustered SIFT keypoints using k-d-b tree. K-d-b tree combines the multidimensional capability of k-d tree and balancing efficiency of B tree. During indexing phase, each cluster center is used to traverse the region pages of k-d-b tree to reach an appropriate point page for insertion. For m cluster centers, m such trees are constructed. Insertion of a node into k-d-b tree is dynamic that generates balanced data structure and incorporates deduplication check as well. For retrieval, range search approach is used which finds the intersection of probe cluster center with each region page being traversed. The iris identifiers on the point page referenced by probe iris image are retrieved. Results are obtained on publicly available BATH and CASIA Iris Image Database Version 3.0. Empirically it is found that k-d-b tree is preferred over state-of-the-art biometric database indexing approaches.

Keywords: SIFT, K-d-b, Range search, Indexing, Iris.

1 Introduction

Iris is one of the most widely used biometric modality for recognition due to its reliability, non invasive characteristic, speed and performance [1]. Attributable to these advantages, the application of iris biometric is increasingly encouraged by various commercial as well as government agencies. Iris is publicly deployed at various airports and border crossings that generates large scale databases. In India, a large scale project (Aadhaar) is undertaken to issue Unique Identification (UID) number to each individual across the country using fingerprint and iris [2]. The idea of UID is deduplication by keeping a check during enrollment that the same citizen is not enrolled more than once. The major concern when working with such large scale databases is the number of false matches which increases with the size of the database [3]. Thus, there is a strong requirement to develop some indexing approach that keeps a check on duplicate identities and simultaneously reduces probe retrieval time, without loss of accuracy. There exists several indexing approaches for iris

D.-S. Huang et al. (Eds.): ICIC 2013, LNAI 7996, pp. 475–484, 2013.

biometrics [4][5] but they fail to perform on iris images taken under noncooperative environment. In this paper, an effort has been made to develop an indexing approach suitable for non-ideal images.

Significant but limited research for indexing biometric databases exists in literature. A multimodal binning and pruning approach has been proposed using signature and hand geometry database [6]. An indexing approach using pyramid technique is proposed in [3]. Feng et al. [4] have proposed iris database indexing approach using beacon guided search (BGS). The authors in [5] make use of two approaches for iris indexing. In the first approach, PCA and k-means clustering is used to partition the database. The second approach is based on statistical analysis of pixel intensities and positions of blocks in iris texture. In [7], the authors have used the concept of hashing to search the templates. Local feature based indexing approach is proposed in [8] using geometric hashing (GH) of Scale Invariant Feature Transform (SIFT) keypoints [9]. In [10], an iris color is used to determine index for reducing the search space. Finally, Speeded Up Robust Features (SURF) [11] are used for matching query with the retrieved set of iris from the database. Recently an effort has been made to further reduce the time required during identification by parallelizing geometric hashing approach [12]. Through asymptotic analysis it is found that there is significant gain in speed in comparison to traditional geometric hashing approach. From the existing literature it is found that the conventional approaches to indexing works using global features. Thus, they are unsuitable for iris images taken under unconstrained environment. To develop a robust indexing approach local features are used. There are few differences between the proposed approach and other indexing approaches that exist in literature. Firstly, existing indexing approaches constructs a single multidimensional tree using *global* features whereas the proposed indexing constructs multiple trees (m) using *local* features. There exists an indexing approach suitable for unconstrained iris identification using geometric hashing of SIFT keypoints [8]. The drawback of geometric hashing approach is that it performs indexing in $O(N^3)$ time for N identities enrolled in the database. Actuated by the need of iris indexing for real time scenarios, an efficient indexing approach is proposed using k-d-b trees [13] constructed using clustered SIFT keypoints. In this paper, the variable number of keypoints, extracted from iris using SIFT, are mapped to fixed number of clusters as presented in Section 2. The application of balanced multidimensional tree for iris database indexing is given in Section 3. Experimental results using the proposed approach are given in Section 4. Conclusions are given at the end.

2   Feature Extraction and Clustering

Prior to feature extraction, the acquired iris image is localized for inner and outer boundary using image morphology [14]. The annular region between the iris circles is considered for feature extraction. In order to eliminate noise due to eyelids, sector based approach is used [8]. Due to expansion and contraction of pupil as a natural phenomenon, the texture pattern of iris undergoes linear deformation. In this paper, SIFT is applied directly to annular iris that provides stable set of features while being less sensitive to local image distortions [9]. The keypoints are detected directly from annular iris image using scale space construction. The scale space is obtained by convolving the annular image with variable scale Gaussian (σ). The Difference of Gaussian (DOG) for two nearby scales is obtained. DOG images are used to detect keypoints located at (x, y), by finding local maxima/minima across different scales. The orientation (ϕ) is obtained for each detected keypoint by forming orientation histogram of gradient orientation. Once orientation has been selected, the feature descriptor is computed as a set of orientation histograms on 4×4 pixel neighborhoods. The orientation histograms are relative to the keypoint orientation. Histogram contains 8 bins each and each descriptor contains an array of 16 histograms around the keypoint. This generates SIFT feature descriptor of 4×4×8 = 128 elements.

The number of keypoints (n) varies across iris images in the database. The traditional approaches to database indexing become unsuitable for such local features. The proposed indexing approach is developed using local feature descriptor. Fuzzy c means (fcm) clustering [15] is used to group the number of keypoints (n), for each iris image, sharing similar descriptor property. The idea is to have transformation from variable number of keypoints (n) to fixed number of clusters (m), ascertained a priori. These fuzzy cluster centers are used to perform indexing using multidimensional k-d-b tree.

3   K-d-b Tree Based Indexing

In this research, a dynamic data structure coined k-d-b tree [13] is used to provide an efficient solution for indexing. The multidimensional capability of k-d tree [16] is combined with organization efficiency of B trees to develop a more robust indexing approach. This approach partitions the search space that consist of k-dimensional point(s) represented as $(x_0, x_1..., x_{k-1})$ to be an element of $domain_0 \times domain_1 \times ... domain_{k-1}$ into mutually exclusive **regions** satisfying

$$min_i \leq x_i < max_i, \ 0 \leq i \leq k-1$$

where min_i and max_i are the maximum and minimum values of all the elements in the region along i^{th} dimension. The k-d-b tree consist of two types of pages

- **Region page:** is the collection of regions and each region is represented by storing (min_i, max_i)
- **Point page:** is the collection of points represented by $(x,$ image id$)$

There are few properties that defines k-d-b tree. The region pages cannot contain null pointer which implies that point pages are the leaf nodes of the tree. All the leaf nodes are at the same level and contains pointer to iris identifiers in the database. The union of two disjoint regions in a region page makes a region. The structure of 2-D k-d-b tree is shown in Fig. 1 with point and region pages.

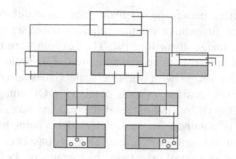

Fig. 1. Structure of k-d-b tree (here k = 2). The shaded regions are not included in region pages and point pages are leaf nodes of the tree.

3.1   Indexing

For indexing iris database, the cluster centers are obtained to construct k-d-b trees. In this approach, the yth cluster center is used to develop the corresponding k-d-b tree. To perform insertion into k-d-b tree if root node of tree does not exist, then a point page is created and the keypoint center with corresponding iris ID is inserted in the point page. If root node exists, then point page is reached to match points with the one to be inserted. If the keypoint center to be inserted has a very close similarity to any point existing in the point page then **duplicate** message is displayed and the insertion stops. The (point, irisID) pair is added to the point page, otherwise. If the page overflows then splitting happens by finding a median (x_i) of all the elements in the point page along i^{th} dimension. This generates left and right children of the point page. A point y lies on left of x_i if $y_i < x_i$ and otherwise on the right. A new region page is created and marked as root node with left and right children assigned. Similarly, if the region page overflows then the splitting is done along x_i to generate the left and right regions of the page. Let the region page be defined as $I_0 \times I_1 \times ... I_{k-1}$, if $x_i \notin I_i$ the region remains unchanged by splitting otherwise let $I_i = [min_i \; max_i)$ and splitting generates two regions

left region $I_0 \times ... \times [min_i \; x_i) \times ... \times I_{k-1}$
right region $I_0 \times ... \times [x_i \; max_i) \times ... \times I_{k-1}$

To find the position of regions for $x_i \notin I_i$ the following scheme is used

$$\text{region} = \begin{cases} x_i < min_i & \text{left} \\ x_i \geq max_i & \text{right} \end{cases}$$

If the region lies to the left of x_i add (region, page id) to the left and otherwise on the right. If splitting is done on root page then new region pages are created with the regions $(domain_0 \times ... \times [min_i \; x_i) \times ... \times domain_{k-1}$, left id) and $(domain_0 \times ... \times [x_i \; max_i) \times ... \times domain_{k-1}$, right id). For splitting page other than root replace in parent of page to be split (left region, left id) and (right region, right id). If this causes the parent page to overflow then the same process is iterated.

The proposed indexing scheme is repeated for m k-d-b trees representing m cluster centers. This approach is *dynamic* and avoids re-indexing of the entire database on insertion of a new record. Further, the proposed indexing structure is *balanced* by design and is even successful in finding *duplicate* identities during enrollment.

3.2   Range Query Retrieval

For retrieving gallery iris images corresponding to probe, range search is used [17]. The intersection of probe cluster center is found for each region on the page and those regions whose intersection with query is non-null are considered further to reach the point page(s). The image identifiers on the point page referenced by probe image are retrieved. This operation is repeated for m k-d-b trees and the identifiers retrieved are combined using set union operation to generate a combined candidate list. Finally, top S matches are obtained by individually comparing the probe iris with candidate irises using SIFT.

4   Experimental Results

The proposed indexing approach is tested on publicly available BATH [18] and CASIA Iris Image Database Version 3.0 (or CASIA-V3 in short) [19]. BATH iris database comprises images from 50 subjects. For each subject both left and right iris images are acquired, each containing 20 images of the respective eyes. CASIA-V3 is acquired in an in-door environment. The images have been captured in two sessions, with an interval of at least one month. The database comprises 249 subjects with total of 2655 images from left and right eyes.

To measure the identification accuracy, each database is divided into two mutually exclusive gallery and probe sets. The gallery set consists of iris templates with known identities. However, probe set consist of iris templates whose identity is to be known. In order to reduce system response time during identification, the gallery set is partitioned into bins. This reduces the number of comparisons required to find the identity of the probe. In this paper the gallery set is partitioned using k-d-b tree based indexing. Some well-known performance measures are used for identification [20]. The rank-k identification indicates the number of correct identities that occur in top k matches. Cumulative match characteristic (CMC) curve shows probability of identification (pi) at various ranks (R). The penetration rate (pr) of probe search is defined as a ratio of expected number of comparisons against total number of elements in the gallery set. Bin miss (bm) occurs when probe is wrongly searched in incorrect bin due to indexing errors. To mark the trade off between pr and bm a new error measure (γ) [21] is used which is defined as

$$\gamma = \sqrt{(1 - pr) \times (1 - bm)} \qquad (1)$$

To justify the strength of the proposed approach, two sets of experiments have been conducted. In *Experiment 1*, results of proposed k-d-b tree based indexing scheme are obtained. In *Experiment 2*, proposed multidimensional tree based approach is compared to existing geometric hashing approach [8].

4.1   Experiment 1: k-d-b Tree Performance

In this experiment results are obtained for multidimensional k-d-b tree based indexing approach on standard databases. Fig. 2(a) shows CMC curves for change in number of clusters on BATH database. In this approach m is varied from 2 to 4 as bm becomes 0 following $m=3$. Table 1 shows the values of pi for change in number of clusters at different ranks. Precisely, the value of pi becomes 1.00 at rank 29 for $m=4$ using the proposed indexing approach. The pi for all ranks is not shown in the table due to space constraints. Fig. 3 shows the plot of penetration rate against change in number of clusters. The plot of bin miss rate using k-d-b tree is shown in Fig. 4. This plot shows that k-d-b tree gives considerably low bin miss rate, $bm = 0$ for $m = 3$. The choice of an indexing approach becomes crucial when both speed (measured in terms of pr) and accuracy (measured in terms of bm) plays significant role. Thus, γ is found to strike balance between pr and bm. Fig. 5 shows γ for change in m using k-d-b tree. The value γ falls after $m=3$ which is due to zero bm and linear rise in pr. The value of m is chosen at optimum point where γ is maximum. When working with identification system the primary objective is to achieve $pi = 1$ which is attained at $m = 4$ (refer Table 1), hence optimum number of cluster for BATH database is 4.

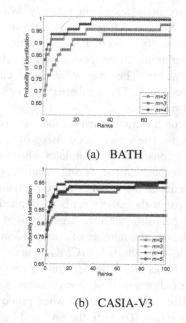

(a)   BATH

(b)   CASIA-V3

Fig. 2. CMC curves of k-d-b tree for change in number of clusters

Similar observations are made for CASIA-V3 database. The CMC curves for k-d-b tree are shown in Fig. 2(b). Here the value of m is varied from 2 to 5. From the curves it can be found the tree based indexing approaches fail to achieve $pi = 1$ for CASIA-V3 database. Table 2 shows identification probabilities for change in ranks. The penetration rate for CASIA-V3 database is shown in Fig. 3. The balanced tree gives considerably low *pr* for CASIA-V3 in comparison to BATH database. The plot of bin miss rate and γ are shown in Fig. 4. and Fig. 5 respectively. As *bm* > 0, the value of *pi* < 1. Thus, maximum value of γ is considered for performance evaluation. The optimum number of cluster for CASIA-V3 database is chosen to be 4.

Table 1. Probability of identification for k-d-b tree based indexing on BATH database

$R{\downarrow}/m{\rightarrow}$	2	3	4
1	0.68	0.70	0.83
2	0.72	0.85	0.87
5	0.78	0.92	0.94
10	0.85	0.92	0.94
20	0.91	0.94	0.96
50	0.94	0.96	**1.00**
100	0.94	0.98	1.00

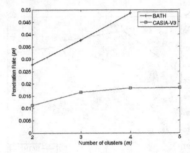

Fig. 3. Penetration rate for BATH and CASIA-V3 databases using k-d-b tree based indexing

Table 2. Probability of identification of k-d-b tree based indexing on CASIA-V3 database

$R{\downarrow}/m{\rightarrow}$	2	3	4	5
1	0.76	0.68	0.75	0.78
2	0.78	0.79	0.80	0.84
5	0.81	0.88	0.86	0.89
10	0.83	0.91	0.95	0.90
20	0.83	0.91	0.95	0.94
30	0.83	0.91	0.95	0.94
50	0.83	0.92	0.95	0.94
100	0.83	0.94	**0.96**	0.95

4.2   Experiment 2: Comparison with Geometric Hashing

The proposed multidimensional tree based indexing approach is compared to geometric hashing based indexing using local features [8]. Fig. 6 shows the identification probabilities of these two approaches for change in ranks on BATH database. From the results it is inferred that the proposed tree based indexing approach performs considerably better in comparison to existing geometric hashing based indexing approach.

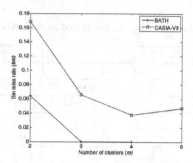

Fig. 4. Bin miss rate for k-d-b tree based indexing on BATH and CASIA-V3 databases

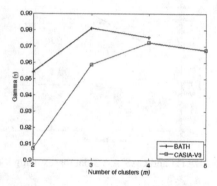

Fig. 5. γ for k-d-b tree based indexing approach with change in number of clusters (*m*)

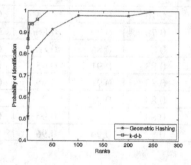

Fig. 6. Identification probabilities of geometric hashing based indexing and proposed k-d-b tree based indexing on BATH database

5 Conclusions

In this paper, local feature based indexing approach is proposed for iris biometric using k-d-b tree. K-d-b tree construction is scalable to new enrollments and the resultant is a balanced index structure. Further, k-d-b tree also finds duplicates during enrollment. The comparison of proposed multidimensional indexing approach with existing geometric hashing approach shows an improvement in performance. The value of γ is significantly high and marks the suitability of proposed k-d-b tree based indexing approach for very large scale iris biometric databases. Further, in this paper m is chosen through extensive experiments. However, in future there is a need to adaptively select the number of clusters (m) based on distribution of features.

Acknowledgements. This work is partially supported by financial assistance received from Department of Science and Technology, Government of India under Women Scientist Scheme (WOS-A) vide sanction order number: SR/WOS-A/ET-142/2011 (G) dated: 9th May 2012.

References

1. Daugman, J.: How iris recognition works. IEEE Transactions on Circuits and Systems for Video Technology 14(1), 21–30 (2004)
2. Unique Identification Authority of India, http://uidai.gov.in/
3. Mhatre, A., Chikkerur, S., Govindaraju, V.: Indexing biometric databases using pyramid technique. In: Kanade, T., Jain, A., Ratha, N.K. (eds.) AVBPA 2005. LNCS, vol. 3546, pp. 841–849. Springer, Heidelberg (2005)
4. Feng, H., Daugman, J., Zielinski, P.: A Fast Search Algorithm for a Large Fuzzy Database. IEEE Transactions on Information Forensics and Security 3(2), 203–212 (2008)
5. Mukherjee, R., Ross, A.: Indexing iris images. In: 19th International Conference on Pattern Recognition (ICPR), pp. 1–4 (2008)
6. Mhatre, A.J., Palla, S., Chikkerur, S., Govindaraju, V.: Efficient search and retrieval in biometric databases. In: SPIE Biometric Technology for Human Identification II, vol. 5779 (2005)
7. Rathgeb, C., Uhl, A.: Iris-Biometric Hash Generation for Biometric Database Indexing. In: 20th International Conference on Pattern Recognition (ICPR), pp. 2848–2851 (2010)
8. Mehrotra, H., Majhi, B., Gupta, P.: Robust iris indexing scheme using geometric hashing of SIFT keypoints. Journal of Network and Computer Applications 33(3), 300–313 (2010)
9. Lowe, D.G.: Distinctive Image Features from Scale-Invariant Keypoints. International Journal of Computer Vision 60(2), 91–110 (2004)
10. Jayaraman, U., Prakash, S., Gupta, P.: An iris retrieval technique based on color and texture. In: 7th Indian Conference on Computer Vision, Graphics and Image Processing (ICVGIP), pp. 93–100 (2010)
11. Bay, H., Ess, A., Tuytelaars, T., Gool, L.V.: Speeded-Up Robust Features (SURF). Computer Vision and Image Understanding 110(3), 346–359 (2008)
12. Panda, A., Mehrotra, H., Majhi, B.: Parallel geometric hashing for robust iris indexing. Journal of Real-Time Image Processing 1–9 (2011)

13. Robinson, J.T.: The K-D-B-tree: A search structure for large multidimensional dynamic indexes. In: ACM SIGMOD International Conference on Management of Data, pp. 10–18 (1981)

14. Bakshi, S., Mehrotra, H., Majhi, B.: Real-time iris segmentation based on image morphology. In: International Conference on Communication, Computing & Security (ICCCS), pp. 335–338 (2011)

15. Dunn, J.: A fuzzy relative of the Isodata process and its use in detecting compact well-separated clusters. Journal of Cybernetics 3(3), 32–57 (1973)

16. Bentley, J.L.: Multidimensional binary search trees used for associative searching. Communications of the ACM 18(9), 509–517 (1975)

17. Agarwal, P.K., Erickson, J.: Geometric range searching and its relatives. Advances in Discrete and Computational Geometry 23, 1–56 (1998)

18. BATH University Database, http://www.bath.ac.uk/elec-eng/research/sipg/irisweb

19. CASIA Iris Image Database Version 3.0, http://www.cbsr.ia.ac.cn/english/Databases.asp

20. Wayman, J.L.: Error rate equations for the general biometric system. IEEE Robotics and Automation Magazine 6(1), 35–48 (1999)

21. Gadde, R., Adjeroh, D., Ross, A.: Indexing iris images using the Burrows-Wheeler Transform. In: IEEE International Workshop on Information Forensics and Security, pp. 1–6 (2010)

Research on Remote Sensing Images Online Processing Platform Based on Web Service

Xiujuan Tian[1] and Binge Cui[2]

[1] School of Materials Science and Engineering,
Shandong University of Science and Technology, 266590 Qingdao, Shandong, P.R. China
txj79@sohu.com
[2] College of Information Science and Engineering,
Shandong University of Science and Technology, 266590 Qingdao, Shandong, P.R. China
cuibinge@gmail.com

Abstract. With the rapid development of remote sensing technology, more and more remote sensing image processing applications were developed. However, most of the legacy applications can only execute on the local host. In this paper, we proposed an approach to encapsulate and publish the legacy DLLs based on Web service. Through our published platform, users can experience each kind of image processing results online and decide which one is most suitable for them. As long as the DLL owners shared their applications, end users can upload image files to the platform and start the image processing. The platform provides an opportunity for the DLL owners to advocate and demonstrate their special programs; in the meantime it also provides more and better choices for end users.

Keywords: Web Service, Legacy Applications, Resource Sharing.

1 Introduction

Today, many of us like to get relaxed and entertained or complete work on the Web, e.g., hearing songs, seeing movies, editing a Word document, etc. The required resource (song, movie, documents) need not have been stored in our local host. Meanwhile, the application (music player, video player, word processer) need not have been installed on our host too. What we need is just a Web browser. Moreover, we can get the latest and most complete resources and services on the Web. We have rarely seen people store hundreds of songs or movies on disk. SaaS (Software as a Service) has been gaining in popularity [1-2].

However, there are still a lot of resources and applications that have not been converted to Web Service [3]. Some of them are made in earlier language and tools, such as C, C++, FORTRAN, etc [4]. Such applications can only be executed or invoked in the local host. How can we access them on the Web? If this becomes feasible, we no longer need to waste time to search, purchase, download and install all applications. We can experience different applications online. Moreover, new applications can be

D.-S. Huang et al. (Eds.): ICIC 2013, LNAI 7996, pp. 485–493, 2013.

added into the Web at any time by DLL owners, and old applications can be maintained or replaced at any time. Only when we want to process a lot of images locally, we should contact the DLL owners to purchase and install them on our local host.

To achieve the above objective, we construct an online remote sensing images processing platform. The platform addresses three key technical problems. The first problem is the dynamic invocation of DLL based on reflection [5]. Because the platform is designed to be extensible, DLL static binding is not applicable for requirement. The second problem is the dynamic construction of user interface. The third problem is the online processing of remote sensing images. We have developed one integrated development toolkit. We also give instructions for the DLL owners to improve the DLL implementation, which enables them to display process of task completion. This feature is particularly useful for the long-time processing task.

2   DLL Improvement and Registration

As usual, DLL is designed and implemented by remote sensing professional programmers. Thus, we won't repeat specific programming details. However, not all of the remote sensing applications will complete in a very short period of time. Some applications take a few minutes or even hours to complete. Let's image what we can do if one Web page is not responding for a few minutes? 99% of people will turn off this page impatiently! Therefore, progress information of the computation is essential for the online remote sensing image processing. Unfortunately, almost no applications can provide information about the running progress. How can we improve the existing DLL to compensate for this flaw?

Before we embed the new code in the DLL, we must adhere to the principle that the original code is not changed. In other words, the new code will not affect the execution result of original code or reduce their efficiency. In order to solve this problem, we set up one monitor for the long running computation process, which is defined as follows:

```
struct Monitor  {
  CString key;
  float progress;   //The process of this computation
  Monitor(CString key="0",float progress=0) {
    this->key=key;
    this->progress=progress;
  }
};
```

We also defined a class Monitor Manager to create and manage all monitors. Its definition is as follows:

```
class MonitorMgr  {
  private:
    list <Monitor> MonitorList;
    CString uniqueKey; // the key for the created monitor
```

```
public:
    void CreateMonitor(CString key);
    CString GetUniqueKey();
    void SetProgress(CString key, float progress);
    float GetProgress(CString key);
    void FreeMonitor(CString key);
};
```

When a long-running computation process starts, one monitor is created to monitor that process with unique key. One statement should be written in the beginning of the function.

```
CString localKey = monitormgr.GetUniqueKey();
```

The rate of progress is written by the application itself. For example, suppose an application will loop 100 times. At the end of the loop, programmers should add such statement:

```
monitorMgr.SetProgress(localKey, counter / 100);
```

So far, DLL improvement is completed. However, some information will usually be missing in the binary DLL, e.g., function names, parameter names, data types, etc. Thus, DLL owners have to re-provide such information during the registration process. For each function, the function name, description, return value type, show progress or not should be provided. For each parameter of the function, the parameter name, description and data type should be provided. Moreover, we defined an extra attribute for parameter: passMode. If one parameter is a data value or a path for the uploaded image file, then passMode is input; if one parameter is a path for the generated result file by the DLL, then passMode is output.

3   DLL Dynamic Invocation and User Interface Design

3.1   DLL Dynamic Invocation

As we know, DLL programs can only execute in the local host. In order to invoke one function in a DLL, we must use the keyword "static" and "extern" to declare that function, and use "DLLImport" to import it. For example, if we want to use the "MessageBoxA" function in "user32.dll", the following codes should be preceded.

```
[DllImport("user32.dll", EntryPoint="MessageBoxA")]
static extern int MsgBox(int hWnd, string msg, string
caption, int type);
```

A typical function call in C# is as follows.

```
MsgBox(0,"Hello World!","Message",0x30);
```

This static function call cannot meet the scalability requirements of our system. Firstly, we don't know what DLL will be added later. Secondly, we don't know what functions is included in these DLLs. Thirdly, we also don't know the parameter number, parameter type of each function. Thus static binding approach is not feasible. We investigated DLL dynamic invocation approach in C#. The detail procedure is stated as follows:

1) All the necessary information for function call, including DLL name, function name, input parameters' name and value, output parameter's name and value, will be organized in a hash table. Each item of the hash table is a name-value pair.

2) Define three arrays: parameter values, parameter types, pass modes. Parameter value can be obtained from the hash table, and parameter type and pass mode can be obtained from the registration file. These three arrays and function return type is the four input parameters of DLL dynamic invocation.

3) Get the DLL handle and the function handle using the external function "LoadLibrary" and "GetProcAddress". The detail code can be obtained via Google.

4) Invoke the target function in C# based on reflection mechanism. The key code is described as follows:

```
AssemblyBuilder MyAssemblyBuilder =
  AppDomain.CurrentDomain.DefineDynamicAssembly(new
  AssemblyName(), AssemblyBuilderAccess.Run);
ModuleBuilder MyModuleBuilder =
  MyAssemblyBuilder.DefineDynamicModule("InvokeDLL");
MethodBuilder MyMethodBuilder =
  MyModuleBuilder.DefineGlobalMethod("MyFun",
  MethodAttributes.Public | MethodAttributes.Static,
  Function Return Type, Parameter Type Array);
ILGenerator IL = MyMethodBuilder.GetILGenerator();
for (int i = 0; i < Parameter Value Array.Length; i++) {
    switch (Pass Mode Array [i]) {
        case PassMode.ByValue: IL.Emit(OpCodes.Ldarg, i);
          break; //Push the parameter value into the stack
        case PassMode.ByRef: IL.Emit(OpCodes.Ldarga, i);
          break; //Push the parameter address into the stack
}}
IL.Emit(OpCodes.Ldc_I4, farProc.ToInt32());
IL.EmitCalli(OpCodes.Calli, CallingConvention.StdCall,
  Function Return Type, Parameter Type Array);
IL.Emit(OpCodes.Ret);
MyModuleBuilder.CreateGlobalFunctions();
MethodInfo MyMethodInfo =
  MyModuleBuilder.GetMethod("MyFun");
return MyMethodInfo.Invoke(null, Parameter Value Array);
```

3.2   DLL Management Web Service

DLL registration information is the prerequisite and basis for the DLL dynamic invocation. Thus, DLL management web service is composed of two main functions, DLL
registration and DLL invocation. The WSDL interface for DLL registration contains
six functions:

```
addDLL(string dllName);
delDLL(string dllName);
addFunction(string dllName, string funcName, string re-
turnType, bool showProgrss, string description);
delFunction(string dllName, string funcName);
addParameter(string dllName, string funcName, string
paramName, string dataType, string passMode, string de-
scription);
delParameter(string dllName, string funcName, string
paramName);
```

The WSDL interface we defined for the DLL invocation is different from that of
the previous work [6]. They generate dynamically one port type for each function of
the DLL. However, we only define one port type for DLL invocation in the WSDL
description. In other words, no matter how many DLLs and functions there are, port
type is the same. The common function is as follows:

```
string invoke(string dllName, string funcName, string pa-
ramValues);
```

The third parameter paramValues is a very long string. It is composed by concate-
nating the value of all parameters using commas. The following statement is one
typical invocation.

```
invoke("Computation", "TMCompute", "1.2378, 0.055158, 2,
3, band6.raw, emissivity.raw, result.raw");
```

The meaning and data type of each part of paramValues can be interpreted through
the DLL registration information. Detail implementation of the function "invoke" is
based on the dynamic invocation approach discussed in Section 4.1.

3.3   User Interface Design

Since the number of parameters, parameter types of each function is not the same,
DLL owners want to design a particular interface for each function. As the development of a visual design environment is too difficult for us, we developed a tree-based
design environment, which is shown in Figure 1.

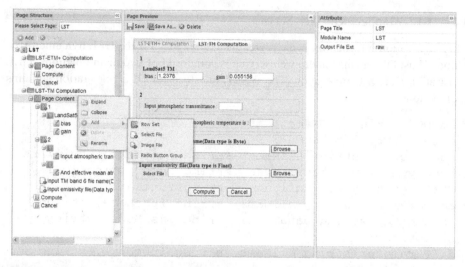

Fig. 1. Tree-based Page Design Environment

The design environment is composed of three parts. The left part is page structure design panel. Right click each node in the tree, one menu will pop up to help DLL owners add controls in the page. There are four types of first-level controls: Row Set, Select File, Image File, Radio Button Group. Row Set can contain second-level controls: Row. Row can contain three-level controls: Text Box and Check Box. Other first-level controls can also be further decomposed. Then central part is page preview panel. When users add one control in the left part, the corresponding control will also appear in the preview panel to show the design result. The right part is attribute panel. When users select one control in the left part, its attributes will appear in the attribute panel. Different controls have different attribute set.

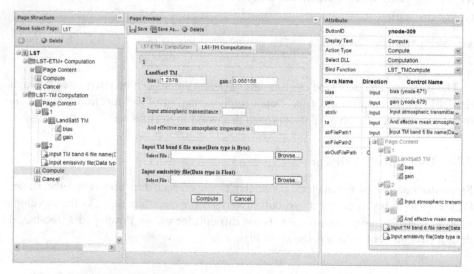

Fig. 2. Build binding relationship between foreground page and background function

The Compute button should tell us the binding relationship between the page and the DLL. Click the button in the design panel, we can see all attributes in the attribute panel, as shown in Figure 2. Display Text can be modified by users. Action Type contains Compute and Cancel. Select DLL is a drop down list box, which contain all registered DLLs. Bind Function is also a drop down list box, which contains all the function of the selected DLL. After the function has been selected, its parameters will appear in the grid at the bottom. Each input should be bound with one control in the page. In other words, when end user inputs one value in the control (or select one file from the disk), the value will be passed to the corresponding parameter of the function. In this way a visual mapping between the foreground page and the background function is built. The output parameter should not be bound with the page control, because its value is generated by the function, not inputted by end users.

After the page design is complete, users can click the "Save" button to save the page as an XML file. Due to space limitations, we omit the content of the page profile file. The saved page can also be re-opened and edited. Through the above design environment, DLL owners can independently design the appropriate user interface without the help of system administrator.

4   Remote Sensing Image Online Processing

Remote sensing image processing is quite different from the general compute programs. It requires one or more image files as input. After computation, it will generate one or more output files. Moreover, these files are usually very large, from several megabytes to several hundreds of megabytes. As the WAN speed is relatively slow, the upload and download files take a long time. To allow users to see the progress of file upload or download, we developed a file transfer ActiveX control based on FTP.

The values of bias and gain are the default values that assigned during the design time. They can be modified by end users. Fill out the value of the other text boxes, select two image files from the local host and click the "Begin Upload" button. The image files will be uploaded from the local host to the platform server. When everything is ready, end users can click the "Compute" button. The underlying DLL function will be invoked and executed, and the computation process will be displayed in the page. After calculating one hyperlink "Download the Results" will appear to allow end users download the result files, as shown in Figure 3.

Through this process, we can see that end users needn't buy and install any programs in their local host. They only need to upload their images and start the process. They do not need a server with high performance. The platform has provided them with all the needed equipment, programs and auxiliary data. In all, it has achieved our goals to share programs and computing resources on the Internet.

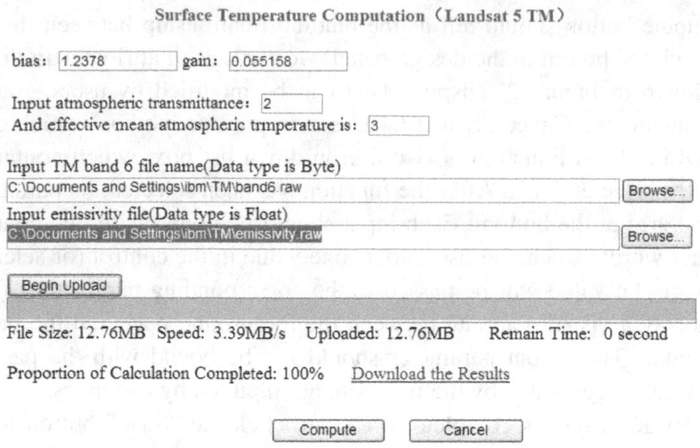

Surface Temperature Computation (Landsat 5 TM)

bias: 1.2378   gain: 0.055158

Input atmospheric transmittance: 2
And effective mean atmospheric trmperature is: 3

Input TM band 6 file name(Data type is Byte)
C:\Documents and Settings\ibm\TM\band6.raw   Browse...

Input emissivity file(Data type is Float)
C:\Documents and Settings\ibm\TM\emissivity.raw   Browse...

Begin Upload

File Size: 12.76MB   Speed: 3.39MB/s   Uploaded: 12.76MB   Remain Time: 0 second

Proportion of Calculation Completed: 100%   Download the Results

Compute   Cancel

Fig. 3. Remote Sensing Image Online Processing

5   Related Works

Wang Bin studied key technologies in gridifying legacy scientific computing resources [7]. He put forward a "Grid-enabled execution environment for computing jobs", which supports invocation and execution of legacy codes under Grid circumvents. He did not solve the problem of monitoring the progress of the computation, which is an important factor for long running task.

Sneed studied integrating legacy software into service oriented architecture [8]. His approach is appropriate when the salvaged code is sufficiently independent of its environment. His proposal requires the larger changes of the code and structure. However, our proposal need not modify or delete any existing source code.

6   Conclusion

In recent years, Web services and cloud computing has been more and more people's attention and recognition. Their common idea is to achieve the share of computing resources and application resources. Our goal is to provide a platform which can help DLL owners share their remote sensing data processing programs on the Web. Our contribution consists of three aspects. Firstly, we designed and implemented a general Web service based on the dynamic invocation technology. Secondly, we provide an IDE for DLL owner to design user interface for each function. Thirdly, the platform can display the designed user interface and support the online processing of remote sensing images.

References

1. Sun, W., Zhang, K., Chen, S.-K., Zhang, X., Liang, H.: Software as a Service: An Integration Perspective. In: Krämer, B.J., Lin, K.-J., Narasimhan, P. (eds.) ICSOC 2007. LNCS, vol. 4749, pp. 558–569. Springer, Heidelberg (2007)

2. Sun, W., Zhang, X.,, G.C.J., Sun, P., Su, H.: Software as a Service: Configuration and Customization Perspectives. In: IEEE Congress on Services Part II, pp. 18–24. IEEE Press, New York (2008)

3. Benatallah, B., Casati, F., Toumani, F.; Hamadi. R.: Conceptual Modeling of Web Service Conversations. In: Eder, J., Missikoff, M. (eds.) CAiSE 2003. LNCS, vol. 2681, pp. 449–467. Springer, Heidelberg (2003)

4. Delaitre, T., Kiss, T., Goyeneche, A., et al.: GEMLCA: Running Legacy Code Applications as Grid Services. Journal of Grid Computing 3(1-2), 75–90 (2005)

5. Zhou, R.Z., Wei, Z.K., Tang, S.G., Kim, J.H.: Study on Application of.NET Reflection in Automated Testing. In: 12th International Conference on Advanced Communication Technology (ICACT), pp. 797–800 (2010)

6. Harry, M.S.: Integrating Legacy Software into a Service Oriented Architecture. In: 10th European Conference on Software Maintenance and Reengineering (CSMR), pp. 293–313 (2006)

7. Wang, B.: A Study of Key Technologies in Gridifying Legacy Scientific Computing Resources. PhD thesis, Peking University (2005) 遗留科学计算资源网格化关键技术研究

8. Sneed, H.M.: Encapsulation of Legacy Software: A Technique for Reusing Legacy Software Components. Annals of Software Engineering 9, 293–313 (2000)

An Advanced Particle Swarm Optimization Based on Good-Point Set and Application to Motion Estimation

Xiang-pin Liu[1], Shi-bin Xuan[1,2,*], and Feng Liu[1]

[1] College of Information Science and Engineering,
Guangxi University for Nationalities, Nanning
[2] Guangxi Key Laboratory of Hybrid Computation and IC Design Analysis, Nanning
sbinxuan@gxun.cn

Abstract. In this paper, an advanced particle swarm optimization based on good-point set theory is proposed to reduce the deviation of the two random numbers selected in velocity updating formula. Good-point set theory can choose better points than random selection, which can accelerate the convergence of algorithm. The proposed algorithm was applied to the motion estimation in digital video processing. The simulation results show that new methods can improve the estimation accuracy, and the performance of the proposed algorithm is better than previous estimation methods.

Keywords: good-point set, particle swarm optimization, motion estimation, block matching algorithm.

1 Introduction

Particle swarm optimization (PSO) is a population based stochastic optimization technique developed by Dr. Eberhart and Dr. Kennedy in [1], because of its good characteristics, PSO algorithm has become a research hotspot of intelligent algorithm.

Good-point set theory was proposed by mathematician Hua Luogeng in [2]. The deviation of selecting points of good-point set is much smaller than randomly selecting. In recent years, some scholars have applied it to the crossover operation in genetic algorithm [3], the simulation results showed that the new algorithm has superiority in speed and accuracy. Then improvements of PSO were proposed, mainly to be used in the construction of the initial population and the crossover PSO [4]. Stimulated by it, an advanced particle swarm optimization based on good-point theory is proposed to reduce the deviation of the two random numbers selected in velocity updating formula and applied to the motion estimation. The simulation results show that the proposed algorithm can improve the global convergence effectively.

---

* Corresponding author.

D.-S. Huang et al. (Eds.): ICIC 2013, LNAI 7996, pp. 494–502, 2013.

2   An Advanced Particle Swarm Optimization Algorithm Based on Good-Point Set

2.1   Good-Point Set

(1) Set G_t is the unit cube in S-dimensional Euclidean space, i.e. $x \in G_t$, $x = (x_1, x_2, \cdots, x_t)$, where $0 \le x_i \le 1$, $i = 1, 2, \cdots, t$.

(2) Set G_t has a point set $P_n(i) = \{x_1^{(n)}(i), \cdots, x_t^{(n)}(i), 1 \le k \le n\}$, $0 \le x_i^{(n)}(i) \le 1$.

(3) For any given point $r = (r_1, r_2, \cdots, r_t)$ in G_t, let $N_n(r) = N_n(r_1, r_2, \cdots, r_t)$ represents the points number of which satisfies the following inequalities in $P_n(i)$: $0 \le x_j^{(n)}(i) < r_j$. If $|r| = r_1 r_2 \cdots r_t$, then point set has deviation $\varphi(n) = \sup_{r \in G_t} \left| \frac{1}{n} N_n(r) - |r| \right|$.

(4) Set $r \in G_t$, the deviation $\varphi(n)$ of $P_n(i) = \{(r_1 \times i, r_2 \times i, \cdots, r_t \times i, i = 1, 2, \cdots n\}$ satisfy $\varphi(n) = C(r, \varepsilon) n^{-1+\varepsilon}$, where $C(r, \varepsilon)$ is a constant only related to r, ε $(\varepsilon > 0)$, then $P_n(i)$ is called good-point set, r is referred to as good point.

(5) Set $r_k = \{2 \cos 2\pi k / p\}$, $1 \le k \le t$, p is the smallest prime number satisfying $(p - t)/2 \ge t$, then r is a good point. Let $r_k = \{e^k\}$, $1 \le k \le t$, then r is a good point.

Theorem (Kai lai Chuang, Kiefer) Set $x_1, x_2, \cdots, x_n$ be a $i.i.d.$ uniform distribution on D_t, $P_n = \{x_1, x_2, \cdots, x_n\}$, then the probability of $D(n, P_n) = O(n^{-1/2}(\log\log n)^{1/2})$ is 1.

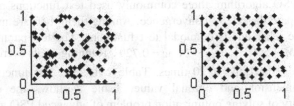

Fig. 1. Distributions of 100 points generated by random and good-point set strategy

In order to illustrate that good-point set sampling is better than random sampling, impression drawings of 100 points using two methods respectively are given in Fig.1. Obviously, the random distribution is scattered, as shown in Fig.1(left), but the good points distribution is relatively uniform, as shown in Fig.1(right).

2.2   Particle Swarm Optimization

PSO is a population based stochastic optimization technique. In PSO algorithm, the potential solutions are called particles, each particle updates its velocity $v_i^{(t+1)}$ and position $x_i^{(t+1)}$ to catch up the best particle g, as follows:

$$v_i^{(t+1)} = wv_i^{(t)} + c_1 * r_1 * (pbest_i - x_i^t) + c_2 * r_2 * (gbest - x_i^t) \qquad (1)$$

$$x_i^{(t+1)} = x_i^{(t)} + v_i^{t+1} \tag{2}$$

Where w is inertia weight, c_1 and c_2 are two position constant, namely, learning factors, usually set to be 2.0. r_1 and r_2 are two random real number within the range of $[0,1]$, which being generated for acceleration toward *pbest* and *gbest* locations. It can be known from Theorem (Kai lai Chuang, Kiefer), the deviation of selecting points at random is large, and this will lead to slow convergence speed of the algorithm. Aiming at this problem, a novel approach to updating the velocity equation based on good-point set in PSO algorithm is proposed. In this work, equation (1) which used to calculate the velocity in standard PSO algorithm is modified to:

$$v_i^{(t+1)} = wv_i^{(t)} + c_1 * gp_1 * (pbest_i - x_i^t) + c_2 * gp_2 * (gbest - x_i^t) \tag{3}$$

Where gp_1 and gp_2 are numbers which generated according to a good-point set. First, constructing the good-point set (n points) in S-dimensional Euclidean space:

$$P_n(i) = \{(\{r_1 \times i\}, \{r_2 \times i\}, \cdots, \{r_s \times i\}), i = 1, 2, \cdots, n\} \tag{4}$$

Where $r_k = 2\cos(2\pi k / p)$, $1 \le k \le S$, p is the smallest prime number satisfying $p \ge 2S + 3$ or $r_k = e^k$, $1 \le k \le t$. $\{a\}$ is decimal part of a. Then, selecting two numbers from $P_n(i)$ randomly, assigning to gp_1 and gp_2.

Aiming to prove the introduction of good-point set theory can accelerate the convergence of PSO algorithm, three commonly used test functions are selected as test cases to compare the rate of convergence. Among them, two are multimodal test functions, and the other one is unimodal test functions. The test parameters setting: population size $M = 500$, $c_1 = c_2 = 2$, $w = 0.729$. Running trials were repeated for each of the chosen function for 20 times. Table 1 lists the test functions and their search space, dimension and optimal value. Table 2 shows the effect on the optimization ability of solving optimization problem of advanced PSO algorithm.

Table 1. Test functions

Function	Space	Dim	val
$f_1(x) = 20\exp(-0.2\sqrt{(1/n)\sum_{i=1}^{n} x_i^2}) - \exp((1/n)\sum_{i=1}^{n}\cos(2\pi x_i)) + 20 + e$	[-32,32]	30	0
$f_2(x) = (1/4000)\sum_{i=1}^{n} x_i^2 - \prod_{i=1}^{n}\cos(x_i/\sqrt{i}) + 1$	[-600,600]	30	0
$f_3(x) = \sum_{i=1}^{n}[x_i^2 - 10\cos(2\pi x_i) + 10]$	[-5.12,5.12]	30	0

Table 2. Comparative tests on the rate of convergence

Func	Good-point set	Iterations	Average (20 times)	optimal values
$f_1(x)$	Yes	110~139	126.2	4.42
	No	130~197	173.4	4.34
$f_2(x)$	Yes	107~130	121.9	9.86e-3
	No	124~170	145.8	1.23e-2
$f_3(x)$	Yes	93~119	105.3	27.04
	No	110~162	129.2	28.20

It can be seen from Table 2, for the function $f_1(x)$, the results of optimization did not reach the optimal value, but the average iterations reduced 47.2 after introduced the good-point set theory. The proposed method accelerates obviously the convergence speed of the algorithm. For the function $f_2(x)$, the accuracy of the optimal solution has reached e-3, and higher than the result which did not use the good-point set theory, and the average iterations reduced 23.9 at the same time. For the function $f_3(x)$, the results of optimization did not reach the optimal value, but the average iterations reduced 23.9 after introduced the good-point set theory. This shows that the introduction of good-point set can significantly improve the search ability of PSO algorithm, and accelerate the convergence speed of the algorithm.

3   Application of the Advanced PSO Based on Good-Point Set to Motion Estimation

Motion estimation is the prediction of the motion vector between the current frame and the reference frame. Because of its simple structure, easy to implement, block matching algorithm BMA[5] has been widely used. Full search algorithm(FS) is a traditional BMA, but its high computational complexity makes it not suitable for real-time implementation, so a number of fast BMAs such as three-step search(TSS)[6], four-step search(FSS)[7], diamond search(DS)[8], adaptive rood pattern search(ARPS)[9] etc, have been proposed which rely on the assumption that Block Distortion Measure(BSM) increases monotonically as the checking point moves away from the global minimum BSM point. These algorithms tend to get trapped in local minimum route of their search of a global minimum, owing to their sub-optimal search strategies. In order to handle this, Genetic algorithm(GA) was applied to BMA in [10]. Although it reduces the computational complexity compared with FS, but the time cost is still much higher than that of fast BMAs. Therefore, PSO was applied to BMA instead of GA in [11], called PSOS, which can improve both computational complexity and estimation precision, but still fail to accomplish an ideal result. In this paper, an advanced PSO based on good-point set was proposed, named GPPSOS.

3.1 Initialization of Population

In BMA, the block is regarded as particle. According to the center of bias of the video sequence, the home position of the current block is taken as the coordinate of the first particle. The four candidate blocks which distance the first particle 2 pixels are taken as the other particles. Besides, a certain number of random block within the search window should be selected in order to maintain the global stochastic of PSO.

3.2 Fitness Function

This paper uses mean absolute difference(MAD) as the fitness function, it is defined:

$$MAD_i(t) = \frac{1}{N_1 N_2} \sum_{x \in 3} |\varphi_2(x+d) - \varphi_1(x)| \tag{5}$$

Where $\varphi_1(x)$ is the pixel value of the current block, $\varphi_2(x+d)$ is the pixel value of the candidate block, N_1 and N_2 are the horizontal and vertical size of the block.

3.3 Update of Velocity and Position

A good-point set with n points is constructed according to equation(4). At each time iteration, two numbers selected randomly from the good-point set are assigned to gp_1 and gp_2 respectively. Then, both the velocity and the position are updated according to equation(3) and equation(2). In equation(3), w usually takes a fixed value, but in this paper w adopts linear decreasing weight strategy, it is defined as follows:

$$w^{(t)} = (w_{ini} - w_{end})(K_{max} - k) / K_{max} + w_{end} \tag{6}$$

Where K_{max} denotes the maximum number of iterations, w_{ini} and w_{end} are the initial inertia weight and the final inertia weight.

3.4 Termination Strategy

Two termination strategies: the maximum number of iterations and the minimum of the fitness difference between the global optimum and the local optimum.

3.5 Algorithm Steps

Step 1. Confirm the current block and initialize the population. Calculate the fitness value of each particle, update the global optimal value and the local optimal value of each particle. Construct the good-point set.

Step 2. Start iteration. Verify whether the termination strategies are satisfied. If satisfied, return the optimal value, or else go to Step 3.

Step 3. Update the inertia weight according to equation(6), select two numbers from the good-point set randomly and assign them to gp_1 and gp_2 respectively. Then, update the velocity and the position according to equation(3) and equation(2).

Step 4. Cross-border judgment on the updated position, if it crosses the border, then limited to the boundary line.

Step 5. Calculate the fitness of each particle updated, update the local optimal value and the global optimal value of the population. Iterations plus 1, back to Step 2.

4   Simulation Results and Performance Comparison

In the simulation experiment, the first 100 frames of the 5 typical standard test sequences: Akiyo(slow), Foreman(middle), Soccer(violent), Football(violent) and Bus(very violent) are used as the test object. Image format is CIF(352×288), frame rate is 30f/s. The block size is fixed at 16×16, search radius is 16. The number of initial population $m=10$, $K_{max}=10$, $c_1=c_2=2$, $w_{ini}=0.9$, $w_{end}=0.4$. In the same test condition, FS, TSS, DS, PSOS and GPPSOS are tested from both the search precision and complexity.

4.1   Search Precision

PSNR(peak signal to noise ratio) is used to evaluate the subjective quality of a reconstructed video sequence, so in this paper, average PSNR is used to evaluate the accuracy of the algorithms. The average PSNR of each algorithm and the difference value between these search methods and FS are shown in Table 3.

Table 3. Average PSNR value comparison of each algorithm

Sequence	Akiyo/dB		Foreman/dB		Soccer/dB		Football/dB		Bus/dB	
Algorithm	PSNR	D-val	PSNR	D-val	PSNR	D-val	PSNR	D-val	PSNR	D-val
FS	37.75	0.00	30.2	0.00	26.5	0.00	22.77	0.00	23.35	0.00
TSS	37.74	(0.01)	29.2	(0.99)	25.1	(1.37)	21.92	(0.85)	20.97	(2.39)
DS	37.75	(0.00)	29.7	(0.55)	23.9	(2.57)	21.16	(1.61)	19.32	(4.03)
PSOS	37.04	(0.71)	29.4	(0.83)	25.4	(1.10)	21.97	(0.80)	21.26	(2.09)
GPPSOS	37.70	(0.05)	29.7	(0.53)	26.0	(0.54)	22.03	(0.74)	21.67	(1.69)

From the analysis of Table 3, for slow and middle motion, difference value of average PSNR between TSS, DS and FS is relatively small, it is because the assumption that BSM increases monotonically is tenable. But for violent motion, difference value of average PSNR is big because of the failure of the monotonicity assumption. PSOS has a good overall, especially for the sequences of violent motion, it can get a good optimization, so the average PSNR is higher than that in TSS, DS. But for the slow motion, PSOS cannot find the optimal value accurately due to its stochastic nature which leads to the slow convergence speed. GPPSOS with fast convergence speed overcomes the defects of PSOS, and its average PSNR is satisfactory. Besides, for the violent motion sequences, the average PSNR of GPPSOS is far higher than other algorithms, further improving the search accuracy.

The PSNR values of Soccer and Bus sequences using TSS, DS, PSOS, GPPSOS respectively are shown in Fig.2. In the figure, circle represents the PSNR value of GPPSOS. As can be seen, for the Soccer sequence, most PSNR value using GPPSOS are higher than other algorithms; for the Bus sequence, the positions of circle are significantly higher than other symbols, it clearly shows the superiority of GPPSOS.

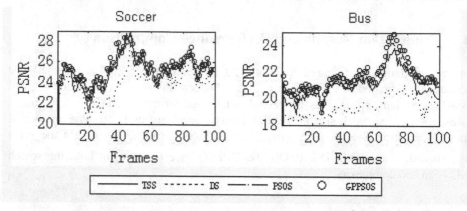

Fig. 2. First 100 frame PSNR value of two kinds of algorithm

Fig. 3. Motion compensation images of Soccer sequence at the third frame

In order to illustrate the superiority of the algorithm proposed in this paper more intuitively, the reconstructed image of the third frame of Soccer sequence is shown in Fig.3. According to the observation of the changing parts in Fig.3, the left leg circle marked, for example, the reconstructed images of TSS, DS and PSOS are incomplete, but the reconstructed image of GPPSOS algorithm is almost as same as that of FS, better than any other algorithm.

4.2   Computational Complexity

This paper takes the average number of candidate blocks which need to be calculated as the criterion of the computational complexity. The average number of candidate blocks of each algorithm and the ratio with that of   FS are shown in Table 4.

Table 4. Average number of candidate blocks comparison of each algorithm

Sequence	Akiyo/dB		Foreman/dB		Soccer/dB		Football/dB		Bus/dB	
Algorithm	block	ratio	block	ratio	block	ratio	block	ratio	block	ratio
FS	886	100	985	100	985	100	985	100	985	100
TSS	28.31	3.20	30.90	3.14	31.48	3.20	31.24	3.17	31.30	3.17
DS	11.42	1.29	17.67	1.79	26.99	2.74	31.15	3.16	22.21	3.16
PSOS	82.04	9.26	101.0	10.26	104.9	10.65	105.79	10.74	102.1	10.6
GPPSOS	53.01	5.98	60.35	6.13	68.65	6.97	53.88	5.47	67.61	6.86

From the analysis of the results in Table 4, the average number of candidate blocks of PSOS algorithm has been greatly decreased compared with FS, but still far more than TSS, DS due to its multiple iterations. GPPSOS algorithm adopts two kinds of termination strategy, so iteration can be terminated ahead the time when any condition is satisfied, therefore the average number of candidate blocks of GPPSOS has been further decreased compared with PSOS.

5   Conclusions

In this paper, an advanced particle swarm optimization algorithm based on the good-point set theory of number theory is proposed. According to Theorem Kai lai Chuang, the deviation of good-point set method is far smaller than that of random method, which speeds up the convergence of the algorithm. The application of the advanced algorithm to the block matching algorithm, not only overcoming the defect of traditional fast search algorithm easily falling into local optimum by using the global stochastic of the particle swarm optimization algorithm, but also accelerating the convergence to the optimal by using the small deviation of good-point set theory. The simulation results show that the performance of the advanced algorithm proposed in this paper is better than any other block matching algorithms.

Acknowledgments. This research is supported by the National Science Foundation Council of Guangxi(2012GXNSFAA053227).

References

1. Kennedy, J., Eberhort, R.: Particle swarm optimization. In: Perth: IEEE International Conference on Neural Networks, pp. 1941–1948 (1995)
2. Luo, H., Yuan, W.: Applications of Number-Theoretic Methods in Approximate Analysis. Science Press (1978)
3. Ling, Z., Bo, Z.: Good point set based genetic algorithms. Chinese J. Computers 24(9), 917–922 (2001)
4. Wei, W., Jia, C., Sheng, H.: Particle swarm algorithm based on good point set crossover. Computer Technology and Development 19(12), 32–35 (2009)
5. Jing, Y., Kai, S.: A survey of block-based motion estimation. Journal of Image and Graphics 12(12), 2031–3041 (2007)
6. Zeng, R., Li, B., Liou, M.L.: A new three-step search algorithm for block motion estimation. IEEE Trans. on Circuits and System for Video Technology 4(4), 438–442 (1994)
7. Po, L.M., Ma, W.C.: A novel four-step search algorithm for fast block motion estimation. IEEE Trans. on Circuits and System for Video Technology 6(6), 313–317 (1996)
8. Zhu, S., Ma, K.K.: A new diamond search algorithm for fast block matching motion estimation. IEEE Trans.on Image Processing 9(2), 287–290 (2000)
9. Xiao, W., Jian, Z.: Adaptive Rood Pattern Search for Fast Block-Matching Motion estimation. Journal of Electronics & Information Technology 27(01), 104–107 (2005)
10. Shen, L., Wei, X., et al.: A Novel Fast Motion Estimation Method Based on Genetic Algorithm. Acta Electronica Sinica 6(28), 114–117 (2000)
11. Du, G.Y., Huang, T.S., Song, L.X., et al.: A novel fast motion estimation method based on particle swarm optimization. In: The Fourth International Conference on Machine Learning and Cybernetics, Guangzhou, pp. 5038–5042 (2005)

Contour Segmentation Based on GVF Snake Model and Contourlet Transform

Xinhong Zhang[1,2], Kun Cai[3], Fan Zhang[1,3], and Rui Li[4]

[1] Institute of Image Processing and Pattern Recognition, Henan University,
Kaifeng 475001, China
[2] Software School, Henan University, Kaifeng 475001, China
[3] College of Computer and Information Engineering, Henan University,
Kaifeng 475001, China
zhangfan@henu.edu.cn

Abstract. A contour segmentation algorithm is proposed based on GVF Snake model and Contourlet transform. Firstly, object contours of images can be obtained based on Contourlet Transform, and those contours will be identified as the initial contour of GVF Snake model. Secondly, GVF Snake model is used to detect the contour edge of human gait motion. Experimental results show that the proposed method can extract the edge feature accurately and efficiently.

Keywords: Gait Recognition, Contour Extraction, GVF Snake models, Contourlet Transform.

1   Introduction

With the rising demand for intelligent monitoring system, the non-contact and long range identification technologies have attracted more and more research interest. Among them, the biometric identification technology identifies people by their inherent physiology or behavior characteristics. As a biometric identification technology, gait recognition identifies authentication by extracting human walking features [1]. Gait recognition is to analyze image sequences containing human motions primarily. Gait recognition mainly consists of three stages: the motion segmentation and classification, feature extraction and description, gait recognition [2]. Gait recognition conquers the limitation of other biological features such as Fingerprint recognition, Iris recognition, Facial recognition *etc*, and has been widely researched in recent years.

Currently, the main methods of motion segmentation include background subtraction method [3], frame difference method and approximate motion field method [4]. The frame difference method is the most commonly used method, which is fast but sensitive to noise. Some researchers have been done and have received good segmentation results through combining the background subtraction method and the frame difference method [5,6]. The feature extraction methods of gait recognition include ellipse model method [7] and graph model method [8] etc. The ellipse model method

D.-S. Huang et al. (Eds.): ICIC 2013, LNAI 7996, pp. 503–508, 2013.

expresses different parts of binary image of human profile using ellipses. The graph model expresses different parts of human body using curves, and it tracks the angle-swing of every part of human body in the image sequence and takes the angles as the feature for gait recognition. The matching algorithms of gait recognition include Dynamic Time Warping (DTW) and Hidden Markov Model (HMM) etc. The DTW can complete pattern matching when the test sequences and reference sequences are not accordant. DTW is widely used in speech recognition fields. Hidden Markov model is a statistical model in which the system being modeled is assumed to be a Markov process with unknown parameters, and the challenge is to determine the hidden parameters from the observable parameters. The extracted model parameters can then be used to perform further analysis, for example for pattern recognition applications.

Feature extraction and description is the key steps for gait recognition. How to extract the motion contour effectively is most important in gait recognition, and it is the main point of this paper. This paper combines the background subtraction method with symmetric differential method to segment the motion human image, and then extracts the contour of motion human with improved GVF Snake model. The experimental results show that the proposed method can extract contour features effectively for the gait recognition.

2   Active Contour Model

Active contour model (snakes) was first introduced in 1988 [9], since then, it has been improved by many researchers. Snake models can move under the influence of internal forces within the curve itself and external forces derived from the image. The internal and external forces are defined so that the snake will conform to an abject boundary or other desired features within an image. Snake is parametric active contour model through curve and surface deformation. There are two key difficulties in Snake models, the first is the location of the initial contour setting, and the second is the bad convergence effect to concave boundary regions of image. On this basis of Snake models, there are many improved models, such as Balloon model, Ziploc Snake model, T-Snake, GVF Snake [10] etc.

Snake models minimize the energy functional to restrain the contour of target object. A traditional snake is a curve $X(s) = (x(s), y(s))$ $s \in [0,1]$, its minimizing energy function is as follows,

$$E = \int_0^1 \frac{1}{2}(\alpha \mid X'(s) \mid^2 + \beta \mid X''(s) \mid^2) + E_{ext}(X(s))ds \qquad (1)$$

where α and β are parameters which control the snake's tension and rigidity respectively. The first term is the internal force, which controls the curve changes, while the second term Eextis is the external force, which pulls the curve to desired features. Different Eextcan be constructed in different models.

To analyze the movement of snake model curve from the aspect of force balance, the minimized E of a snake must satisfy the Euler equation:

$$\alpha X''(S) - \beta X'''(S) - \nabla E_{ext} = 0 \qquad (2)$$

To add additional flexibility to the snake model, it is possible to start from the force balance equation directly, $F_{int} + F_{ext}^1 = 0$, where $F_{int} = \alpha X''(S) - \beta X'''(S)$ and $F_{ext}^1 = -\nabla E_{ext}$.

As the Snake contour is dynamic, the $X(s)$ can be viewed as the function of t and s, then,

$$X_t(s,t) = \alpha X''(S) - \beta X'''(S) - \nabla E_{ext} \qquad (3)$$

Once getting solution of the equation (3), we will find a solution of equation (2).

3   Contourlet Transform

The limitations of commonly used separable extensions of one-dimensional transforms, such as the Fourier and wavelet transforms, in capturing the geometry of image edges are well known. In 2002, Do and Vetterli proposed a 'true' two dimensional transform, contourlet transform, which called pyramidal directional filter bank. The resulting image expansion is a directional multiresolution analysis framework composed of contour segments, and thus is named contourlet. This will overcome the challenges of wavelet and curvelet transform.

Contourlet transform is a double filter bank structure. It is implemented by the pyramidal directional filter bank (PDFB) which decomposes images into directional subbands at multiple scales. In terms of structure the contourlet transform is a cascade of a Laplacian Pyramid and a directional filter bank. In essence, it first uses a wavelet-like transform for edge detection, and then a local directional transform for contour segment detection. The contourlet transform provides a sparse representation for two-dimensional piecewise smooth signals that resemble images.

Contourlet coefficients of natural images exhibit the following properties:

1. Non-Gaussian marginally.
2. Dependent on generalized neighborhood.
3. Conditionally Gaussian conditioned on generalized neighborhood.
4. Parents are (often) the most influential.

The contourlet transform expresses image by first applying a multi-scale transform, followed by a local directional transform to gather the nearby basis functions at the same scale into linear structures. In the first stage of contourlet transform, the Laplacian pyramid (LP) is used for sub-band decomposition. The resulting image is subtracted from the original image to obtain the bandpass image. This process can be iterated repeatedly on the coarser version of the image. The LP decomposition at each step generates a sampled lowpass version of the original image and the difference between the original and the prediction, resulting in a bandpass high frequency image. Directional filter bank (DFB) is used in the second stage to link the edge points into linear structures, which involves modulating the input image and using quincunx filter banks (QFB) with diamond-shaped filters.

4 Contour Extraction Based on GVF Snake and Contourlet Transform

In the GVF field, some points have an important influence to the initial contour setting. When these points within the target, the initial contour must contain these points; and when these points outside the target, the initial contour do not contain these points, otherwise it will not converge to correct results. We call these points the critical points. The creation of critical points has the following factors: (1) critical points are associated with the image gradient Δf. By minimizing the energy functional, Δf will be spread out, the impact of noise generated by the local minima will produce the critical points. (2) critical points are associated with the smoothing coefficient μ. The higher the value μ is, the greater the smoothing effect is. The critical points will reduce, and GVF performance will reduce. Choosing a large-scale Gaussian smoothing and a smaller μ value will get good results. (3) critical points are associated with iterations number of solving equations. When the number of iterations is more than a thousand times, GVF field has changed dramatically. Although we require the final converged solution, but in the actual image processing applications, an intermediate result is acceptable. Therefore, the initial contour should contain the critical points as much as possible. So we can reduce the need of capture and reduce the amount of computation.

Contourlet transform offers a much richer sub-band set of different directions and shapes, which helps to capture geometric structures in images much more efficiently. The Laplacian pyramid (LP) is first used to capture the point discontinuities, and then followed by a direction filter banks (DFB) to link point discontinuities into linear structures. In particular, contourlets have elongate supports at various scales, directions, and aspect ratios. The contourlets satisfy anisotropy principle and can capture intrinsic geometric structure information of images and achieve better a sparser expression image than discrete wavelet transform (DWT).

In order to solve the initial active contour problem of Snake model, In this paper, Contourlet transform is introduced into the GVF Snake model, which will provides a way to set the initial contour, as a result, will improves the edge detection results of GVF Snake model effectively. Firstly, the contours of the object in images can be obtained based on Contourlet Transform, and this contours will be identified as the initial contour of GVF Snake model. Secondly, then GVF Snake model is used to detect the contour edge of human gait motion.

5   Experimental Results

This section shows several experimental results achieved using MATLAB 7.0. Images are from the database of the Biometrics and Security Research Center [11].

In experimental results, we can see that the gradient force field locates in the real contours and points to the real contours in the direct neighborhood of the real contour. The GVF force field smoothly spreads and points to the real contour in the homogeneous regions of image. In the GVF, the center is the neighborhood of real contour, and the force field smoothly diamond diffuses out according to the number of iterations. The direction of diffusion is same as the gradient force field center and the diffusion area is proportional to the number of iterations.

Fig. 1(a) and Fig. 1(b) are the rough contour segmented from the background in a continuous images, and the extracted body contour using the improved GVF Snake model ($\mu= 0.1$, the number of iterations is 80 times).

Experimental results show that using GVF Snake model cannot detect accurately the crotch position of body. While the improved anisotropic GVF Snake model can produce better Concavities fitting effect.

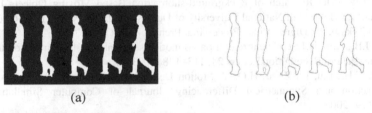

(a)   (b)

Fig. 1. (a) The rough contour segmented from the background. (b) The extracted body contour using the improved GVF Snake model.

6   Conclusions

The background subtraction method and the symmetry differential method can be effectively combined for the moving target detection. The moving target detection is the key step in gait recognition. This paper combines the background subtraction method with symmetric differential method to segment the motion human image, and then extracts the contour of motion human with improved GVF Snake model. The experimental results show that the proposed method can extract contour features effectively for the gait recognition. But the computation based on is complicated and time-consuming, so how to effectively combine GVF Snake model with gait recognition is also an open topic need of further research.

Contourlet transform can be used to captures smooth contours and edges at any orientation. In order to solve the initial active contour problem of Snake model, Contourlet transform is introduced into the GVF Snake model, which will provides a way to set the initial contour, as a result, will improves the edge detection results of GVF Snake model effectively. The multi-scale decomposition is handled by a Laplacian pyramid. The directional decomposition is handled by a directional filter bank.

Firstly, the contours of the object in images can be obtained based on Contourlet Transform, and this contours will be identified as the initial contour of GVF Snake model. Secondly, then GVF Snake model is used to detect the contour edge of human gait motion.

Acknowledgments. This research was supported by the Foundation of Education Bureau of Henan Province, China grants No. 2010B520003, Key Science and Technology Program of Henan Province, China grants No. 132102210133 and 132102210034, and the Key Science and Technology Projects of Public Health Department of Henan Province, China grants No. 2011020114.

References

1. Nixon, M.S., Carter, J.N.: Automatic Gait Recognition, pp. 231–249. Kluwer Academic (1999)
2. Tian, G.J., Zhao, R.C.: Survey of Gait Recognition. Application Research of computers 22(5), 17–19 (2005)
3. Lin, H.W.: The Research of Background-subtraction Based Moving Objects Detection Technology. Journal of National University of Defense Technology 25(3), 66–69 (2003)
4. Murat Tekalp, A.: Digital Video Processing. Prentice-Hall (1996)
5. Kim, J.B., Kim, H.J.: Efficient region-based motion segmentation from a video monitoring system. Pattern Recognition Letters 24, 113–128 (2003)
6. Zhou, X.H., Liu, B., Zhou, H.Q.: A Motion Detection Algorithm Based on Background Subtraction and Symmetrical Differencing. Journal of Computer Simulation 22(4), 117–119 (2005)
7. Lily, L.: Gait Analysis for Classification. R. AI Technical Report 2003-014, Massachusetts Institute of Technology-artificial Intelligence Laboratory (2003)
8. Yoo, J.H., Nixon, M.S., Harris, C.J.: Extracting Gait Signatures Based on Anatomical Knowledge. In: Proceedings of BMVA Symposium on Advancing Biometric Technologies (2002)
9. Kass, M., Witkin, A., Terzopoulos, D.: Snake: Active contour models. Intentional Journal of Computer Vision 1(4), 321–331 (1988)
10. Xu, C., Prince, J.L.: Gradient Vector Flow: A New External Force for Snakes. In: Proceeding of IEEE International Conference on CVPR, pp. 66–71 (1997)
11. Center for Biometrics and Security Research, http://www.cbsr.ia.ac.cn/

Australian Sign Language Recognition Using Moment Invariants

Prashan Premaratne[1], Shuai Yang[1], ZhengMao Zou[2], and Peter Vial[1]

[1] School of Electrical Computer and Telecommunications Engineering,
University of Wollongong, North Wollongong, NSW, Australia
[2] Dalian Scientific and Technological Research Institute of Mining Safety,
No.20 Binhai West Road, Xigang District, Dalian, Liaoning, China
prashan@uow.edu.au

Abstract. Human Computer Interaction is geared towards seamless human machine integration without the need for LCDs, Keyboards or Gloves. Systems have already been developed to react to limited hand gestures especially in gaming and in consumer electronics control. Yet, it is a monumental task in bridging the well-developed sign languages in different parts of the world with a machine to interpret the meaning. One reason is the sheer extent of the vocabulary used in sign language and the sequence of gestures needed to communicate different words and phrases. Auslan the Australian Sign Language is comprised of numbers, finger spelling for words used in common practice and a medical dictionary. There are 7415 words listed in Auslan website. This research article tries to implement recognition of numerals using a computer using the static hand gesture recognition system developed for consumer electronics control at the University of Wollongong in Australia. The experimental results indicate that the numbers, zero to nine can be accurately recognized with occasional errors in few gestures. The system can be further enhanced to include larger numerals using a dynamic gesture recognition system.

Keywords: Australian Sign Language, Dynamic Hand Gestures, moment invariants.

1   Introduction

1.1   Historical Perspective of Auslan

Sign languages in any part of the world have their inception in natural need to communicate especially for children with their parents. Auslan has its roots in British Sign Language (BSL) and Irish Sign Language (ISL) and is known to have been used in early 1800s in large residential schools for the deaf in Australia [1]. ISL was brought to Australia by Irish nuns who established the first school for Catholic deaf children in 1875 [1, 2]. The Irish one-handed alphabet and a tradition of Irish-based signs was kept alive well into the middle of the twentieth century through private Catholic schools that used many Irish signs and one-handed finger spelling, while public schools used Auslan signs (originally BSL) and two-handed finger spelling. Separate

D.-S. Huang et al. (Eds.): ICIC 2013, LNAI 7996, pp. 509–514, 2013.

education systems aside, the two communities mixed freely, with British based signing being undoubtedly the dominant linguistic influence [1, 2].

Schools dedicated for deaf children were first established in Australia in the mid-nineteenth century. The Sydney school for the deaf was established in 1860 by Thomas Pattison. He had his education from the Edinburgh Deaf and Dumb Institution. At the same time another deaf person, Frederick Rose founded the Melbourne School for Deaf who had is formal education at Old Kent Road School in London [1, 2].

Even though Auslan is an offshoot of both BSL and ISL, it has developed some distinct gestures since its inception in Australia in the nineteenth century. New signs developed in the Australian deaf community, particularly in the residential schools for deaf children because signers may have had little contact with deaf communities in other parts of the country [1, 2].

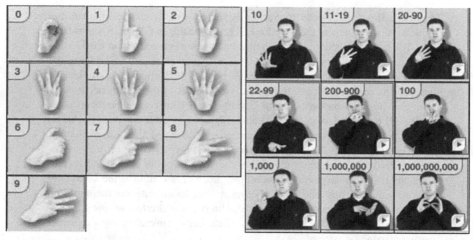

Fig. 1. Numerals of zero to nine and how numerals above 10 are calculated using a gesture sequence. Courtesy of [1].

1.2   Finger Spelling

A number of signs in modern Auslan clearly have their origins in ISL (and through ISL to the French and European signing tradition). Also as a consequence of this mixing and exposure to Irish-based signing, the one-handed alphabet (including its modern American form) does not feel quite so 'alien' to Auslan signers as one might expect. Initialised signs based on one-handed finger spelling have been and continue to be accepted by this linguistic community, even though finger spelling is regularly produced using the two-handed alphabet [2].

1.3   Auslan Evolution

Today Auslan seems to be undergoing a period of rapid change. The enormous expansion of sign language interpreter services, especially in the area of secondary and tertiary education and in the delivery of governmental, legal and medical services, has put great demands on the language by both interpreters and deaf people themselves. These developments have produced three main responses: (i) attempts to standardise usage, (ii) the development of new signs to meet new needs, (iii) the borrowing of signs from other sign languages, particularly from American Sign Language (ASL) [1, 3].

Most members of the deaf community have a personal and political preference for drawing on the internal resources of Auslan to expand and develop its vocabulary. However, some Auslan signers either do not object to ASL borrowings (sometimes they do not even realize that some signs are borrowed from ASL) or are actually willing borrowers (new signs are adopted because they are sometimes seen as more prestigious). The fact that ASL signers also have English as the language of the wider community, as do Auslan signers, may encourage this process. Many borrowed ASL signs are technical and deal with vocabulary used in education and in written English. Nevertheless, many Auslan signers reject any attempts to introduce borrowed ASL signs when a perfectly good and adequate Auslan sign already exists [1, 3].

2   Machine Recognition of Auslan

Currently the Auslan website list 7415 words in its data base. Looking at this figure, it would be almost impossible for a machine to see the subtle variation from one word to the other [1]. With our experience in Hand Gesture Recognition systems for over 7 year, interpreting sign language can be meaningfully attempted only for numerals [4, 5]. Applying hand gesture recognition techniques developed for consumer electronics control will be sufficient to accurately decipher hand gestures used for numerals between 0 and 9. Tens, hundreds, thousands may still need a dynamic hand gesture recognition developed by the authors [6]. Overall, this research will discuss attempt to recognize numerals from 0 to 9 and will ascertain their accuracy.

2.1   Moment Invariants

Image classification is a very mature field today. There are many approaches to finding matches between images or image segments [7, 8]. Starting from the basic correlation approach to the scale-space technique, they offer a variety of feature extraction methods with varying success. However, it is very critical in hand gesture recognition that the feature extraction is fast and captures the essence of a gesture in unique small data set. Neither the Fourier descriptor [9], which results in a large set of values for a given image, nor scale space [10] succeed in this context. The proposed approach of using moment invariants stems from our success in developing the "Wave Controller" [4-6]. Gesture variations caused by rotation, scaling and translation can be circumvented by using a set of features, such as moment invariants, that are invariant to these operations.

The moment invariants algorithm has been recognized as one of the most effective methods to extract descriptive feature for object recognition applications and has been widely applied in classification of subjects such as aircrafts, ships, and ground targets [4, 11]. Essentially, the algorithm derives a number of self-characteristic properties from a binary image of an object. These properties are invariant to rotation, scale and translation. Let f(i,j) be a point of a digital image of size M×N (i = 1,2, …, M and j = 1,2, …, N). The two dimensional moments and central moments of order (p + q) of f(i,j), are defined as:

$$m_{pq} = \sum_{i=1}^{M} \sum_{j=1}^{N} i^p j^q f(i, j)$$

$$U_{pq} = \sum_{i=1}^{M} \sum_{j=1}^{N} (i - \bar{i})^p (j - \bar{j})^q f(i, j)$$

Where

$$\bar{i} = \frac{m_{10}}{m_{00}} \qquad \bar{j} = \frac{m_{01}}{m_{00}}$$

From the second order and third order moments, a set of seven (7) moment invariants are derived as follows [12]: (we are displaying only four here)

$$\phi_1 = \eta_{20} + \eta_{02} \tag{1}$$

$$\phi_2 = (\eta_{20} - \eta_{02})^2 + 4\eta_{11}^2 \tag{2}$$

$$\phi_3 = (\eta_{30} - 3\eta_{12})^2 + (3\eta_{21} - \eta_{03})^2 \tag{3}$$

$$\phi_4 = (\eta_{30} + \eta_{12})^2 + (\eta_{21} + \eta_{03})^2 \tag{4}$$

Where η_{pq} is the normalized central moments defined by:

$$\eta_{pq} = U_{pq} \Big/ U_{00}^r$$

$$r = [(p+q)/2] + 1, \quad p+q = 2,3,...$$

The above four features have achieved accurate gesture recognition for a limited number of gestures in our previous research [4-6]. These moment invariants can be used to train a Support Vector Machine (SVM) approach or Neural Network for Classification.

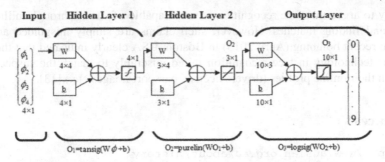

Fig. 2. Structure of the neural network classifier (implementation perspective)

Hand Gesture	Recognition Accuracy %	Hand Gesture	Recognition Accuracy
	100	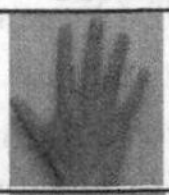	97
	100	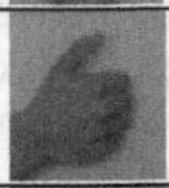	97
	91		90
	89	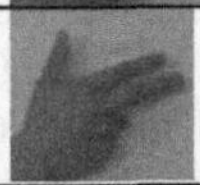	90
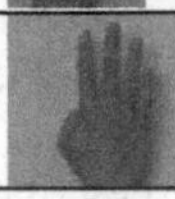	86	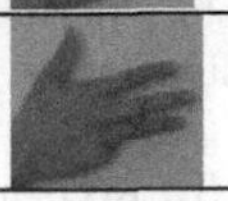	87

Fig. 3. Classification scores for numerals from zero to nine

3   Experimental Results and Conclusion

We developed a database containing 10 images for each gesture and used features extracted using moment invariants for classification. Only the first four moments were used as features similar to our approach in static gesture recognition system [4-6]. The classification was carried out using Neural Network similar to our previous approach for gesture classification as shown in Fig. 2 [4]. Fig. 3 shows the classification scores for the 10 numerals. The results indicate that this approach do have some strength in accurately interpreting 6 gestures and 4 gestures might misclassify 5% to 15% occasionally. Looking at these figures, we conclude that it would almost be an impossible task to decipher sign language using machine learning. Current approaches into gesture recognition and face recognition will never be suitable for accurate interpretation of sign language as it already has 7415 different words and phrases. One might argue that how and why the authors would make such a bold statement. They might try to argue that face recognition is very capable of sifting through millions of images and finding matches. However, such claims are simply misguided and not true. The recent bombing (April 2013) in Boston, USA clearly indicated that the state of the art technology in face recognition failed miserably to match the suspects images with their online profiles (drivers license, passport images etc.) [13].

References

1. http://www.auslan.org.au/about/history
2. Johnston, T.: Signs of Australia: A new dictionary of Auslan, North Rocks. North Rocks Press, NSW (1998)
3. Johnston, T., Schembri, A.: Australian Sign Language: An introduction to sign language linguistics. Cambridge University Press (2007)
4. Premaratne, P., Nguyen, Q.: Consumer Electronics Control System based on Hand Gesture Moment Invariants. IET Computer Vision 1(1), 35–41 (2007)
5. Premaratne, P., Ajaz, S., Premaratne, M.: Hand Gesture Tracking and Recognition System Using Lucas-Kanade Algorithm for Control of Consumer Electronics. Neurocomputing Journal (2012), How to Cite or Link Using DOI,
   http: ==dx:doi:org=10:1016=j:neucom:2011:11:039
6. Premaratne, P., Ajaz, S., Premaratne, M.: Hand gesture tracking and recognition system for control of consumer electronics. In: Huang, D.-S., Gan, Y., Gupta, P., Gromiha, M.M. (eds.) ICIC 2011. LNCS, vol. 6839, pp. 588–593. Springer, Heidelberg (2012)
7. Premaratne, P., Premaratne, M.: New structural similarity measure for image comparison. In: Huang, D.-S., Gupta, P., Zhang, X., Premaratne, P. (eds.) ICIC 2012. CCIS, vol. 304, pp. 292–297. Springer, Heidelberg (2012)
8. Premaratne, P., Premaratne, M.: Image similarity index based on moment invariants of approximation level of discrete wavelet transform. Electronics Letters 48-23, 1465–1467 (2012)
9. Harding, P.R.G., Ellis, T.: Recognizing hand gesture using Fourier descriptors. In: Proceedings of the 17th International Conference on Pattern Recognition, vol. 3, pp. 286–289 (2004)
10. Ho, S., Greig, G.: Space on image profiles about an object boundary. In: Griffin, L.D., Lillholm, M. (eds.) Scale-Space 2003. LNCS, vol. 2695, pp. 564–575. Springer, Heidelberg (2003)
11. Zhongliang, Q., Wenjun, W.: Automatic ship classification by superstructure moment invariants and two-stage classifier. In: ICCS/ISITA 1992 Communications on the Move (1992)
12. Hu, M.K.: Visual Pattern Recognition by Moment Invariants. IRE Trans. Info. Theory IT8, 179–187 (1962)
13. http://spectrum.ieee.org/riskfactor/computing/networks/face-recognition-failed-to-find-boston-bombers

Palmprint Recognition Method Based on a New Kernel Sparse Representation Method

Li Shang

Department of Communication Technology, College of Electronic Information Engineering,
Suzhou Vocational University, Suzhou 215104, Jiangsu, China
sl0930@jssvc.edu.cn

Abstract. To capture the nonlinear similarity of palmprint image features, a new palmprint recognition method utilizing the kernel trick based sparse representation (KSR) algorithm is proposed in this paper. KSR is in fact an essential sparse coding technique in a high dimensional feature space mapped by implicit mapping function, and it can efficiently reduce the feature quantization error and enhance the sparse coding performance. Here, to reduce the time of sparse coding, the fast sparse coding (FSC) is used in coding stage. FSC solves the L_1-regularized least squares problem and the L_2-constrained least squares problem by iterative method, and it has a faster convergence speed than the existing SC model. In test, the PolyU palmprint database used widely in palmprint recognition research is selected. Using the Gauss kernel function and considering different feature dimensions, the task of palmprint recognition obtained by KSR can be successfully implemented. Furthermore, compared our method with general SR and SC under different feature dimensions, the simulation results show further that this method proposed by us is indeed efficient in application.

Keywords: Sparse representation, Fast sparse coding, Kernel function, Classifier; Palmprint recognition.

1   Introduction

In personal recognition, palmprint verification is one of the emerging technologies at present. Compared with other biologic features, those of one's palmprint are stable and remain unchanged throughout the life of a person, so palmprint identification has been applied formally in biologic character identification technology [1-3]. In the task of palmprint recognition, the important issue is how to extract palmprint features, and now many palm recognition methods have been developed, such as ones based on Gabor filters [4-5], wavelet [6-8], independent component analysis (ICA) [9-10], sparse coding (SC) [2, 13-15], etc. In these methods, sparse coding technique has been used widely in many applications, especially in image processing field, due to its state-of-the-art performance [16-18]. And from another angle, SC belongs to in fact one of sparse representation (SR) methods [12, 19-22]. However, existing work based on sparse coding only seeks the sparse representation of the given signal in original

D.-S. Huang et al. (Eds.): ICIC 2013, LNAI 7996, pp. 515–523, 2013.
© Springer-Verlag Berlin Heidelberg 2013

signal space, so the non-linear features of images can't be classified well only by using SC. Recall that kernel trick maps of non-linear separable features into high dimensional feature space, in which features of the same type are easier grouped together and linearly separable, the kernel based sparse representation (KSR) was proposed [17]. This method is the sparse coding in the mapped high dimensional feature space. However, common SC algorithms are very slow in learning feature bases. To reduce learning time of SC stage, here, the fast SC (FSC) algorithm [20-22] is introduced in KSR. FSC algorithm is based on iteratively solving two convex optimization problems, the L_1 -regularized least squares problem and the L_2 - constrained least squares problem. Compared with common SC models, this FSC algorithm can equally model the receptive fields of neurons in the visual cortex in brain of human, but it has a faster convergence speed. In this paper, the Gauss-kernel function is first used to map the feature and basis to a high dimensional feature space. Then, the mapped features and basis are substituted to the formulation of FSC, and a new KSR method is deduced and further discussed in the palmprint recognition task. Here, the PolyU palmprint database used widely in palmprint recognition research is utilized. Considering different feature dimensions, the task of palmprint recognition obtained by our method can be successfully implemented. Furthermore, compared with standard SR and Poly-kernel based SR under different feature dimensions, the simulation results show further that this method proposed by us is indeed efficient in application.

2   Kernel Sparse Representation Algorithm

2.1   The Sparse Representation of Dictionary Bases

Let $X = [x_1, x_2, \cdots, x_C]$ denotes the training sample matrix, where C is the class number of training samples, and x_i ($i = 1, 2, \cdots, C$) is a column vector, which represents the *ith* class of palm images. Let the matrix $D = [d_1, d_2, \cdots, d_K] \in \Re^{N \times K}$ be an over-complete dictionary of K prototype atoms. And then, vector $x_i \in \Re^N$ can be represented as a sparse linear combination of these atoms, namely, x_i can be approximately written as $x_i \approx D_{s_i}$, satisfying $\| x_i - D_{s_i} \|_2 \le \varepsilon$, where $s_i \in \Re^K$ a vector with very few nonzero entries [17-22] is. Thus, the optimized object function of sparse representation for the *ith* class of samples can be written as follows

$$\hat{s}_i = \arg\min_{s_i} \left\| x_i - \sum_i^K D_{s_i} \right\|_2^2 \qquad s.t \ \ \| s_i \|_0 \le \varepsilon \ . \tag{1}$$

where $\hat{s}_i$ is the estimation of s_i, which is the sparse representation coefficient vector of x_i, and ε is the sparse threshold. Here, the l^0 norm represents the nonzero number of coefficient vector s_i, and its physical meanings denote the sparse degree. In the ideal condition, only the coefficient values corresponding to feature bases of a given class of samples are nonzero and other coefficient values corresponding to feature bases of other class of samples are zero. Usually, the optimization solution of the l^0 norm is hardly difficult, therefore, in practical application, the l^1 norm is used to substitute the l^0 norm. Thus, the formula is rewritten as follows

$$\hat{s}_i = \arg\min_{s_i}\left\|x_i - \sum_i^K D s_i\right\|_2^2 + \lambda\|s_i\|_1 .$$ (2)

2.2   The Idea of KSR

KSR is in fact a sparse coding technique. For general sparse coding, it aims at finding the sparse representation under the given basis A ($A \in \mathfrak{R}^{N \times K}$), while minimizing the reconstruction error. Considering the matrix form, it equals to solving the following objective [17]:

$$\min_{A,V} J(A,v) = \|y - Av\|_2^2 + \lambda\|v\|_1 \quad subject\ to : \|a_j\|^2 \leq 1 .$$ (3)

where y denotes an image, v denotes the sparse coefficient vector, parameter λ is positive, and A ($A = [a_1, a_2, \cdots, a_K]$) is the feature basis matrix. The first term of Equation (3) is the reconstruction error, and the second term is the sparsity control function of the sparse codes v.

Suppose there exists a feature mapping function $\phi : \mathfrak{R}^N \rightarrow \mathfrak{R}^K (N < K)$. Utilizing this function $\phi(\cdot)$, the new high dimensional feature space can be obtained, namely, $Y \rightarrow \phi(Y)$, $A = [a_1, a_2, \cdots, a_K] \rightarrow \tilde{A} = [\phi(a_1), \phi(a_2), \cdots, \phi(a_K)]$. Utilizing the mapping relationship, Equation (3) is substituted and the KSR object function can be obtained:

$$\min_{A,v} J(\phi(y), \tilde{A}) = \|\phi(y) - \tilde{A}v\|_2^2 + \lambda\|v\|_1 .$$ (4)

In this paper, the Gaussian kernel is used to be the mapping function, which is written

$$k(y_1, y_2) = \exp(-\gamma\|y_1 - y_2\|^2) .$$ (5)

Note that $k(a_i, a_i) = \exp(-\gamma\|a_i - a_i\|^2) = \phi(a_i)^T \phi(a_i) = 1$, the constraint on a_i can be removed. The goal of KSR is to seek the sparse representation for a mapped feature under the mapped basis in the high dimensional space. Otherwise, note that the Equation (4) is not convex, the optimization problem of the sparse codes v and the codebook $\tilde{A} = [\phi(a_1), \phi(a_2), \cdots, \phi(a_K)]$ is implemented alternatively. When the codebook $\tilde{A}$ is fixed, the Equation (4) can be rewritten as follows

$$\min_v J(\phi(y), \tilde{A}) = k(y, y) + v^T K_{AA} v - 2v^T K_A(y) + \lambda\|v\|_1 = P(v) + \lambda\|v\|_1 .$$ (6)

where $P(v) = 1 + v^T K_{AA} v - 2v^T K_A(y)$, K_{AA} is a $K \times K$ matrix with $\{K_{AA}\}_{ij} = k(a_i, a_j)$, and $K_A(y)$ is a $K \times 1$ vector with $\{K_A(y)\}_i = k(a_i, y)$. The objective is the same as that of sparse coding except for the definition of K_{AA} and $K_A(y)$. So it is easy to extend the FSC algorithm to solve the sparse codes. As for the computational cost, they are the same except for the difference in calculating kernel matrix.

2.3   The Implementation of KSR

In Equation (6), when v is fixed, the feature basis matrix A. Due to the large amount of features, it is hard to use all the features to learn A. Suppose N features are randomly sampled, the objective is rewritten as follows (m, s, t are used to index the column number of A)

$$\min_{A,v} F(A) = \frac{1}{N} \sum_{i=1}^{N} \left[1 + \sum_{s=1}^{K} \sum_{t=1}^{K} v_{i,s} v_{i,t} k(a_s, a_t) - 2 \sum_{s=1}^{K} v_{i,s} k(a_s, y_i) + \lambda \|v_i\|_1 \right] . \tag{7}$$

The derivative of $F(A)$ with respect to a_m is (a_m is the column to be updated)

$$\frac{\partial F}{\partial a_m} = \frac{-4\gamma}{N} \sum_{i=1}^{N} \left[\sum_{t=1}^{K} v_{i,m} v_{i,t} k(a_m, a_t)(a_m - a_t) - v_{i,m} k(a_m, y_i)(a_m - y_i) \right] . \tag{8}$$

To solve the Equation (8) easily, denote the a_m in the nth updating process as $a_{m,n}$, then the equation with respect to $a_{m,n}$ becomes:

$$\frac{\partial F}{\partial a_{m,n}} \cong \frac{-4\gamma}{N} \sum_{i=1}^{N} \left[\sum_{t=1}^{K} v_{i,m} v_{i,t} k(a_{m,n-1}, a_t)(a_{m,n} - a_t) - v_{i,m} k(a_{m,n-1}, y_i)(a_{m,n} - y_i) \right] . \tag{9}$$

3   The FSC Algorithm in Mapped Feature Space

For given one class of training sample y_i, the object function formula of FSC algorithm in mapped feature space is deduced as follows:

$$\min_{A,V} J(\tilde{A}, V) = \|K_{\tilde{Y}\tilde{Y}} - \tilde{A}\tilde{V}\|_2^2 + \lambda \|V\|_1, \quad s.t \ \forall : a_j^T a_j = 1 . \tag{10}$$

where $\tilde{Y} = \left[\phi(y_1^{(i)}), \phi(y_2^{(i)}), \cdots, \phi(y_N^{(i)}) \right] = \left[\tilde{Y}_1^{(i)}, \tilde{Y}_2^{(i)}, \cdots, \tilde{Y}_N^{(i)} \right]$ is the mapped matrix of Y, which is the *ith* class of training set, and obtained by the non-linear mapped function $\phi(\cdot)$. Matrix $\tilde{A} = \left[\phi(a_1), \phi(a_2), \cdots, \phi(a_K) \right]$ is the mapped form of feature matrix A, $\tilde{V} = [\tilde{v}_1, \tilde{v}_2, \cdots, \tilde{v}_N]$ is the sparse representation of $\tilde{Y}$ on the dictionary basis $\tilde{A}$. Here $K_{\tilde{Y}\tilde{Y}}$ is calculated by the following form

$$
K_{\tilde{Y}\tilde{Y}} = \begin{bmatrix} K\left(\tilde{Y}_1^{(i)}, \tilde{Y}_1^{(i)}\right) & K\left(\tilde{Y}_1^{(i)}, \tilde{Y}_2^{(i)}\right) & \cdots & K\left(\tilde{Y}_1^{(i)}, \tilde{Y}_p^{(i)}\right) \\ K\left(\tilde{Y}_2^{(i)}, \tilde{Y}_1^{(i)}\right) & K\left(\tilde{Y}_2^{(i)}, \tilde{Y}_2^{(i)}\right) & \cdots & K\left(\tilde{Y}_2^{(i)}, \tilde{Y}_p^{(i)}\right) \\ \vdots & \vdots & \vdots & \vdots \\ K\left(\tilde{Y}_p^{(i)}, \tilde{Y}_1^{(i)}\right) & K\left(\tilde{Y}_p^{(i)}, \tilde{Y}_2^{(i)}\right) & \cdots & K\left(\tilde{Y}_p^{(i)}, \tilde{Y}_p^{(i)}\right) \end{bmatrix} . \tag{11}
$$

Keeping feature bases fixed, the sparse coefficient matrix $\tilde{V}$ can be solved by optimizing each $\tilde{v}_i$. Referring to the standard FSC algorithm, this learning process of sparse vector v_i is in fact the feature-sign search algorithm, and the detail step of the feature-sign search algorithm can refer to the document [13-14]. This algorithm converges to a global optimum of the object function in a finite number of steps.

Fixed the sparse coefficient vector v_i, the mapped feature matrix $\tilde{A}$ can be learned by $\left\| K_{\tilde{Y}\tilde{Y}} - \tilde{A}\tilde{V} \right\|_2^2$, Subject to $\sum_{i=1}^{k} \tilde{A}_{i,j}^2 \le c$. This is a least squares problem with quadratic constraints. Using a Lagrange dual, this problem can be much more efficiently solved. The Lagrange form of $\left\| K_{\tilde{Y}\tilde{Y}} - \tilde{A}\tilde{V} \right\|_2^2$ is written as follows [13]:

$$
L\left(\tilde{A}, \lambda\right) = trace\left(\left(K_{\tilde{Y}\tilde{Y}} - \tilde{A}\tilde{V}\right)^T \left(K_{\tilde{Y}\tilde{Y}} - \tilde{A}\tilde{V}\right)\right) + \sum_{j=1}^{n} \lambda_j \left(\sum_{i=1}^{k} \tilde{A}_{i,j}^2 - c\right) . \tag{12}
$$

Where each $\lambda_j > 0$ is a dual variable. Minimizing over $\tilde{A}$ analytically, the Lagrange dual can be obtained:

$$
D(\lambda) = \min_{\tilde{A}} L\left(\tilde{A}, \lambda\right) = trace\left(K(\tilde{Y}, \tilde{Y}) - \phi(\tilde{Y})V^T\left(VV^T + \Lambda\right)^{-1}\left(\phi(\tilde{Y})V^T\right)^T - c\Lambda\right) . \tag{13}
$$

where $\Lambda = diag(\lambda)$. According to the gradient of $D(\lambda)$, the optimal basis vectors are deduced as follows:

$$
\tilde{A} = \left(\tilde{Y}V^T\right)\left[\left(VV^T + \Lambda\right)^{-1}\right]^T . \tag{14}
$$

The advantage of solving the dual is that it uses significantly fewer optimization variables than the primal. And note that the dual formulation is independent of the sparsity function, and can be extended to other similar models.

4   Experimental Results and Analysis

4.1   Palm Data Preprocessing

In test, the PolyU palmprint database is used to verify the palmprint recognition method based on KSR proposed in this paper. The PolyU database includes 600 palmprint images with the size of 128×128 from 100 individuals. Several images of two classes of palmprint images were shown in Fig.1. Each person has six images. For each person, the first three images are used as training samples, while the remaining ones are treated as testing samples. Thus the training set X_{train} and the testing set X_{test} have the same size of 300×(128×128). Here, X_{train} and X_{test} are preprocessed to be centered and have zero-mean. To reduce the computational cost, each sub-image is scaled to the size of 64×64, and is converted a column vector. Thus, the size of the training set and test set are both changed to be 300×4096 pixels. Otherwise, to reduce the number of sources to a tractable number and provide a convenient method for calculating representations of test images, PCA is first used to realize data whitening and a dimension reduction. Thus, these sub-images were utilized to represent the original palmprint images in our experiments.

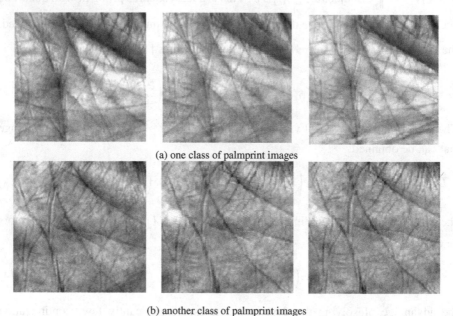

(a) one class of palmprint images

(b) another class of palmprint images

Fig. 1. Two classes of original palmprint images selected randomly from PolyU database

4.2   Recognition Results of Palmprint Images

Referring to the ICA framework described in the document [20-22], and considering different feature dimensions and the complete dictionary basis, the recognition results of palm images using our KSR method, both utilizing the standard FSC algorithm to learn codebook, were discussed in this subsection. And considering the length limitation of this paper, the feature basis images with first 16 dimension obtained by the general SR and those obtained by our KSR method, were respectively shown in Fig.2. From Fig.2, it is clear to see that the feature basis images of KSR behave orientation in space and distinct sparsity as the same as those of SR and SC [2, 11]. Using the Euclidean distance to measure the recognition property in mapped feature space with different feature dimensions, such as 25, 36, 49, 64, 81, 121 and so on, the calculated recognition rates were shown in Table 1. From Table 1, clearly, in despite of recognition method used, the larger the feature dimension is, the larger the recognition rate is.

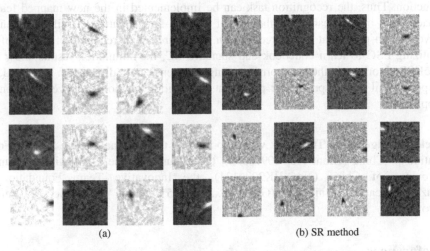

(a)   (b) SR method

Fig. 2. Basis images obtained by general SR and KSR, both using FSC algorithm to learn codebook. (a) Basis images of KSR method. (b) Basis images of general SR method.

Table 1. Recognition of palmprint images using different algorithms and feature dimension

Algorithms Noise variance	SR	KSR	SC
25	78.63	81.18	71.76
36	82.27	86.63	78.51
49	88.62	90.73	82.83
64	91.84	93.26	87.72
81	93.47	95.41	90.68
121	95.68	97.79	94.76

To prove that the recognition method based on KSR is efficient in palmprint recognition task, the comparison results of general SR and SC, also listed in Table 1 were given out here. For given feature dimensions, it is easy to show that the larger the feature dimension, the larger the recognition rate is despite of any algorithm.

As well as, it is easy to see that the recognition rate of KSR is the best, that of SR is the second, and that of SC is the worst. It is noted that KSR and SR both used the FSC algorithm to learn codebook in training.

5   Conclusions

A novel recognition method of palmprint images using the Kernel based sparse representation (KSR) is discussed in this paper. The palmprint database used in test is the PolyU database. The Gauss Kernel function is selected as the non-linear mapped function. Thus, the recognition task can be implemented in the new mapped feature space. This method can solve the task of the non-linear feature classification and is favorable in palmprint image recognition. Compared with recognition results of SR utilizing FSC to learn codebook and standard SC, the simulation results show that theKSR algorithm is the best corresponding to given feature dimension. In a word, the experimental results prove our palmprint recognition method is indeed efficient in application.

Acknowledgement. This work was supported by the National Nature Science Foundation of China (Grant No. 60970058), the Nature Science Foundation of Jiangsu Province of China (No. BK2009131), the Innovative Team Foundation of Suzhou Vocational University (Grant No. 3100125), and the "333 Project" of Jiangsu Province.

References

1. Jain, A.K., Ross, A., Prabhakar, S.: An introduction to biometric recognition. IEEE Transcations on Circuits and Systems for Video Technology 14(1), 4–40 (2004)
2. Olshausen, B.A., Field, D.J.: Emergence of simple-cell receptive field properties by learning a sparse code for natural images. Nature 381, 607–609 (1996)
3. Connle, T., Jin, A.T.B., Ong, M.G.K., et al.: An automated palmprint recognition system. Image and vision computing 23(5), 501–515 (2005)
4. Ghandehari, A., Safabakhsh, R.: Palmprint Verification Using Circular Gabor Filter. In: Tistarelli, M., Nixon, M.S. (eds.) ICB 2009. LNCS, vol. 5558, pp. 675–684. Springer, Heidelberg (2009)
5. Kong, W.K., Zhang, D., Li, W.: Palmprint recognition using eigenpalm features. Pattern Recognition Letters 24(9), 1473–1477 (2003)
6. Strang, G., Nguyen, T.: Wavelet and Filter Banks. Wellesley-Cambridge Press (1997)
7. Li, W.-X., Zhang, D., Xu, Z.-Q., et al.: Palmprint identification by fourier transform. International Journal of Pattern Recognition and Artificial Intelligence 16(4), 417–432 (2002)
8. Han, X., Xu, J.-J.: Palmprint recognition based on support vector machine of combination kernel function. In: The 2nd International Conference on Information Science and Engineering (ICISE), December 15-16, pp. 6748–6751. IEEE Press, Kyiv Ukraine (2010)

9. Yu, P.-F., Xu, D.: Palmprint recognition bsed on modified DCT features and RBF neural network. In: nternational Conference on Machine Learning and Cybernetics, Suntec Singapore, October 12-15, vol. 5, pp. 2982–2986. IEEE Press (2008)

10. Shang, L., Huang, D.-S., Du, J.-X., et al.: Palmprint recogniton using FastICA algorithm and radial basis probabilistic neural network. Neurocomputing 69(13), 1782–1786 (2006)

11. Hyvärinen, A.: Independent component analysis in the presence of gaussian noise by maximizing joint likelihood. Neurocomputing 22, 49–67 (1998)

12. Hyvarinen, A., Oja, E.: Independent component analysis: algorithms and applications. Neural Networks 13, 411–430 (2000)

13. Shang, L.: Image reconstruction using a modified sparse coding technique. In: Huang, D.-S., Wunsch II, D.C., Levine, D.S., Jo, K.-H. (eds.) ICIC 2008. LNCS, vol. 5226, pp. 220–226. Springer, Heidelberg (2008)

14. Wang, Q.-X., Cao, C.-H., Li, M.-H., et al.: A new model based on grey theory and neural network algorithm for evaluation of AIDS clinical trial. Advances in Computational Mathematics and Its Applications 2(3), 292–297 (2013)

15. Pu, J., Zhang, J.-P.: Super-Resolution through dictionary learning and sparse representation. Pattern Recognition and Artificial Intelligence 23(3), 335–340 (2010)

16. Jun, Y., Zhonghua, L., Zhong, J., et al.: Kernel Sparse Repre-sentation Based Classification. Neurocomputing 77(1), 120–128 (2012)

17. Gao, S.-H., Tsang, I.W.-H., Chia, L.-T., et al.: Kernel sparse representation for image classification and face recognition

18. Zeyde, R., Elad, M., Protter, M.: On single image scale-up using sparse-representations. In: Boissonnat, J.-D., Chenin, P., Cohen, A., Gout, C., Lyche, T., Mazure, M.-L., Schumaker, L. (eds.) Curves and Surfaces 2011. LNCS, vol. 6920, pp. 711–730. Springer, Heidelberg (2012)

19. Zu, H.-L., Wang, Q.-X., Dong, M.-Z., et al.: Compressed sensing based fixed point DCT image encoding. Advances in Computational Mathematics and Its Applications 2(2), 259–262 (2012)

20. Ekinci, M.: Palmprint recognition using kernel PCA of Gabor features. In: The 23rd International Symposium on Computer and Information Sciences (ISCIS 2008), Istanbul, Turkey, October 27-29, pp. 27–29. IEEE Press (2008)

21. Connie, T., Teoh, A., Goh, M., et al.: Palmprint recognition with PCA and ICA. Image and Vision Computing NZ 3, 227–232 (2003)

22. Aharon, M., Elad, M., Bruckstein, A., et al.: K-SVD: An algorithm for designing overcomplete dictionaries for sparse representation. IEEE Transactions on Signal Processing 54(11), 4311–4322 (2006)

A Linear Method for Determining Intrinsic Parameters from Two Parallel Line-Segments[*]

Jianliang Tang[1], Jie Wang[2], and Wensheng Chen[1]

[1] College of Mathematics and Computational Science, Shenzhen University
Shenzhen, 518060, China
{jtang,chenws}@szu.edu.cn
[2] Key Laboratory of Multimedia and Intelligent Software
College of Computer Science and Technology, Beijing University of Technology
Beijing 100022, China

Abstract. In this paper, a linear method to determining intrinsic parameters from two parallel line-segments is proposed. Constrains based on the length ratio of line-segments are used to solve the camera calibration problem from images of two parallel line-segments under different conditions. And for each setting, we can get linear solution for intrinsic parameters of a usual camera. Simulated experiments are carried out to verify the theoretical correctness and numerical robustness of our results.

1   Introduction

Camera calibration is a very important and necessary technique in computer vision, especially in 3-D reconstruction. These years, camera calibration based on plane and 1-D object is becoming a hot research topic for its flexibility. Different methods were developed to compute the images of circular points since the images of circular points can provide constraints of the absolute conic. Zhang's calibration method uses a planar pattern board as the calibration object and computes the images of circular points through the world-to-image homnographies [2]. Meng and Hu [3] developed a calibration pattern consisting of a circle and a set of lines through its centers and got the images of circular points by computing the vanishing line and its intersection with the circle in the image plane. Different techniques based on circular points are proposed in papers [4, 5, 6, 7].

Camera calibration methods based on 1-D object have been studied in recent literatures. 1-D object refers to three or more collinear points with known distances. Zhang's work solved the camera calibration problem from images of the 1-D calibration object rotating around a known fixed point [8]. Wu and Hu have also given the different descriptions on camera calibration based on 1-D object in papers [9], [10], and [11]. Recently, Zhao developed a new method based on three non-collinear points, different from three collinear points, constraints based on length and

---

[*] Partially supported by the National Natural Science Foundation of China (No. 61272252, No.61075037)

D.-S. Huang et al. (Eds.): ICIC 2013, LNAI 7996, pp. 524–532, 2013.

their angle of calibration object are obtained by rotating all the three points around center axes and the Euclidean structure then be extracted from the images of the original points and the constructed points [12].

For the calibration based on two parallel line-segments, Zhu's method [9] uses homography from plane- to- image to compute images of circular points and n images of circular points form 2n equations with unknown factors and intrinsic parameters. Then the constraints are obtained by considering the rank of the matrix from 2n equations. However, the constraints are of quadratic and sixth degree if the length ratio of the two parallel segments is known and unknown respectively. Constraints based on the length ratio of line-segment are used to solve the camera calibration problem from images of two parallel line-segments under different conditions. And in each condition, linear solution can be obtained from constraint equations. Simulated experiments are carried out to verify the theoretical correctness and numerical robustness of our results.

The rest of the paper is organized as follows. In section 2, we introduce some preliminaries. In section 3, we give the main result. In section 4, we present the result obtained by the simulated result. In section 5, the conclusions are presented.

2   Preliminaries

In this paper, a 2-D point position in pixel coordinates is represented by $m = (u, v)^T$, a 3-D point in world coordinates is represented by $M = (x, y, z)^T$. We use $\tilde{M} = (x, y, z, 1)^T$ and $\tilde{m} = (u, v, 1)^T$ to denote the homogeneous coordinates of the 3-D point M and the 2-D point m. A camera is modeled by the usual pinhole, then the relationship between M and m is given by

$$S_M \tilde{m} = K[R, t]\tilde{M} \quad \text{with} \quad K = \begin{pmatrix} f_u & s & u_0 \\ 0 & f_v & v_0 \\ 0 & 0 & 1 \end{pmatrix} \quad (1)$$

where S_M is the projection depth of the point M, K is the camera intrinsic matrix with the focal length (f_u, f_v), the principle point (u_0, v_0) and skew factor S; $[R, t]$ is the camera extrinsic matrix, that is, the rotation and translation from the world frame to the camera frame, $P = K[R, t]$ is referred as the camera matrix. The task of camera calibration is to determine these five intrinsic parameters.

Homography is a transformation given by a non-singular matrix, which defines a homogeneous linear transformation from a plane to another plane in projective space. Let π be an arbitrary space plane not passing through the camera center, if we choose π as the oxy plane of the world frame, then homography H from plane π to the image plane can be expressed as

$$H = K[r_1, r_2, t]$$

Where K is the intrinsic matrix, r_1, r_2 are the first two columns of R. Particularly, let π_∞ be the plane at infinity 3-D space, the homography from π_∞ to the image plane, called the infinite homography, is given by $H_\infty = K[r_1, r_2, r_3]$. The absolute conic Ω_∞ is a virtual point conic on the infinite plane π_∞. Points with the homogeneous coordinates $X = (x_1, x_2, x_3, x_4)^T$ on Ω_∞ satisfy

$$x_1^2 + x_2^2 + x_3^2 = 0, x_4 = 0$$

Let C be the image of the absolute conic Ω_∞. From the infinite homography H_∞, then we have

$$C = H_\infty^{-T} \Omega_\infty H_\infty = K^{-T} K^{-1} \qquad (2)$$

Given four collinear points by $X_i = (u_i, v_i, 1)^T (i = 1, 2, 3, 4)$, then the cross ratio of the four points is defined as

$$Cross(X_1 X_2, X_3 X_4) = \frac{u_3 - u_1}{u_3 - u_2} : \frac{u_4 - u_1}{u_4 - u_2} = \frac{v_3 - v_1}{v_3 - v_2} : \frac{v_4 - v_1}{v_4 - v_2}$$

3   Camera Calibration Based on Two Parallel Line Segments

Two arbitrary parallel line-segments determine a plane π in the space, if we choose π as the oxy plane of the world frame, our discussion can base on plane instead of space. For the convenience of writing, the line defined by two points A and B is called line AB and the segment defined by two points A and B is called segment AB.

As shown in Fig. (1), the two arbitrary line-segments are denoted by AB and CD. Point O is the intersection of the line AD and BC. A line through O which parallels to line AB(CD) intersects line AC and BD at E and F respectively. Assume that the length ratio of segment AB and CD is t, the image points of A, B, C, D are $\tilde{a}, \tilde{b}, \tilde{c}, \tilde{d}$ respectively, then the image point of O can be computed by $\tilde{o} = (\tilde{a} \times \tilde{d}) \times (\tilde{b} \times \tilde{c})$.

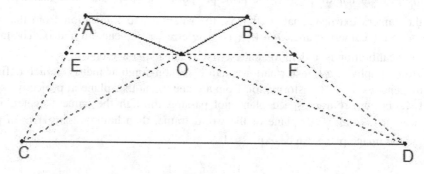

Fig. 1.

From the image plane, we can determine the image vanishing point $\tilde{p}$ of the line AB(CD) by $\tilde{p} = (\tilde{a} \times \tilde{b}) \times (\tilde{c} \times \tilde{d})$. Since AB // CD, $\tilde{p}$ is also the vanishing point of the line EF since EF // AB(CD). Thus the image line of line EF, denoted by $\tilde{l}_{ef}$, can be computed as $\tilde{l}_{ef} = \tilde{p} \times \tilde{o}$. Line EF intersects line AC and BD at E and F, and this holds also for their images in the image plane, hence the image points of E and F, denoted by $\tilde{e}, \tilde{f}$ can be obtained by $\tilde{e} = (\tilde{p} \times \tilde{o}) \times (\tilde{a} \times \tilde{c})$, $\tilde{f} = (\tilde{p} \times \tilde{o}) \times (\tilde{b} \times \tilde{d})$.

Due to the similarity of triangles AOB and COD it holds that the ratio of segment BO and CO is equal to the ratio of segment AO and DO. This may be written in terms of coordinates of points as

$$\tilde{O} = \frac{1}{1+t}\tilde{D} + \frac{t}{1+t}\tilde{A}, \tilde{O} = \frac{1}{1+t}\tilde{C} + \frac{t}{1+t}\tilde{B} \tag{4}$$

Where $\tilde{A}, \tilde{B}, \tilde{C}, \tilde{D}, \tilde{O}$ are homogeneous coordinates of the points A, B, C, D, O. According to (1) and (4), we get

$$S_O \tilde{o} = \mu S_D \tilde{d} + (1-\mu) S_A \tilde{a}, \quad S_O \tilde{o} = \mu S_C \tilde{c} + (1-\mu) S_B \tilde{b} \tag{5}$$

where $\mu = 1/(1+t)$, S_A, S_B, S_C, S_D, S_O are unknown projection depths of the five points. Using the properties of cross product of vectors we have

$$S_A = \frac{S_O}{1-\mu}\frac{(\tilde{o} \times \tilde{d}) \cdot (\tilde{a} \times \tilde{d})}{(\tilde{a} \times \tilde{d}) \cdot (\tilde{a} \times \tilde{d})}, \quad S_D = \frac{S_O}{\mu}\frac{(\tilde{o} \times \tilde{a}) \cdot (\tilde{d} \times \tilde{a})}{(\tilde{d} \times \tilde{a}) \cdot (\tilde{d} \times \tilde{a})} \tag{6}$$

$$S_B = \frac{S_O}{1-\mu}\frac{(\tilde{o} \times \tilde{c}) \cdot (\tilde{b} \times \tilde{c})}{(\tilde{b} \times \tilde{c}) \cdot (\tilde{b} \times \tilde{c})}, \quad S_C = \frac{S_O}{\mu}\frac{(\tilde{o} \times \tilde{b}) \cdot (\tilde{c} \times \tilde{b})}{(\tilde{c} \times \tilde{b}) \cdot (\tilde{c} \times \tilde{b})}. \tag{7}$$

Similar to (6) and (7), since O is the midpoint of segment EF, we have

$$S_E = 2S_O\frac{(\tilde{o} \times \tilde{f}) \cdot (\tilde{e} \times \tilde{f})}{(\tilde{e} \times \tilde{f}) \cdot (\tilde{e} \times \tilde{f})}, \quad S_F = 2S_O\frac{(\tilde{o} \times \tilde{e}) \cdot (\tilde{f} \times \tilde{e})}{(\tilde{f} \times \tilde{e}) \cdot (\tilde{f} \times \tilde{e})}. \tag{8}$$

We chose the following constraints for camera calibration:

$$t^2 \|AE\|^2 = \|EC\|^2, \qquad t^2 \|BF\|^2 = \|FD\|^2 \tag{9}$$

According to (1), we also have

$$t^2(S_A\tilde{a} - S_E\tilde{e})^T K^{-T}K^{-1}(S_A\tilde{a} - S_E\tilde{e}) = (S_C\tilde{c} - S_E\tilde{e})^T K^{-T}K^{-1}(S_C\tilde{c} - S_E\tilde{e}) \tag{10}$$

$$t^2(S_B\tilde{b} - S_F\tilde{f})^T K^{-T}K^{-1}(S_B\tilde{b} - S_F\tilde{f}) = (S_D\tilde{d} - S_F\tilde{f})^T K^{-T}K^{-1}(S_D\tilde{d} - S_F\tilde{f}) \tag{11}$$

Substitute (6), (7) and (8), for $S_A, S_B, S_C, S_D, S_E, S_F$ and eliminate the unknown factor S_o^2, we obtain:

$$t^2 h_3^T K^{-T} K^{-1} h_3 = h_4^T K^{-T} K^{-1} h_4 \tag{12}$$

$$t^2 h_5^T K^{-T} K^{-1} h_5 = h_6^T K^{-T} K^{-1} h_6 \tag{13}$$

Where

$$h_3 = \frac{(1+t)}{t} \frac{(\tilde{o} \times \tilde{d}) \cdot (\tilde{a} \times \tilde{d})}{(\tilde{a} \times \tilde{d}) \cdot (\tilde{a} \times \tilde{d})} \tilde{a} - 2 \frac{(\tilde{o} \times \tilde{f}) \cdot (\tilde{e} \times \tilde{f})}{(\tilde{e} \times \tilde{f}) \cdot (\tilde{e} \times \tilde{f})} \tilde{e} \qquad h_4 = (1+t) \frac{(\tilde{o} \times \tilde{b}) \cdot (\tilde{c} \times \tilde{b})}{(\tilde{c} \times \tilde{b}) \cdot (\tilde{c} \times \tilde{b})} \tilde{c} - 2 \frac{(\tilde{o} \times \tilde{f}) \cdot (\tilde{e} \times \tilde{f})}{(\tilde{e} \times \tilde{f}) \cdot (\tilde{e} \times \tilde{f})} \tilde{e}$$

$$h_5 = \frac{(1+t)}{t} \frac{(\tilde{o} \times \tilde{c}) \cdot (\tilde{b} \times \tilde{c})}{(\tilde{b} \times \tilde{c}) \cdot (\tilde{b} \times \tilde{c})} \tilde{b} - 2 \frac{(\tilde{o} \times \tilde{e}) \cdot (\tilde{f} \times \tilde{e})}{(\tilde{f} \times \tilde{e}) \cdot (\tilde{f} \times \tilde{e})} \tilde{f} \qquad h_6 = (1+t) \frac{(\tilde{o} \times \tilde{a}) \cdot (\tilde{d} \times \tilde{a})}{(\tilde{d} \times \tilde{a}) \cdot (\tilde{d} \times \tilde{a})} \tilde{d} - 2 \frac{(\tilde{o} \times \tilde{e}) \cdot (\tilde{f} \times \tilde{e})}{(\tilde{f} \times \tilde{e}) \cdot (\tilde{f} \times \tilde{e})} \tilde{f}$$

Under this condition, $2n$ constraint equations are obtained from n images of two parallel segments with length ratio known.

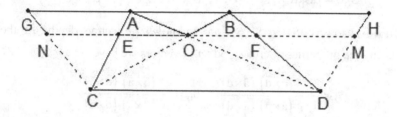

Fig. 2.

As shown in Fig. (2), for different choices of $d(d>1)$, there exists different positions of points G, H on the line AB such that $\|AG\| = \|BH\| = d\|AB\|$. Once d is given, then the position of G and H is fixed. Suppose EF intersects line GC and HD at N and M respectively. Since the cross ratio of four collinear points is invariant under any projective transformation, the image points of G and H, denoted by $\tilde{g}$ and $\tilde{h}$, can be computed from the following equations:

$$\text{Cross}(GB, AP_\infty) = \text{Cross}(\tilde{g}\tilde{b}, \tilde{a}\tilde{p}) = -d, \quad \text{Cross}(AH, BP_\infty) = \text{Cross}(\tilde{a}\tilde{h}, \tilde{b}\tilde{p}) = -1/d$$

Where P_∞ is the point at infinity in the direction of line AB(CD).

Let $\tilde{a} = (a_1, a_2, 1)^T, \tilde{b} = (b_1, b_2, 1)^T, \tilde{g} = (g_1, g_2, 1)^T, \tilde{h} = (h_1, h_2, 1)^T, \tilde{p} = (p_1, p_2, 1)^T$
According to (3), we have

$$\frac{a_1 - g_1}{a_1 - b_1} : \frac{p_1 - g_1}{p_1 - b_1} = \frac{a_2 - g_2}{a_2 - b_2} : \frac{p_2 - g_2}{p_2 - b_2} = -d, \quad \frac{b_1 - a_1}{b_1 - h_1} : \frac{p_1 - a_1}{p_1 - h_1} = \frac{b_2 - a_2}{b_2 - h_2} : \frac{p_2 - a_2}{p_2 - h_2} = -1/d$$

By solving the above equations we obtain

$$g_1 = \frac{(p_1 - b_1)a_1 + d(a_1 - b_1)p_1}{d(a_1 - b_1) + (p_1 - b_1)}, \qquad g_2 = \frac{(p_2 - b_2)a_2 + d(a_2 - b_2)p_2}{d(a_2 - b_2) + (p_2 - b_2)}$$

$$h_1 = \frac{(p_1 - a_1)b_1 + d(b_1 - a_1)p_1}{(p_1 - a_1) + d(b_1 - a_1)}, \qquad h_2 = \frac{(p_2 - a_2)b_2 + d(b_2 - a_2)p_2}{(p_2 - a_2) + d(b_2 - a_2)}$$

Hence, the image points of G and H which satisfy $\|AG\| = \|BH\| = d\|AB\|$ are determined and the image points of M and N, denoted by $\tilde{m}$ and $\tilde{n}$ can be computed by $\tilde{m} = (\tilde{p} \times \tilde{o}) \times (\tilde{h} \times \tilde{d})$, $\tilde{n} = (\tilde{p} \times \tilde{o}) \times (\tilde{g} \times \tilde{c})$ subsequently. Since O is also the midpoint of the segment MN, similar to (8), we have

$$S_M = 2S_O \frac{(\tilde{o} \times \tilde{n}) \cdot (\tilde{m} \times \tilde{n})}{(\tilde{m} \times \tilde{n}) \cdot (\tilde{m} \times \tilde{n})}, \qquad S_N = 2S_O \frac{(\tilde{o} \times \tilde{m}) \cdot (\tilde{n} \times \tilde{m})}{(\tilde{n} \times \tilde{m}) \cdot (\tilde{n} \times \tilde{m})} \qquad (14)$$

Considering $(1+d)^2 \|EO\|^2 = \|NO\|^2$ and $(1+d)^2 \|FO\|^2 = \|MO\|^2$, we get two constraints as follows:

$$(1+d)^2 (S_E \tilde{e} - S_O \tilde{o})^T K^{-T} K^{-1} (S_E \tilde{e} - S_O \tilde{o}) = (S_N \tilde{n} - S_O \tilde{o})^T K^{-T} K^{-1} (S_N \tilde{n} - S_O \tilde{o}) \qquad (15)$$

$$(1+d)^2 (S_F \tilde{f} - S_O \tilde{o})^T K^{-T} K^{-1} (S_F \tilde{f} - S_O \tilde{o}) = (S_M \tilde{m} - S_O \tilde{o})^T K^{-T} K^{-1} (S_M \tilde{m} - S_O \tilde{o}) \qquad (16)$$

Substituting S_E, S_F, S_M, S_N by (8) and (14) gives

$$(1+d)^2 h_7^T K^{-T} K^{-1} h_7 = h_8^T K^{-T} K^{-1} h_8 \qquad (17)$$

$$(1+d)^2 h_9^T K^{-T} K^{-1} h_9 = h_{10}^T K^{-T} K^{-1} h_{10} \qquad (18)$$

with $\quad h_7 = 2 \dfrac{(\tilde{o} \times \tilde{f}) \cdot (\tilde{e} \times \tilde{f})}{(\tilde{e} \times \tilde{f}) \cdot (\tilde{e} \times \tilde{f})} \tilde{e} - \tilde{o}, \qquad h_8 = 2 \dfrac{(\tilde{o} \times \tilde{m}) \cdot (\tilde{n} \times \tilde{m})}{(\tilde{n} \times \tilde{m}) \cdot (\tilde{n} \times \tilde{m})} - \tilde{o}$

$h_9 = 2 \dfrac{(\tilde{o} \times \tilde{e}) \cdot (\tilde{f} \times \tilde{e})}{(\tilde{f} \times \tilde{e}) \cdot (\tilde{f} \times \tilde{e})} \tilde{f} - \tilde{o}, \qquad h_{10} = 2 \dfrac{(\tilde{o} \times \tilde{n}) \cdot (\tilde{m} \times \tilde{n})}{(\tilde{m} \times \tilde{n}) \cdot (\tilde{m} \times \tilde{n})} \tilde{m} - \tilde{o}.$

Under this condition, $2n$ constraint equations can also be obtained from n images of two parallel segments with the length ratio unknown.

4   Experiments

In the simulation experiments, the camera has the following setup: $f_u = f_v = 1200$, $u_0 = 320$, $v_0 = 240$, $s = 1$. The image resolution is 1024×768. If the ratio of them is known, we assume the length ratio of them is 2. We choose d=5 when there is no information about the two parallel segments in section 3.

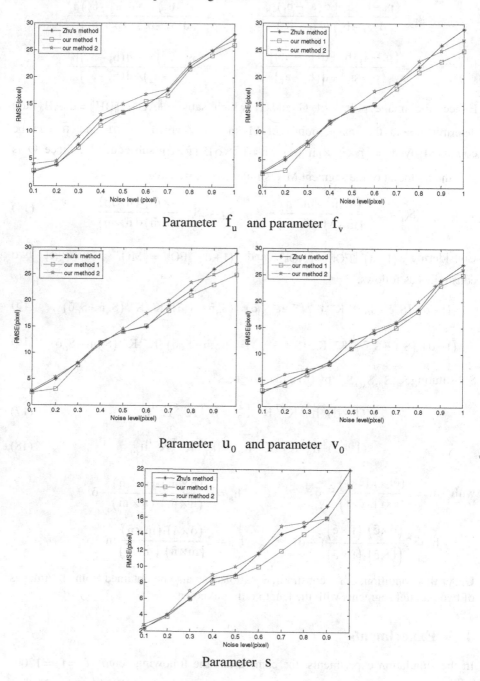

Parameter f_u and parameter f_v

Parameter u_0 and parameter v_0

Parameter s

Fig. 3.

Gaussian noise with 0 mean and σ standard deviation is added to the projected image points. The estimated camera parameters are compared with the ground truth, and RMS errors are measured. We vary the noise level σ from 0.1 to 1 pixel. For each noise level, 50 independent trials are performed, and the results under three conditions are shown in Fig. (3). We can see that RMS errors under two conditions increase almost linearly with the noise level. Our linear method is slightly better than Zhu's method when the length ratio of the two parallel segments is known and worse when the ration is unknown.

5   Conclusion

A linear method to solve camera calibration problem from two parallel line-segments is proposed in this paper. Different priori information of the two parallel line-segments provides different constraints for intrinsic parameters. Simulation experiments show that this linear method can be applied in practical as parallel line-segments are not rare in any man-made scene.

References

1. Abidi, M.A., Chandra, T.: A New Efficient and Direct Solution for Pose Estimation Using Quadrangular Targets: Algorithm and Evaluation. IEEE Transaction on Pattern Analysis and Machine Intelligence 17(5), 534–538 (1995)
2. Zhang, Z.Y.: A Flexible New Technique for Camera Calibration. IEEE Transactions on Pattern Analysis and Machine Intelligence 22(11), 1330–1334 (2000)
3. Meng, X.Q., Hu, Z.Y.: A new easy camera calibration technique based on circular points. Pattern Recognition 36(5), 1155–1164 (2003)
4. Zhu, H.J., Wu, F.C., Hu, Z.Y.: Camera calibration based on two parallel line segments. Acta Automatica Sinica 31(6), 853–864 (2005)
5. Wu, Y., Zhu, H., Hu, Z., Wu, F.C.: Camera calibration from the quasi-affine invariance of two parallel circles. In: Pajdla, T., Matas, J(G.) (eds.) ECCV 2004. LNCS, vol. 3021, pp. 190–202. Springer, Heidelberg (2004)
6. Chen, Q.-A., Wu, H., Wada, T.: Camera calibration with two arbitrary coplanar circles. In: Pajdla, T., Matas, J(G.) (eds.) ECCV 2004. LNCS, vol. 3023, pp. 521–532. Springer, Heidelberg (2004)
7. Gurdjos, P., Sturm, P., Wu, Y.: Euclidean structure from $N \geq 2$ parallel circles: Theory and algorithms. In: Leonardis, A., Bischof, H., Pinz, A. (eds.) ECCV 2006, Part I. LNCS, vol. 3951, pp. 238–252. Springer, Heidelberg (2006)
8. Zhang, Z.Y.: Camera calibration with one-dimensional objects. IEEE Transactions on Pattern Analysis and Machine Intelligence 26(7), 892–899 (2004)
9. Wu, F.C., Hu, Z.Y., Zhu, H.J.: Camera calibration with moving one-dimensional objects. Pattern Recognition 38(5), 755–765 (2005)
10. Wang, L., Wu, F.C., Hu, Z.Y.: Multi-camera calibration with one-dimensional object under general motions. In: Proc. ICCV, pp. 1–7 (2007)
11. Wu, F.C.: Mathematical Method in Computer Vision. Science Press, Beijing (2008)

12. Zhao, Z.J., Liu, Y.C., Zhang, Z.Y.: Camera calibration with three non-collinear points under special motions. IEEE Transactions on Pattern Analysis and Machine Intelligence 17(12), 2393–2400 (2008)
13. Hartley, R.I., Zisserman, A.: Multiple View Geometry in Computer Vision. Cambridge University Press (2000)
14. Hartley, R.: Self-calibration from multiple views with a rotating camera. In: Eklundh, J.-O. (ed.) ECCV 1994. LNCS, vol. 800, pp. 471–478. Springer, Heidelberg (1994)
15. Kim, J., Gurdjos, P., Kweon, I.: Geometric and algebraic constraints of projected concentric circles and their applications to camera calibration. IEEE Transactions on Pattern Analysis and Machine Intelligence 27(4), 637–642 (2005)
16. Gurdjos, P., Kim, J., Kweon, I.: Euclidean structure from confocal conics: Theory and application to camera calibration. In: Proc. CVPR, pp. 1214–1221 (2006)

Real-Time Visual Tracking Based on an Appearance Model and a Motion Mode

Guizi Li, Lin Zhang[*], and Hongyu Li

School of Software Engineering, Tongji University, Shanghai, China
{12guizili,cslinzhang,hyli}@tongji.edu.cn

Abstract. Object tracking is a challenging problem in computer vision community. It is very difficult to solve it efficiently due to the appearance or motion changes of the object, such as pose, occlusion, or illumination. Existing online tracking algorithms often update models with samples from observations in recent frames. And some successful tracking algorithms use more complex models to make the performance better. But most of them take a long time to detect the object. In this paper, we proposed an effective and efficient tracking algorithm with an appearance model based on features extracted from the multi-scale image feature space with data-independent basis and a motion mode based on Gaussian perturbation. In addition, the features used in our approach are compressed in a small vector, making the classifier more efficient. The motion model based on random Gaussian distribution makes the performance more effective. The proposed algorithm runs in real-time and performs very well against some existing algorithms on challenging sequences.

Keywords: object tracking, motion mode, appearance model.

1   Introduction

Object tracking is a challenging problem in computer vision community. Although many researchers [1-10] have proposed numerous algorithms in literature, it is still very difficult to solve it efficiently due to the appearance or motion changes of the object, such as pose, occlusion, or illumination. The movement of objects is unpredictable. Sometimes it moves smoothly and its position is easy to be caught. But when it moves abruptly, catching it will be difficult. To deal with this problem, tracking methods need to be more complex. However, high complexity brings low efficiency. To achieve a real-time tracking algorithm, an effective and efficient appearance is very important.

Under the premise of ensuring an acceptable error rate, to improve the efficiency has become an important issue. There are some methods that have achieved a fast object tracking and detection, such as improving character description (i.e. like integral image [11]), and using a better classifier (i.e. the classifier should balance the performance and the efficiency).

---

[*] Corresponding author.

D.-S. Huang et al. (Eds.): ICIC 2013, LNAI 7996, pp. 533–540, 2013.
© Springer-Verlag Berlin Heidelberg 2013

Zhang *et al.* [12] proposed a compressive tracking method, in which the feature of the object was compressed to a vector. It performed well in feature extraction, making the feature easy to be compared. But it will be useless in the situation that the size of the object has changed.

Kwon *et al.* [13] proposed an algorithm which was based on a visual tracking decomposition scheme for the efficient design of observation and motion models as well as trackers. It was proved to be accurate and reliable when the appearance and motion are drastically changing over time.

However, in these papers, the accuracy and efficiency was not well balanced. Either high efficiency or good accuracy was acquired at the expense of the other one. In this paper, we use motion mode to catch the abrupt move [14, 15], and the image patch is divided into several small pieces to adapt to the slight change of the object, so that the robustness of the algorithm can be assured. Of course, for the sake of efficiency, some time consuming methods will be removed, which, at some extent, might affect the accuracy. But the experimental result shows that the proposed algorithm is well balanced between accuracy and efficiency.

2   Gaussian Perturbation Decomposition

When tracking an object, its moving speed is unknown. So to catch the objects with different moving speeds, we decompose the motion into two kinds of motion mode, smooth and abrupt motion, and use different variances of the Gaussian distributions to represent the different motion mode. When the object moves fast, the distance of the object between two frames will be longer, so the variance of the Gaussian distribution should be large to represent the abrupt moving. On the other hand, when the object moves slow or smoothly, in the next frame, it will appear around the object in the former frame. So the variance should be small. Let's take a bomb blowing up as the example. In the previous frame, the position of the object is the bomb itself, and in the next frame it diffuses to all the direction. When it diffuses fast, in the next frame, it will be far away from the original position. On each direction, the moving distance can be fitted for a Gaussian distribution, and the distribution of the same origin in each direction should be the same.   In 3D version, it looks just like a ring, but consisting with Gaussian distribution.

Since the Gaussian distribution is fair enough to describe the object's moving, we can use it to predict the position of the object in the next frame. It is a priori knowledge of our algorithm, and can be expressed as:

$$p_i(P_t(x, y) \mid P_{t-1}(x, y)) = G(P_{t-1}(x, y), \sigma_i^2), i = 1, ..., s, \tag{1}$$

where G represents the Gaussian distribution with the mean P_{t-1} and variance σ_i^2, $P_{t-1}(x,y)$ means the position of the object at time t-1, and p_i means the probability of each motion mode. In this paper we use both smooth mode and abrupt mode. When the mode is smooth, $p_1(P_t(x,y)|P_{t-1}(x,y))$ is explained with a small σ_i^2, while when the mode is abrupt, a larger σ_i^2 will be used.

Based on the Gaussian distribution, we randomly [17, 18] select some points for tracking. These points are the position we assume the object will move to. And each point has three scale image patches, describing the change of object size. Combined with Bayesian theory, the position of the object can be known.

3   Tracking Using Bayesian Theory

After getting a priori knowledge for Bayesian formulation with Gaussian distribution, we can use Bayesian theory to get the similarity of each patch [21]. Since it is a priori knowledge, we should train it in different situation to get the different variances. For different tracking videos, the variances could be different. But in one video, we can assume that the variance is relatively stable. For instance, when a human is walking on the street, the speed of the human is usually constant.

We use the following Bayesian formulation to combine the prior probability with the likelihood got in the next part:

$$p(P_t(x, y)) = p(P_t(x, y) \mid P_{t-1}(x, y))(L(P_t(x, y)) + \gamma), \tag{2}$$

where $p(P_t(x,y))$ means the probability of the position at which the object will be, $L(P_t(x,y))$ means the likelihood between the position at which the object will be and the object detected in the previous frame, and γ is a weight to balance the importance of the prior probability and the likelihood. Then the Maximum a Posteriori (MAP) estimate is used to choose the best position of the object at the next frame from all the possible positions with all sizes selected by Gaussian distribution. Such a process can be expressed as,

$$P_t(x, y) = \arg \max p(P_t(x, y)), \tag{3}$$

where $P_t(x,y)$ is the position the object at next frame.

4   Classifying All the Motion Patches

After randomly selecting image patches based on Gaussian distribution, we classify all the patches to get the most possible position of the object.

4.1   Reducing Sample Dimensionality

At the first frame, the position of the object should be got from the user, base on which we can get all the samples in the future frames. For all the samples got from the random function, their sizes are not the same. Some are larger than the image patch got in the former frame, while some are smaller. The reason is that when selecting each position, we choose three image patches, a smaller one, a larger one and one as larger as the object patch. And the sizes of the image patches are usually different from our expectation.

For each image patch, we use a vector to represent it. So the sizes of the samples can be reduced and the same with the vector which represent the object patch. Making the samples into the same size is good for the classification [20]. For each sample, to deal with the scale problem, firstly, the patch is decomposed into small patches with same sizes. If the initial size of the patch is larger than forty pixels, then it will be decomposed to sixteen small patches, and for the image patches selected in next frames, their lengths and heights will never be reduced to be smaller than forty pixels. If the initial size of the patch is too small (i.e. either height or width is shorter than forty pixels), it will be decomposed into four small same-sized patches, and the size will never be changed to be smaller than 16 pixels. For each decomposed patch, we extract a feature to represent it. The feature is the mean value of the pixels' values disposed by integral image. All the features are combined to a vector to represent the image patch, so the size of the vector is same for all the image patches in different sizes.

Table 1. Our propoposed tracking algorithm

Input: t-th video frame
1.   Sample a set of image patches base on Gaussian distribution at the t-th frame, and extract the features with dimensionality reduction.
2.   Sample the positive and negative image patches at the $(t-1)$-th frame, and extract the features with dimensionality reduction.
3.   Compute the distance between the each feature extracted in step 1 and the features extracted in step 2. Remove some position not of the object.
4.   Use eq. (2) and (3) to find the best position.
Output: Tracking location P_t.

4.2   Classifying the Features

For each feature vector, we compare it with three kinds of features, namely, not the object, last frame object and initial frame object [19].The feature that is not the object means that when we know the object's position at frame t and want to find the position of the object at the next frame $t+1$, we select four image patches around the object (i.e. each patch should be out of the image boundary). Obviously, the object is not in the four images, and we call it negative part. In the last frame with object, we select five image patches. One patch is at the position of the object, and four patches are at the positions around the object closely. Because sometimes the position we detect might be imperfect, we use the four image patches to fix the small miss, and we call them positive part. The initial object position is the position chosen by the user, which is the perfect position. So definitely, it belongs to the positive part. At this point, for each frame, we have got four negative patches and six positive patches, and each patch has one vector to represent the feature [23].

For each patch for classification, we compute the distance between negative part and positive part. Firstly, we compute the distance between it and the four patches belonging to negative part, and take the average value of them as Na. Then we compute the distance between it and the five patches belonging to the positive part except the patch in initial frame, and take the average value of them as Pa. Finally we compute the distance between it and the patch in the initial frame, and take it as In. The distance we use is Euclidean distance. If the Na is the minimum value of the three values, then we remove the image patch. For the remain image patches, the likelihood $L(P_t(x,y))$ is the minimum value of the Pa and In. Using the eq. (2) and (3), the best position can be found. Main steps for our proposed tracking algorithm are summarized in Table 1.

5   Experiments

We evaluated our tracking algorithm on six challenging sequences which are common for tracking experiments. The four trackers we compare with are the MILTrack algorithm [7], the online AdaBoost method (OAB) [3], the Semi-supervised tracker (SemiB) [6] and the fragment tracker (Frag) [16].

Since all the trackers except for Frag involve randomness,we run them 5 times and report the average result for each video clip. Our trackers is implemented in MATLAB, whic runs at 25frames per second (FPS) on a Pentium Dual 2.2GHz CPU with 2 GB RAM.

5.1   Experimental Setup

When tracking at frame t, we generate 4 negative image patches at frame t-1 around the object's position. For motion mode, we set different values for α and β, where α is the number of the positions randomly selected in smooth mode, and β is the number of the positions randomly selected in abrupt mode. For each position, we generate 3 image patches in different scales. For the variance of the Gaussian distribution, first we setup two different values for the smooth mode and abrupt mode, and the variance of the smooth mode is smaller. When the tracker is working, if the position of the object detected is in the smooth part, the two variances will be set smaller in the next frame, while they will be set larger when the object is detected in abrupt part. Both the variances will vary in a specific range. And the change of the variance also has impact on the values α and β. Obviously, α will become larger when the object detected in smooth mode. All the parameters vary smoothly, like the method in [22].

5.2   Experimental Results

All the video frames are in gray scale and we evaluated the proposed algorithm with other four trackers using success rate. We computed the success rate with the follow equation:

$$score = \frac{area(ROI_T \cap ROI_G)}{area(ROI_T \cup ROI_G)} \tag{4}$$

where the ROI_T is the tracking bounding box and the ROI_G is the ground truth bounding box. If the score is larger than 0.5, the tracking result is considered as a success.

Table 2 shows the quantitative results average 5 times using success rate. The tracking algorithm we proposed achieves the best or second best results in all the sequences in terms of success rate. Furthermore, our tracker runs fastest among all the trackers. Besides, it needs to be noted that all the other trackers involved in our comparison were implemented in C or C++. None of the other algorithms are faster than 11 frames per second (FPS) and ours is 25frames per second (FPS). Figure 1 shows the screenshots of our tracking results.

Table 2. Success rate (SR)(%). Bold numbers indicate the best performance

Sequence	Ours	MILTrack	OAB	SemiB	Frag
Bike	**67**	21	42	62	26
Cliff Bar	**74**	65	23	65	22
David	**73**	68	31	46	8
Face occ2	68	**99**	47	40	52
Girl	**73**	50	71	50	68
Tiger 1	**45**	39	24	28	19

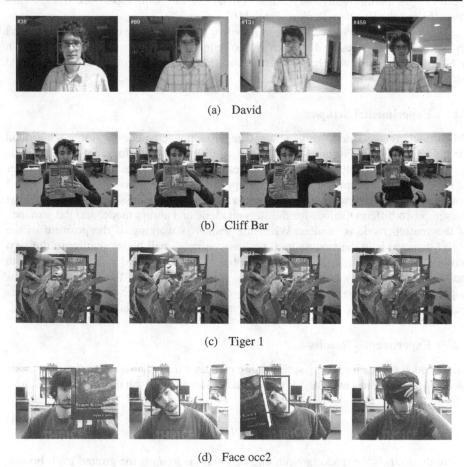

(a)   David

(b)   Cliff Bar

(c)   Tiger 1

(d)   Face occ2

Fig. 1. Screenshots of our tracking results

6   Conclusion

In this paper, we proposed an effective tracking algorithm, which can address the position of the object efficiently and accurately. Specifically, we use the motion mode to catch the abrupt move, and the image patch is divided into several small pieces to adapt to the slight change of the object. The experiments conducted on the benchmark testing video sequences demonstrated that the proposed algorithm could achieve a good balance between accuracy and efficiency. In the future, we will use C++ code to re-implement the algorithm to make it more efficient.

Acknowledgement. This work is supported by the Fundamental Research Funds for the Central Universities under grant no. 2100219033, the Natural Science Foundation of China under grant no. 61201394, and the Innovation Program of Shanghai Municipal Education Commission under grant no. 12ZZ029.

References

1. Black, M., Jepson, A.: Eigentracking: Robust matching and tracking of articulated objects using a view-based representation. IJCV 38, 63–84 (1998)
2. Avidan, S.: Support vector tracking. PAMI 26, 1064–1072 (2004)
3. Grabner, H., Grabner, M., Bischof, H.: Real-time tracking via online boosting. In: BMVC, pp. 47–56 (2006)
4. Jepson, A., Fleet, D., Maraghi, T.: Robust online appearance models for visual tracking. PAMI 25, 1296–1311 (2003)
5. Collins, R., Liu, Y., Leordeanu, M.: Online selection of discriminative tracking features. PAMI 27, 1631–1643 (2005)
6. Grabner, H., Leistner, C., Bischof, H.: Semi-supervised On-Line Boosting for Robust Tracking. In: Forsyth, D., Torr, P., Zisserman, A. (eds.) ECCV 2008, Part I. LNCS, vol. 5302, pp. 234–247. Springer, Heidelberg (2008)
7. Babenko, B., Yang, M.-H., Belongie, S.: Robust object tracking with online multiple instance learning. PAMI 33, 1619–1632 (2011)
8. Li, H., Shen, C., Shi, Q.: Real-time visual tracking using compressive sensing. In: CVPR, pp. 1305–1312 (2011)
9. Mei, X., Ling, H.: Robust visual tracking and vehicle classification via sparse representation. PAMI 33, 2259–2272 (2011)
10. Ross, D., Lim, J., Lin, R., Yang, M.-H.: Incremental learning for robust visual tracking. IJCV 77, 125–141 (2008)
11. Du, W., Piater, J.H.: A probabilistic approach to integrating multiple cues in visual tracking. In: Forsyth, D., Torr, P., Zisserman, A. (eds.) ECCV 2008, Part II. LNCS, vol. 5303, pp. 225–238. Springer, Heidelberg (2008)
12. Zhang, K., Zhang, L., Yang, M.-H.: Real-time compressive tracking. In: Fitzgibbon, A., Lazebnik, S., Perona, P., Sato, Y., Schmid, C. (eds.) ECCV 2012, Part III. LNCS, vol. 7574, pp. 864–877. Springer, Heidelberg (2012)
13. Kwon, J., Lee, K.: Visual tracking decomposition. In: CVPR, pp. 1269–1276 (2010)

14. Kwon, J., Lee, K.M.: Tracking of abrupt motion using wang-landau monte carlo estimation. In: Forsyth, D., Torr, P., Zisserman, A. (eds.) ECCV 2008, Part I. LNCS, vol. 5302, pp. 387–400. Springer, Heidelberg (2008)
15. Li, Y., Ai, H., Yamashita, T., Lao, S., Kawade, M.: Tracking in low frame rate video: A cascade particle filter with discriminative observers of different lifespans. PAMI 30, 1728–1740 (2008)
16. Adam, A., Rivlin, E., Shimshoni, I.: Robust fragments-based tracking using the integral histogram. In: CVPR, pp. 798–805 (2006)
17. Achlioptas, D.: Database-friendly random projections: Johnson-Lindenstrauss with binary coins. J. Comput. Syst. Sci 66, 671–687 (2003)
18. Baraniuk, R., Davenport, M., DeVore, R., Wakin, M.: A simple proof of the restricted isometry property for random matrices. Constr. Approx 28, 253–263 (2008)
19. Liu, L., Fieguth, P.: Texture classification from random features. PAMI 34, 574–586 (2012)
20. Candes, E., Tao, T.: Decoding by linear programming. IEEE Trans. Inform. Theory 51, 4203–4215 (2005)
21. Ng, A., Jordan, M.: On discriminative vs. generative classifier: a comparison of logistic regression and naive bayes. In: NIPS, pp. 841–848 (2002)
22. Raina, R., Battle, A., Lee, H., Packer, B., Ng, A.Y.: Self-taught learning: Transfer learning from unlabeled data. In: ICML, pp. 759–766 (2007)
23. Santner, J., Leistner, C., Saffari, A., Pock, T., Bischof, H.: PROST parallel robust online simple tracking. In: CVPR, pp. 723–730 (2010)

Vanishing Point Based Image Segmentation
and Clustering for Omnidirectional Image

Danilo Cáceres Hernández, Van-Dung Hoang, and Kang-Hyun Jo

Intelligent Systems Laboratory, Graduate School of Electrical Engineering,
University of Ulsan Ulsan 680-749, Korea
{danilo,dungvanhoang}@islab.ulsan.ac.kr, acejo@ulsan.ac.kr

Abstract. Regarding the autonomous of robot navigation, vanishing point (VP) plays an important role in visual robot applications such as iterative estimation of rotation angle for automatic control as well as scene understanding. Autonomous navigation systems must be able to recognize feature descriptors. Consequently, this navigating ability can help the system to identify roads, corridors, and stairs; ensuring autonomous navigation along the environments mentioned before the vanishing point detection is proposed. In this paper, the authors propose solutions for finding the vanishing point in based density-based spatial clustering of applications with noise (DBSCAN).First, the unlabeled data set is extracted from the training images by combining the red channel and the edge information. Then, the similarity metric of the specified number of clusters is analyzed via k-means algorithm. After this stage, the candidate area is extracted by using the hypothetical cluster set of targets. Second, we proposed to extract the longest segments of lines from the edge frame. Third, the set of intersection points for each pair of line segments are extracted by computing Lagrange coefficients. Finally, by using DBSCAN the VP is estimated. Preliminary results are performed and tested on a group of consecutive frames undertaken at Nam-gu, Ulsan, South Korea to prove its effectiveness.

Keywords: Vanishing point, Lane marking, Clustering, Omnidirectional camera.

1 Introduction

Autonomous ground navigation is still facing important challenges in the field of robotics and automation due to the uncertain nature of the environments, moving obstacles, and sensor fusion accuracy. Therefore, for the purpose of ensuring autonomous navigation and positioning along the environments aforementioned, a visual based navigation process is implemented by using an omnidirectional camera. The camera must operate alongside a variety of portable sensing devices such as Global Positioning System (GPS), Inertial Measurement Unit (IMU) and Laser Range Finder (LRF) and online interactive maps. Based on the perspective drawing theory, one VP is a point in which a set of parallel lines converge and disappear into the horizon. Ultimately, by using VP for indoor and outdoor in 2D images, roads as well as corridors can be described as a set of orthogonal lines due to the set of parallel lines that

D.-S. Huang et al. (Eds.): ICIC 2013, LNAI 7996, pp. 541–550, 2013.
© Springer-Verlag Berlin Heidelberg 2013

typically exhibit these structures in 3D scenes, see figure 1(a). In the field of autonomous navigation systems, efficient navigation, guidance and control design are critical in averting current challenges. In this sense, referring to the need for estimation of rotation angle for automatic control, one VP plays an important role at this stage. By detecting VP in the 2D image, autonomous unmanned systems are able to navigate towards the detected VP. Several approaches for VP detection have been used, for example, Hough transforms (HT), RANSAC algorithm, dominant orientation, and more recently equivalent sphere projection. In the case of HT, for estimating lines into the 2D images [1] propose the randomized HT to estimate the lane model based VP voting, [2] propose a VP detection method based HT and K-mean algorithm mainly based on the straight lines orientation given by the edges of corridors. In [3] authors use the RANSAC approach to describe a parametric model of the lane marking into the image. The clear examples of dominant orientation approaches are [4] [5] in which authors detected the VP by using a bank of 2D Gabor wavelet filters with a voting process. In case of spherical representation, [6] [7] authors used the 3-D line RANSAC approach in real time. In the case of omnidirectional scenes, parallel lines are projected as curves that converge and disappear into the horizon, see figure 2(b). Therefore, in order to address the challenges of VP detection in omnidirectional scenes, the authors decided to present an iterative VP detection based on DBSCAN. To this end, the main contributions of the presented method are:

- Implementation of real time polynomial algorithm for VP detection,
- Implementation of DBSCAN, due that the number of cluster in the image are not specified.

Fig. 1. Perspective projections of 3D parallel lines are projected as straight but diagonal lines or as curves in the 2D space. (a) VP perspective in single CCD images. (b) VP perspective in omnidirectional images. Lines in red represent orthogonal lines set towards the vanishing point. The point in blue represents the vanishing point on the horizon where all the orthogonal lines converge.

The rest of this paper is structured as follows: (2) Proposed Method, (3) Experimental Result, (4) Conclusions and Future Works.

2   Proposed Method

Essentially, the proposed method consists mainly in extracting the information surrounding the ground plane. Frames are extracted in short time intervals that started just before the earliest detection from the video capture sequence. In this section, the proposed algorithm for Iterative Vanishing Point Estimation Based on DBSCAN for Omnidirectional Image has three steps: (1) lane marking segmentation, (2) extracting line segments, (3) polynomial curve intersection, (4) cluster extraction by DBSCAN.

2.1   Lane Marking Segmentation.

In the field of Computer Vision, image segmentation plays an important task in image processing based application [8]. Segmentation can be defined as the process that divides an image into segments. As a result of the segmentation it is possible to identify regions of interest and objects in the 3D environment. The data set of lane marking imagery were extracted from a group of consecutive frames. For the training process we used 40x10 image resolution, see figure 2(a). The color images are decomposed in their respective RGB channels. In order to extract the foreground mask image a threshold selection method from gray level histograms is applied to the red channels (Otsu's method), see figure 2(b), (c). Then, by using the foreground binary mask the next step consists in extracting from the color image the RGB pixel information, see Fig. 2(d). Finally, the unlabeled training data is used in the k-means algorithm to determine the best cluster area can be used for the image segmentation step. K-means clustering is a method of cluster analysis which aims to partition n observations into k clusters in which each observation belongs to the cluster with the nearest mean. In other words, this algorithm is simple, straightforward and is based on the firm foundation of analysis of variances. The k-means clustering is defined as follows:

$$\arg\min \sum_{j=1}^{k} \sum_{i \in C_j} \left| x_i - \mu_j \right|^2 \tag{1}$$

where x_i is the given data set of intersection points, μ_j is the mean of the points in C_j, k is the number of cluster.

2.2   Extracting Line Segments

Road scenes can be described as structures which contain lane marking, soft shoulders, gutters, or a barrier curb. Therefore, in 2D images these features are represented as a set of connected points. The main idea of this section consisted of extracting the longest line segments around the ground by applying the canny edge detector from a group of consecutive frames. After the edge detection step, the subsequent task was to remove the smallest line segments by extracting the longest line segments after applying canny edge detection. This was achieved by using a 3 chain rule for each possible curve candidate. From the tracking algorithm the authors were able to extract the basic information such as: length (l), number of point per line segments as well as the pixel position location of the endpoints (P1, P3) and midpoint (P2) of line segments into the image plane. As a result, extraction from an image sequence the set of longest line segments was completed.

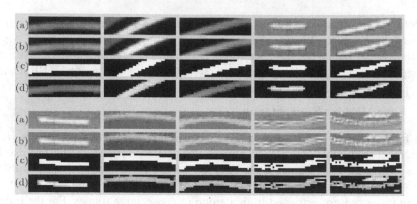

Fig. 2. Steps of the training process. (a) Training data set. (b) Red channels image. (c) Binary image after applying Otsu's thresholding method. (d) RGB color extraction.

Fig. 3. Steps of the lane color segmentation. (a) Frames from different sequence positions. (b) Otsu's method result. (c) Result after applying the color segmentation .(d) Binary image results after applying Otsu's method.

2.3   Polynomial Curve Intersection

At this point the set of line segments in the image plane are known. Hence, the new task of defining the polynomial function for each line segment. This means, finding the function which goes exactly through the points P1, P2 and P3. In numerical analysis, there are various approaches for solving a polynomial system. For example, in [9] authors use an Auzinger and Stetter method, which is a polynomial approach to estimating VP. The main contribution of the authors work is to estimate VP in real time processing, required by autonomous navigation. To this end, the approach used to achieve this is based on an algebraic property of Lagrangian interpolation polynomial approximation. Given the data set of points $(x1, y1 = f(x1)), (x2, y2 = f(x2)), ..., xn$, $yn = f(xn)$) the interpolation polynomial is defined as follows:

$$f_n(x) = \sum_{j=1}^{n} L_j(x) f(x_j) \tag{2}$$

$$L_j(x) = \prod_{\substack{k=1 \\ k \neq j}}^{n} \frac{x - x_k}{x_j - x_k} \tag{3}$$

Thus, the given 3 points of the interpolation polynomial equate to:

$$f(x_1) = y_1 \frac{(x - x_2)(x - x_3)}{(x_1 - x_2)(x_1 - x_3)} \tag{4}$$

$$f(x_2) = y_2 \frac{(x - x_1)(x - x_3)}{(x_2 - x_1)(x_2 - x_3)} \tag{5}$$

$$f(x_3) = y_3 \frac{(x - x_1)(x - x_2)}{(x_3 - x_1)(x_3 - x_2)} \tag{6}$$

$$f_2(x) = f(x_1) + f(x_2) + f(x_3) \tag{7}$$

Once the polynomials for each line segments have been disclosed, the next step consist in finding or each pair of polynomial curve the intersection point, see figure 4.

2.4   Cluster Extraction by DBSCAN

Given the data set of intersection point from the previous step, the algorithm should be able to extract clusters. From figure 4, it is clear that the projected data into the image plane give us vital information about the data set points, as follows:

• The data does not depict a well-defined shape.

- Presence of noise in the data due to the lack of a pre-processing model for road analysis. To compensate, all line segments are considered as a part of the road.
- The number of cluster could not be described in advance.

Considering the above mentioned occurrences, among the various clustering algorithms it was proposed to use the DBSCAN algorithm [10]; an unsupervised method, due to the algorithm having achieved a good performance with respect to some other algorithms. In essence, the main idea of DBSCAN is that for each point of a cluster, the neighborhood of a given radius has to contain a minimum number of points. The DBSCAN algorithm depends mostly on two parameters:

- Eps: number of points within a specified radius
- MinPts: minimum number of points belonging to the same cluster.

After the candidate clusters are formed, the idea is to define the cluster which contains the largest amount of points. Then, the centroids of the cluster are calculated, see figure 5.

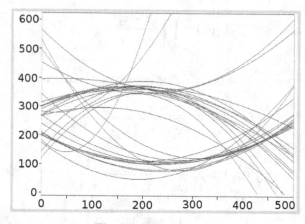

Fig. 4. Polynomial curve

3   Experimental Result

In this section, the ending results of the experiment will be introduced. All the experiments were done on Pentium Intel Core 2 Duo Processor E4600, 2.40 GHz, 2 GB RAM and the implementation was done in C++ under Ubuntu 12.04. The algorithm used a group of 4,355 consecutive frames taken at Nam-gu, Ulsan, South Korea with a frame size of 160x146 pixels. The depicted results are to the implementation of the lane marking segmentation, the Lagrange polynomial curve fitting and DBSCAN. The consecutive frame consists on urban road scene images. The experimental result of the proposed method in different frame sequences are shown in the cluster with the largest amount of points. Points in blue show the cluster with the lowest amount of points. The map in fig 8 shows the chosen trajectory, it consist of approximately 3 km drive from A to B. Figure 7 shows the frame-rate image sequences at a resolution of 160x146 pixels.

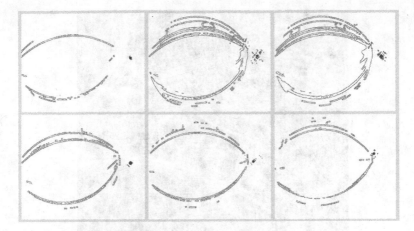

Fig. 5. Result from a different frame sequences. DBSCAN result. The point in green shows the estimated VP result. Points in red show the set of intersection points. Points in celestial blue show cluster with the largest amount of points. Points in blue show the cluster with the lowest amount of points.

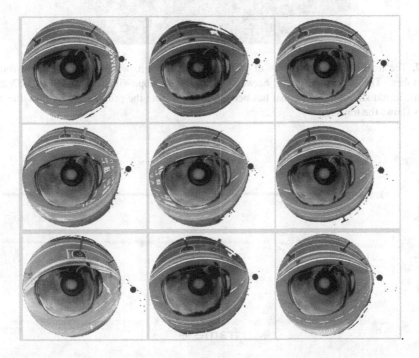

Fig. 6. Experimental results for a frame sequences. Points in red and green represent the extracted clusters. Points in blue represent the VP candidate.

Fig. 7. The map shows the location of Nam-gu area in Ulsan, where the consecutive frames were taken. The average distance between A and B is approximately 187 mi (3.02 km). The line in red shows the route that has been used for testing the proposed algorithm, the blue arrows shows the trajectory.

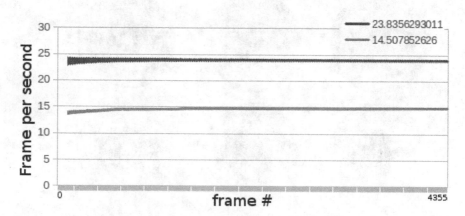

Fig. 8. Frame-rate image sequences at a resolution of 160x146. Data in blue represent the video acquisition. Data in red represent the performance of the algorithm.

4   Conclusions and Future Works

The preliminary results of the proposed method presented relevant information for finding the VP estimation in road scenes. Figure 6 shows the result of VP estimation. This result reinforces the point of usage a set of sensor (GPS, IMU,LRF, and online interactive maps) for dealing with the autonomous navigation problem in real time. As a result, the proposed algorithm is able to determine the VP in the 3D space based on DBSCAN. The preliminary results demonstrated its effectiveness and competitiveness. However, due to the illumination distribution, strong sunlight, and dark shadows problems there will be improvements to the performance and the experiment by estimating the likely illumination conditions of the scene.

Acknowledgements. This work was supported by the MKE (The Ministry of Knowledge Economy), Korea, under the Human Resources Development Program for Convergence Robot Specialists support program supervised by the NIPA (National IT Industry Promotion Agency) (NIPA-2012-HI502-12-1002).

References

1. Samadzadegan, F., Sarafraz, A., Tabibi, M.: Automatic Lane Detection in Image Sequences for Vision-Based Navigation Purposes. In: Proceeding of the International Society for Photogrammetry and Remote Sensing (ISPRS), Dresden, Germany (2006)
2. Ebrahimpour, R., Rassolinezhad, R., Hajiabolhasani, Z., Ebrahimi, M.: Vanishing point detection in corridors: using Hough transform and K-means clustering. IET Computer Vision J. 6, 40–51 (2012)
3. López, A., Cañero, C., Serra, J., Saludes, J., Lumbreras, F., Graf, T.: Detection of Lane Markings based on Ridgeness and RANSAC. In: Proceeding of the 8th International IEEE Conference on Intelligent Transportation Systems (ITSC 2005), Vienna, Austria (2005)
4. Miksik, O., Petyovsky, P., Zalud, L., Jura, P.: Detection of Shady and Highlighted Roads for Monocular Camera Based Navigation of UGV. In: IEEE International Conference of Robotics and Automation (ICRA 2011), Shanghai, China (2011)
5. Kong, H., Audibert, J.-Y., Ponce, J.: Vanishing point detection for road detection. In: IEEE Computer Society Conference on Computer Vision and Patter Recognition (CVPR 2009), Miami Beach, USA (2009)
6. Bosse, M., Rikoski, R., Leonard, J., Teller, S.: Vanishing Point and 3D Lines from Omnidirectional Video. In: Proceeding of the International Conference on Image Processing (ICIP 2002), New York, USA (2002)
7. Bazin, J.-C., Pollefeys, M.: Vanishing Point and 3D Lines from Omnidirectional Video. In: IEEE/RSJ International Conference on Intelligent Robots and Systems (IROS 2012), Algarve, Portugal (2012)
8. Velat, S.J., Lee, J., Johnson, N., Crane III, C.D.: Vision Based Vehicle Localization for Autonomous Navigation. In: roceeding of the International Symposium on Computational Intelligence in Robotics and Automation (CIRA 2007), Jacksonville, FL, USA (2007)

9. Mirzaei, F.M., Roumeliotis, S.I.: Optimal Estimation of Vanishing Points in a Manhattan World. In: IEEE International Conference on Computer Vision (ICCV 2011), Barcelona, Spain (2011)
10. Ester, M., Kriegel, H.-P., Sander, J., Xu, X.: A Density-Based Algorithm for Discovering Clusters in Large Spatial Databases with Noise. In: Proceeding of 2nd International Conference of Knowledge Discovery and Data Mining (KDD 1996), Oregon, USA (1996)

Secure SMS Based Automatic Device Pairing Approach for Mobile Phones

Shoohira Aftab[1], Amna Khalid[1], Asad Raza[2,*], and Haider Abbas[1,3]

[1] National University of Sciences and Technology, Islamabad, Pakistan
[2] Faculty of Information Technology, Majan University College, Muscat, Oman
asad.raza@majancollege.edu.om
[3] Center of Excellence in Information Assurance, King Saud University, Saudi Arabia
hsiddiqui@ksu.edu.sa, haidera@kth.se

Abstract. Bluetooth is recognized as one of the efficient way of data transmission at short distances wirelessly. Despite the fact that Bluetooth is widely accepted wireless standard for data transmission, there are also some security issues associated with Bluetooth. The pairing of Bluetooth and secure sharing of the Bluetooth pairing key is the challenge faced by the Bluetooth infrastructure today. This paper proposes the use of Out Of Band (OOB) channel for key exchange; hence proposing the use of security strengths of Global System for Mobiles (GSM) to securely communicate the encryption keys between the mobile devices, even if the users are not in the visibility range of each other. Once users accept the request to pair their devices, keys are exchanged between them and are used to set a secure communication channel between the mobile devices. The keys are exchanged via Short Message Service (SMS a service provided by GSM) and are automatically delivered to the Bluetooth pairing procedure by eliminating the need to manually communicate the Keys.

Keywords: Mobile Device Pairing, Out Of Band (OOB) Channel, Global System for Mobiles (GSM), Short Messaging Service (SMS).

1   Introduction

Mobile cell phones are nowadays equipped with Bluetooth and Wi-Fi technology that is used to share confidential data like pictures and documents among various users. Due to broadcast nature of Bluetooth and Wi-Fi it is quiet easy to intercept the communication between the devices. Therefore there is a need for a secure channel to be established between the devices for such communication. As these devices lack high computation power so the protocol and the algorithm need not to be cumbersome and at the same time secure enough to protect against attacks.

Pairing of two mobile phones requires a trusted channel to exchange keys. Most common approach is to enter the PIN(Personal Identification Number) on both devices and then that PIN is used to derive session keys. This mechanism of authentication

---

* Corresponding author.

D.-S. Huang et al. (Eds.): ICIC 2013, LNAI 7996, pp. 551–560, 2013.

through PIN; requires both the users to share the PIN through verbal communication. Another method is to use a cable as Out Of Band (OOB) channel to secure paring. Sharing the PIN through verbal communication can be easily eavesdropped by the attacker in range. Medium range for Bluetooth is 10m but to transfer the key by speaking, user needs to be less than 2-3 meter [8], otherwise it would be similar to broadcasting the key. Attacker could also intercept the key if one user is telling the other at the max distance. Using cable as OOB to sure exchange of PIN is comparatively a better solution but it affects the sole purpose of Bluetooth communication which is based on wireless. It also requires the users to be at a closer distance. Hence, both the options have drawbacks for establishing a secure channel.

Motivated by the fact that cost of SMS, provided by GSM network has been reduced drastically with time [15]. All mobile phone has the capability of SMS. The paper proposes solution to use SMS to transfer the key securely to other device. The process would require minimum user interaction .General overview is described in Fig. 1.

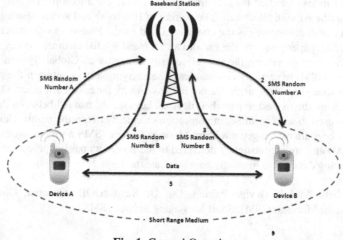

Fig. 1. General Overview

The device pairing mechanism depicted in Fig.1 has the following contributions:

- SMS based automatic key transfer method. This method does not require comparison or entering number while most of the methods require this.
- This proposed method assumes that GSM network offers secure / encrypted communication [10].
- This method doesn't require user to be at closer distance in audio visual range of other mobile device user.

The paper is organized as follows. Section 2 provides an overview of the cryptographic protocols and device pairing methods, which are relevant to this approach. Section 3 explains the problem and requirements associated with the area. Section 4 describes our approach to use GSM network for secure key exchange.

Section 5 explains the algorithm of the proposed approach. Section 6 describes the GSM packet and the proposed packet format. Section 7 shows the result of the survey based on this approach. Section 8 provides conclusions and future work.

2   Background and Related Work

This section presents an overview of the relevant key cryptographic protocols, exchange methods, device pairing methods and security of GSM network.

2.1   Cryptographic Protocols

Balfanz [3] uses public key cryptography to secure the key exchange. The protocol works in the following fashion. Public key of A (Pka) and Public Key of B (Pkb) are exchanged over the insecure channel and their respective hashes H(Pka) and H(Pkb) over OOB channel to make sure that the public keys are not forged [3]. This protocol requires hash function to be collision resistant and requires at least 80 bits of data to be transferred over OOB channel in each direction. MANA protocol [4] use k bits OOB data and limiting the attacker success probability to 2^{-k}. However they doesn't cater for the possibility that attacker can delay or replay OOB messages.

Another approach is to use Short Authentication String (SAS) protocol presented by F. Stajano and R. J. Anderson in [5].The protocol prevent the attack probability to 2^{-k} for k bit of OOB message. SAS protocols are usually based on two things i.e. 1). Use of OOB channel to transmit SAS from one device to another, or 2). the user compare two values generated by the devices involved in pairing.

Password Authenticated Key exchange (PAK) protocols [1] require users to generate or select a secret random number and somehow enter in both devices. Authenticated key exchange is then performed using the secret random number as authentication [2].

2.2   Device Pairing Methods

Based on many proposed cryptographic protocols, a number of pairing methods have been proposed. The methods are proposed by keeping in mind the wide variety of devices involved in the process of pairing. Based on different limitation of devices methods have been proposed. These methods use different available OOB channel and thus offer varying degree of security, reliability and usability.

A method named "Resurrecting Duckling" is proposed to counter the Man in The Middle (MiTM) attack [5]. It uses the standardized physical interfaces and cables. Due to wide variety of devices available today this method is now almost obsolete. Another reason is that use of the cable voids the advantages of wireless communication. "Talking to Stranger" [6] was another one of the initial approaches to counter MiTM attack. It uses infrared (IR) communication as OOB channel to authenticate and establish a secure channel. This method requires line of sight for communication via infrared which not only raises the issue of close distance but it is also prone to MiTM attack.

Another approach is based on synchronization of audio visual signal [7]. This method uses blink or beep to synchronize two devices. This method was suitable for devices which lack the display and input capability. As it require only two LEDs or a basic speaker. However all of them require users comparing audio/visual patterns. "Loud and clear" [8] uses audio OOB channel along with vocalized sentence as a digest of information that is transmitted over wireless channel.

Although the methods explained in this section provide certain level of trust but they also limit in terms of OOB, distance and MiTM. On the contrary, the paper proposed an approach which uses the GSM security to handle some problems identified in existing solutions.

3   Problem Setting

The problem is to establish a secure channel between two mobile phones that require transferring key to both devices. After successful transfer of key the device can use that to derive session keys to provide confidentiality of data. Requirement for the solution are as follows:

- Reliability: User can rely on the process to establish a secure channel.
- Usability: User should find the pairing process as easy-to-use.
- Automation: The process for secure key exchange should not require any manual configurations. Once the secret key is exchanged securely, secure communication channel should be established between devices.

4   SMS Based Key Transfer

This section presents SMS based key exchange mechanism. The proposed solution takes the advantage of security provided by GSM networks. By using SMS service both the parties are authenticated to each other. The proposed scheme overcomes limitations of existing schemes. This section will present complete algorithm and packet format for this scheme.

The possession of symmetric secret key confirms the authentication process. For data transfer the key can be used to provide confidentiality of data. The only information known to both the user is the cellular mobile number so that they can exchange the key. The proposed solution is as follows:

- User selects the object to be sent via Bluetooth.
- A random bit string of variable length is generated. Variable length key means that key length of desired security can be generated. 256 bit key is considered to provide enough protection.[14]
- User enters the other user's cellular number or can select from phonebook. As in most of the cases, one has already the number of the person, he wants to share data with. Even if the Cellular Identification (ID) needs to be told it is considered as a public entity and doesn't affect the security. There is a possible scenario that if someone doesn't want to share his/her mobile number with other user. Later in

this paper we have conducted a survey to further support this approach that most of the users have no problem in sharing their cellular ID with the people   they wish to exchange data/information via Bluetooth.

- Using SMS; these random numbers are transferred to the other device.
- By combining both numbers first initiator and then acceptor a common secret is securely exchanged between mobile phones. Now this key can be used directly to encrypt the data and provide integrity check or can be used to generate further keys for this process. By combining both number reduces the chance of guessing. Even if one user's random numbers are guessed by the attacker, even then the other user's random number are to be find out.

This scheme also facilitate the discovery process even if the device is made invisible to everyone a secure connection can establish without turning it visible to all. This would avoid denial of service attack in which the attacker tried to exhaust the device by sending repeated connection requests. There is also an optional feature to automatically switch on the Bluetooth device on receiving data. This option is kept optional as the survey results shows that quite a large percentage of people want to switch on Bluetooth manually, although higher percentage preferred this process to be automatic. The sequences of events are shown in Fig. 2.

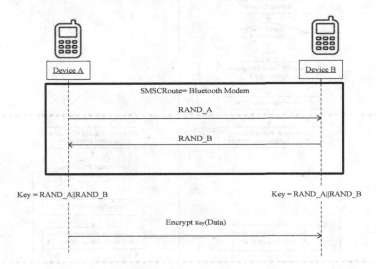

Fig. 2. Sequence Diagram

5   Algorithm

The algorithm for SMS based automatic device pairing is described in this section. Fig. 3 explain the algorithm sequence of events, when device A wants to share data with device B.

- Device A selects an object to be sent via Bluetooth and select Cellular ID of Device B.
- Device A generates a random bit string RAND_A (256 bit long) and composes a GSM Packet which contains RAND_A and pairing request identifier.
- Device A→ Device B (Secure GSM network)
- Device B receive request to pair via SMS.
- Device B accepts Device A's request to pairing by conforming the Cellular number to Device A's ID.
- Device B generates another Random bit string RAND_B (256 bit) and compose a GSM Packet which contain RAND_B and pairing conform identifier.
- Device B → Device A (secure GSM network)
- Device B calculate key as RAND_A||RAND_B
- Device A receives RAND_B with conform identification
- Device A calculate key RAND_A||RAND_B

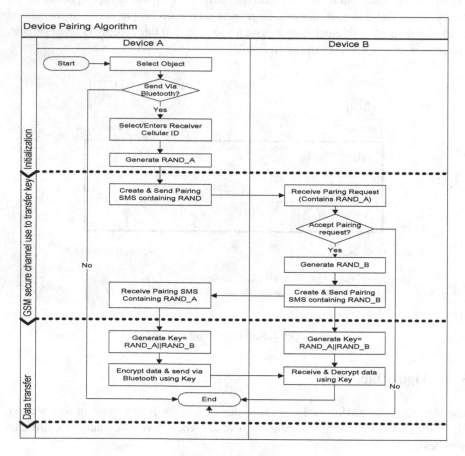

Fig. 3. Algorithm

6   Proposed SMS Packet

GSM is the second generation protocol used to transmit and receive digitally modulated voice and data signals. In this paper, the concern is with the Short Message Service provided by GSM. It was considered to be the OOB channel for the key exchange during the Bluetooth pairing procedure. GSM provides sufficient security on the SMS packet. It provides A3, A5 and A8 security features. Our objective is to send the random number via SMS and it is received in the Bluetooth bucket. For this matter it was not needed to suggest changes in the GSM SMS format however it is needed to add new parameters to the GSM SMS invoking and delivering Uniform Resource Locator. Since every SMS is sent via URL; these URLs define the destination, port and the type of the data in the SMS packet.

The problem in automating Bluetooth pairing procedure by an SMS was to redirect an SMS packet to the Bluetooth modem, because if the normal SMS is sent , it will be delivered to the inbox port of the receiving device and the user have to manually transfer the Random number from the received message (in the inbox) to the Bluetooth pairing procedure. However in order to automate, it should be received at the Bluetooth modem. Here it is needed to add changes into the SMS URL not the SMS packet frame. Each SMS is delivered via URLs. These URL describe parameters through the network. For example

"http://127.0.0.1:8800/?PhoneNumber=xxxxxxxx&..."

127.0.0.1 is the gateway, 8800 is the port number and after '?' are the parameters. As in example there is a Phone number; that cannot be written as +923214004499 rather following the syntax and write it as %2B923214004499 [15].

Here is another example URL that can send a text message to the desired Phone number provided the sender knows the IP address of the gateway it is connected through [15].

http://127.0.0.1:8800/?PhoneNumber=xxxxxxxx&Text=this+is+my+message

This is the basic URL format and the parameters define: where to send the packet, either acknowledgement is required or not, what is the type of packet. All these questions are answered in the SMS URL parameters. The Short Message Service Center (SMSC) is the body that reads these parameters and route the SMS to the path described in the parameters. There are certain HTTP parameters that route the packet through a specific SMSC. The HTTP interface supports SMSCRoute parameter, identified by a route name (Route name can be added by making amendments to SMSGW.IN) for example Bluetooth modem [5]. By defining the SMSCRoute to Bluetooth modem the particular SMS would be delivered on the Bluetooth modem. Therefore the message containing the random number is directly received by the Bluetooth hence making it automated because user is neither involved in random number generation nor in number transmission.

7   Results

The results of this paper are survey based. A Questionnaire based survey was initiated on a social networking platform (Facebook) and the people from different backgrounds (IT and non-IT professionals) polled for the answers. These people were not influenced by any social or political pressures from any individual or organization having any concerns with the writing or reading of this paper. Following questions Q1,Q2,Q3 were asked as shown in Fig.4, Fig.5 and Fig.6:

Q.1. The people you share data with (on Bluetooth) are listed in your phonebook?

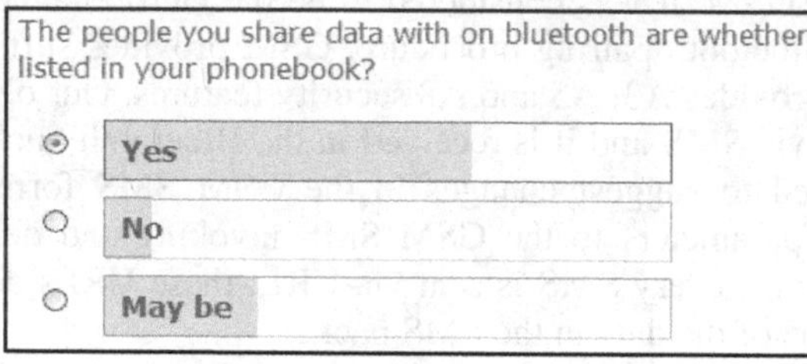

Fig. 4. 64.8% Yes+8.1% No+24.3% May be

Out of a total of 74 people; 48 people said 'Yes', they already have the phone numbers of the people they usually share data with. 6 people said 'No', they don't have the cell numbers of the people they are sharing data with and 18 people said May be i.e., sometimes they have the cell number sometimes don't.

Q.2. Do you mind sharing your cell number with the people you share data (on Bluetooth) with?

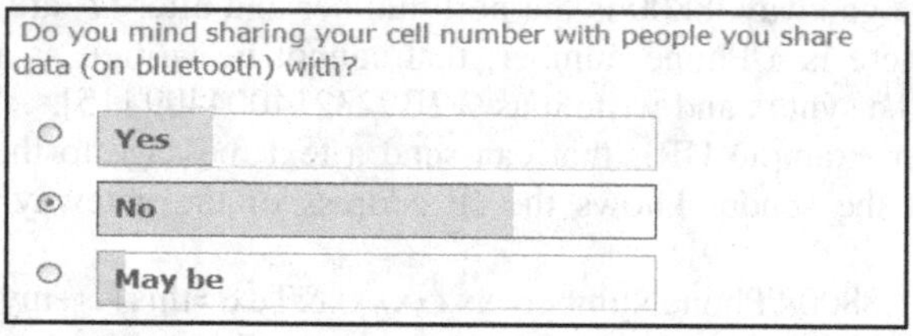

Fig. 5. 12.6% Yes+ 83.78% No+ 4.02% May be

Out of a survey of 74 people 9 said Yes they would mind sharing their cell numbers with people they are sharing data on Bluetooth with. 62 said No, They don't have any objection on sharing the cell number and 3 people said sometime, they may object sharing cell number sometime they may not.

Q.3. Would you like to automatically receive the data after accepting the pairing request?

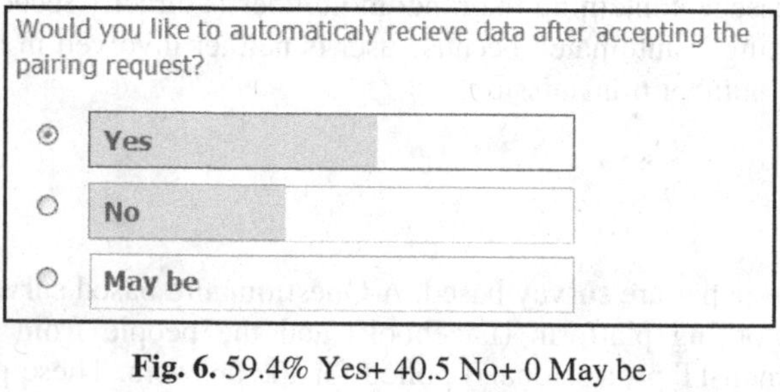

Fig. 6. 59.4% Yes+ 40.5 No+ 0 May be

When this question was presented to 74 different people the result came out to be; 44 people said Yes, we would like to get rid of that frequent acceptance of every file. 30 said No, it should ask me before transferring a file.

According to the survey people usually have the cell numbers already listed in their phonebooks with whom they are going to share data with, and most of the people don't mind sharing their cell number and most of the people want to get rid of the per file acceptance decision notification.

8   Conclusion

The paper presents a possible solution to transfer the PIN between the two mobile devices without extra effort of exchanging pairing keys manually through SMS. It inherits the GSM technology's security needed for the initial key derivation between the devices. Once the secure channel is set, the comunication can be carried out privately. There are some limitations associated with the approach which can be further improved. This method is dependent on GSM security. If GSM network is not available then this method can't function properly. The paper also presents the survey results of using SMS as Bluetooth key exchange to get an opinion of the users. The survey results show that most of the people are willing to communicate the cell number (already a public entity) instead of the private entities like pairing code. The future direction of this research is to develop a mechanism to encrypt the key with extra shield using Bluetooth technology.

References

1. Boyko, V., MacKenzie, P.D., Patel, S.: Provably Secure Password-Authenticated Key Exchange Using Diffie-Hellman. In: Preneel, B. (ed.) EUROCRYPT 2000. LNCS, vol. 1807, pp. 156–171. Springer, Heidelberg (2000)
2. Alfred, K., Rahim, S., Gene, T., Ersin, U., Wang, Y.: Serial hook-ups: a comparative usability study of secure device pairing methods. In: Proceedings of the 5th Symposium on Usable Privacy and Security (SOUPS 2009) (2009)
3. Balfanz, D., Smetters, D.K., Stewart, P., Wong, H.C.: Talking to strangers: Authentication in ad-hoc wireless networks. In: Proc. NDSS 2002: Network and Distributed Systems Security Symp. The Internet Society (February 2002)
4. Gehrmann, C., Mitchell, C.J., Nyberg, K.: Manuall authentication for wireless devices. RSA CryptoBytes 7(1), 29–37 (2004)
5. Stajano, F., Anderson, R.J.: The resurrecting duckling: Security issues for ad-hoc wireless networks. In: Security Protocols Workshop (1999)
6. Perrig, Song, D.: Hash visualization: a new technique to improve real-world security. In: International Workshop on Cryptographic Techniques and E-Commerce (1999)
7. Margrave, D.: GSM Security and Encryption, http://www.hackcanada.com/blackcrawl/cell/gsm/gsm-secur/gsm-secur.html (last Accessed on September 12, 2012)

8. Ameen, Y.: Pakistan has the lowest SMS cost In The World: Research Report,
   `http://www.telecomrecorder.com/2011/01/14/pakistan-has-the-`
   `lowest-sms-cost-in-the-world-research-report/`
   (last Accessed on May 12, 2012)
9. Stepanov, M.: GSM Security : Available at, Helsinki University of Technology,
   `http://www.cs.huji.ac.il/~sans/students.../GSM%20Security.ppt`
   (last Accessed on May 12, 2012)
10. Brute force attack, `http://en.citizendium.org/wiki/Brute_force_`
   `attack/Draft` (last Accessed on May 18, 2012)
11. SMS URL Parameters for HTTP, `http://www.nowsms.com/doc/`
   `submitting-sms-messages/url-parameters` (last Accessed on July 17, 2012)
12. Suominen, M.: GSM Security, Helsinki University of Technology,
   `http://www.netlab.tkk.fi/opetus/s38153/k2003/Lectures/`
   `g42GSM_security.pdf` (last Accessed on April 17, 2012)
13. Brookson, C.: GSM & PCN Security Encryption,
   `http://www.kemt.fei.tuke.sk/predmety/KEMT414_AK/_materialy/`
   `Cvicenia/GSM/PREHLAD/gsmdoc.pdf` (last Accessed on April 14, 2012)
14. Hwu, J.S., Hsu, S.F., Lin, Y.B., Chen, R.J.: End-to-end Security Mechanisms for SMS.
   International Journal of Security and Networks 1(3/4), 177–183 (2006)
15. Routing SMS messages to a Specific SMSC Route, `http://www.nowsms.com/`
   `routing-sms-messages-to-a-specifc-smsc-route`
   (last Accessed on February 14, 2012)

Optimization Algorithm Based
on Biology Life Cycle Theory

Hai Shen[1,*], Ben Niu[2], Yunlong Zhu[3], and Hanning Chen[3]

[1] College of Physics and Technology, Shenyang Normal University, Shenyang, China
drshenhai@gmail.com
[2] College of Management, Shenzhen University, Shenzhen, China
drniuben@gmail.com
[3] Laboratory of Information Service and Intelligent Control, Shenyang Institute of Automation, Chinese Academy of Sciences Shenyang, Shenyang, China
{ylzhu,chenhanning}@sia.cn

Abstract. Bio-inspired optimization algorithms have been widely used to solve various scientific and engineering problems. Inspired by biology life cycle, this paper presents a novel optimization algorithm called Lifecycle-based Swarm Optimization. LSO algorithm simulates biologic life cycle process through six optimization operators: chemotactic, assimilation, transposition, crossover, selection and mutation. Experiments were conducted on 7 unimodal functions. The results demonstrate remarkable performance of the LSO algorithm on those functions when compared to several successful optimization techniques.

Keywords: Bio-inspired Optimization Technique, Lifecycle, Lifecycle-based Swarm Optimization.

1   Introduction

Currently, the bio-inspired optimization techniques possessing abundant research results include Artificial Neural Networks (ANN) [1], Genetic Algorithm (GA) [2], Particle Swarm Optimization (SI) [3], ant Ant Colony Optimization (ACO) [4], Artificial Bee Colony (ABC) [5], Bacterial Colony Optimization (BCO) [6] and Artificial Immune Algorithm (AIA) [7] and so on. All bio-inspired optimization techniques have bionic features, the ability of highly fault tolerance, self-reproduction or cross-reproduction, evolution, adaptive, self-learning and other essential features. At present, the bio-inspired optimization algorithms have been widely applied [8-11].

All living organisms have life cycle. Four stages including birth, growth, reproduction, and death comprise the biologic life cycle [12]. This process repeated continuously made the endless life on earth, and biologic evolution become more and more perfecting. Borrowing the biology life cycle theory, this paper presents a Lifecycle-based Swarm Optimization (LSO) technique.

The rest of this paper is organized as follows. Section 2 describes the proposed Lifecycle-based Swarm Optimization (LSO) technique. Sections 3 and 4 present and discuss computational results. The last section draws conclusions and gives directions of future work.

D.-S. Huang et al. (Eds.): ICIC 2013, LNAI 7996, pp. 561–570, 2013.
© Springer-Verlag Berlin Heidelberg 2013

2   Lifecycle-Based Swarm Optimization

2.1   Population Spatial Structure

For the first stage of biologic life cycle, we consider in this paper not the how an organism is born, organism's size after birth, and the time of birth, but the distributed position state of all organisms in a population in their living space, which is population spatial distribution. Clumped distribution is the most common type of population spatial structure found in nature. It can be simulated by normal distribution of statistics [13]. Normal distribution is a continuous bell-shaped distribution. It is symmetric around a certain point. Variables fall off evenly on either side of this point. It is described as follows:

$$f(x) = \frac{1}{\sqrt{2\pi}\,\sigma} e^{-\frac{(x-\mu)^2}{2\sigma^2}}, \quad -\infty < x < +\infty \tag{1}$$

where parameters μ and σ^2 are the mean and the variance. The frequency distribution of variables which obey the normal distribution was completely determined by the μ and σ^2. Each possible value of μ and σ^2 defines a specific normal distribution.

(1) The μ specify the position of the central tendency of the normal distribution. It can be any value within the search space.

(2) The σ^2 describes the discrete level of variables distribution, and it must be positive. The larger σ, the more scattered the data; otherwise, the smaller σ, the more concentrated the data.

2.2   Six Operators

***Definition 1* Chemotaxis Operator:** based on the current location, the next movement will toward the better places. Chemotaxis operator forage strategy is chosen only by optimal individual of population.

Since 1970's, a large number of biologic model simulation explained that the chaos is widespread exist in biologic systems. Borrowing chaotic theory [14], chemotactic operator which was employed by the optimal individual performs chaos search strategy. The basic idea is introducing logistic map to optimization variables using a similar approach to carrier, and generate s set of chaotic variables, which can shown chaotic state [15, 16]. Simultaneously, enlarge the traversal range of chaotic motion to the value range of the optimization variables. Finally, the better solution was found directly using chaos variable.

The logistic map equation is given by equation (2):

$$x_{i+1} = rx_i(1 - x_i) \tag{2}$$

where r (sometimes also denoted μ)is driving parameter, sometimes known as the "biotic potential", and is a positive constant between 0 and 4. Generally, r=4. x_i represent the current chaos   variable, and x_{i+1} represent the next time's.

Definition 2 **Assimilation Operator:** by using a random step towards the optimal individual, individuals will gain resource directly from the optimal individual.

$$X_{i+1} = X_i + r_1(X_p - X_i) \tag{3}$$

where $r_1 \in R^n$ is a uniform random sequence in the range $(0,1)$. X_P is the best individual of the current population. X_i is the position of an individual who perform assimilation operator and X_{i+1} is the next position of this individual.

Definition 3 **Transposition Operator:** individuals will random migrates within their own energy scope.

$$ub_i = \frac{X_p}{X_i} \cdot \Delta \quad ; \quad lb_i = -ub_i \tag{4}$$

$$\varphi = r_2(ub_i - lb_i) + lb_i \tag{5}$$

$$X_{i+1} = X_i + \varphi \tag{6}$$

where φ is the migration distance of X_i; $r_2 \in R^n$ is a normal distributed random number with mean 0 and standard deviation 1; ub_i and lb_i is the search space boundary of the i_{th} individual; Δ is the range of the global search space.

Definition 4 **Crossover Operator:** exchange of a pair of parent's genes in accordance with a certain probability, and generate the new individual. In LSO, the crossover operator selects single-point crossover method.

Definition 5 **Selection Operator:** In this algorithm, the selection operator performs elitist selection strategy. A number of individuals with the best fitness values are chosen to pass to the next generation, and others will die.

Definition 6 **Mutation Operator:** randomly changes the value of individual. In this algorithm, the mutation operator performs dimension-mutation strategy. Every individual $X_i \in R^n$, $X_i = (x_{i1}, x_{i1}, ..., x_{in})$, one dimension of an individual selected according to the probability will re-location in search space.

Mutations are changes in a genomic sequence. It is an accident in the development of the life, and is the result of adaptation to the environment, and is important for the evolution, no mutation, and no evolution.

2.3   Algorithm Description

Lifecycle-based Swarm Optimization is a population-based search technique. In beginning, population spatial structure meet the clumped distribution, and evaluation the fitness function, then establishes an iterative process through implementation six operators proposed above. The Pseudo-code of the LSO is as follows:

Initialization:	Initialization parameters.
Born:	(1) Initialize the population with a normal distributed.
	(2) Compute the fitness values of all individuals.
Growth:	(1) The best individual of population executes the chemotaxis operator via the chaos searching.
	(2) A number of individual selected social foraging strategies will perform assimilation operator.
	(3) The rest individuals would execute the transposition operator.
Reproduction:	(1) Randomly select a pair of individuals to implement single-point crossover operation. All individuals generated by crossover operation comprised the offspring-population, which is called *SubSwarm*.
	(2) Compute the fitness values of the *SubSwarm*.
Death:	(1) Sort all individuals of *Swarm* and *SubSwarm*.
	(2) The *S* individuals with the better fitness were selected and others die.
Mutation:	Execute dimension-mutation operation based on the mutation probability.

3   Experiments

To evaluate the performance of the LSO algorithm, we employed 7 unimodal benchmark functions [17]. Detailed description of these functions, including the name, the boundary of search space, the dimensionalities, and the global optimum value of every function are all given in Table 1.

We compared the optimization performance of LSO with the well-known algorithm: PSO, GA and GSO [18]. Each algorithm was tested with all numerical benchmarks. Each of the experiments was repeated 30 times, and the max iterations in a run T_{max} =3000. The same population size S=50. In LSO, the probability using for decision the individual's foraging strategy P_f=0.1; the number of chaos variables S_c =100; crossover probability P_c=0.7; mutation probability P_m=0.02. The PSO and GA we used in this paper are the standard algorithms. In PSO, the acceleration factors $c1$=1, and $c2$=1.49; and a decaying inertia weight w starting at 0.9 and ending at 0.4 was used. In GA, crossover probability P_c and mutation probability P_m is respectively 0.7 and 0.05; selection operation is roulette wheel method. The parameter setting of the GSO algorithm can refer to the paper «A novel group search optimizer inspired by animal behavioral ecology» listed in the references [18].

Table 1. Unimodal Benchmark Functions

Functions	$\{N, \text{SD}, f_{min}\}$
$f_1(x) = \sum_{i=1}^{n} x_i^2$	$\{30, [-100,100]^n, f_1(\vec{0}) = 0\}$
$f_2(x) = \sum_{i=1}^{n} \|x_i\| + \prod_{i=1}^{n} \|x_i\|$	$\{30, [-10,10]^n, f_2(\vec{0}) = 0\}$
$f_3(x) = \sum_{i=1}^{n} (\sum_{j=1}^{i} x_j)^2$	$\{30, [-100,100]^n, f_3(\vec{0}) = 0\}$
$f_4(x) = max_i\{\|x_i\|, 1 \le i \le n\}$	$\{30, [-100,100]^n, f_4(\vec{0}) = 0\}$
$f_5(x) = \sum_{i=1}^{n-1} (100(x_{i+1} - x_i)^2) + (x_i - 1)^2)$	$\{30, [-30,30]^n, f_5(\vec{1}) = 0\}$
$f_6(x) = \sum_{i=1}^{n} (x_i + 0.5)^2$	$\{30, [-100,100]^n, f_6(\vec{0}) = 0\}$
$f_7(x) = \sum_{i=1}^{n} i x_i^4 + random[0,1)$	$\{30, [-1.28,1.28]^n, f_7(\vec{0}) = 0\}$

Table 2. Results for all algorithms on 7 benchmarks functions

Functions		LSO	GSO	PSO	GA
f_1	Mean Best	**6.13E-12**	1.51E-07	1.57 E-07	1.25E+05
	Std	**(2.07E-12)**	(3.89E-08)	(4.74E-05)	(1.22E+04)
f_2	Mean Best	**4.79E-07**	4.06E-04	5.01E+01	3.47E+26
	Std	**(1.14E-07)**	(1.30E-04)	(1.51E+01)	(1.07E+26)
f_3	Mean Best	**7.70E-08**	4.68E+02	2.16E+04	1.31E+07
	Std	**(1.98E-08)**	(1.40E+02)	(6.39E+03)	(3.26E+06)
f_4	Mean Best	**3.02E-06**	6.54E+00	3.02E-01	9.64E+00
	Std	**(7.74E-07)**	(2.11E+00)	(7.18E-02)	(4.10E-01)
f_5	Mean Best	2.78E+01	**1.80E+01**	9.01E+04	3.64E+09
	Std	**(1.26E-01)**	(1.34E+01)	(2.31E+04)	(8.43E+08)
f_6	Mean Best	**0**	**0**	1.40E+01	1.22E+05
	Std	**(0)**	**(0)**	(4.18E+00)	(1.15E+04)
f_7	Mean Best	**1.23E-03**	9.82E-01	3.22E+01	9.06E+02
	Std	**(3.35E-04)**	(2.74E-01)	(1.12E+01)	(1.97E+02)

4   Results and Discussion

Experimental results were recorded and are summarized in Table 2. All algorithms has a consistent performance pattern on functions $f_1 \sim f_4$. LSO is the best, GSO is almost as good and PSO and GA are failed. On function f_5, all of these algorithms can't find the optimum. Function f_6 is the step function and consists of plateaus, which has one minimum and is discontinuous. It is obvious that finding the optimum solution by LSO and GSO is easily, but is difficulties for PSO and GA. Function f_7 is a noisy quartic function, where random $[0, 1)$ is a uniformly distributed random variable in $[0, 1)$. On this function, LSO can find the optimum, other algorithms can't.

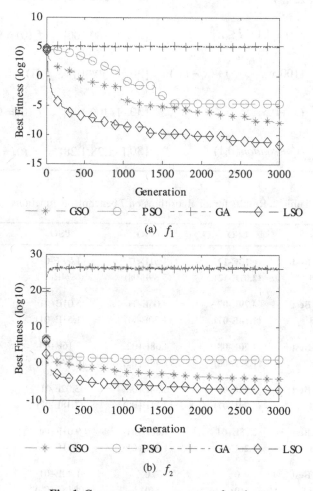

Fig. 1. Convergence curves on test functions

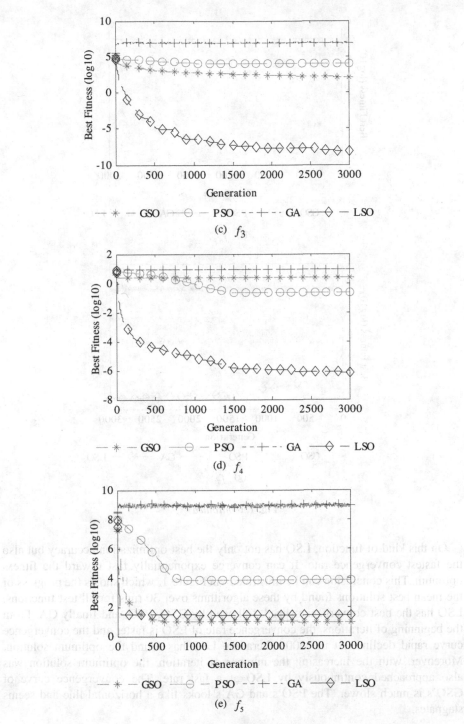

Fig. 1. (*continued*)

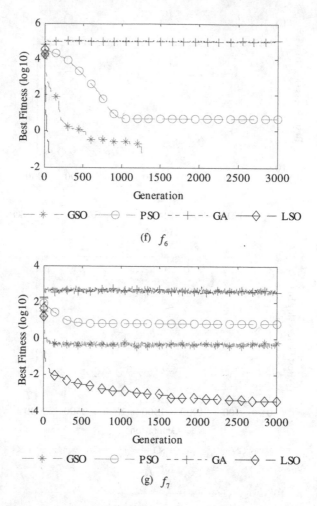

Fig. 1. (*continued*)

On this kind of function, LSO has not only the best optimization accuracy but also the fastest convergence rate. It can converge exponentially fast toward the fitness optimum. This conclusion can be illustrated via Figure 1, which shows the progress of the mean best solutions found by these algorithms over 30 runs for all test functions. LSO has the best convergence speed, followed by GSO, PSO, and finally GA. From the beginning of iterations, the convergence rate of LSO is faster and the convergence curve rapid decline. At the 500th iteration, LSO has found the optimum solution. Moreover, with the increasing the number of iteration, the optimum solution was also approached continuously by LSO at a fast rate. The convergence curve of GSO's is much slower. The PSO's and GA's looks like a horizontal line and seems stagnates.

5   Conclusion

This paper proposed a novel optimization algorithm: LSO, which is based on biologic life cycle theory. Each biologic must go through a process from birth, growth, reproduction until death, this process known as life cycle. Based on these features of life cycle, LSO designed six optimization operators: chemotactic, assimilation, transposition, crossover, selection and mutation. This paper uses 7 benchmark functions to test LSO, and compare it with GSO, PSO and GA, respectively. LSO appeared to be an overpowering winner compared to the GSO, PSO and GA in terms of effectiveness, as well as efficiency. In our future work, LSO could be studied and tested on multimodal benchmark functions and real-world problems.

Acknowledgment. This project is supported by the National Natural Science Foundation of China (Grant Nos. 61174164, 51205389, 61105067, 71001072, 71271140, 71240015), and the National Natural Science Foundation of Liaoning province of China (Grant No. 201102200), and the Natural Science Foundation of Guangdong Province (Grant Nos. S2012010008668, 9451806001002294).

References

1. Hebb, D.O.: The Organization of Behavior. John Wiley, New York (1949)
2. Holland, J.H.: Adaptation in Natural and Artificial Systems. The University of Michigan Press, USA (1975)
3. Kennedy, J., Eberhart, R.: Particle Swarm Optimization. In: IEEE International Conference on Neural Networks, pp. 1942–1948. IEEE Press, New York (1995)
4. Dorigo, M., Blum, C.: Ant Colony Optimization Theory: A Survey. Theoretical Computer Science 344(2-3), 243–278 (2005)
5. Karaboga, D., Akay, B.: A Comparative Study of Artificial Bee Colony Algorithm. Applied Mathematics and Computation 214(1), 108–132 (2009)
6. Niu, B., Wang, H., Chai, Y.J.: Bacterial Colony Optimization. Discrete Dynamics in Nature and Society 2012, article ID, 698057, 1–28 (2012)
7. Dasgupta, D.: Artificial Immune Systems and Their Applications. Springer, Germany (1999)
8. Niu, B., Fan, Y., Xiao, H., et al.: Bacterial Foraging-Based Approaches to Portfolio Optimization with Liquidity Risk. Neurocomputing 98(3), 90–100 (2012)
9. Donati, A.V., Montemanni, R., Casagrande, N., et al.: Time Dependent Vehicle Routing Problem with a Multi Ant Colony System. European Journal of Operational Research 185(3), 1174–1191 (2008)
10. Parpinelli, R.S., Lopes, H.S., Freitas, A.A.: Data Mining with An Ant Colony Optimization Algorithm. IEEE Transactions on Evolutionary Computation 6(4), 321–332 (2002)
11. Chandrasekaran, S., Ponnambalam, S.G., Suresh, R.K., et al.: A Hybrid Discrete Particle Swarm Optimization Algorithm to Solve Flow Shop Scheduling Problems. In: IEEE Conference on Cybernetics and Intelligent Systems, pp. 1–6 (2006)
12. Roff, D.A.: The Evolution Of Life Histories: Theory And Analysis. Chapman and Hall, New York (1992)

13. Berman, S.M.: Mathematical Statistics: An Introduction Based on the Normal Distribution. Intext Educational Publishers, Scranton (1971)
14. Lorenz, E.N.: Deterministic Non-Periodic Flow. Journal of the Atmosphéric Sciences 20, 130–141 (1963)
15. Verhulst, P.F.: Recherches Mathématiques Sur la loi d'accroissement de la population. Nouv. mém. de l'Academie Royale des Sci. et Belles-Lettres de Bruxelles 18, 1–41 (1845)
16. Verhulst, P.F.: Deuxième mémoire sur la loi d'accroissement de la population. Mém. de l'Academie Royale des Sci., des Lettres et des Beaux-Arts de Belgique 20, 1–32 (1847)
17. Yao, X., Liu, Y., Lin, G.M.: Evolutionary Programming Made Faster. IEEE Transactions on Evolutionary Computation 3(2), 82–102 (1999)
18. He, S., Wu, Q.H., Saunders, J.R.: A Novel Group Search Optimizer Inspired By Animal Behavioral Ecology. In: IEEE International Conference on Evolutionary Computation, pp. 1272–1278 (2006)

An Idea Based on Plant Root Growth for Numerical Optimization

Xiangbo Qi[1,2], Yunlong Zhu[1], Hanning Chen[1], Dingyi Zhang[1], and Ben Niu[3]

[1] Shenyang Institute of Automation Chinese Academy of Sciences 110016, Shenyang, China
[2] University of Chinese Academy of Sciences 100039, Beijing, China
[3] College of Management, Shenzhen University 518060, Shenzhen, China

Abstract. Most bio-inspired algorithms simulate the behaviors of animals. This paper proposes a new plant-inspired algorithm named Root Mass Optimization (RMO). RMO simulates the root growth behavior of plants. Seven well-known benchmark functions are used to validate its optimization effect. We compared RMO with other existing animal-inspired algorithms, including artificial bee colony (ABC) and particle swarm optimization (PSO). The experimental results show that RMO outperforms other algorithms on most benchmark functions. RMO provides a new reference for solving optimization problems.

Keywords: Root mass optimization, artificial bee colony algorithm, particle swarm optimization algorithm.

1 Introduction

In order to solve tough optimization problems, many researchers have been drawing inspiration from the nature and many meta-heuristic algorithms inspired by biology behavior are proposed. Genetic Algorithm (GA) mimics the process of natural evolution[1]. Particle Swarm Optimization (PSO) algorithm simulates the swarm behavior of birds and fish[2]. Firefly algorithm (FA) simulates the bioluminescent communication behavior of fireflies[3]. Bacterial Foraging Optimization (BFO) simulates the foraging behavior of bacteria[4]. BCO is based on a lifecycle model that simulates some typical behavior of E. coli bacteria[5].Artificial Fish Swarm Algorithm (AFSA) simulates prey, swarm and follow behavior of a school of fish[6]. Ant Colony Optimization(ACO) algorithm modeled on the actions of an ant colony[7]. Artificial Bee Colony (ABC) algorithm inspired by the foraging behavior of a swarm of bees is proposed by Karaboga[8].

Can be seen from above, most bio-inspired algorithms simulate some behaviors of animals. Algorithms simulating the growth behavior of plants are rarely seen. However, plants also have 'brain-like' control[9, 10]. Some researchers proposed a plant growth simulation algorithm (PGSA) simulating plant growth [11]. Every algorithm has its advantages and disadvantages. "No free Lunch "theorems [12, 13] suggests one algorithm impossibly shows the best performance for all problems. Many strategies including improving existed algorithms or studying new algorithms can get better

D.-S. Huang et al. (Eds.): ICIC 2013, LNAI 7996, pp. 571–578, 2013.

optimization effect. Inspired by the root growth behavior of plants, this paper proposes a new algorithm named Root Mass Optimization (RMO) algorithm.

The remainder of the article is organized as follows. Section 2 introduces some researches about root growth model. Section 3 proposes Root Mass Optimization (RMO) algorithm and gives the pseudo code. Section 4 gives the experiment process in detail, presents the experimental results and gives the analysis. Section 5 discusses a similarity in form between PSO and RMO. Finally, section 6 gives the conclusions.

2   Research of Root Growth Model

To understand the biological process of plant roots, many researchers built all kinds of models as a means of simulating the growth behaviors of plants. Different models show different purposes. In order to describe the process of plant growth, some mathematical models were constructed and were useful for the investigation of the effects of soil, water usage, nutrient availability and many other factors on crop yield [14]. There is increasing evidence that root–root interactions are much more sophisticated [15]. Roger Newson characterized the root growth strategy of plants in a model [16]: (1) Each root apex may migrate downwards (or sideways) in the substrate. (2) Each root apex, as it migrates, leaves behind it a trail of root mass, which stays in place.(3) Each root apex may produce daughter root apices.(4) Each root apex may cease to function as above, and "terminally differentiate" to become an ordinary, on-migrating, on-reproducing piece of root mass.

From the above point of view, the meaning of each root apex growth contains two aspects. One is root apex itself grows. The other is producing branch roots. These two kinds of growth may stop for some reasons. Existing root system models can be divided into pure root growth models, which focus on describing the root system's morphology, and more holistic models, which include several root-environment interaction processes, e.g. water and nutrient uptake[17]. However, we don't want to pay attention to root system's biological significances and agricultural significances. Plant roots are fascinating as they are able to find the best position providing water and nutrient in soil depending on their growth strategy. These strategies include gravitropism, Hydrotropism, chemotropism, and so on. We link the root growth process to the optimizing process for an objective function.

3   Root Mass Optimization Algorithm

Root growth offers a wonderful inspiration for proposing a new optimization algorithm. The objective function is treated as the growth environment of plant roots. The initial root apices forms a root mass. Each root apex can be treated as the solution of the problem. Roots turn to the direction that provides the optimal soil water and fertilizer conditions, so they may proliferate. That process can be simulated as an optimizing process in the soil replaced with an objective function. In view of this, we proposed the root mass optimization (RMO) algorithm. Some rules are made to idealize the root growth behaviors in RMO:(1)All root apices forms a root mass. Two operators including root regrowing and root branching are needed to idealize the root growth behavior. Each root apex grows using one of these two operators. (2)Root mass are divided into three groups according to the fitness. The group with better fitness is called regrowing group. The group with worse fitness called stopping group stops growing. The rest of root mass is called branching group.

The meanings of two operators including root regrowing and root branching are listed below.

(a) Root regrowing: this operator means that the root apex regrows along the original direction. The root apex may migrate downwards (or sideways) the best position which provides the optimal soil water and fertilizer conditions. This operator is formulated using the expression (1).

$$n_i = x_i + r + (g_{best} - x_i) \tag{1}$$

Where r is a random vector each element of which is between $[-1, 1]$. n_i is the new position of the ith root apex. x_i is the original position of the ith root apex. g_{best} is the root apex with the best fitness in each generation.

(b) Root branching: this operator means that the root apex produces a new growth point instead of regrowing along the original direction. The growth point may be produced at a random position of the original root with a random angle β. It is worth noting that, a growth point is seen as a root apex in this paper. This operator is formulated using the expression (2).

$$n_i = \beta \alpha x_i \tag{2}$$

Where α is a random number between $(0, 1)$. n_i is the new position of the ith root apex. x_i is the original position of the ith root apex. β is calculated using the expression (3).

$$\beta = \lambda_i / \sqrt{\lambda_i^T \lambda_i} \tag{3}$$

Where λ_i is a random vector.

The pseudo code of RMO algorithm is listed in Table 1. In each generation, sort the root apices in descending order according to the fitness. The selection of root apices participating in the next generation employs the linear decreasing way according to the expression (4). This way makes the root apices with better fitness perform root regrowing or root branching and makes the worse ones stop going on growing. In the selected part, select a percentage of the root apices from the front and let these root apices (growing group) regrow using the operator root growing; the rest of root apices (branching group) branch using the operator root branching.

$$ratio = sRatio - (sRatio - eRation)\frac{eva}{mEva} \tag{4}$$

Where eva is the current function evaluation count and mEva is the maximum function evaluation count. sRatio is the initial percentage and eRatio is the last percentage.

Table 1. Pseudo Code of RMO

1.	Initialize the position of root apices to form a root mass and Evaluate the fitness values of root apices
2.	While not meet the terminal condition
3.	Divide the root apices into regrowing group , branching group and stopping group
4.	Regrowing phase
5.	For each root apex in regrowing group
	Grow using the operator root regrowing
	Evaluate the fitness of the new root apex
	Apply greedy selection
	End for
7.	Branching phase
8.	For each root apex in branching group
	Produce two growing point using the operator root branching
	Evaluate the fitness
	Apply greedy selection
	End for
9.	Rank the root apices and memorize the current best root apex
10.	End while
11.	Postprocess results

4   Validation and Comparison

In order to test the performance of RMO, PSO and ABC were employed for comparison.

4.1   Experiments Sets and Benchmark Functions

The max evaluations count is 10000. In PSO, inertia weight varied from 0.9 to 0.7 and learning factors c1 and c2 were set 2.0. The population size of three algorithms was 40. Each algorithm runs for 30 times and takes the mean value and the standard deviation value as the final result. In RMO, the number of root apices in regrowing group is thirty percent of the selected root apices in each generation. sRatio is 0.9 and eRatio is 0.4. Seven well-known benchmark functions which are widely adopted by other researchers[18] are listed as follows.

Sphere function: $\qquad f_1 = \sum\limits_{i=1}^{n} x_i^2$ $\qquad\qquad\qquad\qquad x \in [-5.12, 5.12]$

Quadric function: $\qquad f_2 = \sum\limits_{i=1}^{n} \left(\sum\limits_{j=1}^{i} x_j \right)^2$ $\qquad\qquad\qquad x \in [-10, 10]$

Rosenbrock function: $\quad f_3 = \sum\limits_{i=1}^{n} (100(x_i^2 - x_{i+1})^2 + (1-x_i)^2)$ $\qquad x \in [-15, 15]$

Rastrigin function: $\qquad f_4 = \sum\limits_{i=1}^{n} (x_i^2 - 10\cos(2\pi x_i) + 10)$ $\qquad x \in [-15, 15]$

Schwefel function: $\qquad f_5 = n * 418.9829 + \sum\limits_{i=1}^{n} (-x_i \sin(\sqrt{|x_i|}))$ $\qquad x \in [-500, 500]$

Ackley function: $\qquad f_6 = 20 + e - 20\exp(-0.2\sqrt{\dfrac{1}{n}\sum\limits_{i=1}^{n} x_i^2}) - \exp(\dfrac{1}{n}\sum\limits_{i=1}^{n} \cos(2\pi x_i))$ $\quad x \in [-32.768, 32.768]$

Griewank function: $\qquad f_7 = \dfrac{1}{4000}(\sum\limits_{i=1}^{n} x_i^2) - (\prod\limits_{i=1}^{n} \cos(\dfrac{x_i}{\sqrt{i}})) + 1$ $\qquad x \in [-600, 600]$

4.2   Experiment Results and Analysis

As can be seen in Table2, on function with dimension of 2, ABC performs better than RMO and PSO on f1, f2, f4, f5 and f6. PSO shows the best performance on function f3. RMO not only gets the best result on f7 but also gets satisfactory accuracy on f1, f2, and f4. From Table4 and Table5, we can see that RMO performs much better than ABC and PSO on most functions except f5. From Table3, RMO performs much better than ABC and PSO on most functions except f3 and f5. On most benchmark functions with multiple dimensions, RMO is much superior to other algorithms in terms of accuracy.

In view of the above comparison, we can see RMO is a very promising algorithm. It has a very strong optimizing ability on test functions with multiple dimensions.

Table 2. Results of ABC, RMO and PSO on benchmark functions with dimension of 2

Function		ABC	RMO	PSO
f_1	Mean	**2.67680e-018**	6.31869e-017	3.13912e-014
	Std	**2.30179e-018**	1.15724e-016	5.99525e-014
f_2	Mean	**2.21443e-017**	1.57823e-016	4.18295e-013
	Std	**1.94285e-017**	1.78346e-016	1.03217e-012
f_3	Mean	2.44453e-002	2.48765e-002	**2.17246e-010**
	Std	2.90649e-002	3.54920e-002	**7.02026e-010**
f_4	Mean	**0**	7.56728e-014	4.28023e-010
	Std	**0**	1.11558e-013	1.60122e-009
f_5	Mean	2.54551e-005	2.24088e+002	7.83010e+001
	Std	**1.37842e-020**	1.27893e+002	6.31680e+001
f_6	Mean	**8.88178e-016**	6.38552e-008	3.00013e-006
	Std	**2.00587e-031**	6.33102e-008	3.06260e-006
f_7	Mean	6.01118e-007	**1.18805e-013**	2.25163e-003
	Std	3.28120e-006	**2.56555e-013**	3.61174e-003

Table 3. Results of ABC, RMO and PSO on benchmark functions with dimension of 15

Function		ABC	RMO	PSO
f_1	Me	4.99706e-007	**3.17457e-017**	3.00736e-004
	Std	1.24507e-006	**3.16538e-017**	2.55414e-004
f_2	Me	3.87111e+001	**4.33235e-017**	5.60171e-002
	Std	1.48006e+001	**3.33955e-017**	3.01647e-002
f_3	Me	**5.75084e+000**	1.39190e+001	3.39054e+001
	Std	**5.25142e+000**	3.52324e-002	4.30184e+001
f_4	Me	2.16707e+000	**0**	2.79516e+001
	Std	1.22010e+000	**0**	1.11010e+001
f_5	Me	**3.43993e+002**	4.46660e+003	2.73225e+003
	Std	**1.08184e+002**	3.50720e+002	5.48055e+002
f_6	Me	3.38283e-002	**1.08484e-010**	1.73043e+000
	Std	2.37433e-002	**1.13511e-010**	8.12113e-001
f_7	Me	3.67391e-002	**2.96059e-017**	2.18996e-001
	Std	2.13001e-002	**4.99352e-017**	9.42148e-002

Table 4. Results of ABC, RMO and PSO on benchmark functions with dimension of 30

Function		ABC	RMO	PSO
f_1	Me	6.12616e-003	**3.09257e-017**	2.93943e-002
	Std	1.10377e-002	**3.06840e-017**	1.14802e-002
f_2	Me	2.55908e+002	**3.02047e-017**	6.67699e+000
	Std	5.10078e+001	**2.89243e-017**	2.76913e+000
f_3	Me	1.66525e+002	**2.89378e+001**	1.62023e+002
	Std	2.46282e+002	**4.43782e-002**	6.00192e+001
f_4	Me	2.77714e+001	**0**	1.04094e+002
	Std	8.58554e+000	**0**	1.76451e+001
f_5	Me	**2.07710e+003**	1.02135e+004	6.17883e+003
	Std	**3.14194e+002**	5.28514e+002	8.70712e+002
f_6	Me	3.83016e+000	**1.70717e-011**	3.54363e+000
	Std	7.92153e-001	**2.68388e-011**	6.32969e-001
f_7	Me	5.82595e-001	**2.96059e-017**	1.09786e+000
	Std	3.01591e-001	**4.99352e-017**	4.47926e-002

5   Discussion

Eberchart & Kennedy (1995) firstly presented PSO algorithm (PSO) which imitated the swarm behavior of birds and fish [2]. In the mathematical models of PSO, particle swarm optimizer adjusts velocities by the following expression:

$$v_i^{t+1} = v_i^t + c_1 rand1()(pbest_i^t - x_i^t) + c_2 rand2()(gbest^t - x_i^t) \tag{5}$$

$i : 1,2,...,N$, N :population size; t :iterations; c_1 and c_2 :positive constants; $rand1()$ and $rand2()$:uniform distribution in [0, 1].The expression (5) means that the new velocity of a particle is adjusted by the original velocity,own learning factor and social learning factor. $pbest_i^t$ is the best previous position of the ith particle. $gbest^t$ is the best position among all the particles in the swarm. In RMO algorithm, the expression (1) is simialr in form to the expression (5) .However, the expression (1) has its own biological significance for roots growth in essence. A root apex regrows depending many factors including gravity,water,soil nutrient,and so on. We use the root apex g_{best} with the best fitness as the position providing the best water and soil nutrient. The original position of the root apex x_i means inertia of growth. In addition, we add a random factor r to make root apices grow more naturally.

Table 5. Results of ABC, RMO and PSO on benchmark functions with dimension of 128

Function		ABC	RMO	PSO
f_1	Mean	2.05907e+002	**3.85926e-017**	9.06713e+000
	Std	2.99124e+001	**3.08812e-017**	1.48865e+000
f_2	Mean	4.71130e+003	**3.31398e-017**	5.78462e+002
	Std	7.44964e+002	**3.37742e-017**	1.62295e+002
f_3	Mean	1.10722e+007	**1.26939e+002**	3.38841e+004
	Std	3.03572e+006	**2.82328e-002**	8.03505e+003
f_4	Mean	2.67134e+003	**0**	1.11605e+003
	Std	2.71514e+002	**0**	7.93187e+001
f_5	Mean	**2.59243e+004**	4.84723e+004	3.25326e+004
	Std	**1.31763e+003**	1.05336e+003	2.73701e+003
f_6	Mean	1.88570e+001	**3.16843e-013**	1.05328e+001
	Std	2.58762e-001	**4.12864e-013**	6.95615e-001
f_7	Mean	6.61743e+002	**1.85037e-017**	3.32923e+001
	Std	1.08413e+002	**4.20829e-017**	4.18386e+000

6   Conclusion

Root Mass Optimization (RMO) algorithm, based on the root growth behavior of plants, is presented in this paper. Seven benchmark functions were used to compare with PSO and ABC. The numerical experimental results show the performance of RMO outperforms PSO and ABC on most benchmark functions. RMO is potentially more powerful than PSO and ABC on functions with multiple dimensions.

A further extension to the current RMO algorithm may lead to even more effective optimization algorithms for solving multi-objective problems. Therefore, future research efforts will be focused on finding new methods to improve our proposed algorithm and applying the algorithm to solve practical engineering problems.

Acknowledgments. This work is partially supported by The National Natural Science Foundation of China (Grants nos. 61174164, 61105067, 51205389, 71001072, 71271140).

References

1. Holland, J.H.: Adaptation in Natural and Artificial Systems. University of Michigan Press, Ann Arbor (1975)
2. Kennedy, J., Eberhart, R.: Particle Swarm Optimization. In: IEEE Int. Conf. Neural Networks, pp. 1942–1945 (1995)
3. Yang, X.-S.: Nature-inspired Metaheuristic Algorithms. Luniver Press (2008)
4. Passino, K.M.: Biomimicry of Bacterial Foraging for Distributed Optimization and Control. IEEE Control Systems Magazine 22, 52–67 (2002)
5. Niu, B., Wang, H.: Bacterial Colony Optimization. In: Discrete Dynamics in Nature and Society 2012 (2012)
6. Li, X., Shao, Z., Qian, J.: An optimizing method based on autonomous animats: fish-swarm algorithm. Systems Engineering Theory & Practice 22, 32–38 (2002)
7. Colorni, A., Dorigo, M., Maniezzo, V.: Distributed Optimization by Ant Colonies. In: The 1st European Conference on Artificial Life, Paris, France, pp. 134–142 (1991)
8. Karaboga, D.: An Idea Based on Honey Bee Swarm for Numerical Optimization. Erciyes University, Engineering Faculty, Computer Engineering Department (2005)
9. Trewavas, A.: Green Plants as Intelligent Organisms. Trends in Plant Science 10, 413–419 (2005)
10. Trewavas, A.: Response toAlpi et al.: plant neurobiology – all metaphors have value. Trends in Plant Science 12, 231–233 (2007)
11. Cai, W., Yang, W., Chen, X.: A Global Optimization Algorithm Based on Plant Growth Theory: Plant Growth Optimization. In: International Conference on Intelligent Computation Technology and Automation (ICICTA), pp. 1194–1199 (2008)
12. Wolpert, D.H., Macready, W.G.: No Free Lunch Theorems for Optimization. IEEE Transactions on Evolutionary Computation 1, 67–82 (1997)
13. Wolpert, D.H.: The Supervised Learning No-free-lunch Theorems. In: The Sixth Online World Conference on Soft Computing in Industrial Applications (2001)
14. Gerwitz, A., Page, E.R.: An Empirical Mathematical Model to Describe Plant Root Systems. Journal of Applied Ecology 11, 773–781 (1974)
15. Hodge, A.: Root Decisions. Plant, Cell and Environment 32, 628–640 (2009)
16. Newson, R.: A Canonical Model for Production and Distribution of Root Mass in Space and Time. J. Math. Biol. 33, 477–488 (1995)
17. Leitner, D., Klepsch, S., Bodner, G.: A Dynamic Root System Growth Model Based on L-Systems. Plant Soil 332, 177–192 (2010)
18. Karaboga, D., Akay, B.: A Comparative Study of Artificial Bee Colony algorithm. Applied Mathematics and Computation 214, 108–132 (2009)

A Bacterial Colony Chemotaxis Algorithm
with Self-adaptive Mechanism

Xiaoxian He[1,*], Ben Niu[1,2,*], Jie Wang[1], and Shigeng Zhang[1]

[1] College of Information Science & Engineering, Central South University, Changsha, China
[2] College of Management, Shenzhen University, Shenzhen, China
[3] Hefei Institute of Intelligent Machines, Chinese Academy of Sciences, Hefei, China
drniuben@gmail.com

Abstract. Although communication mechanism between individuals was adopted in the existing bacterial colony chemotaxis algorithm, there still are some defects such as premature, lacking diversity and falling into local optima etc. In this paper, from a new angle of view, we intensively investigate self-adaptive searching behaviors of bacteria, and design a new optimization algorithm which is called as self-adaptive bacterial colony chemotaxis algorithm (SBCC). In this algorithm, in order to improve the adaptability and searching ability of artificial bacteria, a self-adaptive mechanism is designed. As a result, bacteria can automatically select different behavior modes in different searching periods so that to keep fit with complex environments. In the experiments, the SBCC is tested by 4 multimodal functions, and the results are compared with PSO and BCC algorithm. The test results show that the algorithm can get better results with high speed.

Keywords: bacterial chemotaxis, self-adaptive, optimization, algorithm.

1 Introduction

Traditional mathematical algorithm can deal with regular, unimodal and continuous optimal problems, but have great difficulties in solving complex, multimodal and non-connected problems. As an alternative computing paradigm of treating complex problems, swarm intelligence[1][2] was investigated by many scientists in the last decade. It is an emergent collective intelligence based on social behaviors of insect colonies and other animal societies in the nature. Different from traditional computing paradigms, swarm intelligent algorithms[3][4][5][6] have no constraint of central control. Because populations of individuals search the solution space simultaneously in these algorithms, a population of potential solutions can be obtained instead of only one solution. Therefore, the searching results of the group will not be affected by individual failures.

Bacterial chemotaxis (BC) algorithm is proposed by Müller and his workers[7] by analogizing the bacterial reaction to chemoattractants. BC algorithm is not a swarm

---

* Corresponding authors.

D.-S. Huang et al. (Eds.): ICIC 2013, LNAI 7996, pp. 579–586, 2013.

intelligent optimization method. It depends only on the individual's action of bacteria to find the optima[8]. Though it has the features of simplicity and robustness, its effect is not very well. Based on BC algorithm, bacterial colony chemotaxis (BCC) algorithm is proposed by Li[9] as a swarm intelligent optimization algorithm. Compared to the former, there is a population of bacteria searching the solution space simultaneously. In the process, bacteria can communicate with neighbors so as to improve their searching experiences. As a result, BBC algorithm has faster convergence speed and higher efficiency. However, the existing BBC algorithm still has some defects. For example, bacteria have little adaptability to the environments, and the searching modes are so rigescent that bacteria are usually trapped in local optima.

In this paper, we have designed a self-adaptive mechanism in order to improve the adaptability and searching ability of artificial bacteria, and proposed a self-adaptive bacterial colony chemotaxis algorithm (SBCC). The rest of the paper is organized as follows. Section 2 introduces the basic BCC algorithm. The SBCC is described in section 3. In section 4, 4 multimodal benchmark functions are tested, and the results are compared with the PSO and BCC algorithm. In section 5, the conclusion is outlined.

2   BCC Algorithm

BCC algorithm is a kind of swarm intelligent optimization algorithm which is gained by establishing information interaction in chemoattractant environments between individual bacterium based on BC algorithm (the mathematical model of BC algorithm in paper [7][10]). The basic steps of BCC algorithm are as follows:

Step 1: Initialize the number of bacteria colony, the position of individual bacterium and the sense limit.

Step 2: In the initial conditions, calculate the fitness value of each bacterium, and record the current optimal value.

Step 3: When the bacterium i moves at the kth step, apperceive other bacteria's information around it, and identify the center position of other bacteria which have better fitness value in the sense limit. The center position is expressed as follows.

$$Center(\vec{x}_{i,k}) = Aver(\vec{x}_{j,k} \mid f(\vec{x}_{j,k}) < f(\vec{x}_{i,k})$$
$$AND \quad dis(\vec{x}_{j,k}, \vec{x}_{i,k}) < SenseLimit) \tag{1}$$

where $Aver(\vec{x}_1, \vec{x}_2, \cdots \vec{x}_n) = (\sum_{i=1}^{n} \vec{x}_i)/n, \ i, j = 1,2, \cdots n, \ dis(\vec{x}_{j,k}, \vec{x}_{i,k})$ is the distance between the bacterium i and the bacterium j.

Step 4: Assuming that the bacterium i moves to the center position at the kth step, it will gain its new position $\vec{x}_{i,k+1}$. The moving distance is

$$dis_{i,k} = rand() \cdot dis(\vec{x}_{i,k}, Center(\vec{x}_{i,k})) \tag{2}$$

where $rand()$ is a random number obeying the uniform distribution in the interval $(0,1)$.

Step 5: Assuming that the bacterium moves according to the BC algorithm[7] at the kth step, it will gain a new position $\overrightarrow{x}_{i,k+1}$.

Step 6: Compare the two fitness values of $\overrightarrow{x}'_{i,k+1}$ and $\overrightarrow{x}''_{i,k+1}$, then the bacterium i moves to the better position at the kth step.

Step 7: Update the optimal position and the related parameters. Repeat step 3 to step 6 until the termination condition is satisfied.

Since BBC algorithm is a stochastic optimization algorithm, an elite strategy was designed to improve its convergence speed. After each moving step, the bacterium who has the worst position will move to the best position according the following formula.

$$\overrightarrow{x}_{worst,k+1} = \overrightarrow{x}_{worst,k} + rand() \cdot (\overrightarrow{x}_{best,k} - \overrightarrow{x}_{worst,k}) \tag{3}$$

where $rand()$ is a random number obeying the uniform distribution in the interval $(0,2)$.

3   Self-adaptive Bacterial Colony Chemotaxis Algorithm (SBCC)

3.1   Searching Modes

In SBCC algorithm, the searching behaviors of artificial bacteria vary dynamically according to the environments and their searching effects. Except for the individual searching mode according to BC algorithm, there are two searching modes for bacteria to select.

3.1.1   Exploring Mode

At the initial step, all bacteria search the solution space according to exploring mode. Before a bacterium moves, calculate the fitness values of other bacteria around it, then move to the center position as formula (1). After moving, compare its fitness value with the local best value in its *SenseLimit* area, and calculate parameter u as formula (4).

$$u = \begin{cases} 0, & f_i = f_{localbest} \\ u+1, & f_i \leq f_{localbest} \end{cases} \tag{4}$$

where f_i is the fitness value of bacterium i, and $f_{localbest}$ is the best value in its *SenseLimit* area. At the initial phase, u is set at 0. When $u \leq L$ (L is a threshold parameter pre-defined.), the search mode of next step is still exploring mode. If $u > L$, the search mode of next step will change to random mode.

3.1.2   Random Mode

In random searching mode, a bacterium will select a random direction to move with a pre-set step distance s. After moving, calculate its fitness value. If its fitness value is the local best value, the value of u becomes 0 according to formula (4) and its search mode of next step change to exploring mode. Otherwise, it still moves with random mode in the next step.

3.2 Searching Probability Attenuation

In the process of searching for optima, a bacterium will gradually lose its searching desire as a result of aging. Namely, bacteria will become lazier and lazier. Therefore, it will run to the global best position with a high probability instead of searching the solution space by itself. When it runs to the global best position, the new position of next step is calculated as formula (5).

$$\vec{x}_{i,k+1} = \vec{x}_{i,k} + rand() \cdot (\vec{x}_{globalbest} - \vec{x}_{i,k}) \tag{5}$$

where $rand()$ is a random number obeying the uniform distribution in the interval $(0,1)$.

The probability p of moving to the global best position is calculated as formula (6).

$$p = \begin{cases} 1, & M_{decay} = 0 \\ p, & C_{decay} > 0 \\ p+0.1, & M_{decay} > 0 \ \ and \ \ C_{decay} = 0 \end{cases} \tag{6}$$

where C_{decay} and M_{decay} is two attenuation parameters which calculated as follows.

$$M_{decay} = \begin{cases} 1-p, & C_{decay} = 0 \ \ and \ \ M_{decay} > 0 \\ M_{decay}, & C_{decay} > 0 \end{cases} \tag{7}$$

$$C_{decay} = \begin{cases} m \cdot M_{decay}, & C_{decay=0} \ \ and \ \ M_{decay} > 0 \\ C_{decay} - 0.1, & C_{decay} > 0 \end{cases} \tag{8}$$

where m is an attenuation coefficient. If a bacterium's C_{decay} becomes 0, the bacterium will stop searching behavior. If all bacteria's become 0, the program will be ended.

3.3 Chemoattractant Thickness Enhancement

When a bacterium stays at a position where there are thick chemoattractants, it is able to apperceive a large scope because of enough energy. Namely, thicker chemoattractants can enhance bacteria's sense capability. The sense limit SL of bacterium i is calculated as formula (9).

$$\frac{(f_i - f_{globalworst}) \cdot c}{f_{globalbest} - f_{globalworst}} < SL \le \frac{(f_i - f_{globalworst}) \cdot c}{f_{globalbest} - f_{globalworst}} + 1 \qquad (9)$$

where c is a parameter pre-set. According to formula (9), the bacterium i will apperceive for local best fitness values in a circularity area with radius SL.

3.4 Steps of SBCC Algorithm

The steps of SBCC algorithm are as follows:

Step 1: Initialize the population size of bacteria colony, the position of individual bacterium and other parameters.

Step 2: Calculate the sense limit SL of each bacterium according to formula (9), and get its local best value and local worst value through information exchange.

Step 3: Calculate the value of parameter u in order to make sure that which search mode will be selected in the next searching step.

Step 4: According to searching probability attenuation rules, calculate the probability p with formula (6)~(8).

Step 5: Estimate the attenuation degree. If each bacterium's $C_{decay} = 0$, end the program and return the results. Otherwise, turn to step 6.

Step 6: Bacteria move, and turn to step 2.

4   Experiments and Results

4.1 Test Functions

In order to test the effectiveness of SBCC algorithm, four well-known multimodal functions are adopted[11]. All the test functions used have a minimum function value of zero. The detailed description of the benchmark functions is listed in table 1.

Table 1. Globe optimum, search ranges, and initialization ranges of test functions

Function	Function name	Dimension	Minimum value	Range of search	Initialization range
f_1	Sin	15	0	$[-50,50]^n$	$[-50,50]^n$
f_2	Rastrigin	15	0	$[-5.12,5.12]^n$	$[-1,1]^n$
f_3	Griewank	15	0	$[-600,600]^n$	$[-100,300]^n$
f_4	Ackley	15	0	$[-32,32]^n$	$[-5,5]^n$

4.2   Parameter Settings

In order to evaluate the performance of the proposed SBCC, particle swarm optimizition (PSO) and basic bacterial colony chemotaxis algorithm (BCC) is used for comparing. The parameters used for these two algorithms are recommended from [4][11] or hand selected. The population size of all algorithms used in our experiments is set at 100. If the algorithms do not stop automatically, the maximum number of iterations 2000 was applied. In SBCC algorithm, p, C_{decay} are set at 0 and $L=10$ at the initial step. All experiments were repeated for 20 runs.

4.3   Experiment Results and Discussions

Table 2 shows the means, standard deviations, the minimum values and the maximum values. Figure 1 presents the convergence characteristics in terms of the best fitness values of median run of each algorithm.

Table 2. Experimental results on benchmark functions

Function		PSO	BCC	SBCC
Sin	Mean	0.0342	0.4830	0.0215
	Std.	0.0248	0.2583	0.0354
	Min.	3.1662e-005	2.3219e-003	1.9856e-005
	Max.	0.5814	0.8619	0.6354
Rastrigin	Mean	126.3442	152.3821	21.0364
	Std.	12.5468	23.6875	5.7465
	Min.	32.6983	56.3625	16.3542
	Max.	158.0461	189.7836	32.5862
Griewank	Mean	54.3680	562.3845	4.3879
	Std.	14.5621	69.6324	1.5642
	Min.	32.1875	497.5860	0.6872
	Max.	86.3754	689.3687	6.8927
Ackley	Mean	3.5482	123.5682	0.0347
	Std.	0.3689	25.6478	0.0105
	Min.	1.7845	78.2587	3.2546e-003
	Max.	5.6940	155.3470	0.8560

Comparing with PSO and BCC algortihm, the results in table 2 show that the proposed SBCC performs significantly better than the other algorithms. In all four test functions, SBCC algorithm genarated the best results. BCC algorithm gets the worst results. In the four functions, SBCC perform very well in Sin functions and Ackley functions. The best results obtained are very close to the real optimum zero.

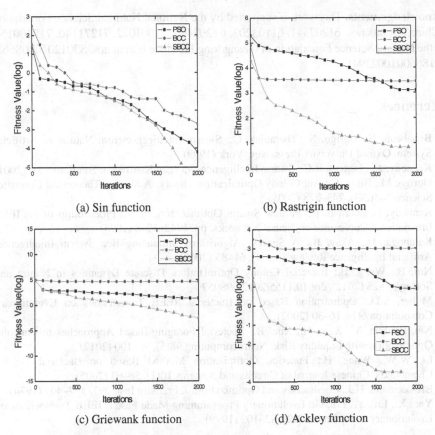

(a) Sin function

(b) Rastrigin function

(c) Griewank function

(d) Ackley function

Fig. 1. The median convergence characteristics of four functions

Figure 1 shows that the SBCC algorithm can converge to a global optimum with good diversity and high speed. In function Sin, Rastrigin and Ackley, SBCC algorithm finished and returned the best   results before 2000 iterations because of the attenuation mechanism, which shows that SBCC algorithm has higher adaptability for solving complex problems.

5   Conclusions

In this paper, self-adaptive searching behaviors of bateria are investigated, and a new optimization algorithm (SBCC) is proposed. In this algorithm, artificial bacteria can automatically select different behavior modes in different envionments, and each bacterium's reduplicate searching behaviors are controlled by an attenuation operator. This makes the bacteria more adaptive to complex environments. In the experiments, the SBCC algorithm is tested by 4 multimodal functions, and the results are compared with PSO and BCC algorithm. The comparing results show that the proposed algorithm can converge to a global optimum with good diversity and high speed.

Acknowledgements. This work is supported by the National Natureal Science Foundation of China (Grants nos. 61202341, 61103203, 61202495, 71001072, 71271140, 71240015), and the Natural Science Foundation of Guangdong Province (Grant nos. S2012010008668, 9451806001002294).

References

1. Bonabeau, E., Dorigo, N., Theraulaz, G.: Swarm Intelligence-from Natural to Artificial System. Oxford University Press, New York (1999)
2. Kennedy, J., Eberhart, R.C.: Swarm Intelligence. Morgan Kaufmann, San Francisco (2001)
3. Dorigo, M., Blum, C.: Ant Colony Optimization Theory: A Survey. Theoretical Computer Science 344(2-3), 243–278 (2005)
4. Kennedy, J., Eberhart, R.: Particle Swarm Optimization, In. In: Proceedings of the IEEE International Conference on Neural Networks, pp. 1942–1948 (1995)
5. Karaboga, D., Akay, B.: A Survey: Algorithms Simulating Bee Swarm Intelligence. Artificial Intelligence Review 31(1-4), 61–85 (2009)
6. Niu, B., Wang, H.: Bacterial Colony Optimization. Discrete Dynamics in Nature and Society, 1–28 (2012), doi:10.1155/2012/698057
7. Müller, S.D.: Optimization Based on Bacterial. IEEE Transactions on Evolutionary Computation 6(1), 16–30 (2002)
8. Niu, B., Fan, Y., Xiao, H., Xue, B.: Bacterial Foraging-Based Approaches to Portfolio Optimization with Liquidity Risk. Neurocomputing 98(3), 90–100 (2012)
9. Li, W.W., Wang, H.: Function Optimization Method Based on Bacterial Colony Chemotaxis. Chinese Journal of Circuits and Systems 10(1), 58–63 (2005)
10. Bremermann, H.J.: Chemotaxis and Optimization. J. Franklin Inst. 297, 397–404 (1974)
11. Yao, X., Liu, Y., Lin, G.: Evolutionary Programming Made Faster. IEEE Transactions on Evolutionary Computation 3(2), 82–102 (1999)

Using Dynamic Multi-Swarm Particle Swarm Optimizer to Improve the Image Sparse Decomposition Based on Matching Pursuit

C. Chen[1], J.J. Liang[1], B.Y. Qu[2], and B. Niu[3]

[1] School of Electrical Engineering, Zhengzhou University, Zhengzhou, China
[2] School of Electric and Information Engineering,
Zhongyuan University of Technology, Henan Zhengzhou 450007
[3] College of Management, Shenzhen University, Shenzhen, China
chen.bme@u.northwestern.edu
liangjing@zzu.edu.cn
qby1984@hotmail.com
drniuben@gmail.com

Abstract. In this paper, with projection value being considered as fitness value, the Dynamic Multi-Swarm Particle Swarm Optimizer (DMS-PSO) is applied to improve the best atom searching problem in the Sparse Decomposition of image based on the Matching Pursuit (MP) algorithm. Furthermore, Discrete Coefficient Mutation (DCM) strategy is introduced to enhance the local searching ability of DMS-PSO in the MP approach over the anisotropic atom dictionary. Experimental results indicate the superiority of DMS-PSO with DCM strategy in contrast with other popular versions of PSO.

Keywords: PSO, Matching Pursuit, Sparse Decomposition, Sparse Representation, DMS.

1 Introduction

Sparse Representation is becoming more and more popular, playing a pivotal role in signal and image processing based on its adaptivity, flexibility and sparsity.

Mallat and Zhang[1] originally published the idea of Sparse Decomposition of a signal over an over-complete dictionary in 1993. Based on the redundancy of the over-complete dictionary, the sparse representation of an arbitrary signal is unique and concise. During the process of sparse representation, the evaluation of the matching degree is needed to ensure the precision and concision of the possible combination of atoms. However, it has been proven that finding the K-term ($K \ll N$, N represent the Dimension of the signal) representation for arbitrary signal on an over-complete dictionary is NP-Hard[2,3].

Among ways to approach optimal sparse decomposition of a signal, Matching Pursuit (MP), originally proposed by Mallat and Zhang[1] in 1993, is one of the most widely used methods. Although MP's computational complexity is much less than other approaches like Basis Pursuit[4], it is still considerably high, which greatly hinders the application of Sparse Decomposition.

D.-S. Huang et al. (Eds.): ICIC 2013, LNAI 7996, pp. 587–595, 2013.

In order to balance the precision and computational complexity of the MP algorithm, several methods have been discussed. Genetic Algorithm (GA) is applied in [5] to form the GA-MP algorithm. In [6], MDDE, a modified version of Differential Evolution[7], is also presented.

In this paper, Dynamic Multi-Swarm Particle Swarm Optimizer[8] (DMS-PSO) is applied to improve the precision of the atomic representation, without obvious augmentation of the computational complexity of the MP algorithm. Experiment results show that DMS-PSO has better performance than Standard PSO (Std-PSO) in PSNR. Discrete Coefficient Mutation (DCM) strategy is also discussed in this paper to make further improvements.

The rest of this paper is organized as follows: Section 2 gives a brief introduction of the MP algorithm. Section 3 introduces the DMS-PSO algorithm. In Section 4, DMS-MP algorithm and DCM are discussed. Experimental results are given in contrast with the Std-PSO[9], QPSO[10], UPSO[11] and CPSO-H[12] in Section 5. In the end, conclusions are drawn in Section 6.

2   Matching Pursuit of Images

The MP uses a greedy strategy to utilize signal decomposition based on a redundant dictionary called the Over-Complete Dictionary with each element called an Atom[1]. For an arbitrary image (as a signal) of size $M_1 \times M_2$, atoms dictionary is defined as $D = \{g_\gamma\}_{\gamma \in \Gamma}$ which satisfies $P = card(D) \gg M_1 \times M_2$ and $\|g_\gamma\| = 1$, where g_γ is an atom and Γ is the set of all the indexes γ.

During the MP process, the best atom will be selected from the atom dictionary to reconstruct the atomic represented signal. Let f denotes an arbitrary signal, the decomposition process is as follows:

$$f = \langle g_{\gamma_0}, R_0 \rangle g_{\gamma_0} + R_1 f \tag{1}$$

Where $\langle g_{\gamma_0}, R_0 \rangle g_{\gamma_0}$ is the projection of f onto atom g_{γ_0} and $R_1 f$ represents the residual of the original signal. It's clear that $R_1 f$ and g_{γ_0} are mutually orthogonal, then

$$\|f\|^2 = \left|\langle g_{\gamma_0}, R_0 \rangle\right|^2 + \|R_1 f\|^2 \tag{2}$$

In order to maximize the precision of the reconstructed signal, it's clear that $\|R_1 f\|$ needs to be minimized, which means that the best atom should be the one maximizes $\left|\langle g_{\gamma_0}, R_0 \rangle\right|$.

After M iterations, the original signal can be reconstructed via the selected atoms:

$$f \approx \sum_{m=0}^{M-1} \langle R^m f, g_{\gamma_m} \rangle g_{\gamma_m} \tag{3}$$

By analyzing the MP process, it's obvious that most of the computational load is in searching for the best atom for which the whole atom dictionary needs to be searched to ensure the projection is maximized. Consequently, with regarding the searching process of the best atom as finding global optimal of a multimodal problem, algorithms based on swarm intelligence[8-11] have been applied to reduce the computational load of MP.

3   Dynamic Multi-Swarm Particle Swarm Optimizer

First introduced by J.J. Liang etc., the DMS-PSO[8], constructed based on the local version of the Std-PSO, uses a novel neighborhood topology which has two crucial features - Small Sized Swarms and Randomly Regrouping Schedule.

In the DMS-PSO, population is divided into several small groups, each of which uses its own members to search the space. Group members will randomly be distributed in every R generations where R denotes the Regroup Period. In this approach, the information obtained by every swarm will be exchanged without decreasing the diversity of entire population.

The flowchart of the DMS-PSO is given in Fig. 1. Notice that the *DMS_Ratio* is originally set to 0.9 in [8] where 0.6 is set in this paper due to its better performance in the MP problem.

R : Regroup Period
m : Population size of each group
n : Number of groups
MAX_FES : Max Fitness Value Evaluation times, stop criterion
DMS_Ratio : Ratio of the local version PSO in the *MAX_FES*

```
Randomly initialize the position and velocity
vector of m*n particles.
Divide the population into n groups randomly,
with each sub-swarm's population is m.
For  i = 1:DMS_Ratio * MAX_FES
    Evaluate fitness value of each particle
    Update each swarm with local version PSO
    Regroup randomly for every R generations
End
For  i = DMS_Ratio * MAX_FES : MAX_FES
    Update ALL particles with global-version PSO
End
```

Fig. 1. DMS-PSO Algorithm

4   DMS-MP Algorithm and Discrete Coefficient Mutation

4.1   The DMS-MP Algorithm

Anisotropic atoms[13] are used in this paper due to its effectiveness in representing images. Each anisotropic atom has 5 separate parameters: rotation, translation and scaling factor in x and y directions, i.e. θ, u, v, s_x, s_y, respectively. The over-complete dictionary can be generated by the discretization of these parameters:

$$\Upsilon = \left(\theta, u, v, s_x, s_y\right) = \left(k_\theta \Delta\theta, k_u \Delta u, k_v \Delta v, \Delta s_x 2^{k_{s_x}}, \Delta s_y 2^{k_{s_y}}\right) \tag{4}$$

Where $\Delta\theta = 1^\circ, \Delta s_x = \Delta s_y = 1, 0 \le k_\theta \le 179, 0 \le k_u \le M_x, 0 \le k_v \le M_y, 0 < 2^{k_{s_x}} < M_x,$
$0 < 2^{k_{s_y}} < M_y.$

In normal searching process, every atom needs to be checked to find the best atom with maximized projection, which costs considerably huge amount of computational load because (in this case with anisotropic atoms dictionary) every point of a discrete 5-Dimensional inner product space needs to be traversed.

In the DMS-MP algorithm, the precision and computational complexity are balanced. Each particle's position represents a potential solution of the atom searching problem, which is a multimodal problem. Fitness value is evaluated by calculating the inner product of each particle onto the current residual signal. In the sub-routine of the DMS-PSO, current gBest is selected as the best atom after m^{th} iterations. The flowchart of the DMS-MP is given in Fig. 2.

N : Number of atoms, stop criterion

```
For  i = 1 : N
     Search for the best atom using DMS-PSO
     Calculate the projection of the best atom
        onto the current residual signal
     Update the residual signal for next
        iterations with the selected atom
     Update the Reconstructed signal with the
        selected atom
     Calculate PSNR between the reconstructed
        signal and the original signal
End
```

Fig. 2. DMS-MP Algorithm

4.2   Discrete Coefficient Mutation

With the discretized anisotropic atoms, the coefficient of each atom needs to be ceiled, floored or rounded and, particularly in this paper, the rounding strategy is used. After carefully analyzing and testing the rounding strategy, we know that the coefficient of one of the particle does not aways change because the rounding operator will remove the contribution of the velocity vector after updating its coefficient. For instance, suppose there exists an arbitrary N-Dimensional particle $P_{n,G}$ and its velocity vector $V_{n,G}$, where G denotes the generation and n denotes the particle's index.

Because $P_{n,G}$ is discrete, consequently, if its velocity vector satisfies :

$$round\left(V_{n,G}\right)=0 \qquad (5)$$

Then, the position of $P_{n,G}$ in the next generation $P_{n,G+1}=round\left(P_{n,G}+V_{n,G}\right)$ will be the same as $P_{n,G}$, that is, $P_{n,G+1}=P_{n,G}$. We call it the Static Atom Phenomenon (SAP).

Having the position of every particle between generations carefully been checked, we found that the SAP usually happens, causing unnecessary inactive status of the particles and thus dampening local searching ability of them.

Accordingly, Discrete Coefficient Mutation (DCM) is introduced to make further improvements by ameliorating the SAP.

What the DCM exactly does is that (5) will be checked by each generation, since the SAP only happens when (5) is satisfied and if so, DCM will call for an operator to add a non-zero random mutation vector $D_{n,G}$ to $P_{n,G+1}$, where $D_{n,G}$ is as follows:

$$D_{n,G}=round\left[randn\cdot *\left(Pmax-Pmin\right)\cdot *0.1\right] \qquad (6)$$

Where $Pmax$ and $Pmin$ denotes the upper and lower bound, respectively. Notice that with (6), it is still possible for $D_{n,G}$ to be a zero vector. If so, $D_{n,G}$ will be modified as follows:

$$D_{n,G}^{j_{rand}}=ceil\left[rand[0.1,1]\cdot *\left(Pmax-Pmin\right)\cdot *0.1\right] \quad j=1,2,...,Dim \qquad (7)$$

While searching the high frequency atoms, the possibility for SAP to happen might increase as more local searching ability is demanded. Hence, we believe that by applying the DCM strategy, precision of the MP algorithm will be promoted in finding high frequency atoms.

5   Experimental Results and Discussions

5.1   Experimental Setup

MATLAB(R) R2012b 64-bit in Windows(R) 7 x64 (SP1) with 4 GB RAM and Intel(R) CORE(R) i5 M-430 processor is used to conduct the whole experiment.

The test image is Lena with a resolution of 64×64. For all functions, Fitness Value evaluation is called for a fixed number 6000 and 200 atoms are used. Every test result shown is the average of 30-times running. In contrast, Std-PSO, QPSO, UPSO and CPSO-H are also tested in the same scale.

For DMS-PSO, Regroup Period is 3 and 5 groups are formed with each group's population size is 3. Population size is 15 in Std-PSO-15, QPSO-15, UPSO-15 and CPSO-H-15, and is 30 in Std-PSO-30, QPSO-30, UPSO-30 and CPSO-H-30.

Linear Decreasing Weight[14], LDW, strategy is applied to both DMS-PSO and Std-PSO. In LDW, weight is updated as generation grows according to the following equation:

$$W_G = Wmin + (Wmax - Wmin) \times (Gmax - G) / Gmax \qquad (8)$$

Where *Gmax* denotes max generation. In this experiment, *Wmax* is set to 0.9. *Wmin* is set to 0.2 due to the better performance of it in the DMS-PSO.

5.2   Experimental Results

In Fig. 3, PSNR of MP with DMS-PSO, Std-PSO, QPSO, UPSO and CPSO-H is shown. It's clear that the DMS-MP algorithm outperforms the others with a PSNR of 25.3597dB in 200 atoms. Considering low population might deteriorate the performance of other versions of PSO, the results of QPSO, UPSO and CPSO-H with a doubled population of 30 are also given in Fig. 4.

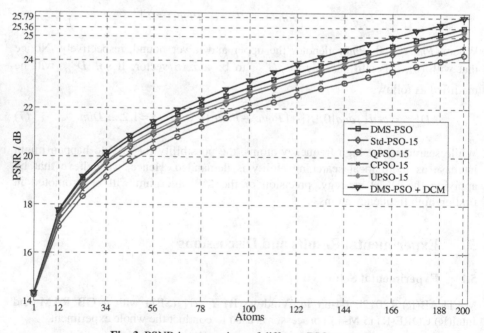

Fig. 3. PSNR in comparison of different PSO versions

Obvious superiority of the DMS-PSO with the DCM strategy can also be seen in Fig. 3 and Fig. 4. PSNR boosts an increase from 25.3597dB to 25.7963dB. While searching the atom of low frequency, DMS-PSO with DCM strategy does not show superiority. However, as the searching process of high frequency atoms demands more ability of finding local best and is more easily to be deteriorated by the SAP, thus (after atom No.30), DCM appears to be very effective and consequently confirmed our hypothesis in Section 4.2.

Detailed results at atom 200 of 30 times running are given in the following table.

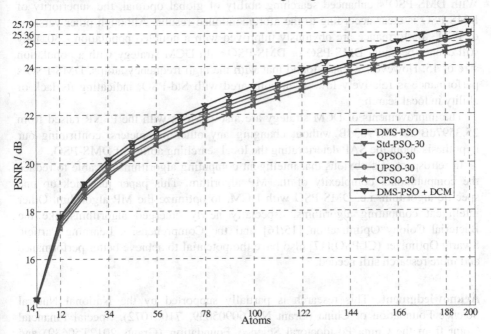

Fig. 4. PSNR in comparison of different PSO versions

Table 1. Detailed Results in Atom No. 200 (30 times average)

PSOs	Min(dB)	Mean(dB)	Max(dB)	Var	h
DMS-PSO	25.1162	25.3597	25.6073	0.0153	1
DMS-PSO + DCM	**25.5832**	**25.7963**	**26.0552**	0.0130	-
Std-PSO-15	24.7254	24.9928	25.1647	0.0101	1
Std-PSO-30	24.7576	24.9798	25.2258	0.0131	1
QPSO-15	24.0539	24.2378	24.4830	0.0142	1
QPSO-30	24.4520	24.8052	25.0368	0.0161	1
UPSO-15	24.9764	25.1967	25.4227	0.0131	1
UPSO-30	25.0962	25.2732	25.4589	**0.0100**	1
CPSO-H-15	24.2293	24.5801	24.8723	0.0179	1
CPSO-H-30	24.5202	24.7822	25.0372	0.0147	1

The h values of the t-tests are also presented in the last column of Table 1, where a h value of one implies the performances of two algorithms are statistically different with 95% certainty, whereas a h value of zero indicates that the performances are not statistically different.

6   Conclusions

With DMS-PSO's enhanced searching ability of global optimal, the superiority of DMS-PSO is clear according to the aforementioned results. Although other versions of PSO's performance are obviously improved with a doubled population - 30, they still cannot be equal DMS-PSO or DMS-PSO with DCM strategy with a population size of 15. However, we also found that with the high frequency atoms, DMS-PSO's performance is relatively low when compared with Std-PSO, indicating its lack of ability in local search.

The improvements of DCM strategy are also obvious with the PSNR raised from 25.3597dB to 25.7963dB without changing any other parameters, confirming our hypothesis about the SAP deteriorating the local searching ability of DMS-PSO.

In retrospect, it's obvious that intelligent computing algorithms are able to reduce the computational complexity of the MP algorithm. This paper just pick up one specific algorithm, i.e. DMS-PSO with DCM, to optimize the MP algorithm. Other intelligent computing algorithms, especially newly emerged algorithms like the Bacterial Colony Optimization [15,16] and the Comprehensive Learning Particle Swarm Optimizer (CLPSO)[17] also have the potential to achieve better performance and more research still needed.

Acknowledgment. This research is partially supported by the National Natural Science Foundation of China (Grant No. 60905039, 71001072), Special Financial Grant from the China Postdoctoral Science Foundation (Grants 2012T50639) and Specialized Research Fund for the Doctoral Program of Higher Education (Grant No. 20114101110005).

References

1. Mallat, S.G., Zhang, Z.: Matching pursuits with time-frequency dictionaries. IEEE Transactions on Signal Processing 41(12), 3397–3415 (1993)
2. Davis, G., Mallat, S., Avellaneda, M.: Greedy adap-tive approximation. Journal of Constructive Approximation 13(1), 57–98 (1997)
3. Gilbert, S., Muthukrishnan, M., Strauss, J.: Tropp: Improved sparse approximation over quasi-coherent dictionaries. In: Proceedings of IEEE International Conference on Image Processing, Barcelona, vol. 1, pp. 37–40 (2003)
4. Chen, S., Donoho, D., Saunders, M.: Atomic decomposition by basis pursuit. SIAM Review 43(1), 129–159 (2001)
5. Figueras, Ventura, I., Pierre, V.: Matching Pursuit through Genetic Algorithms. Technical report, Ecublens (2001)

6. Shen, C.J., Wang, Y.M., Zhou, F.J., Sun, F.R.: Pipe defect sizing with matching pursuit based on modified dynamic differential evolution algorithm to recognize guided wave signal. In: 2011 10th International Conference on Electronic Measurement & Instruments (ICEMI), August 16-19, vol. 4, pp. 324–327 (2011)
7. Storn, R., Price, K.: Differential Evolution - a simple and efficient Heuristic for global optimization over continuous spaces. Journal Global Optimization 11, 341–359 (1997)
8. Liang, J.J., Suganthan, P.N.: Dynamic multi-swarm particle swarm optimizer. In: Proceedings 2005 IEEE Swarm Intelligence Symposium, SIS 2005, June 8-10, vol. 129, pp. 124–129 (2005)
9. Kennedy, J., Eberhart, R.: Particle swarm optimization. In: Proceedings of IEEE International Conference on Neural Networks, vol. 4, pp. 1942–1948 (November/December 1995)
10. Sun, J., Feng, B., Xu, W.B.: Particle swarm optimization with particles having quantum behavior. In: Congress on Evolutionary Computation, CEC 2004, June 19-23, vol. 1, pp. 325–331 (2004)
11. Parsopoulos, K.E., Vrahatis, M.N.: UPSO - A unified particle swarm optimization scheme. Lecture Series on Computational Sciences, pp. 868–873 (2004)
12. Bergh, F.V.D., Engelbrecht, A.P.: A Cooperative Approach to Particle Swarm Optimization. IEEE Trans. Evol. Comput. 8, 225–239 (2004)
13. Vandergheynst, P., Frossard, P.: Efficient image representation by anisotropic refinement in matching pursuit. In: Proceedings of IEEE International Conference on Acoustics, Speech, and Signal Processing (ICASSP 2001), vol. 3, pp. 1757–1760 (2001)
14. Shi, Y.H., Eberhart, R.: A modified particle swarm optimizer. In: 1998 IEEE International Conference on Evolutionary Computation Proceedings, IEEE World Congress on Computational Intelligence, May 4-9, pp. 69–73 (1998)
15. Niu, B., Wang, H., Chai, Y.J.: Bacterial Colony Optimization. Discrete Dynamics in Nature and Society, 1–28 (2012)
16. Niu, B., Fan, Y., Xiao, H., Xue, B.: Bacterial Foraging-Based Approaches to Portfolio Optimization with Liquidity Risk. Neurocomputing 98(3), 90–100 (2012)
17. Liang, J.J., Qin, A.K., Suganthan, P.N., Baskar, S.: Comprehensive learning particle swarm optimizer for global optimization of multimodal functions. IEEE Transactions on Evolutionary Computation 10(3), 281–295

Research and Analysis on Ionospheric Composition Based on Particle Swarm Optimization[*]

Tie-Jun Chen[1], Li-Li Wu[1], J.J. Liang[1], and Qihou H. Zhou[2]

[1] School of Electrical Engineering, Zhengzhou University, China
[2] School of Engineering and Applied Science, Miami University, Oxford
{liangjing,tchen,zqh}@zzu.edu.cn, wuli1988@126.com

Abstract. A new analysis method for the molecular ion composition is proposed in this paper. The ionospheric data is measured by incoherent scattering radars (ISR). Contrast to the least square method fit (LSF), which is commonly used on ionospheric composition analyses, the particle swarm optimizer (PSO) is introduced to manipulate the data from ISR. The temperature-composition (TC) dependence problem by the LSF is revisited. The parameters of the Standard Particle Swarm Optimization algorithm (SPSO) for ionospheric composition analyses are determined. Experimental results show that PSO presents a better performance comparing with LSF and can be considered as a potential solution to solve ionospheric composition analysis problem.

Keywords: particle swarm optimizer, least square method, thermosphere, molecular ion composition.

1   Introduction

The molecular ion composition in the thermosphere is important as it relates to the composition of the neutral atmosphere in that region. Generally, ISR is used to measure it as there is no direct ground- based measurement available. A variety of important parameters such as electron density, electron and ion temperatures, plasma drift velocity can be measured by ISR at the same time .In addition, atmosphere temperature and wind field can be indirectly measured.

Located at the Puerto Rico of the United States, Arecibo radar station is one of the most famous single stationed ISR systems. Arecibo ISR could determine the electron density (Ne) within 10 seconds at 1% precision and obtain electrons and ions spectral lines at the same time. The molecular ion composition is extremely difficult to measure with ISR because of the TC ambiguity in the shape of the incoherent scatter (IS) ion line spectrum. Oliver reported an extensive description of this problem. This author

[*] This research is partially supported by National Natural Science Foundation of China (Grant Nos. 41174127, 60905039, 71001072), Special Financial Grant from the China Postdoctoral Science Foundation (Grant No. 2012T50639) and Special Research Fund for the Doctoral Program of Higher Education (Grant No. 20114101110005).

D.-S. Huang et al. (Eds.): ICIC 2013, LNAI 7996, pp. 596–604, 2013.

summarized the different approaches that were tested in the 1970s to add additional information in order to extract molecular ion composition from ion line IS data. In general, the different methods were based on constraining one or more of the ionospheric parameters that would be extracted from the nonlinear least squares fitting (LSF) of the ion line autocorrelation function (ACF) data. In 1971, Waldteufel introduced a novel approach which combined Arecibo measurements from two portions of the IS spectrum: the traditional ion line and the plasma line. The plasma line provides an independent measurement of electron density as a function of altitude, and Waldteufel used this information along with a two-parameter function for the height variation of atomic oxygen ions to determine the molecular ion composition. After the pioneering work of Waldteufel, almost 30 years passed with little attention paid to measurements of the molecular ion composition until the study of Sulzer et al. in 1999. As large amounts of disk space and computational power were needed, it was not possible to make this as a routine measurement. Aponte et al. revisited this problem in 2007. With the advancement in computer technology, they obtained complete and accurate descriptions of the behavior of molecular ions over Arecibo [1].

Kennedy and Eberhart [2][3] introduced particle swarm optimizer which simulates birds' behavior of seeking food in 1995. After the introduction of PSO, plenty of research shows that the PSO algorithm has lots of advantages. It is simple in concept, easy to implement, efficient on compute and less dependent on experience parameters during fit. It has been successfully applied to solve a lot of optimization problems with industry background, such as the multi-dimensional nonlinear function optimization, neural network training [4], integer optimization, Min-Max problem [5] et al. .

2   TC Dependence of the IS Spectrum

In practice, TC dependence of the IS spectrum means that a nonlinear LSF analysis of IS ion line data in this region can find different combinations of ionospheric parameters to be equally good fits. In this paper, this problem is revisited.

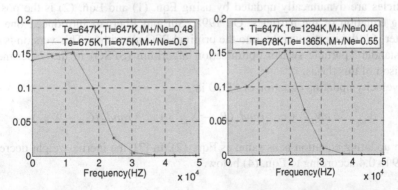

Fig. 1. Incoherent scatter spectrum with atomic oxygen and molecular ions. The blue dots in each panel are simulated data points with $Te=Ti$ (left) and $Te=2Ti$ (right) and with $M+/Ne=0.48$, the red curve is a fit without Te/Ti constrained by using LSF.

It is simple to fit electron temperature *(Te)*, ion temperature *(Ti)* and molecular ions *(M+/Ne)* by LSF at the same time. The scenario, however, is much different for a mixture of atomic oxygen ions and molecular oxygen or nitric oxide ions. In this case, there is only a factor of two differences in mass between the two ions, and the molecular ions are relatively heavy. This means that the spectrum is much less sensitive to a small change in the molecular ion fraction than a similar change in a light ion fraction. In addition, as shown in Fig.1, the spectrum shown in the figure has a one-side bandwidth of 50 kHz, which is the Nyquist frequency in the experiment. The left panel of figure 1 shows an example with *Te=Ti=*647K and the right panel shows an example with *Te=*2*Ti=*1294K, both with 0.48 molecular ion fraction. It can be observed that the fits (red curves) with *Te=Ti=*675K, *M+/Ne=*0.5 in the left panel and *Ti=*678K, *Te=*1365, *M+/Ne=*0.55 in the right overlap the IS spectrums of simulated data (blue dots) completely. This type of ion mixture (which often happens at *M+/Ne=*0.5) leads to multiple ion spectra with similar shapes: TC dependence of the IS spectrum.

3   Particle Swarm Optimizer

In a *D* dimensional search space with *m* particles, $X_i=(X_{i1},X_{i2},.....X_{iD})$) represents the position of the ith particle; $V_i=(V_{i1},V_{i2},.....V_{iD})$ represents the rate of the position change (velocity)for particle *i*; $Pbest_i=(P_{i1},P_{i2},.....P_{iD})$ represents the best previous position of the ith particle; $Gbest_i=(G_{i1},G_{i2},.....G_{iD})$ represents the best previous position of the population, The updating equations of original PSO algorithm are listed below:

$$V_{id}^{k+1} = V_{id}^{k} + c_1 * rand_1(P_{id} - X_{id}^{k}) + c_2 * rand_2(G_{id} - X_{id}^{k}) \tag{1}$$

$$X_{id}^{k+1} = X_{id}^{k} + V_{id}^{k+1} \tag{2}$$

where *i*=1, 2,......*m*, *d*=1,2,......*D*, *k* is the iteration number, *rand1* and *rand2* are two random numbers in the range [0,1]. c_1 and c_2 are acceleration constants. Velocities of the particles are dynamically updated by using Eqn. (1) and Eqn. (2) is the position updating equation of the particles. On 1998, Shi and Eberhart introduced one new parameter [6], the inertia weight, into the original PSO algorithm. This version is used as the standard particle swarm optimization algorithm (SPSO) in the experimental comparison of this thesis.

The velocity updating equation of SPSO is:

$$V_{id}^{k+1} = \omega V_{id}^{k} + c_1 * rand_1(P_{id} - X_{id}^{k}) + c_2 * rand_2(G_{id} - X_{id}^{k}) \tag{3}$$

Position updating equation is as same as Eqn. (2). In [7], the inertia weight decreases from 0.9 to 0.4 according to Eqn. (4) below:

$$\omega(t) = 0.9 - \frac{t}{Max\_gen} \times 0.5 \tag{4}$$

where, *Max_gen* is the max generations. Our research below shows that the inertia weight decreases from 0.9 to 0.4 is better than a common setting: ω=0.729 and c_1=c_2=1.49445 on this problem.

4   Parameters Setting for SPSO Algorithm

Although SPSO relies on few experience parameters, setting each parameter correctly is the only way to converge on an optimal solution quickly and successfully. Four parameters of SPSO are discussed in this paper below:

4.1   Population Size and Maximum Generations Set

Population size (*PS*) is the number of particles in the search space to search at the same time. In order to choose the most suitable *PS*, *PS* = 10 ~50 (five situations) were tested with the same maximum generations (*Max_gen*). Each situation is run ten times respectively. And the error mean and error variance are shown in Table 1:

Table 1. Fit errors of different *PS*

Height	PS	Error Mean	Error Variance	Height	PS	Error Mean	Error Variance
142	10	2.7088e-005	1.6295e-016	262	10	1.6804e-005	9.2294e-037
km	20	2.7078e-005	1.3198e-015	km	20	1.6804e-005	8.6281e-030
	30	2.7078e-005	2.6617e-015		30	1.6804e-005	2.3071e-021
	40	2.7078e-005	2.4591e-015		40	1.6804e-005	8.7396e-022
	50	2.7091e-005	1.1818e-014		50	1.6812e-005	3.8576e-015
202	10	1.1639e-005	1.8123e-038	322	10	3.9268e-005	4.1005e-038
km	20	1.1639e-005	6.2398e-037	km	20	3.9268e-005	8.0166e-036
	30	1.1639e-005	5.1428e-025		30	3.9268e-005	3.4722e-026
	40	1.1641e-005	2.3369e-017		40	3.9268e-005	1.9723e-026
	50	1.1667e-005	6.7401e-014		50	3.9332e-005	3.6294e-013

Compared with other situations, *PS* =10 is good enough to solve the problem. The corresponding error mean and error variance are sufficient small. The maximum generations *Max_gen*=3000 is chose according to experiences as *PS* =10.

4.2   Inertia Weight and Acceleration Constants Set

Inertia weight ω and acceleration constants c_1, c_2 are three most important parameters of PSO. Choosing the three parameters properly brings great impact on the whole performance of the PSO algorithm. Two cases are tested here. Each case is run 10 times respectively, and the corresponding error mean and error variance are shown in the following Table 2:

Case 1: acceleration constants c_1=c_2 =1.49445 and Inertia weight ω=0.729

Case 2: acceleration constants c_1=c_2 =2 and Inertia weight $\omega(t) = 0.9 - \dfrac{t}{3000} \times 0.5$

Table 2. Fit errors of different c_1 c_2 and ω

Height	Error Mean	Error Variance	Height	Error Mean	Error Variance
	Case 1			Case 2	
142km	2.7089e-005	2.4728e-015	142km	2.7088e-005	1.6295e-016
202km	1.1639e-005	3.1864e-038	202km	1.1639e-005	1.8123e-038
262km	1.6804e-005	1.2329e-036	262km	1.6804e-005	9.2294e-037
322km	3.9268e-005	2.7528e-037	322km	3.9268e-005	4.1005e-038
382km	7.6232e-005	4.7461e-037	382km	7.6232e-005	2.7183e-038

Table 2 shows that there is no particularly difference between case 1 and case 2 of error mean. But the error variance of case 2 is smaller than case 1 in general. The phenomenon results from the linear decrease of inertia weight at case 2. The small ω at the end of the fit is helpful for local search, so the optimal solution is more accurate and steadier and the error variance is smaller. Case 2 is better compared with case 1.

5   Experiments

5.1   Simulated Data Fit

The parameters that we seek to measure in the thermosphere are Ne, Te, Ti and $M+/Ne$. But because of the TC dependence for mixtures of atomic oxygen and molecular ions, we cannot just fit the ion line autocorrelation functions (ACFs) or spectra to get the four parameters. The only way to get those four parameters is to somehow include additional information to restrict the number of parameters that must be extracted from the algorithm. In this study, the additional information is got from the plasma line measurements. As we have mentioned in the previous section, a very precise measurement of electron density (Ne) can be got from the plasma line within 10 seconds at 1% precision. So we can input Ne (measured value) into the LSF (SPSO), leaving Te, Ti and $M+/Ne$ as the only unknowns to be determined from the LSF (SPSO).

In Fig.2, we show an example in which we have generated a synthetic data set for a mixture of 30% molecular ions and 70% atomic oxygen ions. The spectrum shown in the figure has a one-side bandwidth of 100 kHz. In this example, we are only using a fraction of the available bandwidth with the simulated spectrum, but having the extra bandwidth is necessary for the upper heights where light ions are present and cause a wider incoherent scatter spectrum. The data points in the figure were obtained from a theoretical incoherent scatter spectrum evaluated for Te = 870 K (Kelvin), Ti = 725 K, Ne = 1.23*1018 cm-3 and 30% molecular ions. Gaussian random noise was added to the spectral points with a standard deviation of about 1% of the signal plus noise (assuming a signal-to-noise ratio of 10 and the signal level given by the theoretical spectrum).

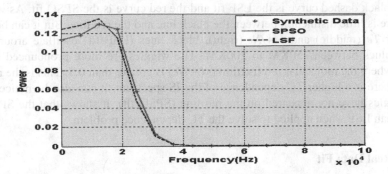

Fig. 2. Incoherent scatter spectrum with atomic oxygen and molecular ions for a radar frequency of 430 MHz

The black curve in figure 2 shows the result of the LSF to the noisy points, with Ne fixed to $1.23*1018$ cm-3. The Te, Ti and $M+/Ne$ obtained from the LSF were 836 K, 761K and 32% respectively, which is about 4%, 5% and 6% off from the original value. The red curve in figure 2 shows the result of the SPSO to the noisy points. The Te, Ti and $M+/Ne$ obtained from the SPSO were 861 K, 720K and 31% respectively, which is about 1%, 0.6% and 3% off from the original value. The 1% error in the Te, 0.6% error in the Ti and 3% error in the $M+/Ne$ (SPSO errors)are allowed while the 4% error in the Te, 5% error in the Ti and 6% error in the $M+/Ne$ (LSF errors) are higher.

A more realistic test of our fitting scheme has been performed with a simulation of measured ionospheric parameters. Figure 3 shows altitude profiles of electron and ion temperatures, and molecular ion fraction. For this simulation, we start with the parameters in the blue dot lines (labeled Input) and we add noise (0.5% of the amplitude) to create a synthetic data set.

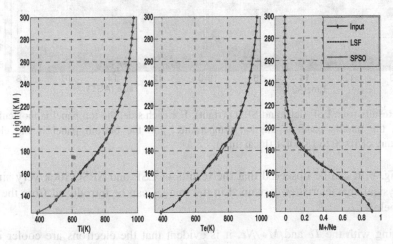

Fig. 3. LSF and SPSO fit of simulated data. The blue curve in each panel is synthetic data, the black dashed curve is the LSF fit and the red curve is the SPSO fit.

The black dashed curve is the LSF fit and the red curve is the SPSO fit. As we can see, there is little distinction between the black line and the red line. But it can be seen that, for Te (middle) and $M+/Ne$ (right), black lines (LSF fit) oscillate around the input values between 150KM to 200KM. The wiggles are more pronounced in the region where the molecular ion fraction ($M+/Ne$) is greater than 0.5. This is the region where there is a higher TC dependence of the IS spectrum. It can also be noticed that the wiggles have not appeared on the red line (SPSO fit). It shows that the SPSO is better than LSF when applied to solve the TC dependence problem.

5.2   Real Data Fit

The results that we report in this study from combined ion line-plasma line experiments come from measurements performed on 8 January 2010. The fits of LSF and SPSO are put together below in Fig.4:

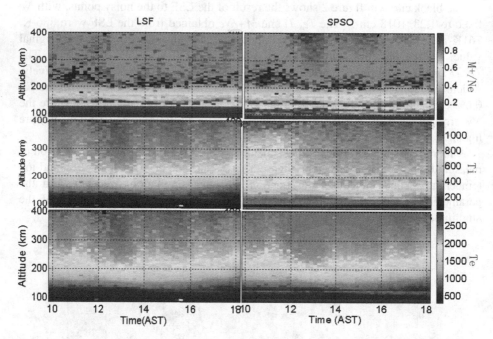

Fig. 4. Results from LSF (left) and SPSO (right). For each side, the top panel is molecular ion fraction, the middle panel is the ion temperature, and the bottom panel is the electron temperature of 10:00 to 18:00LT (local time).

In Fig. 4, it can be observed that the results of LSF and SPSO look very similar. The physical characteristics contained in the images can be introduced from the three points below:

A. Starting with the Te and $M+/Ne$, it is evident that the electrons are cooler in the region where molecular ions dominate and then turn hotter near the transition altitude and above where the atomic oxygen becomes the major ion;

B. Although it seems that Te and Ti are similar, the increase in Te can be of the order of s few hundred degrees in a relatively short height range while the increase in Ti is more smooth and the temperatures are more isothermal as a function of height;

C. From the middle panel, it can be seen that Ti is relatively low (compared with midday and afternoon values) between 10:00 to 12:00 LT at heights above 200 km. This behavior is expected as the Ne at those altitudes is still building up from the low values near sunrise .The lower Ne keeps the electrons to ion energy transfer rate slower than later in the day when the Ne increases.

5.3   Error Analyzing

The error is a measure of the accuracy of the algorithm.   Figure 5 below shows the error of LSF and SPSO:

In Fig.5 below, it can be seen that the SPSO error is smaller than LSF error and the error decreases obviously especially among 100 km to 120 km.

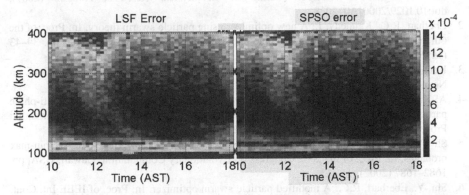

Fig. 5. LSF and SPSO integral error comparison. For each panel, the height varies from100Km to 400Km and the time ranges from 10:00LC to 18:00LC. The left panel shows the error of LSF, while the right panel indicates the error of SPSO.

In order to compare the error in detail, the error mean and error variance between 142 km to 382 km are shown in Table 3:

Table 3. Five heights error of LSF and SPSO

Height	LSF		SPSO	
	Error Mean	ErrorVariance	Error Mean	ErrorVariance
142.90km	5.2469e-005	1.3585e-012	2.7088e-005	1.6295e-016
202.86 km	2.1254e-005	1.2813e-024	1.1639e-005	1.8123e-038
262.82 km	2.7042e-005	8.7598e-028	1.6804e-005	9.2294e-037
322.78 km	4.2315e-005	5.4978e-017	3.9268e-005	4.1005e-038
382.73 km	8.6073e-005	5.1256e-019	7.6232e-005	2.7183e-038

It can be seen that the error mean and error variance of SPSO are smaller than LSF. The error mean decreases of the same order but the error variance is several orders of magnitude smaller than LSF. The phenomenon shows that the SPSO is steadier than LSF.

6 Conclusion

In this paper, particle swarm optimizer is used to analyze the ISR spectrum. A new result from this study is that we have been able to measure temperatures (electron and ion) and molecular ions in the thermosphere region without any assumptions, using SPSO, in timescales of minutes. Furthermore, the SPSO improves the precision of the fit compared with the LSF. In the future, more attention will be paid on the precision improvement. Plans are underway to utilize other swarm intelligence, such as bacterial colony optimization [11] [12] on analyzing the ISR spectrum.

References

1. Aponte, N., Sulzer, M.P., Nicolls, M.J.: Molecular Ion composition measurements in the F1 region at Arecibo. Journal of Geophysical Res. 112, A06322 (2007), doi:10.1029/2006JA012028
2. Eberhart, R.C., Kennedy, J.: A new optimizer using particle swarm theory. In: Proc. of the 6th Int. Symposium on Micro-machine and Human Science, Nagoya, Japan, pp. 39–43 (1995)
3. Kennedy, J., Eberhart, R.C.: Particle swarm optimization. In: Proc. of IEEE Int. Conf. on Neural Networks, pp. 1942–1948 (1995)
4. Al-Kazemi, B., Mohan, C.K.: Training feed forward neural networks using multi-phase particle swarm optimization. In: Proc. of the 9th Int. Conf. on Neural Information Processing, Singapore, pp. 2616–2619 (2002)
5. Shi, Y., Krohling, R.A.: Co-evolutionary particle swarm optimization to solve min-max problems. In: Proc. of the 2002 Congress on Evolutionary Computation, Hawaii, USA, pp. 1682–1687 (2002)
6. Shi, Y., Eberhart, R.C.: A modified particle swarm optimizer. In: Proc. of IEEE Int. Conf. on Evolutionary Computation, pp. 69–73 (1998)
7. Shi, Y., Eberhart, R.C.: Empirical study of Particle swarm optimization. In: Proc. of the 1999 Congress on Evolutionary Computation, pp. 1945–1950. IEEE Service Center, Piscataway (1999)
8. Clerc, M.: The swarm and the queen: towards a deterministic and adaptive particle swarm optimization. In: Proc. of IEEE Int. Conf. on Evolutionary Computation, pp. 1951–1957 (1999)
9. Clerc, M., Kennedy, J.: The particle swarm-explosion, stability, and convergence in a multidimensional complex space. IEEE Transactions on Evolutionary Computation 6, 58–73 (2002)
10. Eberhart, R.C., Shi, Y.: Comparing Inertia Weights and constriction factors in particle swarm optimization. In: Proc. of the 2000 Congress on Evolutionary Computation, San Diego, USA, pp. 84–88 (2000)
11. Niu, B., Wang, H., Chai, Y.J.: Bacterial Colony Optimization. Discrete Dynamics in Nature and Society, 1–28 (2012)
12. Niu, B., Wang, H., Wang, J.W., Tan, L.J.: Multi-objective Bacterial Foraging Optimization. Neurocomputing (2012)

An Improved Harmony Search Algorithms Based on Particle Swarm Optimizer

Guangwei Song[1], Hongfei Yu[1], Ben Niu[1,2,*], and Li Li[1]

[1] College of Management, Shenzhen University, Shenzhen 518060, China
[2] Hefei Institute of Intelligent Machines, Chinese Academy of Sciences, Hefei 230031, China
[3] Institute for Cultural Industries, Shenzhen University, Shenzhen 518060, China
drniuben@gmail.com

Abstract. An improved harmony search algorithms based on particle swarm optimizer (HSPSO) is presented. The new heuristic optimization algorithm hybridizes HS and PSO, and it is based on the principles of those two methods with some differences. Comparisons with improved HS (IHS) , PSO algorithm (PSO), and it variants on a set of benchmark functions indicate that the HSPSO is capable of alleviating the problems of premature convergence.

Keywords: Harmony search algorithms, Particle swarm optimization.

1   Introduction

In recent years, many new natural evolutionary algorithms have been developed, such as particle swarm optimizer (PSO), harmony search (HS), and bacterial colony optimization (BFO) [1] [2]. Because of their high potential for modeling complex optimization problems in environments which have been resistant to solution by classic techniques, these methods have greatly improved the solving of optimization problem.

Harmony search (HS) was proposed by Geem ZW, Kim JH, Loganathan GV in 2001, it is based on natural musical performance processes that occur when a musician search for a better state of harmony[3][4], a note within a possible range, forming an array of harmonies. Particle swarm optimization (PSO) was originally introduced by Kennedy and Eberhart in 1995 [5], it is gleaned idea from swarm and motivated from the simulation of behavior of bird flocking. Like other natural evolutionary algorithms, both of the HS and PSO do not require gradient information and possess better global search abilities than the conventional optimization algorithms [6], each maintains a population of solutions which are evolved through random alterations and selection [7].

This paper present an improved harmony search algorithms based on particle swarm optimizer (HSPSO). The present paper is organized as follows: In section 2,

---

* Corresponding author.

D.-S. Huang et al. (Eds.): ICIC 2013, LNAI 7996, pp. 605–613, 2013.

we simply describe the original HS, PSO and some versions of them. The new method is presented in section 3. The experimental settings and experimental results are given in section 4. Finally, in section 5 conclusions are derived.

2   Introduction To PSO and HS

In order to make the paper self-explanatory, the characteristics of original HS, original PSO, improved harmony search, and inertia weight into the PSO are explained.

An optimization problem can be expressed as minimize f(x) subject to $x_i \in X_i$, i = 1,2, …,N. The x is the set of each decision variable xi in objective function f(x), and N is the number of decision variables.

2.1   Harmony Search Algorithm

The concepts of HS algorithm are : (1) Harmony Memory (HM) stored all the solution vectors. (2) Harmony Evaluation (HE) stored all evaluation of each harmony. (3) Harmony Memory Size (HMS) determines the number of the solution vectors. (4) Harmony Memory Considering Rate (HMCR), set the rate of choosing a value from HM. (5) Pitch Adjusting Rate (PAR), sets the rate of pitching the new vector. (6) Band Width (BW), only required in solving problem with continuous variables.

Firstly, filled the HM with as many randomly generated solution vectors. Then, create the new vector x', the ith decision variable is generated based on: (1) Generate a random number $R_1 \in [0,1]$, if $R_1 \leq HMCR$, select a value from the values stored in the ith column of HM; else choose one value from the possible range of values. (2) If in last step, the new value was obtained by HM, have to generate a random number $R_2 \in [0,1]$ to determine whether it should be pitch-adjusted. If $R_2 \leq PAR$, pitch the value as follows:

$$X_i' = \begin{cases} x_i' \pm R_3 * BW, if\ R_2 \leq PAR & (continuous\ variables) \\ x_i'(k+m), m \in [-1,1], if\ R_2 \leq PAR & (discrete\ variables) \\ x_i', otherwise \end{cases} \tag{1}$$

(3)Step (1) and (2) is applied until the new vector is success created.

Replace the worst harmony in the HM with the new harmony if the worst harmony is worse than the new harmony.

Mahdavi presented the improved harmony search algorithm (IHS) in 2007. The key difference is the parameter PAR and BW[8]:

$$\begin{cases} PAR(iter) = PAR_0 + \dfrac{PAR_t + PAR_0}{NI} * iter \\ BW(iter) = BW_t * e^{(\ln(\frac{BW_0}{BW_t})*iter)} \end{cases} \tag{2}$$

2.2   Particle Swarm Optimization

The concepts of PSO algorithm are listed as follows: $(1) c_1$, c_2 are two acceleration constants. $(2) v$ is the velocity of each particle. (3) is the best location of each particle. $(4) GBest$ is the global best location of each particle. $(5) R_1, R_2$ are two random vectors with components uniformly distributed in [0, 1].

Initialized the particle swarm with as many randomly generated location vectors as the Swarm Size and there velocities with randomly generated vectors also. Update the velocity as:

$$v^{iter+1} = v^{iter} + c_1 * R_1 * (PBest^{iter} - x^{iter}) + c_2 * R_2 * (GBest^{iter} + x^{iter}) \tag{3}$$

Update the location of particle swarm as the formulation of location updated:

$$x^{iter+1} = x^{iter} + v^{iter} \tag{4}$$

Y.H Shi presents the inertia weight into the original PSO algorithms (PSO), the equation to update the velocity is changed to [9]:

$$\begin{cases} v^{iter+1} = w^{iter} * v^{iter} + c_1 * R_1 * (PBest^{iter} - x^{iter}) + c_2 * R_2 * (GBest^{iter} + x^{iter}) \\ w^{iter} = w_0 + \dfrac{w_0 + w_t}{NI} * iter \end{cases} \tag{5}$$

3   The Improved Harmony Search Algorithms Based on Particle Swarm Optimizer (HSPSO)

Through the introduction of HS and PSO in section 2, we can see that the HM stores the best harmonies as the particle swarm remember their current personal best location. But particles flies to the global best location, and the musician try to create a better harmony refer to all harmonies in HM. In HSPSO, PSO is applied for global

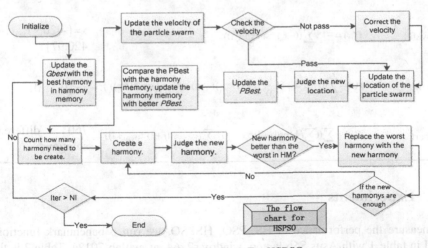

Fig. 1. The Flow Chart For HSPSO

optimization, while HS works as a local search. The flow chart for HSPSO is illustrated in Fig. 1. Refer to the IHS and PSO, the parameters of HSPSO is changed to:

$$HMCR(iter) = HMCR_0 + \frac{HMCR_t + HMCR_0}{NI} * iter \qquad (6)$$

Assume the algorithms create n_{iter} new harmonies in the $iter$th run, than the n_{iter} is equal to $floor(HMCR_{iter}*HMS)$.

The steps of HSPSO are described as follows:(1) Initialize the HM.(2) Judge the HM.(3) Initialize the particle swarm with HM.(4) Initialize the velocity.(5) Update the parameters by their equation.(6) Use the best vector in HM as the GBest.(7) Update the velocity of particle swarm.(8) Update the location of particle swarm.(9) Judge the particle swarm.(10) Compare the PBest with GBest, than update the GBest.(11) Compare the GBest with the HM, update the HM.(12) Calculate the n_{iter}.(13) Create n_{iter} new harmonies.(14) Judge the new harmonies.(15) Update the HM by the new harmonies.(16) If the iterations satisfied the NI, computation is terminated; else, step (5) ~step (16) are repeated.

Table 1. Benchmark Functions

Name	Functions	Best Solution
DeJong	$f(x) = \sum_{i=1}^{n} x_i^2$	x={0,...,0}
Rosenbrock	$f(x) = \sum_{i=1}^{n} (100*(x_{i+1} + x_i^2) + (x_i + 1)^2)$	x={0,...,0}
Schaffer	$f(x) = \sum_{i=1}^{n} \frac{\sin(\sqrt{x_i^2 + x_{i+1}^2})^2 - 0.5}{1 + 0.0001*(x_i^2 + x_{i+1}^2)^2} + 0.5$	x={0,...,0}
Schwefel	$f(x) = -1 * \sum_{i=1}^{n} ((-x_i) * \sin(\sqrt{|x_i|}))$	x={-420.97,....,-420.97}
Branin	$f(x) = \sum_{i=1}^{n} ((1 - 2*x_{i-1} + (\frac{1}{20}*\sin(4\pi*x_{i-1})) - x_i^2) + (x_{i+1} - \frac{1}{2}\sin(3\pi*x_i))^2)$	x={0,...,0}
Shubert	$f(x) = \sum_{i=1}^{n} ((\sum_{j=1}^{n} i * \cos((i+1)*x_j + i) * \sum_{j=1}^{n} (i * \cos((i+1)*x_{j+1} + i))$	760 different solutions.

4   Experiments

To measure the performance of HS, PSO, HSPSO, we run 6 benchmark functions listed in table 1 with Asus S56 laptop, windows7x64, at matlab 2012a. Table 2 is the parameters of each algorithm. The results listed in table 3 were averaged in 50 runs. Figure 2~7 shows the average best fitness curves with 50 independent run.

4.1   Analysis

From figure 2 ~ 7 we can found that the HSPSO converges quickly under all cases of 6 benchmark functions. The curves of DeJong, Rosenbrock, Schaffer, Branin, and Shubert shows faster converges in the last half of the iterations, that means the improvement we suggest has some effective to some of the optimization problem.

Table 2. Parameters of HS, PSO, HSPSO

Parameters	value	Parameters	value	Parameters	value
HMS	10	C_1	2	Dimension	10
$HMCR_0$	0.3	C_2	2	Iterations	1000,5000
$HMCR_T$	0.95	V_{max}	100	L-bounds	-500
PAR_0	0.3	V_{min}	-100	U-bounds	500
PAR_T	0.95	W_0	0.4	W_t	0.9
BW_0	0.000001	BW_T	1		

Table 3. Results

Function	Algorithms	Mean best fitness(Generation)		Standard deviation	
		1000	5000	1000	5000
	IHS	8.7893e+03	706.7554	4.7525e+03	395.3126
DeJong	PSO	3.3341e-08	1.2816e-04	2.3575e-07	8.6166e-04
	HSPSO	1.6815e-14	3.2271e-09	9.4436e-14	2.2461e-08
	IHS	4.0372e+09	852.1815	3.9431e+09	412.4797
Rosenbrock	PSO	379.0625	21.2710	1.3232e+03	56.0759
	HSPSO	172.1365	10.2771	392.5290	6.9739
	IHS	3.0434	1.7146	0.3885	0.4384
Schaffer	PSO	1.9370	1.4979	0.6398	0.7753
	HSPSO	0.3694	0.1156	0.2571	0.0939
	IHS	-3.8982e+03	-4.1538e+03	127.4210	32.4438
Schwefel	PSO	-2.3270e+03	-2.4335e+03	325.4701	292.5127
	HSPSO	-4.2122e+03	-4.2119e+03	58.5848	51.1705
	IHS	3.7437e+04	8.6696e+03	1.9303e+04	4.7040e+03
Branin	PSO	196.0759	149.7126	729.8310	372.8973
	HSPSO	32.2131	12.8479	95.8704	18.3714
	IHS	-662.8991	-884.9700	80.2533	35.5971
Shubert	PSO	-584.9698	-840.2870	138.4033	112.8200
	HSPSO	-905.3241	-932.1326	19.6699	1.0250

From figure 2, 6, PSO is performed better than HSPSO in earlier iterations, but the HSPSO converge quickly after about half iteration, and got better fitness at the end, that shows the HSPSO can improve the fine search around a local optimum. From figure 4, HSPSO show greater performance than the IHS and PSO, that shows the HSPSO has the capability to act better. From figure 3, 5, 7 we can see the HSPSO can inherits the capability of both IHS and PSO, and get better fitness.

Compare the standard deviation presented in table 2, we can easily found out that the stabilization of best fitness of HSPSO has greatly improved compare with IHS and PSO. With the larger iterations, experiment of Branin、 Shubert show greater ability of local search because of the pitch adjusting of HS has a better performance with algorithms runs.

All of the results has improved compare to HS, PSO expect DeJong, I think it is due to DeJong is a very simple unimodal function, exploitation is not stable. Generally, HSPSO converge more quickly and more stable than HS and PSO, prove that the new algorithm is propitious to better result.

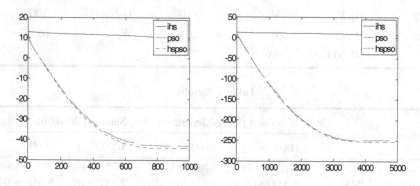

Fig. 2. Convergence Curves Of DeJong

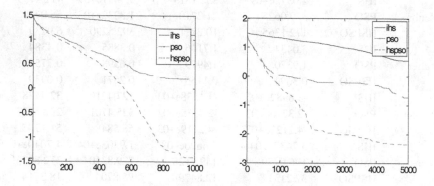

Fig. 3. Convergence Curves Of Schaffer

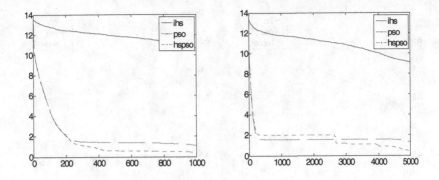

Fig. 4. Convergence Curves Of Branin

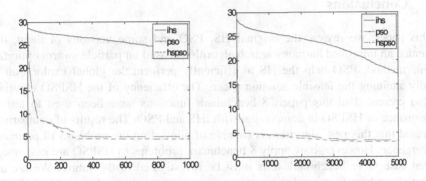

Fig. 5. Convergence Curves Of Rosenbrock

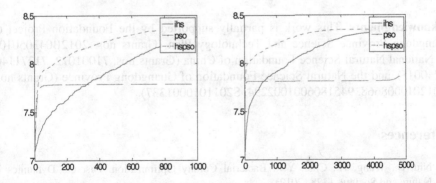

Fig. 6. Convergence Curves Of Schwefel

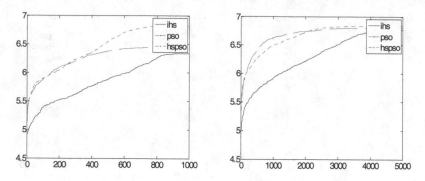

Fig. 7. Convergence Curves Of Shubert

5   Conclusions

In this paper, we review the original HS, PSO and some versions of them, then presented an improved harmony search algorithms based on particle swarm optimizer. In this method, PSO help the HS to efficiently perform the global exploration for rapidly attaining the feasible solution space. The efficiency of the HSPSO algorithm is also presented in this paper. 8 benchmark functions have been used to test the performance of HSPSO in comparison with IHS and PSO. The results of comparisons indicated that this new algorithm is capable of alleviating the problems of premature convergence. However, only apply 8 benchmark problems to HSPSO are not enough, in that case more benchmark tests must be investigated in the future. We are also setting about how to control the parameters to improve the performance of HSPSO. Then, we plan to apply the new algorithms to some research like image retrievalt[10], neural network[11].

Acknowledgments. This work is partially supported by the Foundation Project of Guangdong Province Science and Technology Plan (Grants nos. 2012B04305010), the National Natural Science Foundation of China (Grants nos. 71001072, 71271140, 71240015), and the Natural Science Foundation of Guangdong Province (Grants nos. S2012010008668, 9451806001002294, S2011010001337).

References

1. Niu, B., Wang, H., Chai, Y.J.: Bacterial Colony Optimization. Discrete Dynamics in Nature and Society, 1–28 (2012)
2. Niu, B., Wang, H., Wang, J.W., Tan, L.J.: Multi-objective Bacterial Foraging Optimization. Neurocomputing (October 2012)
3. Geem, Z.W., Kim, J.H., Loganathan, G.V.: A New Heuristic Optimization Algorithm: Harmony Search. Simulation 76, 60–68 (2001)

4. Lee, K.S., Geem, Z.W.: A New Meta-heuristic Algorithm for Continuous Engineering Optimization: Harmony Search Theory and Practice. Computer Methods in Applied Mechanics and Engineering 194, 3902–3933 (2005)
5. Eberhart, R., Kennedy, J.: New Optimizer Using Particle Swarm Theory. In: Proceedings of the International Symposium on Micromechatronics and Human Science, pp. 39–43. IEEE, Piscataway (1995)
6. Coello, C.: Theoretical and Numerical Constraint-handling Techniques Used with Evolutionary Algorithms: A Survey of the State of The Art. Computer Methods in Applied Mechanics and Engineering 191, 1245–1287 (2002)
7. Mahdavi, M., Fesanghary, M., Damangir, E.: An Improved Harmony Search Algorithm for Solving Optimization Problems. Applied Mathematics and Computation 188, 1567–1579 (2007)
8. Shi, Y., Eberhart, R.: Empirical Study of Particle Swarm Optimization. In: Proceedings of the 1999 Congress on Evolutionary Computation, pp. 1945–1950. IEEE, Piscataway (1999)
9. Eberhart, R., Shi, Y.: Comparison between Genetic Algorithms and Particle Swarm Optimization. In: Porto, V.W., Waagen, D. (eds.) EP 1998. LNCS, vol. 1447, pp. 611–616. Springer, Heidelberg (1998)
10. Zhao, Z.Q., Glotin, H.: Diversifying Image Retrieval by Affinity Propagation Clustering on Visual Manifolds. IEEE Mutimedia 16, 34–43 (2009)
11. Zhao, Z.Q.: A Novel Modular Neural Network for Imbalanced Classification Problems. Pattern Recognition Letters 30, 783–788 (2009)

An Emergency Vehicle Scheduling Problem with Time Utility Based on Particle Swarm Optimization

Xiaobing Gan[1,*], Yan Wang[1], Ye Yu[1], and Ben Niu[1,2,*]

[1] College of Management, Shenzhen University, Shenzhen 518060, China
[2] Hefei Institute of Intelligent Machines, Chinese Academy of Sciences, Hefei 230031, China
drniuben@gmail.com

Abstract. In this paper, utility function is introduced to the emergency vehicle scheduling problem. An exponential utility function of time is designed as an indicator of operational efficiency. A mathematic model for the emergency vehicle scheduling problem is constructed. A particle swarm optimization algorithm is designed for the proposed problem. In the PSO algorithm, $N + K - 1$ dimensions encoding scheme are introduced. Finally, we study an experiment. PSO algorithm and MCPSO algorithm are applied. Comparing with the results of the MCPSO algorithm, the PSO algorithm performs better to solve the problem.

Keywords: Emergency Vehicle Scheduling Problem, Time Utility, Particle Swarm Algorithm.

1 Introduction

Emergency vehicle scheduling problem (EVSP), as a NP hard problem, has been researched widely in recent years. EVSP is a special scheduling problem in logistics system for its rigidly limited time. Generally, EVSP is described as choosing route for limited vehicles to rescue a group of disaster areas in limited time. And the object of this problem is to minimize the cost, distance and time of all vehicles traveled. Chen et al. [1] established a general mathematical model for EVSP, and solved the problem by artificial immune algorithm. Zhang et al. [2] established an emergency logistics distribution VSP model with multiple objectives. Zhang et al. [3] constructed an EVSP model to minimize the system time and proposed a FSACO algorithm to deal with this problem. Considering the different relief distribution service priorities of affected areas, Ji and Zhu [4] introduced weighted coefficient to reflect the impact of each disaster area.

However, the objectives of all those models can not exactly mirror the real request of this problem. In emergency situation, time is a kind of rare resource. The key of EVSP is appropriate distribution of time. As each disaster area has respective

---

[*] Corresponding authors.

D.-S. Huang et al. (Eds.): ICIC 2013, LNAI 7996, pp. 614–623, 2013.

sensitivity of time, we introduce time utility into EVSP. Time utility has been taken seriously for many years. In 1994, Spekman, Salmond and Kamauff [5] indicated that waiting time is a measure of service level. In recent year, time has got more attention in the range of services. Li and Su [6] proposed a priority queuing model based on time utility in the out-patient waiting room. The time utility was introduced to describe patients' waiting satisfaction and the utility function was developed based on queue length. Kazutomo et al. [7] dealt with the priority control of contents delivery networks from the viewpoint of utility. Through a subjective evaluation experiment, the exponential utility function of time was proved to be reasonable. Ma et al. [8] defined the time satisfaction function and raised the Time-Satisfaction-Based Maximal Covering Location Problem. Qian [9] brings the customers satisfaction function of time to the vehicle scheduling problem. Kahneman and Tversky [10] testified that the value function is normally convex for losses. While, time utility hasn't got much attention in emergency logistics operations.

In this paper, time utility is introduced. The objective of the proposed EVSP is maximizing utility. In order to solve this problem, PSO algorithm is used. In PSO, $N + K - 1$ particle encoding method is applied. In addition, we set fitness function by penalty function method.

The rest of this paper is arranged as follows. Section 2 describes the EVSP, Section 3 describes the PSO for the proposed problem, Section 4 presents experiment study, and section V contains the conclusion.

2 Description of the Emergency Vehicle Scheduling Problem

The Emergency Vehicle Scheduling Problem (EVSP) can be defined as choosing route for limited number of vehicles to rescue a group of disaster areas in the hard time windows. Each vehicle which starts from the Rescue Center and goes back to the Rescue Center has a limited capacity. Each disaster area has an index of relative importance and should be rescued exactly once. And the vehicle's arrived time must not be later than the deadline of the disaster area.

2.1 Utility Function

In this paper, the concept of utility is introduced. Firstly, the emergency vehicle scheduling is a non-economic behavior. The Rescue Center is concerned more about the victims than the cost of scheduling. So we employ utility to describe the effect of the rescue behavior. Secondly, the rescue material, like perishable goods, is time-sensitive. Even though the material will not rot, the effect can drop off over time. The longer transport time, the less utility the certain area achieved. Thirdly, the ith disaster area's utility U_i is related to the index of importance of each disaster area β_i and the time utility coefficient $\dfrac{U_0}{LT_i}$. The time utility coefficient is defined as the utility of unit time.

According to Kahneman and Tversky [10], the utility function is convex. Based on reference [7], we design an exponential utility function of time for the proposed EVSP. While the base power of the utility function is set to $\beta_i \times \dfrac{U_0}{LT_i}$, which means that the bigger of β_i and $\dfrac{U_0}{LT_i}$, the greater of utility U_i. Formula (1) expresses the time utility function. Which is consistent with the reality that people initially overreact in case of emergency, but they accommodate it later.

$$U_i = U_0 \times (\beta_i \frac{U_0}{LT_i})^{-t_i} \qquad (1)$$

2.2   Mathematical Model for EVSP

Before describe the mathematical formulation of the VSP, we define some Parameters, which are shown as follows.

K :	The number of vehicles
n :	The number of disaster areas
k :	The index of the vehicles
i, j :	The index of the nodes
q_k :	The capacity of vehicle k
v_k :	The velocity of vehicle k
c_k :	The unit freight of the vehicle k
d_{ij} :	The distance between node i and j
t_{ij} :	The time of transportation between node i and j
t_i :	The time when the vehicle arrives node i

β_i :   The index of relative importance of disaster area i

LT_i :   The deadline of disaster area i to accept the rescue goods

UT_i :   The unload time at node i

U_i :   The utility of node i, when its demands are satisfied

U_0 :   The utility of each disaster area if they get the goods at time 0

$\dfrac{U_0}{LT_i}$:   The time utility coefficient

b :   The cost utility coefficient

D_i :   The demand of disaster of node i

x_{ijk} :   The decision variable, which represents whether the vehicle k travels

from i to j or not.

$$x_{ijk} = \begin{cases} 1 & \text{if the vehicle k travels from i to j} \\ 0 & \text{else} \end{cases}$$

Then the description of the Emergency Vehicle Scheduling Problem is as follows. n Disaster areas need a kind of material very urgently. And the Rescue Center is in charge of the rescue mission. Each disaster area i $(i = 1, 2, ..., n)$ has a

demand D_i and an index of relative importance β_i $(\sum_{i=1}^{n} \beta_i = 1)$. The Rescue Center

only has K vehicles. And each vehicle k $(k = 1, 2, ..., K)$ has a capacity q_k. The time window of disaster area i is $[0, LT_i]$, which means the vehicle must arrive before LT_i. And the unload time of each disaster area is UT_i. What's more, we suppose that the demand of each disaster is lower than the capacity of each vehicle. That is, $\max(D_1, D_2, ..., D_n) \le \min(q_1, q_2, ..., q_k)$. The mathematical model is as follows.

$$\max U = \sum_{i=1}^{n} U_i - b \sum_{k=1}^{K} \sum_{j=1}^{n} \sum_{i=0}^{n} c_k D_i d_{ij} x_{ijk} \tag{1}$$

s.t.

$$\begin{cases} t_j = \sum x_{ijk}(t_i + d_{ij}/v + UT_i) \qquad (t_0 = 0, UT_0 = 0) & (2) \\[2mm] \sum_{j=1}^{n}\sum_{k=1}^{K} x_{0jk} = \sum_{i=1}^{n}\sum_{k=1}^{K} x_{i0k} = K & (3) \\[2mm] \sum_{j=1}^{n}\sum_{k=1}^{K} x_{ijk} = 1 \qquad\qquad (i = 0,1,\ldots,n) & (4) \\[2mm] \sum_{i=0}^{n}\sum_{k=1}^{K} x_{ijk} = 1 \qquad\qquad (j = 1,2,\ldots,n) & (5) \\[2mm] \sum_{j=1}^{n} D_j \sum_{i=0}^{n} x_{ijk} \le q_k \qquad (k = 1,2,\ldots,K) & (6) \\[2mm] 0 \le t_i \le LT_i \qquad\qquad (i = 1,2,\ldots,n) & (7) \\[2mm] U_i = U_0(\beta_i * \dfrac{U_0}{LT_i})^{-t_i} & (8) \\[2mm] D_i \le q_k & (9) \\[2mm] x_{ijk} = \begin{cases} 1 & \text{if the vehicle } k \text{ travels from } i \text{ to } j \\ 0 & \text{else} \end{cases} & (10) \end{cases}$$

In the model, formula (1) is the goal of the problem, which is maximum the utility. Formula (2) calculates the transport time from the Rescue Center to disaster j. Constrain (3) calculates the number of vehicles, which are from and back to the Rescue Center. Constrain (4) and (5) represent that each disaster area is rescued exactly once. Constrain (6) means that the load of each vehicle should not exceed its capacity. Constrain (7) represents the time window, that is, arrived time should not be later than the deadline. Formula (8) describes the utility of each disaster area. Formula (9) defines the decision variable.

3 PSO for EVSP

3.1 PSO

In 1995, inspired by the social behavior of animals, Kennedy and Eberhart proposed the Particle Swarm Optimization (PSO). In particle swarm optimization algorithm, each member is regard as a particle, and each particle represents a potential solution to the problem. The particle keeps memory of its previous locations, and evolves from one generation to another. What's more, the global best particle can get the other particles' experiences. So PSO optimum search is speeded up owning to the communication mechanism.

Suppose there are n particles searching for the best location in D dimensional space. When one particle finds a better location, it will change its location and velocity as the following formulas. Formula (11) is used to calculate particle's velocity, which is related to the particle's previous velocity, the distance between the particle's best and current position, and the distance between swarm's best position and the particle's current position. Formula (12) represents the new location of the particle.

$$v_{id}^{k+1} = wv_{id}^k + c_1 \times rand_1 \times (pbest_{id}^k - x_{id}^k) + c_2 \times rand_2 \times (gbest_{id}^k - x_{id}^k) \qquad (11)$$

$$x_{id}^{k+1} = x_{id}^k + v_{id}^k \qquad (12)$$

Where v_{id}^{k+1} means the velocity of the ith particle and x_{id}^{k+1} represents the location of the ith particle. The superscripts, $k \& k+1$, are the number of iterations. Note that c_1 and c_2 are accelerating constant, w is inertial weight, while $rand_1$ and $rand_2$ are uniform random sequences in the range [0, 1].

3.2 Particle Encoding Scheme

The appropriate expression of particles in PSO algorithm is an important issue for the proposed EVSP. Determining the best rescue sequence is the essence of EVSP. The object of the proposed problem is to maximum the sum of all utility. For the EVSP with N disaster areas and K vehicles, $N+K-1$ dimensions encoding scheme is applied. Each particle has $N+K-1$ dimensions, and the numeric sort of all dimensions represents the disaster areas' rescue sequence. Take an example, table 1 shows one particle's position vector for EVSP with 3 vehicles and 7 disaster areas.

Number 0 represents the Rescue Center. We sort all dimensions of the particle in an ascending order. The vector $X = (5, 6, 3, 0, 1, 7, 2, 0, 4)$ represents the rescue sequence, which is shown in table 2.

Table 1. The Particle's Position Vector

Nodes	1	2	3	4	5	6	7	0	0
X	4.2	5.7	3.2	6.4	2.0	2.7	5.3	4.0	5.8

Table 2. The Rescue Scheduling

The number of vehicles	Rescue sequence
1	0→5→6→3→0
2	0→1→7→2→0
3	0→4→0

3.3   Design of Fitness Function

In our proposed EVSP, time window and vehicle's capacity are basic constrains. We adjust the object of this problem by penalty function method, so that both of constrains are mirrored in the fitness function. Formula (13) calculates the fitness function, where M is an infinite penalty coefficient and N is vehicle's loads.

$$fitness = \max Z - M \max(ssupport - q, 0) - M \max(t_i - LT_i, 0) \qquad (13)$$

4   Experimental Study

To measure the performance of the PSO algorithm, in this section, we set the following situation. The Rescue Center deliveries relief goods to 8 disaster areas. Each disaster area's index of relative importance, demands, unloading time and time window is shown in table 3. Rescue Center has 3 vehicles. And all the vehicles have a capacity of 8 units and velocity of 60 units. Table 4 lists the distance between nodes. In addition, other parameters are set as follows: $c_k = 1, b = 0.1, U_0 = 200$.

Based on Matlab 7.2, we apply PSO and MCPSO algorithms [11] to solve this EVSP. In both of the two algorithms, the number of particles is 40, and the number of iterations is 200. What's more, $c1 = c2 = 2, w = 0.8$.

The results of MCPSO and PSO are shown in table 5 and table 6. In table 5, the arrival time of each disaster area is within the time window. Each vehicle's routing is shown in table 6. Compared the results of the two algorithms, we can see that PSO algorithm obtain the higher utility than MCPSO. From figure 1, it is obvious that the

Table 3. Basic Information of Disaster Areas

Disaster areas	Demands/t	Unload time/h	Time window/h	Index of relative importance
1	1	0.2	[0,6]	0.18
2	2	0.3	[0,6]	0.15
3	4	0.5	[0,3]	0.1
4	3	0.4	[0,7]	0.14
5	1.5	0.2	[0,6]	0.12
6	1	0.5	[0,5]	0.13
7	2.5	0.4	[0,7]	0.1
8	3	0.4	[0,9]	0.08

Table 4. Distance between Nodes

	0	1	2	3	4	5	6	7	8
0	0	20	80	65	70	80	40	80	60
1		0	18	35	50	50	40	70	60
2			0	75	40	60	75	75	75
3				0	30	50	90	90	150
4					0	20	75	75	100
5						0	70	90	75
6							0	70	100
7								0	100
8									0

Table 5. The Arrival time of Each Disaster Area

Disaster areas	1	2	3	4	5	6	7	8
PSO	0.33	0.83	2.70	1.80	4.03	0.67	2.33	1.00
MCPSO	0.33	0.83	2.70	1.80	4.23	0.67	2.33	1.00

Table 6. The Route and Utility

Vehicle	Route of vehicle	Utility
PSO	0→1→2→4→3→5→0 0→6→7→0 0→8→0	440.9027
MCPSO	0→1→2→4→3→0 0→6→7→5→0 0→8→0	438.2221

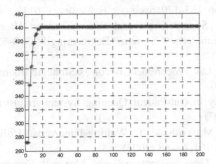

Fig. 1. The convergence graph for PSO

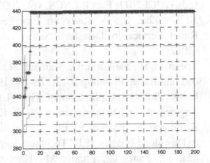

Fig. 2. The convergence graph for MCPSO

best solution 440.9027 is obtained by PSO algorithm at 18th iteration. However, MCPSO algorithm achieves 438.221 as the best solution, which is smaller than 440.9027.

5 Conclusions

In this paper, time utility, which is very significant for emergency vehicle scheming, is introduced to the traditional EVSP. We construct a mathematic model for the extended EVSP, where utility is regard as the objective of the proposed problem. To solve the proposed problem, PSO algorithm is designed. A new encoding method is proposed, which may be somewhat benefit to deal with other problems. Furthermore, the effectiveness of PSO and MCPSO algorithm for the proposed problem is verified through an experimental study. But compared with MCPSO, PSO algorithm is at least better for our proposed problem. In the future, we consider a Bacterial colony Optimization algorithm [12-13] to solve this problem.

Acknowledgements. This work is supported by the National Natureal Science Foundation of China (Grants nos. 71001072, 71271140, 71240015), and the Natural Science Foundation of Guangdong Province (Grant nos. S2012010008668, 9451806001002294), and the Science and Technology Planning Project of Guangdong Province (Grants nos. 2012B040305010).

References

1. Chen, M., Li, Y., Luo, Y.: Research on Emergency Logistics Distribution Vehicle Routing Problems. Computer Engineering and Applications 45(24), 194–197 (2009)
2. Zhang, G., Zhang, Y.: Emergency Logistics Distribution VRP Based on Particle Swarm Optimization. Value Engineering 30(34), 9–10 (2011)
3. Zhang, L.-Y., Fei, T., Liu, T., Zhang, J., Li, Y.: Emergency Logistics Routing Optimization Algorithm Based on FSACO. In: Deng, H., Miao, D., Lei, J., Wang, F.L. (eds.) AICI 2011, Part I. LNCS, vol. 7002, pp. 163–170. Springer, Heidelberg (2011)
4. Ji, G., Zhu, C.: A Study on Emergency Supply Chain and Risk Based on Urgent Relief Service in Disasters. Safety and Emergency Systems Engineering 5, 313–325 (2012)
5. Spekman, R., Salmond, D., Kamauff, J.: At Last Procurement Becomes Strategic. Journal of Long-Range Planning 27(2), 76–84 (1994)
6. Li, X., Su, Q.: Application of Time Utility in Out-Patient Queuing System. Industrial Engineering and Management 6, 26–29 (2007)
7. Kazutomo, N., Kyoko, Y., Eiji, T., Takumi, M., Yoshiaki, T.: Waiting Time versus Utility to Download Images. IEIC Technical Report 101, 63–68 (2001)
8. Ma, Y., Yang, C., Zhang, M., Hao, C.: Time-Satisfaction-Based Maximal Covering Location Problem. Chinese Journal of Management Science 14(2), 45–51 (2006)
9. Qian, H.: Research for the VRP in Agricultural Products Based on Genetic Algorithm. Logistics Sci-Tech. 9, 106–110 (2012)
10. Kahneman, D., Tversky, A.: Prospect Theory: An Analysis of Decision under Risk. Econometrica 47(2), 263–293 (1979)

11. Gan, X.B., Wang, Y., Li, S.H., Niu, B.: Vehicle Routing Problem with Time Windows and Simultaneous Delivery and Pick-Up Service Based on MCPSO. Mathematical Problems in Engineering (2012)
12. Niu, B., Wang, H.: Bacterial colony Optimization. Discrete Dynamics in Nature and Society, 1–28 (2012)
13. Niu, B., Fan, Y., Xiao, H., Xue, B.: Bacterial Foraging-Based Approaches to Portfolio Optimization with Liquidity Risk. Neurocomputing 98(3), 90–100 (2012)

DEABC Algorithm for Perishable Goods Vehicle Routing Problem

Li Li[1], Fangmin Yao[1], and Ben Niu[1,2,*]

[1] College of Management, Shenzhen University
Shenzhen 518060, China
[2] Hefei Institute of Intelligent Machines, Chinese Academy of Sciences
Hefei 230031, China
drniuben@gmail.com

Abstract. In this paper, a hybrid DEABC algorithm is used to solve the perishable food distribution. This particular problem is formulated as a vehicle routing problem with time windows and time dependent (VRPTWTD) by considering the randomness of perishable food distributing process. The DEABC algorithm is implemented in a reliable coding scheme. The performance of proposed algorithm is demonstrated by using modified Solomon's problems. We also analyze the important parameters in perishable food VRPTWTD and their influences have been verified.

Keywords: Vehicle routing problem, perishable food delivery, DEABC algorithm.

1   Introduction

Perishable goods, which including food products, vegetables, blood, often deteriorate during the distribution processes. Therefore perishable goods vehicle routing problem has become the key factor to achieve profit maximization for suppliers.

As a combinatorial optimization problem, perishable goods vehicle routing problem is studied from different perspectives. And some heuristic algorithms have been proposed to solve this problem in recent years.

Hsu C. I. et al [1-2] constructed a perishable goods VRP model and used a heuristic method which extended time-oriented nearest-neighbor heuristic to solve the problem. Osvald A. et al [3] considered the impact of the perishability and adopted a heuristic approach which based on the tabu search to solve the problem. Chen H.K. et al [4] proposed a mathematical model which considered production scheduling and vehicle routing with time windows. A solution algorithm composed of the constrained nelder-mead method and a heuristic for the vehicle routing with time windows was used.

---

* Corresponding author.

D.-S. Huang et al. (Eds.): ICIC 2013, LNAI 7996, pp. 624–632, 2013.

Gong W.W. et al [5] applied two-generation ant colony optimization with ABC customer classification (ABC-ACO) to solve the perishable food distribution problem.

However, many mathematical models have been proposed for the perishable goods vehicle routing problem (VRP) in these papers. Most of them assume the travel speeds are constant, but in the reality travel speeds can change. In this paper, based on this reason, a DEABC algorithm is presented for solving perishable goods vehicle routing problem with time windows and time dependent (VRPTWTD). The rest of paper is organized as follows: section 2 formulates the VRP model for perishable food, the section 3 gives a briefly introduce to the DEABC algorithm. Section 4 is comparative computational result. And we analyze the important parameters in this part. A few conclusions are given in Section 5.

2   Perishable Goods Vehicle Routing Problem with Time Windows and Time Dependent

As perishable goods often deteriorate, their value will decay. So the costs perishable goods supplier undertakes consist of fixed cost, transportation cost, inventory cost, energy cost. And if vehicle violates time windows, suppliers will undertake penalty cost. In addition, most VRP problems assume the travel speeds are constant between different locations. This assumption may be far from the reality because of variable traffic conditions in the city. So in this paper, soft time windows and time dependent are considered.

The perishable goods vehicle routing problem formulation is presented by Hsu C.I., Hung S.F. and Li H.C. [2]. We present the formulation again for taking into account time dependence and time windows. So the perishable goods VRPTWTD objective function can be formulated as follows:

$$\min \sum_{k=1}^{m} f + \sum_{i=0}^{n}\sum_{j=0}^{n}\sum_{k=1}^{m} C * x_{ij}^{k} + \sum_{j=0}^{n}\sum_{k=1}^{m} P * z_{j}^{k} * \overline{b}_{j} + \sum_{k=1}^{m}\left[\int_{t_{s}^{k}}^{t_{f}^{k}} \alpha \Delta H(t)dt\right] + \sum_{i=1}^{n} C_{i}(t_{i}) \qquad (1)$$

While f is the fixed cost for dispatching vehicle. C is transportation cost, which is generally proportional to driving distance. The third term represents inventory cost, P is the cost per item of perishable food, if vehicle k serves customer i, then $Z_{j}^{k} = 1$; otherwise $Z_{j}^{k} = 0$. $\overline{b}_{j}$ represents the expected loss which from customer $(j-1)$ to customer j. The forth term denotes energy cost, α is coefficient, reflecting thermal load. $\Delta H(t) = H(t) - H_{0}$ represents the difference in temperature between the outside and inside of the vehicle. $\sum_{i=1}^{n} C_{i}(t_{i})$ is the penalty cost, as equation (2) below:

$$C_i(t_i) = \begin{cases} M, & t_i \prec ET_i \\ a(et_i - t_i), & ET_i \le t_i \prec et_i \\ 0, & et_i \le t_i \le lt_i \\ a(t_i - lt_i) + \eta P q_i (t_i - lt_i)^\omega, & lt_i \prec t_i \le LT_i \\ M, & t_i \succ LT_i \end{cases} \tag{2}$$

While ET is the earliest allow time, LT is the latest allow time. In equation (2), a is penalty rate, η, ω are parameters, relating to perishable goods decay. q_i is quantity for customer i.

The equation (3) is used to calculate the time point to customer j, $F(v_{ij})$ represents travel speeds function, which proposed by Ichoua S. et al [6]. s_i is the service time.

$$t_j = \sum x_{ij}^k \left(t_i + d_{ij} / F(v_{ij}) + s_i \right) \tag{3}$$

Equations (4) to (13) are constraints as described in the perishable goods VRP formulations. Equation (4) ensures every customer is served by one vehicle and each route starts and ends at the depot.

$$\sum_{k=1}^m z_i^k = \begin{cases} m & i = 0 \\ 1 & i = 1, 2, ..., m \end{cases} \tag{4}$$

The flow conservation constraints are shown as Equations (5) and (6). Equations (7) to (9) make sure the arrival times of any two customers wouldn't conflict with each other. Equation (10) is the capacity constraint. Equations (11) and (12) count the loss amount of decay food.

$$\sum_{i=0}^n x_{ij}^k = z_j^k \quad j = 0, ..., n \quad k = 1, ..., m \tag{5}$$

$$\sum_{j=0}^n x_{ij}^k = z_i^k \quad i = 0, ..., n \quad k = 1, ..., m \tag{6}$$

$$t_j \ge t_i + s_i + t_{ij}^k - (1 - x_{ij}^k) M \quad i = 1, ..., n \quad j = 1, ..., n \quad k = 1, ..., m \tag{7}$$

$$t_i \ge t_s^k + t_{0i}^k - (1 - x_{0i}^k) M \quad i - 1, ..., n \quad k = 1, ..., m \tag{8}$$

$$t_f^k \ge t_j + s_j + t_{j0}^k - (1 - x_{j0}^k) M \quad j = 1, ..., n \quad k = 1, ..., m \tag{9}$$

$$L^k = \sum_{i=1}^{n} z_i^k q_i \le Q \quad k=1,...,m \tag{10}$$

$$x_{0i}^k \overline{b}_i = x_{0i}^k L^k * \left[F\left(t_i - t_s^k + s_i\right) + G\left(q_i\right) \right] \tag{11}$$

$$x_{ij}^k \overline{b}_j = x_{ij}^k L^k * \left[F\left(t_j - t_s^k + s_j\right) - F\left(t_i - t_s^k + s_i\right) + G\left(q_i\right) \right] \tag{12}$$

$$x_{ij}^k = \begin{cases} 1 & \text{if vehicle k departures from demand i to demand j} \\ 0 & \text{otherwise} \end{cases} \tag{13}$$

3   DEABC Algorithm

To solve the problems, we have extended our previous DEABC algorithm in Li L. et al [9]. DEABC algorithm, like bacterial colony optimization [10-11], has successfully solved the continuous optimization problems. To make DEABC algorithm available for discrete problems, we construct $2N$ dimensional space vector X_{2N}, which N represents the number of customers in VRP problems [12]. X_{2N} is divided to two N dimensional space vector X_v and X_r. X_v represents the customer of the corresponding service vehicle, X_r represents the service sequence.

To understand the coding scheme of DEABC algorithm, as Table 1 shown, there is a VRP problem with 5 customers. $X_v = (2,1,3,2,3)$ means 3 vehicles will be used. $X_r = (1.5, 2.0, 1.7, 1.3, 0.8)$ represents the service sequence for the 3 vehicles. For the vehicle 1, there is an only customer 2 need to service. For the vehicle 2, the service sequence is 1, 4. For the vehicle 3, the service sequence is 3, 5.

Table 1. Coding scheme for DEABC algorithm

N	1	2	3	4	5
X_v	2	1	3	2	3
X_r	1.5	2.0	1.7	1.3	0.8

This coding scheme ensures every customer would receive the service only from one vehicle. So in the new DEABC algorithm, fitness value refers to the minimum cost of the solution. DEABC generates two populations, one generated by the ABC and the other by DE. When they are executed, the ABC algorithm obtains the information of global best individual through scouts. The global best individual is chosen according to the last global best positions from the two populations in mutation operator for DE algorithm.

The pseudo code of the proposed DEABC algorithm could be shown as follows:

step 1   Initialize DE, ABC population. Evaluate the fitness value of each individual.
step 2   Compare the fitness value of DE and ABC, and memorize the best individual.
step 3   Perform employed bees search and onlookers selection, search processes.
step 4   Execute scouts search process. The new search points should be determined according to a given probability, whether are randomly produced or obtained from the best positions.
step 5   Update $\vec{x}_{best}^{t}$ and execute mutation, crossover and selection operators.
step 6   If the termination criterion is not met, go to step 2. Otherwise, output the best solution and the global best fitness.

4   Experiment Study

The modified Solomon's problems are used to measure the performance of DEABC algorithm. We assume the service time of vehicle is from 7:00 -11:00 a.m. According to the distance of the two customers, the travel speed matrix is shown in the Table 2.

Table 2. Time-dependent travel speed

	$d \leq 15$	$15 < d \leq 30$	$d > 30$
Speed from 7:00 to 8:00	2.2	2.4	2.6
Speed from 8:00 to 9:00	1.4	1.7	1.9
Speed from 9:00 to 10:00	1.9	2.1	2.5
Speed from10:00 to 11:00	2.3	2.8	3.0

The values of other parameters are shown in Table 3. Based on time windows of Solomon's problems, we assume $ET = \max(et - 120, 0)$, $LT = \min(240, lt + 120)$. The initial temperature outside the vehicle is 26°C, and the outside temperature every hour rises 1°C.

Table 3. Values of other parameters

Parameters	Value	Parameters	Value
f	200	H_0	0°C
C	2.1	a	0.5
P	50	η	0.1
α	0.05	ω	1.2
$F()$	$\dfrac{\Delta t}{1440}$	$G(q_i)$	$\dfrac{s_i}{720}$

In every experiment, the DE parameters are $F = 0.8$ and $CR = 0.5$. The parameter of ABC is as follows: limit is 100. For DEABC, the parameters $F, CR,$ limit are all the same with those defined in ABC and DE. The designated probability S of sharing information is 0.5. A total of 10 runs for each experimental setting are conducted.

4.1 Experiment I: The Comparison of GA, DE, ABC and DEABC

GA, ABC, DE and DEABC were tested to solve the problems using the settings presented in the previous paragraph. Figs 1-4 show the convergence graphs of GA, DE, ABC and DEABC to the problems of R105.25, R105.50, RC101.25 and RC101.50. Table 4 lists the comparison results including the minimum cost, the maximum cost, the mean cost and standard deviations.

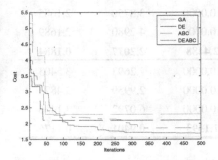

Fig. 1. 25 customers of R105(R105.25)

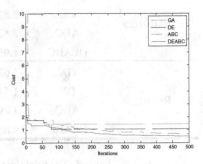

Fig. 2. 50 customers of R105(R105.50)

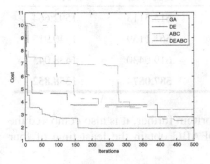

Fig. 3. 25 customers of RC101(RC101.25)

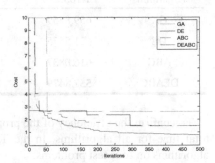

Fig. 4. 50 customers of RC101(RC101.50)

It is shown from Fig.1 to Fig.4 that ABC has a fast convergence except for R105.25 problem. Table 4 presents the summary results of the computational experiments of the four algorithms. The experimental results show that the DEABC algorithm outperforms the GA, DE, ABC algorithm with the same discrete mechanism.

Table 4. Results for all algorithms on test problems

Problem set	Algorithms	Minimum	Maximum	Mean	Std
R105.25	GA	2.2864	3.8313	3.2678	0.4825
	DE	2.1038	5.3820	2.6808	0.9581
	ABC	1.8516	2.2050	2.0169	0.1173
	DEABC	**1.6859**	**2.0294**	**1.7969**	**0.1145**
R105.50	GA	1.4564	1.8979	1.6864	0.1426
	DE	1.0800	2.2256	1.3031	0.3327
	ABC	0.6582	0.7706	0.7116	0.0400
	DEABC	**0.5115**	**0.6277**	**0.5615**	**0.0342**
RC101.25	GA	3.3701	5.8346	4.6715	0.9432
	DE	2.8277	10.0000	3.8750	2.1685
	ABC	2.5235	10.0000	4.2980	2.1689
	DEABC	**1.9570**	**2.4488**	**2.2077**	**0.1884**
RC101.50	GA	2.6240	10.0000	7.2693	3.5403
	DE	1.5123	10.0000	2.9959	2.4822
	ABC	1.4917	10.0000	6.0792	4.1456
	DEABC	**0.8020**	**1.1393**	**0.9401**	**0.1176**

Table 5. The total time for all algorithms on test problems

Algorithms	R105.25	R105.50	RC101.25	RC101.50
GA	659.8459	1314.5414	615.5113	1592.664
DE	621.5575	1222.5762	661.1159	1440.915
ABC	616.0837	1246.7638	619.6430	1438.043
DEABC	**557.8479**	**1175.391**	**587.0874**	**1358.852**

To confirm the efficiency of the proposed algorithm further, it is also compared the total time for all algorithms. In the Table 5, DEABC obtain the least time of four algorithms on four test problems.

The results generated by DEABC are robust. This illustrates that DEABC makes full use of the best individuals of DE population and bee colony into the evolution.

4.2   Experiment II: The Influence of Parameters: $H(t)$

To confirm the influence of some parameters in the model, two experiments have been done separately in this section. As the perishable goods need to be preserved in a fixed temperature, the difference in temperature between the outside and inside of the vehicle is the influential factor in total delivery costs.

So in this part we assume the initial temperature outside the vehicle changes from 24°C to 26°C. DEABC is used to solve the problems.

Table 6. The results of different initial temperature outside the vehicle

Problem set	24°C	25°C	26°C	27°C	28°C
R105.25	1.8473	1.8896	1.9319	1.9743	2.0166
R105.50	4.3719	4.4827	4.5935	4.7043	4.8151
RC101.25	1.6661	1.7072	1.7484	1.7895	1.8307
RC101.50	6.4562	6.6334	6.8105	6.9877	7.1649

The results which at the same service sequence of customers are presented in Table 6. As the Table 6 shown, the total cost of perishable goods rises when the initial temperature increases. The changes in the R105.50 and RC101.50 are obvious. This illustrates the initial temperature outside the vehicle is an influential factor in total costs. And in the same initial temperature, the more customers to service, so the cost is bigger also.

Secondly, we assume the initial temperature outside the vehicle is 26°C. But every hour changes are different. The changes and results which at the same service sequence of customers are presented in Table 7.

Table 7. The Results of changes in outside temperature

Problem set	Every hour drop 2°C	Every hour drop 1°C	No change	Every hour rise 1°C	Every hour rise 2°C
R105.25	1.8207	1.8577	1.8948	1.9319	1.9690
R105.50	4.0903	4.2581	4.4258	4.5935	4.7613
RC101.25	1.5914	1.6437	1.6960	1.7484	1.8007
RC101.50	5.7408	6.0974	6.4540	6.8105	7.1671

It is shown in Table 7 that the changes of temperature is different, the total cost varies. Because of the perishability, perishable goods distribution is different to the conventional goods. It is more sensitive on the outside temperature change.

5   Conclusions and Future Work

In this paper, a perishable food vehicle routing problem with time windows and time dependent (VRPTWTD) is formulated. For solving the problem, an integrated swarm intelligent approach DEABC algorithm is presented. Four modified Solomon's problems are used to evaluate the performance of the DEABC algorithm. The results indicate that DEABC algorithm is efficient. And we analyze the important parameters. The experiments results show that perishable food is sensitive on the outside temperature change. Future study can analyze other parameters such as speed and time windows in the model. On the other hand, the DEABC algorithm can compare with other swarm intelligent algorithm.

Acknowledgements. This work is supported by National Natural Science Foundation of China (Grants nos. 71001072, 71271140, 71240015), and the Natural Science Foundation of Guangdong Province (Grant nos. S2012010008668, 9451806001002294, S2011010001337).

References

1. Hsu, C.I., Hung, S.F.: Vehicle Routing Problem for Distributing Refrigerated Food. Journal of the Eastern Asia Society for Transportation Studies 5, 2261–2272 (2003)
2. Hsu, C.I., Hung, S.F., Li, H.C.: Vehicle Routing Problem with Time-Windows for Perishable Food Delivery. Journal of Food Engineering 80, 465–475 (2007)
3. Osvald, A., Stirn, L.Z.: A Vehicle Routing Algorithm for the Distribution of Fresh Vegetables and Similar Perishable Food. Journal of Food Engineering 85, 285–295 (2008)
4. Chen, H.K., Hsueh, C.F., Chang, M.S.: Production Scheduling and Vehicle Routing with Time Windows for Perishable Food Products. Computers & Operations Research 36, 2311–2319 (2009)
5. Gong, W.W., Fu, Z.T.: ABC-ACO for Perishable Food Vehicle Routing Problem with Time Windows. In: The 2010 International Conference on Computational and Information Sciences (ICCIS), pp. 1261–1264. IEEE Press, Chengdu (2010)
6. Ichoua, S., Gendreau, M., Potvin, J.Y.: Vehicle Dispatching with Time-Dependent Travel Times. European Journal of Operations Research 144, 379–396 (2003)
7. Karaboga, D.: An Idea Based on Honey Bee Swarm for Numerical Optimization. Technical Report (TR06), Computer Engineering Department, Erciyes University, Turkey (2005)
8. Price, K.V.: An Introduction to Differential Evolution. In: Corne, D., Dorigo, M., Glover, F., Dasgupta, D., Moscato, P., Poli, R., Price, K.V. (eds.) New Ideas in Optimization, pp. 79–108. McGraw-Hill Ltd. (1999)
9. Li, L., Yao, F., Tan, L., Niu, B., Xu, J.: A Novel DE-ABC-Based Hybrid Algorithm for Global Optimization. In: Huang, D.-S., Gan, Y., Premaratne, P., Han, K. (eds.) ICIC 2011. LNCS, vol. 6840, pp. 558–565. Springer, Heidelberg (2012)
10. Niu, B., Wang, H.: Bacterial Colony Optimization. Discrete Dynamics in Nature and Society, 1–28 (2012)
11. Niu, B., Fan, Y., Xiao, H., Xue, B.: Bacterial Foraging-Based Approaches to Portfolio Optimization with Liquidity Risk. Neurocomputing 98, 90–100 (2012)
12. Li, N., Zou, T., Sun, D.B.: Particle Swarm Optimization for Vehicle Routing Problem. Journal of Systems Eneineering 19, 596–600 (2004)

BFO with Information Communicational System Based on Different Topologies Structure

Qiwei Gu[1], Kai Yin[1], Ben Niu[1,2,*], Kangnan Xing[1], Lijing Tan[4], and Li Li[1]

[1] College of Management, Shenzhen 518060, China
[2] Department of Industrial and System Engineering,
Hong Kong Polytechnic University, Hong Kong
[3] Management School, Jinan University, Guangzhou 510632, China
drniuben@gmail.com

Abstract. Bacterial foraging optimization (BFO) is a swarm intelligent algorithm which draws inspiration from the foraging behavior of Escherichia coli. This paper improves BFO by introduced information communicational system in which bacteria share information according to neighbor topologies to slow down the premature convergence. The effects of full connected topology, ring topology, star topology and Von Neumann topology on the BFO are systematically investigated, and the new BFO algorithms are named as BFO-FC, BFO-R, BFO-S, and BFO-VM, respectively. Experimental results on four benchmark functions validate the effectiveness of the proposed algorithms.

Keywords: BFO, Topology, Information communication, Swarm intelligence.

1 Introduction

In 2002, Passino proposed a novel optimization algorithm known as Bacterial Foraging Optimization (BFO) algorithm based on the foraging strategies of the E.Coli bacteria cells [1]. BFO displayed good performance for solving optimization problems due to its strong optimization capability, so, more and more researchers began to analyze and improve the algorithm. On the one hand, researchers improved its efficiency by redesigned the mechanism of the algorithm, e.g., Niu et al. developed a bacterial behaviour model called BCO which includes Chemotaxis, Communication, Elimination, Reproduction and Migration [2]. On the other hand, researchers developed the hybridization of BFO with another algorithm, e.g., Chu et al. provided a hybrid system consisting of PSO and BFO [3], Kim et al. presented a novel bacterial swarming algorithm combined BFO and GA [4]. BFO and its variants have successfully been applied to many kinds of complex optimization problems, such as RFID network planning problem [5], portfolio optimization problem [6].

Although, researchers have carried out considerable research about refinement of the BFO algorithm, most of them considered little about the information communicational system of bacteria, especially from the perspective of topology

---

* Corresponding author.

D.-S. Huang et al. (Eds.): ICIC 2013, LNAI 7996, pp. 633–640, 2013.

structure. This paper provides the way of improving the performance of BFO algorithm based on exploring the effect of different neighbor topologies structure.

The rest of the paper is organized as follows. Section 2 systematically discusses neighbor topologies for BFO and describes its implementation details. Section 3 tests the algorithms on the benchmark functions, and gives out the results and analysis. Finally, Section 4 concludes the paper and mentions directions for future work.

2   BFO with Different Neighbor Topologies

BFO includes three kinds of operations, namely chemotaxis, reproduction and elimination-dispersal [1]. Fig.1 shows the flowchart of BFO algorithm.

The bacteria colony becomes stagnated because BFO considers little about the information communicational of bacteria, especially from the perspective of topology structure, which results in converge prematurely. To deal with the problem, this paper introduces information communicational system based on different neighbor topologies into chemotaxis step to explore the effect of the BFO algorithm.

Neighbor topologies system is considered effective in the BFO because topology could regulate the information flow among bacteria individuals. The topology of a bacteria colony, which indicates how information is communicated between bacteria, typically represents an undirected graph, where each vertex is a bacteria. And, the set of vertices with shared edges show the neighbor relationship of a bacteria colony [7]. This paper focuses on static topologies where connections were undirected, unweighted, and do not vary over the course of a trial.

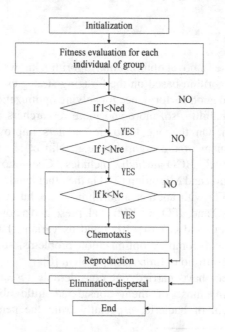

Fig. 1. The flowchart of BFO

Table 1. Variable declaration

Variable	Declaration
$\theta^i(j,k,l)$	position for i^{th} bacteria at j^{th} chemotactic, k^{th} reproductive and l^{th} elimination-dispersal step
$P^i(j,k,l)$	the historical best position for i^{th} bacteria at j^{th} chemotactic, k^{th} reproductive and l^{th} elimination-dispersal step
S	size of bacteria colony
ω	inertia weight
c	learning factor
r	random number
i_left	left neighbor for i^{th} bacteria
i_right	right neighbor for i^{th} bacteria
i_above	above neighbor for i^{th} bacteria
i_blow	blow neighbor for i^{th} bacteria
i_hub	the hub of a bacteria colony in star topology

Various types of neighbor topologies are presented and studied in previous literatures [7, 8]. The neighbor topologies which are considered and investigated in this paper are: A). Full connected topology. B). Ring topology. C). Star topology. D). Von Neumann topology. And, the new BFO algorithms are denoted as BFO-FC, BFO-R, BFO-S, and BFO-VM, respectively. The variable declaration shows in Table 1.

2.1   Full Connected Topology

Fig.2 (A) illustrates the full connected topology which treated the entire bacteria colony as the individual's neighborhood. In this model, the bacteria tumble in chemotaxis step using equation (1):

$$\theta^i(j+1,k,l) = \omega*\theta^i(j,k,l) + \sum_{n=1}^{S} c_n r_n (P^i(j,k,l) - \theta^i(j,k,l)) \tag{1}$$

2.2   Ring Topology

Fig.2 (B) illustrates the ring topology which connects each bacteria to its left and right neighbors. In this model, the bacteria tumble in chemotaxis step using equation (2):

$$\theta^i(j+1,k,l) = \omega*\theta^i(j,k,l)$$
$$+ c_1 r_1 (P^{i-left}(j,k,l) - \theta^i(j,k,l)) + c_2 r_2 (P^{i-right}(j,k,l) - \theta^i(j,k,l)) \tag{2}$$

2.3   Star Topology

Fig.2 (C) illustrates the star topology, one bacteria is selected as a hub, which is connected to all other bacteria in the colony. However, all the other bacteria are only connected to the hub. The bacteria tumble in chemotaxis step using equation (3):

$$\theta^i(j+1,k,l) = \omega*\theta^i(j,k,l) + c_1 r_1 (P^{i-hub}(j,k,l) - \theta^i(j,k,l)) \tag{3}$$

2.4   Von Neumann Topology

Fig.2 (D) illustrates the Von Neumann topology, bacteria are connected using a grid network (2-dimensional lattice) where each bacteria is connected to its four neighbor bacteria (above, below, right, and left). In this model, the bacteria tumble in chemotaxis step using equation (4):

$$\begin{aligned}
\theta^i(j+1,k,l) &= \omega*\theta^i(j,k,l) \\
&+ c_1 r_1 (P^{i-left}(j,k,l) - \theta^i(j,k,l)) + c_2 r_2 (P^{i-right}(j,k,l) - \theta^i(j,k,l)) \\
&+ c_3 r_3 (P^{i-above}(j,k,l) - \theta^i(j,k,l)) + c_4 r_4 (P^{i-blow}(j,k,l) - \theta^i(j,k,l))
\end{aligned} \tag{4}$$

Table 2. The pseudo-code of BFO with different neighbor topologies

---

`INITIALIZE.`
`Set parameters` Dim `,` S `,` N_s `,` N_c `,` N_{re} `,` N_{ed} `,` Ped `,` c `,` R `,etc.`
`Randomize positions and evaluate` $Pbest$ `,` $fPbest$ `and` $fGbest$ `.`
`WHILE` `(the termination conditions are not met)`
`FOR` `(` $l=1:N_{ed}$ `) Elimination-Dispersal loop`
`FOR` `(` $k=1:N_{re}$ `) Reproduction loop`
`FOR` `(` $j=1:N_c$ `) Chemotaxis loop`
`Update position and cost function:`
`Tumble using equation (1) or (2) or (3) or (4). Running.`
`Evaluate cost function for each bacteria.`
`Update` $Pbest$ `,` $fPbest$ `and` $fGbest$ `.`
`END FOR` `Chemotaxis loop end`

`Select highest` J^i_{health} `bacteria using` $J^i_{health} = \sum_{j=1}^{N_c} J^i(j,k,l),$ `Reproduce.`

`END FOR` `Reproduction loop end`
`With probability` Ped `, eliminates and disperse each bacteria.`
`END FOR` `Elimination and Dispersal loop end`
`END WHILE`

So, tumbling step is accompanied with information communication between bacteria in the chemotaxis step. Then, running step is operated which guided by tumbling direction. Finally, the reproduction and elimination-dispersal step are sequentially operated. Table.2 shows the pseudo-code of BFO with different neighbor topologies.

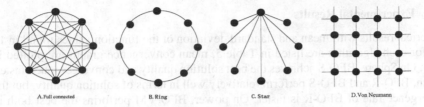

Fig. 2. The neighbor topologies

3   Experiments and Results

3.1   Benchmark Functions

To test the effectiveness of the proposed BFO algorithms, we have done experiments with four benchmark functions. These functions which include a unimodal function (f_1) and three multimodal functions (f_2 to f_4) are formulated as follow. Table 3 shows the parameters setting of those algorithms on four benchmark functions.

Table 3. Globe optimum, search ranges of test functions

Name	Function	Minimum Value	Range of Search
Sphere	$f_1 = \Sigma_{i=1}^{n} x_i^2$	0	$[-100, 100]^n$
Power	$f_2 = \sum_{i=1}^{n} \mid x_i \mid^{i+1}$	0	$[-1,1]^n$
Sin	$f_3 = \dfrac{\pi}{n}\left\{10\sin^2 \pi x_1 + \sum_{i=1}^{n-1}(x_i -1)^2(1+10\sin^2 \pi x_{i+1}) + (x_n -1)^2\right\}$	0	$[-50, 50]^n$
Ackley	$f_4 = -20\exp(-0.2\sqrt{\dfrac{1}{30}\sum_{i=1}^{n} x_i^2}) - \exp(\dfrac{1}{30}\sum_{i=1}^{n}\cos 2\pi x_i) + 20 + e$	0	$[-32, 32]^n$

3.2   Parameters Setting

Comparisons are made among five algorithms of BFO, namely BFO, BFO-FC, BFO-R, BFO-VN and BFO-S. The corresponding parameters setting of BFO variants are showed in Table 4. And all the experiments in this paper are conducted 10 times. Maximum number of iterations is chosen as 5000.

Table 4. The corresponding parameters setting of BFO variants

Parameter	Dim	S	N_s	N_c	N_{re}	N_{ed}	P_{ed}
Value	15	100	40	500	2	5	0.75

3.3   Experimental Results

Numerical results with mean and standard deviation of the function values found in 10 runs for each algorithm are listed in Table 5, mean convergence graphs are plotted in Fig. 3. On Sphere, BFO-S achieves the best solution quality and converges much faster. On Sin, BFO-R and BFO-S perform relatively well in terms of solution quality, but the convergence rate of BFO-R is faster. On power, BFO-VM performs the best both in terms of solution quality and convergence rate. On Ackley, the four BFO variants almost achieve the same solution quality.

Table 5. Results for all algorithms on benchmark functions $f_1 - f_4$

	Sphere	Power	Sin	Ackley
BFO	1.005e-004 ± 2.002e-004	2.742e-004 ± 1.567e-004	4.006e-001 ± 9.009e-002	1.805e+001 ± 5.020e-001
BFO-FC	4.206e-069 ± 3.775e-069	4.653e-010 ± 2.683e-010	4.884e-011 ± 2.611e-011	**1.332e-014 ± 5.599e-015**
BFO-R	**2.726e-072 ± 1.304e-073**	2.332e-140 ± 3.107e-140	**3.141e-032 ± 1.487e-033**	2.664e-015 ± 7.993e-016
BFO-S	**5.577e-073 ± 1.245e-073**	7.615e-064 ± 1.027e-063	3.192e-032 ± 1.154e-033	4.440e-015 ± 2.512e-015
BFO-VM	1.424e-020 ± 4.884e-020	**3.296e-192 ± 5.034e-193**	1.227e-027 ± 2.415e-028	**2.664e-015 ± 1.679e-015**

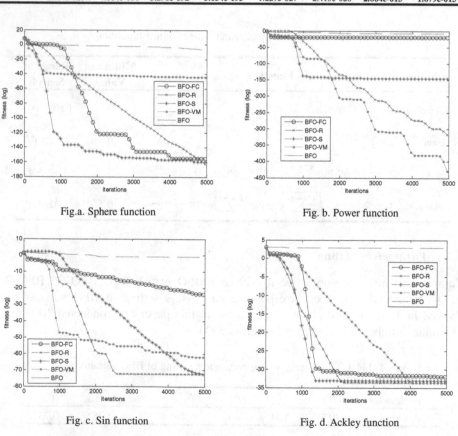

Fig.a. Sphere function   Fig. b. Power function

Fig. c. Sin function   Fig. d. Ackley function

Fig. 3. The mean convergence characteristics of 15-D test functions

We observe that: 1) BFO based on different topologies achieve the better solution quality and converges much faster than the standard BFO. 2) Some BFO variants based on topology worked well on some benchmark functions may perform poorly for another. 3) Bacteria colony with fewer connections might perform better on highly multimodal problems because bacteria are able to "flow around" local optima and to explore different regions, while highly interconnected colony converge quickly on problem solutions but has a weakness for becoming trapped in local optima.

4 Conclusions

We have proposed four new variants on BFO by introduced information communicational system based on four different topologies structure. Each of these neighborhood topologies can be a model of a BFO implementation in different contexts. The performance of the proposed algorithm was compared to the standard BFO with a set of unimodal functions and multimodal functions. The results showed that, the new algorithms almost performed better fitness and has faster convergence rate comparing to the standard one. Given its robust and good performance on different types of problems, the proposed algorithms have a good trade-off between exploration and exploitation capability.

Future work will focus on extending these ideas to dynamic neighborhood topology and hybrid model of BFO based on the multi-types of neighborhood topologies. Another direction will be multi-colony cooperative using neighborhood topologies.

Acknowledgments. This work is partially supported by the National Natural Science Foundation of China (Grants nos. 71001072, 71271140, 71210107016, 71240015), The Hong Kong Scholars Program 2012 (Grant no. G-YZ24), China Postdoctoral Science Foundation (Grant nos. 20100480705, 2012T50584), and the Natural Science Foundation of Guangdong Province (Grant nos. S2012010008668, 94518060010022 94, S2011010001337) .

References

1. Passino, K.M.: Biomimicry of Bacterial Foraging for Distributed Optimization and Control. IEEE Control Systems Magazine 22(3), 52–67 (2002)
2. Niu, B., Wang, H., Chai, Y.J.: Bacterial Colony Optimization. Discrete Dynamics in Nature and Society, 1–28 (2012)
3. Chu, Y., Mi, H., Liao, H., Ji, Z., Wu, Q.H.: A Fast Bacterial Swarming Algorithm for High-Dimensional Function Optimization. In: Proceedings of IEEE World Congress on Computational Intelligence, pp. 3135–3140 (2008)
4. Kim, D.H.: Hybrid GA-BF Based Intelligent PID Controller Tuning for AVR System. Applied Soft Computing 11(1), 11–22 (2011)
5. Chen, H.N., Zhu, Y.L.: RFID Networks Planning Using Evolutionary Algorithms and Swarm Intelligence. In: Proceedings of 4th IEEE International Conference on Wireless Communications, Networking and Mobile Computing, pp. 1–4 (2008)

6. Niu, B., Fan, Y., Xiao, H., Xue, B.: Bacterial Foraging-Based Approaches to Portfolio Optimization with Liquidity Risk. Neurocomputing 98(3), 90–100 (2012)
7. Kennedy, J., Mendes, R.: Population Structure and Particle Swarm Performance. In: Proceedings of the IEEE Congress on Evolutionary Computation, pp. 1671–1676 (2002)
8. McNabb, A., Gardner, M., Seppi, K.: An Exploration of Topologies and Communication in Large Particle Swarms. In: Proceedings of the IEEE Congress on Evolutionary Computation, pp. 712–719 (2009)

PSO-Based SIFT False Matches Elimination
for Zooming Image

Hongwei Gao[1,3,*], Dai Peng[1], Ben Niu[2], and Bin Li[3]

[1] School of Information Science & Engineering, Shenyang Ligong University,
Shenyang, 110159 China
ghw1978@sohu.com
[2] College of Management, Shenzhen University, Shenzhen, 518060 China
drniuben@gmail.com
3 State Key Laboratory of Robotics, Shenyang Institute of Automation,
Chinese Academy of Sciences, Shenyang, 110016 China

Abstract. According to the numerous false matches of SIFT feature attribution of zooming image, false matches elimination algorithm, combined with geometric constraint of zooming image, is proposed in this paper. It aims to optimize square sum function of distance from point to corresponding polar line and adopt PSO to do iterative optimization that false matches points could be eliminated. The experimental results prove that the proposed algorithm is efficient and stable.

Keywords: Zooming image, SIFT matching, PSO, False matches elimination.

1 Introduction

As a basic question in research of computer vision, depth estimation of zooming image plays a key role in the image understanding and can be applied in robotics, scene understanding and 3D reconstruction. Zooming image, as depth cue of monocular vision[1], is widely used in many fields, such as visual surveillance, visual tracking and environmental perception and map building of robot. Ma and Olsen[2] put forward the method which used the zoom lens to realize depth estimation. And the results showed that zooming image can provide information of depth theoretically. Lavest, etc[3,4], did precise study to optical properties of zoom lens and came up with thick lens model to describe the zoom lens. According to the actual structure of lens, Asada and Baba, etc[5], presented the zoom lens model which had three parameters: zoom, focus and aperture. Fayman, etc[6], used depth estimate for vision tracking field and put forward a active vision technology of zoom tracking. However, most of these studies, lack of automatic matching algorithms, focused on 3D model reconstruction of zooming image. How to select the image features is the key technique in matching. The classic operator includes Harris corner detection operator[7], Susan

---

* Corresponding author.

D.-S. Huang et al. (Eds.): ICIC 2013, LNAI 7996, pp. 641–648, 2013.
© Springer-Verlag Berlin Heidelberg 2013

corner detection operator[8] and SIFT detection operator[9],etc. Because of insensitivity with light, rotation and zooming, SIFT operator is widely researched. In this paper, SIFT feature detection algorithm of zooming image is researched and an iterative false matches elimination algorithm which is based on distance from point to polar line and PSO is put forward. The experimental results show that false matches can be eliminated efficiently.

2   Feature Matching of SIFT of Zooming Image

In 1999, David G.Lowe, a professor of British Columbia, summarized feature detection methods which were based on invariant technology and put forward a local feature descriptor of image. It was what we call SIFT (scale invariant feature transform) which was improved in 2004[10]. A image can be mapped as a local feature vector set finally. This feature vector is invariant for translation, scaling and rotating and has invariance for change of illumination, affine transformation and projection transformation in a sense. There are three parts in SIFT algorithm for matching. First, detect extreme the value point and extract the key point under the scale space of image. The feature information of key point, including position, scale and direction, should be calculated. Second, divide regions around key point into pieces and calculate gradient histogram of every piece. Then, feature vector of descriptor about key point can be created. Last, matching could be completed by comparison of key point descriptor. Euclidean distance is generally used for similarity measurement of key point descriptor.

Fig. 1. Result of traditional SIFT matching

In figure 1, there is a pair of zooming image for matching. Experimental result is shown. It is convenient that every pair of matching point is connected by straight line. In the left image, there are 1478 feature points which are calculated by algorithm and 732 feature points in right image. 132 pairs of matching point are got by initial matching. Obviously, there are so many false matches that we should think of how to eliminate false matches. The primary problem is how to get the ideal matching points which are needed for depth estimation of zooming image.

3   Geometric Constraint of Zooming Image

In the research of depth estimate of zooming image, there is a basic assumption. The radial slopes of matching points are same when look zoom center as the origin of the image coordinate system ideally. (the matching points p1 and p2 in Fig.2). Obviously, having the same radial slope is a necessary condition for correct matching point. So we can use this condition to get rid of false matches points in the matching results.

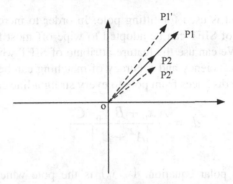

Fig. 2. The ideal matching point and actual matching point of zooming image

Actually, due to the influence of the distortion of the imaging, even if the correct match point, radial slope could not be completely the same. (the matching points p1' and p2' in Fig.2).Therefore, it's needed to give a reasonable tolerance to screen the ideal match point, and try to eliminate those false match by matching algorithm. But don't feel free to give permissible error. The best way is experiment.

4   Iterative False Matches Elimination Algorithm Based on Distance From Point to Polar Line

4.1   Distance from Point to Polar Line

A few pairs of false matches point can be eliminated by geometric constraint of zooming image. But the remaining pairs of matching point is too little that it goes against following display of 3D reconstruction. In this paper, iterative false matches elimination algorithm which is based on distance from point to polar line and PSO is proposed. This algorithm not only meet the demand for a lot of matching points but also get high accuracy matching.

Actually, zooming image is a special kind of translation image. In ideal condition, connecting lines of matching points shall intersect at a common pole as shown in figure 3. So, a pole can be fitted by polar of matching point directly. Then, we can use the distance from pole to polar to eliminate the abnormal polar lines. Meanwhile, the false matches points could also be eliminated.

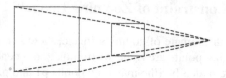

Fig. 3. Schematic drawing of pole

Least square method is used for fitting pole. In order to increase the accuracy of pole, feature attribute of SIFT can be adopted to wipe off most false matches points before pole is fitted. We can use the feature attribute of SIFT when doing key points matching. So that the efficiency and accuracy of matching can be improved. According to equation (1), the distances from pole to every straight line can be calculated.

$$d_i = \frac{A_i x_0 + B_i y_0 + C_i}{\left| A_i^2 + B_i^2 \right|} \tag{1}$$

A_i, B_i, C_i signifies the polar equation. (x_0, y_0) is the pole which is fitted by least square method.

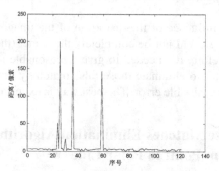

(a) Distribution from pole to every euclidean distance of polar line

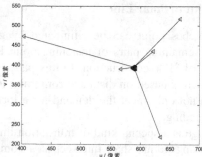

(b) Vertical distribution from pole to polar line

Fig. 4. Distance from pole to polar line

Distribution from pole to every euclidean distance of polar line is shown in figure 4(a). And vertical distribution from pole to polar line is shown in figure 4(b). Length of vertical is the euclidean distance from pole to polar line. The picture shows that a few poles which are far from polar lines should be eliminated. Suppose the first i distance of polar line is $d(i), i \in [1, N]$. The average rate of every distance of polar line is:

$$\mu_d = \frac{1}{N} \sum_{i=1}^{N} d(i) \tag{2}$$

Standard deviation is

$$\sigma_d = \sqrt{\frac{1}{N} \sum_{i=1}^{N} (d(i) - \mu_d)^2} \tag{3}$$

Presume that distance of polar line obeys normal distribution. Abnormal matching points can be eliminated by setting a confidence interval. The confidence interval of polar line distance of matching point is $[\mu_d - k_d \sigma_d, \mu_d + k_d \sigma_d]$. The k_d is the standard deviation factor of polar line distance of matching point. The k_d can be taken in smaller value because of application of SIFT feature attribution. Here, $k_d = 0.5$.

The question of how to eliminate false matches points can be transformed to solve minimum value if the $d(i)$ is solved.

$$D = \min \sum_{i=1}^{n} (d_i^2), i \in [1, N] \tag{4}$$

d_i^m signifies the value of first i polar line distance after first m operation. N^m is the amount of polar line. The size of D determines the value of m. The smaller the value of D, the higher the accuracy of pole fitting. The fitting pole is the zoom center if the zooming image uses the identical coordinate system.

4.2   PSO Based on False Matches Elimination for Iteration

The PSO (particle swarm optimization), derived from the prey behavior of birds, is a kind of optimization tools[11] based on iterative just like genetic algorithm. The system is initialized to a random solutions and searches the optimal value by iterative. PSO is a effective optimization tool for nonlinear continuous optimization problem, combinatorial optimization problem and mixed integer nonlinear optimization problems. This algorithm has been used widely in function optimization, neural network training, fuzzy system control, and other application of genetic algorithm. And this paper attempts to applied it to false matches elimination.

PSO is initialized as a group of random particles (random solution). The particle update itself by tracking two extreme value points in each iteration. The first one is the best solution, called the individual extreme value point (the location is expressed as $pbest_{id}$).The other extreme value point of PSO is the best solution in the entire population, called the global extreme value point (the location is expressed as $gbest_d$).After this, according to the following update equation (5) and (6), the particles update their own speed and position.

$$v_{id}^{k+1} = v_{id}^k + c_1 rand_1^k \left(pbest_{id}^k - x_{id}^k \right) + c_2 rand_2^k \left(gbest_d^k - x_{id}^k \right) \tag{5}$$

$$x_{id}^{k+1} = x_{id}^k + v_{id}^{k+1} \tag{6}$$

v_{id}^k is d-D velocity of the particle i in iteration of the first k ; c_1 and c_2 are acceleration coefficient(or learning factor) to adjust the maximum step of global best particle and the individual best particle direction of flight; $rand_{1,2}$ is a random number in [0 1]; x_{id}^k is the current position of the particle i in iteration k ; $pbest_{id}$ is the position of the individual extreme value point of the particle i in d-D; $gbest_d$ is the position of the global extreme value point of the whole group in d-D.

Parameter optimization based on PSO is shown as follows:

(1) Random initialization of population. The location of initial search point and its speed are usually generated randomly in initial value of neighborhood space. The initial number of particle is 200, d=30. Calculate the corresponding individual extreme value. Global extreme value is the best among individual extreme value.

(2) Evaluation of each particle in the population. The fitness value of particle can be calculated by equation (4). Update individual extreme value if it is better than the current individual extreme values of the particle. If the best individual extreme value of all the particles is better than the current global extreme value, update the global extreme value.

(3) Updated the population according to equation (5) and (6).

(4) The operation should stop if the termination condition is met. Then return the best individual combination or else turn to step 2).

5   Experimental Results and Analysis

Using the method to process another zooming images and results are still good. The experimental results are shown in figure 5 and 6. There are 146 matching points in figure 5(a). In figure5(b), there are 119 matching points which accounting for 81.51% of the total matching points after false matches elimination. And there is no false-matches basically. In the second experiment, because of onefold scene, less matching points are got by SIFT. In figure 6(a), there are 34 matching points. In figure 6(b), there are 22 matching points which accounting for 64.71% of the total matching points after false matches elimination. By the above zooming image experiment of different scene, it can be seen that the proposed algorithm is effective. It can get high accuracy matching points and has good robustness.

(a) Direct matching result of SIFT algorithm

(b) Matching results of false matches elimination algorithm

Fig. 5. False matches elimination results of flowerpot

(a) Direct matching result of SIFT algorithm

(b) Matching results of false matches elimination algorithm

Fig. 6. False matches elimination results of computer

6   Conclusions

In this paper, a false match elimination algorithm, based on distance from points to polar line is put forward. This method combines epipolar constraint and SIFT characteristic attribute. It also uses PSO to optimize. The experimental results show that the high accuracy and strong robustness matching points can be got. We also want to use the proposed method to solve some other image processing problems, such as image retrieval [12], image annotation [13].

Acknowledgement. This work is supported by National Natural Science Foundation of China (Grant nos. 71001072, 71271140), State Key Laboratory of Robotics Foundation (Grant No. 2012017), education department project of Liaoning province (Grant No.L2011038), and the Natural Science Foundation of Guangdong Province (Grants nos. S2012010008668, 9451806001002294, S2011010001337).

References

1. Wang, J., Wang, Y.Q.: A Monocular stereo vision algorithm based on bifocal imaging. Robot 33(6), 935–937 (2007)
2. Ma, J., Olsen, S.I.: Depth from zooming. Journal of the Optical Society of America 7(10), 1883–1890 (1990)
3. Lavest, J.M., Rives, G., Dhome, M.: Three-dimensional reconstruction by zooming. IEEE Transactions on Robotics and Automation 9(2), 196–207 (1993)
4. Lavest, J.M., Delherm, C., Peuchot, B., Daucher, N.: Implicit reconstruction by zooming. Computer Vision and Image Understanding 66(3), 301–315 (1997)
5. Baba, M., Oda, A., Asada, N., Yamashita, H.: Depth from Defocus by Zooming Using Thin Lens-Based Zoom Model. lectronics and Communications in Japan (89), 53–62 (2006)
6. Fayman, J.F., Sudarsky, O., Rivlin, E., Rudzsky, M.: Zoom tracking and its applications. Machine Vision and Applications 13(1), 25–37 (2001)
7. Smith, S.M., Brady, J.M.: SUSAN-a new approach to low level image processing. International Journal of Computer Vision, 45–78 (1997)
8. David, G., Lowe, D.G.: Distinctive Image Features from Scale-Invariant Key Points. International Journal of Computer Vision 60(2), 91–110 (2004)
9. Ke, Y., Sukthankar, R.: PCA-SIFT: A more distinctive representation for local image descriptors. In: IEEE Conf. on Computer Vision and Pattern Recognition, vol. (2), pp. 506–513 (2004)
10. Krystian, M., Cordelia, S.: A performance evaluation of local description. IEEE. Transactions on Pattern Analysis and Machine Intelligence 27(10), 1615–1630 (2005)
11. Niu, B., Zhu, Y.L., He, X.X., Wu, Q.H.: MCPSO: A multi-swarm cooperative particle swarm optimizer. Applied Mathematics and Computation 185(2), 1050–1062 (2007)
12. Zhao, Z.Q., Glotin, H.: Diversifying image retrieval by affinity propagation clustering on visual manifolds. IEEE Mutimedia 16, 34–43 (2009)
13. Zhao, Z.Q., Glotin, H., Xie, Z., Gao, J., Wu, X.: Cooperative sparse representation in two opposite directions for semi-supervised image annotation. IEEE Transactions on Image Processing 21, 4218–4231 (2012)

Object Tracking Based on Extended SURF and Particle Filter

Min Niu[1], Xiaobo Mao[1], Jing Liang[1], and Ben Niu[2]

[1] School of Electrical Engineering, Zhengzhou University, Zhengzhou, China
angel12318@126.com, {mail-mxb,liangjing}@zzu.edu.cn
[2] College of Management, Shenzhen University, Shenzhen, China
drniuben@gmail.com

Abstract. Under complex environment, it is difficult to track target successfully by single feature. To solve this problem, the paper propose a novel object tracking approach which fuses color and SURF(Speeded Up Robust Features) in the frame of particle filter. SURF remain invariant for illumination, scale and affine. Add color to make up for the shortcoming(SURF is based on image gray scale information.). It not only maintains the characteristics of SURF, but also makes use of the image color information. The experimental results prove that the proposed method is real-time and robust in different scenes.

Keywords: SURF, Color Feature, Particle Filter, Object Tracking.

1 Introduction

Video target tracking is an important research subject in the field of machine vision. Many scholars has carried on the extensive research to target tracking, a variety of algorithms are proposed, but most algorithms are based on single feature, which are effective in a particular environment. It may result in decline in accuracy even failure when the target or environment changes. For the situation, multiple features have been widely used in tracking systems. However, the performance still depend on single feature even the fusion[1].

Compared with color, shape and texture etc features, SIFT (scale invariant feature transform) proposed by D.G. Lowe in 2004 is more capable to remain invariable for illumination, scale and affine [2-4]. SIFT is the local characteristics of the image, and although it has more advantages over other features, the matching algorithm based on SIFT is obliged to deal with complex computing problems and long time consuming. Even though Grabner etc put forwards an idea that increases the calculating velocity by integral image at cost of the superiority [5]. In 2006, Bay etc provided SURF (speeded up robust features) [6] algorithm on the foundation of SIFT. SURF has good adaptability for zooming, small perspective changes, noise, and brightness changes, like SIFT. In addition, it can take short time to accomplish SURF matching algorithm. So SURF has excellent performance not only in speed but also in accuracy [7-10].

D.-S. Huang et al. (Eds.): ICIC 2013, LNAI 7996, pp. 649–657, 2013.

However, SURF only uses the gray information of images, ignores color information, so it is difficult to identify and match the target with similar texture. The paper fuses color vector and build extend SURF descriptor to solve the problem.

When tracking object, it needs to process a large of invalid information to determine the optimal matching position if we calculate all the pixels in images directly, and it is heavy computation and time-consuming. Therefore, it can improve the efficiency of tracking algorithm if we adopt a certain method to reduce the search area of candidate target. Due to the particle filter can predict target location in the next frame image and realize the state estimation of nonlinear non-Gaussian systems, it is commonly used in tracking algorithm at present. The paper uses extend SURF descriptor to set up the target model and particle filter to search candidate target location and the novel method is robust and real-time.

2   Particle Filter

For particle filter, its fundamental is Monte Carlo method. It uses a set of particles with weights $\{(x^i_k, \omega^i_k), i=1...n\}$ to estimate the posterior density $p(x_k \mid y_{1:k})$. x^i_k describes the particle, namely the target state. ω^i_k represents the weight, and $\sum_{i=1}^{N} \omega^i_k = 1$.

In ideal conditions, all the weights should be $1/N$, the particles should be random sampled from the posterior density, but it is impossible in the actual situation. So sampling new particles from the proposed density $\pi (x^i_k \mid x^i_{k-1}, y_{1:k})$ which is similar with posterior density, and computing weights again to make up the difference between the proposed density and the posterior density. Then normalizing the weights, the posterior density is computed as:

$$p(x_k|y_{1:k}) \approx \sum_{i=1}^{N} \omega^i_k \delta(x_k - x^i_k) \tag{1}$$

Where, $\delta(*)$ is DE carat function. Weights updating formula as follows:

$$\omega^i_k \propto \omega^i_{k-1} \frac{p(y_k|x^i_k)p(x^i_k|x^i_{k-1})}{\pi(x^i_k|x^i_{k-1}, y_{1:k})} \tag{2}$$

Where, it is significant how to choose the proposed density $\pi (x^i_k \mid x^i_{k-1}, y_{1:k})$. It is easy to make $\pi (x^i_k \mid x^i_{k-1}, y_{1:k}) = p(x^i_k \mid x^i_{k-1})$ for practical application. Weights updating is computed as:

$$\omega^i_k \propto \omega^i_{k-1} p(y_k|x^i_k) \tag{3}$$

3   SURF Algorithm

3.1   Feature Point Detection

Image pyramid is used to express scale space usually in the field of vision. It is variable for image size when using the traditional way to conduce a scale space. Next, reuse Gaussian filer to smooth each layer image. In order to speed up, we adopt the box filters increasing gradually which approximates second-order Gaussian filter and the integral image to make the convolution to form different scales of image pyramid.

SURF algorithm detects the extreme points in images by computing the Hessian matrix. To a point (x,y) in image I, at the scale σ the Hessian matrix is defined as follows:

$$H(x,y,\sigma) = \begin{bmatrix} L_{xx}(x,y,\sigma) & L_{xy}(x,y,\sigma) \\ L_{xy}(x,y,\sigma) & L_{yy}(x,y,\sigma) \end{bmatrix} \tag{4}$$

Where, $L_{xy}(x,y,\sigma)$ is the convolution between Gaussian second order derivative $\dfrac{\partial}{\partial x^2} g(x,y,\sigma)$ and the pixel value $I(x,y)$ in image I. Similarly for L_{xy} and L_{yy}. Because the box filter approximate second order Gaussian filter, for convenience of calculation, we use the convolution of $I(x,y)$ with box filter instead, the results are described by D_{xx}, D_{xy}, D_{yy}. Introduce weight to reduce the error between the approximation and the accurate value. So the determinant of Hessian matrix is computed as follows:

$$\det H = D_{xx}D_{yy} - (0.9D_{xy})^2 \tag{5}$$

The extreme point obtained by calculating the determinant of Hessian matrix, 8 pixel points at the same scale and 18 pixel ones at upper and lower adjacent scales form a 3*3*3 neighborhood. If the extreme point larger or smaller than the remaining 26 pixel ones, it is the feature point. Then localize feature point and over scale by interpolating it in scale and image space with the method proposed by Brown et al.

3.2   SURF Feature Descriptor

Firstly, calculating the Harr wavelet response (Harr response side length is 4σ, σ is scale at which the feature point.) in X and Y direct in a circular neighborhood of radius 6σ around the feature point, then the responses are weighted with a Gaussian (2σ) centered on the feature point. Next computing the weighted sum of all responses within a sliding orientation window covering 60° angle to form a new vector, the longest vector over all windows is defined as orientation of the feature point, the process is shown in figure 1.

Constructing a square region centered on the feature point and oriented along the orientation confirmed before, with the size 20σ.The region is divided into 4*4 square sub-regions evenly. In each of sub-regions, we compute the Harr responses at 5σ scale in orientation selected before and its vertical direction, which are represented by d_x and d_y respectively. In order to increase the robustness, d_x and d_y are weighted with a Gaussian(3.3σ) centered on the feature point. Every sub-region has a four-dimensional descriptor vector.

$$V = \left(\sum d_x, \sum d_y, \sum |d_x|, \sum |d_y| \right) \qquad (6)$$

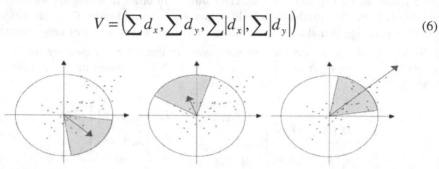

Fig. 1. Diagram for confirming orientation of the feature point

For every feature point, its descriptor vector is 64-dimensional.Figure 2 shows the descriptor vector building.

3.3   Extend SURF Descriptor

SURF descriptor only uses the gray information of images, ignores color information, so it is difficult to identify and match the target with similar texture. So the paper adopts a new kind of descriptor based on color. Namely, fuses color vector to build extend SURF descriptor to solve the problem.

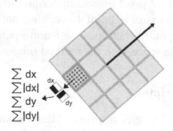

$\sum dx$
$\sum |dx|$
$\sum dy$
$\sum |dy|$

Fig. 2. Diagram for building descriptor vector

Because SURF has good adaptability for zooming, small perspective changes, noise, and brightness changes, and color descriptor can not change structure of SURF descriptor, so extend SURF one is more robust. To reduce the influence of illumination change during target tracking, the paper uses HSV color model [11]. Where, H is hue, S is saturation, they are not sensitive to light. V is value and opposite. And three components are independent mutually. So we only compute the sum of H and S to obtain Σh and Σs. Therefore, every sub-region has a 2-dimensional color vector.

$$C = \left[\sum h \quad \sum s \right] \tag{7}$$

16 sub-regions form 32-dimensional color descriptor vector. Add color vector C to SURF descriptor vector V to construct the extend SURF descriptor vector with 96-dimensional.

3.4   Feature Point Matching

The paper adopts the nearest neighbor matching vector method to match feature point [12]. Given A and B are the set of feature points in two images I_1 and I_2, a_j is a point in A, compute the Euclidean distances between a_j and all the points in B, call d_1 and d_2 the nearest distance and next nearest distance regularly, corresponding to two feature points b_j and b_j^*, if $d_1 \leqslant \mu d_2$ (μ=0.65), a_j and b_j match. Otherwise, we discard the feature point. Traversing all the feature points in A can find out all possible matching feature point pairs.

4   Target Tracking

4.1   Target Motion Model

Supposing object keeps a constant velocity, its movement is small for the same object of adjacent frames. The paper adopts a rectangle to represent the object area, $X=(x,y,v_x,v_y,h_x,h_y)^T$ is described as motional state of an object, (x,y) are the center coordinates of the object, (v_x,v_y) is velocity of movement in x and y direction, (h_x,h_y) are height and width of object area. The dynamic equation of the system is shown in formula 8.

$$X_k = AX_{k-1} + V_k \tag{8}$$

$$A = \begin{bmatrix} 1 & 1 & 0 & 0 & 0 & 0 \\ 0 & 1 & 0 & 0 & 0 & 0 \\ 0 & 0 & 1 & 1 & 0 & 0 \\ 0 & 0 & 0 & 1 & 0 & 0 \\ 0 & 0 & 0 & 0 & 1 & 0 \\ 0 & 0 & 0 & 0 & 0 & 1 \end{bmatrix}$$

Where, A is state transfer matrix of the system, v_k is system noise.

4.2   Similarity Measurement

The paper adopts the extend SURF feature to track object, we draw a conclusion that similarity between reference target and candidate one is related to the number of matching feature point pairs and the Euclidean distance between matching feature points. Supposing m is the number of matching feature point pairs between candidate target S_k and reference one S_0 at time k, dis_t is the Euclidean distance between a matching feature point pair, then the average Euclidean distance between two objects is defined as:

$$D_k = \sum_{t=1}^{m} dis_t / m \tag{9}$$

We judge the similarity degree between targets in two frames by combining the number and Euclidean distance, and observation likelihood function is defines as:

$$p\left(y_t \mid x_t\right) = 1 - \exp\left(-mD_k\right) \tag{10}$$

4.3   Algorithm Description

(1) In the initial frame, we choose the object tracking area manually, calculate the extend SURF feature vector contained in the object as reference template, and sample a set of particles with weights $\{(x^i_0, 1/N), i=1...N\}$ from the proposed density $\pi(x^i_k \mid x^i_{k-1}, y_{1:k})$, for $k=0$. According to the actual conditions, set the initialization parameter, such as target coordinates, velocity and rectangular frame size;

(2) Input next frame image, exploit formula 7 to acquire a new particle collection for time k;

(3) Compute the extend SURF feature vector in image for time k.

(4) Utilize formula 3 and 9 to calculate the weight of each particle and to normalize it as follows:

$$\omega^i_k = \omega^i_k / \sum_{i=1}^{N} \omega^i_k \tag{11}$$

(5) If $1/\sum_{i=1}^{N} \left(\omega^i_k\right)^2 < 2N/3$, resample to generate a new set of particles: $\{(x^i_k, 1/N), i=1...N\}$.

(6) Output the state estimation: $x_k = \sum_{i=1}^{N} \omega^i_k x^i_k$;

(7) $k = k+1$, if go on tracking, switch(2); otherwise end.

5   Experiment Analysis

To validate the superiority of the algorithm above, several video of different scenes are tested, and the results are compared with SURF algorithm. The environment of experiment is personal computer, AMD Athlon, 2.8GHz, 2GB Memory, OS Windows XP, and the proposed algorithm is programmed by VC++6.0 and OpenCV.

The first video sequences are people tracking on a simple scenes, the experiment result is as in figure 3. We choose the 120th frame, the 150th frame, the 170th frame, the 480th frame to analyze the tracking performance. During tracking, he human body target selected is interfered with similar color can, moreover the distance between target and vidicon is bigger gradually when human is walking. So some changes have taken place in scale with target appearing in the lens. The experiment result shows the extend SURF algorithm overcomes the interference with similar color, and retains invariant for scale, achieves good tracking performance.

Fig. 3. Human tracking on a simple scene

(a) Human tracking based on SURF

(b) Human tracking based on the algorithm proposed

Fig. 4. Human tracking under occlusions

The second video is human tracking with occlusion. Two persons are running in the playground, a person in brown was occluded by the other one in blue now and then. From the tracking results in figure 4, we can see that the algorithm base on SURF is distracted seriously and lose the target when occlusion occurs, but the algorithm based on the extend SURF works well throughout the whole sequence. Reasons for the case were analyzed basing on experimental phenomenon. SURF feature only uses the gray information of images, ignores color information, so it is so difficult to identify and match the target with similar texture that miss real target. Extend SURF feature not only maintains the characteristics of SURF, but also makes use of color information, so the proposed algorithm can track the object selected accurately when the above situation exists.

The third video is car tracking in a complex environment, the light is dark and there are some other ones near the object. We analyze the 210th frame, the 232nd frame, the 301st frame, the 340th in video as shown in figure 5. The experiment result indicates the proposed algorithm can track object accurately.

Fig. 5. Car tracking under the complex environment

6   Conclusions

The paper proposed a novel object tracking algorithm by integrating color and SURF (Speeded Up Robust Features) in the frame of particle filter. The proposed algorithm has been tested on different scenes and proved stability and robustness. The results show it is stable and robust. In our future work, we also want to use some new swam intelligence based algorithms [13,14] to solve the problem of object tracking.

Acknowledgment. This research is partially supported by the Specialized Research Fund for the Doctoral Program of Higher Education of China (Grant No. 20114101110005), Science and technology research plan funded key projects province department of education of Henan (Grant No. 12A410002), the National Natural Science Foundation of China(Grants nos.71001072, 71271140, 71240015), and the Natural Science Foundation of Guangdong Province (Grants nos. S2012010008668, 9451806001002294).

References

1. Wang, H., Wang, J.T., Ren, M.W.: A new robust object tracking algorithm by fusing multi-features. Journal of Image and Graphics 14, 489–495 (2009)
2. David, G.L.: object recognition from local scale-invariant features. In:7th IEEE International Conference on Computer Vision, vol. 2, pp. 1150–1157 (1999)

3. David, G.L.: Distinctive image features from scale-invariant keypoints. International Jornal of Computer Vision 60, 91–110 (2004)
4. Zeng, L., Wang, Y.Q., Tan, J.B.: Improved algorithm for SIFT feature extraction and matching. Optics and Precision Engineering 19, 1392–1397 (2011)
5. Tong, R.Q., Huang, Y.Q., Tian, R.J.: SURF algorithm and its Detection effect on object tracking. Journal of Southwest University of Science and Technology 26, 63–67 (2011)
6. Grabner, M., Grabner, H., Bischof, H.: Fast approximated SIFT. In: Narayanan, P.J., Nayar, S.K., Shum, H.-Y. (eds.) ACCV 2006. LNCS, vol. 3851, pp. 918–927. Springer, Heidelberg (2006)
7. Bay, H., Ess, A., Tuytelaars, T.: Speeded-up robust features(SURF). Computer Vision and Image Understanding 110, 404–417 (2008)
8. Dong, H., Han, D.Y.: Research of image matching algorithm based on SURF features. In: International Conference on Computer science and Information Processing, pp. 1140–1143 (2012)
9. Li, J.G., Wang, T., Zhang, Y.M.: Face detection using SURF cascade. In: IEEE International Conference on Computer Vision Workshops, pp. 2183–2190 (2011)
10. Sun, X., Fu, K., Wang, H.J.: High resolution remote sensing image understanding, 68–74 (2011)
11. Comaniciu, D., Meer, P.: Mean shift: a robust approach toward feature space analysis. Pattern Analysis and Machine Intelligence 24, 603–619 (2002)
12. Marius, M., David, G.L.: Fast approximate nearest neighbors with automatic algorithm configuration. In: International Conference on Computer Vision Theory and Applications, VISAPP, pp. 331–340 (2009)
13. Niu, B., Wang, H.: Bacterial colony optimization. Discrete Dynamics in Nature and Society, 1–28 (2012)
14. Niu, B., Fan, Y., Xiao, H., Xue, B.: Bacterial foraging-Based approaches to portfolio optimization with liquidity risk. Neurocomputing 98(3), 90–100 (2012)

First Progresses in Evaluation of Resonance in Staff Selection through Speech Emotion Recognition

Vitoantonio Bevilacqua[1,*], Pietro Guccione[1,*], Luigi Mascolo[1],
Pasquale Pio Pazienza[1], Angelo Antonio Salatino[1], and Michele Pantaleo[2]

[1] Dip. di Ingegneria Elettrica e dell'Informazione, Politecnico di Bari, via Orabona 4,
70125 Bari, Italy
[2] AMT Services s.r.l., Viale Europa, 22 – Bari, Italy
{bevilacqua,guccione}@poliba.it

Abstract. Speech Emotion Recognition (SER) is a hot research topic in the field of Human Computer Interaction. In this paper a SER system is developed with the aim of providing a classification of the "state of interest" of a human subject involved in a job interview. Classification of emotions is performed by analyzing the speech produced during the interview. The presented methods and results show just preliminary conclusions, as the work is part of a larger project including also analysis, investigation and classification of facial expressions and body gestures during human interaction. At the current state of the work, investigation is carried out by using software tools already available for free on the web; furthermore, the features extracted from the audio tracks are analyzed by studying their sensitivity to an audio compression stage. The Berlin Database of Emotional Speech (EmoDB) is exploited to provide the preliminary results.

Keywords: Emotional Speech Classification, Emotion Recognition, Acoustic Features Extraction.

1   Introduction and Motivations

Automatic Speech Emotion Recognition is a very active research topic, having a wide range of applications. It can be used to detect the customers dissatisfaction in automatic remote call centres, to monitor the mental level of attention of a pilot in an aircraft cockpit, the trend of depressive symptoms in patients with mood disorders, the level of captivation skill of a teacher during a lesson in order to enhance the quality of the lesson, and in many other contexts [28]. In this paper, speech prosodic features are analyzed to the aim of finding reciprocal interest, empathy and agreement during a job interview between the examiner(s) and the candidate. This suit of emotional states can be defined *resonance*. The ultimate goal of such analysis is to infer, through meta-language characteristic (face expressions, speech, body position and gestures), the level of *synergy* between a set of actors in a specific situation (for example a job interview). To this aim speech emotion recognition plays a fundamental role, as the

---

* Corresponding authors.

D.-S. Huang et al. (Eds.): ICIC 2013, LNAI 7996, pp. 658–671, 2013.
© Springer-Verlag Berlin Heidelberg 2013

interview is basically a spoken process between two or more actors. Emotion recognition in speech is a relatively new field of research that has been explored with valuable results (see for example [24], [25], [11]). In literature there are several works that explore the performances of different classifiers, such as linear discriminant analysis, k-nearest neighbour, hidden Markov models, artificial neural networks and support vector machines, on different databases of utterances [29]; other works explore the possibility to select innovative features as the short time log frequency power coefficients [22] or the Teager energy operator [18] and so on. However, the case of emotional state recognition in speech during an interview involves specific challenges that have not been sufficiently explored in literature, as the need to work in near real time, in presence of several subjects and in different environmental conditions. The framework is new and suggestive for its potentials; currently companies spend a considerable part of their budget in staff selection. Anyway, the proposed system has not been thought of as a fully automated software tool that could be able to substitute the role of psychologists, in future. Rather, it can be seen as a helpful support to human evaluation and it is potentially opened to new applications (e.g. judiciary questioning). The specific contribution of the paper in the field of speech emotion recognition is mainly in the novel organization of the existing stuff (basically algorithmic methods, class labels definitions, freely available software tools and databases), to the aim of providing a new and integrated framework for a decision support tool. The paper is organized as follows. In Section 2 the system model and the framework are outlined; in Section 3 the system implementation is described and in Section 4 and 5 some results and preliminary conclusions are drawn.

2   Model Framework

The main purpose of the emotion recognition system here described is to provide a response to the level of positive (or negative) resonance of a candidate during a job interview. Some simplifying hypotheses have been assumed (that shall be removed in future, with the project progress). It is supposed that the job interview is carried out between a single candidate and one or more examiners and it is, for the most, a talk (the presence and the interaction between more candidates is hence excluded). During the interview the candidate speech, facial expressions and body gestures are recorded, as the final decision could be taken by integrating the information from these three sources[1]. The system must be able to work both in an offline (i.e. after the interview) and an online (i.e. during the interview) context. The last condition implies the ability of the software tool to be able to work in (near) real time. Finally the system must be robust to several possible impairments, as the environmental noise during recording, the level of (audio/video) compression, the speaker gender or age, the language and so on. Another impairment, very important and for now not accounted for, is the superimposition of many audio sources: for example the voice of the examiners and that of the candidate. For this reason, Blind Source Separation (BSS) techniques (as

---

[1]   The legal implications of the recording step are not accounted for here.

Independent Component Analysis) should be exploited to separate the different sources before the emotion recognition process. The main parts needed to set up such system are basically a human computer interface (i.e. a system to acquire and extrapolate the useful audio/video information), a software tool, which performs the processing and the classification of the achieved information, and a suitable way to present the output to the examiners (see Fig. 1). We currently focused the analysis just on the software tool, specifically on the algorithmic choices needed for an effective system response, leaving the first and the third parts to future developments. In absence of any experimental set up, we used one of the databases freely available on the web, namely EmoDB [9]. As the majority of speech emotion data bases, EmoDB classifies the emotional states grounding on the studies and theories originally proposed by Ekman [26], for which from six to seven universal emotions can be recognized, regardless of the culture and language involved. In these DBs actors usually play a set of sentences with little or no coupling with the emotional state they are required to interpret. For details on EmoDB see Section 4.1.

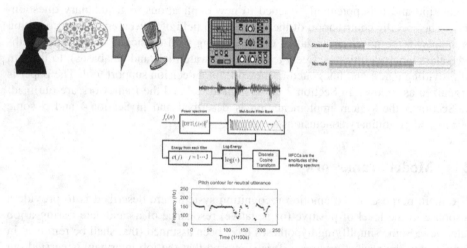

Fig. 1. A schematic impression of the main parts of the system

3 Description of the Algorithm

The main steps of the algorithm are sketched in Fig. 2. The algorithm has three basic parts: feature extraction, dimensionality reduction and classification, each expounded in the following sections.

3.1 Feature Extraction

Voice features are instantaneous or time averaged characteristics that jointly describe some property of the speech. Among the others, the pitch, i.e. the fundamental voice frequency, is the most important. Properties related to energy, energy distribution

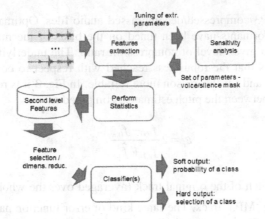

Fig. 2. Block scheme of the algorithm proposed

throughout the frequencies, rate of the speech and spectrum shape, are extracted on a frame-by-frame basis, as the speech is usually considered a non-stationary random process. Since the process must work in near real time, a lag T_L has been fixed which represents the maximum tolerable delay to wait before having an overall system response/classification. This lag must include the time needed to ingest the current part of the input track, to extract the features and to take a decision about the class label. It is intended that $T_L \gg T_f$, being T_f the frame duration, since in the time interval of a lag the features collected in the frames are analyzed to extract a number of statistics. It is for this reason that T_L must be sufficiently long. Indeed, in order to infer statistics as mean, standard deviation, percentiles and so on it is necessary to arrange a suitable number of frames. In future the lag could be time-variant; currently its duration is set as fixed. For example, if we fix $T_f = 0.06$ s (a suitable frame size for the pitch estimation), $T_L = 1$ s and a superimposition of the frames of 25% (so every frame starts 15 ms after the previous one), we achieve a reasonable number of about 64 frames to get the statistics (in the hypothesis of "all voiced" frames). It is well known that a too narrow or a too large set of features can impair the performance of a classifier (see [27] and the discussion in Sec. 4.2). For this reason we started the analysis using a reasonable (not too large) set of features, illustrated in Table 2. An important step of a feature extraction algorithm is the tuning of the parameters. As an example, for the pitch extraction algorithm here adopted [6], it is important to set the silence and voiced/unvoiced thresholds (to correctly recognize the frames) as well as a reasonable research pitch interval; for the formant extraction [13] the number of formants and the ceiling frequency have to be decided; for the Mel-frequency Cepstrum Coefficients (MFCC) [8] it is necessary to set up the position and distance between the filters, and so on. As a practical way to tune the parameters, we compared the features extracted from the original track with the corresponding ones extracted from the same track passed through a compression/decompression step. Audio compression is commonly adopted for disk space saving and is considered in this context a possible impairment, as it is usually a loss compression, when offline processing is carried out. For this reason, a comparison has been performed between the features extracted from the original and the MP3-compressed/decompressed audio files. Optimal sets of parameters for pitch and formants have been tuned on the basis of the minimization of an error function, for a given level of compression ratio. The underlying hypothesis for such optimization is that the robustness reached with respect to compression makes the estimates stable and for this reason more reliable. In Fig. 3 it is reported the distribution of the error between the pitch estimates, computed as:

$$\varepsilon_i = 2 \frac{f_{0,i} - \hat{f}_{0,i}}{f_{0,i} + \hat{f}_{0,i}} \tag{1}$$

where $f_{0,i}$ is the pitch of the original track (averaged over the whole track) and $\hat{f}_{0,i}$ is the pitch over the 'MP3' track. The same kind of error function has been applied to find the best ceiling frequency for the formants extraction algorithm. The silence and voiced/unvoiced thresholds have been tuned within suitable intervals until the standard deviation of the error reached a minimum. To evaluate the goodness of such minimum we referenced to previous studies in which it is stated that the MP3 compression/decompression steps can be accounted for as an additive noise effect. In [6] it has been estimated that adding a Gaussian noise with SNR = 20 dB to the audio signal produces an impairment in the pitch estimation whose error has a stddev < 1%, which is in line with our analysis. After parameter tuning and feature extraction, each feature has been normalized by zeroing the mean and then dividing the result by its standard deviation.

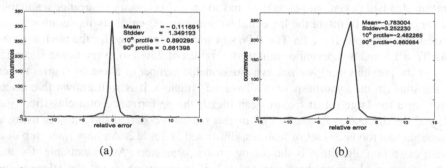

(a)   (b)

Fig. 3. Histogram of the relative error (in %) between the pitch estimates in the original tracks and the compressed/decompressed tracks. Compression ratio is 4 in (a) and 16 in (b).

3.2   Dimensionality Reduction

There are several good reasons why the number of features should be reduced when a classifier is trained: (i) satisfying the *generalization* property of the classifier and overcome the risk of *overfitting*; (ii) reducing the computational burden; (iii) improve the performance of the classification system (*peaking phenomenon*) [27]. In this preliminary design of the classification system, two feature ranking techniques have been explored, followed by an empirical selection. The first ranking criterion is the

Information Gain (IG) measure. The IG of an attribute A is defined as the expected reduction in entropy caused by partitioning the data collection S according to this attribute:

$$IG(S,A) = H(S)-H(S|A), \tag{2}$$

where H denotes the information entropy [21]. The first term in (2) is the entropy of the original collection S and the second term is the expected value of the entropy after having partitioned S using attribute A. More specifically:

$$IG(S,A) = H(S) - \sum_{v \in Values(A)} \frac{|S_v|}{|S|} H(S_v), \tag{3}$$

where $Values(A)$ is the set of all possible values for attribute A, and S_v is the subset of S for which attribute A has value v (i.e., $S_v = \{s \in S \mid A(s) = v\}$). The second criterion exploits a linear discriminant classifier to rank the features. In order to explain the idea, let us consider a two-class classification problem. In [16] it has been suggested that some given feature ranking coefficients $\mathbf{w}$ can be used as weights of a linear discriminant classifier:

$$D(\mathbf{x}) = \mathbf{w} \cdot (\mathbf{x} - \mathbf{\mu}), \tag{4}$$

where $D(\cdot)$ is the scalar decision function, $\mathbf{x}$ is a d-dimensional input pattern and $\mathbf{\mu}$ is a vector of means. Inversely, the weights multiplying the inputs of a linear discriminant classifier can be used as feature ranking coefficients. To this aim, the following iterative procedure, called *Recursive Feature Elimination*, has been proposed [17]:

1. Train the classifier (optimize the weights w_i with respect to a cost function J).
2. Arrange the features according to a ranking criterion (for example according to the square of the weights, w_i^2).
3. Remove the feature with the smallest ranking.

A linear SVM has been used as classifier. The multi-class classification problem is faced by separately ranking the features for each class using a one-vs-all method in which a *winner-takes-all* strategy is adopted: the classifier with the highest output function assigns the class. The use of linear transformations for dimensionality reduction, as Principal Component Analysis [19], has been investigated too. However such methods require the evaluation of all the original features, thereby preventing the gain of the reduced computational complexity. A straight PCA approach to all the available features has been applied for dimensionality reduction, but the method led to very poor classification performances (here not shown). Given the IG and the SVM-based rankings, the feature selection for the classifier design has been performed by choosing part of the top ranked features. The *keeping threshold* has been empirically set on the performance of a linear SVM classifier.

3.3   Classification

The performances of two classification methods have been investigated: Artificial Neural Networks and Support Vector Machines. In this section a short description of both algorithms is given. Learning procedures have been omitted as more details can be found in literature [5]. Artificial Neural Networks (ANNs) have turned out to be a fundamental and flexible tool in the context of pattern recognition and classification [1], [2], [4]. ANNs are made up of simple elements, called neurons, with basic computational properties. Each neuron activates its output or not, accordingly to a nonlinear function (usually a sigmoid-like function) fed by a linear combination of the inputs. The number of layers of the network and the number of neurons in each layer are chosen according to data and design considerations; the weights of the combination are the unknowns to be found by means of a suitable training stage and exploiting labelled data samples. The training comes to a halt when the network output matches the right targets (in the sense of an error function minimization). The network is then able to properly classify new data samples, which were not used in the training stage. In this work several Multi-Layer Perceptron Neural Networks (MLP) have been trained by the Error Back Propagation algorithm and their performances compared [5]. Support vector machines (SVMs) are supervised learning models in which the decision boundary is optimized to be the one for which the margin, i.e. the smallest distance between the decision boundary and any of the samples, is maximized [10], [5], [3]. A polynomial kernel has been adopted in the SVMs and the Sequential Minimal Optimization (SMO) algorithm has been used to train the classifier [23]. The multi-class problem has been faced by using pairwise classification, according to Hastie and Tibshirani [15]. In Sec. 4.2 the performances of both ANNs and SVMs are compared under different conditions, by using the 10-fold cross-validation technique.

4   Preliminary Results

4.1   Dataset, Working Environment and Feature Extracted

In EmoDB 10 speakers of both genders played 10 different sentences in German language using 7 emotional states (see Table 1, first two columns). Each sentence is repeated up to 4 times. A total of 535 tracks are publicly available. Each track has a mean duration of 2 s and has been recorded as a mono audio signal @16 kHz, 8 bps.

Table 1. Re-classification of EmoDB emotions

Index	Emotion	Classification
1	Anxiety/Fear	-
2	Disgust	-
3	Happiness/Surprise	+
4	Boredom	-
5	Neutral	+
6	Sadness	-
7	Anger	-

For the most of the feature extraction procedures, the algorithms developed in Praat have been used [7]; the fine tuning of parameters for each feature extraction algorithm has been implemented as described in Section 3.1. Feature ranking and selection and classification have been implemented in Weka [14], which is commonly known to provide the environment to develop and explore a large set of machine learning algorithms on data mining problems. According to the purpose of the paper (i.e. identifying the positive / negative resonance between a candidate and the interviewers), we arbitrarily re-classified the emotions included in EmoDB in two classes, grouping them on the basis of what can be considered positive or negative to the aim of a job interview. The grouping of EmoDB emotions in the positive resonance (+) and negative resonance (-) classes is reported in the third column of Table 1. It is worthwhile that we decided to assign the neutral behaviour in the + class as we deemed that during an interview this state should be considered a sign of serenity shown by the candidate and so ultimately positive.

Table 2. Descriptors and their statistics

#	Feature Description
1-5	Pitch (mean, std, min, max, range)
6-10	1st formant (mean, std, min, max, range)
21-25	4th formant (as above)
26-30	Average Intensity (mean, std, min, max, range)
31-35	1st Mel-freq coeff. (mean, std, min, max, range)
86-90	12th Mel-freq coeff. (as above)
91-95	1st Linear Predicting coeff. (mean, min, max, std, range)
166-170	16th Linear Predicting coeff. (as above)
171-175	Harmonic-to-noise ratio (mean, min, max, std, range)
176	Voiced frames in pitch
177	Pitch envelope slope
178-179	Zero crossing rate (mean, std)

The features extracted from the tracks are reported in Table 2. To simplify, we have initially supposed to compute the features by using a time lag T_L equal to the track length and considering the simulation as an offline analysis. The values, extracted on a frame-by-frame basis, for (1) the pitch, (2) the formants (f_i, i=1,…,4), (3) the average intensity, (4) the harmonic-to-noise ratio (HNR), (5) the mel-frequency cepstrum coefficients (MFCCs) and (6) the linear prediction coefficients (LPCs) have been used to get some statistics for the whole track: the mean, the standard deviation, the minimum and maximum (actually computed as the 10% and 90% percentiles) and the range (i.e. max-min) have been extracted for each feature. The voiced frames in pitch is the ratio between the number of frames where pitch has been taken for the analysis divided by the total number of frames. The pitch envelope slope is the slope (with sign and magnitude) of a LMS linear fitting over the voiced frames, i.e. those for which the pitch has been estimated. Finally, the zero crossing rate represents the number of crossings by zero in a frame. For this parameter just mean and standard deviation over frames have been evaluated.

4.2   Preliminary Results and Discussions

The classification for both the 2 classes (+ / -) and the 7 classes (the emotional states of EmoDB) problems are here presented as preliminary results, in order to test the effectiveness of the ranking methods, the performances of the classifiers and their sensibility to the number of features. In a future work such results will be integrated with those coming from different sources, i.e. facial expressions and body gestures recognition. To this aim, it will be considered the opportunity of using soft outputs, i.e. the posterior probabilities, provided by alternative classification algorithms like the *Relevance Vector Machine* (RVM), for example. Experiments have been carried out with two different classification algorithms: a Feed Forward Neural Network (NN, from now on) and a Support Vector Machine (SVM) with a linear kernel (with fixed parameter $C = 1$, $\varepsilon = 1.0E - 12$). Both classifiers have been shortly described in Section 3.3. Two different NN have been trained by keeping fixed the number of hidden layers (1 hidden layer) and the nodes activation functions (*logsig* for the hidden layer, *purelin* for the output layer), while varying the number of neurons: one with 20 neurons (NN-20) and one with $a = (N_d + N_c)/2$ neurons (NN-a) in the hidden layer, being N_d the number of features and N_c the number of classes. N_c neurons have been fixed in the output layer (2 or 7). In both cases, 500 epochs have proved sufficient to get the best performance.

Table 3. Average performances using the complete set of extracted features

Classifier	# of classes	TP rate	FP rate	Precision	F-measure
SVM	7	0.778	0.040	0.779	0.777
NN-a	7	0.804	0.036	0.803	0.802
NN-20	7	0.796	0.037	0.795	0.795
SVM	2	0.753	0.369	0.752	0.753
NN-a	2	0.804	0.325	0.798	0.800
NN-20	2	0.806	0.312	0.801	0.803

All the results are the outcomes of the 10-fold cross-validation technique. Firstly, the classifier performances have been evaluated using all the 179 extracted features. The results of the analysis are shown in the previous Tab. 3. For each trial the true positive rate (or recall), the false positive rate, the precision and the F-measure (harmonic mean of precision and recall) have been reported. Each number refers to the weighted average over the 7 or 2 classes (using as weights their relative abundance).
In both the 7-class and the 2-class problems, the NN classifier outperforms the SVM, as shown by the F-Measure scores. Furthermore, it seems that the number of neurons in the hidden layer is not crucial for the NN performances.As mentioned in Sec. 3.2, exploiting all the features may entail a decay of the classification performances.

Therefore, both the Information Gain (IG) and the SVM-Recursive Feature Elimination (SVM-RFE) ranking techniques have been applied to cut the number of features. In order to choose the optimal number of retained features for each ranking algorithm, the SVM classifier has been trained and validated several times, with an increasing number of input features, by following the feature rankings. The *keeping threshold* has been set as the one which has led to the best SVM classifier performances in terms of *F*-measure. The accuracies obtained by such analysis and by varying the number of features from 10 to 90 are shown in Fig. 4.

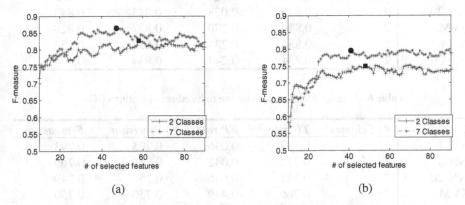

(a)   (b)

Fig. 4. Accuracy (in term of F-Measure) of the SVM classification algorithm vs. the number of selected features in case of (a) SVM-RFE and (b) IG-based rankings. Both the 7 class (red dotted line with star markers) and the 2 class cases (blue solid line with plus markers) have been represented. The black markers address the maximum of each curve, which corresponds to the optimal number of retained attributes.

Table 4. Optimal number of selected features

Ranking algorithm	7 classes	2 classes
Information Gain	41	48
SVM-RFE	47	59

It can be noticed that the accuracy of the classifier gains a maximum in correspondence of a certain number of features and does not improve anymore by adding further features, which can be supposed to act as nuisances. The selection results are summarized in the previous Tab. 4. It is noteworthy that the two ranking methods produce agreeing results: it seems that in order to distinguish between a smaller number of classes it is necessary to retain more features. This could be explained as the consequence of a high overlap between the defined positive and negative resonance classes, so that a larger dimensionality is required to discriminate samples in the data space. For example, the *surprise* emotion class could probably overlap with the *anger* class while the *neutral* emotion class could overlap with the *boredom* class. Probably, to achieve better results, it could be better to split again the resonance classes in the 7 emotions so that, the *hypersurface* describing the data space in a smaller dimensionality becomes enough precise to better distinguish the classes.

Table 5. Average performances on the reduced set of features (SVM-RFE)

Classifier	# of classes	TP rate	FP rate	Precision	F-measure
SVM	7	0.860	0.024	0.864	0.861
NN-*a*	7	0.864	0.026	0.863	0.862
NN-20	7	0.843	0.028	0.843	0.843
SVM	2	0.830	0.270	0.827	0.828
NN-*a*	2	0.854	0.236	0.852	0.853
NN-20	2	0.834	0.244	0.834	0.834

Table 6. Average performances on the reduced set of features (IG)

Classifier	# of classes	TP rate	FP rate	Precision	F-measure
SVM	7	0.793	0.036	0.795	0.793
NN-*a*	7	0.770	0.042	0.769	0.769
NN-20	7	0.751	0.045	0.751	0.749
SVM	2	0.766	0.449	0.750	0.750
NN-*a*	2	0.742	0.402	0.737	0.739
NN-20	2	0.746	0.408	0.738	0.741

The classifier performances have been evaluated using only the selected number of features as in Tab. 4. The results for the SVM-RFE and the IG ranking methods are shown in Tab. 5 and Tab. 6, respectively. As expected, the feature reduction step has improved the results for all the classifiers, gaining up to 8.4% in the case of the SVM (7 class case). Smaller increases in performance have been observed by exploiting the IG-based feature selection technique, compared to the SVM-RFE one. This is an expected conclusion if we compare the classification curves of the *F*-Measure score as a function of the number of (ranked) features shown in Fig. 4. For both the 7 and the 2 class problems, the IG curves lay under the ones produced by SVM-RFE. On the other hand, the SVM-RFE method is of course more time consuming and computationally expensive compared to IG. The best classification performance (accounted by the *F*-measure) has been provided by the NN-*a* classifier in both the 7-class (0.862%) and the 2-class (0.853%) problems, by the SVM-RFE based selection technique. This result is a valuable improvement compared to what reported in literature (compare, for example [22], [29], [20]), though it is important to highlight the differences of the analysis conditions. In [22] the short time log-frequency power coefficients have been used as features and the Hidden Markov Models to classify 6 primary emotions on utterances (not on complete sentences) of two different languages (chinese mandarin and burmese), getting an average accuracy of 78%. In [29] an average classification rate of 51.6% has been obtained on a Danish emotional speech database, distinguishing between 5 emotions and combining different classifiers with a feature selection method. In [20], finally, just syllables (from which 32 features were selected) have been used to distinguish between 5 emotions, reaching a 55% of classification accuracy. For the best results the confusion matrices have been reported in Tab. 7 and in Tab. 8. By inspecting the confusion matrix for the 7 class problem, the following misclassifications have proved frequent:

— anger is easily misclassified as happiness;
— happiness, besides being misclassified as anger, is also misclassified as fear or disgust;
— neutral is sometimes misclassified as either sadness, boredom or fear.

Table 7. Confusion matrix for the NN-*a* classifier with 7 classes (SVM-RFE Selection)

True class / Classified as →	1	2	3	4	5	6	7
1	64	0	3	1	0	0	2
2	0	37	4	1	2	1	1
3	4	3	48	1	0	1	14
4	2	0	0	72	5	2	0
5	3	0	1	4	68	3	0
6	1	1	0	4	0	56	0
7	1	0	8	0	0	0	117

Table 8. Confusion matrix for the NN-*a* classifier with 2 classes (SVM-RFE Selection)

True class / Classified as →	+	-
+	106	44
-	34	351

5   Conclusions

A speech emotional classifier has been presented. The algorithm suite has been implemented starting from the analysis of the optimal parameter settings for feature extraction; feature ranking and selection have been evaluated by means of the SVM classifier; finally SVM and NN have been exploited to classify both the 7-emotions and the 2-resonance cases providing valuable results, sensibly better than those in literature, in both situations. A preliminary conclusion is that in order to get more accuracy it is better to distinguish between more classes (the available emotions of the dataset) and then to perform the class merging in a second step. Anyway the work represents just a preliminary analysis of the system perspective, as room for further improvements exists. In detail, other features as well as other ways to select the features should be explored; furthermore more accurate methods to classify should be investigated. For example we did not faced the problem of the data resampling, needed to balance the number of samples when the classes are joined. Another open problem is that of testing the implemented system on larger datasets, possibly with different languages or emotion labelling, in order to make the system as data independent as possible and to generalize its performances. Clearly, facing such open problems will probably provide different results. As the aim is to improve the system performances as much as possible, the presented results must be taken with caution and shall be used as a basis for comparison. Moreover the algorithms should be developed in a single software environment in order to collect the processing times and validate the feasibility of the near real time case. Finally, soft decision on classes, provided as posterior probabilities, will be considered as the real outcome of the system in future implementations.

Acknowledgements. This study was supported by the Italian PON FIT Project called "Sviluppo di un sistema di rilevazione della risonanza (SS-RR) N° B01/0660/01-02/X17" - Politecnico di Bari and AMT Services s.r.l. – Italy.

References

1. Bevilacqua, V., et al.: A face recognition system based on Pseudo 2D HMM applied to neural network coefficients. Soft Computing 12(7), 615–621 (2008)
2. Bevilacqua, V., et al.: 3D nose feature identification and localization through Self-Organizing Map and Graph Matching. Journal of Circuits Systems and Computers, 191–202 (2010)
3. Bevilacqua, V., Pannarale, P., Abbrescia, M., Cava, C., Paradiso, A., Tommasi, S.: Comparison of data-merging methods with SVM attribute selection and classification in breast cancer gene expression. BMC Bioinformatics 13(7), S9 (2012), doi:10.1186/1471 2105-13-S7-S9
4. Bevilacqua, V.: Three-dimensional virtual colonoscopy for automatic polyps detection by artificial neural network approach: New tests on an enlarged cohort of polyps. Neurocomputting, 0925–2312 (2012)
5. Bishop, C.M.: Pattern Recognition and Machine Learning Information Science and Statistics (2006)
6. Boersma, P.: Accurate short-term analysis of the fundamental frequency and the harmonics-to-noise ratio of a sampled sound. Proceedings of the Institute of Phonetic Sciences 17, 97–110 (1993)
7. Boersma, P.: Praat: a system for doing phonetics by computer. Glot International 9(10), 341–345 (2001)
8. Bou-Ghazale, S.E., Hansen, J.H.L.: A comparative study of traditional and newly proposed features for recognition of speech under stress. IEEE Transactions on Speech and Audio Processing 8(4), 429–442 (2000)
9. Burkhardt, F., Paeschke, A., Rolfes, M., Sendlmeier, W., Weiss, B.: A database of german emotional speech. In: Proceedings on Interspeech 2005, pp. 1517–1520 (2005)
10. Cortes, C., Vapnik, V.: Support-Vector Networks. Mach. Learn 20(3) (1995)
11. Dellaert, F., Polzin, T., Waibel, A.: Recognizing Emotion in Speech. Proceedings on Spoken Language 3, 1970–1973 (1996)

12. Eyben, F., Wollmer, M., Schuller, B.: Introducing the Munich open-source emotion and affect recognition toolkit. In: 3rd International Conference on Affective Computing and Intelligent Interaction and Workshops (2009)
13. Fant, G.: The Acoustic Theory of Speech Production. Mouton, The Hague (1960)
14. Hall, M., Frank, E., Holmes, G., Pfahringer, B., Reutemann, P., Witten, I.: The WEKA Data Mining Software: An Update. SIGKDD Explorations 11, 1 (2009)
15. Hastie, T., Tibshirani, R.: Classification by Pairwise Coupling. In: Advances in Neural Information Processing Systems (1998)
16. Golub, T.R.: Molecular classification of cancer: class discovery and class prediction by gene expression monitoring. Science 286, 531–537 (1999)
17. Guyon, I., Weston, J., Barnhill, S., Vapnik, V.: Gene Selection for Cancer Classification using Support Vector Machines. Mach. Learn. 46(1-3) (2002)
18. Jabloun, F.: Teager Energy Based Feature Parameters for Speech Recognition in Car Noise. IEEE Signal Processing Letters 6(10) (1999)
19. Jolliffe, I.T.: Principal Component Analysis, 2nd edn. Springer Series in Statistics. Springer, NY (2002)
20. McGilloway, S., Cowie, R., Douglas-Cowie, E., Gielen, S., Westerdijk, M., Stroeve, S.: Approaching automatic recognition of emotion from voice: a rough benchmark. In: ISCA Tutorial and Research Workshop (ITRW) on Speech and Emotion (2000)
21. Mitchell, T.M.: Machine Learning, 1st edn. McGraw-Hill, New York (1997)
22. New, T.L., Foo, S.W., De, S.L.C.: Speech emotion recognition using hidden Markov models. Speech Communication 41(4), 603–623 (2003)
23. Platt, J.C.: Fast Training of Support Vector Machines using Sequential Minimal Optimization. In: Schoelkopf, B., Burges, C., Smola, A. (eds.) Advances in Kernel Methods - Support Vector Learning. MIT Press (1998)
24. Scherer, S., Hofmann, H., Lampmann, M., Pfeil, M., Rhinow, S., Schwenker, F., Palm, G.: Emotion Recognition from Speech: Stress Experiment. In: Proceedings of the Sixth International Language Resources and Evaluation (LREC 2008) European Language Resources Association, ELRA (2008)
25. Scherer, K.R., Banse, R., Wallbott, H.G., Goldbeck, T.: Vocal cues in Emotion Encoding and Decoding. Motivation and Emotion 15, 123–148 (1996)
26. Scherer, K.R., Johnstone, T., Klasmeyer, G.: Vocal expression of emotion. Oxford University Press, New York (2003)
27. Theodoridis, S., Koutroumbas, K.: Pattern Recognition, 4th edn. Academic Press (2008)
28. Ververidis, D., Kotropoulos, C.: Emotional speech recognition: Resources, features, and methods. Speech Communication 48(9), 1162–1181 (2006)
29. Ververidis, D., Kotropoulos, C., Pitas, I.: Automatic emotional speech classification. In: Proceedings of IEEE International Conference on Acoustics, Speech, and Signal Processing (ICASSP 2004), vol. 1, pp. I-593. IEEE (2004)

A Mass Spectra-Based Compound-Identification Approach with a Reduced Reference Library[*]

Zhan-Li Sun[1], Kin-Man Lam[2], and Jun Zhang[1]

[1] School of Electrical Engineering and Automation,
Anhui University, China
[2] Department of Electronic and Information Engineering, Hong Kong
Polytechnic University, Hong Kong
zhlsun@foxmail.com

Abstract. In this paper, an effective and efficient compound identification approach is proposed based on the frequency feature of mass spectrum. A nonzero feature-retention strategy, and a correlation based-reference library reduction strategy, are designed in the proposed algorithm to reduce the computation burden. Further, a frequency feature based-composite similarity measure is adopted to decide the chemical abstracts service (CAS) registry numbers of mass spectral samples. Experimental results demonstrate the feasibility and efficiency of the proposed method.

Keywords: Spectrum matching, similarity measure, discrete Fourier transform, feature selection.

1   Introduction

The similarity based-spectrum matching is one widely used chemical compound identification method. So far, various spectrum-matching algorithms have been developed for chemical compound identification, e.g. cosine correlation, probability-based matching, and normalized Euclidean distance, etc. Although they are the most informative ions for compound identification, the peak intensities of fragment ions with large mass-to-charge (m/z) values in a GC-MS mass spectrum tend to be smaller. Therefore, the performance of compound identification can be improved by increasing the relative significance of the large fragment ions via weighing more on their peak intensities. In [1], a classical composite similarity measure is proposed by computing the cosine correlation with weighted intensities. Information in a frequency domain is useful in pattern recognition [2-4]. Instead of the ratios of peak pairs [1], a dot product of the frequency features, which are obtained via discrete Fourier transform (DFT), is adopted in [5] to measure the similarity of the query

---

[*] The work was supported by grants from National Science Foundation of China (Nos. 60905023 and 61271098) and a grant from National Science Foundation of Anhui Province (No. 1308085MF85).

D.-S. Huang et al. (Eds.): ICIC 2013, LNAI 7996, pp. 672–676, 2013.

sample and the reference sample. Further, skewness and kurtosis of mass spectral similarity scores among spectra in a reference library are considered in [6] to design the optimal weight factors. In [7], a retention index is utilized to assist the determination of the CAS index of unknown mass spectra. Partial and semi-partial correlations, along with various transformations of peak intensity, are proposed in [8] as mass spectral similarity measures for compound identification.

Even though various methods have been developed and demonstrated to be feasible and effective for compound identification, how to decrease the computation burden is a prominent and intractable problem because the sizes of the query library and the reference library become larger and larger. In this paper, an effective and efficient compound identification approach is proposed based on the frequency feature of mass spectrum. Considering the sparsity of mass spectrum, a nonzero feature-retention strategy is proposed to decrease the feature space dimension of mass spectral samples. Moreover, for the mass spectrum with less features, a two-stage similarity measure scheme is designed further to reduce the computation burden. In the first stage, a lower-accuracy but more efficient similarity measure, Pearson's linear correlation coefficient, is computed and used to select a few most correlated mass spectral samples of the reference library. In the second stage, a high-accuracy but less efficient similarity measure, the frequency feature based-composite similarity measure [5], is adopted to perform the spectrum matching on the reduced reference library. Experimental results on the standard mass spectral database demonstrate the feasibility and efficiency of the proposed method.

The remainder of the paper is organized as follows. In Section 2, we present our proposed algorithm. Experimental results and related discussions are given in Section 3, and concluding remarks are presented in Section 4.

2   Methodology

Let a $m \times p$ matrix X_0 denote the mass spectrum data of a reference library, which includes m samples and p features. The element X_{ij}^0 of X^0 is the peak intensity of the jth fragment ion m/z value in the ith sample. Each row X_i ($X_i = [X_{i1}^0, \ldots, X_{ip}^0], i = 1, \ldots, m$) of X_0 denotes a mass spectral sample. Similarity, denote a $n \times p$ matrix Y^0 as an unknown mass spectral library with n query samples. In a reference spectral library, each mass spectrum is attached with a CAS registry number, which is used to denote the compound category. The task of chemical compound identification is to assign a query mass spectrum with a correct CAS index by using a spectrum-matching algorithm.

There are three main parts to the proposed method: decrease feature dimension, select most correlated reference samples and constitute a new reference library, extract frequency feature and compute similarity measure. A detailed description of these three parts is presented in the following subsections.

2.1   Nonzero Feature Selection

After observation, we found that mass spectral signals are very sparse. In the computation of similarity measure, these zeros elements of the mass spectral signals cannot bring any benefit to the computation accuracy. On the contrary, a large feature dimension will increase the computation time. Therefore, those features with zero peak intensity values can be removed, while the recognition performance of the similarity measure will not be changed significantly. In the proposed method, the nonzero indices of all samples are found at first, and then collected together. The repetitive indices are then removed from the index collection. Finally, the samples with the retained nonzero features constitute a new query library Y and a new reference library X.

2.2   Correlation Based-Reference Library Reduction

As the computation burden is mainly caused by the reference library size, a two-stage similarity measure computation strategy is designed in the proposed algorithm to alleviate this problem. In the first stage, due to its simplicity and computation efficiency, Pearson's linear correlation coefficient is adopted to select the most correlated mass spectral samples of the reference library.

For each query sample, we compute the Pearson's linear correlation coefficients $(r_1, \ldots, r_m)$ between it and all the samples of the reference library. Assume that the number of selected samples is d, after sorting the r_i values, d most correlated samples are chosen from the reference library X, and constitute a new reference library X'. As the computation time of Pearson's linear correlation coefficient is far less than most relative complicated similarity measures, the compaction time of the whole spectrum matching algorithm can be reduced significantly.

2.3   Frequency Feature Based-Similarity Measure

For a reference sample $x (x \in X')$ and a query sample y $(y \in Y')$, the weighted peak intensities can be given by

$$x_j{}^w = (z_j)^\alpha (x_j)^b, j = 1, \ldots, q \tag{1}$$

and

$$y_j{}^w = (z_j)^\alpha (y_j)^b, j = 1, \ldots, q \tag{2}$$

where z_j is the m/z value of jth intensity, the optimized weight factors a and b are set at 3 and 0.5, respectively [1]. The DFT $\mathbf{x}^f (\mathbf{x}^f = (x^f, \cdots, x^f))$ of $\mathbf{x}$ can be calcuated as

$$x_k{}^f = \sum_{d=0}^{q-1} x_d e^{-j\frac{2\pi}{q}kd}$$

$$= x_d (\cos \frac{2\pi}{q} kd + j \sin \frac{2\pi}{q} kd), k = 0, \ldots, q-1 \qquad (4)$$

Given x, y, x^{fr} and y^{fr}, the composite similarity can be defined as [5]:

$$S_{wr}(x, y, x^{fr}, y^{fr}) = \frac{N_x S_w(x, y) + N_{x \wedge y} S_w(x^{fr}, y^{fr})}{N_x + N_{x \wedge y}} \qquad (6)$$

where N_x and $N_{x \wedge y}$ are the numbers of nonzero peak intensities existing in the unknown query and in both the reference and the unknown query spectrum, respectively. For each query mass spectral sample X, the composite similarity between X and all samples of the reference library are computed and sorted. The CAS index of the query sample is considered to be that of the mass spectrum of the reference library with the largest composite similarity value.

3   Experimental Results

3.1   Databases and Experimental Set-Up

In order to verify the effectiveness of the proposed method (denoted as TS-RL WDFTR), in the experiments, the well-known NIST mass database is chosen as a reference library, and the repetitive library is selected as the query data. The compound identification performance is evaluated by the top one recognition accuracy. Considering the accuracy and the efficiency, the sample number Ns of the new reference library is set as 400 for all the following experiments.

3.2   Experimental Comparisons of Different Compound Identification Methods

As a comparison, we present the experimental results on the same data with three typical similarity measures, including weighted cosine correlation (denoted as WC), Stein and Scott's composite similarity (denoted as SSM), and DFT-based similarity measure (denoted as WDFTR) [5]. Moreover, the experimental results of the proposed method without feature reduction (denoted as RL-WDFTR) are also given here in order to investigate its function. Table 1 shows the recognition rates (RR) and the computation times $(CT, \times 10^4)$ of five methods. We can see that the recognition rate of WDFTR is higher than those of WC and SSM. And the computation times of these three methods are close. Further, it can be seen that the recognition rate of TS-RL-WDFTR is slightly lower than that of WDFTR. Nevertheless, the computation time of the former is only about 7% of the later. Considering the accuracy and computation time, we can conclude that the proposed method has a competitive compound identification performance to three existing methods, but with a far less computation burden than them. Moreover, the experimental results of RL-WDFTR and TS-RL-WDFTR indicate that the nonzero feature-retention strategy can effectively reduce the computation time.

Table 1. Comparisons of the recognition rates (RR) and the computation times (CT,$\times 10^4$) for five methods

Method	WC	SSM	WDFTR	RL-WDFTR	TS-RL-WDFTR
RR	0.8016	0.8050	0.8351	0.8326	0.8334
CT	3.8691	6.8560	5.7504	1.0643	0.4134

4　Conclusions

An effective and efficient compound identification approach is proposed in this paper based on the frequency feature of mass spectrum. Experimental results indicate that, the nonzero feature-retention strategy in the preprocessing stage can effectively decrease the feature dimension, while does not deteriorate the algorithm performance. Moreover, a correlation based-filtering strategy is verified to be able to provide an optimized reference library. In general, the proposed algorithm has been demonstrated to be an efficient spectrum matching method.

References

1. Stein, S., Scott, D.R.: Optimization and Testing of Mass Spectral Library Search Algorithms for Compound Identification. Journal of the American Society 5(9), 859–866 (1994)
2. Oliva, A., Torralba, A.: Modeling the Shape of the Scene: A Holistic Representation of the Spatial Envelope. International Journal of Computer Vision 42(3), 145–175 (2001)
3. Sun, Z.L., Rajan, D., Chia, L.T.: Scene Classification Using Multiple Features in A Two-Stage Probabilistic Classification Framework. Neurocomputing 73(16-18), 2971–2979 (2010)
4. Sun, Z.L., Lam, K.M., Dong, Z.Y., Wang, H., Gao, Q.W., Zheng, C.H.: Face Recognition with Multi-Resolution Spectral Feature Images. PLOS One (2013), doi:10.1371/journal.pone.0055700
5. Koo, I., Zhang, X., Kim, S.: Wavelet- and Fourier-transformations-based Spectrum Similarity Approaches to Compound Identification in Gas Chromatography/Mass Spectrometry. Analytical Chemistry 83(14), 5631–5638 (2011)
6. Kim, S., Koo, I., Wei, X.W., Zhang, X.: A Method of Finding Optimal Weight Factors for Compound Identification in Gas Chromatography-Mass Spectrometry. Bioinformatics 8(28), 1158–1163 (2012)
7. Zhang, J., et al.: A Large Scale Test Dataset to Determine Optimal Retention Index Threshold Based on Three Mass Spectral Similarity Measures. Journal of Chromatography A 1251, 188–193 (2012)
8. Kim, S., Koo, I., Jeong, J., Wei, X.W., Shi, X., Zhang, X.: Compound Identification Using Partial and Semipartial Correlations for Gas Chromatography-Mass Spectrometry Data. Analytical Chemistry 84(15), 6477–6487 (2012)

Role of Protein Aggregation and Interactions between α-Synuclein and Calbindin in Parkinson's Disease

M. Michael Gromiha[1], S. Biswal[2], A.M. Thangakani[3], S. Kumar[4],
G.J. Masilamoni[5], and D. Velmurugan[3]

[1] Department of Biotechnology, Indian Institute of Technology Madras,
Chennai 600 036, Tamilnadu, India
gromiha@iitm.ac.in
[2] Georgia Institute of Technology, Atlanta, USA
[3] Department of Crystallography and Biophysics, University of Madras, Chennai 600025,
Tamilnadu, India
[4] Biotherapeutics Pharmaceutical Sciences, Pfizer Inc., JJ230-D Chesterfield Parkway West,
Chesterfield, MO 63017 USA
[5] Department of Neurology, Emory University, USA

Abstract. Parkinson's disease is one of the neurodegenerative diseases caused by protein aggregation. It has been reported that the proteins, α-synuclein and calbindin are related with Parkinson's disease. However, the interactions between these proteins and their relationship with protein aggregation prone regions have not yet been explored. In this work, we have systematically analyzed the characteristic features of amino acid residues in α-synuclein and calbindin, and obtained a structural model for the complex using protein docking procedures. The structural model of calbindin:α-synuclein complex structure was used to identify the binding site residues based on distance and energy based criteria. The aggregation prone regions in these proteins were identified using different computer programs and compared with binding site residues. Specific residues, which participate in aggregation and preventing calbindin-α-synuclein complex formation were explored and these residues may be main causes for Parkinson's disease. Further, we have carried out mutational analysis and estimated the energetic contributions of the aggregation prone regions towards stability. The results obtained in the present work would provide insights to understand the integrative role of protein-protein interactions and protein aggregation in Parkinson's disease and lead to new directions for inhibiting this disease.

Keywords: protein aggregation, α-synuclein, calbindin, Parkinson's disease, docking, binding sites.

1   Introduction

Parkinson's disease (PD) is a progressive neurological disorder that leads to tremor, difficulty in walking, slowness of movements and co-ordination. Nerve cells use a brain chemical called dopamine to help control muscle movement. PD occurs when the

D.-S. Huang et al. (Eds.): ICIC 2013, LNAI 7996, pp. 677–684, 2013.
© Springer-Verlag Berlin Heidelberg 2013

cells that produce dopamine slowly get destroyed, leading to the failure of sending signals and loss of muscle function. PD is mainly sporadic and several cases have been found to carry genetic form of the disease [1]. Further, the early onset of PD might be due to several mutations. On the other hand, accumulation of protein aggregates in Lewy bodies is also a major cause for PD among the elderly patients [2].

The experimental and computational analysis of α-synuclein showed that it is a major fibriliar component of Lewy bodies found in brains of PD patients. In addition, single point mutations in α-synuclein are correlated to early onset of Parkinson's disease and it is one of the proteins associated with neurodegenerative disorders that have high propensity to aggregate [3]. The end product of α-synuclein aggregation is the formation of heavily insoluble polymers of the protein, known as fibrils. This fibriliar α-synuclein is the building blocks of Lewy bodies.

Calbindin is a calcium binding protein, which is widely expressed in calcium transporting tissues such as epithelial–absorptive tissues of intestine in the nervous system. Calbindin was also implicated in Parkinson's disease [4]. Hence, it is possible that interaction of calbindin with α-synuclein may inhibit the aggregation of α-synuclein. It has been shown that Calbindin-D28K acts as a calcium-dependent chaperone suppressing α-synuclein fibrillation [5].

The data on known and predicted protein-protein interactions available in STRING database [6] showed that α-synuclein and calbindin are interacting with several other proteins. However, the information on direct interaction between these two proteins are not available in literature. In this work, we have analyzed the potential interactions between α-synuclein and calbindin and compared our results with potential aggregation prone regions in α-synuclein. We found that the potential aggregation prone regions in α-synuclein are not part of the regions interacting with calbindin. Our result suggests that proper mutation at the aggregation prone regions in α-synuclein to enhance the tendency of interaction with calbindin, via allostery, would help to inhibit PD.

2   Materials and Methods

2.1   Three-Dimensional Structures of α-Synuclein and Calbindin

The structural data for α-synuclein is available in Protein Data Bank, PDB [7] and we used 1XQ8 for the present work. It contains 140 amino acid residues with three domains, (i) amphipathic N-terminal (1-60 residues), highly amyloidogenic middle (61-95 residues) and acidic C-terminal (96-140 residues). Human calbindin has 261 amino acid residues, which is homologous to rat calbindin with the sequence identity of 98% (2G9B). We used this structure as template and carried out homology modeling to get the final structure. The programs I-TASSER [8], Swiss-model [9] and Modeller [10] have been used for obtaining the structure and the RMSD between the structures obtained with any of the two programs is less than 0.3Å. The experimental structure of 1XQ8 and the modeled structure of calbindin have been used for docking to identify the interacting residues. The programs Hex [11], Rosetta-dock [12], Clus-pro [13] and patch-dock [14] were used to obtain the α-synuclein-calbindin complex structure.

2.2   Identification of Interacting Residues and Visualization

We have identified the interacting residues between α-synuclein and calbindin using distance and energy based criteria. In the distance based criteria, two residues are considered to interact if the distance between any of the heavy atoms in α-synuclein and calbindin is less than 5Å [15].

In the energy based criteria, we computed the interaction energy between all the atoms in α-synuclein and calbindin and summed up for each residue. The residue pairs, which have the interaction energy of lower than -1kcal/mol were considered as interacting [16]. The interactions between α-synuclein and calbindin were visualized using the software, Cytoscape [17].

2.3   Prediction of Aggregation Prone Regions

Recently, we have developed a program to identify aggregation prone regions in proteins [18] and used the same in the present study. In addition, other methods such as TANGO [19] and Aggrescan [20] have been used to predict the aggregation prone regions.

2.4   Structural Analysis of Interacting Residues/Aggregation Prone Regions

We have computed various structure based parameters such as solvent accessibility obtained with DSSP [21], surrounding hydrophobicity [22] and long-range order [23] for the interacting residues in α-synuclein and calbindin. In addition, we have made an attempt to mutate different residues to evaluate the stability using PopMuSiC [24].

3   Results and Discussion

3.1   Generating   α-Synuclein   and   Calbindin Complex Structure

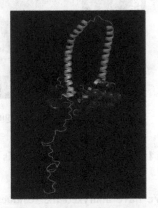

We have used different docking programs such as hex, Rosetta-dock, Clus-pro and Patch-doc to obtain the complex structure of α-synuclein and calbindin. Figure 1 shows the complex structure obtained with Rosetta-dock. The terminals of α-helices in α-synuclein are interacting with calbindin in the complex. This observation is consistent with the complex structures obtained with all the docking methods.

Fig. 1. α-synuclein (green)-calbindin (blue) complex

3.2   Interacting Residues in α-Synuclein-Calbindin Complex

We have utilized both distance and energy based criteria to identify the interacting residues in α-synuclein-calbindin complex. The interacting residues in calbindin are Asp26, Gly27, Ser28, Gly29, Tyr30, Glu32, Pro56, Glu57, Thr60, Phe61, Val62, Gln64, Tyr65, Asp70, Gly71, Lys72, Glu77, His80, Val81, Arg93, Gln96, Asp155, Ser156, Asn157, Arg169, Leu179, Lys180, Gln182, Gly183, Ile184, Lys185, Cys187, Gly188 and Lys189. The interacting residues in α-synuclein are Met1, Asp2, Val3, Phe4, Met5, Leu8, Gln79, Val82, Glu83, Gly84, Ala85, Gly86, Ser87, Ile88, Ala89, Ala90, Thr92, Phe94, Val95, Lys96, Asp98 and Gln99. The interactions between these two proteins can be pictorially seen using Cytoscape. Figure 2 shows the pattern of interactions in α-synuclein-calbindin complex.

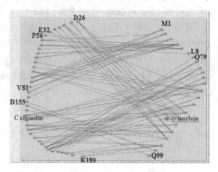

Fig. 2. Interaction between α-synuclein and calbindin

3.3   Aggregation Prone Regions in α-Synuclein

We have computed the aggregation prone regions in α-synuclein using our in-house program as well as with TANGO and AGGRESCAN. The aggregation profile for α-synuclein is shown in Figure 3. We noticed that the region with residues 69-77 has high propensity for aggregation. This result is consistent with all the three prediction methods.

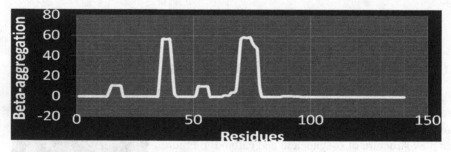

Fig. 3. Aggregation prone regions in α-synuclein

3.4   Comparative Analysis of Residues Involved in Aggregation and Interaction

We have analyzed the residues in α-synuclein, which are prone to aggregation and interacting with calbindin. The results are shown in Table 1. Interestingly, the residues, which are aggregating in α-synuclein are not interacting with calbindin. This analysis shows that the mutation of aggregation prone regions to make the residues with high probability to interact with calbindin may help to prevent aggregation and inhibit PD. Disruption of APRs via mutation should improve solubility of α-synuclein. This would, in turn, make more of α-synuclein available for binding with calbindin and other proteins.

Table 1. Aggregating residues in α-synuclein and their tendency to interact with calbindin

Residue	Aggregating	Interacting
Val37	Yes	No
Tyr39	Yes	No
Val40	Yes	No
Ala69	Yes	No
Val70	Yes	No
Val71	Yes	No
Thr72	Yes	No
Gly73	Yes	No
Val74	Yes	No
Thr75	Yes	No
Ala76	Yes	No

3.5   Structural Features of Residues Involved in Interaction

We have computed the solvent accessibility of residues in α-synuclein and calbindin using the program DSSP [21] and obtained the plot with ASAView [25]. The normalized solvent accessibility values of all the residues in α-synuclein is shown in Figure 4. We noticed that all the interacting residues in both the proteins are located at the surface.

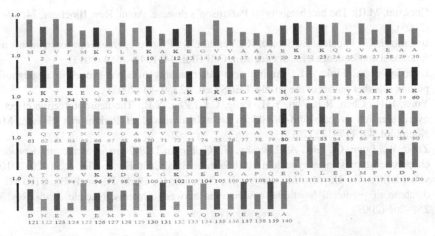

Fig. 4. Solvent accessibility of a-synuclein obtained with ASAView

Further, the computed surrounding hydrophobicities are in the range of 17-20 kcal/mol and the long-range order lies between 0.015 and 0.020. These results show that the residues make fewer long-range contacts and are less hydrophobic as compared to stabilizing residues. The point mutations of these residues indicates that the proteins are stabilized with an average energy of 0.5 kcal/mol upon mutations. Further detailed analysis on aggregation and the structural aspects are on progress.

4   Conclusions

We have analyzed the interplay between protein interactions and aggregation for the proteins α-synuclein and calbindin, which are involved in Parkinson's disease. We have modeled the complex structure using docking programs with the experimental structure of α-synuclein and modeled structure of calbindin. The interacting residues in the complex have been identified with distance and energy based criteria. The aggregation prone regions have been identified using different programs and compared the aggregation prone regions in α-synuclein with the residues, which are interacting with calbindin. Interestingly, the aggregating residues in α-synuclein are not interacting with calbindin. Further, we have analyzed the structural parameters of residues, which are involved in binding. These residues are highly accessible to solvent, less hydrophobic and have less number of long-range contacts.

Acknowledgments. This research was supported by Indian Institute of Technology Madras (BIO/10-11/540/NFSC/MICH) and the Department of Biotechnology research grant (BT/PR7150/BID/7/424/2012).

References

1. Cookson, M.R.: The biochemistry of Parkinson's disease. Annu. Rev. Biochem. 74, 29–52 (2005)
2. Invernizzi, G., Papaleo, E., Sabate, R., Ventura, S.: Protein aggregation: mechanisms and functional consequences. Int. J. Biochem. Cell Biol. 44, 1541–1554 (2012)
3. Uversky, V.N., Fink, A.L.: Amino acid determinants of alpha-synuclein aggregation: putting together pieces of the puzzle. FEBS Lett. 522, 9–13 (2002)
4. Yuan, H.H., Chen, R.J., Zhu, Y.H., Peng, C.L., Zhu, X.R.: The neuroprotective effect of overexpression of calbindin-D(28k) in an animal model of Parkinson's disease. Mol. Neurobiol. 47, 117–122 (2013)
5. Zhou, W., Long, C., Fink, A., Uversky, V.: Calbindin-D28K acts as a calcium-dependent chaperone suppressing α-synuclein fibrillation in vitro. Cent. Eur. J. Biol. 5, 11–20 (2010)
6. von Mering, C., Huynen, M., Jaeggi, D., Schmidt, S., Bork, P., Snel, B.: STRING: a database of predicted functional associations between proteins. Nucleic Acids Res. 31, 258–261 (2003)

7. Rose, P.W., Bi, C., Bluhm, W.F., Christie, C.H., Dimitropoulos, D., Dutta, S., Green, R.K., Goodsell, D.S., Prlic, A., Quesada, M., Quinn, G.B., Ramos, A.G., Westbrook, J.D., Young, J., Zardecki, C., Berman, H.M., Bourne, P.E.: The RCSB Protein Data Bank: new resources for research and education. Nucleic Acids Res. 41(Database issue), D475–D482 (2013)

8. Roy, A., Kucukural, A., Zhang, Y.: I-TASSER: a unified platform for automated protein structure and function prediction. Nature Protocols 5, 725–738 (2010)

9. Arnold, K., Bordoli, L., Kopp, J., Schwede, T.: The SWISS-MODEL Workspace: A web-based environment for protein structure homology modelling. Bioinformatics 22, 195–201 (2006)

10. Eswar, N., Marti-Renom, M.A., Webb, B., Madhusudhan, M.S., Eramian, D., Shen, M., Pieper, U., Sali, A.: Comparative Protein Structure Modeling With MODELLER. Current Protocols in Bioinformatics 15(suppl.), 5.6.1–5.6.30 (2006)

11. Ritchie, D.W., Venkatraman, V.: Ultra-fast FFT protein docking on graphics processors. Bioinformatics 26, 2398–2405 (2010)

12. Lyskov, S., Gray, J.J.: The RosettaDock server for local protein-protein docking. Nucleic Acids Research 36(Web Server Issue), W233–W238 (2008)

13. Kozakov, D., Hall, D.R., Beglov, D., Brenke, R., Comeau, S.R., Shen, Y., Li, K., Zheng, J., Vakili, P., Paschalidis, I.C., Vajda, S.: Achieving reliability and high accuracy in automated protein docking: Cluspro, PIPER, SDU, and stability analysis in CAPRI rounds 13–19. Proteins: Structure, Function, and Bioinformatics 78, 3124–3130 (2010)

14. Schneidman-Duhovny, D., Inbar, Y., Nussinov, R., Wolfson, H.J.: PatchDock and SymmDock: servers for rigid and symmetric docking. Nucl. Acids. Res. 33, W363–W367 (2005)

15. Li, W., Keeble, A.H., Giffard, C., James, R., Moore, G.R., Kleanthous, C.: Highly discriminating protein-protein interaction specificities in the context of a conserved binding energy hotspot. J. Mol. Biol. 337(3), 743–759 (2004)

16. Gromiha, M.M., Yokota, K., Fukui, K.: Energy based approach for understanding the recognition mechanism in protein-protein complexes. Mol. Biosyst. 5, 1779–1786 (2009)

17. Shannon, P., Markiel, A., Ozier, O., Baliga, N.S., Wang, J.T., Ramage, D., Amin, N., Schwikowski, B., Ideker, T.: Cytoscape: a software environment for integrated models of biomolecular interaction networks. Genome Res. 13(11), 2498–2504 (2003)

18. Thangakani, A.M., Kumar, S., Velmurugan, D., Gromiha, M.M.: Distinct position-specific sequence features of hexa-peptides that form amyloid-fibrils: application to discriminate between amyloid fibril and amorphous β- aggregate forming peptide sequences. BMC Bionformatics (in press).

19. Fernandez-Escamilla, A.M., Rousseau, F., Schymkowitz, J., Serrano, L.: Prediction of sequence-dependent and mutational effects on the aggregation of peptides and proteins. Nat. Biotechnol. 22, 1302–1306 (2004)

20. Conchillo-Sole, N.S., de, G.A.F.X., Vendrell, J., Daura, X., Ventura, S.: AGGRESCAN: a server for the prediction and evaluation of "hot spots" of aggregation in polypeptides. BMC Bioinf., 65 (2007)

21. Kabsch, W., Sander, C.: Dictionary of protein secondary structure: pattern recognition of hydrogen-bonded and geometrical features. Biopolymers 22, 2577–2637 (1983)

22. Manavalan, P., Ponnuswamy, P.K.: Hydrophobic character of amino acid residues in globular protein. Nature 275, 673–674 (1978)

23. Gromiha, M.M., Selvaraj, S.: Comparison between long-range interactions and contact order in determining the folding rates of two-state proteins: application of long-range order to folding rate prediction. J. Mol. Biol. 310, 27–32 (2001)
24. Gilis, D., Rooman, M.: PoPMuSiC, an algorithm for predicting protein mutant stability changes: application to prion proteins. Protein Eng. 13(12), 849–856 (2000)
25. Ahmad, S., Gromiha, M., Fawareh, H., Sarai, A.: ASAView: database and tool for solvent accessibility representation in proteins. BMC Bioinformatics 1(5), 51 (2004)

Cognitive Models of Peer-to-Peer Network Information of Magnanimity

Tian Tao[1], Yin Yeqing[1], and Li Yue[2]

[1] Business School, Guangxi University for Nationalities, Nanning, Guangxi 530006, China
aehousman@126.com
[2] School of Foreign Studies, Guilin University of Electronic Technology, Guilin,
Guangxi 541004, China

Abstract. The worldwide application of the P2P system results in an explosive expansion of information on its peer-to-peer networks, and it becomes a very realistic issue of considerable significance as how to help people effectively select and utilize valuable data. Combining to the information filtering technology, this paper builds a cognitive model designed for peer-to-peer network with an attempt to provide quick access to valuable information, and in the meanwhile, trying to eliminate those irrelevant data.

Keywords: P2P, Magnanimous information, Data filtering.

1 Introduction

In recent years, the information-based P2P application system is developed rapidly worldwide, which greatly expands the access to information channels. The P2P application system constructs a peer-to-peer network which allows its users around the world to be both data providers and recipients, containing information so magnanimous that people cannot imagine. With such magnanimous information, it becomes a focus issue for both academic and business circles to help people effectively select and utilize valuable data, shield those irrelevant ones, and protect personal privacy on information selection.

In the face of the growing mass information on the P2P network, it is far from enough to collect and organize information relying on the manual labor. Therefore, it becomes new challenges and opportunities for the development of information industry to automatically collect and organize all types of data needed. According to different application backgrounds and purposes of its usage, information processing technology has evolved to information retrieval, information filtering, information classification, and so forth.

Information filtering is based on users' needs of information. In the dynamic flow of information, it can help its users to search valuable data and shield other useless ones. For example, its application background includes subscription of electronic books and technical documentations, shielding of pornography, violence,

D.-S. Huang et al. (Eds.): ICIC 2013, LNAI 7996, pp. 685–690, 2013.

unauthorized dissemination of information, and automatically obtaining of information, and the like. In this context, information filtering technology emerges and becomes an important branch in the field of information processing.

Currently, the popular filtration products on the market are basically those characterized with keyword matching technology, the processing targets of which are focused on the Internet. Most information adopts centralized storage on the Internet, with which the source of information is easy to obtain, while the P2P application system is characterized by its complete distributed control, high dynamic node and autonomy, which results in difficult access to information. With the semantic-based information filtering model on the Internet, this paper constructs a semantic recognition model on P2P network in an attempt to help people to quickly obtain valuable information in such a P2P network.

2   P2P Network Cognitive Model Architecture

The Peer-to-peer network cognitive model system include P2P network, internal network, message filtering module, P2P application system message identification module, the P2P application system communication protocol identification module, and semantic parsing filter model, which can be shown as follows.

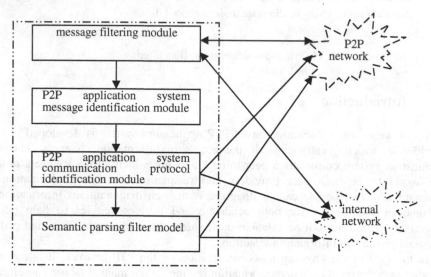

Fig. 1. Peer-to-peer network cognitive model architecture

The part surrounded by the dotted lines in fig.1. is the main body of the cognitive model, in which the semantic parsing filter agent is the core of the peer-to-peer information processing. By semantic parsing filtering proxy, the following target text information will be filtered in the process according to preset rules and user templates.

Peer-to-peer network: the network is not a physical presence on the network, but a logical network constituted by the large number of computers running P2P application system, each node of which has a dual role: a resource provider and resource requester. The most important peculiarity in the P2P network is the height of discreteness and, dynamics, the nodes of which can leave or plug in the network any time, thus resulting in an extreme instability in the access to information.

Message filter module: it is used to judge the category of the requesting information, to identify the contents of data communication, and determine whether the requested information is a message of peer-to-peer application. If the requested information is judged not to belong to the P2P areas, it will be delivered directly to the internal network, otherwise it will be submitted to the next processing, the message library of which is used to store message mode and composition information of the P2P application system.

Message identification module of P2P application system: it is used to determine the types of P2P applications to which they belong in current communication. For different P2P applications, the communication messages vary in their designs. It needs the support of the knowledge base message library and the P2P systems of knowledge. The P2P system knowledge base is used to store characteristics, categories, message formats (such as join, quit, query, and data interaction request syntax), and communication protocol in their different P2P application systems.

Communication protocol identification module of the P2P application system: It is used to deal with the contents of the request of a P2P application, including the split information request, and information analysis, so as to gain the contents requests, and then submit for further processing. It includes a logical selection mechanism between normal request (to join, quit, yet do not include the text content, and so on) to be released unconditionally.

Semantic parsing filter module: Its work includes two aspects, namely the intranet template and the text representation of the rule templates with text-matching techniques. According to the intranet template and text-matching techniques, the model can effectively identify and filter network information.

Based on Figure 1.1, the message filtering module is used to deal with the logical judgment. For instance, when the information of the external network is sent to the internal network, such data must be judged to which it belongs—if the information is a P2P message, it will be delivered to the parsing process, otherwise, the message will be submitted directly to the internal network; When the information of internal network is sent to the external network, the information may also need to go through the same logical judgment.

3   Semantic Parsing Filter Model

The semantic parsing filter model is the core of the peer-to-peer network cognitive model, which is used to identify and filter the information passed on peer-to-peer networks. Identification and filtering behavior is implemented by an intranet demand template and a text representation.

The Chinese text has its own characteristics compared to other language texts. Before conducting such a filter, first of all, the problem to be solved includes splitting words, unknown words and part-of-speech tagging. Secondly, it includes such a problem of the construction of stop-words table as the construction of the concept dictionary, with a semantic analysis of the problem. The above problems lead to a complexity in the Chinese text filtering. With the help of references of the corresponding foreign filtration system, and combined with the characteristics of the Chinese text processing, a Chinese text filtering model is built as illustrated in Figure 2.

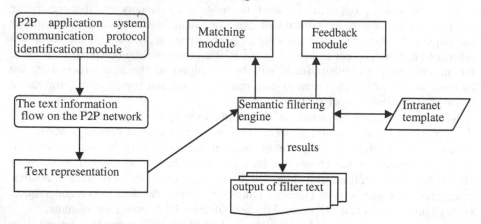

Fig. 2. Chinese text filtering model

It is obvious that, combined with characteristics of Chinese text information in a peer-to-peer network, the model has added a variety of Chinese processing for representation of information needs and the text such as splitting words, concept labeling, extension of the concept, stop-word processing, and so on.

Representation of the template of the internal network is the keyword tables and sample text.

The output of the filter text represents the text processed by semantic filtering engine. Matching module provides a similarity measure algorithms and sorting algorithms, in which the feedback module is used to improve the function of the filter algorithm.

The semantic filtering engine is the core of the system's scheduling, which gets the text from flow of information from the peer-to-peer network; Then, the text information will be identified, filtered according to the template with text match. The result will be sent to its users in the internal network.

In filtering models, the need of the internal network and documentation is expressed by the vector space model, while the demand and document matching mechanism use similarity computation to express.

4   Internal Network Template

In the text filter, this paper uses the method of the framework to describe the information needs of the intranet network. The theoretical framework is a structured method of knowledge representation, and it applies to natural language understanding. Theoretical framework: human understanding of the real world things similar framework structure is stored in memory. When faced with a new thing, people begin to search a matching framework in memory. In accord with the actual situation, we can get the details of the framework to be updated and supplemented. And finally, it is formed in the brain on the identification of new things. The frame theory provides a good resolution mechanism for knowledge representation of the internal network template.

In the text recognition, we have to deal with two aspects.

1. The information needs of the internal network, which represents a restrictive text needed in the internal network, can be limited the ability to interact with internal network information through demand
2. The understanding of the text, which means to understand the content of text passed in P2P network, theme expressed and the method of expression, and an analysis of the semantics of the text and its characteristics.

The internal network managers can describe the demand by determined template and some formal agreement. Understanding of the text can not be described by simple rules or information extraction, which must use complex understanding technology of natural language. The filtering system based on the statistical text use the feature items of the text to describe the text, while the basis of the matching is quantity of common items and concept owned by template and the text. Because of an emphasis on quantitative analysis of this type of system, and the lack of semantics constraints, the obtained results are often less than ideal. To solve this problem, we can only identify each feature item's role and functions in the text by semantics analysis, join this semantic information to expressed and text representation of the intranet templates, and improved the shortcomings of statistical methods.

The natural language understanding is currently unable to achieve a comprehensive understanding of the main ideas of the text, and draw a clear analytical expression at present. So in recent years, computational linguistics has a new development trend that real text processing uses the part of the analysis.

Generally the text is consisted of multiple sentences, and sentences of the composition are not a simple sentence superimposed. Have some logical relationship between sentences, and reflect a certain level of relations and sentence epitaxial determines the text epitaxy, then, an understanding of the sentences form the basis for text understanding.

According to the CD theory, language comprehension is the process of mapping to a conceptual base of the sentence, the sentence is considered reflected the relationship between one concept and another, and the formation of conceptual structures. Contain the meaning of a sentence, not the grammatical structure of a sentence, nor the semantic structure of a sentence, but the concept of the structure. We may have the following roles in the structure.

Intranet template represents a restrictive intranet managers define information needs, awareness about objective things, is a predefined pattern. Analogy of the text as an intranet template, fill an expression to a corresponding framework, according to a certain way, for analog operation, the process completed by computing the similarity. Therefore, text filtering, dynamic framework is an ongoing process to match with the corresponding intranet framework.

Intranet templates based on semantic is semantic relations expressed in the templates, template is not only features a simple list, but should include the interrelationship between characteristics, so as to better reflect the restrictive intranet information needs. To achieve this objective, the user template and concept definition must be assigned a role. According to requirements, the template needs to be defined in terms of a role model in filling, forming a comparable framework.

In accordance with the understanding of the law of the human, to analysis of a thing, should define its subject, object, time, place, scope, frequency and so on. Habitual or frequent will form a certain pattern in the human mind, triggering of the appropriate information, and you will associate it with their patterns, and in accordance with non-complete information to predict the missing information, visible intranet templates to match text with incomplete information.

Relationship between things in the real world is complex; it is hard to using simple models to analyze all aspects of a thing. Thus, logical combinations of the models must be strengthened, associated with logical relations pattern sequences can be used to express the relationships between things individual properties. Therefore, intranet template should have a logical computing capacity.

5   Conclusions

The article attempts to use cognitive model of natural language to filter information in a peer-to-peer network. It can improve people's efficiency to access information in a peer-to-peer network, and avoid interference from irrelevant information. Appling semantics parsing filter model applications in internal network and external network and setting according to intranet template, it can help to effectively control legal flow of information and avoid the spread of unauthorized information.

Acknowledgement. This research was supported by Guangxi Natural Science Foundation, Nos. 2010GXNSFB013053.

References

1. Schank, R.: Conceptual Information Processing. North Holland, Amsterdam (1975)
2. Schank, R.: Identification of conceptualizations underlying natural language. In: Schank, R., Colby, K. (eds.) Computer Models of Thought and Language (1973)
3. Cullingford, R.E.: Natural Language Processing, Rowman & Littlefield (1986)
4. Qulian, M.R.: Semantic memory. In: Minsky, M. (ed.) Semantic Information Processing (1968)
5. Schank, R.C., Abelson, R.P.: Scripts, Plans, Goals and Understanding (1977)
6. Cullingford, R.: Sam. In: Schank, R.C., Riesbeck, C.K. (eds.) Inside Computer Understanding (1981)
7. Woods, W.A.: Progress in natural language understanding: An application to lunar geology. In: Proc. AFIPS Conference, p. 42 (1973)
8. Woods, W.A.: Transition network grammar for natural language analysis. Communications of the ACM 13, 591–606 (1970)
9. Chomsky, N.: Aspects of the theory of syntax. MIT Press, Cambridge
10. Burton, R.R.: Semantic grammar: An engineering technique for constructing natural language understanding system. Technical Report 3453, Bolt Beranek and Newman (1976)

Author Index